T0190103

Facilities Operations & Engineering Reference

AFE Association for FACILITIES ENGINEERING

John Wiley & Sons, Inc.

RSMeans

Facilities Operations & Engineering Reference

A Technical & Management Handbook for:

- Planning & Analyzing Projects
- Complying with Codes & Standards

The Certified Plant Engineer Reference

AFE Association for FACILITIES ENGINEERING

RSMeans

This book is printed on acid-free paper.

Published by John Wiley & Sons, Inc., Hoboken, New Jersey

Published simultaneously in Canada

Managing Editor: Mary Greene. Editors: John Marchetti and Marla Marek. Production Manager: Michael Kokernak. Production Coordinator: Marion Schofield. Composition: Paula Reale-Camelio. Proofreader: Wayne Anderson. Graphics coordination and scanning: Jonathan Forgit. Book and cover design: Norman R. Forgit.

For general information on our other products and services, or technical support, please contact our Customer Care Department within the United States at 800-762-2974, outside the United States at 317-572-3993 or fax 317-572-4002.

Wiley also publishes its books in a variety of electronic formats. Some content that appears in print may not be available in electronic books. For more information about Wiley products, visit our web site at www.wiley.com.

Library of Congress Cataloging-in-Publication Data:

ISBN: 978-0-87629-462-8

Printed in the United States of America

10 9 8 7

Table of Contents

Foreword

Beginning in 1993, the Association for Facilities Engineering [AFE] and Westinghouse Electric Company entered into a collaborative effort to provide quality reference materials for facilities engineering and maintenance professionals. It began with the development of supporting materials for each of the component technical and management disciplines included in a training program designed to help Westinghouse employees prepare for the certifying examination for Plant Engineers. In the original effort, each of the participating Westinghouse training instructors and reviewers compiled materials as hand-outs to equip the Certified Plant Engineer [CPE] candidates with the requisite background information on engineering concepts, formulas and other data needed to successfully complete the CPE program. These hand-outs were assembled into bound booklets in each of these disciplines: Civil Engineering; Electrical Engineering; Mechanical Engineering; Engineering Economics; Environmental and Safety Engineering; Maintenance; and Management.

Westinghouse made these booklets available to AFE for use by all of the candidates striving to achieve the Certified Plant Engineer credential. The booklets served as the starting point and the foundation for this *Facilities Operations & Engineering Reference,* which has been co-produced by the Association for Facilities Engineering and the R.S. Means Company, Inc. Participating Westinghouse employees and AFE volunteers made a personal commitment and dedicated a great deal of time and effort to additional review and updating of this material over the past six years. Means' editors worked with these individuals to further expand and enhance the materials, with the goal of creating a reliable technical reference that would address essential engineering, management and construction fundamentals for facilities professionals.

The completed work represents the essence of the extensive body of knowledge with which a successful facilities or plant engineer must be conversant, including not only the originally covered disciplines previously listed, but also Heating, Ventilation and Air-Conditioning; Energy Efficiencies; and Instrumentation and Controls. The comprehensive nature of the book illustrates the depth of knowledge required of a facilities or plant engineer and stands as a monument to the significance of the industry's investment in this critical profession. It forms a reference base for anyone who would aspire to the significant and often daunting responsibility for the engineering, design, operation or maintenance of the physical assets of any corporation or business enterprise.

J. Bruce Medaris, CPE

Executive Director, AFE - The Association for Facilities Engineering

Acknowledgments

We would like to acknowledge the outstanding supporting efforts of the Westinghouse Electric Company, and their contributing authors for their untiring efforts and personal dedication to the production of this work. Included in this group are Chuck Baker, who focused on Electrical and Mechanical; Irwin Dobrushin on Environmental and Safety; Tom Hart on Maintenance; Alan Mochnick on Mechanical; Rick Sovic on Civil Engineering and Engineering Economics; and Rick Wolfe on Management. Without their early and continuing contributions, this work would not have been possible. *(See "Contributors to this Publication" following the Introduction for more on these individuals.)*

In addition, the editors would like to express appreciation to the following individuals and organizations who assisted with this publication.

Alan Vautrinot of Vautrinot Land Surveying, Inc. in Plympton, Massachusetts, contributed site plans used to illustrate site planning and surveying documentation in Chapter 3, "Civil Engineering and Construction Practices." Mr. Vautrinot, a registered Professional Land Surveyor, also offered expert guidance in state-of-the-art surveying techniques and equipment.

Francis McDonald, CPE, James McDonald, CPE, and William Haynes, CPE of MHM Engineering Services in Osterville, Massachusetts, identified some publications and provided text that was helpful in developing several chapters of this book.

Finally, we would like to acknowledge the authors of several R.S. Means publications that served as primary references for many of the chapters:

- William H. Rowe, III, AIA, PE, author of *HVAC: Design Criteria, Options, Selection;*
- Harold Colen, PE, author of *HVAC Systems Evaluation*;
- Richard W. Seivert, Jr., author of *Total Productive Facilities Management;*
- Kweku K. Bentil, AIC, author of *Fundamentals of the Construction Process;*
- Richard Ringwald, PE, author of *Means Heavy Construction Handbook;*
- Reinhold A. Carlson, PE and Robert A. DiGiandomenico, authors of *Understanding Building Automation Systems*, and
- Wayne DelPico, author of *Plan Reading and Material Takeoff*.

Introduction

Facility engineers and managers come from a variety of backgrounds and different levels of technical expertise. Some, such as the AFE Certified Plant Engineers, are trained in mechanical, electrical or civil engineering, and are often Registered Professional Engineers as well. Others have concentrated on business and finance. Whatever their academic training, facilities managers and engineers require a tremendous range of knowledge to effectively manage a facility's assets. This book is designed to help fill the information needs of facilities professionals at all levels of knowledge, and to improve communications by bridging the gaps between technical experts, managers and administrators.

The book is a compilation of technical information on the major disciplines of engineering, management, economics and construction. The Management chapter reviews basic skills, communications and documentation, team building, decision-making, problem-solving and some project management techniques. Chapter 2 provides guidance on planning, budgeting and investment issues, including funding new projects and performing life cycle cost analysis. Chapter 3, "Civil Engineering and Construction Practices," covers a tremendous range of information, from construction delivery methods, contract documents and CPM scheduling, to essentials of structural design, including foundations, steel structures and concrete practices; site planning, surveying and soils design; grading and paving. A good background knowledge of the various engineering and construction disciplines makes it possible to consider the appropriate alternatives and their ramifications, and to ask the right questions—from the first conceptual discussion through selection of the design team, the design process, construction, commissioning, move-in and daily operations.

Chapter 4, "Maintenance," provides extensive practical guidance for establishing and maintaining a maintenance program that not only meets the organization's needs, but actually performs as a profit center by contributing in enhanced output and savings to the bottom line. The author of this chapter shows how to establish maintenance program goals and organizational structure; evaluate the financial impacts and opportunities for improvement; define training requirements; manage effective work order systems; control purchasing and inventory; and plan and schedule maintenance. He also covers preventive and predictive maintenance and management reporting.

"Energy Efficiencies" (Chapter 5) addresses energy costs; explains how to create a facility's energy profile; explores new energy alternatives and the effects of deregulation; and outlines energy savings opportunities in HVAC, lighting, water usage, the building envelope, windows and doors. The chapter that follows focuses on HVAC alone, with a comprehensive treatment of essential issues including heating and cooling loads; fan, ductwork, and exhaust ventilation design; refrigeration; heat recovery; psychrometrics and computer programs. Chapter 7, "Mechanical Engineering," reviews principles, such as statics and dynamics, thermodynamics, hydraulics and fluid mechanics, that apply to the design and operation of building systems. This chapter includes sections on steam, pumps and air compressors that may be useful in conjunction with the HVAC chapter.

Chapter 8, "Instrumentation & Controls," defines the terms and explains the key criteria in selecting, operating and troubleshooting controls for HVAC and other equipment. Chapter 9 addresses electrical engineering issues, including power factors and quality, motors, lighting, overcurrent protection, the National Electrical Code and electrical costs. Chapter 10 covers environmental, health and safety issues, including indoor air quality, air emissions, wastewater and leaking underground storage tanks, hazardous materials, and federal environmental and safety regulations, including OSHA and ADA requirements.

Clearly, one book cannot capture an entire field of engineering, construction, economics or management in depth, much less provide comprehensive coverage of the ten major fields addressed in this book. Our editors have therefore focused on presenting key data with the most common applications in managing today's facilities operations. Much of the material in this publication is reprinted with permission from authoritative publications that are known and widely used in these fields, and from industry associations that set standards (adopted by national building codes) for design and construction. These texts are particularly recommended when full-scale information is required. The editors recommend consulting the additional resources listed at the end of each chapter for more detailed information, as needed, to support particular projects. These are currently available publications, related associations, and Internet web sites.

This book is intended to support facilities engineers and managers, and maintenance managers in playing an active, informed role in the design and development of new facility projects and in the efficient management of day-to-day operations. It is designed to provide essential technical information in a form that can be accessed quickly, without the need to search through technical manuals or identify the appropriate industry source. Our hope is that this publication will help facility professionals work more effectively with financial representatives, consulting engineers, architects, and construction personnel, as well as their own internal staffs. Talking to these individuals in their own "language" saves time in the planning and execution of projects, prevents misunderstandings, and leads to a more satisfied owner organization.

As noted in the foreword by Bruce Medaris, Executive Director of the Association for Facilities Engineering (formerly AIPE), Facilities Operations & Engineering Reference *was edited by members of that organization, to ensure a final product that would meet the fundamental reference needs of participants in AFE's Certified Plant Engineer program and exam. The book serves as the primary text for the CPE program.*

Contributors to This Publication

Assembling this book involved the efforts and cooperation of a team of experts, including leaders in the fields of facilities management and engineering, several of whom are prominent members of the Association for Facilities Engineering (AFE). Many of these individuals served as contributing authors and editors—writing, selecting and assembling materials from other sources and adding their own experience-based editorial comments. Following is a list of contributors and brief biographical descriptions.

Richard Sovic, PE, CPE, CFEP, is Manager of Project Management and Technical Services in the Facilities Department of the Westinghouse Electric Company. He holds a Masters degree in Engineering Management and is a registered Professional Engineer, an AFE Certified Plant Engineer, and an IFMA Certified Facility Manager. Mr. Sovic has over 20 years of professional experience including construction management, civil engineering, and facilities engineering design projects. He is the Chairman of the AFE (Association for Facilities Engineering) Certification Board and a member of the AFE Board of Directors, and has served as a past president of his AFE chapter.

Terry Wireman, an internationally recognized expert in facilities maintenance, has assisted hundreds of clients in North America and abroad in improving their maintenance management practices. Currently the Director of Advanced Services for Indus International (formerly TSW International), he determines clients' return on investment for improvement programs, and also conducts technical maintenance management seminars and workshops for universities, trade associations, and other clients. Previously, Mr. Wireman assisted clients in developing Best Practices methodologies for asset care and maintenance management. Mr. Wireman is the Maintenance Editor for "Engineer's Digest" magazine and the author of numerous articles and nine textbooks on maintenance topics. He is a member of the Institute of Industrial Engineers, the Society of Manufacturing Engineers, the Association for Facilities Engineering, and the International Maintenance Institute, and a frequent speaker for the National Plant Engineering and Maintenance Conference.

Charles Armbruster, CPE is Manager of Plant Engineering for the J.M. Smucker Company. He has practiced facilities engineering in a variety of capacities for over 20 years. Mr. Armbruster has also held positions at the national level of AFE's Planning and Resources Group and Certification Board.

Charles F. Baker, PE, CPE is Senior Electrical Engineer in Facilities Engineering at Westinghouse Electric Company in Monroeville, Pennsylvania, where he is responsible for electrical projects at the Westinghouse Energy Center Complex. He is a registered Professional Engineer in Pennsylvania and a Certified Plant Engineer. Mr. Baker is a member of AFE, the Institute of Electrical and Electronics Engineers, and the National Fire Protection Association. He has been an instructor in the Electrical and Energy AFE CPE national review courses.

Todd Brown, PE, CPE is the Director of Industrial Services for Tighe & Bond Consulting Engineers in Westfield, Massachusetts. He specializes in industrial wastewater treatment, remediation system design and regulatory compliance. In his ten years of membership in AFE, Mr. Brown has served as a chapter president, group director and member of the Certification Board. He frequently conducts sessions on environmental regulatory compliance for audiences around the country.

John Carroll, PE, CPE is a Vice President with the MetroHealth System. Located in Cleveland, Ohio, this facility includes a 740-bed medical center, two nursing homes, and community clinics. His administrative responsibilities include construction and renovation, food service operation, environmental services, transportation, clinical engineering, security and safety, and laundry. Prior to joining MetroHealth, Mr. Carroll held similar positions with the University of Chicago Hospitals and University Hospitals of Cleveland. He also teaches at Cleveland State University, and has been a member of AFE for 20 years.

Richard Cheslock, PE, CPE has over 25 years experience in facilities engineering/management. He is currently a Facilities Maintenance Program Engineer at the U.S. Department of State, Office of Foreign Buildings Operations. Previously, Mr. Cheslock held engineering and management positions with the Alexandria Hospital, the Norwalk Hospital, and the Houston Manned Spacecraft Center.

George O. Glavis, PE, CPE, CFEP is Environmental Maintenance and Program Manager at the U.S. Department of State. He is a member of the AFE Certification Board.

Richard Kosian, PE, CPE is Vice President and Director of Engineering at Beals and Thomas, Inc., a Westborough, Massachusetts environmental consulting and design firm. He is a graduate of the University of New Hampshire, and a registered Professional Engineer in 12 states, as well as a Certified Plant Engineer. Mr. Kosian has over 20 years of professional experience, which includes land surveying, field engineering, construction management, and civil engineering design projects. He has been a member of AFE since 1991 and has instructed the Civil Engineering session of AFE's CPE review course at the chapter level and nationally on numerous occasions.

Naren Mehta, CPE is Plant Services Manager for the Boeing Company. He has 21 years of plant engineering and maintenance experience, and a BS in both Electrical Engineering and Finance. Mr. Mehta is currently managing Boeing's Anaheim facility, where his responsibilities include energy management, maintenance and minor rearrangement of this 1.6 million square foot industrial site.

David Millay, CPE has been active in plant engineering for over 26 years. He is currently the Assistant Director of Facilities Management at the University of Delaware, and is past national president of AFE. Mr. Millay is a Certified Plant Engineer and holds a BS in Industrial Technology. Prior to his position with the University of Delaware, Mr. Millay was Senior Manager at the Massachusetts Institute of Technology. In 1990, he was voted Plant Engineer of the Year by the Boston chapter of AFE for his many contributions to the profession. Mr. Millay has served as an instructor for the CPE review program and is a frequent lecturer.

J. Alan Mochnick, PE, CPE, a Senior Mechanical Engineer in the Facilities Department at Westinghouse Electric Company, has 27 years of experience in plant and facilities engineering. He is a graduate of the Case Institute of Technology and a registered Professional Engineer. Mr. Mochnick is a long-time member of the AFE, and is a member of both the American Society of Heating, Refrigeration and Air-Conditioning Engineers (ASHRAE) and the American Society of Mechanical Engineers (ASME).

Vincent A. Rojas, PE, CPE is Facilities Engineer at Los Alamos National Laboratory. He has 35 years experience in plant and facilities management and is a registered Professional Engineer in New Mexico. Mr. Rojas is a member of the Certification Board of AFE and the Chairman of AFE's Education Committee.

James Armstrong, currently an Applications Engineer for Trigen: Boston Energy Corporation, has 20 years experience in facilities maintenance, engineering and project management. Prior to Trigen, he served as an energy consultant for Energy Vision, Inc., conducting facility audits and designing solutions for system efficiencies. Mr. Armstrong also was the Manager of Property and Utilities for MassDevelopment, overseeing the development of Fort Devens Air Force Base in Massachusetts in its transition to an industrial center. In this capacity he worked with the Massachusetts Dept. of Utilities to develop an unbundled utility rate structure. Earlier in his career, Mr. Armstrong served as Facilities and Operations Director for institutions including Colby Sawyer College and the Boston Museum of Science, and as Maintenance Supervisor for Anderson Power Products and the Massachusetts Institute of Technology.

Barbara Balboni is currently an Engineer/Editor with R.S. Means Company, where she is responsible for the content of the *Square Foot Costs, Assemblies* and *Interior Cost Data* publications. Ms. Balboni also presents seminars on square foot and unit price estimating. Prior to her position at Means, she held positions as Senior Designer and Project Manager for several architectural firms in Massachusetts.

John Chiang, PE is a registered Professional Engineer who holds degrees in both Electrical and Industrial Engineering. He is the editor of *Means Electrical Estimating Methods, Electrical Cost Data,* and *Electrical Change Order Cost Data.* He has over 15 years experience in electrical design and construction project management on projects including manufacturing and industrial, hotels and apartments, and medical and office buildings.

Jesse Page, Senior Engineer/Editor at R.S. Means Company, has held project management and estimating positions at large construction firms, where his duties included quality control systems management on several large Army Corps of Engineers projects. Mr. Page's responsibilities at Means include research and development of new cost data line items, and preparation of budgets and detailed cost estimates for Means Consulting Services clients.

Phillip Waier, PE is a registered Professional Engineer in Massachusetts and Senior Editor/Engineer at R.S. Means Company. He served for 16 years as project manager and chief engineer for numerous general and industrial construction projects, during which time he was also responsible for the structural design of chemical process, sewerage treatment, and printing plants. Mr. Waier is responsible for some of Means' largest cost databases and cost data publications, including *Facilities Maintenance & Repair Cost Data* and *Building Construction Cost Data.* He also manages Means' custom database projects for national firms and presents professional development seminars on construction estimating and project management.

Chapter 1

Management

Chapter 1 Management

We live in a time of great change. Nowhere is this more evident than in the workplace. Most facilities and plant managers agree that employer/employee relationships, wants, needs, and expectations have changed more in the past two decades than over the last hundred years, with even more dramatic changes predicted early in the next century. Today's facility and plant managers are faced with greater process and personnel challenges, including a globalized economy and fierce competition. Concepts like workplace diversity, synergy, reengineering, and the now infamous downsizing have all demanded far more interpersonally from today's facility and plant managers.

The field of management is as broad, complex, ambiguous, and diverse as any subject that has ever been presented in an academic format. Management can best be thought of in terms of either a process or people. While the two overlap, they can also be approached as separate skills. Under the rubric of "managing a process" are all of the technical skills a manager must possess, including life-cycle costing, just-in-time logistics, project management, cost estimating and budgeting, evaluating alternatives, computerized maintenance management, and preventive maintenance. The last two topics are addressed in Chapter 4. For in-depth coverage of these and the other process management topics, one should review *The Facilities Manager's Reference* by Harvey Kaiser and *Total Productive Facilities Management* by Richard W. Sievert, Jr. (both published by R.S. Means Company, Inc.).

Managing people includes all of the personal, interpersonal, and group skills. Most facility and plant managers will not dispute the fact that there is never a shortage of good ideas. What there is a shortage of is people with the talent, drive and "people skills" necessary to make these good ideas an integral part of an organization's value system. To reduce the time gap between good ideas and accepted practice in any organization, the exceptional manager must have good communication and problem-solving skills, and be able to manage conflict, motivate employees, delegate properly, and work in groups. He or she must also have enough power within the organization to perform these tasks.

Most facility and plant managers would also agree that it is not the technical problems that consume their time and give them the biggest headaches. The biggest challenge is dealing effectively with their people. Much research has been done that shows personnel problems, and more explicitly, interpersonal problems, cause the greatest concern among managers and are the greatest detriment to employee productivity.

The field of "people skills" is rooted in behavioral psychology. It is a wide field with a long list of texts that grows exponentially each year. For facility and plant managers who want to delve further into the field of people skills, we recommend the fourth edition of *Developing Management Skills* by David Whetten and Kim Cameron (Addison-Wesley Educational Publishers, Inc.). The book addresses issues like personal skills, problem-solving, and motivating employees. The first section of this chapter outlines some of the book's principles that are most relevant to the facility manager.

The remainder of the chapter focuses on the communications and team-building aspects of management. It is excerpted from *Total Productive Facilities Management* by Richard W. Sievert, Jr. (R.S. Means Co., Inc.). This recommended reference also includes information on evaluating and improving facilities performance, value engineering projects and cost management, scheduling, and contracting and procurement methods.

Management Skills

Developing Management Skills is a practical introduction to business and management skills. Using a series of interactive exercises and examples, Whetten and Cameron place a strong emphasis on self-assessment and self-awareness.

Self-awareness is a difficult concept to grasp and an even more difficult state to achieve. With a heightened sense of who you are, however, the task of working with subordinates, peers, and superiors becomes easier and more rewarding. Most researchers agree that self-awareness is a function of four discrete attributes:

- One's values
- One's cognitive style (i.e., how one gathers and processes information)
- One's adaptability and attitude toward change
- One's interpersonal style

Ironically, some of us may not be as aware of our values and style as we think. Self-awareness is a lifelong process. Knowing and understanding yourself should manifest itself in more effective interaction with others. See Figure 1.6 later in this chapter for a form that may be used as a self-awareness exercise.

Managing Yourself

The best managers know that to lead effectively, they must first manage themselves. They start by drawing a realistic picture of their style and their place in their organization, then look for ways to improve their own efficiency.

Assessing Your Managerial Type: As you work to better understand yourself, try to determine your managerial style and how you might improve it. According to Whetten and Cameron, there are essentially four types of managers. The *indulging manager* stresses employee happiness and satisfaction at the expense of employee productivity. The *imposing manager* stresses employee productivity at the expense of employee happiness and satisfaction. The *ignoring manager* stresses neither employee satisfaction nor employee productivity. Finally, the *integrative manager* stresses both employee productivity and employee satisfaction.

Not surprisingly, Whetten and Cameron claim the best managers are integrative ones. This managerial style can only be developed through applied practice, but the important point is understanding that employee productivity and employee satisfaction are not mutually exclusive variables. In fact, it is actually the opposite that is true.

Determining Your Power Level: Another important aspect in increasing self-awareness involves determining your true level of power in your organization. Counterintuitively, one's salary is not a good predictor of his or her power in an organization. The causal independent variables that are the best predictors of power include a person's personal attributes and the characteristics of the position they hold.

Whetten and Cameron emphasize the following personal attributes as the most relevant in influencing others:

- professional expertise
- "likability" and personal attractiveness
- the amount of effort put forth on behalf of the organization
- legitimacy within the organization's value system

The specific characteristics of your position are also strong predictors of your power. The following five areas should help you define your position's power.

- *Criticality*: How critical is your job? If you were missing for a week, what would happen?
- *Centrality*: How central is your job—both horizontally and vertically—to the communication flow in your organization?
- *Flexibility*: How much discretion do you have to make decisions?
- *Visibility*: Managing people will always make you more powerful in an organization than managing tasks. For example, those who write speeches seldom get the same recognition as those who deliver them because they are not as visible.
- *Relevance*: How much does your position affect the bottom line of the company?

An excellent first step to enhance your position in a company is to take on tasks that cross departmental lines. In this way, you increase your visibility and put yourself directly in the flow of information. In addition, those performing cross-departmental tasks are often given more latitude in decision-making and are likely to be critical to the company's bottom line.

Managing Your Time: Efficient time management is essential for managers, as most agree that the one thing they do not have enough of is time. Most people that excel in time management do so by:

- Prioritizing tasks and creating "to do" lists,
- Doing many trivial jobs simultaneously, and
- Using discretion when reading material and skimming for the important points.

Meetings can be the time-conscious manager's worst enemy. To keep them running efficiently, Whetten and Cameron suggest the following:

- Schedule routine meetings near the end of the day. Productivity is highest in the morning and this time should not be wasted on routine meetings.
- Start meetings on time
- Set time limits and stick to them.
- All meetings should have an agenda and clear-cut objectives.
- Whenever possible, hold short meetings standing up. This will help keep the meeting short.
- Minutes of a meeting should be promptly prepared and distributed.

Managing Others

The effective manager has the ability to react appropriately to an employee's strengths and weaknesses while bolstering the employee's confidence and commitment to the organization.

Diagnosing Poor Performance: One of the greatest mistakes a manager can make is misdiagnosing an employee's poor performance.

Coaching and Counseling:

When an employee's productivity suffers because of a lack of information or technical skills, there exists a coaching problem. A manager should assess it as such, explain to the employee that the company requires more from him or her, and take steps to eliminate the gap in knowledge. When an employee's productivity drops due to emotional problems or friction with other employees, however, a manager must recognize it as a counseling problem and respond accordingly. A skilled manager will always approach this kind of situation by first convincing the subordinate that, in fact, there *is* a problem.

Ability and Motivation:

A manager may also be called on to determine whether an employee's unsatisfactory performance is due to a limited ability or to a lack of motivation. Motivation can be thought of as the product of an employee's desire to complete a task multiplied by his or her commitment to the task. The variable ability can be further broken down into a person's knowledge, how much specific training he or she has had, and how much access he or she has to certain resources. According to Whetten and Cameron, the best predictor of work performance is the product of an employee's ability multiplied by his or her motivation.

Delegating: The ability to skillfully delegate tasks is vital to a successful manager. When properly implemented, a delegated task can give an employee confidence and a sense of importance. It tells the employee that you trust and believe in him or her. The result is an employee who is more committed to the organization and a manager with more time to focus on the big picture. Whetten and Cameron offer the following tips on delegating.

- Managers must avoid the practice of delegating only when they are completely overloaded. Likewise, they should be careful not to delegate just the unpleasant tasks.
- Whenever possible, managers should delegate a task completely. A sure formula for an unhappy subordinate is to give them all the responsibility and then limit their authority.
- Tasks should be delegated to the lowest hierarchic level possible.
- Subordinates should be involved in the delegating process. People are far more likely to "buy into" something they helped create.
- A good manager also knows how to avoid the infamous "upward delegation," a phenomenon that occurs when managers allow subordinates to dictate tasks to them. The classic example is when a subordinate comes to a manager with a problem and asks the manager to "get back to me" with the solution. In this case, the worker has shifted the responsibility for the task to the manager. The best managers avoid "upward delegation" by insisting that, when subordinates come to them with problems, they bring along some possible solutions to the problems as well.

The following section is reprinted from *Total Productive Facilities Management*, by Richard W. Sievert, Jr., R.S. Means Co., Inc.

Communication is a basic skill that is needed to establish and maintain productive relationships. A high percentage of the friction, confusion, frustration, disputes, and inefficiencies in our working relationships are traceable to poor communication. Facility management work is especially susceptible to communication problems because of the broad and overlapping areas of responsibility, and the multidisciplinary and sometimes complex nature of facility-related projects. Facility managers and facility engineers must be aware of the importance of communication in a building services environment to ensure people are working together effectively to meet organizational objectives. And, of course, facility managers need to develop better oral and written communication skills themselves.

Communication problems are an enormous threat to profitability. In almost every case, the misinterpretation of a customer's requirements, failure to execute a plan or carry out instructions, or a missed delivery date is a result of a breakdown in communications. Applying some fundamental principles and techniques to this area will lead to more effective use of available time, improved cooperation and coordination of efforts, fewer disputes, and a reduction in associated costs.

Why do communication breakdowns exist in our Age of Information Technology? Messages and images can be transmitted instantaneously almost anywhere in the world. Through use of the Internet combined with high-speed computers and reproduction equipment, we can distribute reports and other information, in real time, almost anywhere at any time.

Communication is a process. The key is that you have to receive and transmit the *right* information on a *timely* basis in order to maintain control. The fax and the computer sometimes overload us with information, and we have to make judgments about what is truly useful. Effective communication, versus mere sending of data, means the person receiving the message must *understand* the message and be *motivated* to take the action recommended by the sender. The cycle must be completed for communication to be successful. Transmitting information is the easy part. The problem is reception—more precisely, intelligent reception—of the information.

Also, the receiver must be able to secure clarification and additional information. The sender, in turn, must have feedback enabling him to assess the degree of understanding and compliance. Through feedback, the sender determines requirements for new or follow-up communication. Successful communication, therefore, lies in making it a two-way process: downward (from sender to receiver) and upward (from receiver to sender). Distributing established schedules and status reports are examples of one-way communication, while project team planning meetings, status review meetings, value engineering workshops, and post-project review meetings are examples of interactive two-way communication.

The only way to have feedback is to include an aspect of communication often overlooked—listening. Communication cannot be two-way unless we listen actively and with sensitivity. In the best working environments, people feel free to express their views, knowing that their opinions will be considered, and their suggestions and ideas recognized.

There is an old maxim: "If you talk too much, you can't hear what others are saying!" Experience demonstrates that by listening long enough, one starts to get answers. One hears others define the problems and suggest answers. This is considerably more helpful than evaluating only what one person thinks from a vantage point that may be far removed from the problem. It is also important to try to understand the other person's point of view.

Listening can be disturbing because it sometimes forces you to recognize unexpected problems. Often it is more comfortable not to listen, and to ignore warning signs that will require involvement in solving a problem. Another obstacle to effective listening is ego, which prevents one from hearing another point of view. For example, an architect's mission is not to design a "trophy" building for his or her own accolades, but rather to fulfill the customer's needs. The grand entrance or spiral staircase may not always be cost-justifiable. The focus should not be on "me," but on the customer's perspective.

Often people will withhold communication of information (such as a potential problem) if they fear somebody will not respond favorably to it. Remember the old maxim, "Don't kill the messenger." The fact is that major problems can often be avoided if warning signs are recognized and reported while there are still opportunities for corrective actions. Many problems can be minimized or avoided if somebody has the courage to communicate bad news. Facility managers should instruct their staff to listen carefully, be succinct, and follow through by taking the appropriate action after communicating with the right individuals in a timely fashion.

Project Communications

Early warning signs that a project could be heading for trouble include:

- Delays (schedule slippage, request for time extensions)
- Breakdown in communications (Common causes include incomplete, inaccurate, or untimely transmittal of shop drawings, and cost and schedule reports.)
- Slow payments to consultants, equipment vendors, and contractors.
- Substantial increase in change order requests and claims
- Inefficient crew sizes
- Complaints from consultants, contractors, and vendors
- Quality defects
- Abnormal number of contractor requests for substitutions
- Increasing number of contractor requests for information
- Deteriorating supervision

Part of good communication is learning your customers' culture and what they need. Be a counselor to your customers and support them constantly throughout your working relationship. Emphasize the two-way communication process: They need to be informed, and you need to request their feedback constantly in the form of suggestions and advice. You can keep them informed through vehicles such as regularly scheduled briefing sessions and routine reports and written plans.

At times, facility managers may get involved with a project only to discover that team members are reluctant to share information that may be crucial to successful completion. This is a normal protective human trait. Managers who sensitively convince their staff that they are not trying to assign blame will be more apt to obtain information about conditions that may impede progress. Perhaps a safety hazard needs to be removed or more efficient equipment installed. When people are preoccupied with personal defensiveness rather than organizational objectives, they are not working together effectively to complete their assignments.

To gain cooperation, managers should share their own experiences with similar situations to create an open environment where people can freely share their ideas. Staff members, consultants, or contractors who are continually criticized for sharing their ideas will stop sharing them. Despite your time constraints, it is important to demonstrate consideration for new ideas. Sometimes what appears at first glance to be the most ridiculous idea, turns out to be the best one.

Competent executives, managers, and staff specialists often diminish their effectiveness by maintaining only one-way interpersonal contact. They issue hurried or unclear instructions. They initiate change orders, but fail to ask whether important points are grasped. They assume that others use and understand precisely the same terminology, and often ignore suggestions. With one-way directives, they are actually talking without listening. Managers must take time to communicate thoughtfully—both orally and in writing.

"Communication" is a term seldom found written into contract provisions. Nevertheless, you can write better contracts with a complete statement of the work to be done, and a definition of the roles and relationships between the parties to the contract, lines of communication, frequency of team meetings, and relevant reporting and report formats. Everyone involved in a project should be provided with a project directory containing a list of important telephone numbers. The project manager should have the home as well as the work and pager or cell phone numbers for key project players. The home phone number is important to permit quick answers or exchange of information that cannot wait until the next day.

Written Communications

Communication is at its best when it is clear and to the point, particularly when customers are not experienced or familiar with industry jargon. Moreover, complicated quantitative data is often more understandable to readers when presented in graphic form. Much of the material included in reports is more easily understood when the written information is accompanied by graphic aids such as tables, bar charts, pie charts, and graphs.

Important messages delivered orally must be followed up in writing. Meeting minutes and memoranda must be given to project team members to ensure documentation of important information. Keep project team members informed by regularly sending copies of correspondence, progress reports, calculations, and other significant project documentation.

Shop Drawings

Construction drawings and specifications communicate in graphic and written form the designers' expectations for the work to be performed by the contractor(s). The construction documents prepared by the design professionals do not show all details that affect the constructed product to be put in place. Also, design professionals typically do not communicate means, methods, sequence of construction, or related job safety procedures which are typically the contractors' responsibility. The contract between the owner and contractor(s) should state that the contractor(s) submit more detailed drawings, diagrams, and schedules for review and approval by the design professional (architect and/or engineer) prior to beginning construction. These shop drawings or submittals serve as important feedback from the construction or installation contractors, signifying that they understand the design intent and the contract documents.

Meetings, Meetings, Meetings

Meetings are required to complete—on-time and on-budget—projects that require the involvement of a multi-disciplined team. Meetings offer a sure and fast way to share information among a group of individuals because they provide instant two-way communication. Every participant obtains an immediate response to a question, or clarification of unclear points. Properly run, meetings also save considerable time that otherwise would be spent sending memos, waiting for responses to the memos, or answering letters.

Successful meetings must be planned. Many meetings are too long or attempt to accomplish too much. Meetings should be conducted only when necessary and planned with a clear objective in mind. To properly plan, conduct and control productive meetings, you need to define the objectives and prepare an agenda. Preparing an agenda not only provides you with a tool to control the meeting, but also demonstrates that you have done your research and are prepared to make the most productive use of other people's time.

Project orientation meetings enable management to review the project requirements and obtain the cooperation of consultants and contractors. Project orientation and "kick-off" sessions should be designed to establish a "we" attitude, rather than the "me" outlook that might be prevalent with the involvement of numerous independent consultants and contractors. A "we" attitude transforms a group of people working on a job into a working team doing a job. Regular meetings for clients and contractors, for instance, could include a complete, but brief, report of job progress. Decision-making meetings may solve specific problems such as a productivity issue, pointing out the required corrective action.

Project Record Filing System

Documentation is an essential aspect of communication. Records of design criteria and scope of work serve as a basis for identifying, documenting, and reporting changes. A record file containing sketches, charts, telephone conversations, notes, proposals, estimates, schedules, and other important communications should be maintained for a complete record of the project's evolution. Each item should be dated, and minutes included so that meetings and agreements can be recorded. Not only does careful documentation protect your interests for the current project, but it also provides a storehouse of knowledge for application to future projects.

A structured project record filing system should be maintained in a central location. Duplicates of contracts should be kept in several locations for protection in the event of fire or other losses and as a defense against claims, project overruns, and delays. Incoming and outgoing correspondence should be differentiated. Responsibility for the maintenance and security of files should be delegated to one person who can help prevent their disappearance by controlling file removal and access. Figure 1.1 is a sample procedural outline for a project record filing system.

- Project Description
- Budget
- Equipment List
- Contracts
- Meeting Minutes and Telephone Memorandums
- Correspondence and Transmittals
- Engineering Information
- Estimates
- Specifications
- Request for Bid Packages
- Bid Analysis Documents
- Contractor Proposals
- Insurance Certificates
- Submittals
- Schedules
- Permits
- Reports
- Change Orders
- Utility Agreements
- Notes

Photographs provide excellent records and should be taken (and dated) of site conditions, including the conditions of nearby buildings and site characteristics, prior to the start of construction projects. Foundations, wall systems, parking lots, and roads should be included. Photographs serve as important records in disputes.

A transmittal letter (see Figure 1.2) should accompany all important documents and submittals so that the sender has a record of what was sent. Owners should ensure that minutes are kept, and validate their understanding of what occurred at meetings based on the recorded minutes. If owners receive minutes with which they disagree, they should issue written clarifications. Minutes are also used to communicate the team's understanding of the project requirements.

Following these guidelines for an organized documentation system will improve productivity, as it saves project personnel from wasting time trying to locate reference material or duplicate previous efforts.

Procedures for Project Record Filing

A job information sheet is filled out by the Project Manager and given to the Accounting Department, listing:

A. Client
B. Job Name
C. Location
D. Client Contact
E. Type of Contract
F. Contract Price if Applicable
G. Estimate if Applicable
H. Large or Small Job

The Accounting Department will complete the Contract Data Sheet, issue a job number, fill out an estimate/ contract sheet, and enter necessary information to the "Jobs in Process" binder.

Accounting Department will collect all new jobs and enter them into the cost system the following week.

Files are prepared as follows:

Job Files contain:

1. Green pendaflex indicating job number and identity.
2. Manila folders will be made up with the following tabs:
   a. Contracts and Approved Change Orders
   b. Quotations, Estimates and Proposals
   c. Correspondence
   d. Engineering
   e. Purchase Orders
   f. Job Information
   g. A separate file is made up for the Accounts Payable/Billing in Process drawer for bookkeeping and billing procedures.
   h. Invoices
3. Extra work order (a field-generated change in contract amount due to unanticipated work).
   a. Rough draft is prepared by Project Manager and given to Accounting Department listing the following:
      1. Description of work
      2. Estimated hours worked on job
      3. Materials
      4. Equipment
   b. The Accounting Department will compile labor dollars and extend figures into final cost.
   c. Finished rough draft is reviewed with Project Manager and, when approved, is initialed.
   d. Extra work is then prepared in final form.
   e. Contract data sheet is updated by Accounting.
4. Change Order—An office-generated change order in contract amount due to change in job. (Processing is the same as noted above for the extra work order.)
5. Sections providing information on the overall project generally contain the following files:

R.S. Means Co., Inc., *Total Productive Facilities Management*

Figure 1.1

a. Specifications
b. Architects/Engineers
c. Contractors
d. Contracts/Approved Change Orders
e. Correspondence
f. Transmittals ("Incoming" and "Outgoing" arranged chronologically)
g. Design Criteria
h. Estimates
i. Field Notes
j. General
k. Insurance
l. Job Information
m. Memoranda
n. Meeting Minutes, Telephone Logs, Field Observation Reports
o. Permits
p. Project Criteria
q. Purchase Orders
r. Change Orders
s. Schedules
t. Pertinent Documentation
u. Shop Drawings
v. Transmittal Copies (Alphabetical)
w. Vendor Quotations and Proposal Evaluations
x. Zoning
y. Contract Close-out Documents

6. For large projects, files may be set up by contractor work packages or classified by CSI MasterFormat Divisions:

Div. 1 - General Requirements
Div. 2 - Site Work
Div. 3 - Concrete
Div. 4 - Masonry
Div. 5 - Metals
Div. 6 - Wood and Plastics
Div. 7 - Thermal and Moisture Protection
Div. 8 - Doors and Windows
Div. 9 - Finishes
Div. 10 - Specialties
Div. 11 - Equipment
Div. 12 - Furnishings
Div. 13 - Special Construction
Div. 14 - Conveying Systems
Div. 15 - Mechanical
Div. 16 - Electrical

R.S. Means Co., Inc., *Total Productive Facilities Management*

Figure 1.1 cont.

LETTER OF TRANSMITTAL

FROM ______________________

DATE ______________________

PROJECT ______________________

LOCATION ______________________

ATTENTION ______________________

RE: ______________________

Gentlemen:

WE ARE SENDING YOU ☐ HEREWITH ☐ DELIVERED BY HAND ☐ UNDER SEPARATE COVER

VIA __________ THE FOLLOWING ITEMS:

☐ PLANS ☐ PRINTS ☐ SHOP DRAWINGS ☐ SAMPLES ☐ SPECIFICATIONS

☐ ESTIMATES ☐ COPY OF LETTER ☐ __________

COPIES	DATE OR NO.	DESCRIPTION

THESE ARE TRANSMITTED AS INDICATED BELOW:

☐ FOR YOUR USE ☐ APPROVED AS NOTED ☐ RETURN ______ CORRECTED PRINTS

☐ FOR APPROVAL ☐ APPROVED FOR CONSTRUCTION ☐ SUBMIT ______ COPIES FOR ______

☐ AS REQUESTED ☐ RETURNED FOR CORRECTIONS ☐ RESUBMIT ______ COPIES FOR ______

☐ FOR REVIEW AND COMMENT ☐ RETURNED AFTER LOAN TO US ☐ FOR BIDS DUE ______

REMARKS ______________________

IF ENCLOSURES ARE NOT AS INDICATED, PLEASE NOTIFY US AT ONCE.

SIGNED ______________________

R.S. Means Co., Inc., *Total Productive Facilities Management*

Figure 1.2

Handling Change Order Communications

Change orders that occur during the life cycle of a construction project also demand timely and accurate communication. They represent a change in the cost and time, from that originally planned and budgeted. Depending on the scope of the change, these situations can be of considerable concern to a project owner.

When changes occur, the project manager must move quickly to provide the customer with adequate documentation concerning the cause of the change. Changes can be the result of a design omission, an unexpected occurrence, or a change in the customer's preference. Cost estimates should be provided in a timely fashion, and the necessary work to implement the change should be scheduled. Regular progress meetings and update reports will help prevent time and budget surprises, and will enhance the owner/project manager relationship.

Time is of the essence not only for change orders, but for other project communications. With fax, email, and voice mail systems readily available, customers not only expect, but demand, timely communications. Reporting must be prompt. They identify potential problems, thereby making it possible to take corrective measures to prevent small problems from turning into big ones. Modern project management software systems allow for relatively easy updates.Variance analysis can also be used to identify problems and to determine the reasons for the variance.

On-Site Communication

The owner's site representative or contractor's field superintendent plays a special communication role. This is a leadership position that goes far beyond policing contractors to make sure they are working and that construction materials are delivered on time. The owner's site representative's most important task is to serve as the company's liaison with the "workers in the trenches." A variety of craftsmen skillfully make the designer's drawings and specifications come to life with their building skills. The site representative has to deal with a wide range of personalities and situations daily. Each worker comes to the job with his own unique worries, long-standing attitudes, prejudices, possible health and other issues. Each person working on the project draws conclusions, expresses emotions, and is often influenced by inaccurate reports from the "grapevine." It is important to work with the whole individual, not just that person's engineering, carpentry, or administrative competencies.

Regardless of whether the worker is a corporate employee, an on-site laborer or a consultant, that individual's freedom to contribute opinions on the project allows him or her to feel involved and important. Whether the work is new construction or ongoing maintenance, managers benefit from the suggestions of workers closest to the project. Managers must learn to hear the unspoken, or that which may not be explicitly stated. For example, when asked, "How many times will we have to repair this broken-down equipment?" the response should not be merely a number. The manager should ask if they are trying to say that the equipment has become worn to a point that it is interfering with productivity. It is important to listen for the meaning behind certain communications.

A Word About Advanced Communications Technology

Owners and facility managers can utilize proven new communication technologies to reengineer facility management practices. Communicating via fax and e-mail reduces the time period for delivering a project. Collaborative work technologies are producing major changes in the way construction projects are procured and facilities are designed, built, and managed.

In the past, long distance business relationships were relatively scarce and costly. It was not productive for designers and engineers, who were geographically separated, to collaborate on projects. Much time was wasted waiting for information that was sent via U.S. mail (snail mail), and it was difficult to process or decipher information sent on a piecemeal basis. Today, communications technologies enable organizations that may be geographically remote to work together on projects simultaneously.

Video conferencing enables face-to-face teamwork over distances. Internet access and groupware programs provide instantaneous electronic access to complex information from remote locations. Portions of a project can be assigned to businesses located across the globe in order to work on a project 24 hours a day. The services of lower-cost design and technical consultants from any locale can be retained to minimize design costs without compromising quality and performance. When consultants are required to travel by air, it is much more convenient and affordable than in prior years.

Capitalizing on developments in communication technologies provides a way to get more use out of less space. It is now a generally accepted practice for sales and consulting personnel who frequently work outside the main office to telecommute. By working primarily out of their homes, they reduce office space needs and the associated cost. When necessary they can reserve space at the main office. Facility managers who need to avoid unnecessary costs can take advantage of these same arrangements.

Computer-Aided Facility Management (CAFM) programs can be used to store, organize, and process large amounts of facilities data in a variety of ways. Computerizing facility management functions enables companies to systematize their operations, handle more inquiries, and establish uniform standards for maximum efficiency.

Computer-aided building management systems can be installed to monitor and control the condition of systems and equipment from remote locations. For example, in 1987, the resident operating engineer was required to be on the premises to monitor operation of the physical plant. Now the computer can perform many of these functions, and it will summon the mechanic when needed. When equipment is operating outside a control set point, the system automatically contacts the mechanic and other designated personnel via beeper or a communication alternative. Remote adjustments can be made through a laptop or other computer. Automating facility management functions can simultaneously improve performance and reduce overhead. Fewer personnel are required to operate and maintain an automated facility than are needed for a plant that is not automated.

The merging of computer and communication technology enables facility managers to maximize productivity of the work environment. This increase in automation is costly at first. It would be unreasonable to expect an immediate payback from an investment in automation, as the new systems must operate alongside the old ones for a period of time. However, the long-run benefits from productivity improvement may far outweigh the initial cost. Careful analysis is required before investing substantial funds in automating a facility management operation.

Team Building

Effective communication is the backbone of a productive, team-based work environment. Communication clearly plays an important role in team building and facility management. Make a rule of encouraging team members to participate in the planning and implementation of work or events that will affect them. There are three benefits to this approach: you may get some valuable ideas, team members will better understand the reasons for the decisions and actions taken, and they will see you are sensitive to their needs and motivations.

Assembling Contractors and Consultants for the Project Team

The way in which the team is formed is important if you want to optimize a team approach throughout a project. Consultants, contractors, and suppliers should be carefully selected and managed under the terms and conditions of well-defined contracts. When you retain these services, you are engaging in a partnership. You and they are partners working together to accomplish the common goals of the project.

Select the best qualified consultants, rather than emphasizing only low cost. By recognizing the value of qualified designers, engineers, construction managers, contractors, and suppliers, the owner reinforces the team approach, which serves the project's objectives.

Selection of consultants or contractors should be based on professional and technical qualifications. You may want to consider forming a committee to help you develop qualification requirements, and to prepare written requests for proposal forms. If there is any question about why the consultant was selected, you will have a basis for justifying your decision.

Owners should prequalify contractors and consultants, and require them to fill out standard questionnaires that request technical and financial information. References from their previous projects are invaluable. AIA Document A305, "Contractor's Qualification Statement," is a good form to pre-qualify bidders. It can be obtained from The American Institute of Architects' headquarters in Washington, D.C. Careful evaluation of consultants and construction firms prior to entering into a contract is a major factor in ensuring quality and performance, and can determine the owner's ability to obtain the best facility for the money.

In evaluating consultants, contractors, and suppliers, keep in mind that it is time-consuming and expensive for them to prepare proposals. You should not subject a consultant, contractor, or supplier to the hassle of preparing a proposal if they have little chance of getting the job. Some criteria for selecting consultants include:

- Experience
- Education
- References
- Subconsultants
- Project approach/new ideas
- Chemistry or fit with staff
- Financial health
- Insurance
- Familiarity with applicable facility systems and equipment
- Size of staff
- Years in business

Techniques for Team Decision-Making and Problem-Solving

Once the team is assembled and the project under way, there is an ongoing need for decisions, and for solutions to the problems that inevitably arise. Consensus decision-making is one of the most powerful tools at your disposal. Team decisions can minimize mistakes and disputes, and vastly increase productivity. It is a good idea to keep a few simple rules you can follow to diffuse conflicts between team members. For example:

- Affirm the opinions of the conflicting team members before expressing your perspective.
- Attack the problem or process, not the person's character.
- Speak the truth in a kind and respectful way.
- Separate facts from emotions.

The Function Analysis System Technique

The Function Analysis System Technique (FAST) applied in the value engineering process is a great example of team problem-solving. It gets a multi-disciplinary group of people together as a team to solve problems based on the analysis of functions. Function Analysis System Technique helps build a consensus among team members on what the problem is, how the problem will be solved, and why the selected corrective actions are being taken. This technique answers not only *how* something is being done, but *why*. Function Analysis System Technique is also a powerful tool to determine and rank priorities.

When this technique is applied in the value engineering process, customers can understand where their funds are being spent and can therefore decide which areas are most important when it comes to planning future uses for available funds. When an owner participates in the value engineering process, he or she will often seek trade-offs and set priorities that comply with budget parameters. This up-front communication helps avoid criticism when the project is under way or upon project completion, when "I wish I had done this or that" hindsight is common.

Cause and Effect Diagram

A practical tool for generating ideas and making decisions by consensus about a problem or issue is a Cause and Effect diagram, created by the project team.

This is often called a "fishbone" diagram because its lines resemble the skeleton of a fish. The basic problem, issue, or desired effect becomes the "head" of the fish. Then the team identifies *causes* behind the basic problem or issue. Frequently the major causes are grouped into four categories: *manpower, methods, materials*, and *machinery*. However, teams may choose their own categories to identify and classify the major causes. During the process of creating the fishbone diagram, the team uses creativity techniques such as brainstorming to identify causes within each of the four categories, using a minimum number of words, which then become additional "bones" in each category. Sometimes teams include words that describe the environment; these become the "tail" of the fish. Use of the fishbone diagram can help organizations find new ways to improve and fine-tune their operations. An example is shown in Figure 1.3.

Affinity Diagram

An Affinity Diagram is another useful tool for team members to gather and classify shared ideas. An Affinity Diagram is effective in collecting information on a specific problem, as shown in Figure 1.4. This format can also be used with larger groups or when the topic offers a wide variety of choices that require grouping. Follow these steps to create an Affinity Diagram:

1. Define the issues the team is to consider.
2. Generate ideas individually, writing clearly on slips of paper or special sticky-back note paper.
3. Use a brief description for each idea.
4. Sort recorded ideas (silently) into related groupings by spreading them out on a table or posting them on a wall.
5. Create new categories using new or existing ideas.
6. Look for patterns and reach a team consensus on the highest ranking ideas.

Positive/Negative Forces Analysis

A two-column Positive/Negative Forces Analysis can also be used to stimulate thinking. For example, the format may be used to receive the team's input regarding the positive and negative aspects (or costs and benefits) of proposed alternative solutions. Other uses for this technique include analyzing the positive and negative aspects of a meeting at its conclusion, or to communicate what team members want, and do not want, to occur on a project. Figure 1.5 is an example of a positive/negative forces analysis used to determine the feasibility of various site locations for a new facility.

Post-Mortem Team Review

At the end of every significant project, it is helpful to assemble project team members to conduct a post-project evaluation. Ask yourselves "Knowing what we know now about the project, if we had to do it over again, what would we do differently?" The Positive/Negative Forces Analysis is a useful tool for collecting this information.

In addition to significant planned projects, the post-mortem team review should also be applied to any project that requires emergency response from a contractor or in-house personnel. Taking a proactive stance and avoiding the

Figure 1.3

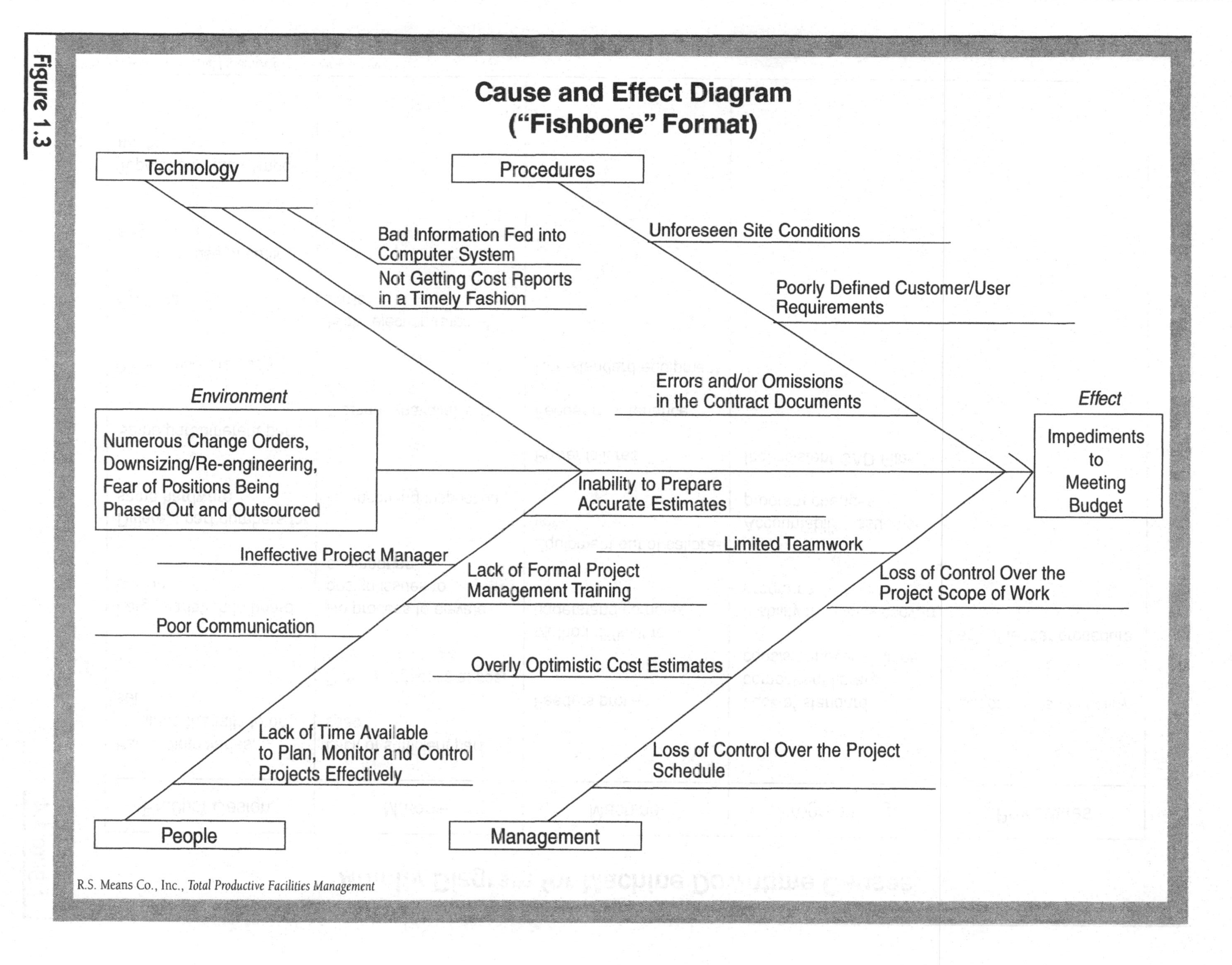
Cause and Effect Diagram
("Fishbone" Format)
Technology
Procedures
Bad Information Fed into Computer System
Not Getting Cost Reports in a Timely Fashion
Unforeseen Site Conditions
Poorly Defined Customer/User Requirements
Errors and/or Omissions in the Contract Documents
Environment
Numerous Change Orders, Downsizing/Re-engineering, Fear of Positions Being Phased Out and Outsourced
Effect
Impediments to Meeting Budget
Inability to Prepare Accurate Estimates
Limited Teamwork
Ineffective Project Manager
Lack of Formal Project Management Training
Loss of Control Over the Project Scope of Work
Poor Communication
Overly Optimistic Cost Estimates
Lack of Time Available to Plan, Monitor and Control Projects Effectively
Loss of Control Over the Project Schedule
People
Management

R.S. Means Co., Inc., *Total Productive Facilities Management*

Figure 1.4

Affinity Diagram for Machine Downtime Causes

Product Design	Material	Machine	Programming	Procedures
Panel width varies, requiring the rails to be set	Lack of sufficient part spec	Feeders broken	Lack of standard component library consistent over all lines	Lack of escalation policy
Large variation in board design	Purchasing looks at cost only	Method difficult to understand/complex	Inability to quickly convert programs	Lack of feeder procedure
Different part numbers for same hardware	No process to elevate design issues to engineering	Equipment out of calibra-tion	Accountability issues on program changes	
Same part/different part	No incoming inspection	Power failures	Inconsistent CAD Files	
0 ohm resistors used	Material unavailable to run	Feeder maintenance		
No DFM	High rejection rates of parts	Non-standard equipment		
Extensive use of cutting edge				
Top/bottom side compo-nents				

R.S. Means Co., Inc., *Total Productive Facilities Management*

necessity for work done under these circumstances represents a clear opportunity for savings because this kind of work often involves premium labor and equipment cost, and possible disruption of productivity.

Characteristics of Team Leaders

Decision-making and problem-solving skills are essential for productive facility management. Beyond this, managers need to make a lifelong commitment to cultivating teamwork and developing their own leadership abilities. In the pursuit of their profession, facility managers affect the quality of other people's lives. Conducting their work in an ethical manner is one of the fundamental ways they earn and maintain the confidence of team members, supervisors, peers, subordinates, suppliers, customers, and the general public.

The facility manager or project manager must not hide problems, acknowledging errors promptly, so that progress can be made toward a solution. Feeding bad information or withholding important information to try to control the situation is a good way to destroy credibility for the manager and his organization. Customers are entitled to accurate and timely facts so informed decisions can be made. Leaving out relevant information in status reports or misrepresenting potential costs to get a project approved can reduce the manager's credibility and effectiveness.

Positive/Negative Forces Analysis for New Facility

	+	–
Alternate 1	**Pole Building Site** Ease of truck access Distant from utilities "Flat" site	Requires extensive demolition Visibility No expansion capability Eliminates expansion capability for adjacent building Distance to parking Requires new storage space
Alternate 2	**Parking Lot Site** Ease of truck access Adjacent parking Visibility Allows space for future expansion "Flat" site	Consumes existing parking space Distant from utilities
Alternate 3	**Addition to Existing Building Site** Part of campus Proximity to utilities Interconnects buildings More aesthetically pleasing	Distance to parking More difficult truck access More difficult construction access Expansion limited by setback requirements Relocation of high voltage feeders may be required Eliminates planned expansion for other operations

R.S. Means Co., Inc., *Total Productive Facilities Management*

Figure 1.5

It is important to remember that the purpose of a team is to achieve an organization's objectives. A group of people might be inclined to choose someone with a dominant personality to become their team leader, with the idea that a take-charge type will get the job done. On the other hand, a domineering, authoritarian leader may lead the team down the wrong path. The best leaders want to serve and equip others.

Leaders who possess character value the people they lead and serve. Leaders of character think of themselves as servants, setting aside their own egos and self interests and inspiring other team members to act on behalf of the organization's greater goals. They demonstrate an appreciation for other team members' perspectives. They find ways to protect the interests of their followers and still achieve the collective goals of the organization.

Leaders should foster open, two-way communication. Commitment from team members is elicited by enabling them to participate jointly in analyzing problems and offering solutions. When a leader gives directions, he or she should provide opportunities for team members to ask questions in order to clarify what is expected of them. An effective leader also asks questions to uncover and resolve problems. He or she also reads body language to recognize signs of impatience, approval or disapproval, hesitance, confusion, or understanding.

Leaders must stay ahead of the game, including the competition. They strive to learn as much as they can from each personal and business situation. They understand the ultimate goal in learning is the application of that knowledge to overcome obstacles that will confront them in the future.

Effective leaders work to develop their own skill repertoire and empower others to expand theirs. One way to empower team members is by noticing their efforts and potential and encouraging them to persevere. Acknowledge each individual's accomplishments with a reward appropriate for his or her unique needs. Everyone needs to know that the work they do is appreciated and has merit.

To be an effective leader, one must have a vision of what needs to be accomplished. Leaders are goal-oriented. They plan their work and work their plan. Team members are often required to undertake projects that are unpleasant or difficult. They like to know what steps need to be taken to reach the objectives and how they will be rewarded for their efforts. Leaders motivate people to work toward a common goal.

Leaders must be good facilitators and use every meeting as an opportunity for team building. Sharing of experiences should be encouraged so team members can learn from and build on the knowledge of others. Leaders can tap into the knowledge and skills available through other people, empowering those individuals by initiating and facilitating team building workshops. Workshops may include group problem-solving activities such as Cause and Effect Analysis exercises, brainstorming, FAST Diagramming, Affinity Diagramming, and Positive/Negative Forces Analyses.

Effective leaders are aware of their own strengths and weaknesses, as well as those of other team members. Even if every person on a team should possess similar skills, the differences in timing, behavioral patterns, communication preferences, and motivation affect each person's ability to work productively in various environments.

Complete the form in Figure 1.6 to become more aware of your unique passions, work preferences, skills, and habits. Share the completed form with key members of your team, for their impressions. Repeat this process with all team members periodically. By looking closely at yourself and others, you will take a leadership role in developing a more productive team. Each team member needs to understand how he or she helps or hinders the team as a whole. The process enables each team member to understand individual differences and to identify and set dates for making changes that will encourage the effectiveness of the individual and the work team.

Power is not necessarily bestowed upon those with the most money, brains, or academic degrees. Rather it abides in the individual or organization people turn to for competent advice and help. Facility managers will achieve power and recognition by helping others do their jobs better and faster—and by facilitating the teamwork necessary for productive and safe working environments.

The first step is to commit to a process of ongoing learning and improvement. To provide the basis for development of strategies that will benefit both the facility manager and his or her organization, it is necessary to clarify the greater goals of the company on a continuing basis. Assess facilities performance from the perspectives of internal and external customers and compare your performance against your competition and industry best practices.

Business situation analyses, customer attitude surveys, space utilization studies, flow charts, benchmarking, and value engineering are practical methods for collecting, organizing, and analyzing data; identifying performance requirements; and determining the developmental changes needed to keep the entire organization healthy and vigorous. These methods incorporate multidisciplinary teams, cross-functional systems thinking, and creativity techniques to foster cooperation and consensus decision-making. Remember to evaluate facility management's contribution to corporate goals on a routine basis, because the performance targets and priorities are always changing.

Meticulously apply value engineering to define the functions of facilities, projects, and services. This will help determine lower cost ways to reliably provide the required functions. Customers pay for functions. Value engineering determines which functions and services are the most valuable and how much they are worth from the customers' perspective. Facility management organizations should be prepared to justify the value of their projects and services at all times. If the facility manager does not initiate the value studies, someone else in upper management may.

What do you like? (List work activities or areas that you enjoy doing or have a passion for.)

What do you dislike? (Identify those work activities that are less appealing or that you tend to avoid.)

Solicit input from your team members on the following items. Be fair, honest, and diplomatic in this exchange.

What are you good at? (List special knowledge, talents, and skills.)

What skills do you need to learn, improve, or develop?

Skills	Target Completion Date:

What are your best work habits?

What present work habits do you need to change, or what new habits do you need to develop?

Habits	Target Completion Date:

Team Member Awareness Form

R.S. Means Co., Inc., *Total Productive Facilities Management*

Figure 1.6

What are your working style tendencies? (Use the rating scale below for assistance in clarifying your style. Circle the number on the scale that best describes you.)

Structured									Unstructured
1	2	3	4	5	6	7	8	9	10
Decisive									Indecisive
1	2	3	4	5	6	7	8	9	10
Proactive									Reactive
1	2	3	4	5	6	7	8	9	10
Patient									Impatient
1	2	3	4	5	6	7	8	9	10
Organized									Disorganized
1	2	3	4	5	6	7	8	9	10
Concise									Expansive
1	2	3	4	5	6	7	8	9	10
Reserved									Outspoken
1	2	3	4	5	6	7	8	9	10
Detail-Oriented									Concept-Oriented
1	2	3	4	5	6	7	8	9	10
Impetuous									Deliberate
1	2	3	4	5	6	7	8	9	10
People-Oriented									Technically Oriented
1	2	3	4	5	6	7	8	9	10
Time-Oriented									Results-Oriented
1	2	3	4	5	6	7	8	9	10
Aggressive									Passive
1	2	3	4	5	6	7	8	9	10

Team Member Awareness Form cont.

R.S. Means Co., Inc., *Total Productive Facilities Management*

Figure 1.6 cont.

The next step is to create a project-driven facility management organization and employ a structured project management approach for implementing the desired changes. Business and financial conditions, regulatory requirements, technology developments, and competition influence the type of projects selected. In a competitive, resource-constrained environment, there will always be emphasis on projects that help improve the organization's cost, quality, and cycle-time management.

Matrix organizational structures, decision charts, work breakdown structures, programming and scope documents, cost estimates, CPM schedules, value engineering, responsibility charts, life cycle cost analysis and selection of the appropriate contracting and procurement methods are essential for effective project management. Supervision, use of feedback systems and reports, and visible, easily observable milestones are necessary to measure and control project progress. Some of the benefits of structuring an organization around the management of projects include: efficient allocation and control of limited resources, shorter development times, minimum cost, and improved quality and performance.

The final step is to improve productivity through communications and teamwork. Many organizations can dramatically improve effectiveness and productivity by making improvements in these areas. Management alone cannot keep a company competitive anymore. Everyone needs to share a sense of responsibility for the survival and success of the company.

Today, most organizations are unwilling to empower a single individual to make high risk decisions that impact the competitiveness and return on expensive facility assets. The high stakes, increased number of stakeholders, and the complexity of required tasks, skills, technology and knowledge require the involvement of a diverse mix of people from inside and outside the organization.

The broad task of the facility manager or engineer is to facilitate the creation of flexible teams that can manage projects and respond effectively to new workplace challenges. Managers need to work continuously to improve team effectiveness, share ideas, and develop esprit de corps. Many individuals who have been informed that they are critical components of a team do not feel that they are really working together as a team to solve shared problems in the most efficient and effective manner. Changing current methods of working and organizing company resources takes teamwork, team building, team incentives, and shared information. Conduct team-building workshops using tools such as cause and effect diagrams, affinity diagrams, and positive and negative forces analyses.

A fluid and efficient business organization requires timely and effective communication. Through telephones, pagers, e-mail, facsimiles, and express deliveries, distance is practically eliminated as a barrier to communication. Dissemination of the right information to the right individuals at the right time improves the probability that a project will be successful.

The best facility engineers and managers are equipped with a variety of methods which they continually fine-tune and adapt to fit the specific needs of each situation. They have control over both the cost and productivity of facility resources and foster efficient and effective working environments in which their team will flourish and function well.

Resource Publications

Drucker, Peter F., *Managing for the Future*. New York, NY: Penguin Group, 1992.

Jablonski, Joseph R., *Implementing TQM*. Albuquerque, NM: Technical Management Consortium, Inc., 1994.

Sievert, Richard W., *Total Productive Facilities Management*. Kingston, MA: R.S. Means Co., Inc., 1998.

Stuckenbruck, Linn C., *Team Building for Product Managers*. Drexel Hill, PA: Project Management Institute, 1988.

Whetten, David and Kim Cameron, *Developing Management Skills, Fourth Edition*. Reading, MA: Addison-Wesley Publishing, Inc., 1998.

For Additional Information

American Management Association
1601 Broadway
New York, NY 10019-7420
212-586-8100
www.amanet.org

The American Management Association is a non-profit, membership-based organization that assists individuals and enterprises in the development of organizational effectiveness. It publishes more than 80 business-related books a year, and also issues educational materials in audio, video, and electronic formats.

Association of Higher Education Facilities Officers (APPA)
1643 Prince Street
Alexandria, VA 22314-2818
703-684-1446
www.appa.org

APPA is an international association dedicated to maintaining and promoting the quality of educational facilities. APPA conducts research, presents educational programs, produces publications, and serves as a central information source for its members.

International Facility Management Association
One E. Greenway Plaza, Suite 1100
Houston, TX 77046-0194
713-623-4362
www.ifma.org

IFMA represents and supports facility management as a profession. It forecasts trends in management and provides workplace education through self-study, seminars, and an on-line education program.

Chapter 2

Engineering Economics

Chapter 2 Engineering Economics

Engineering Economy, or 'Engineering Economics,' as it is more commonly called, uses mathematical and economical concepts to analyze alternative economic courses of action utilizing an engineering framework. The underlying assumption to the subject is that money has added value over time (TVM). Money having added value over time is not a new concept. Business majors have studied this exact subject for many years under the title "Principles of Finance," wherein the economic investment alternatives are choices between fixed income (bonds) and equity (stocks) rather than tangible assets, like centrifugal compressors and reciprocating compressors. In either case, the investor/plant manager is constantly evaluating alternatives, over time (t), with an interest rate (i), under the assumption that that same interest rate (i) is available over the complete life of the investment.

Although the word "economics" is part of the title of this subject, there is very little economics associated with this subject. This terminology came about because there was a growing demand to have facility engineering managers skilled in the formulas of finance. A more accurate term might be "engineering investments."

Interest Rates

It should also be pointed out that the term "interest rate" has taken on many different names and acronyms (with slightly different connotations) in the world of finance and economics, and some of these names and acronyms have slightly different connotations. Some of the many terms and definitions for "interest rate" are as follows:

<table>
<tr><th>Terms:</th><th>Meaning:</th></tr>
<tr><td>Interest Rate</td><td>The interest rate an investment earned</td></tr>
<tr><td>Rate of Return</td><td>The interest rate an investment earned</td></tr>
<tr><td>Internal Rate of Return</td><td>The interest rate which makes the net present value of an investment equal zero</td></tr>
<tr><td>The Discount Rate</td><td>The "risk free interest rate" offered by lenders or the interest rate that the federal reserve bank loans money to other banks</td></tr>
<tr><td>Yield</td><td>The interest rate earned on an investment, usually a fixed-income investment, such as bonds</td></tr>
<tr><td>Yield to Maturity (YTM)</td><td>The interest earned on an investment, usually a fixed income investment, such as bonds</td></tr>
<tr><td>The Risk Free Rate</td><td>The "risk-free interest rate" offered by lenders</td></tr>
<tr><td>Return</td><td>The interest rate an investment earned or the dollar amount an investment earned[1]</td></tr>
<tr><td>Capitalization Rate</td><td rowspan="6">The market rate of interest offered by lenders</td></tr>
<tr><td>Going Rate of Interest</td></tr>
<tr><td>Cost of Money</td></tr>
<tr><td>Cost of Capital</td></tr>
<tr><td>Discounted Cost of Capital</td></tr>
<tr><td>Market Rate of Interest</td></tr>
<tr><td>Minimum Attractive Rate of Return (MARR)</td><td>A predetermined rate of return arbitrarily set by the investor(s)</td></tr>
<tr><td>Incremental Rate of Return</td><td>The difference in the rate of return between one investment and the next more expensive investment</td></tr>
</table>

The ability to compare alternatives with differing cost characteristics is essential if we are to select the option that has the most favorable financial impact. The key to making such comparisons is the ability to convert present, future, and repeated cash flows into one equivalent value at a certain point in time (present value or future value), or over time (an annuity). Through the use of financial theory—mainly the time value of money—and engineering economic concepts, financial equivalency can be determined and alternatives compared effectively.

All cash flows can be expressed as present, future, or repeated expenditures. To determine equivalency you need to know the applicable **interest rate**, the **number of time periods** over which that rate will be applied, and the appropriate **conversion formula.** It is also very convenient to draw a horizontal time line showing cash receipts as upward vertical vectors (arrows) and cash disbursements as downward vertical vectors (arrows). Each would denote the end of the period in which the cash receipt or disbursement occurred, with the length of the vector (arrow) proportionate to the cash amount.

Choosing between alternatives is most often done with the assumption that money has added value over time. There are many methods of money management which companies use that employ a TVM conceptual framework. Some of these are:

1. Value engineering
2. Project justification
3. Life cycle cost analysis
4. Capital asset management
5. Capital budgeting.

If TVM methods are not employed to optimize decision-making, companies may use "simple payback analysis" where the cost of each alternative is divided by the benefit per year. Thus, payback analysis does not incorporate the time value of money; it simply tells the investor the number of years before the investment has "paid for itself."

Example 1

A new, more efficient electric motor is purchased to replace an older, less efficient motor. The cost of the motor is $10,000, and the savings per year (due to the higher efficiency) are $2,000: Simply dividing $10,000 by $2,000/year gives five years as the period of time it takes to repay the investment. Using simple payback analysis then, the motor would "pay for itself" in five years. Most engineering economists agree that simple payback analysis is a very rough estimator, and that using TVM analysis is a much more effective tool.

In solving problems with the following formulas, many choose to use the tables rather than work out the problems. While the formulas are not very easy to apply, the use of the tables can simplify the math significantly. However, the tables usually do not cover all interest rates or all periods. Thus, if you want to use the tables for all problems, certain problems with odd interest rates (like 7.22 percent), can only be solved via interpolation. Fortunately, technology is readily available and relatively inexpensive. Menu-driven calculators, with built-in "time value of money" and "cash flow" functions can accommodate any interest rate calculations and enable you to quickly and easily perform "what if-type" reviews. They also allow you to easily solve for other variables, such as the interest rate or the number of periods, where often, the only alternative method is an iterative, arduous process. (Figures 2.1–2.4 are examples of compounding annual interest. See Appendix for additional examples and instructions on creating a loan amortization spreadsheet.)

Table E.9 4% Interest Factors for Annual Compounding Interest

n	Single Payment: Compound-Amount Factor	Single Payment: Present-Worth Factor	Equal-Payment Series: Compound-Amount Factor	Equal-Payment Series: Sinking-Fund Factor	Equal-Payment Series: Present-Worth Factor	Equal-Payment Series: Capital-Recovery Factor	Uniform Gradient-Series Factor
	To Find F Given P F/P, i, n	*To Find P Given F* P/F, i, n	*To Find F Given A* F/A, i, n	*To Find A Given F* A/F, i, n	*To Find P Given A* P/A, i, n	*To Find A Given P* A/P, i, n	*To Find A Given G* A/G, i, n
1	1.040	0.9615	1.000	1.0000	0.9615	1.0400	0.0000
2	1.082	0.9246	2.040	0.4902	1.8861	0.5302	0.4902
3	1.125	0.8890	3.122	0.3204	2.7751	0.3604	0.9739
4	1.170	0.8548	4.246	0.2355	3.6299	0.2755	1.4510
5	1.217	0.8219	5.416	0.1846	4.4518	0.2246	1.9216
6	1.265	0.7903	6.633	0.1508	5.2421	0.1908	2.3857
7	1.316	0.7599	7.898	0.1266	6.0021	0.1666	2.8433
8	1.369	0.7307	9.214	0.1085	6.7328	0.1485	3.2944
9	1.423	0.7026	10.583	0.0945	7.4353	0.1345	3.7391
10	1.480	0.6756	12.006	0.0833	8.1109	0.1233	4.1773
11	1.539	0.6496	13.486	0.0742	8.7605	0.1142	4.6090
12	1.601	0.6246	15.026	0.0666	9.3851	0.1066	5.0344
13	1.665	0.6006	16.627	0.0602	9.9857	0.1002	5.4533
14	1.732	0.5775	18.292	0.0547	10.5631	0.0947	5.8659
15	1.801	0.5553	20.024	0.0500	11.1184	0.0900	6.2721
16	1.873	0.5339	21.825	0.0458	11.6523	0.0858	6.6720
17	1.948	0.5134	23.698	0.0422	12.1657	0.0822	7.0656
18	2.026	0.4936	25.645	0.0390	12.6593	0.0790	7.4530
19	2.107	0.4747	27.671	0.0361	13.1339	0.0761	7.8342
20	2.191	0.4564	29.778	0.0336	13.5903	0.0736	8.2091
21	2.279	0.4388	31.969	0.0313	14.0292	0.0713	8.5780
22	2.370	0.4220	34.248	0.0292	14.4511	0.0692	8.9407
23	2.465	0.4057	36.618	0.0273	14.8569	0.0673	9.2973
24	2.563	0.3901	39.083	0.0256	15.2470	0.0656	9.6479
25	2.666	0.3751	41.646	0.0240	15.6221	0.0640	9.9925
26	2.772	0.3607	44.312	0.0226	15.9828	0.0626	10.3312
27	2.883	0.3468	47.084	0.0212	16.3296	0.0612	10.6640
28	2.999	0.3335	49.968	0.0200	16.6631	0.0600	11.9909
29	3.119	0.3207	52.966	0.0189	16.9837	0.0589	11.3121
30	3.243	0.3083	56.085	0.0178	17.2920	0.0578	11.6274
31	3.373	0.2965	59.328	0.0169	17.5885	0.0569	11.9371
32	3.508	0.2851	62.701	0.0160	17.8736	0.0560	12.2411
33	3.648	0.2741	66.210	0.0151	18.1477	0.0551	12.5396
34	3.794	0.2636	69.858	0.0143	18.4112	0.0543	12.8325
35	3.946	0.2534	73.652	0.0136	18.6646	0.0536	13.1199
40	4.801	0.2083	95.026	0.0105	19.7928	0.0505	14.4765
45	5.841	0.1712	121.029	0.0083	20.7200	0.0483	15.7047
50	7.107	0.1407	152.667	0.0066	21.4822	0.0466	16.8123
55	8.646	0.1157	191.159	0.0052	22.1086	0.0452	17.8070
60	10.520	0.0951	237.991	0.0042	22.6235	0.0442	18.6972
65	12.799	0.0781	294.968	0.0034	23.0467	0.0434	19.4909
70	15.572	0.0642	364.290	0.0028	23.3945	0.0428	20.1961
75	18.945	0.0528	448.631	0.0022	23.6804	0.0422	20.8206
80	23.050	0.0434	551.245	0.0018	23.9154	0.0418	21.3719
85	28.044	0.0357	676.090	0.0015	24.1085	0.0415	21.8569
90	34.119	0.0293	817.983	0.0012	24.2673	0.0412	22.2826
95	41.511	0.0241	1012.785	0.0010	24.3978	0.0410	22.6550
100	50.505	0.0198	1237.624	0.0008	24.5050	0.0408	22.9800

Table E.10 5% Interest Factors for Annual Compounding Interest

n	Single Payment: Compound-Amount Factor	Single Payment: Present-Worth Factor	Equal-Payment Series: Compound-Amount Factor	Equal-Payment Series: Sinking-Fund Factor	Equal-Payment Series: Present-Worth Factor	Equal-Payment Series: Capital-Recovery Factor	Uniform Gradient-Series Factor
	To Find F Given P F/P, i, n	*To Find P Given F* P/F, i, n	*To Find F Given A* F/A, i, n	*To Find A Given F* A/F, i, n	*To Find P Given A* P/A, i, n	*To Find A Given P* A/P, i, n	*To Find A Given G* A/G, i, n
1	1.050	0.9524	1.000	1.0000	0.9524	1.0500	0.0000
2	1.103	0.9070	2.050	0.4878	1.8594	0.5378	0.4878
3	1.158	0.8638	3.153	0.3172	2.7233	0.3672	0.9675
4	1.216	0.8227	4.310	0.2320	3.5460	0.2820	1.4391
5	1.276	0.7835	5.526	0.1810	4.3295	0.2310	1.9025
6	1.340	0.7462	6.802	0.1470	5.0757	0.1970	2.3579
7	1.407	0.7107	8.142	0.1228	5.7864	0.1728	2.8052
8	1.477	0.6768	9.549	0.1047	6.4632	0.1547	3.2445
9	1.551	0.6446	11.027	0.0907	7.1078	0.1407	3.6758
10	1.629	0.6139	12.587	0.0795	7.7217	0.1295	4.0991
11	1.710	0.5847	14.207	0.0704	8.3064	0.1204	4.5145
12	1.796	0.5568	15.917	0.0628	8.8633	0.1128	4.9219
13	1.866	0.5303	17.713	0.0565	9.3936	0.1065	5.3215
14	1.980	0.5051	19.599	0.0510	9.8987	0.1010	5.7133
15	2.079	0.4810	21.579	0.0464	10.3797	0.0964	6.0973
16	2.183	0.4581	23.658	0.0423	10.8378	0.0923	6.4736
17	2.292	0.4363	25.840	0.0387	11.2741	0.0887	6.8423
18	2.407	0.4155	28.132	0.0356	11.6896	0.0856	7.2034
19	2.527	0.3957	30.539	0.0328	12.0853	0.0828	7.5569
20	2.653	0.3769	33.066	0.0303	12.4622	0.0803	7.9030
21	2.786	0.3590	35.719	0.0280	12.8212	0.0780	8.2416
22	2.925	0.3419	38.505	0.0260	13.1630	0.0760	8.5730
23	3.072	0.3256	41.430	0.0241	13.4886	0.0741	8.8971
24	3.225	0.3101	44.502	0.0225	13.7987	0.0725	9.2140
25	3.386	0.2953	47.727	0.0210	14.0940	0.0710	9.5238
26	3.556	0.2813	51.113	0.0196	14.3752	0.0696	9.8266
27	3.733	0.2679	54.669	0.0183	14.6430	0.0683	10.1224
28	3.920	0.2551	58.403	0.0171	14.8981	0.0671	10.4114
29	4.116	0.2430	62.323	0.0161	15.1411	0.0661	10.6936
30	4.322	0.2314	66.439	0.0151	15.3725	0.0651	10.9691
31	4.538	0.2204	70.761	0.0141	15.5928	0.0641	11.2381
32	4.765	0.2099	75.299	0.0133	15.8027	0.0633	11.5005
33	5.003	0.1999	80.064	0.0125	16.0026	0.0625	11.7566
34	5.253	0.1904	85.067	0.0118	16.1929	0.0618	12.0063
35	5.516	0.1813	90.320	0.0111	16.3742	0.0611	12.2498
40	7.040	0.1421	120.800	0.0083	17.1591	0.0583	13.3775
45	8.985	0.1113	159.700	0.0063	17.7741	0.0563	14.3644
50	11.467	0.0872	209.348	0.0048	18.2559	0.0548	15.2233
55	14.636	0.0683	272.713	0.0037	18.6335	0.0537	15.9665
60	18.679	0.0535	353.584	0.0028	18.9293	0.0528	16.6062
65	23.840	0.0420	456.798	0.0022	19.1611	0.0522	17.1541
70	30.426	0.0329	588.529	0.0017	19.3427	0.0517	17.6212
75	38.833	0.0258	756.654	0.0013	19.4850	0.0513	18.0176
80	49.561	0.0202	971.229	0.0010	19.5965	0.0510	18.3526
85	63.254	0.0158	1245.087	0.0008	19.6838	0.0508	18.6346
90	80.730	0.0124	1594.607	0.0006	19.7523	0.0506	18.8712
95	103.035	0.0097	2040.694	0.0005	19.8059	0.0505	19.0689
100	131.501	0.0076	2610.025	0.0004	19.8479	0.0504	19.2337

Figure 2.1-2.2

Figure 2.3-2.4

Table E.13 8% Interest Factors for Annual Compounding Interest

n	Single Payment		Equal-Payment Series				Uniform
	Compound-Amount Factor	Present-Worth Factor	Compound-Amount Factor	Sinking-Fund Factor	Present-Worth Factor	Capital-Recovery Factor	Gradient-Series Factor
	To Find F Given P F/P, i, n	To Find P Given F P/F, i, n	To Find F Given A F/A, i, n	To Find A Given F A/F, i, n	To Find P Given A P/A, i, n	To Find A Given P A/P, i, n	To Find A Given G A/G, i, n
1	1.080	0.9259	1.000	1.0000	0.9259	1.0800	0.0000
2	1.166	0.8573	2.080	0.4808	1.7833	0.5608	0.4808
3	1.260	0.7938	3.246	0.3080	2.5771	0.3880	0.9488
4	1.360	0.7350	4.506	0.2219	3.3121	0.3019	1.4040
5	1.469	0.6806	5.867	0.1705	3.9927	0.2505	1.8465
6	1.587	0.6302	7.336	0.1363	4.6229	0.2163	2.2764
7	1.714	0.5835	8.923	0.1121	5.2064	0.1921	2.6937
8	1.851	0.5403	10.637	0.0940	5.7466	0.1740	3.0985
9	1.999	0.5003	12.488	0.0801	6.2469	0.1601	3.4910
10	2.159	0.4632	14.487	0.0690	6.7101	0.1490	3.8713
11	2.332	0.4289	16.645	0.0601	7.1390	0.1401	4.2395
12	2.518	0.3971	18.977	0.0527	7.5361	0.1327	4.5958
13	2.720	0.3677	21.495	0.0465	7.9038	0.1265	4.9402
14	2.937	0.3405	24.215	0.0413	8.2442	0.1213	5.2731
15	3.172	0.3153	27.152	0.0368	8.5595	0.1168	5.5945
16	3.426	0.2919	30.324	0.0330	8.8514	0.1130	5.9046
17	3.700	0.2703	33.750	0.0296	9.1216	0.1096	6.2038
18	3.996	0.2503	37.450	0.0267	9.3719	0.1067	6.4920
19	4.316	0.2317	41.446	0.0241	9.6036	0.1041	6.7697
20	4.661	0.2146	45.762	0.0219	9.8182	0.1019	7.0370
21	5.034	0.1987	50.423	0.0198	10.0168	0.0998	7.2940
22	5.437	0.1840	55.457	0.0180	10.2008	0.0980	7.5412
23	5.871	0.1703	60.893	0.0164	10.3711	0.0964	7.7786
24	6.341	0.1577	66.765	0.0150	10.5288	0.0950	8.0066
25	6.848	0.1460	73.106	0.0137	10.6748	0.0937	8.2254
26	7.396	0.1352	79.954	0.0125	10.8100	0.0925	8.4352
27	7.988	0.1252	87.351	0.0115	10.9352	0.0915	8.6363
28	8.627	0.1159	95.339	0.0105	11.0511	0.0905	8.8289
29	9.317	0.1073	103.966	0.0096	11.1584	0.0896	9.0133
30	10.063	0.0994	113.283	0.0088	11.2578	0.0888	9.1897
31	10.868	0.0920	123.346	0.0081	11.3498	0.0881	9.3584
32	11.737	0.0852	134.214	0.0075	11.4350	0.0875	9.5197
33	12.626	0.0789	145.951	0.0069	11.5139	0.0869	9.6737
34	13.690	0.0731	158.627	0.0063	11.5869	0.0863	9.8208
35	14.785	0.0676	172.317	0.0058	11.6546	0.0858	9.9611
40	21.725	0.0460	259.057	0.0039	11.9246	0.0839	10.5699
45	31.920	0.0313	386.506	0.0026	12.1084	0.0826	11.0447
50	46.902	0.0213	573.770	0.0018	12.2335	0.0818	11.4107
55	68.914	0.0145	848.923	0.0012	12.3186	0.0812	11.6902
60	101.257	0.0099	1253.213	0.0008	12.3766	0.0808	11.9015
65	148.780	0.0067	1847.248	0.0006	12.4160	0.0806	12.0602
70	218.606	0.0046	2720.080	0.0004	12.4428	0.0804	12.1783
75	321.205	0.0031	4002.557	0.0003	12.4611	0.0803	12.2658
80	471.955	0.0021	5886.935	0.0002	12.4735	0.0802	12.3301
85	693.456	0.0015	8655.706	0.0001	12.4820	0.0801	12.3773
90	1018.915	0.0010	12723.939	0.0001	12.4877	0.0801	12.4116
95	1497.121	0.0007	18701.507	0.0001	12.4917	0.0801	12.4365
100	2199.761	0.0005	27484.515	0.0001	12.4943	0.0800	12.4545

Table E.14 9% Interest Factors for Annual Compounding Interest

n	Single Payment		Equal-Payment Series				Uniform
	Compound-Amount Factor	Present-Worth Factor	Compound-Amount Factor	Sinking-Fund Factor	Present-Worth Factor	Capital-Recovery Factor	Gradient-Series Factor
	To Find F Given P F/P, i, n	To Find P Given F P/F, i, n	To Find F Given A F/A, i, n	To Find A Given F A/F, i, n	To Find P Given A P/A, i, n	To Find A Given P A/P, i, n	To Find A Given G A/G, i, n
1	1.090	0.9174	1.000	1.0000	0.9174	1.0900	0.0000
2	1.188	0.8417	2.090	0.4785	1.7591	0.5685	0.4785
3	1.295	0.7722	3.278	0.3051	2.5313	0.3951	0.9426
4	1.412	0.7084	4.573	0.2187	3.2397	0.3087	1.3925
5	1.539	0.6499	5.985	0.1671	3.8897	0.2571	1.8282
6	1.677	0.5963	7.523	0.1329	4.4859	0.2229	2.2498
7	1.828	0.5470	9.200	0.1087	5.0330	0.1987	2.6574
8	1.993	0.5019	11.028	0.0907	5.5348	0.1807	3.0512
9	2.172	0.4604	13.021	0.0768	5.9953	0.1668	3.4312
10	2.367	0.4224	15.193	0.0658	6.4177	0.1558	3.7978
11	2.580	0.3875	17.560	0.0570	6.8052	0.1470	4.1510
12	2.813	0.3555	20.141	0.0497	7.1607	0.1397	4.4910
13	3.066	0.3262	22.953	0.0436	7.4869	0.1336	4.8182
14	3.342	0.2993	26.019	0.0384	7.7862	0.1284	5.1326
15	3.642	0.2745	29.361	0.0341	8.0607	0.1241	5.4346
16	3.970	0.2519	33.003	0.0303	8.3126	0.1203	5.7245
17	4.328	0.2311	36.974	0.0271	8.5436	0.1171	6.0024
18	4.717	0.2120	41.301	0.0242	8.7556	0.1142	6.2687
19	5.142	0.1945	46.018	0.0217	8.9501	0.1117	6.5236
20	5.604	0.1784	51.160	0.0196	9.1286	0.1096	6.7675
21	6.109	0.1637	56.765	0.0176	9.2923	0.1076	7.0006
22	6.659	0.1502	62.873	0.0159	9.4424	0.1059	7.2232
23	7.258	0.1378	69.532	0.0144	9.5802	0.1044	7.4358
24	7.911	0.1264	76.790	0.0130	9.7066	0.1030	7.6384
25	8.623	0.1160	84.701	0.0118	9.8226	0.1018	7.8316
26	9.399	0.1064	93.324	0.0107	9.9290	0.1007	8.0156
27	10.245	0.0976	102.723	0.0097	10.0266	0.0997	8.1906
28	11.167	0.0896	112.968	0.0089	10.1161	0.0989	8.3572
29	12.172	0.0822	124.135	0.0081	10.1983	0.0981	8.5154
30	13.268	0.0754	136.308	0.0073	10.2737	0.0973	8.6657
31	14.462	0.0692	149.575	0.0067	10.3428	0.0967	8.8083
32	15.763	0.0634	164.037	0.0061	10.4063	0.0961	8.9436
33	17.182	0.0582	179.800	0.0056	10.4645	0.0956	9.0718
34	18.728	0.0534	196.982	0.0051	10.5178	0.0951	9.1933
35	20.414	0.0490	215.711	0.0046	10.5668	0.0946	9.3083
40	31.409	0.0318	337.882	0.0030	10.7574	0.0930	9.7957
45	48.327	0.0207	525.859	0.0019	10.8812	0.0919	10.1603
50	74.358	0.0135	815.084	0.0012	10.9617	0.0912	10.4295
55	114.408	0.0088	1260.092	0.0008	11.0140	0.0908	10.6261
60	176.031	0.0057	1944.79	0.0005	11.0480	0.0905	10.7683
65	270.846	0.0037	2998.28	0.0003	11.0701	0.0903	10.8702
70	416.730	0.0024	4619.22	0.0002	11.0845	0.0902	10.9427
75	641.191	0.0016	7113.23	0.0002	11.0938	0.0902	10.9940
80	986.552	0.0010	10950.57	0.0001	11.0999	0.0901	11.0299
85	1517.932	0.0007	16854.80	0.0001	11.1038	0.0901	11.0551
90	2335.527	0.0004	25939.18	0.0001	11.1064	0.0900	11.0726
95	3593.497	0.0003	39916.63	0.0000	11.1080	0.0900	11.0847
100	5529.041	0.0002	61422.67	0.0000	11.1091	0.0900	11.0930

Example 2

An individual deposits $10,000 dollars and wants to withdraw $2,000 dollars at the end of every year for nine years, such that at the end of nine years the balance will be zero. What interest rate must the $10,000 earn for this to happen? To solve this by hand, or using the tables, you must guess and then guess again until you close in on the answer. The correct calculator iterates for you until it arrives at the answer, which is 13.7%. Certain spreadsheet software programs like Excel™ and Lotus® have built-in financial functions for more complicated engineering economic analysis.

Time Value of Money Analysis

The subject of engineering economics has one essential concept: that money (i.e., costs or investments) has added value over time. The simple question, "would you rather have a dollar today or a dollar tomorrow?" is always answered by the rational investor as "today." Because the rational investor knows that they can invest today's dollar and have more than one dollar by tomorrow. Similarly, to compare any cost or investment, the facility manager must compare them at the same point or points in time. Using three fundamental forms of "Time Value of Money" analyses:

1. The Present Worth or Present Value method
2. The Uniform Series method, often referred to by many other names, such as:
    - Equivalent Uniform Annual Cost (EUAC)
    - Equivalent Uniform Annual Benefit (EUAB)
    - Net Uniform Series (NUS)
    - Payment (PMT)
    - An Annuity (A)
3. The Future Worth or Future Value method

When people borrow money, they must pay "rent" on what they borrowed. The amount of rent is determined by the interest rate per period that the lender charges. In most cases, this is an annual effective rate.

It should also be noted that the all receipts and disbursements (except for initial project costs) in engineering economic analysis take place at the *end* of the period in which they occur. Usually, this is at the end of the year, but it could be the end of the day, the week, or month. There are formulas that assume the "beginning of the period" analysis, but these have been omitted from most engineering economics textbooks. Other assumptions that are often made in engineering economic analysis problems are:

- Asset replacement costs remain constant over time
- No inflation
- Income taxes are not included in the analysis
- The money for either alternative is available
- Any excess money continues to earn interest at the given rate
- All **initial** costs or receipts occur at the *beginning* of the first period
- Nonquantifiable measures are not included in the analysis.

Cash Flows

In solving engineering economic problems it is helpful to represent the cash flows graphically.

Example 3

A plant manager purchases a new processing machine for $1,600 that generates a maintenance savings at the end of year one of $400. This maintenance savings increases by $400 each year for the next four years. The five-year cash flow time line can be shown graphically (see Figure 2.5). Notice that the x-axis represents time (n), or more specifically, compounding-interest time periods. Cash receipts are shown as upward arrows, while cash disbursements are shown as downward arrows.

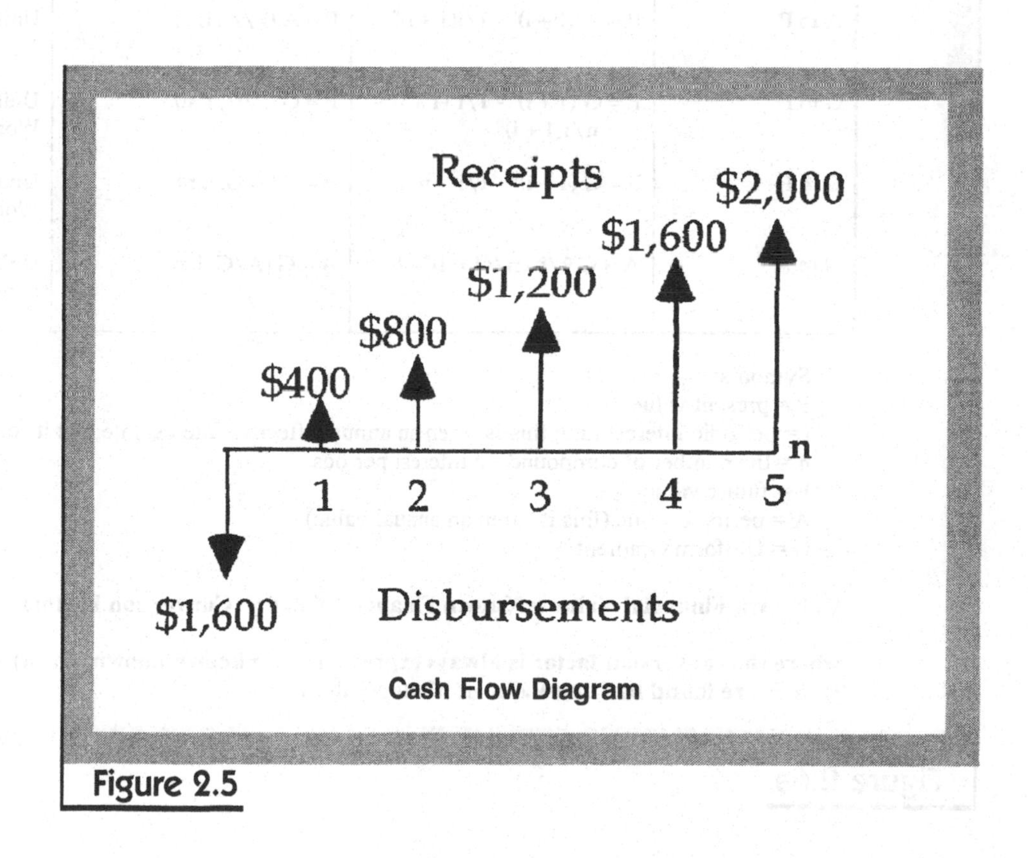

Figure 2.5

Functional Notation

The formulas for discrete compounding are given in Figure 2.6a. Figure 2.6b puts these formulas to use. One can see that the formulas can be quite cumbersome. Therefore, most engineering economic textbooks have adopted a functional notation that provides discounting factors to help solve the problem using tables rather than calculating the formula.

General Financial Formulas

Converts	Long-Hand Formula	Functional Formula	Conversion Factor Name
P to F	$F = P(1 + i)^n$	F = P (F/P, i, n)	Single Payment Compound Amount
F to P	$P = F(1 + i)^{-n}$	P = F (P/F, i, n)	Single Payment Present Worth
F to A	$A = F\, i/(1 + i)^n - 1$	A = F (A/F, i, n)	Uniform Series Sinking Fund
P to A	$A = P\, i(1 + i)^n/(1 + i)^n - 1$	A = P (A/P, i, n)	Capital Recovery
A to F	$F = A\,(1 + i)^n - 1/i$	F = A (F/A, i, n)	Uniform Series Compound Amount
A to P	$P = A\,(1 + i)^n - 1/i(1 + i)^n$	P = A (P/A, i, n)	Uniform Series Present Worth
G to P	$P = G\,(1 + i)^n - 1/i^2(1 + i)^n - n/i(1 + i)^n$	P = G (P/G, i, n)	Uniform Gradient Present Worth
G to F	$F = G\,(1 + i)^n - 1/i^2 - n/i$	F = G (F/G, i, n)	Uniform Gradient Future Worth
G to A	$A = G\, 1/i - n/(1 = i)^n - 1$	A = G (A/G, i, n)	Uniform Gradient Uniform Series

Symbols:
P = present value
i = periodic interest rate; this is often an annual effective interest rate, but it could be any periodic rate
n = the number of compounding interest periods
F = future value
A = periodic value (this is often an annual value)
G = Uniform Gradient

Unknown Financial Value = Known Financial Value × (Conversion Factor)

where the conversion factor is always expressed as (unknown/known, i%, n) and the conversion factors are found in Time Value of Money Tables.

Figure 2.6a

Using Financial Formulas

1) How much would you have at the end of 8 years if you invested $1000 today at 5%?

Solution: Find the future worth of a present amount

$F = P \times (F/P, 5\%, 8 \text{ years}) = \$1000 \times 1.4775 = \$1{,}477.50$

2) How much do you need to invest today at 5% to have $10,000 at the end of 10 years?

Solution: Find the present worth of a future amount

$P = F \times (P/F, 5\%, 10 \text{ years}) = \$10{,}000 \times .6319 = \$6{,}139$

3) If you invest $100,000 today at 7%, how much can you withdraw each year for the next 20 years?

Solution: Find the annuity from a present amount

$A = P \times (A/P, 7\%, 20 \text{ years}) = \$100{,}000 \times .0944 = \$9{,}440$

4) You are searching for a new house. You have determined that you can pay $10,000/year toward a mortgage. If you take a 30-year loan, how much can you borrow if the interest rate is 8%?

Solution: Find the present worth of an annuity

$P = A \times (P/A, 8\%, 30 \text{ years}) = \$10{,}000 \times 11.2578 = \$112{,}578$

5) If you need $100,000 in 15 years to fund your children's education, how much do you have to invest each year at 5% to accumulate that amount?

Solution: Find the annuity for a future amount

$A = F \times (A/F, 5\%, 15 \text{ years}) = \$100{,}000 \times .0463 = \$4{,}630$

6) If you won the lottery and will receive $10,000/year (after taxes) for the next 20 years. If you invest all of this money at 6%, how much will you have at the end of 20 years?

Solution: Find the future worth of an annuity

$F = A \times (F/A, 6\%, 20 \text{ years}) = \$10{,}000 \times 36.7856 - \$367{,}856$

Figure 2.6b

Example 4

A facility engineering manager is contemplating the purchase of a new pump. The vendor says that the buyer can purchase the pump for either $10,000 today or $14,000 at the end of five years. Assuming that the risk-free annual effective interest rate is 6%, what method of payment should the manager choose? (See Figure 2.7, an illustration of the two cash flow choices.)

Answer:

The solution to the problem is to choose the method which minimizes the present cost. This problem could actually be answered by utilizing any of the three methods mentioned earlier. Solving it by the present worth method is as follows:

Present cost of the first payment method is: $P_1 = \$10{,}000$

Present cost of the second payment method is: $P_2 = \$14{,}000/(1.06)^5$

Or using functional notation and the tables: $P_2 = \$14{,}000(P/F, i\%, n)$

In either case: $P_2 = \$14{,}000(.7472)$

Thus: $P_2 = \$10{,}461$

The facility manager should choose the first payment method because it minimizes the present cost. The manager would rather pay $10,000 today than pay $10,461 today. Because $10,461 today is *equivalent* to $14,000 five years from now if the effective annual interest rate is 6%.

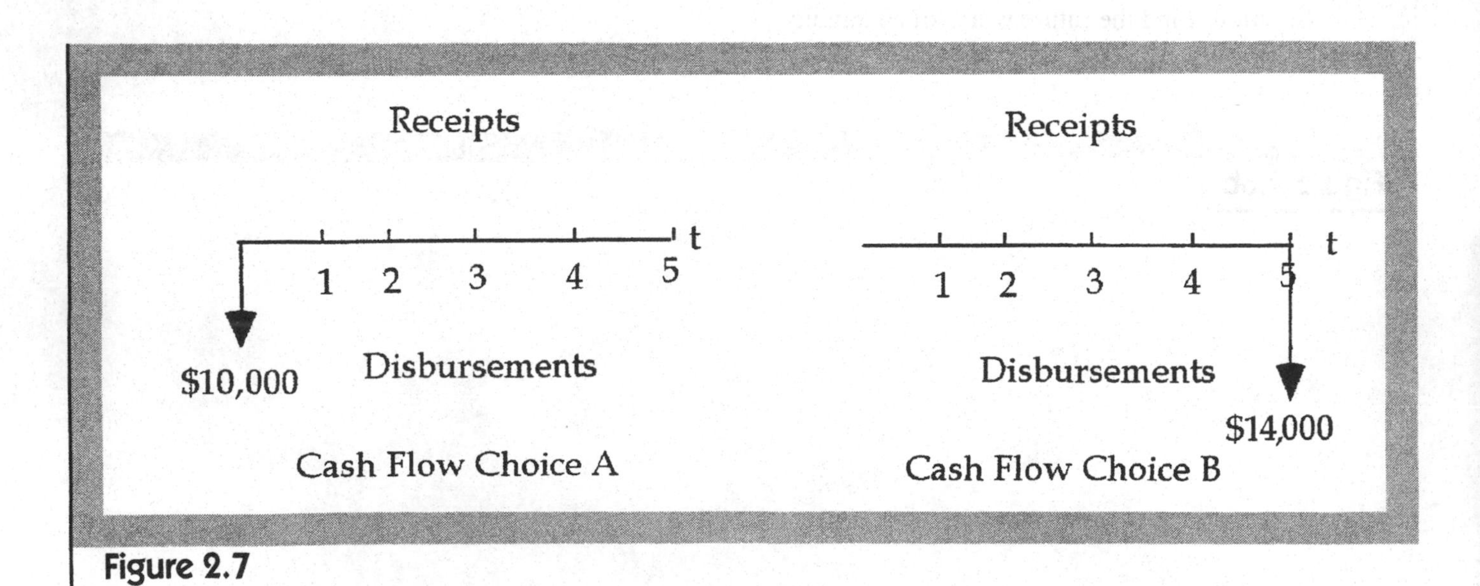

Figure 2.7

Future Value

This same problem could also have been solved using the future value method.

Longhand Formula $F = P(1+i)^n$
Functional Formula $F = P(F/P, i\%, n)$

Answer:

The solution to the problem is to choose the method that minimizes the future cost (which would occur at the end of Year 5).

Future cost of the first payment method is:

$F_1 = P(F/P, i\%, n)$ Looking up the single payment compound amount factor F/P @ 6% for 5 years in the tables gives 1.3382.

Thus $F_1 = \$10,000\,(1.3382)$ and $F_1 = \$13,382$

Future cost of the second payment method is: $F_2 = 14,000$

The facility manager should choose the first payment method because it minimizes the future cost. The manager would rather pay $13,382 in Year 5 dollars than pay $14,000 at the end of Year 5. $13,382 at the end of Year 5 is *equivalent* to $10,000 today, if the effective annual interest rate is 6%. Using the formula to solve the problem or using the factor from the tables to solve the problem and computing the answer lead to the same conclusion.

Annual Value

This same problem could also have been solved using the annual value method.

Longhand Formulas: $A = F\,[i/(1+i)^n - 1]$ $A = P\,\{i(1+i)^n/[(1+i)^n - 1]\}$
Functional Formulas: $A = F(A/F, i\%, n)$ $A = P(A/P, i\%, n)$

Answer:

The solution to the problem is to choose the method that minimizes the annual cost.

Annual cost of the first payment method is:

$A_1 = \$10,000\,\{.06(1.06)^5/(1.06)^5 - 1\}$ or $A_1 = \$10,000(A/P, i\%, n)$
$A_1 = \$2,374$ $A_1 = \$10,000(.2374)$

Annual cost of the second payment method is:

$A_2 = \$14,000\,\{.06/(1.06)^5 - 1\}$ or $A_2 = \$14,000(A/F, i\%, n)$
$A_2 = \$2,483$ $A_2 = \$14,000(.1774)$

The facility manager should again choose the first payment method because it minimizes the annual cost. The manager would rather pay $2,374 each year than $2,483 each year.

Using either method of analysis gives us the correct choice because the period of five years was chosen for either method of payment. It should be noted that whenever the number of periods is not equal, the method of analysis that is easiest to use is the annual method, because other analysis techniques require a cumbersome adjustment to reach common analysis periods for alternatives.

Interest Rate and Time Periods

These two items are discussed together because there is a critical relationship between them that must be recognized if we are to avoid errors. In the preceding problem, the interest rate was six percent **per year**, and the number of periods was five. Therefore, the interest rate and the number of periods were matched. This is not always the case.

Example 5

The facility manager of a startup company borrows $100,000 to purchase a new cooling tower. The lender has advertised an annual effective interest rate of 7% for a 15-year loan, the payments on which are made monthly. What will be the periodic payments made by the borrower?

Answer:

To calculate the monthly payment you must first compute the *monthly interest rate*. This can be tricky.

Formula: $i = (1 + r/m)^m - 1$ where

i = the effective interest rate (often an annual rate)

r = the nominal interest rate (often an annual rate)

m = the number of compounding subperiods per interest period (normally, interest periods are one year, but do not have to be)

n = the number of interest periods

r/m = the periodic interest rate

solving for $r = 6.78$ percent

solving for $r/m = .5654$ percent

Therefore, the monthly interest rate is .5654%, or .005654. Now that we have matched the interest rate and the time period, we can easily solve the problem. The appropriate formula is:

$$A = P\{i(1+i)^n/[(1+i)^n - 1]\}$$

Solving for $A = 100{,}000\{.005654(1.005654)^{180}/[(1.005654)^{180} - 1]\}$

Thus $A = \$886.84$

In this case, one can see where the use of the tables, functional notation, and the compound amount factors should make the problem much simpler. Notice, however, that the tables most likely will not include the interest rate of .005654. Therefore, as the problems become more complicated, the use of a financial calculator or a computerized spreadsheet is highly recommended.

Compounding

When a person does not pay the interest on their debt the debt begins to grow at a greater rate. This is due to the effect of compounding.

Example 6

*A person borrows $1 from a credit card company whose nominal annual interest rate is 18% compounded monthly. If the person pays nothing on their debt for the entire year, what will he or she owe at the end of each month or at the end of the year? What is the annual **effective** interest rate that the credit card company is charging? What is the monthly periodic rate that the credit card company is charging?*

Answer:

First, compute the periodic interest rate. Because the nominal rate (r) was given the monthly interest rate is 18/12 = 1.5%. Next, compute the payment table for each of the 12 months of the year.

Month	Cash Flow
0	-1.00000
1	1.015000
2	1.030225
3	1.045678
4	1.061364
5	1.077284
6	1.093443
7	1.109845
8	1.126493
9	1.143390
10	1.160541
11	1.177949
12	1.195618

One can see by the table that if no payments are made for the year, the borrower will owe $1.1956 at the end of the twelfth month. Notice that the answer is not $1.18. This is due to the effect of compounding, i.e., charging interest on the interest. Also note that although the credit card company is allowed to advertise their rate of 18%, the annual effective rate is actually 19.56%. Many institutions that offer credit subscribe to this method of lending using compound interest rate accrual. The annual effective interest rate could also have been calculated directly by the formula referred to earlier:

$$i = (1 + r/n)^m - 1$$

The following section is reprinted from *Contractor's Business Handbook*, by Michael S. Milliner, R.S. Means Co., Inc.

Capital Asset Management and the Timing of Cash Flows

Capital assets are those that will provide service to the construction firm for years into the future. The distinction separating current from noncurrent (capital) assets is that current assets are consumed or turned over within the year, while noncurrent assets provide service beyond the one-year period.

Since the service period for capital assets is so long, the decision to acquire a noncurrent asset carries a more prolonged risk element than that associated with the acquisition of a current asset. *Risk* generally refers to the probability of the *actual* outcome of an event being different than the *expected* outcome of the event. The greater the uncertainty, the greater the risk. Most of us feel more comfortable forecasting business conditions over the next 6 to 12 months than we would the next 12 to 24 months. By their very nature, capital assets acquired today will be in service during these future periods. Thus, the decisions made today have a significant impact on the contractor's positioning and risk in future business operations.

The following discussion of capital asset management is divided into two sections. The first section focuses strictly on the economics of capital asset acquisition, while the second deals with the more general managerial procedures concerning capital assets.

Financial Analysis of Capital Assets

Typical capital asset acquisition decisions for the contractor include:

- **Equipment expansion and replacement decisions.** Should old equipment be replaced or expanded upon now or in the future?
- **Plant expansion decisions**. Should the building and other plant facilities be expanded, and will the benefits gained by the expansion offset the cost?
- **Purchase vs. lease decisions.** Is it more cost effective in the long run to lease or purchase heavy equipment?

Clearly, acquiring capital assets involves the commitment of funds in the current period, with returns being generated largely in future periods. The problem becomes one of comparing the *costs* of a capital asset with the *benefits* to be generated by the asset. The most immediate problem is that there is a difference in the timing of the cash outflows and the cash inflows. No discussion of decisions related to capital asset acquisition, referred to as *capital budgeting*, can begin without a thorough understanding of the time value of money.

Time Value of Money

History shows that it is very rare when inflation is not a factor in our economy. A dollar received today is, therefore, worth more than a dollar received in the future. This is true for three reasons. First, the purchasing power of a dollar deteriorates over time. Consequently, more dollars are needed to purchase the same quantity of product in the future. Second, the future holds uncertainty. The further one plans into the future, the greater the uncertainty. Therefore, as the benefits to be derived from an investment in a capital asset are received further in the future, they become more uncertain. Third, a dollar received today can be invested and become an income producing liquid asset for the firm.

Two basic concepts are important to the discussion of the time value of money. These are *future value* and *present value*. Future value represents the future earnings of the dollar invested today. For example, if $100 is invested today in an interest-bearing account paying 5% annually, the account will grow to the following future values at the end of each of the next three years:

Original investment	**$100.00**
Interest for the first year:	
$100.00 x 0.05	5.00
Balance at the end of year one	**$105.00**
Interest for the second year:	
$105.00 x 0.05	5.25
Balance at the end of year two	**$110.25**
Interest for the third year:	
$110.25 x 0.05	5.51
Balance at the end of year three	**$115.76**

The interest for each year is greater than the interest for the previous year. This increase is due to the compounding of the interest from the previous year. *Compounding* means the earning of current period interest on interest that has already been earned in past periods. The more often interest is compounded, the faster the money grows.

It would be cumbersome to continue this procedure to discover the value at the end of the tenth year. Fortunately, tables have been established to simplify this calculation. Figure 2.8 shows the factors used to calculate the future value of an amount invested today. The table shows the interest rates in columns, and time periods in rows. For our previous example, the applicable factor would be based on an interest rate of 5% and three periods. Since, in our example, the interest is compounded annually, the periods would be equal to the number of years, times one. If our interest is compounded semi-annually, each year would represent two periods; quarterly would be four periods, etc.

In Figure 2.8, the factor represented by 5% and three periods is 1.1576. To use this factor we employ the formula:

Future Value = Present Value x Future Value Interest Factor

From this point, the following abbreviations are used:

Future Value	= F
Present Value	= P
Future Value Interest Factor	= (F/P, i%, n)

Filling in the formula using $100 for P, we have

F $100 (5%, 3 periods)	= $100 x (F/P, 5%, 3)
	= $100 x 1.1576
	= $115.76

Multiplying $100 by the factor 1.1576, we find the future value is $115.76, the same answer as originally calculated.

The time value of money is further illustrated in the following examples, using our hypothetical contracting company, Eastway Contractors.

Example 1

Eastway Contractors has recently completed a successful project that generated a cash surplus of $20,000. This surplus is now available for investment. Eastway realizes that it may be necessary to expand plant facilities in the next two years and is, therefore, interested in accumulating cash to help fund the expansion. How much cash will Eastway have available at the end of two years if the $20,000 is invested at 8%, compounded annually, for the entire two-year period?

Solution:

F $20,000 (8%, 2 periods)	= $20,000 x (F/P, 8%, 2)
	= $20,000 x 1.1664
	= $23,328

According to this computation, Eastway will have $23,328 available at the end of the second year.

If Eastway chooses a financial institution that compounds the investment semi-annually, there would be four compounding periods over the two years. Figure 2.8 can again be used for this calculation, but the number of periods must be doubled and the interest rate cut in half, to 4%. (The interest rate must be divided by two because the compounding is taking place every six months instead of once each year.)

The new solution for semi-compounding would be:

F $20,000 (4%, 4 periods)	= $20,000 x (F/P, 4%, 4)
for semi -	= $20,000 x 1.1699
compounding	= $23,398

The use of semi-annual compounding results in an additional $70 of interest income over the two years.

Annuities: In the prior examples, the investment is a one-time amount. Sometimes, an investment is made in a series of consecutive periods. The term *annuity* is used to describe a series of investments, or cash flows. To illustrate, assume that Eastway Contractors invested $5,000 in an interest-bearing account each year for the next ten years earning an average rate of return of 8%. To determine the value of the account at the end of the ten-year period involves the use of the future value of an annuity calculation. This calculation is similar to the *future value* calculation used previously and is stated as follows:

Future Value of an Annuity =
Annuity x Future Value Interest Factor for an Annuity

The abbreviations for the variables are as follows:

Future Value of an Annuity	= A (F/A, i%, n)
Annuity	= A
Future Value Interest Factor for the Annuity	= (F/A, i%, n)

Future Value of $1

Periods	2%	3%	4%	5%	6%	7%	8%	10%	12%
1	1.0200	1.0300	1.0400	1.0500	1.0600	1.0700	1.0800	1.1000	1.1200
2	1.0404	1.0609	1.0816	1.1025	1.1236	1.1449	1.1664	1.2100	1.2544
3	1.0612	1.0927	1.1249	1.1576	1.1910	1.2250	1.2597	1.3310	1.4049
4	1.0824	1.1255	1.1699	1.2155	1.2625	1.3108	1.3605	1.4641	1.5735
5	1.1041	1.1593	1.2167	1.2763	1.3382	1.4026	1.4693	1.6105	1.7623
6	1.1262	1.1941	1.2653	1.3401	1.4185	1.5007	1.5869	1.7716	1.9738
7	1.1487	1.2299	1.3159	1.4071	1.5036	1.6058	1.7138	1.9487	2.2107
8	1.1717	1.2668	1.3686	1.4775	1.5938	1.7182	1.8509	2.1436	2.4760
9	1.1951	1.3048	1.4233	1.5513	1.6895	1.8385	1.9990	2.3579	2.7731
10	1.2190	1.3439	1.4802	1.6289	1.7908	1.9672	2.1589	2.5937	3.1058
11	1.2434	1.3842	1.5395	1.7103	1.8983	2.1049	2.3316	2.8531	3.4785
12	1.2682	1.4258	1.6010	1.7959	2.0122	2.2522	2.5182	3.1384	3.8960
13	1.2936	1.4685	1.6651	1.8856	2.1329	2.4098	2.7196	3.4523	4.3635
14	1.3195	1.5126	1.7317	1.9799	2.2609	2.5785	2.9372	3.7975	4.8871
15	1.3459	1.5580	1.8009	2.0789	2.3966	2.7590	3.1722	4.1772	5.4736
16	1.3728	1.6047	1.8730	2.1829	2.5404	2.9522	3.4259	4.5950	6.1304
17	1.4002	1.6528	1.9479	2.2920	2.6928	3.1588	3.7000	5.0545	6.8660
18	1.4282	1.7024	2.0258	2.4066	2.8543	3.3799	3.9960	5.5599	7.6900
19	1.4568	1.7535	2.1068	2.5270	3.0256	3.6165	4.3157	6.1159	8.6128
20	1.4859	1.8061	2.1911	2.6533	3.2071	3.8697	4.6610	6.7275	9.6463
25	1.6406	2.0938	2.6658	3.3864	4.2919	5.4274	6.8485	10.835	17.000
30	1.8114	2.4273	3.2434	4.3219	5.7435	7.6123	10.063	17.449	29.960
40	2.2080	3.2620	4.8010	7.0400	10.286	14.974	21.725	45.259	93.051
50	2.6916	4.3839	7.1067	11.467	18.420	29.457	46.902	117.39	289.00
60	3.2810	5.8916	10.520	18.679	32.988	57.946	101.26	304.48	897.60

Figure 2.8

Figure 2.9 provides the factors used in Future Value of an Annuity calculations.

The calculation used to determine the value of Eastway's account after depositing $5,000 per year for 5 years is as follows.

FVA $5,000 (8%, 5 periods) = $5,000 x (F/A (8%, 5)
= $5,000 x 5.8666
= $29,333

The factor, 5.8666, is based on the interest rate of 8% and the number of annual payment periods, five.

The account balance after the fifth deposit of $5,000 is $29,333. After ten deposits of $5,000, the account balance is $72,435 ($5,000 x 14.487).

Future Value of an Annuity of $1

Periods	2%	3%	4%	5%	6%	7%	8%	10%	12%
1	1.0000	1.0000	1.0000	1.0000	1.0000	1.0000	1.0000	1.0000	1.0000
2	2.0200	2.0300	2.0400	2.0500	2.0600	2.0700	2.0800	2.1000	2.1200
3	3.0604	3.0909	3.1216	3.1525	3.1836	3.2149	3.2464	3.3100	3.3744
4	4.1216	4.1836	4.2465	4.3101	4.3746	4.4399	4.5061	4.6410	4.7793
5	5.2040	5.3091	5.4163	5.5256	5.6371	5.7507	5.8666	6.1051	6.3528
6	6.3081	6.4684	6.6330	6.8019	6.9753	7.1533	7.3359	7.7156	8.1152
7	7.4343	7.6625	7.8983	8.1420	8.3938	8.6540	8.9228	9.4872	10.089
8	8.5830	8.8923	9.2142	9.5491	9.8975	10.260	10.637	11.436	12.300
9	9.7546	10.159	10.583	11.027	11.491	11.978	12.488	13.579	14.776
10	10.950	11.464	12.006	12.578	13.181	13.816	14.487	15.937	17.549
11	12.169	12.808	13.486	14.207	14.972	15.784	16.645	18.531	20.655
12	13.412	14.192	15.026	15.917	16.870	17.888	18.977	21.384	24.133
13	14.680	15.618	16.627	17.713	18.882	20.141	21.495	24.523	28.029
14	15.974	17.086	18.292	19.599	21.015	22.550	24.215	27.975	32.393
15	17.293	18.599	20.024	21.579	23.276	25.129	27.152	31.772	37.280
16	18.639	20.157	21.825	23.657	25.673	27.888	30.324	35.950	42.753
17	20.012	21.762	23.698	25.840	28.213	30.840	33.750	40.545	48.884
18	21.412	23.414	25.645	28.132	30.906	33.999	37.450	45.599	55.750
19	22.841	25.117	27.671	30.539	33.760	37.379	41.446	51.159	63.440
20	24.297	26.870	29.778	33.066	36.786	40.995	45.762	57.275	72.052
25	32.030	36.459	41.646	47.727	54.865	63.249	73.106	98.347	133.33
30	40.568	47.575	56.085	66.439	79.058	94.461	113.28	164.49	241.33
40	60.402	75.401	95.026	120.80	154.76	199.64	259.06	442.59	767.09
50	84.579	112.80	152.67	209.35	290.34	406.53	573.77	1163.9	2400.0
60	114.05	163.05	237.99	353.58	533.13	813.52	1253.2	3034.8	7471.6

Figure 2.9

Example 2

Eastway Contractors has entered into a financing arrangement with a lending institution which agrees to provide $250,000 in debt financing to the firm. Repayment of the principal amount will occur at the end of five years with interest-only payments being made each quarter until the principal payoff. Eastway wants to provide for the payoff by depositing an equal amount into a bank account each year for the next five years so that after the final deposit, the account balance is equal to the required $250,000. Deposits into the account will begin one year from today and are expected to earn an average return of 8% over the five years.

In this example, the future value of the annuity is known. The unknown is the amount of the annual deposit necessary to fund the future payoff of the bank note's principal balance. This unknown amount can be calculated using the Future Value of an Annuity formula.

Two of the three variables within the formula are known. The Future Value is equal to $250,000 and the Future Value Interest Factor is determined to be 5.8666 based on an 8% interest rate and five periods. By plugging these known variables into the formula, the solution is:

$$F = A(F/A, 8\%, 5)$$

$$\$250{,}000 = A \times 5.8666$$

$$A = \frac{\$250{,}000}{5.8666}$$

$$A = \$42{,}614$$

Solving for A involves isolating it from the other variables. This is accomplished by dividing each side of the equation by 5.8666.

While calculating the future value of a single investment or a series of investments is useful for the contractor, the primary focus in capital budgeting is on the *present value of future cash flows*. It is the present value calculations that enable a contractor to evaluate the future benefits of an investment against its current cost.

Capital Budgeting

Capital budgeting involves committing funds in the current period with the expectation of receiving a desired return on those funds in the future. For the contractor, this means investing funds in business assets that are expected to provide a positive contribution to overall business operations and profitability over a period of years.

In most cases, the contractor has a limited budget within which to operate and is, therefore, eager to commit funds to the assets that promise the highest expected rate of return. Generally, the measure of expected return is, at best, only an indicator of the *relative merit* of the project, since the capital budgeting process tends to be subjective. This subjectivity is due to the following variables.

1. Forecasting the firm's cost of capital in the future.
2. Forecasting the amount and timing of a project's cash inflows and outflows.

3. Anticipating the possible technical obsolescence of the acquired equipment, as well as its useful life.
4. Potential changes in the firm's corporate tax rate.

The fact that it is a subjective process does not invalidate the usefulness of capital budgeting. On the contrary, requiring the manager's involvement ensures that steps are taken to effectively plan the acquisition and use of capital assets. This budgeting exercise creates a benchmark against which actual asset performance can be measured. In this way, the relative effectiveness of the assets employed by the firm can be measured.

A number of methods are used to evaluate alternatives for investing in capital assets. The most useful methods are those employing *discounted cash flow*. Since the contractor is weighing today's investment against future benefits, it is necessary to compare dollars in terms of purchasing power. For example, let us assume that Eastway Contractors has an opportunity to invest $50,000 in either of two mutually exclusive pieces of heavy equipment, such as a track loader or a backhoe. *Mutually exclusive* means that the pieces of equipment will provide similar benefits and only one or the other will be acquired. Each asset will provide labor cost savings over a five-year period. At the end of three years, the asset will have no salvage value. When evaluating investments of this type, Eastway Contractors uses a discount rate of 12%. This percentage is decided based on (a) a rate of return commensurate with the risk of the investment and (b) a rate that will cover the firm's cost of capital, which is the rate that the contractor would have to pay when financing the asset.

Projecting the net cash flows associated with each investment alternative is the first step in evaluating the asset acquisitions. The projected net cash flows for the track loader and backhoe, due largely to labor and rental savings, are as follows:

		Track Loader	Backhoe
Cash inflow	(period 1)	20,000	14,200
Cash inflow	(period 2)	18,000	14,200
Cash inflow	(period 3)	14,000	14,200
Cash inflow	(period 4)	10,000	14,200
Cash inflow	(period 5)	8,000	14,200
Total net cash inflows		**70,000**	**71,000**

The backhoe produces total net cash inflows $1,000 greater than the net cash inflows produced by the track loader. Does this mean that the backhoe should be accepted over the track loader? The answer is not readily apparent since the timing of the cash inflows is different for each of these investments.

To convert future cash flows to present dollars, the following equation is used:
Present Value = Future Value x Present Value Interest Factor.

Abbreviations for the formula variables are as follows:

Present Value	= P
Future Value	= F
Present Value Interest Factor	= (P/F, i%, n)

Figure 2.10 provides the factors used to calculate the present value of a single sum to be received in the future. The structure of the Present Value table is the same as the Future Value tables. It should be noted that no factor on the table is greater than .9901. This is due to the fact that unless the assumed discount factor is 0%, no future dollar will be worth as much as a present day dollar. In order to properly evaluate the investment in the track loader or backhoe, the future cash inflows must be discounted back to their present value. The schedule in Figure 2.11 shows the Present Value calculations for the future cash flows generated by both investments.

For both investments, the present value of the net cash inflows are greater than the initial outlay of $50,000. This tells Eastway Contractors that both investments generate a return greater than the 12% required rate. Since the present value of the cash inflows is greater for the track loader than the

Present Value of $1

Periods	1%	2%	3%	4%	5%	6%	7%	8%	10%	12%
1	.99010	.98039	.97087	.96154	.95238	.94340	.93458	.92593	.90909	.89286
2	.98030	.96117	.94260	.92456	.90703	.89000	.87344	.85734	.82645	.79719
3	.97059	.94232	.91514	.88900	.86384	.83962	.81630	.79383	.75131	.71178
4	.96098	.92385	.88849	.85480	.82270	.79209	.76290	.73503	.68301	.63552
5	.95147	.90573	.86261	.82193	.78353	.74726	.71299	.68058	.62092	.56743
6	.94205	.88797	.83748	.79031	.74622	.70496	.66634	.63017	.56447	.50663
7	.93272	.87056	.81309	.75992	.71068	.66506	.62275	.58349	.51316	.45235
8	.92348	.85349	.78941	.73069	.67684	.62741	.58201	.54027	.46651	.40388
9	.91434	.83676	.76642	.70259	.64461	.59190	.54393	.50025	.42410	.36061
10	.90529	.82035	.74409	.67556	.61391	.55839	.50835	.46319	.38554	.32197
11	.89632	.80426	.72242	.64958	.58468	.52679	.47509	.42888	.35049	.28748
12	.88745	.78849	.70138	.62460	.55684	.49697	.44401	.39711	.31863	.25668
13	.87866	.77303	.68095	.60057	.53032	.46884	.41496	.36770	.28966	.22917
14	.86996	.75788	.66112	.57748	.50507	.44230	.38782	.34046	.26333	.20462
15	.86135	.74301	.64186	.55526	.48102	.41727	.36245	.31524	.23939	.18270
16	.85282	.72845	.62317	.53391	.45811	.39365	.33873	.29189	.21763	.16312
17	.84438	.71416	.60502	.51337	.43630	.37136	.31657	.27027	.19784	.14564
18	.83602	.70016	.58739	.49363	.41552	.35034	.29586	.25025	.17986	.13004
19	.82774	.68643	.57029	.47464	.39573	.33051	.27651	.23171	.16351	.11611
20	.81954	.67297	.55368	.45639	.37689	.31180	.25842	.21455	.14864	.10367
25	.77977	.60953	.47761	.37512	.29530	.23300	.18425	.14602	.09230	.05882
30	.74192	.55207	.41199	.30832	.23138	.17411	.13137	.09938	.05731	.03338
40	.67165	.45289	.30656	.20829	.14205	.09722	.06678	.04603	.02209	.01075
50	.60804	.37153	.22811	.14071	.08720	.05429	.03395	.02132	.00852	.00346
60	.55045	.30478	.16973	.09506	.05354	.03031	.01726	.00988	.00328	.00111

Figure 2.10

backhoe, Eastway Contractors should select the track loader, all other factors (delivery time of equipment, maintenance costs, etc.) being equal. Prior to the present value analysis, the backhoe appears to yield the greatest return. However, when the time value of money is considered, the track loader is clearly the better investment.

Figure 2.12 provides the discount factors for the present value of an annuity. Since the cash inflows generated by the backhoe are consistent, this table can be used to determine the appropriate discount factor, in this case, 3.6048. This factor is the sum of the Present Value factors used above. The formula for the present value of an annuity is:

Present Value of an Annuity =
Annuity x Present Value Interest Factor of an Annuity

Abbreviations for the formula variables are as follows.

Present Value of an Annuity	= P
Annuity	= A
Present Value Interest Factor of an Annuity	= (P/A, i%, n)

Investment A

Track loader

Period	Cash Inflow	Present Value Factor	Present Value
1	$20,000	.8929	$17,858
2	18,000	.7972	14,358
3	14,000	.7118	9,965
4	10,000	.6355	6,355
5	8,000	.5674	4,539
	70,000		**Total** **53,067**

Investment B

Backhoe

Period	Cash Inflow	Present Value Factor	Present Value
1	$14,200	.8929	$12,679
2	14,200	.7972	11,320
3	14,200	.7118	10,108
4	14,200	.6355	9,024
5	14,200	.5674	8,057
	71,000		**Total** **51,188**

Figure 2.11

For Investment B, the solution is:

P $14,200 (12%, 5 periods)

= $14,200 x PVIFA (12%, 5 periods)

= $14,200 x 3.6048

= $51,188

These exercises are simple examples of discounted cash flow analysis used to evaluate the acquisition of capital assets. In the next section, we will discuss a variety of other factors that affect the discounted cash flow analysis and the capital budgeting process.

Other Capital Asset Investment Factors

The acquisition of capital assets generally aims for an asset that will provide cash flow benefits to the firm for a number of years. In addition to the purchase price, the asset may require periodic cash investments, as in the case of equipment overhaul and maintenance. Whether the asset is financed with debt or by equity, there is still a cost for the financing. For *debt*, the cost is the explicit interest charged for the use of the funds. For *equity,* the cost is the opportunity cost of not having the funds available for other investments. Any investment of funds must generate a return sufficient to cover the cost of the funds.

Present Value of an Annuity of $1 per Period

Number of Payments	2%	3%	4%	5%	6%	7%	8%	10%	12%
1	.98039	.97087	.96154	.95238	.94340	.93458	.92593	.90909	.89286
2	1.9416	1.9135	1.8861	1.8594	1.8334	1.8080	1.7833	1.7355	1.6901
3	2.8839	2.8286	2.7751	2.7232	2.6730	2.6243	2.5771	2.4869	2.4018
4	3.8077	3.7171	3.6299	3.5460	3.4651	3.3872	3.3121	3.1699	3.0373
5	4.7135	4.5797	4.4518	4.3295	4.2124	4.1002	3.9927	3.7908	3.6048
6	5.6014	5.4172	5.2421	5.0757	4.9173	4.7665	4.6229	4.3553	4.1114
7	6.4720	6.2303	6.0021	5.7864	5.5824	5.3893	5.2064	4.8684	4.5638
8	7.3255	7.0197	6.7327	6.4632	6.2098	5.9713	5.7466	5.3349	4.9676
9	8.1622	7.7861	7.4353	7.1078	6.8017	6.5152	6.2469	5.7590	5.3282
10	8.9826	8.5302	8.1109	7.7217	7.3601	7.0236	6.7101	6.1446	5.6502
11	9.7868	9.2526	8.7605	8.3064	7.8869	7.4987	7.1390	6.4951	5.9377
12	10.575	9.9540	9.3851	8.8633	8.3838	7.9427	7.5361	6.8137	6.1944
13	11.348	10.635	9.9856	9.3936	8.8527	8.3577	7.9038	7.1034	6.4235
14	12.106	11.296	10.563	9.8986	9.2950	8.7455	8.2442	7.3667	6.6282
15	12.849	11.938	11.118	10.380	9.7122	9.1079	8.5595	7.6061	6.8109
16	13.578	12.561	11.652	10.838	10.106	9.4466	8.8514	7.8237	6.9740
17	14.292	13.166	12.166	11.274	10.477	9.7632	9.1216	8.0216	7.1196
18	14.992	13.754	12.659	11.690	10.828	10.059	9.3719	8.2014	7.2497
19	15.678	14.324	13.134	12.085	11.158	10.336	9.6036	8.3649	7.3658
20	16.351	14.877	13.590	12.462	11.470	10.594	9.8181	8.5136	7.4694
25	19.523	17.413	15.622	14.094	12.783	11.654	10.675	9.0770	7.8431
30	22.396	19.600	17.292	15.372	13.765	12.409	11.258	9.4269	8.0552
40	27.355	23.115	19.793	17.159	15.046	13.332	11.925	9.7791	8.2438
50	31.424	25.730	21.482	18.256	15.762	13.801	12.233	9.9148	8.3045
60	34.761	27.676	22.623	18.929	16.161	14.039	12.377	9.9672	8.3240

Figure 2.12

The discounted cash flow technique most often used to evaluate capital asset acquisitions is *Net Present Value* analysis. In this procedure, the present value of cash inflows is compared to the present value of cash outflows over the life of the project. If the present value of the inflows exceeds the present value of the outflows, the acquisition would be accepted. The net present value (NPV) of the track loader in the previous example is $3,067; for the backhoe, it was $1,188. Since the projects were mutually exclusive with equivalent lives, the track loader should be chosen for acquisition.

Our discussion of capital budgeting has focused on *cash flows* rather than *accounting net income.* The reason for this focus is that it is important to understand not only the *dollar amount,* but also the *timing* of the cash flows. Typical cash inflows and outflows related to investments are as follows.

Cash Inflows:

- Additional revenue
- Labor savings
- Material savings
- Salvage value
- Release of working capital
- Tax savings from depreciation

Cash Outflows:

- Acquisition costs
- Repairs/Improvements
- Working capital commitment
- Taxes on profits

The contractor must make a good faith effort to forecast each of the above cash flow items in order to develop a good capital budgeting model.

Example 3

Eastway Contractors is considering the investment of $180,000 in a new piece of equipment. Purchase of the equipment will allow Eastway to avoid some equipment rental costs and will provide the opportunity to generate some additional revenue for the firm. The equipment is expected to last for eight years and have a salvage value of $25,000 at the end of the eight-year period. Eastway's controller is interested in whether or not the equipment will generate sufficient cost savings and additional revenue to cover the cost of purchase and operation. Before beginning the analysis, the controller forecasted the following.

Expected savings in annual equipment rental	$20,000
Annual revenue generated by the equipment	$40,000
Cash operating cost of the new equipment	($20,000)
Net increase in annual cash income	**$40,000**
Equipment overhaul in the fourth year	$10,000
Depreciation method - Straight Line	8 years
Eastway Contractors projected tax rate	30%
Discount Factor used to evaluate projects	**12%**

The following headings suggest a format useful for structuring the capital budgeting analysis:

Event Years Amount PV Factor Present Value

These headings are briefly defined below.

Event: The transaction that causes a cash outflow or inflow.

Years: The number of annual periods affected by the transaction. Some events (e.g., the initial purchase) occur only once, while others (e.g., annual cost savings) occur over a period of years.

Amount: The after-tax cash inflow or outflow associated with the transaction.

Present Value Factor: The factor, as specified in Figures 2.10 and 2.12, which converts the future cash inflow or outflow to present value dollars.

Present Value: The product obtained by multiplying the Amount by the PV Factor.

Figure 2.13 depicts the completed capital budgeting analysis. The initial event, equipment purchase, represents the major cash outflow of the project. It occurs in Period 0 (the current period) and represents an outflow of $180,000. For simplicity, it is assumed in this example that the firm pays the entire amount, up front, using working capital. In reality, it is much more likely that the firm would make an initial downpayment only, with additional monthly payments scheduled over a period of several years.

Capital Budgeting Analysis

Event	Years	Amount	PV Factor	Present Value
Equipment Purchase	0	($180,000)	1.000	($180,000)
Annual Cash Net Income $40,000 (1–.30)	1–8	28,000	4.9676	139,093
Depreciation Tax Savings $22,500 (.30)	1–8	6,750	4.9676	33,531
Year 4 Overhaul $10,000 (1–.30)	4	(7,000)	0.6355	(4,449)
Year 8 Salvage Value $25,000 (1–.30)	8	17,500	0.4039	7,068
			Net Present Value	($4,757)

Figure 2.13

Annual Cash Income

The next event is the annual increase in cash net income due to the acquisition of equipment. It is important to note that the contractor does not have access to the entire increase in cash net income. This is because the increase in net income causes an increase in the income taxes that must be paid. The increase in cash net income must, therefore, be offset by the increase in taxes. On the schedule, this tax adjustment has been calculated by multiplying the annual increase in cash net income by one, minus the firm's tax rate, which is 30% (1–30% = 70%). The result is an after-tax increase in net income of $28,000. In essence, Eastway keeps 70% of the profits and the taxing authorities receive the other 30%. This annual income represents a cash inflow for the firm for the next eight years and, therefore, is multiplied by the Present Value Interest Factor of an Annuity (eight periods at 12% shown in Figure 2.12) of 4.9676. The result is a present value of cash inflows of $139,093.

Depreciation

Depreciation is the allocation of an asset's cost over its useful life. Depreciation is not a cash expense; therefore, it does not require a direct cash disbursement. The entire cost of the equipment has already been considered as a cash outflow. The indirect effect of depreciation is as a deduction on the firm's tax return, where it provides a tax savings and therefore a cash savings.

The annual depreciation amount for our example is $22,500 (180,000 divided by 8). This deduction on the tax return will save the firm $6,750 (22,500 x .30) in income taxes. Since this amount is a very real cash savings, it represents a cash inflow to the firm over the next eight years. Using the same Present Value Factor as that applied to the annual cash net income, a Present Value of $33,531 (6,750 x 4.9676) is calculated for the cumulative depreciation tax savings over 8 years at 12%.

Overhaul: The next event noted is the overhaul of equipment, scheduled to take place at the end of the fourth year. The $10,000 overhaul represents a cash outflow for the firm. It is, however, a tax deduction (assuming a full write-off of the amount) and, therefore, will generate tax savings. The net amount of cash outflow resulting from the overhaul is calculated by multiplying the cost of the overhaul by one, minus the firm's tax rate, which is 30%. The result is a net cash cost of $7,000 (10,000 x .70). This amount multiplied by .6355 (the Present Value Interest Factor $1 for Period 4 at 12% shown in Figure 2.10) results in a present value for the cost of the overhaul of $4,449.

Salvage Value: The final event affecting the investment in equipment is the salvage value of the equipment at the end of the eighth year. For purposes of analysis, it is assumed that the equipment will be sold at the end of the eighth year for the $25,000 salvage value. Since the equipment is fully depreciated at that point, the entire $25,000 is a taxable gain. Since the firm's tax rate is 30%, the remaining 70% is available to the firm as a cash inflow in the eighth year. The $17,500 ($25,000 x .70) cash inflow is discounted back to the present using a factor of .4039. This factor is the Present Value Interest Factor for an amount received in Period 8 at 12% (Figure 2.10) and results in a present value of $7,068.

Deciding on the Investment: The net present value of the cash inflows and outflows is a negative $4,757. This figure indicates that the investment does not generate the required 12% return and, therefore, should not be accepted unless other considerations warrant it. The analysis does not disclose the *actual rate of return* on the investment. The actual rate can be approximated by substituting a lesser rate of return in the foregoing analysis and, by trial and error, discovering the rate of return closest to making the net present value of the project equal 0. There are mathematical procedures that can be used to determine actual rate of return, but the net present value approach yields adequate results for the purpose of comparing alternative investments.

Other Approaches to Capital Budgeting Decisions: Some managers prefer to use one of the other general approaches to capital budgeting. The reason is usually ease of use. One of these other approaches is the *Payback Method.* This method focuses on the length of time it takes an investment to recover its initial cost. For example:

Eastway Contractors needs a new machine and is considering two options, Machine X and Machine Y. Machine X costs $20,000 and Machine Y costs $18,000. Machine X will generate annual cash inflows of $5,000 and Machine Y will generate annual cash inflows of $4,600. The calculation of the payback period for each machine is as follows:

Machine X	$20,000 (initial cost)	= 4.0 years to recover cost
	$5,000 (annual cash inflow)	
Machine Y	$18,000 (initial cost)	= 3.9 years to recover cost
	$4,600 (annual cash inflow)	

Since Machine Y has a faster payback, it would be selected based on the payback method criteria.

The payback method has two primary flaws. First, it does not consider cash flows after the payback period and, second, it does not address the time value of money. For short term or very risky projects, the payback method has particular merit and should be used in conjunction with the discounted cash flow method previously discussed.

Capital budgeting techniques represent a *quantitative* process, helpful to the manager in making investment decisions. There are many options for the contractor's investment dollar. It is the manager's goal to direct available dollars to the investment that is expected to produce the *most significant benefit* for the firm. While none of us can predict the future with absolute certainty, careful consideration of the different benefits and costs of each alternative will result in a better ultimate decision.

General Management Procedures: One of the major factors in the decision to acquire equipment is the projection of need for the piece of equipment. The contractor must be able to separate a "nice to own" piece of equipment from a "need to own" piece of equipment. After considering all factors, the contractor may decide that hiring subcontractors with their own equipment or renting the equipment is more economically feasible than purchasing the equipment. The final decision should be the result of a *quantitative analysis* and *general management review.*

The acquisition of equipment brings with it other associated costs that must be considered by the contractor in the decision-making process. For example, *storage and insurance costs* will be incurred on the equipment even if it is not in use. Certain pieces of equipment may also require skilled operators. In any event, the cost of "down time" must be considered as an associated cost of equipment purchase. Other major considerations include:

1. The age, maintenance costs, and efficiency of equipment currently owned.
2. Warranties provided with new equipment.
3. The outlook for new projects and the economy in general.
4. *All* costs of ownership. Have they been considered?
5. Alternative arrangements. What terms are offered by other leasing agencies?
6. Inflation. Is it likely to affect the future purchase price of the equipment?
7. Future tax changes. Are there any revisions on the horizon that will affect the net return on investment?

The purchase of a capital asset is only the beginning of the asset management process. Since capital asset expenditures represent a significant obligation of funds, proper care and monitoring of the use of the asset must be ensured.

It is important to maintain detailed records regarding the acquisition of capital assets. The records should include the following.

- Asset Description
- Cost
- Purchase Date
- Model/Serial Number
- Warranty Information
- Seller

This information is needed to maintain current records of the assets owned by the contractor, for such purposes as:

- Proper depreciation expense calculations
- Internal control over equipment security
- Proper calculation of gain or loss upon disposal
- Security liens by financial institutions

Asset management focuses on the effective use of the contractor's assets in meeting the objectives of the firm. Each asset plays an important part in the overall success of the firm, and thus requires (to varying degrees) management's attention. Cash, receivables, inventory, and plant and equipment are the major assets employed by the contractor and, therefore, merit their own specific policies and procedures.

Current Asset Management

Current asset management concerns those assets that will be converted into cash or used up within one year. The primary emphasis in current asset management is the efficient and timely turnover of the asset. Problems such as old accounts receivable and slow moving inventory should be minimized.

The contractor's accounts receivable management should be supported by a sound credit policy. The credit policy should cover the following key areas.

1. Credit Standards
2. Credit Terms
3. Credit Account Tracking
4. Collection
5. Timely Billing Procedures

Capital Asset Management

Capital asset management involves the management of assets that will provide the firm with benefits for periods beyond one year. Typical capital asset management issues include:

1. Equipment Expansion/Replacement Decisions
2. Plant Expansion/Improvement Decisions
3. Purchase vs. Lease Decisions

Since capital asset decisions involve the acquisition of an asset that will affect future cash flows, either in a single period or a series of periods, the contractor must consider the *time value of money.* Because of the erosion of the purchasing power of the dollar, a dollar received today is worth more than a dollar received in the future. Thus the contractor must consider not only the amount of the cash flows, but their timing.

The time value of money methodology is often used to calculate when and how to replace capital assets. This is often referred to as a **Life Cycle Cost Analysis**.

The following section is reprinted from *Total Productive Facilities Maintenance,* by Richard W. Sievert, Jr., R.S. Means Co., Inc.

Life Cycle Cost Analysis Output/ Results

The results of a Life Cycle Cost Analysis (LCCA) can be described in a variety of units or terms, e.g., *Discounted Payback Period (DPP), Return on Investment (ROI), Present Worth of Expenditures (PWE), or Equivalent Annual Charges (EAC).* The units that you will use to describe your project will probably be the ones with which the approving authority is most comfortable. With the exception of DPP, the superior project will always compare favorably regardless of the units used. DPP can be used, but it should never be used as the sole criteria for approving one proposal over another because there are some circumstances where DPP can be shorter for an option that otherwise is inferior to its competition. If Option "A" has a better ROI than Option "B," it will also have a better PWE and EAC. The bottom line is that your choice of units for describing your project will be driven by your target audience. Choose the terms that they are familiar with and expecting.

Examples

Sometimes selecting between options is easy...

I.	Option "A"	Option "B"
Initial Cost	$5,000	$8,000
Operating Expenses	$2,000/yr.	$2,500/yr.

Here, it is obvious that "A" is the superior plan since both first costs and ongoing operating expenditures are less. However, the choice would not be as clear if the Operating Expense values were reversed.

II.	Option "A"	Option "B"
Initial Cost	$5,000	$8,000
Operating Expenses	$2,500/yr.	$2,000/yr.

To identify the better plan, you would need to know the duration of the installation and your cost of money or discount rate. For the purpose of this example, let's use ten years and 8%, and solve for the present worth (PW) of the expenditures.

Option "A"
PW (Option "A") = PW (Initial Cost) + PW (Annual Costs)

We define the "initial cost" as occurring at the start of the project or in the "present" time frame. Therefore, the Present Worth of the initial cost is equal to the initial cost; no conversion is required. The Operating Expenses are expressed as an annual expenditure which therefore must be converted to a present value by the use of the following formula:

PW (Annual Costs) = (Op. Exp.) x (P/A) @ 8%, 10 Yrs.

Where (P/A) is the Present Worth of an Annuity factor that can be found in the Figure 2.3. For this problem we will use the 8% table and use the P/A factor for 10 years (6.7101).

PW (Op. Exp.) = $2,500 x 6.7101 = $16,755.50

We now return to our original formula to solve the PW (Option "A"):

PW (Option "A") = PW (Initial Cost) + PW (Annual Costs)
= $5,000 + $16,755 = $21,775

which is the equivalent present worth of the expenditures associated with Option "A." Said another way, if you received $21,775 today and could earn 8% on money held for future use, you could fund the initial $5,000 investment and the $2,500/year operating expenses for ten years! Don't let the fact that your company wouldn't operate in this manner throw you off course. Remember that the purpose of the conversion is to get all cost elements into a common format—present, future, or annual costs—for the purpose of comparison. It does not indicate how you would actually operate.

Option "B"

PW (Option "B") = PW (Initial Cost) + PW (Annual Costs)
= $8,000 + $2,000 (P/A) @ 8%, 10 Yrs.
= $8,000 + $13,420 = $21,420

Now that we have both Options "A" and "B" expressed in a common format, Present Worth, we can compare the two and identify the one that will cost the least over the life of the installation.

	Option "A"	Option "B"
Present Worth	$21,775	$21,420

When the closeness of the values makes the decision a virtual toss-up, you will need to consider other factors, such as inflation. If you were doing a revenue requirements type of analysis, it would be essential that all costs be fully accounted for, and you would have to calculate the impact of inflation. However, in simple comparison situations, it is often sufficient to simply apply logic. In this example, we can be pretty confident that operating expenses will rise over time. However, if we assumed that inflation will affect each option in the same way, we can logically conclude that the Option "A" expenses will increase more than those for "B." Since "B" already has the economic advantage, inflation will only serve to increase that advantage. Here no further calculations would be necessary. Accounting for inflation will be illustrated in the next example.

Finally, even when there is a clear "winner," your choice may be influenced by the availability of capital. If the budget is tight, you may not be able to fund the more economical option even if it is clearly a far superior choice.

The previous example is extremely simplistic, dealing only with initial and annual operating costs, and expressing those costs in "lump sum" terms. In the following example, we will add a third type of expenditure (a future cost), break the other cost elements into their component parts, and partially take inflation into account.

III. The furnace serving the administrative offices in your plant needs to be replaced. You have narrowed the decision down to options (A), an 80% efficient unit, and (B), a more expensive, 95% efficient unit. The estimated costs and study factors are as follows:

	"A"	"B"
Hardware Cost	$1,500	$2,200
Installation Labor (using in-house staff)	300	300
Annual Servicing/Tune-up	50	75
Annual Energy Use	1,200	1,000
Igniter Replacement (every 10 years; option "A" only)	120*	—
Vent Damper Motor Replacement (every 12 years: option "B" only)	—	200*

General Assumptions:

Study period = 20 years
Investment rate = 8%
Inflation rate = 4%

* Based on industry experience, you expect to have to replace these components at specific times in the future. While you don't know what the future costs will be, you do know what the work would cost if done today. To improve the accuracy of our study, we will estimate those future costs by applying an inflation factor.

As with the previous example, we will convert all expenditures to their equivalent present worth. The PW for either option is:

PW (Option) = Initial Costs + PW (Annual Costs) + PW (Future Costs)

Option "A"

Initial Costs = Hardware + Installation Labor
= $1,500 + $300 = **$1,800**

PW (Annual Costs) = PW (Annual Service) + PW (Energy)
= $50 x (P/A) @ 8%, 20 years + $1,200 (P/A) @ 8%, 20 years
= $50 x (9.8182) + $1,200 x (9.8182) = **$12,273**

So far the process has been virtually identical to that used in Example II. You have probably noticed that the annual cost elements could have been combined before multiplying by the conversion factor, which raises the question, "Why separate the costs?" The purpose of separating the cost components is to allow you to see that all relevant factors have been included, and that the estimate for each is reasonable. After documenting the various components, it is perfectly acceptable to combine them numerically when performing this conversion.

The final element in our study is the cost for the future replacement of the igniter. To make our results a little more accurate, we will use the Time Value of Money tables to estimate what the actual charges will be in the future, then convert those future costs to their equivalent present value. We know that it would cost $125 to replace the igniter if we did it today. Based on historical trends for this type of service, we estimate that this cost will increase approximately 4% per year. To calculate a future value from a present amount:

FW (Replacement) = Present Cost x (F/P) @ 4%, 10 years
= $125 x (1.480) = $185

At this point we are halfway home. We have just calculated what we expect to pay for this work when it is done in ten years. But remember that our goal is to convert all costs to present value, which means we now must convert these future dollars back to their present equivalent. Keep in mind that we are dealing with a hypothetical situation in which we are calculating the amount we would have to invest today to have the funds necessary to pay this $185 bill when it comes due in ten years. Since we are dealing with the amount we would invest, we must use the investment rate (8%) when we do the conversion, not the inflation value!

PW (Replacement) = Future Cost x (P/F) @ 8%, 10 years
= $185 x (.4632) = **$86**

We now have all of the costs to plug into our original formula:

PW (Option "A") = Initial Costs + PW (Annual Costs) + PW (Future Costs)
= $1,800 + $12,723 + $86 = **$14,609**

Following the pattern used in solving for "A", solving for "B" is quite easy...

Option "B"

Initial Costs = Hardware + Installation Labor
= $2,200 + $300 = **$2,500**

PW (Annual Costs) = PW (Annual Service) + PW (Energy)
= $75 x (P/A) 8%, 20 years + $1,000 (P/A) @ 8%, 20 yrs.
= ($75 + $1,000) x (9.8182) = **$10,555**

FW (Replacement) = Present Cost x (F/P) @ 4%, 12 yrs.
= $200 x (1.601) = **$320**

PW (Replacement) = Future Cost x (P/F) @ 8%, 12 yrs.
= $320 x (.3971) = **$127**

PW (Option "B") = Initial Costs + PW (Annual Costs) = PW (Future Costs)
= $2,500 + $10,555 + $127 = **$13,182**

With all costs converted to a common base—Present Worth—it is easy to compare the two choices and see that Option "B" costs less than "A."

	Option "A"	**Option "B"**
Present Worth	$14,609	$13,182

Besides Present Worth, it is fairly common to convert all cost factors to an *Equivalent Annual Charge.* Many managers are more comfortable with this frame of reference because it indicates how much revenue must be collected to pay the actual and equivalent annual expenses associated with the project. Using the Time Value of Money tables we can convert the Present Worth results of Example III to an Equivalent Annual Charge.

EAC = PW x (A/P) @ i%, n years

EAC (Option "A")	= PW ("A") x (A/P) @ 8%, 20 years
	= $14,609 x (0.1019) = $1,489
EAC (Option "B")	= PW ("B") x (A/P) @ 8%, 20 years
	= $13,182 x (0.1019) = $1,343

As expected, Option "B" is still less expensive than Option "A," with ownership and operating cost being $145/year less.

Figures 2.1–2.4 are tables that allow you to calculate Present Worth.

Effective cost management requires carefully prepared feasibility studies, a cost estimate, and a responsive cost monitoring and reporting system based on an established work breakdown structure. The design and budget stage of the cost management process may include a Life Cycle Cost Analysis. The relative simplicity or complexity of the analysis will depend on your situation; there is no single approach. For example, one organization may be chiefly interested in payback, while another is in a position to commit to a long-term capital expenditure. An established history of good cost analysis and control procedures may be the factor that tips the scale in getting your plans approved over competing corporate investment alternatives.

Capitalized Cost

In some cases, a manager must estimate the present worth of an *infinite* annual cost. This type of analysis is referred to as a capitalized cost.

Example 7

A person wants to withdraw $1,000 at the end of each year forever. The investment pays 5% annual effective interest each year. How much must be deposited in order to withdraw $1,000 each year forever? The savvy engineering economist would refer to this as the capitalized cost *of the investment.*

Capitalized Cost Formula: **P = A/i**

Note that there is no functional notation for this due to the simplicity of the formula.

Answer:

P = A/i Thus P = 1000/.05 P = $20,000

Uniform Gradient

In some cases the cash flows of an investment are an increasing uniform series. The formulas and functional notation for a uniform gradient are given in Figure 2.15.

Example 8

A process machine cost $1,600 and generates a profit at the end of Year 1 of $400, at the of Year 2 of $800, at the end of Year 3 of $1,200, at the end of Year 4 of $1,600, and at the end of Year 5 of $2,000. If the cost of capital is 8%, what is the net present value of this machine?

Answer:

The first step is to draw the cash flow diagram, as shown in Figure 2.14.

Then resolve the cash flow into two components as shown in Figure 2.16:

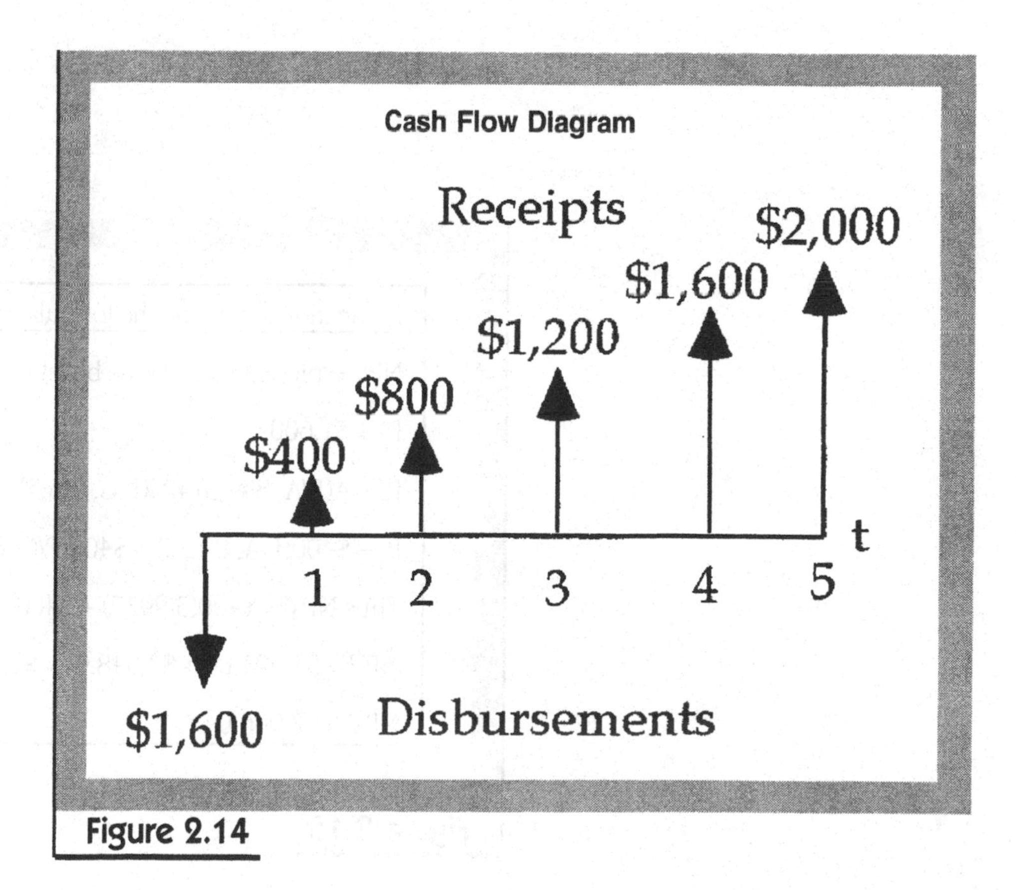

Figure 2.14

Notice that the gradient does not begin until the end of Year 2. Although this is the case the formula was calculated using the number of periods, not the number of cash flows. Thus, in either case, n will equal 5.

It is important to note that this problem could also have been solved by simply repeating the single payment present worth formula for each of the five cash flows and subtracting it from the present value of the cost, $1,600. The functional notation for this would be:

NPV = F(P/F, 8%, 1) + F(P/F, 8%, 2) + F(P/F, 8%, 3) + F(P/F, 8%, 4) + F(P/F, 8%, 5) -$1,600

NPV = $400(.9259) + $800(.8573) + $1,200(.7938) + $1,600(.7350) + $2,000(.6806) - $1,600

NPV = $370.36 + $685.84 + $952.56 + $1,176 + $1,361.20 - $1,600

NPV = 2945.96 (the numbers are not exact due to rounding)

In functional notation the formula would be:

NPV = present value of the benefit - present value of the cost

P_C = $1,600

P_B = A(P/A, 8%, 5) + G(P/G, 8%,5)

P_B = $400(P/A, 8%, 5) + $400(P/G, 8%,5)

Thus NPV = $400(3.9927) + $400(7.3724) - $1,600

NPV = $1,597.08 + $2,948.96 - $1,600

NPV = $2,946.04.

Figure 2.15

Rate of Return Criterion

As stated in example problem 2 calculating the rate of return on an investment whose cash flow is periodic involves an iterative process and should be done with a financial calculator or a computerized spreadsheet.

Example 9

Let's look at the cash flow from Example 2. First in tabular format:

Year	Cash Flow
0	-$10,000
1	$2,000
2	$2,000
3	$2,000
4	$2,000
5	$2,000
6	$2,000
7	$2,000
8	$2,000
9	$2,000

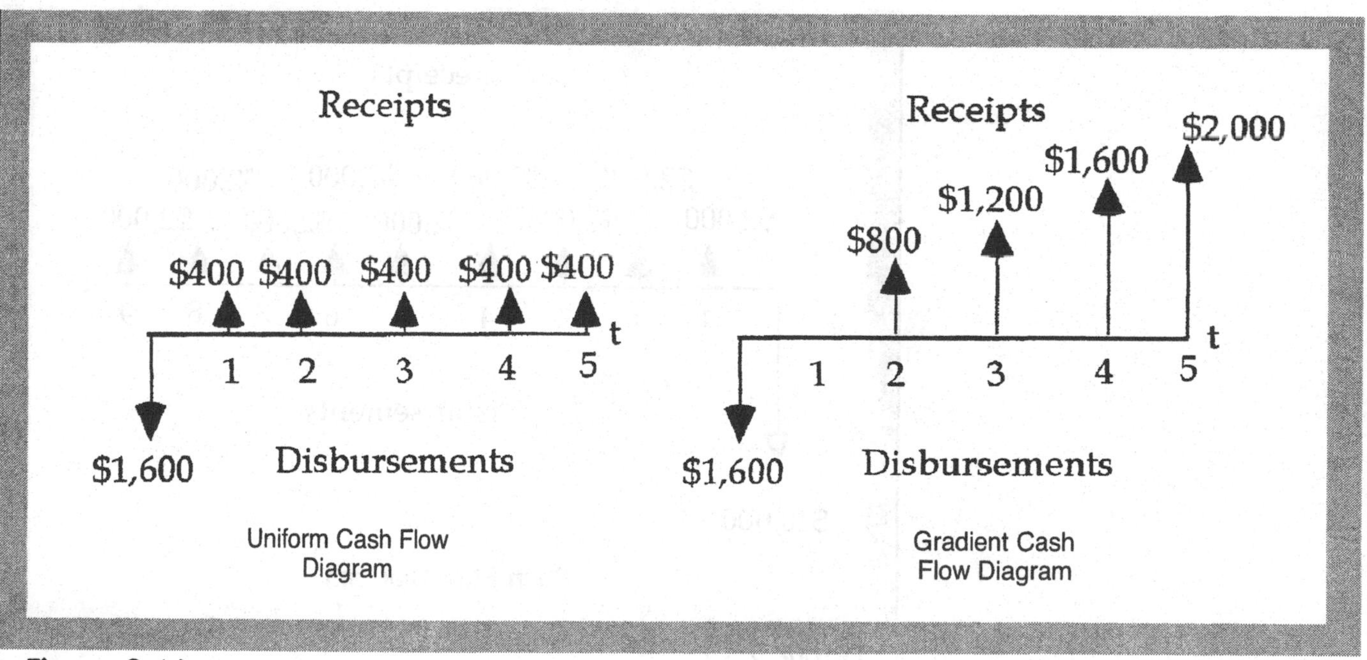

Figure 2.16

See Figure 2.17 for a graphic representation.

Answer:

The rate of return that solves this problem is the rate that makes the net present value equal to zero. In the financial world this is referred to as the "internal rate of return."

The internal rate of return is the rate in which;

a. NPV = 0
b. PW of benefits = PW of costs
c. EUAC = EUAB
d. PW of benefits – PW of costs = 0.

Solving for this using functional notation:

A(P/A, i%, n) – $10,000 = 0

$2,000 (P/A, i%, 9) – $10,000 = 0.

At this point, you must go to the tables and pick an interest rate, find the uniform series present worth factor for nine years, plug it back into the equation, and see how close you are. Then, keep repeating the process until you solve the problem. Clearly, the better method is to use either a financial calculator or a computerized spreadsheet. The internal rate of return for this problem was 13.7%.

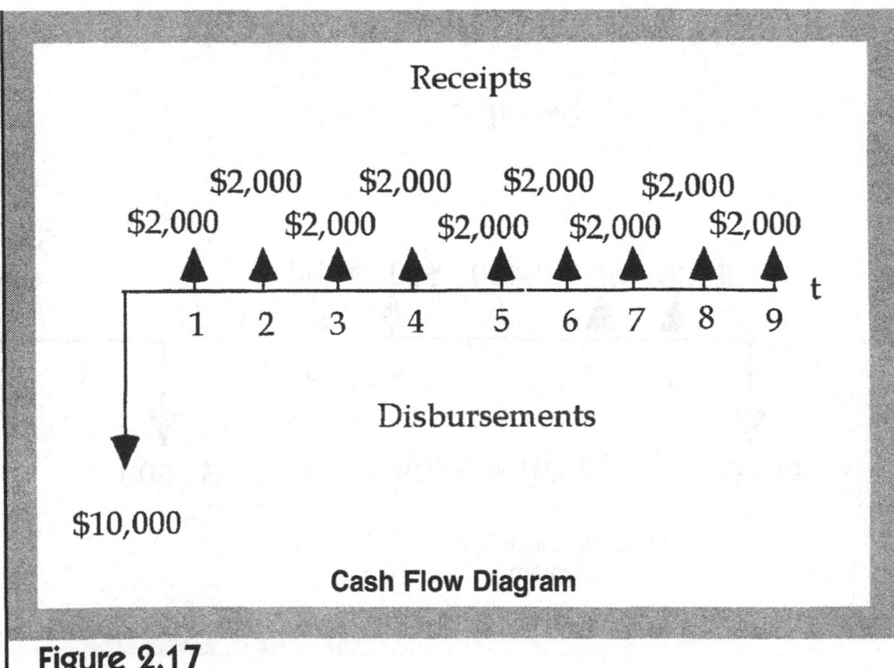

Figure 2.17

Minimum Attractive Rate of Return (MARR) and Incremental Rate of Return

Rate of return is often used as an economic estimator. The common rule-of-thumb is to choose the investment with the highest rate of return. Although this usually makes good financial sense, there are certain cases where it is **not** true. When a facility manager has a choice between two mutually exclusive alternatives, it may be wiser to choose the alternative with the lower ROR, as long as the incremental rate of return is above the company's minimum attractive rate of return (MARR).

Example 10

A facility plant manager needs to purchase a new air compressor. She has narrowed her choices down to two: a reciprocating compressor and a rotary-screw compressor. The company uses 5% as its minimum attractive rate of return (MARR). The cash flows for each compressor, due to maintenance savings, are as follows:

Reciprocating Comp		Rotary Comp	
Year	Cash Flow	Year	Cash Flow
0	($10,000)	0	($16,000)
1	$4,000	1	$6,300
2	$4,000	2	$6,300
3	$4,000	3	$6,300

Solving for the internal rate of return for each:

IRR(recip) = 9.7% IRR(Rotary) = 8.8%

Although it appears that the easy choice is the reciprocating compressor, because it has a higher internal rate of return, this is not the case. First, the plant manager must compute the **incremental** rate of return of the difference between the alternatives. The incremental rate of return is simply the internal rate of return for the cash flow of the difference between the alternatives.

Reciprocating Comp		Rotary Comp		Diff. between Alternatives	
Year	Cash Flow	Year	Cash Flow	Year	Cash Flow
0	($10,000)	0	($16,000)	0	($6,000)
1	$4,000	1	$6,300	1	$2,300
2	$4,000	2	$6,300	2	$2,300
3	$4,000	3	$6,300	3	$2,300
IRR	9.70%	IRR	8.80%	Inc. IRR	7.30%

Notice that the incremental ROR is 7.3%, which is above the company's MARR. Therefore, the plant manager should choose the rotary compressor, even though it is the alternative with the higher initial cost and lower rate of return. This is true because the rotary compressor will earn the company more dollars over time.

Another way to approach this solution is that the company has $16,000 and has two mutually exclusive options. It can either choice:

(a) invest the $16,000 in a rotary compressor or

(b) invest $10,000 in a reciprocating compressor and the other $6,000 in a bank earning 5% interest.

Choice (a) earns the company $6,300 each year for three years.

Choice (b) earns the company $4,000 each year + the capital recovery of the $6,000 @ 5%.

Choice (b) earns $4,000 + $2,203 which equals $6,203.

Thus, choice (a), the rotary compressor, is preferred. The incremental rate of return analysis is the optimal method when choosing from two or more mutually exclusive alternatives.

Benefit/Cost Ratio

Many municipal works projects and state and federal agencies often use the benefit/cost ratio form of analysis. This analysis is very similar to the incremental rate of return analysis.

The formula is B/C ratio = PW of the benefits/PW of the costs with the idea that any ratio greater than 1 is good. But in the case of two alternatives both with a B/C greater than 1, the correct answer is to choose the higher cost alternative if the **incremental** B/C is greater than 1.

Example 11

Solve example 10 using B/C ratio.

Reciprocating Comp		Rotary Comp		Diff. between Alternatives	
Year	Cash Flow	Year	Cash Flow	Year	Cash Flow
0	($10,000)	0	($16,000)	0	($6,000)
1	$4,000	1	$6,300	1	$2,300
2	$4,000	2	$6,300	2	$2,300
3	$4,000	3	$6,300	3	$2,300
B/C ratio	1.089	B/C Ratio	1.072	B/C Ratio	1.044

Compute the B/C ratios using a 5% cost of capital.

B/C ratio of reciprocating compressor = PW of the benefits/PW of the costs

B/C = $4,000(P/A, 5%, 3)/$10,000

B/C_{recipr} = **1.089**.

B/C ratio of rotary compressor = $6,000(P/A, 5%, 3)/$16,000

B/C_{rotary} = **1.021**.

B/C ratio of difference between alternatives = $2,300(P/A, 5%, 3)/$6,000

B/C_{diff} = **1.044.**

Because the incremental B/C ratio is still greater than 1 the correct choice is the rotary compressor even though it costs more and has a **lower** B/C ratio.

1. The dollar amount could take into account the time value of money or it could ignore it. Most standard accounting methods ignore the time value of money.

2. There are also geometric or exponential gradient formulas that are beyond the scope of this text.

Resource Publications

Finkler, Steven A., *Finance and Accounting for Non-Financial Managers*. Paramus, NJ: Prentice Hall, 1996.

Milliner, Michael S., *Contractor's Business Handbook*. Kingston, MA: R.S. Means Co., Inc., 1988.

Muro, Vincent, *Handbook of Financial Analysis for Corporate Managers, Revised Edition*. New York, NY: AMACOM, 1998.

Palmer, William J., William E. Coombs, and Mark A. Smith, *Construction Accounting and Financial Management*. New York, NY: McGraw-Hill, Inc., 1995.

Sievert, Richard W., *Total Productive Facilities Management*. Kingston, MA: R.S. Means Co., Inc., 1998.

Chapter 3

Civil Engineering & Construction Practices

Chapter 3 Civil Engineering & Construction Practices

This chapter takes the facility engineer through the entire construction process, from initial surveys and site plans to construction management and scheduling. Along the way, it reviews basic soils and foundation properties, hydraulics, pipe and storm water design, and the essential principles of structural design, concrete and pavement applications.

While some of the chapter's topics may fall outside the facility engineer's immediate jurisdiction, a general understanding of the subject matter will allow for more effective management of operations. In short, this information lets you know what to expect with civil engineering projects, including primary methods of project delivery and the role of all important team members.

Surveying

Surveying is the measurement of distances, elevations, or angles of the earth's surface, including both natural and man-made features. Plant engineers most often use surveys, which are used to build at proper elevation and location, with new construction. They may also use site surveys to evaluate issues involving utilities, drainage, and site expansions. This section offers help in interpreting topographical maps and summarizes general surveying practices.

Types of Surveys

Surveys are usually conducted for the preparation of a plan or map, and most often fall into one of the following categories:

- **Control:** determines the horizontal or vertical positions of arbitrary points
- **Construction:** determines distances and angles to locate construction lines for buildings, infrastructure, etc., or locates engineering works
- **Hydrographic:** determines the configuration of a surface underwater

- **Property:** determines the positions of boundary lines, and the area of tracts within those lines
- **Topographic:** determines the configuration of the ground and locates its natural and artificial (man-made) features

Topographic surveys are used by engineers to find the most desirable and economical location for buildings, pipelines, utility lines, and infrastructure. The method used to represent hills, depressions, and ground-surface undulations on a topographical map involves the use of *contours*, which are lines connecting points of equal elevation. Most contours are irregular lines, like the closed loops that represent a hill or similar elevation. The vertical difference between the level surfaces that form contours is referred to as *contour interval.*

The following are the characteristics of contours that are most fundamental to understanding a topographical map:

- Contours that increase in elevation represent hills. Valleys are represented with contours of decreasing elevation. Elevations are shown on the uphill side of lines or in breaks in the line.
- Uniform slopes are represented by evenly spaced contours.
- Wide spacing between contours denotes flat slopes; narrow spacing indicates steep ones.
- Smooth contour lines signify gradual changes, while irregular ones indicate rough country.
- Contours never meet, unless they are representing a wall or cliff. They cross only in those rare cases where elevations overlap, as in a cave or overhanging cliff.
- All contours must close upon themselves since the earth is a continuous surface.

Field Data

Survey calculations may be carried out in the field or in the office. Most are now done in the latter by computer. This innovation and the use of a direct digital reading total station instrument has revolutionized the practice of surveying.

Some types of surveys require only a few simple calculations that can be carried out by hand, while others are best tackled with the help of a computer program. Either way, surveying requires a familiarity with field operational techniques and the mathematics applied in surveying data.

Field readings, whether electronic or written, become the record of the survey. For this reason, an exacting approach to taking measurements is critical. The majority of field data is conducted on devices called **total station instruments**, which provide for the measurement and recording of horizontal and vertical angles and distances. Horizontal distance measurements are displayed to one thousandth of a foot, while angles are direct readings in degrees, minutes, and seconds. The surveyor should clearly state the time, location, and purpose for the survey in the field data.

The following section is reprinted with permission from *Surveying*, by Harry Bouchard and Francis Moffitt, International Textbook Company.

Basic Definitions

In order to gain a clear understanding of the procedures for making surveying measurements on the earth's surface, you must be familiar with the meanings of certain basic terms. The terms discussed here have reference to the actual figure of the earth.

A *spheroid* is an ellipse rotated on its shorter axis. The ideal figure of the earth is a spheroid having the actual axis of rotation as the shorter axis. Because of its relief, the earth's surface is not a true spheroid. However, an imaginary surface representing a mean sea level extending over its entire surface very nearly approximates a spheroid. This imaginary surface is used as the figure on which surveys of large extent are computed.

A *vertical line* at any point on the earth's surface is the line which follows the direction of gravity at that point. It is the direction which a string will assume if a weight is attached to the string and the string is suspended freely at the point. At a given point there is only one vertical line. The earth's center of gravity cannot be considered to be located at its geometric center, because vertical lines passing through several different points on the surface of the earth do not intersect in that point. In fact, all vertical lines do not intersect in any common point. A vertical line is not necessarily normal to the surface of the earth, nor even to the spheroid.

A *horizontal line* at a point is any line which is perpendicular to the vertical line at the point. At any point there are an unlimited number of horizontal lines.

A *horizontal plane* at a point is the plane which is perpendicular to the vertical line at the point. There is only one horizontal plane through a given point.

A *vertical plane* at a point is any plane which contains the vertical line at the point. There are an unlimited number of vertical planes at a given point.

A *level surface* is a continuous surface which is at all points perpendicular to the direction of gravity. It is exemplified by the surface of a large body of water at complete rest (unaffected by tidal action).

A *horizontal distance* between two given points is the distance between the points projected onto a horizontal plane. The horizontal plane, however, can be defined at only one point. For a survey the reference point may be taken as any one of the several points of the survey.

A *horizontal angle* is an angle measured in a horizontal plane between two vertical planes. In surveying, this definition is effective only at the point at which the measurement is made or at any point vertically above or below it.

A *vertical angle* is an angle measured in a vertical plane.

The *elevation* of a point is its vertical distance above or below a given level surface.

The *difference in elevation* between two points is the vertical distance between two level surfaces containing the two points.

Plane surveying is that branch of surveying wherein all distances and horizontal angles are assumed to be projected onto one horizontal plane. A single reference plane may be selected for a survey where the survey is of limited extent.

Geodetic surveying is that branch of surveying wherein all distances and horizontal angles are projected onto the surface of the spheroid which represents mean sea level on the earth.

The surveying operation of leveling takes into account the curvature of the spheroidal surface in both plane and geodetic surveying. The leveling operation determines vertical distances and hence elevations and differences of elevation.

Units of Measurement

In English-speaking countries, the linear unit most commonly used is the foot, and the unit of area is the acre, which is 43,560 sq. ft. In Europe and South America, distances are expressed in meters. This unit is also used by the United States Coast and Geodetic Survey, although the published results of leveling are expressed in feet.

For purposes of computation and plotting, decimal subdivisions of linear units are the most convenient. Most linear distances are therefore expressed in feet and tenths, hundredths, and thousandths of a foot. The principal exception to this practice is in the layout work on a construction job, where the plans of the structures are dimensioned in feet and inches. Tapes are obtainable graduated either decimally or in feet and inches.

Volumes are expressed in either cubic feet or cubic yards. Angles are measured in degrees (°), minutes ('), and seconds ("). One circumference = 360°; 1° = 60'; 1' = 60".

Leveling

Leveling is the operation in surveying performed to determine and establish elevations of points, to determine differences in elevation between points, and to control grades in construction surveys. The basic instrument is a level, which establishes a horizontal line of sight.

Direct Differential Leveling: The purpose of differential leveling is to determine the difference of elevation between two points on the earth's surface. The most accurate method of determining differences of elevation is with the level and a rod, in the manner illustrated in Figure 3.1. It is assumed that the elevation of point A is 976 ft. and that it is desired to determine the elevation of point B. The level is set up at some convenient point so that the plane of the instrument is higher than both A and B. A leveling rod is held vertically at the point A, which may be on the top of a stake or on some solid object, and the telescope is directed toward the rod. The vertical distance from A to a horizontal plane can be read on the rod where the horizontal cross hair of the telescope appears to coincide. If the rod reading is 7.0 ft., the plane of the telescope is 7.0 ft. above the point A. The elevation of this horizontal plane is 976 + 7 = 983 ft. The leveling rod is next held vertically at B and the telescope is directed toward the rod. The vertical distance from B to the same horizontal plane is given by the rod reading with which the horizontal cross hair appears to coincide. If the rod reading at B is 3.0 ft., the point B is 3.0 ft. below this plane and the elevation of B is 983 – 3 = 980 ft. The elevation of the ground at the point at which the level is set up need not be considered.

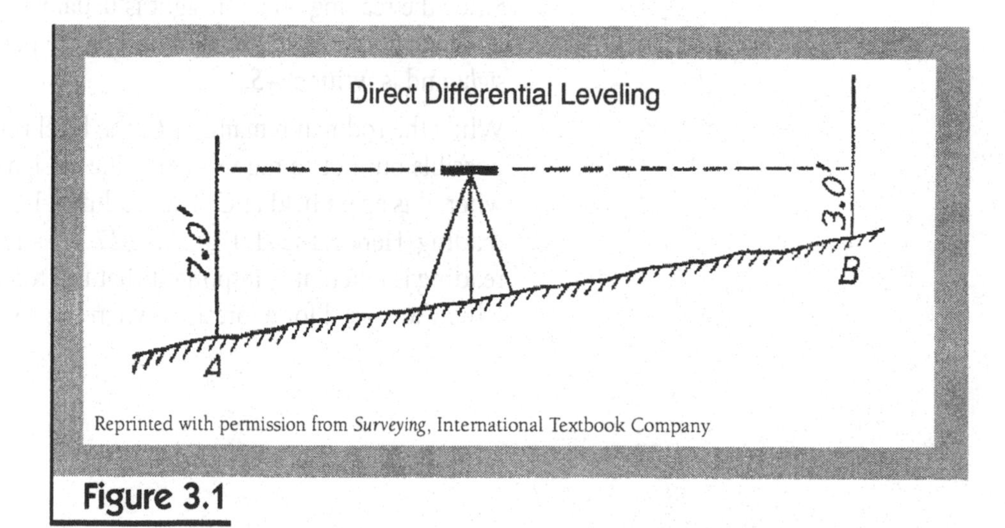

Reprinted with permission from *Surveying*, International Textbook Company

Figure 3.1

The same result may be obtained by noting that the difference in elevation between A and B is 7 – 3 = 4 ft., and that B is higher than A. The elevation of B equals the elevation of A plus the difference of elevation between A and B, or 976 + 4 = 980 ft.

Running a Line of Levels: In the previous example of direct leveling, it was assumed that the difference of elevation between the two points considered could be obtained by a single setting of the level. This will be the case only when the difference in elevation is small and when the points are relatively close together. In Figure 3.2, rods at the points *A* and *K* cannot be seen from the same position of the level. If it is required to find the elevation of point *K* from that of *A*, it will be necessary to set up the level several times and to establish intermediate points such as *C, E,* and *G.* These are the conditions commonly encountered in the field and may serve as an illustration of the general methods of direct leveling.

Let the elevation of the bench mark, abbreviated B. M., at *A* be assumed as 820.00 ft. The level is set up at *B*, near the line between *A* and *K*, so that a rod held on the B. M. would be visible through the telescope; the reading on the rod is found to be 8.42 ft. The *height of instrument*, abbreviated H. I., is the vertical distance from the datum to the plane of sight. Numerically, the H. I. is found by adding the rod reading taken on the bench mark. Thus, the H. I. of the level at *B* is 820.00 + 8.42 = 828.42 ft. The rod reading taken at *A* by directing the line of sight toward the start of the line is called a *backsight reading*, or simply a *backsight*, abbreviated B. S. The backsight is the rod reading that is taken on a point of known elevation to determine the height of instrument. Since a backsight is usually added to the elevation of the point on which the rod is held, it is also called a *plus sight*, written + S.

After the reading has been taken on a rod at *A*, a point *C* is selected which is slightly below the line of sight, and the reading is taken on a rod held at *C*. If this reading is 1.20 ft., the point *C* is 1.20 ft. below the line of sight, and the elevation of *C* is 828.42 – 1.20 = 827.22 ft. The reading on *C* is called a *foresight reading*, or *foresight*, abbreviated F. S. It is taken on a point of unknown elevation in order to determine the elevation from the height of instrument. Since the reading on a foresight is usually subtracted from the height of instrument to obtain the elevation of the point, a foresight is also called a *minus sight* and is written – *S*.

While the rodman remains at *C*, the level is moved to *D* and set up as high as possible but not so high that the line of sight will be above the top of the rod when it is again held at *C*. The reading 11.56 ft. is taken as a backsight or + S reading. Hence, the H. I. at *D* is 827.22 + 11.56 = 838.78 ft. When this reading is taken, it is important that the rod be held on exactly the same point which was used for a foresight when the level was at *B*. The point *C* should be

Figure 3.2

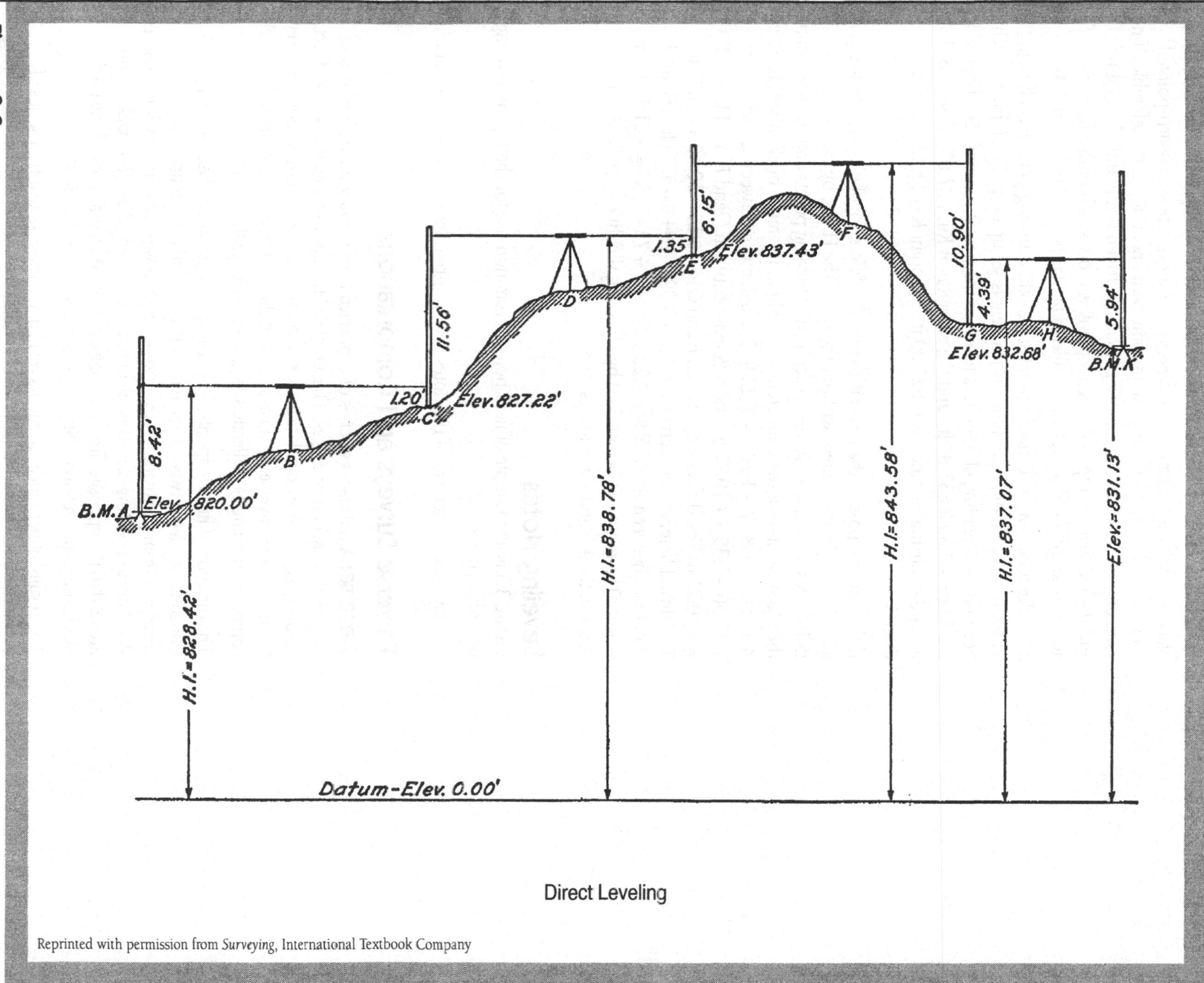

Direct Leveling

Reprinted with permission from *Surveying*, International Textbook Company

some stable object, so that the rod can be removed and put back in the same place as many times as may be necessary. For this purpose, a sharp-pointed rock or a well-defined projection on some permanent object is preferable. If no such object is available, a stake or peg can be driven firmly into the ground and the rod held on top of it. Such a point as *C* on which both a foresight (– S) and a backsight (+ S) are taken, is called a *turning point*, abbreviated T. P.

After the backsight on *C* has been taken, another turning point *E* is chosen. This process continues until this instrument is set up on point H, from which position a + S reading of 5.94 ft. is taken on the new B. M. at *K*. The final H. I. is 832.68 + 4.39 = 837.07 ft., and the elevation of *K* is 837.07 – 5.94 = 831.13 ft. As the starting elevation was 820.00 ft., the point *K* is 11.13 ft. higher than *A*.

If the only purpose of the levels is to find the difference of elevation between *A* and *K*, it can be computed, without the use of the H. I.s, by calculating the differences of elevation between each pair of turning points and then obtaining the algebraic sum of these differences. The first difference of elevation, between *A* and C, is + 8.42 – 1.20 = + 7.22 ft. The difference between *C* and *E* is + 11.56 – 1.35 = + 10.21 ft. The difference between *C* and *E* is + 11.56 – 1.35 = + 10.21 ft. The difference between *E* and *G* is + 6.15 – 10.90 = – 4.75 ft. The final difference between *G* and *K* is + 4.39 – 5.94 = – 1.55 ft. The total difference between *A* and *K* is + 7.22 + 10.21 – 4.75 – 1.55 = + 11.13 ft., as before. The result can be checked by obtaining the difference between the sums of the + S and – S readings.

Leveling Notes

Figure 3.3 is an example of field notes in common use for the type of leveling described above.

The layout/arrangement of the notes is very important to simplify calculations.

Traverse Surveys and Computations

Traverse: A traverse is a series of connected lines of known length related to one another by known angles. The lengths of the lines are determined by direct measurement of horizontal distances. The angles at the traverse stations between the lines of the traverse are measured with the total station instrument. The angles can be interior angles, deflection angles, or angles to the right.

The direction of lines is specified by *bearings* and *azimuths*. The former are horizontal angles between a reference direction and a given line. They are measured from north and south, toward east and west, and are never greater than 90°. Forward bearings are those read in the advancing direction; back bearings are read in the opposite direction. *Azimuths* are horizontal angles measured clockwise from a reference line, most often north, to an object.

The results of field measurements related to a traverse will be a series of connected lines whose lengths and azimuths, or whose lengths and bearings, are known. The lengths are horizontal distances; the azimuths or bearings are true, magnetic, assumed, or grid.

In general, traverses are of two classes. One is an open traverse, and the second is a closed traverse, which can be described in any one of the following three ways: 1) It originates at an assumed horizontal position and terminates at that same point; 2) it originates at a known horizontal position with respect to a horizontal datum and terminates at that same point; 3) it originates at a known horizontal position and terminates at another known horizontal position. A known horizontal position is defined by its geographic latitude and longitude, by its Y- and X-coordinates on a grid system, or by its location on or in relation to a fixed boundary.

Traverse surveys are made for many purposes and types of projects, some of which follow:

- To determine the positions of existing boundary markers.
- To establish the positions of boundary lines.
- To determine the area encompassed within the confines of a boundary.
- To determine the positions of arbitrary points from which data may be obtained for preparing various types of maps, that is, to establish *control* for mapping.
- To establish ground control for photogrammetric mapping.
- To establish control for gathering data regarding earthwork quantities in railroad, highway, utility, and other construction work.
- To establish control for locating railroads, highways, and other construction work.

Level Notes

FORM A

1960-4-10. Leveling, B.M.A. to B.M.K.
Washtenaw Ave. Sewer Project.
Ann Arbor, Mich.

Level #4096. Level & Rec. J. Brown
Rod #18. Rodman F. Smith

Sta.	+S (B.S.)	H.I.	−S (F.S.)	Elev.		
B.M.A.	8.42	828.42		820.00		Top of iron pipe, S.E. cor. Washtenaw & Hill Sts.
T.P.1	11.56	838.78	1.20	827.22		
T.P.	6.15	843.58	1.35	837.43		
T.P.	4.39	837.07	10.90	832.68		
B.M.K.	+30.52		5.94	831.13		Top of iron pipe, N.W. cor. Washtenaw & Oxford Rd.
	−19.39		19.39			
	+11.13					820.00 B.M.A.
						+11.13 Diff. Elev.
						831.13 B.M.K. Check.
						J. Brown

Reprinted with permission from *Surveying*, International Textbook Company

Figure 3.3

Open Traverse: An open traverse is usually run for exploratory purposes. There are no other arithmetical checks on the field measurements. Since the figure formed by the surveyed lines does not close, the angles cannot be summed up to a known mathematical condition.

Closed Traverse: A traverse that closes on itself immediately affords a check on the accuracy of the measured angles, provided that the angle at each station has been measured and recorded. A traverse that closes on itself gives an indication of the consistency of measuring distances as well as angles by affording a check on the position closure of the traverse.

Interior-Angle Traverse: An interior-angle traverse is shown in Figure 3.4. The azimuth or bearing of the line *AP* is known. The lengths of the traverse lines are measured to determine the horizontal distances. With the total station instrument at *A*, the angle at *A from P* to *B* is measured to determine the azimuth of line *AB*. The angle at *A* from *E* to *B* is also measured, as this is one of the interior angles in the figure. Magnetic bearings to *P, E*, and *B* are observed to provide a rough check on the values of the measured angles. The instrument is then set up at *B, C, D*, and *E* in succession, and the indicated angles are measured and magnetic bearings are observed.

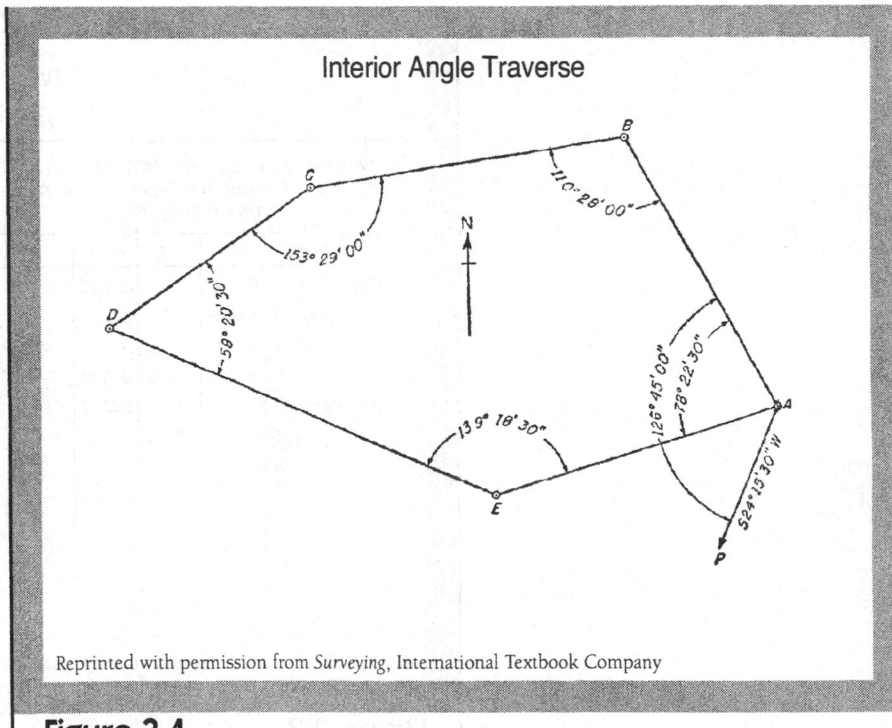

Reprinted with permission from *Surveying*, International Textbook Company

Figure 3.4

To test the angular closure, the interior angles are added and their sum is compared to $(n - 2)$ 180°, which in this example is (5 – 2) 180° = 540°. The total angular error, or the closure, is 1' 30".

After the adjusted angles are computed, they should always be added to see whether their sum is, in fact, the proper amount. A mistake in arithmetic either in adding the measured angles or in applying the corrections will become apparent.

One very important point to observe in the preceding example of an interior-angle traverse is that there is no check on the angle at *A* from *P* to B. If this angle is in error, then the error affects the azimuth of each line in the traverse. To avoid the possibility of making a mistake in measuring this angle, the clockwise angle at *A* from *B* to *P* should also be measured.

Latitudes and Departures

Closure of a traverse is checked by computing the *latitude* and *departure* of each line. The latitude of any line is its projection on a north and south line, and is often called a *northing* or a *southing*. The departure of a line, or its *easting* or *westing*, is its projection on an east or west line. Departures and latitudes are the X- and Y-components found in algebra and mechanics.

North latitudes and east departures are considered positive or "+"; south latitudes and west departures are negative or "–." A line with a northeast bearing would have a positive latitude and departure. Likewise, a line with a southwest bearing would have negative latitude and departure. For closure in a traverse, all sum of all latitudes must equal zero and the sum of all departures must equal zero.

Stated in the form of a general rule applicable to any line,

Latitude = length x cosine bearing angle

Departure = length x sine bearing angle

To determine either, then, one must first determine the bearing and length of the lines, followed by the natural sines and cosines of the bearing angles. Traverse tables are also available that provide for various bearing angles the latitudes and departures for various lengths of line.

Site Planning

Site planning can be simply defined as the act of arranging structures on a site and shaping the spaces between and around them. It involves the location of structures and activities within a specific environment, and defines how these elements relate to and interact with their surroundings. Good planning involves a thorough understanding of the project in relationship to its environment. It is in many ways an art, linking the fields of architecture, engineering, landscape architecture, and city planning.

The goal is to locate a new structure into an existing setting in such a way that the program will be achieved while enhancing the local environment.

There is a wide range of concerns relating to the development or redevelopment of a site for a specific planned purpose. Site planning is the systematic process of identifying these concerns and addressing them. The major design concerns include:

- Zoning requirements
- Environmental concerns and approvals
- Placement of building(s) on site
- Circulation: pedestrian and vehicular
- Drainage and disposal of surface runoff
- Landscape design; new and existing plants
- Soil bearing capacity; placement of foundation and below grade construction
- Utilities on site; connection to off-site services
- Interface with neighboring sites and streets
- Site lighting, natural and artificial
- Sound control
- Air quality
- Fire safety; accessibility by fire department
- Security for people and building contents

A successful site plan must address all these issues. The process is complex. It is generally led by the principal designer of the project (architect, landscape architect, or civil engineer), but involves input from many parties, including the owner, contractor, engineering professionals, and civic authorities.

Sequence of Planning Tasks

Site planning usually follows a logical sequence of tasks. There are no sharp boundaries between these tasks, and frequently they are interrelated and overlapping. The sequence includes the following steps.

Problem Definition: Identifying project name and type, site location, expectations of owner, and available resources.

Site and User Analysis: Includes site visit by planner for familiarity of the site and its environs; collection of site information; and analysis of site for suitability for defined project.

Program: A specific statement of design requirements, stating objectives and behavioral information (i.e., building use, timing, intensity, connections between settings, expected management and service support).

Schematic Plan: Definition of patterns of activity, circulation, and building elements. This is usually presented in sketches of plans, sections, elevations, and diagrams showing interrelationships of the various elements of the project. Several schemes may be developed. At the end of the process, building form and location will be defined in relation to site circulation, outdoor activity, and general landscaping. A rough estimate will be calculated, and adjustments made to the project, if necessary, to meet budget requirements.

Detailed Plan and Contract Documents: After approval of the schematic plan by the owner, detailed plans and contract documents are produced. The plans detail the location of the building(s) on the site, all roads and paved surfaces, drainage structures, utilities, site improvements, and plantings. All changes to site contours are detailed. The specification details the general conditions of the contract, specific requirements of all materials on the project, and a complete description of the project and any alternates. Plans and specifications form the basis for bid documents. At this point a final detailed project estimate is produced for the owner's approval.

Bidding: When the contract documents are complete and approved by the owner, the project is put out to bid by contractors.

Construction, Supervision and Occupation: After the bid has been awarded and construction begins, the site planner reviews the construction periodically to ensure compliance with the plans and to provide assistance if changes are made.

The Planning Team

Many professionals are involved in the process of designing the building and its site. The exact number and type of professionals involved varies by project. In general, the following tasks are performed as indicated.

Surveys and Investigations: The architect researches zoning limitations including setbacks, easements, and other legal constraints on the site. The architect will also identify and document existing site features and seismic conditions as identified by building code authorities. An environmental consultant will be responsible for environmental studies, including identification of hazardous waste. The site engineer will identify and document the existence of surface water, soil conditions, and subsurface conditions, including materials and geology.

Site Engineering: The site engineer will identify all existing topography and design surface contouring, site utilities, drainage, and surface stabilization if required. In addition, the site engineer is concerned with all site construction, including retaining walls and roads, and the interface with existing streets and utilities.

Landscape Design: A landscape architect is responsible for the design of plantings and the location of site elements including drives, parking lots, walkways and terraces.

Foundation Design: The structural engineer is responsible for the building foundation and below-grade construction.

Design for Site Construction: The site engineer and structural engineer are responsible for foundation construction, including excavation, shoring, de-watering, etc.

Generally, the architect oversees the entire site design process. In reality this is a collaborative process, with the work of each professional dependent on and linked to the work of all the others. In small projects where there is not the full complement of professionals, the tasks will be assumed by another professional. For instance, if no landscape architect is involved, parking, walkways, and plantings may be designed by the architect or site engineer. See Figure 3.5 for a diagram of site planning team interaction.

No site can be considered in isolation. Site design must involve the site in relation to its surroundings, with the site's boundaries presenting significant constraints to the designer. Any re-contouring, for instance, must be accomplished completely within the site. Site drainage must be designed so that there are no negative consequences to neighboring sites or the environment. Pedestrian and vehicular access must be provided to the site, either using the existing infrastructure, or by developing new infrastructure with the approval of neighboring landowners. Likewise, utilities and sewers must tie into the existing infrastructure. Building and site orientation, including orientation to the sun and weather exposure, will be affected by the surrounding properties.

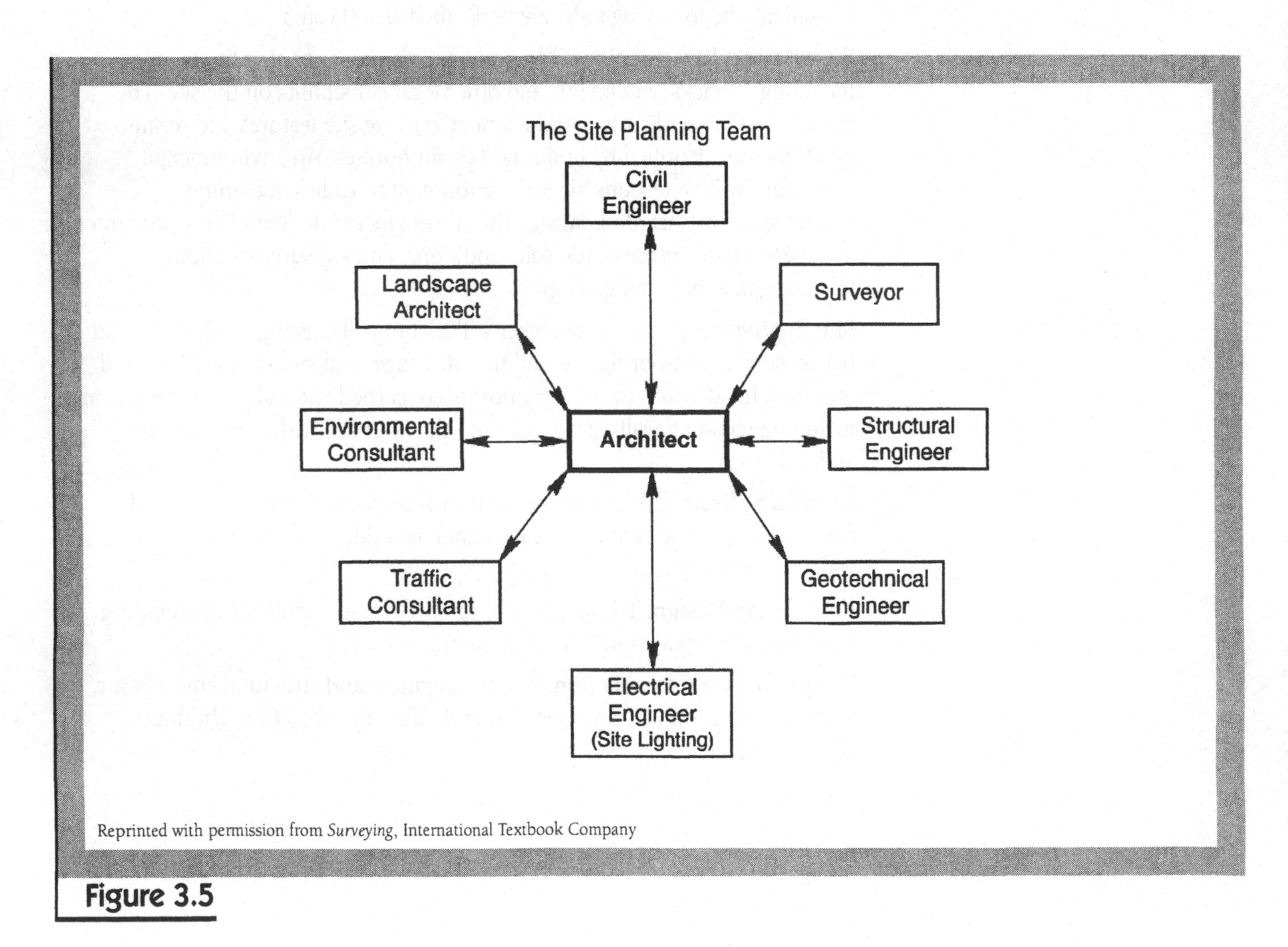

Reprinted with permission from *Surveying*, International Textbook Company

Figure 3.5

The Site Plan

Once preliminary information is gathered and a project is deemed feasible and progresses to the preliminary design phase, detailed information is gathered. A site survey is performed, and a registered land surveyor will create a site plan. This plan becomes a part of the legal description of the property. The survey will indicate site boundaries, locations of adjacent streets, easements for utilities, and the location and description of major site features. Figure 3.6 shows a site plan for a small retail location.

Additional information may be gathered, including general area maps, aerial surveys, geotechnical surveys, and geographic statistics. The amount of information gathered depends on the complexity of the project and the requirements of the owners and their funding sources.

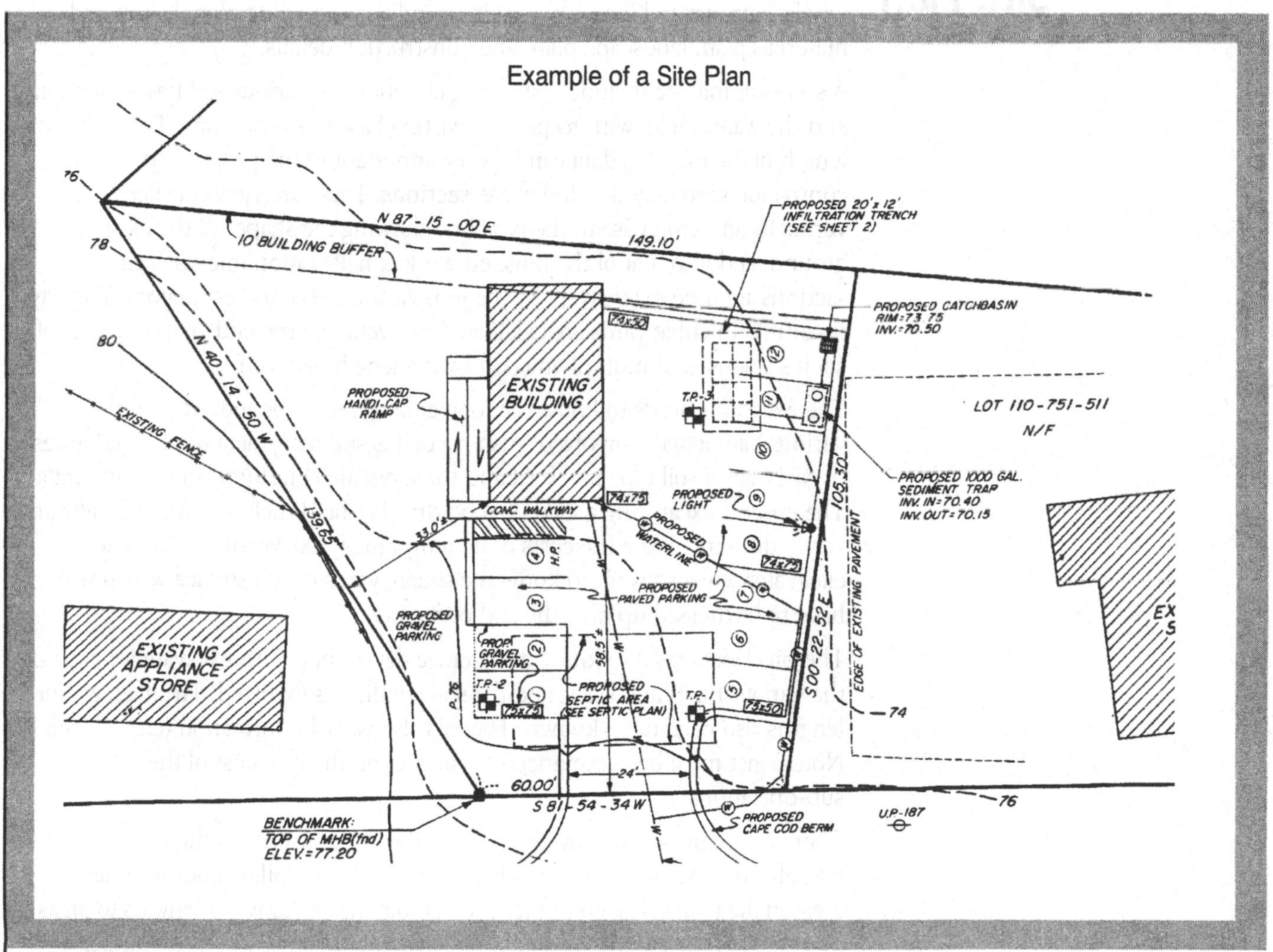

Figure 3.6

The site designer uses the information gathered and the requirements of the program to develop a site development plan. This plan combines the features of the existing site with the proposed changes to the site, including the location of proposed buildings, roads, parking, walkways, retaining walls, etc. Any contour changes and site grading specifications will also be included on this plan. (See Figure 3.7.)

During the construction document phase, additional detailed drawings are produced detailing site contours and grades, site utilities, site structures, plantings, and construction details. These drawings become a part of the contract documents, defining the exact requirements of the site work portion of the construction contract.

The following section is reprinted from *Means Heavy Construction Handbook*, by Richard C. Ringwald, R.S. Means Co., Inc.

Understanding the Site Plan

Plan Features

There is usually a cover page with general information about the job as a whole: title, general location, existing conditions, grading plan, layout and materials plan, landscape plan, and construction details.

A soil plan may be included, showing elevations of various soil types, or strata, and the water table, with respect to existing base line elevations. Through the length of the job, this data can be very important to the project earthmoving contractor, who may develop **cross sections**. These are views of slices "cut" vertically and crosswise to the base line showing the shapes of the existing ground, and also that of the finished work at half-station intervals. Cross sections are used extensively by the job's earthmovers and estimators. The pay quantities in cubic yards are calculated by averaging the end areas of a pair of such sections, and multiplying that by the length between.

The excavator needs to know the total cubic yards to be moved, and the various haul lengths involved. Also needed is summary data on the quantities of each major soil type involved and the moisture conditions in various strata. The costs of excavating a load of good dry clay, hard rock, or saturated silts are vastly different, as are the costs of hauling a load 500' versus 5000'. The excavator will also want to know the square yards of soil surface which will have to be dressed up near the end of the job.

The pipelayers need to know the average depth they must excavate for each of the various types and sizes, and the soil conditions involved. The average line lengths also need to be known. There is also verbal information (e.g., General Notes) that must be summarized for impact on the unit cost of the sub-operations.

There are many software systems that can eliminate some of the drudgery of takeoff work. Most of these involve using a stylus to follow contour lines or area on the plans. The computer then automatically converts lengths to areas or volumes as needed. Computer takeoff systems can reduce the number of errors arising from fatigue. Alas, they can also be misused by fledgling

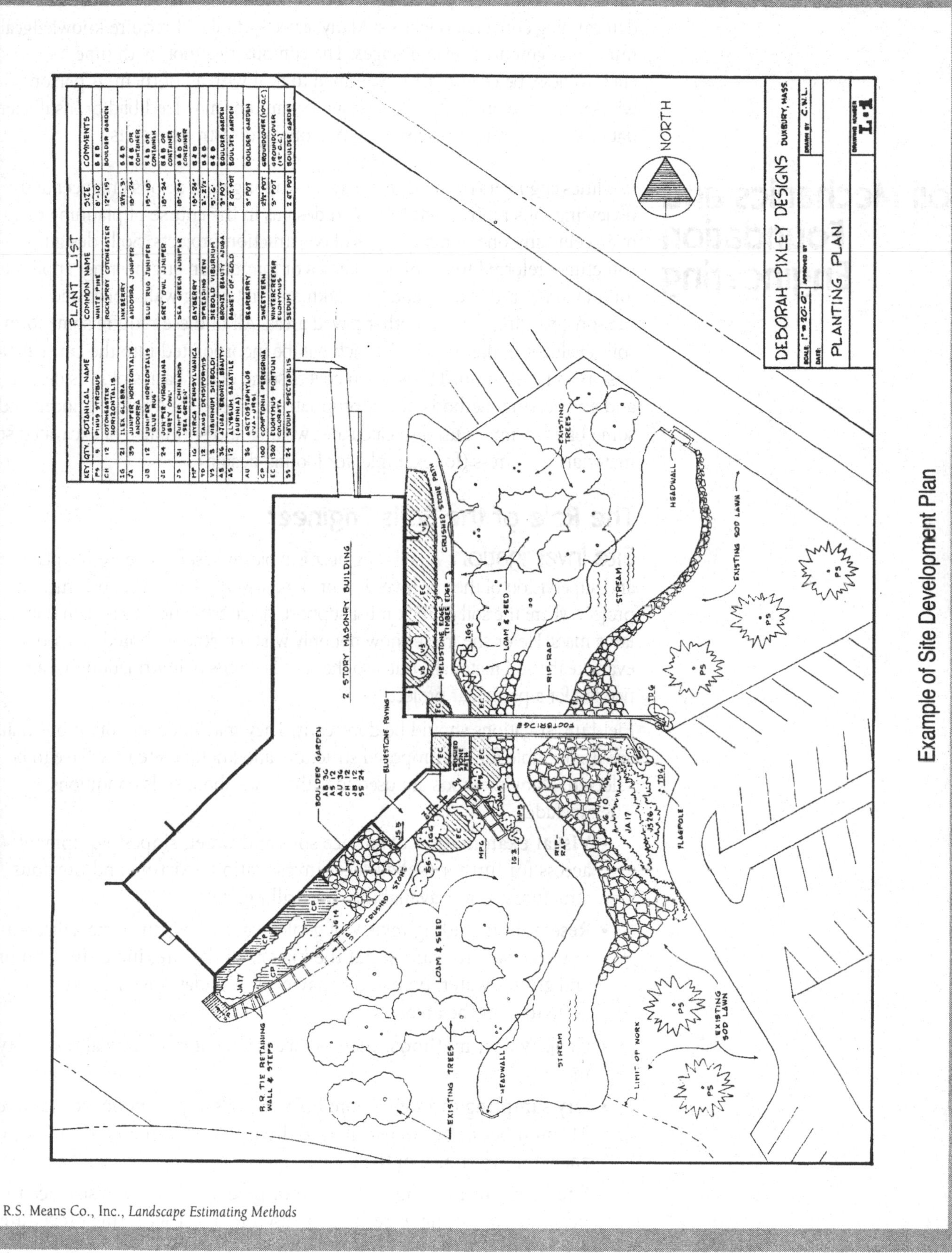

Example of Site Development Plan

R.S. Means Co., Inc., *Landscape Estimating Methods*

Figure 3.7

estimators who fail to give the job sufficient forethought, or who are careless in determining conversion factors. Many aspects of takeoff require knowledgeable human judgment at several stages. The computer cannot, with time effectiveness, be correctly programmed to deal with all of them. A person whose takeoff work is flawed without a computer may feed bad or insufficient data into a computer system, with potentially disastrous results.

Soil Mechanics and Foundation Engineering

Facilities engineers or managers may be working with a soils engineer and reviewing soils reports and proposed designs in the course of planning or overseeing any one of many types of construction projects. Soils design, sometimes referred to as soil mechanics or foundation engineering, applies to foundations, earth slopes and embankments, retaining walls, piles and caissons, parking lots, and other paved areas. All of these require some form of soils evaluation, the results of which will be incorporated into the construction design and execution. This section covers some of the basics of soils design and engineering methods, the appropriate experience of a soils engineer, and some basic formulas used to calculate swelling and shrinkage of excavated soil materials and stress (for example, for footing design).

The Role of the Soils Engineer

Site Investigation: A soils engineer is typically responsible for the planning and supervision of the site investigation, a necessary step in any construction project where the soil or rock must support the construction or remain stable in large mass. He or she must know not only which methods should be used to evaluate the soil materials, but also the specific kinds of information that are needed for a particular project.

Field investigations should be done early. They may uncover conditions that are undesirable for the proposed structure, and another site may have to be selected. Many methods are used to gather data about soils conditions. They include:

- **Visual examination** of surface soils and water, slopes, equipment access for further (subsurface) investigation, existing and previous structures, water availability for drilling, etc.
- **Research**, including review of geological surveys by a specialist—a low-cost way to learn about the history of the site, including flooding and groundwater, type and behavior of foundations of existing structures in the area, etc.
- **Geophysical methods**, such as measurement of electrical resistivity in soil.
- **Dry sampling** through a bore hole, using sample spoons, or samplers. The number of blows required to drive a steel casing is recorded to show the resistance of the soil.
- **Sounding**, or probing, with a rod or pole, to measure resistance to penetration. Jetting water may be used to assist with this procedure.

- **Test pits**, dug by hand or with a backhoe or excavator.
- **Water boring.**
- **Rotary drilling.**
- **Load tests.** Load tests may be performed on a small area of soil in a pit dug to the level of the proposed foundation. Settlements are measured at intervals, when increased load is applied to a bearing plate by jacks or weights. The American Society for Testing and Materials (ASTM) offers a procedure for these tests; building codes may also specify procedures. The problem with field load tests is that they may not indicate an underlying weak stratum that may be unable to resist the stresses of the structure. Field tests such as these should as closely as possible replicate actual foundation conditions. They should be performed at the actual locations where significant loads will be supported. *Ed. Note: See "Soils Properties Related to Foundation Design" later in this section for more on soil bearing capacity.*
- **Laboratory tests**, carried out in order to identify soil types and properties, and likely behavior under the proposed structure's load. Laboratory tests include:
  - mechanical analyses to determine particle size (more about this later in this section)
  - moisture content
  - permeability
  - compression
  - consolidation
  - direct shear, to determine the bearing capacity of the soil and the stability of an embankment.
- **Compaction tests**, like the Proctor and American Association of State Highway and Transportation Officials (AASHTO) tests, are employed to determine the density of soil. When the maximum soil density is ascertained, a job standard can be established to make sure that soil compacted in the field will have minimum performance characteristics. Percentage compaction is 100 times the ratio of the unit dry weight of a section of fill to the job standard. Normally, 90% to 100% compaction is specified. The results of compaction tests are most often used to ensure that all lifts of earth in a fill are meeting specifications.

Foundations: In the design of structural foundations, a soils engineer may help select the correct type of foundation for the existing conditions and structural requirements. He or she should be well acquainted with all kinds of structural foundations, and able to describe the conditions that will occur during construction and after the structure is in place, based on a chosen foundation design.

Ed. Note: See the section on substructure later in this chapter for an overview of foundations.

A soils engineer may also design a plan to improve existing site conditions by methods such as compaction, preloading, drainage or surcharging.

Finally, a soils engineer may be called upon to evaluate existing structures that show evidence of excessive settlement or other problems. Such situations may be the result of unexpected changes in the existing soils (bearing materials) conditions over time. Solutions may involve structural underpinning.

Retaining Walls and Slopes: For retaining walls, sheeting, anchors, or other structures that must resist lateral earth pressure, the soils engineer must determine any soils characteristics that will affect such pressure, and the degree of anticipated pressure. This is done with loading diagrams. Unretained earth slopes depend on earth cuts and embankments that will properly distribute stresses within a given slope and in consideration of other conditions such as seepage of ground water.

Earthworks: The soils engineer may also be involved in planning earthworks, such as cuts and fills for site preparation or road construction. An engineer experienced in earthworks should be familiar with the various types and appropriate uses of heavy equipment. He or she should also have experience in writing specifications and instructing construction personnel.

Cost Estimating: Soils engineers should be able to estimate the costs of the aforementioned procedures. Records of prior projects with similar circumstances can be useful in determining costs of proposed new work. Published cost data, such as *Means Heavy Construction Cost Data* and *Site Work & Landscape Cost Data*, are also useful tools in preparing estimates of work that might be designed and supervised by a soils engineer.

Borings for Heavy Structures

Borings for heavy structures are conducted to determine the characteristics of underlying soils. Borings must be made to greater depths for heavy structures, and extensive sampling of the undisturbed undersoil must be performed.

Bore holes for a heavy structure should be close enough to provide complete information regarding the thickness and extent of the various soil strata. Drilling through rock should continue deep into the rock, with cores taken to not only determine the rock's character but to discover whether the rock is a boulder or thin ledge with minimal load-carrying capacity.

Deeper borings required for heavy structures may require more elaborate drilling equipment than that used for soil surveys. A well-drilling rig may be necessary. This type of boring should be performed by drilling contractors with appropriate equipment and experience. The cost of drilling operations is generally based on a unit price per foot of hole plus a sum for each soil sample.

Penetration Tests and Sampling

Undisturbed sampling requires certain precautions, equipment, and expense. A different approach is to load or drive a sampler and evaluate the soil's strength from its resistance to penetration. Samples acquired in this way are disturbed and should be used only for purposes of identifying strata. Penetration resistance is derived from the number of blows from a standard-sized hammer that cause a certain degree of penetration. The hammer may be lifted by a drill rig or a hand-operated winch and tripod.

Examination of Soil Texture: Fractionation

Soil behavior can, to a large extent, be determined by examining the soil's texture, including particle size and shape. These characteristics, together with moisture content and consistency, are known as index properties. A soils engineer should be familiar with these properties, and their significance. Particle size may be recorded in terms of its diameter (e.g., 4mm.), determined by the particles passing through a sieve. When it is sufficient to determine particles' size ranges, or the relative amounts of material comprised of particles of certain size ranges, a Number 4 sieve (four mesh openings per square inch) is usually adequate. Material that passes through this size sieve is called the minus four fraction (or binder). Material that will not pass through the Number 4 sieve is known as the plus four fraction (or stone). A 200-mesh sieve may be used to determine fine-grain materials. Particles that can pass through this sieve may be referred to as subsieve-size. Materials with still finer particles may be analyzed by sedimentation using a hydrometer test.

The following section is reprinted from *Means Heavy Construction Handbook*, by Richard C. Ringwald, R.S. Means Co., Inc.

Excavation: Formulas and Calculations

The soil mass that supports a structure must be sufficiently firm that it will yield negligibly under load. This condition is usually accomplished by bringing the soil particles into such a tight configuration that they resist still tighter association. The process that causes this close consolidation is called compaction. (More exotic methods that rely on more than just compaction to effect an unyielding surface are called *stabilization*.) All compaction methods employ machines that run over layers ("lifts") of soil, imparting various types of forces to do the work.

There are three soil states involved in the process of excavating, hauling and backfilling earth: bank, loose, and compacted. Bank earth is undisturbed soil, and is of medium density relative to the other states. Loose earth is that which lies in the hauling vehicle or in an unconsolidated lift on the embankment. It is the least dense of the states. After consolidation, the lift is in the compacted, most dense state. (An exception is solid rock which—after moving—can never be compacted as tightly as it existed in the bank state.)

The earthmoving quantity unit is the cubic yard. In these pages, we will use the terms BCY (bank run cubic yard), LCY (loose cubic yard), and CCY (compacted cubic yard) in referring to quantities in the bank, loose, and compacted states, respectively. Of course, one BCY will make one-plus LCY, which will then become less than one CCY after compaction. The term *swell factor* refers to the percent increase in volume of a BCY when it is loaded loose into a vehicle. The term *shrinkage factor* is the percent decrease in volume of one BCY after it has finally been compacted.

BCY is the unit of preference in discussing earthwork of any kind. On heavy construction jobs, the cubic yard figure for which a contractor is paid a unit price is almost always in BCY. Nothing extra is paid for the loose or compacted states occupied by the same BCY throughout the course of the job.

On heavy construction sites, we speak of *cuts* (the earth removed) and *fills* (the new embankment made from the removed earth). BCY are hauled (in loose form) to the fill, where they later become CCY. Ideally, all the BCY of cut on a project would be exactly sufficient to produce the (reduced) CCY of fill required. This does not happen often, however. If there is an excess of cut BCY over fill needs, that excess is called *waste*, and is dumped somewhere off site. If there is not enough BCY in the job's cut to supply fill needs, the shortage is made up with off project cut which is called *borrow*. The relationship of BCY of cut to the CCY of fill it will produce is expressed by (1–Sh F/100), where Sh F means *shrinkage factor*.

The shrinkage factor can be related to the densities of the two states by:

$$\left[1 - \frac{SHF}{100}\right] \frac{CCY}{BCY} = \frac{\text{Bank Density (Dry)}}{\text{Compacted Density (Dry)}}$$

which you can readily prove for yourself. (If no change in soil moisture content is required for the compaction process, the natural densities can be used in this formula.)

When soils are excavated, they increase in volume, or swell, because of an increase in voids:

$$V_b = V_L L = \frac{100}{100 + \%\ \text{swell}} \quad V_L$$

where:

V_b = original volume, yd^3, or bank yards
V_L = loaded volume, yd^3, or loose yards
L = load factor

When soils are compacted, they decrease in volume:

$V_c = V_b S$

where

V_c = compacted volume, yd^3

S = shrinkage factor

Bank yards moved by a hauling unit equals weight of load, lb., divided by density of the material in place, lb. per bank yard.

Figure 3.8 lists various soils and rock materials with their swell percentages and load factors.

The hourly work of a compactor in CCY/HR converting earth from a loose state can be derived from an expression of width times length times compacted lift thickness. The compacted length must be average compactor speed (S) times one hour divided by the number of passes (N) required to achieve final compaction. The final formula shown in Figure 3.9 for CCY/HR of work completed merely uses this formula (MPH/N) in lieu of length in the width times length times thickness expression, then combines conversion factors so that the most convenient units can be used. If you can pass over 3 miles (17,840') in an hour with just one pass, 3 passes are required, you can compact just *one* mile (5280') of lift in one hour.

The formula for gallons of water involved in any compaction process moisture content change is also shown in Figure 3.9. No proof of this formula is needed for this simple expression. It should be noted that moisture content is always expressed in terms of the weight of water itself in a soil sample, divided by the weight of the soil only in the sample (e.g., in a 130 lb. soil sample with 30% moisture, 30 lbs. is water alone, 100 lbs. is soil alone.)

Soils Properties Related to Foundation Design

Footings–Strength Design: Footings are structural elements that transmit loads to the soil beneath. The footing spreads the load over an area large enough to prevent damage to the weaker supporting soil. Footing design involves the two steps. The first is to size the footing with respect to contact area to avoid overstressing the soil below. Secondly, the footing is sized in terms of thickness and reinforcement to avoid failing.

Types of footings include *wall footings, isolated spread footings*, and *combined footings*. A strip of reinforced concrete, a wall footing is wider than the wall and thus distributes the wall load over a larger area. Isolated spread footings transmit a load from columns to the supporting soil. If the soil is weak or the column loads heavy, the isolated spread footings need to be larger. When spread footings are large enough that they merge, combination footings are used. They can be rectangular or trapezoidal, depending on column load.

Figure 3.10 lists soil types with their maximum allowable presumptive bearing values in tons per square foot.

Soil Characteristics and Volume Conversion Factors

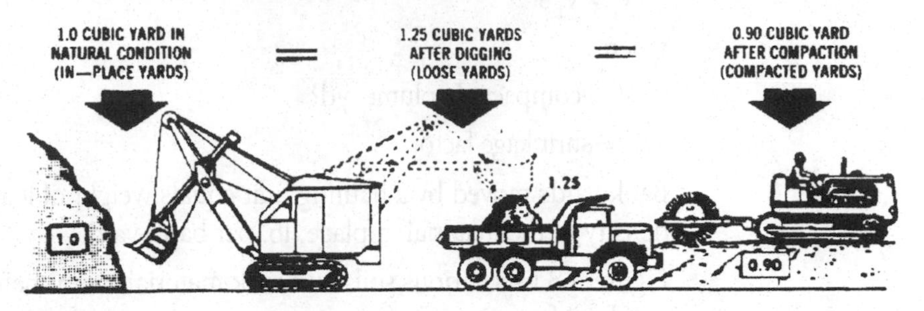

Approximate Material Characteristics*

Material	Loose (lb/cu yd)	Bank (lb/cu yd)	Swell (%)	Load Factor
Clay, dry	2,100	2,650	26	0.79
Clay, wet	2,700	3,575	32	0.76
Clay and gravel, dry	2,400	2,800	17	0.85
Clay and gravel, wet	2,600	3,100	17	0.85
Earth, dry	2,215	2,850	29	0.78
Earth, moist	2,410	3,080	28	0.78
Earth, wet	2,750	3,380	23	0.81
Gravel, dry	2,780	3,140	13	0.88
Gravel, wet	3,090	3,620	17	0.85
Sand, dry	2,600	2,920	12	0.89
Sand, wet	3,100	3,520	13	0.88
Sand and gravel, dry	2,900	3,250	12	0.89
Sand and gravel, wet	3,400	3,750	10	0.91

*Exact values will vary with grain size, moisture content, compaction, etc. Test to determine exact values for specific soils.

Typical Soil Volume Conversion Factors

Soil Type	Initial Soil Condition	Converted to: Bank	Loose	Compacted
Clay	Bank	1.00	1.27	0.90
	Loose	0.79	1.00	0.71
	Compacted	1.11	1.41	1.00
Common earth	Bank	1.00	1.25	0.90
	Loose	0.80	1.00	0.72
	Compacted	1.11	1.39	1.00
Rock (blasted)	Bank	1.00	1.50	1.30
	Loose	0.67	1.00	0.87
	Compacted	0.77	1.15	1.00
Sand	Bank	1.00	1.12	0.95
	Loose	0.89	1.00	0.85
	Compacted	1.05	1.18	1.00

R.S. Means Co., Inc., *Means Heavy Construction Handbook*

Figure 3.8

Cut and Fill Calculations

Name		Bank	Loose	Compacted
Unit:		BCY (Bank Cu Yds)	LCY (Loose Cu Yds)	CCY (Compacted Cu Yds)
State:		Undisturbed Original Ground	In Vehicle or Loose on Fill	In Fill after Compaction
Other Names:		• Cut • Borrow • In Situ • Pay Yards		• Fill • Embankment
Density	Soil	Medium	Lightest	Heaviest
	Rock	Heaviest	Lightest	Medium

S_HF = *Shrinkage Factor:* % reduction in volume of BCY after compaction (negative for rock).

$$(1 - S_HF/100) = CCY/BCY$$

S_WF = *Swell Factor:* % increase in volume of BCY after loosening: $(1 + S_WF/100) = LCY/BCY$

Note: $$\left(1 - \frac{S_HF}{100}\right)\frac{CCY}{BCY} = \frac{(\text{Dry Density LB/BCF}) \times 27\ \text{BCF/BCY}}{(\text{Dry Density LB/CCF}) \times 27\ \text{CCF/CCY}} = \frac{\text{Bank Dry Dens}}{\text{Comp Dry Dens}}$$

Cut & Fill Calculations:

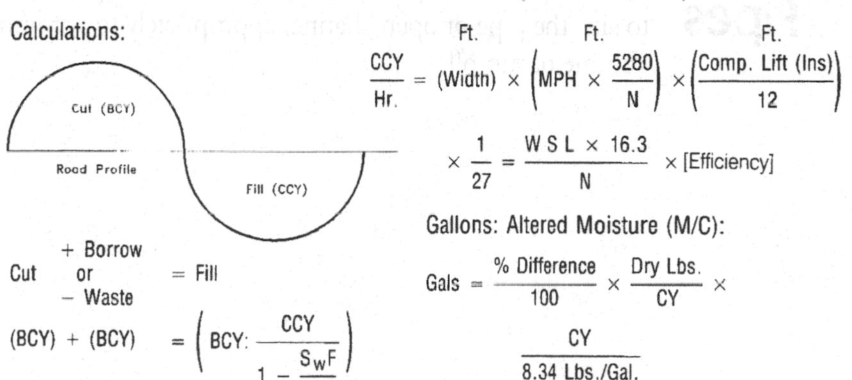

$$\text{Cut} \begin{matrix} + \text{Borrow} \\ \text{or} \\ - \text{Waste} \end{matrix} = \text{Fill}$$

$$(BCY) + (BCY) = \left(BCY: \frac{CCY}{1 - \frac{S_WF}{100}}\right)$$

Hourly Work of a Compactor:

$$\frac{CCY}{Hr.} = (\text{Width})_{Ft.} \times \left(MPH \times \frac{5280}{N}\right)_{Ft.} \times \left(\frac{\text{Comp. Lift (Ins)}}{12}\right)_{Ft.} \times \frac{1}{27} = \frac{W\,S\,L \times 16.3}{N} \times [\text{Efficiency}]$$

Gallons: Altered Moisture (M/C):

$$\text{Gals} = \frac{\%\ \text{Difference}}{100} \times \frac{\text{Dry Lbs.}}{CY} \times \frac{CY}{8.34\ \text{Lbs./Gal.}}$$

Soil Types

Non-Cohesive:
- Cobbles – 3″ and up
- Gravel – ⅛″ – 3″
- Sand – .002″ – ⅛″, feels grainy
- Silt – Very fine granular < .002″; settles in water; feels smooth

Cohesive:
- Clay – Super-fine, submicroscopic grains bond together; suspends in water

R.S. Means Co., Inc., *Means Heavy Construction Handbook*

Figure 3.9

Soil Pressure at a Depth in the Strata Caused By the Weight of the Footing: The increased pressure on the soil at a depth in the strata below the foundation member can be measured in various ways. The simplest measure is to use a 2:1 slope. A stress zone can also be defined by an angle (for example 30° to 45°) with the vertical. If the stress zone is defined by the 2:1 slope, the change in the pressure, Δq, at a depth z beneath the footing is:

$$\Delta q = \frac{V}{(B + z)(L + z)}$$

For a square footing, the calculation is:

$$\Delta q = \frac{V}{(B + z)^2}$$

Where:

V = total load applied to foundation member

B, L = footing dimensions, ft.

z = depth from footing base to soil elevation, where stress is increased

Design of Drainage Pipes

A facility engineer may be involved in drainage design for a parking lot or other impervious surface. Applying the following formulas allows the engineer to size the pipe or open channel appropriately to accommodate the anticipated volume of run-off.

Presumptive Bearing Value

Material	Maximum Allowable Presumptive Bearing Value, tons/sq.ft.
Clay, medium soft	1 – 1.5
Clay, dense	2
Clay, hard	5
Sand, loose	1
Sand, fine	2
Sand, coarse, dense	3
Gravel, dense	4 – 6
Rock, sound and hard	60
Rock, soft	12

Figure 3.10

Kutter's Formula

Kutter's formula is used for estimating open-channel flows. Values were established for the Kutter roughness coefficient, *n*, applicable to sewers. Kutter's formula (in English units) is:

$$V = \left[\frac{\frac{1.81}{n} + 41.67 + \frac{0.0028}{S_6}}{1 + \frac{n}{\sqrt{R}}\left(41.67 + \frac{0.0028}{S_6}\right)} \right] \sqrt{RS}$$

V is the mean velocity of flow, *R* is the hydraulic radius, S_6 is the slope of energy grade line, and *n* is the coefficient of roughness. The $0.0028/S_6$ is sometimes omitted since it was based on data now known to be inaccurate.

The Manning Formula

The Manning formula is simpler than Kutter's formula, and its *n* value is nearly equal to Kutter's *n* for types of pipe commonly used in sewer construction. The Manning formula is most often used for storm and sanitary sewer design. The equation is:

$$V = \frac{1}{n}\left(R^{2/3}S^{1/2}\right) \text{ (Metric Units)}$$

$$V = \frac{1.49}{n}\left(R^{2/3}S^{1/2}\right) \text{ (English Units)}$$

Figure 3.11 is a table listing the values of *n* for various pipe, canal, ditch, stream, and channel surfaces for use with the Manning formula.

The Kutter and Manning formulas are used for pipes and conduits of various shapes, flowing full or partly full. The graphs used to solve these equations are typically compiled for pressure-conduit flow only. Hydraulic-Elements graphs or tabular values are used to determine the flow characteristics at less than full flow. Alignment charts are available that allow the user to solve the Manning equation based on discharge (in gallons and cubic feet), pipe diameter and roughness coefficient, velocity and slope.

Storm Water Runoff

Runoff for Minor Hydraulic Structure

Determining runoff for minor hydraulic structures is typically done using the rational formula:

Q = CIA

where

Q = peak discharge, $ft.^3/s$

C = runoff coefficient = percentage of rain that appears as direct runoff

I = rainfall intensity, in./h

A = drainage areas, acres

The rational formula is based on a set of assumptions that apply to areas with drainage facilities of fixed dimensions and hydraulic characteristics.

Manning Equation n Values

Values of *n* To Be Used With the Manning Equation

Type of Surface	Ideal	Good	Fair	Poor
PIPES:				
Coated cast-iron pipe	0.011	0.012	0.013	X
Commercial wrought-iron pipe, black	0.012	0.013	0.014	0.015
Commercial wrought-iron pipe, galvanized	0.013	0.014	0.015	0.017
Concrete pipe	0.012	0.013	0.015	0.016
Smooth brass and glass pipe	0.009	0.010	0.011	0.013
Uncoated cast-iron pipe	0.012	0.013	0.014	0.015
Vitrified clay sewer pipe	0.011	0.013	0.015	0.017
MISC. SURFACES:				
Cement mortar	0.011	0.012	0.013	0.015
Cement-rubble	0.017	0.020	0.025	0.030
Common clay drainage tile	0.011	0.012	0.014	0.017
Concrete-lined channels	0.012	0.014	0.016	0.018
Dressed-ashlar	0.013	0.014	0.015	0.017
Dry-rubble	0.025	0.030	0.033	0.035
Neat cement	0.010	0.011	0.12	0.013
Semicircular metal flumes, corrugated	0.023	0.025	0.028	0.030
Semicircular metal flumes, smooth	0.011	0.012	0.013	0.015
CANALS AND DITCHES:				
Canals with stony beds, weedy banks	0.025	0.030	0.035	0.040
Dredged earth channels	0.025	0.028	0.030	0.033
Earth bottom, rubble sides	0.028	0.030	0.033	0.035
Earth, straight and uniform	0.017	0.020	0.023	0.025
Rock cuts, smooth	0.025	0.030	0.033	0.035
Rock cuts, irregular	0.035	0.040	0.045	X
Winding canals	0.023	0.025	0.028	0.030
NATURAL STREAM CHANNELS:				
Clean straight bank, no rifts or pools	0.025	0.028	0.030	0.033
Clean straight bank, some weeds and stones	0.030	0.033	0.035	0.040
Winding, some pools	0.033	0.035	0.040	0.045
Winding with lower stages	0.040	0.045	0.050	0.055
Winding with weeds and some stones	0.035	0.040	0.045	0.050
Winding with shoals, many stones	0.045	0.050	0.055	0.060
Weedy with deep pools, weedy	0.050	0.060	0.070	0.080
Very weedy	0.075	0.100	0.125	0.150

Figure 3.11

The rational formula expresses runoff as a fraction of rainfall (instead of rainfall minus losses resulting from all of the factors that affect runoff) in a single coefficient. Using a more complex formula is not justified for minor hydraulic structures.

Figure 3.12 lists the runoff coefficients for various types of drainage.

Rainfall intensity *I* may be determined from any of a number of formulas or from a statistical analysis of rainfall data if enough are available.

Runoff Coefficients

Drainage Area	Runoff Coefficient C
Business:	
Downtown areas	0.70-0.95
Neighborhood areas	0.50-0.70
Industrial:	
Light areas	0.50-0.80
Heavy areas	0.60-0.90
Unimproved Areas	0.10-0.30
Streets:	
Asphalt	0.70-0.95
Concrete	0.80-0.95
Brick	0.70-0.85
Drives and walks	0.75-0.85
Roofs	0.75-0.95
Parks, cemeteries	0.10-0.25
Playgrounds	0.20-0.35
Railroad-yard areas	0.20-0.40
Lawns:	
Sandy soil, flat, 2%	0.05-0.10
avg, 2-7%	0.10-0.15
steep, 7%	0.15-0.20
Heavy soil, flat, 2%	0.13-0.17
avg, 2-7%	0.18-0.22
steep, 7%	0.25-0.35

Figure 3.12

Runoff for Major Hydraulic Structures

The unit-hydrograph method is widely accepted for determining runoff for major hydraulic structures. It can be used to calculate the complete runoff hydrograph from rainfall after the unit hydrograph has been established for a given area.

The unit hydrograph is based on a unit storm (nearly constant rainfall intensity for its duration termed a *unit period*), and a runoff volume of 1" (water with a depth of 1" over a unit area, usually 1 acre). Thus, a unit storm may have a 2"/hour effective intensity lasting ½ hour or a 0.2"/hour effective intensity lasting 5 hours. The volume is less critical than the constancy of the intensity. It is possible to adjust for situations where the runoff volume is other than 1". Corrections for highly variable rainfall rates cannot be made. The unit hydrograph is summarized by the formula:

Effective rain x unit hydrograph = runoff

The unit hydrograph is the link between rainfall and runoff. It integrates various factors that affect runoff and can be derived from rainfall and stream-flow data or from stream-flow data alone.

Strength of Materials

Simple Stress

When a construction process is designed, the architect and/or engineer must select the appropriate materials, positioned correctly and in the correct quantity, to support a structure or building element. To make the proper choices, it is necessary to know or determine a material's strength, stiffness, and other properties.

The unit strength of a material is expressed based on its load capacity. For example, a bar capable of supporting 1000 lbs. has a unit strength of:

$$S_2 = \frac{1000}{1} = 1000 \text{ psi (pounds per square inch)}$$

The unit strength of a material is usually defined as stress, and can be expressed as follows:

$$S = \frac{P}{A}$$

S is the stress (or force) per unit area, P is the applied load, and A is the cross-sectional area.

Ed. Note: Stress or total stress are sometimes referred to as load or force, or stress or stress intensity.

Shearing Stress

Shearing stress is the result of forces acting parallel to the area resisting the forces. Tensile and compressive stresses are caused by forces that are perpendicular to the areas resisting the force. Tensile and comprehensive stresses may be referred to as *normal stresses*, while shearing stress may be called *tangential stress*.

If shearing stress were distributed uniformly over an area, it would be possible to determine it with the following equation:

$$S_s = \frac{V}{A}$$

Where

V = shearing force

A = the cross-sectional area

Since this is almost never the case, however, this equation most often provides the *average shearing stress*.

This equation provides the *maximum shearing stress at a point*:

$$\text{Max } v = \frac{f_1 - f_2}{2}$$

Where f_1 is the maximum principal stress and f_2 is the minimum.

Basic Structural Engineering Principles

This section presents some fundamentals of structural engineering, including loads and stresses, types of support, statics, and shear and moment diagrams. Provided first are some basic definitions of beam and column supports, types of loads, and stresses that affect them, along with the formula for flexure. Listed next are the beam diagrams and formulas of greatest interest to facilities managers.

Beam Diagrams and Formulas

Beam diagrams and formulas require a basic understanding of the associated nomenclature. There are two main types of formulas:

- **M_{max} Formulas** compute the maximum moment for the beam. The location of maximum moment is provided in parentheses.
- **R, R_1, R_2, and R_3 formulas** compute the end beam reactions.

The facility engineer may also encounter the following:

- **Total Equivalent Uniform Load Equations**, which compute the equivalent total load from this loading condition on the beam.
- **V_x equations**, which compute vertical shear at any point along the beam.
- **M_x equations**, which compute moments at any point along the beam.

Figure 3.13 lists common symbols used in formulas.

Nomenclature

BEAM DIAGRAMS AND FORMULAS

Nomenclature

E = Modulus of Elasticity of steel at 29,000 ksi.
I = Moment of Inertia of beam, $in.^4$.
L = Total length of beam between reaction points ft.
M_{max} = Maximum moment, kip in.
M_1 = Maximum moment in left section of beam, kip-in.
M_2 = Maximum moment in right section of beam, kip-in.
M_3 = Maximum positive moment in beam with combined end moment conditions, kip-in.
M_x = Moment at distance x from end of beam, kip-in.
P = Concentrated load, kips
P_1 = Concentrated load nearest left reaction, kips.
P_2 = Concentrated load nearest right reaction, and of different magnitude than P_1, kips.
R = End beam reaction for any condition of symmetrical loading, kips.
R_1 = Left end beam reaction, kips.
R_2 = Right end or intermediate beam reaction, kips.
R_3 = Right end beam reaction, kips.
V = Maximum vertical shear for any condition of symmetrical loading, kips.
V_1 = Maximum vertical shear in left section of beam, kips.
V_2 = Vertical shear at right reaction point, or to left of intermediate reaction point of beam, kips.
V_3 = Vertical shear at right reaction point, or to right of intermediate reaction point of beam, kips.
V_x = Vertical shear at distance x from end of beam, kips.
W = Total load on beam, kips.
a = Measured distance along beam, in.
b = Measured distance along beam which may be greater or less than a, in.
l = Total length of beam between reaction points, in.
w = Uniformly distributed load per unit of length, kips/in.
w_1 = Uniformly distributed load per unit of length nearest left reaction, kips/in.
w_2 = Uniformly distributed load per unit of length nearest right reaction and of different magnitude than w_1, kips/in.
x = Any distance measured along beam from left reaction, in.
x_1 = Any distance measured along overhang section of beam from nearest reaction point, in.
Δ_{max} = Maximum deflection, in.
Δ_a = Deflection at point of laod, in.
Δ_x = Deflection at any point x distance from left reaction, in.
Δ_{x1} = Deflection of overhang section of beam at any distance from nearest reaction point, in.

Figure 3.13

Beams & Columns - Some Basics

Simple Beam: beam supported by two end supports

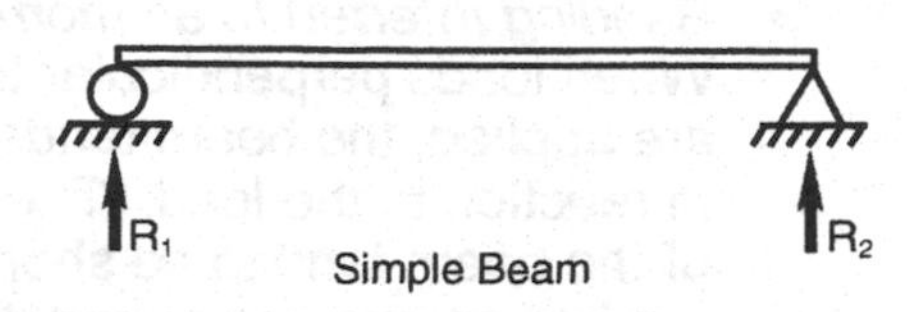

Types of Supports:

- *Hinged* - restrain vertical and horizontal movement, but allow rotation.

- *Roller* - restrain vertical movement, but allow rotation.

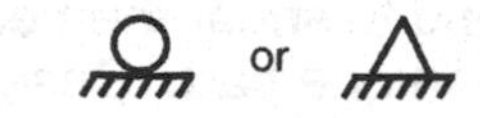

- *Fixed* - restrain vertical, horizontal and rotational movement.

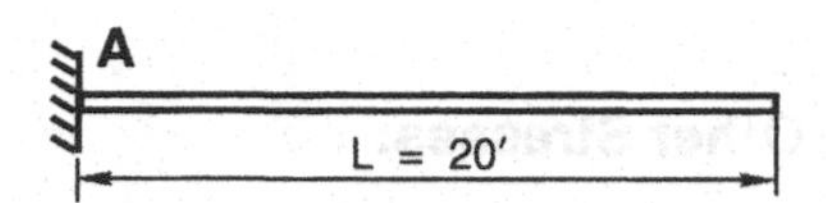

Simply Supported Beams: supported by a hinge on one end and a roller on the other.

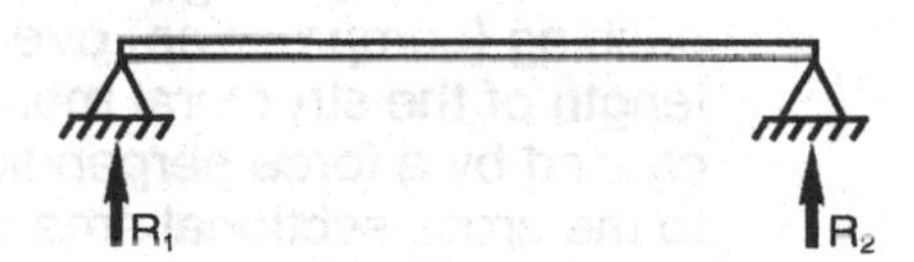

Types of Loads:

- *Concentrated* - acts over so small an area that it can be assumed to act at a point.

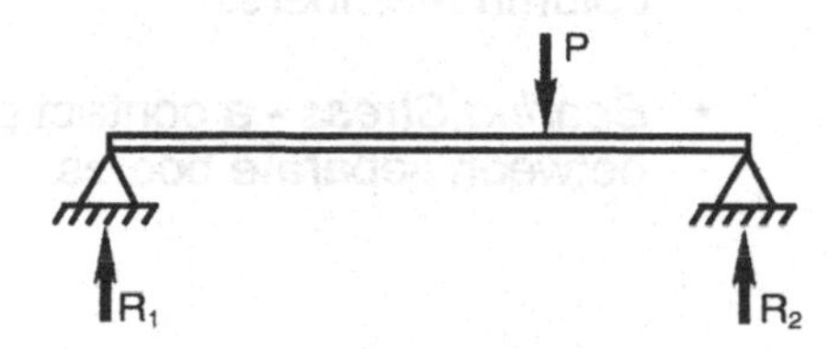

- *Distributed* - acts over a considerable length of the beam or structure.

Note: Loads are also classified as dead, live, moving or impact loads. Refer to "Classification of Loads" in the following pages for further explanation on types of loads.

Stresses on Beams:

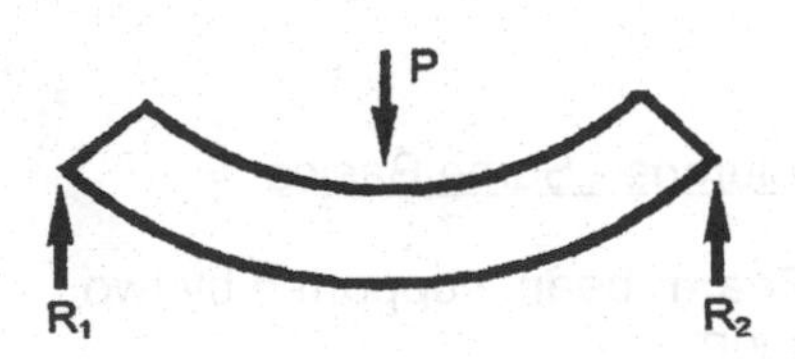

- *Bending referred to as moment* - When loads perpendicular to a beam are applied, the beam tends to bend in reaction to the load. The top fibers of the beam tend to be shortened or put into compression, and the bottom fibers tend to be lengthened and put into tension. Designated as M.

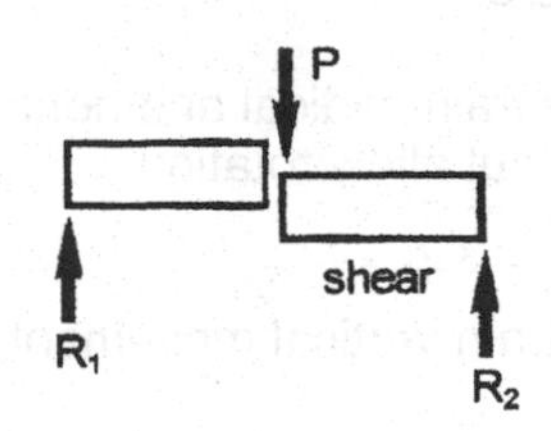

- *Shear Stress* - Stress caused by the vertical loads and reactions which tend to shear the beam into more than one part. Designated with V.

Other Stresses:

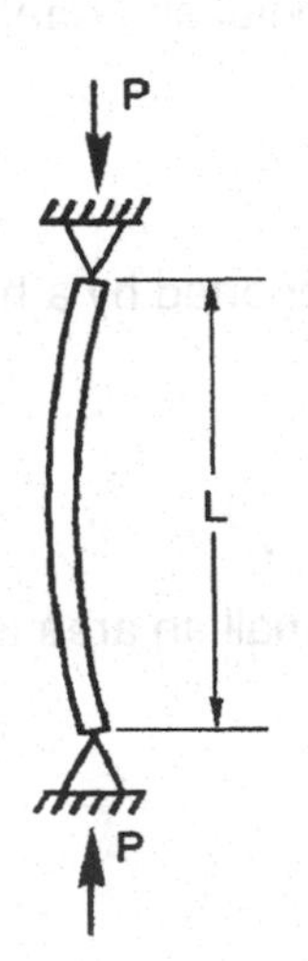

- *Axial Stress* - pulling (tension) or pushing (compression) over the length of the structural member caused by a force perpendicular to the cross-sectional area of the member. This stress occurs in column members.
- *Bearing Stress* - a contact pressure between separate bodies.

Relationships Between Load, Shear and Moment in Beams:

- Sum of forces in y direction = 0
- Sum of forces in x direction = 0
- Sum of moments = 0

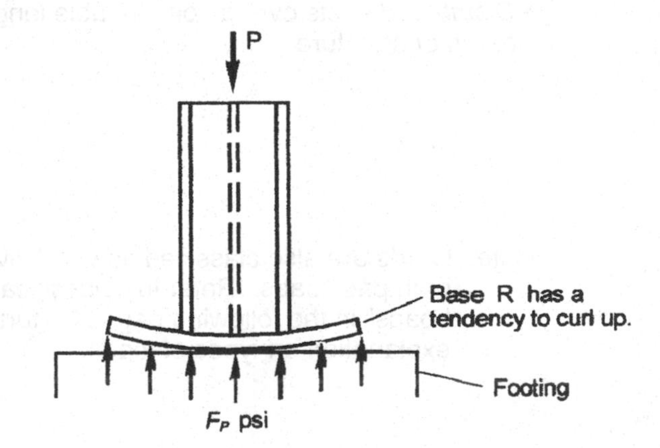

Other Notes:

- Beams usually fail from bending, not shear.
- Points of maximum bending stress occur when shear equals zero.
- Flexure formula

$$f = \frac{M}{S}$$

- Where: f = bending stress
  M = moment
  S = section modulus = I/C
- Shear and moment caused by individual loads are additive.

<u>Trusses</u>

Forces Through Members at an Angle:

Forces transmitted through the structural members at an angle can be separated into forces in the x and y direction. These forces in the x and y direction are proportional to the geometry of the truss structure.

S = axial force through member A-B
V = y component of S force
H = x component of S force
L = length of truss member A-B
v = length of member A-B in y direction
h = length of member A-B in x direction

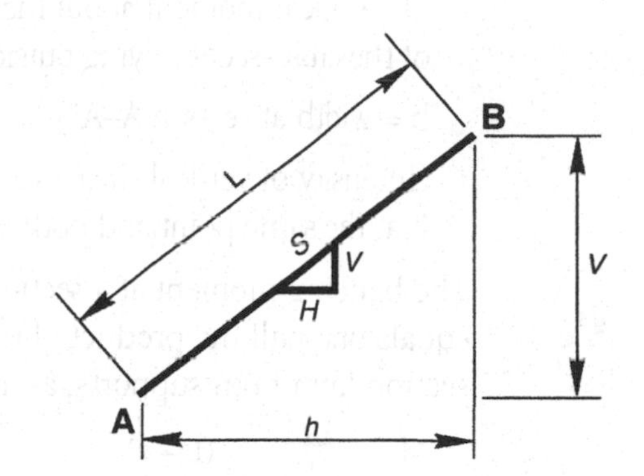

$$\frac{S}{H} = \frac{L}{h}$$

$$\frac{S}{V} = \frac{L}{v}$$

$$S = \sqrt{H^2 + V^2}$$

$$\frac{S}{V} = \frac{L}{v}$$

Frequently Used Diagrams and Formulas: The formulas given below are frequently required in structural design.

Flexural stress at extreme fiber:

$f = Mc/I = M/S$

Flexural stress at any fiber:

$f = My/I$

Where y = distance from neutral axis to fiber

Average vertical shear:

$v = V/A = V/dt$ (for beams and girders)

Horizontal shearing stress at any section A–A:

$v = VQ/I\ b$

Where

Q = statical moment about the neutral axis of the entire section of that portion of the cross-section lying outside section A–A.

b = width at section A–A

(Intensity of vertical shear is equal to that of horizontal shear acting normal to it at the same point and both are usually a maximum at mid-height of beam.)

The bending moment at a section of simply supported, uniformly loaded beam equals one-half the product of the load per linear foot and the distances to the section from both supports, as in this equation:

$$M = \frac{w}{2} x (L - x)$$

The slope of the bending-moment curve at any point on a beam equals the shear at that point:

$$V = \frac{dM}{dx}$$

Where

V = shear

M = moment

x = distance along the beam

Slope and deflection at any point:

$$EI = \frac{d^2y}{dx^2} = M$$

Where

x and y are abscissa and ordinate respectively of a point on the neutral axis, referred to axes of rectangular coordinates through a selected point of support.

(First integration gives slopes; second gives deflections. Constants of integrations must be determined.)

Figures 3.14 – 3.16 represent various static loading conditions.

Simple Beam – Uniformly Distributed Load

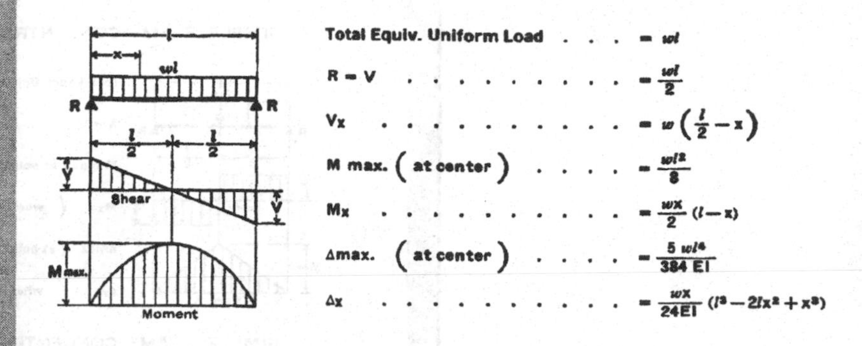

Total Equiv. Uniform Load . . . . $= wl$
R = V . . . . $= \frac{wl}{2}$
Vx . . . . $= w\left(\frac{l}{2} - x\right)$
M max. (at center) . . . . . $= \frac{wl^2}{8}$
Mx . . . . $= \frac{wx}{2}(l - x)$
Δmax. (at center) . . . . . $= \frac{5wl^4}{384EI}$
Δx . . . . $= \frac{wx}{24EI}(l^3 - 2lx^2 + x^3)$
wl
R
R
Shear
Moment
M max.

Reprinted with permission from *Manual of Steel Construction*, American Institute of Steel Construction, Inc.

Figure 3.14

Simple Beam – Uniform Load Partially Distributed

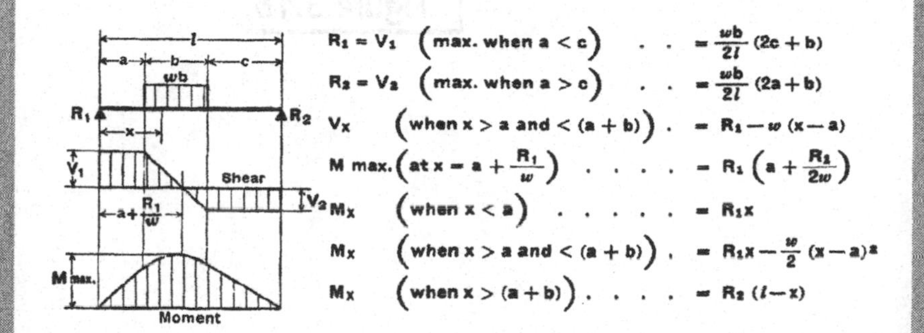

$R_1 = V_1$ (max. when $a < c$) . . $= \frac{wb}{2l}(2c + b)$
$R_2 = V_2$ (max. when $a > c$) . . $= \frac{wb}{2l}(2a + b)$
Vx (when $x > a$ and $< (a + b)$) . $= R_1 - w(x - a)$
M max. (at $x = a + \frac{R_1}{w}$) . . . . . $= R_1\left(a + \frac{R_1}{2w}\right)$
Mx (when $x < a$) . . . . . . $= R_1 x$
Mx (when $x > a$ and $< (a + b)$) . $= R_1 x - \frac{w}{2}(x - a)^2$
Mx (when $x > (a + b)$) . . . . $= R_2 (l - x)$
wb
R1
R2
Shear
Moment
M max.

Reprinted with permission from *Manual of Steel Construction*, American Institute of Steel Construction, Inc.

Figure 3.15

Simple Beam – Concentrated Loads

SIMPLE BEAM—CONCENTRATED LOAD AT CENTER

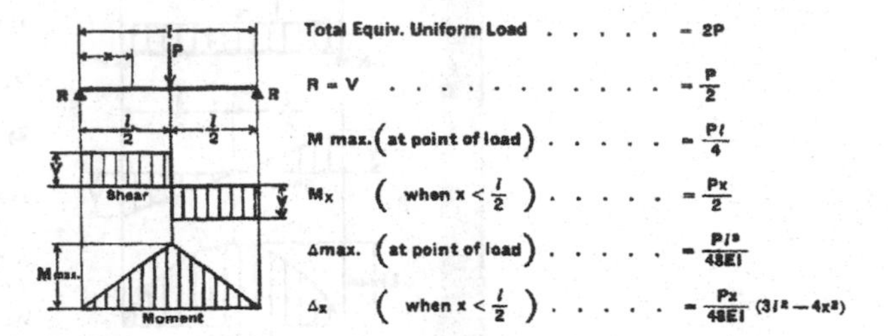

Total Equiv. Uniform Load . . . . . . $= 2P$

$R = V$ . . . . $= \frac{P}{2}$

M max. (at point of load) . . . . . . $= \frac{Pl}{4}$

M_x (when $x < \frac{l}{2}$) . . . . . . $= \frac{Px}{2}$

Δmax. (at point of load) . . . . . . $= \frac{Pl^3}{48EI}$

Δ_x (when $x < \frac{l}{2}$) . . . . . . $= \frac{Px}{48EI}(3l^2 - 4x^2)$

SIMPLE BEAM—CONCENTRATED LOAD AT ANY POINT

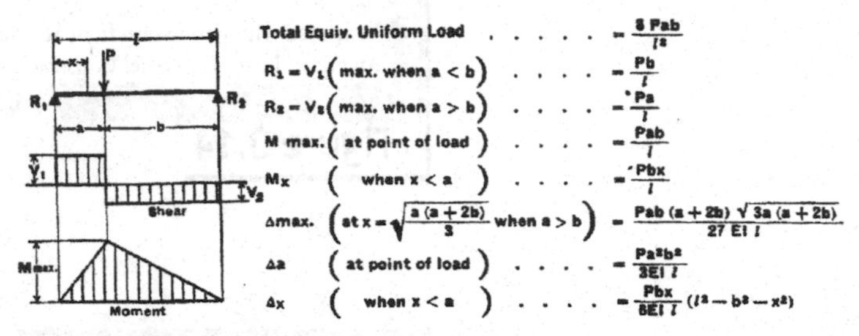

Total Equiv. Uniform Load . . . . . . $= \frac{8Pab}{l^2}$

$R_1 = V_1$ (max. when $a < b$) . . . . . $= \frac{Pb}{l}$

$R_2 = V_2$ (max. when $a > b$) . . . . . $= \frac{Pa}{l}$

M max. (at point of load) . . . . . $= \frac{Pab}{l}$

M_x (when $x < a$) . . . . . $= \frac{Pbx}{l}$

Δmax. $\left(\text{at } x = \sqrt{\frac{a(a+2b)}{3}} \text{ when } a > b\right) = \frac{Pab(a+2b)\sqrt{3a(a+2b)}}{27EIl}$

Δa (at point of load) . . . . . $= \frac{Pa^2b^2}{3EIl}$

Δx (when $x < a$) . . . . . $= \frac{Pbx}{6EIl}(l^2 - b^2 - x^2)$

Reprinted with permission from *Manual of Steel Construction*, American Institute of Steel Construction, Inc.

Figure 3.16

Load Classification

In designing the structure of a building, an engineer must estimate the loads that will be applied to it over the life of the building. Loads are classified as follows:

- *Dead load*, the structural frame's own weight, together with the weight of any other loads that are permanently attached to the frame, do not change in weight or position. Dead load elements include the frame, walls, roof, floors, plumbing and fixtures.
- *Live loads* are all other loads. They do not remain in a fixed position, and may increase or decrease.
- *Moving loads* include trucks, people and other loads that move themselves, under their own power.
- *Movable loads* include anything that may be moved, such as snow, supplies, and furniture.
- *Snow loads* are a significant factor in colder climates. Snow load factors range from 10 to 40 pounds per square foot, depending on the slope of the roof and the type of surface material.
- *Impact loads* refer to the stress caused by the vibration of moving or movable loads.
- *Lateral loads* are primarily caused by wind and earthquake.

Ed. Note: Loads are also described in the context of foundations in "Substructures," the next section of this chapter.

Statics

A body at rest is in equilibrium. The external load (the sum of the horizontal forces and the sum of the vertical forces) on a body at rest is zero. The equations of statics, listed below, apply to structural elements such as beams, rigid frames, or trusses, as these elements are intended to remain in equilibrium:

$$\Sigma H = 0$$

$$\Sigma V = 0$$

$$\Sigma M = 0$$

Where H represents horizontal components, V the vertical ones, and M the moments of the components about any point in the plane.

Shear and Moment Diagrams

These diagrams enable the engineer to determine the values of shear and moment at any point in a beam. Shear and bending moment are the actions of the external loads on a structure. Shear is an algebraic sum of the external forces to the right or left of a section, perpendicular to the beam's axis. Calculations are made to the right and to the left to determine that both sides have the same result. Shear is positive if the sum of the forces to the left is up, or the sum of the forces to the right is down. Bending moment is obtained by adding the moments of all external forces to the left or right of a section. The moments are taken around an axis through the centroid of the cross section. A positive bending moment occurs when the moment to the left is clockwise, or the moment to the right is counterclockwise.

In a shear diagram, vertical lines are drawn to show the quantity and direction of the force for each concentrated load. Horizontal lines are drawn to show that there is no change in shear. A straight, inclined line shows that shear is changing at a constant rate per foot where uniform loads exist.

Moment diagrams show the moments at various points in a structure. The change in moment between two points on a structure equals the shear between those points, times the distance between them:

$$dM = V\,dx$$

Thus, the change in moment equals the area of the shear diagram between the points.

Understanding the relationship between shear and moment simplifies the task of creating moment diagrams. By computing the total area under the shear curve (to the left or right of a section), in consideration of the algebraic signs of the shear curve segments, one can calculate the moment at any particular section.

The following section is reprinted from *Fundamentals of the Construction Process*, by Kweku K. Bentil, R.S. Means Co., Inc.

Ed. Note: See the section on soils mechanics earlier in this chapter for soils properties related to foundations. "Basic Engineering Principles," the section immediately preceding this one, reviews loads and stresses on beams and structures. Finally, a later section on concrete includes information on concrete mixes, drying time, strength, and related areas.

Substructures

The substructure is the portion of a building that is at ground level, and below, and transfers structural loads from a building safely into the supporting soil. The scope of work associated with the substructure portion of a building depends on the ability of the soil at the site to support the building, and the size of the building loads. The substructure, for most buildings, consists of a slab-on-grade, and a foundation.

Slab-On-Grade

A concrete floor placed directly on the ground is known as a slab-on-grade. The slab is placed on compacted granular fill such as gravel or crushed stone. The stone, in turn, is covered with sheets of polyethylene, commonly known in the industry as "visqueen." Polyethylene acts as a vapor barrier to prevent moisture or vapor in the ground from passing into the slab and eventually to the floor above. Slab-on-grade is illustrated in Figure 3.17.

The thickness of the slab depends primarily on the loads acting on the floor, and secondarily on the bearing capacity of the subsoil upon which it is placed.

A concrete slab-on-grade most often is reinforced with welded steel wire fabric—often referred to as "wire mesh." If heavier reinforcement is required, it consists of one or two layers of reinforcing bars, each running both ways (at right angles to each other). Any reinforcing, light or heavy, must have adequate concrete coverage to bond it to the concrete and protect it from corrosion. Supports or "chairs" placed at intervals serve as supports until the concrete cures.

Concrete slabs are subject to cracking, as shrinkage occurs during the curing process. Cracking is controlled through the use of joints in the slab. Examples of four types of joints are illustrated in Figure 3.18.

Isolation joints occur at the perimeter of the slab and where there are openings to allow for columns. It is a full thickness joint filled with a compressible material that allows the slab to move independently of the wall or column.

Construction joints occur in the slab because of the limits placed on the size of the pour (to control shrinkage). These are also full thickness and are bounded by removable or permanent forms. Where differential (vertical) settlement is anticipated across the joint, dowels can be inserted.

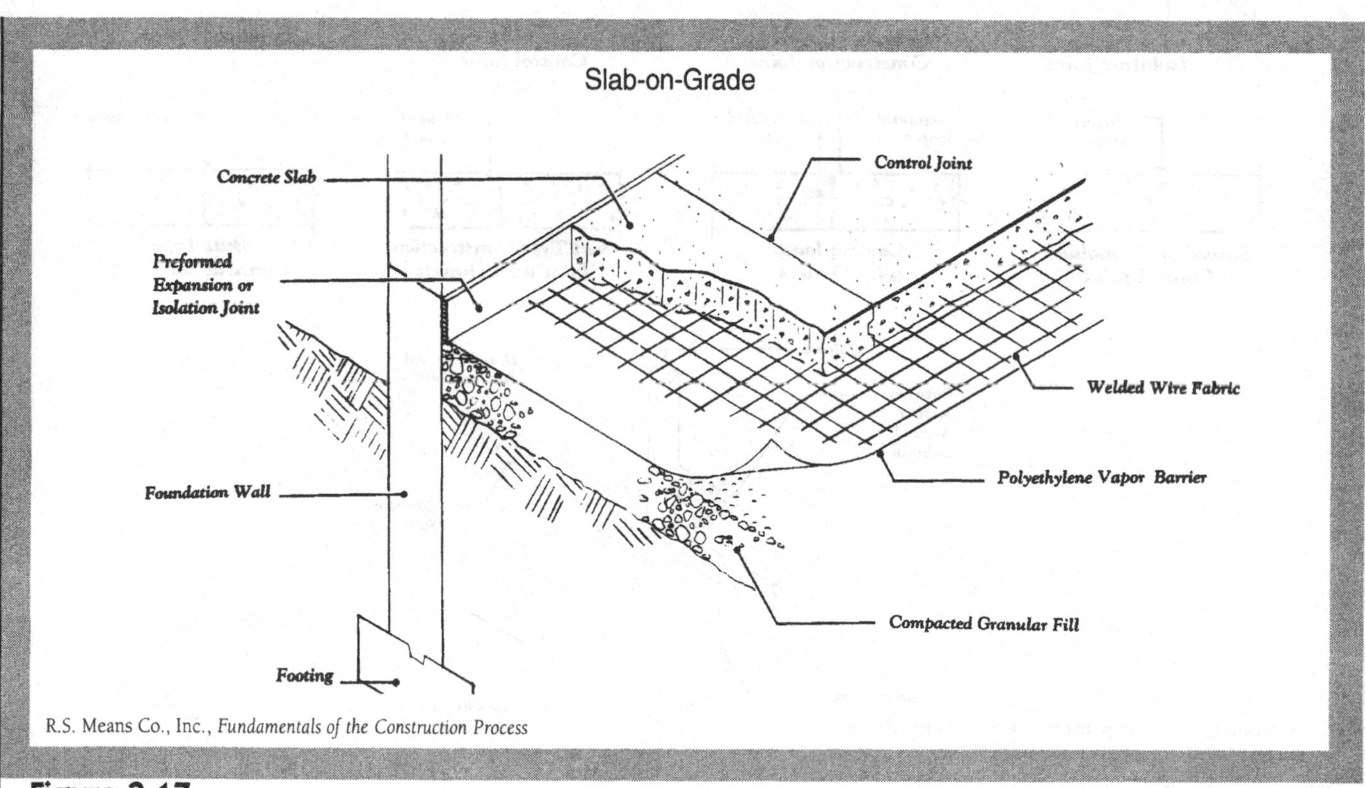

R.S. Means Co., Inc., *Fundamentals of the Construction Process*

Figure 3.17

A **control joint,** or **contraction joint**, is a sawed, tooled, or formed groove in the top part of a concrete slab. Its main purpose is to create a weakened plane, thereby predetermining and regulating the location of natural cracking caused by shrinkage as the concrete cures.

An **expansion joint** is generally used to isolate two structurally independent portions of a building, such as a low rise wing adjacent to a high rise tower. The joint is formed with the slab and, once set, filled with an elastomeric material.

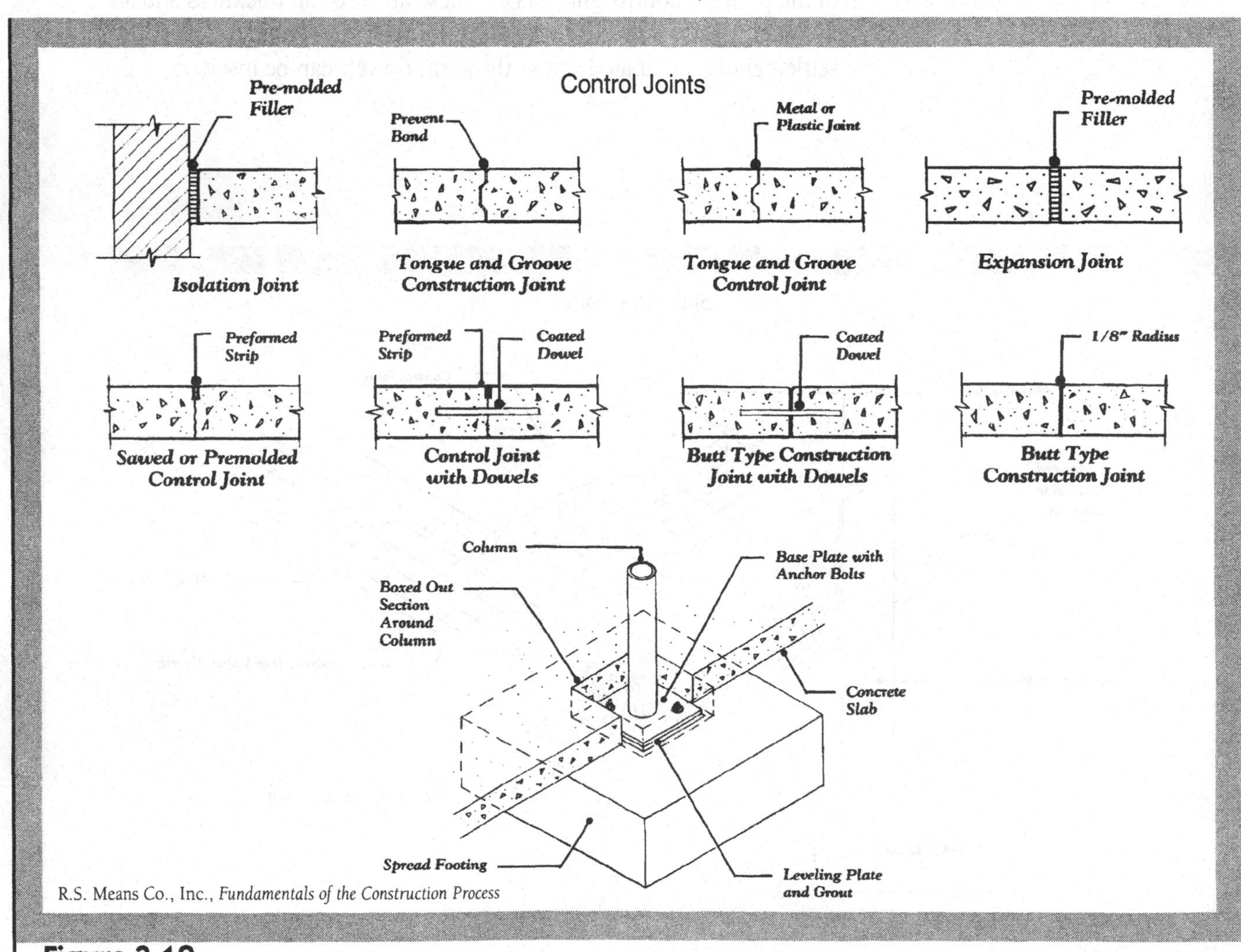

Figure 3.18

The Foundation

Foundations transfer many kinds of loads from a building to the earth below. These include: **dead and live loads**, **wind loads** and **earthquake loads**. The dead load is the total weight of the entire building—the structural frame, walls, floors, roof, and foundation. Live load is the combined weight of people, furniture, furnishings, equipment, and any other items expected to occupy the building, as well as the weight of any snow that may accumulate on the roof. Columns carry the dead and live loads down to the foundation.

The structure carries wind and earthquake loads to the foundations by additional strengthening of the columns or by the introduction of stiffening walls into the structure, usually at stairwells or elevator shafts.

There are two main types of foundations: **shallow** and **deep**. Shallow foundations for walls or columns are called footings, and transfer loads to the earth below provided they possess sufficient bearing capacity. Deep foundations transfer loads from columns to bedrock when the subgrade lacks sufficient bearing. Bedrock or soils with improved bearing are reached by driving piles or drilling caissons through upper layers of unstable soil.

Shallow Foundations: A footing is a type of foundation that spreads and transmits loads directly to the earth (soil). The weaker the bearing capacity of the soil, the wider the footing must be. Footings are usually cast-in-place concrete. Concrete may be placed by pumping, direct chute, buggies, wheelbarrows, or a crane and bucket. When access permits, the most economical method is the direct chute method.

Under favorable soil conditions, the sides of the excavation can hold a vertical slope, and formwork is not necessary. This is known as **neat excavation** or "excavating to neat lines." However, formwork is required where there is an angle of repose (sloped sides) or the footing is shaped irregularly. In most cases, formwork is required and the excavation is extended to provide working space.

The two most common types of shallow footings are **spread footings** and **strip footings**. The important characteristics of each type are discussed below.

Spread Footings

Spread footings are used to distribute column loads to the soil. These are common types of footings because columns are often used to carry building loads. Common configurations of spread footings are rectangular or square. They may be used to support single or multiple column loads. The size of spread footings is governed by the load and the bearing capacity of the soil. Pads, isolated footings, or pier footings are some alternate names for spread footings. Spread footings may be formed with dimension lumber and/or prefabricated plywood. Figure 3.19 illustrates the formwork used to create spread footings with dimension lumber.

Strip Footings

Under walls of concrete, block, brick, or stone, strip footings distribute loads evenly to the supporting soil. Strip footings also act as a leveling pad to facilitate the erection of formwork for the walls. Strip footings are also referred to as **continuous footings** or **wall footings**. An example of strip footing formwork is shown in Figure 3.20.

The bottom of a strip footing is placed on undisturbed soil—usually 12 inches below the deepest frost penetration. The sides may be formed with either dimension lumber or prefabricated panels. Strip footings may have temperature reinforcement running longitudinally. Reinforcing bars may be needed laterally when the width of the footing increases. When reinforced, strip footings should have a minimum of six inches of concrete above the reinforcing. When not reinforced, the footings should be at least eight inches thick.

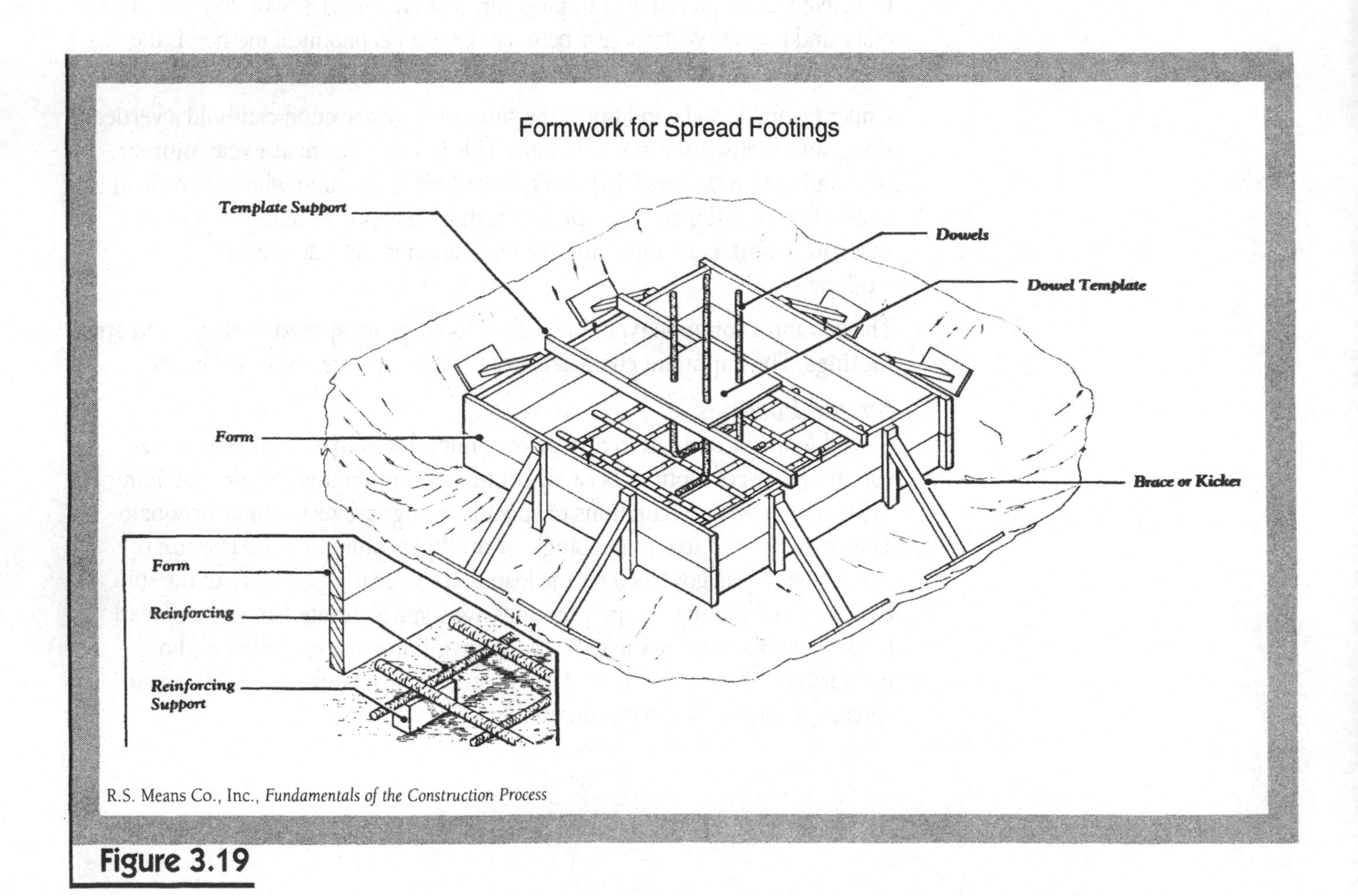

R.S. Means Co., Inc., *Fundamentals of the Construction Process*

Figure 3.19

Structures are commonly built on sloping terrain, or designed with differences in the top of footing elevations. To accommodate such transitions, steps are required in strip footings. Figure 3.21 illustrates a typical stepped footing.

Step footings are generally excavated by hand. Steps should measure at least two feet horizontally, and each vertical step should measure no greater than three-quarters of the horizontal distance between the steps. Vertical risers should be at least six inches thick, and match the footing's width. In comparison to strip and spread footings, stepped footings are time consuming and costly to construct.

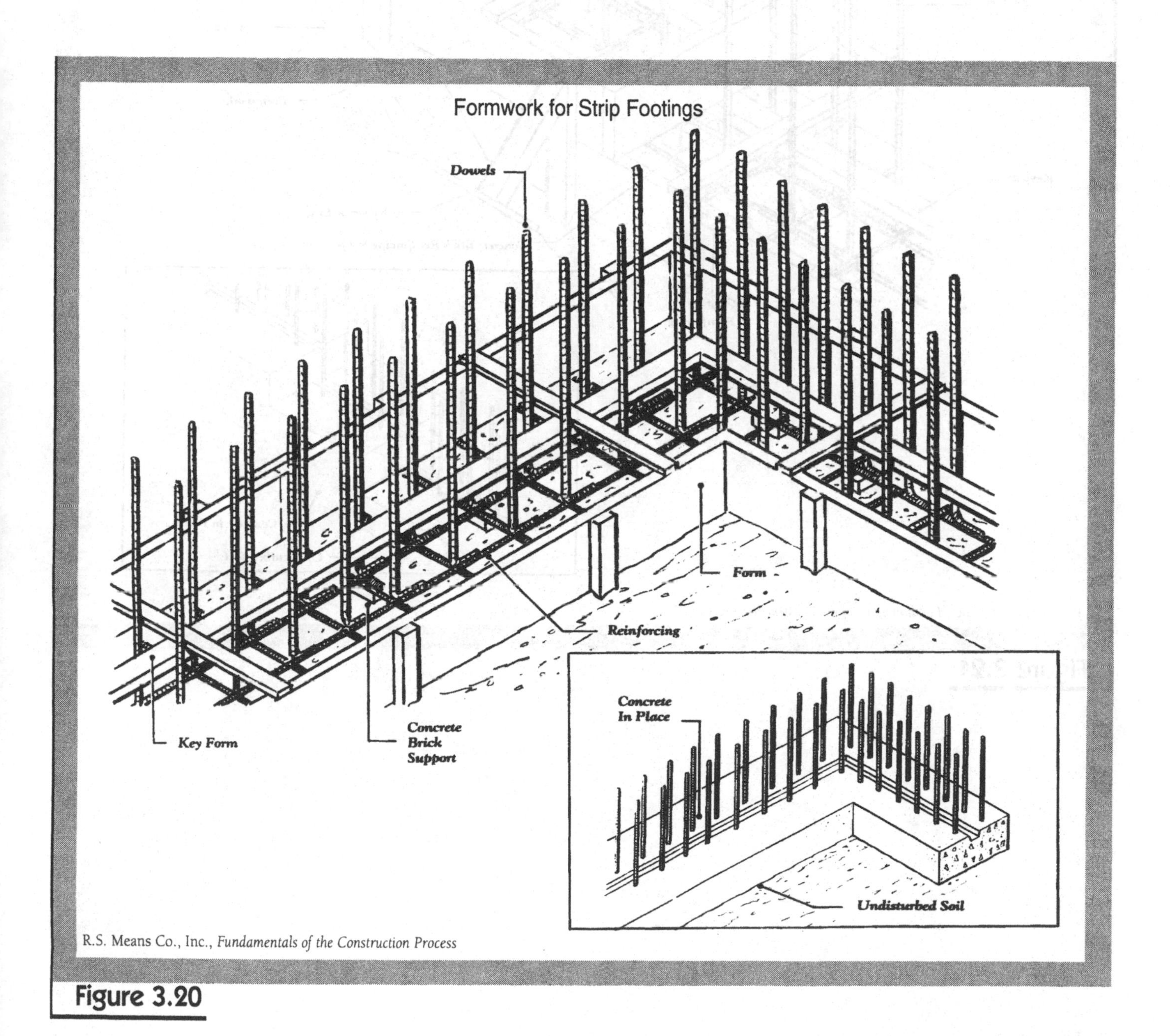

R.S. Means Co., Inc., *Fundamentals of the Construction Process*

Figure 3.20

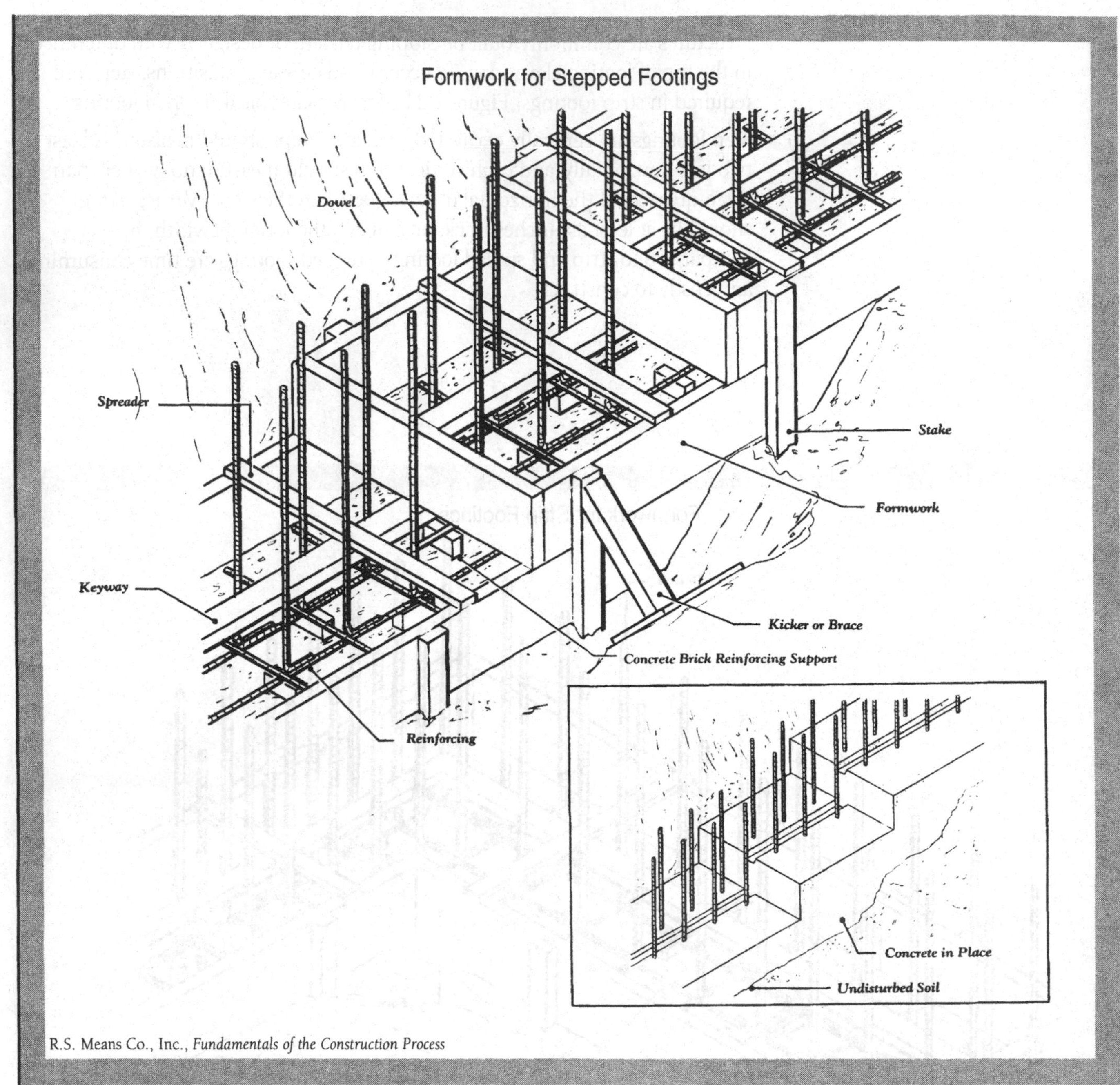

R.S. Means Co., Inc., *Fundamentals of the Construction Process*

Figure 3.21

Foundation Walls

Cast-in-place concrete or concrete block are most often used to construct foundation walls. Height and thickness of the walls depends on the height of backfill the wall must retain, the depth of frost penetration, and whether there is a full basement below grade. Plywood forms, steel concrete forming systems, or a combination of the two are typical methods of forming foundation walls. Examples of the two methods are shown in Figure 3.22.

Although there are several methods for erecting foundation wall formwork, the process can be broken down into the following four sequential steps.

1. Build the formwork for one side of the wall.
2. Tie the steel into the wall or place the reinforcing.
3. Form the opposite side wall.
4. Align, adjust, and straighten the entire forming system prior to placing the concrete.

Grade Beams

Grade beams are used to carry loads to the spread footings. Grade beams are economical where there are heavy column loads and light wall loads, in comparison to the cost of other foundation systems. Typical grade beam construction is illustrated in Figure 3.23.

Waterproofing

Ground water must be prevented from seeping through a foundation wall. The foundation wall is waterproofed and, if necessary, underdrains—drain pipes below the slab-on-grade and near the footings—are installed, as shown in Figure 3.24.

The foundation wall is waterproofed by applying a protective coating to the exterior surface—bituminous coating, troweled onto the wall, applied by brush, or sprayed on. Other materials used for waterproofing include asphalt protective board, mastic, or a waterproofing membrane.

The underdrain pipes are usually porous and have openings to facilitate the collection of water. Asphalt-coated corrugated metal, porous concrete, vitrified clay, or Polyvinyl Chloride (PVC) pipes are commonly used.

Coarse rock or gravel surrounds the pipes to create a channel of least resistance within the fill material. Of course, the pipes should be laid with sufficient slope to give the water adequate velocity to the discharge area.

Deep Foundations: The two most commonly used types of deep foundations are **piles** and **caissons**. A pile is a wood, steel, concrete, or composite shaft driven into place, while a caisson is a large shaft that is drilled into place, and filled with concrete.

Piles:

A pile is a column made of timber, concrete, or steel. The pile is driven into the ground to transfer loads from the pile cap through poor soil layers to deeper stable soil (with adequate strength and acceptable settlement) or rock.

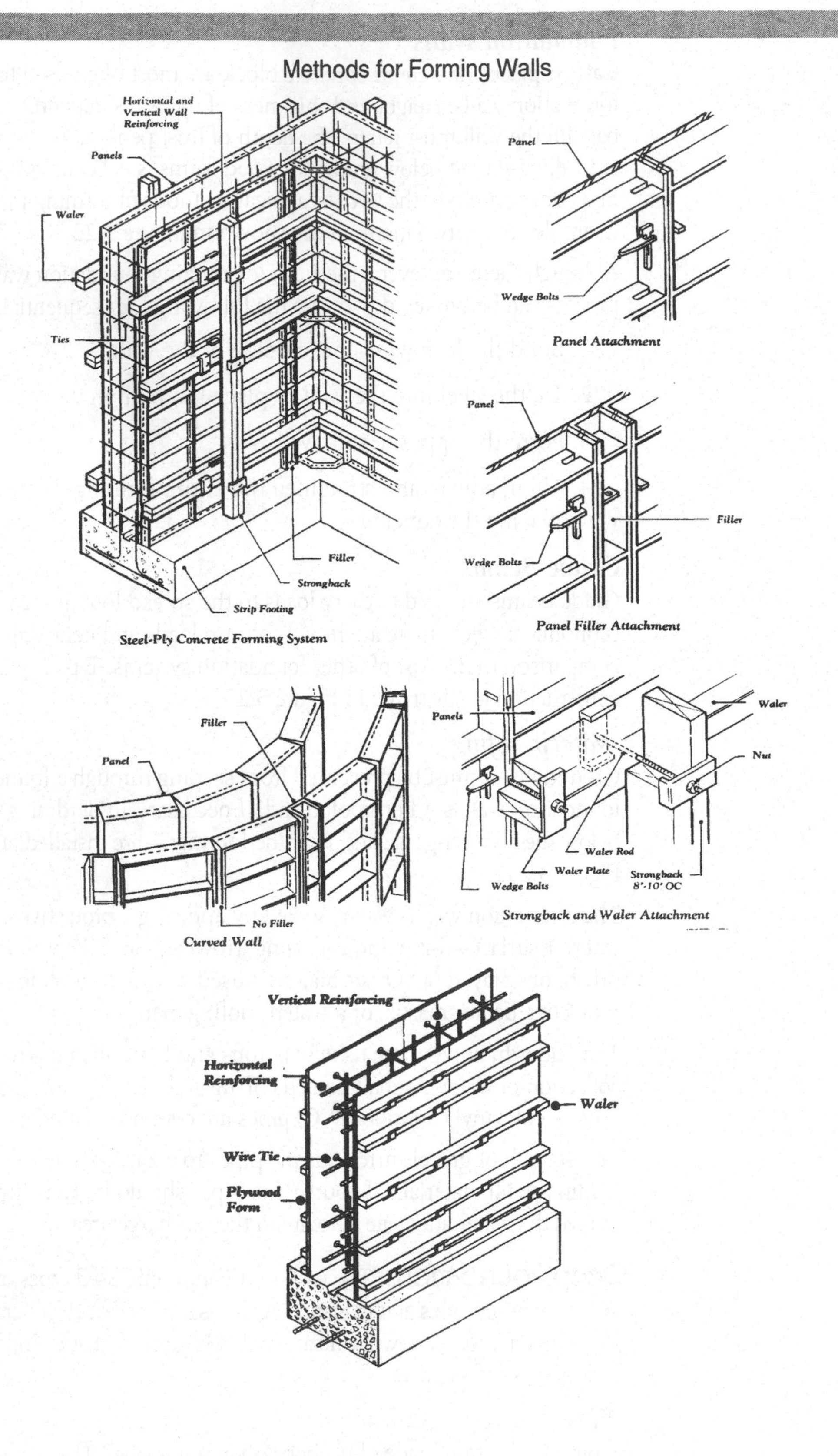

R.S. Means Co., Inc., *Fundamentals of the Construction Process*

Figure 3.22

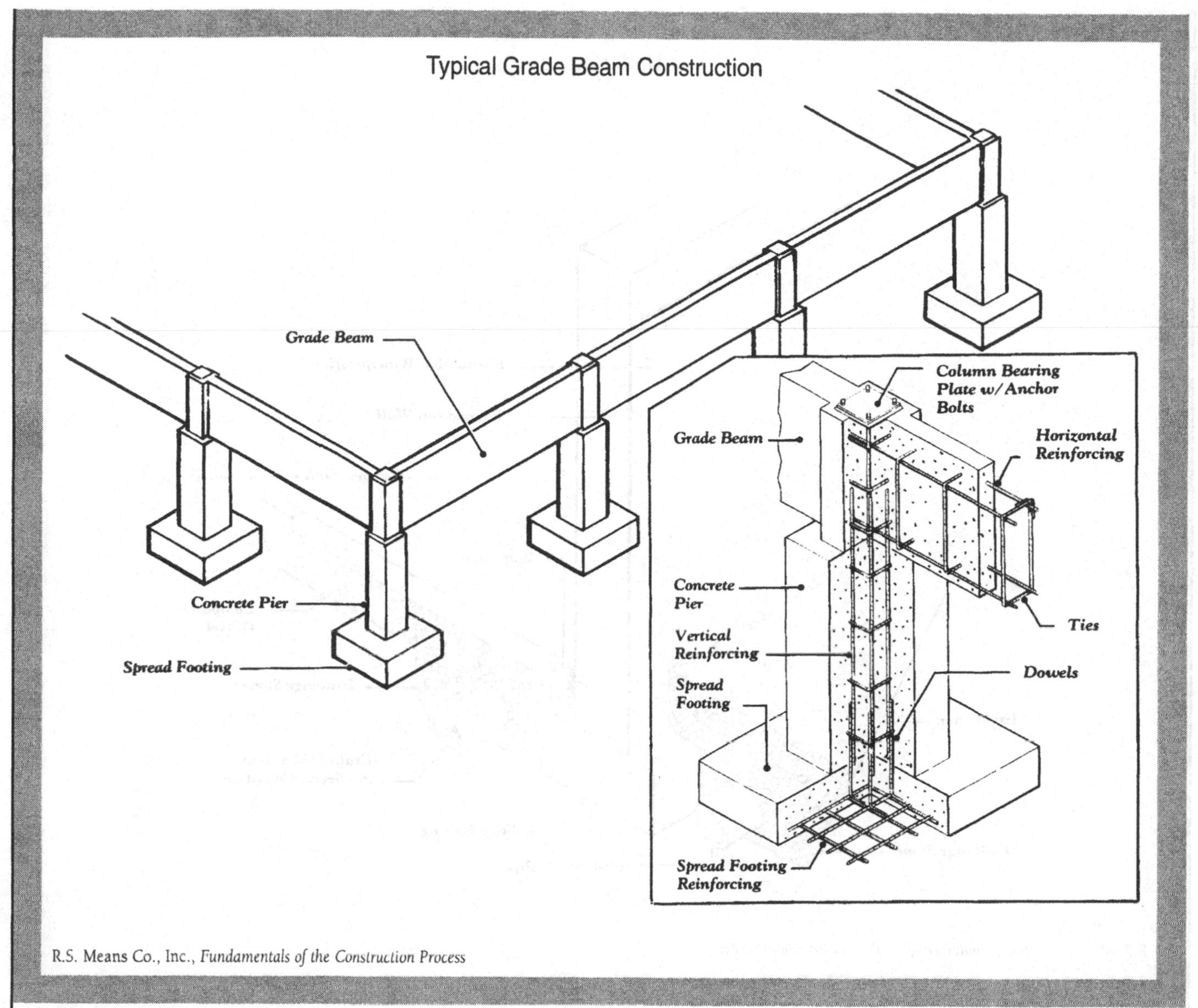

R.S. Means Co., Inc., *Fundamentals of the Construction Process*

Figure 3.23

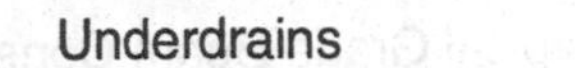
Underdrains

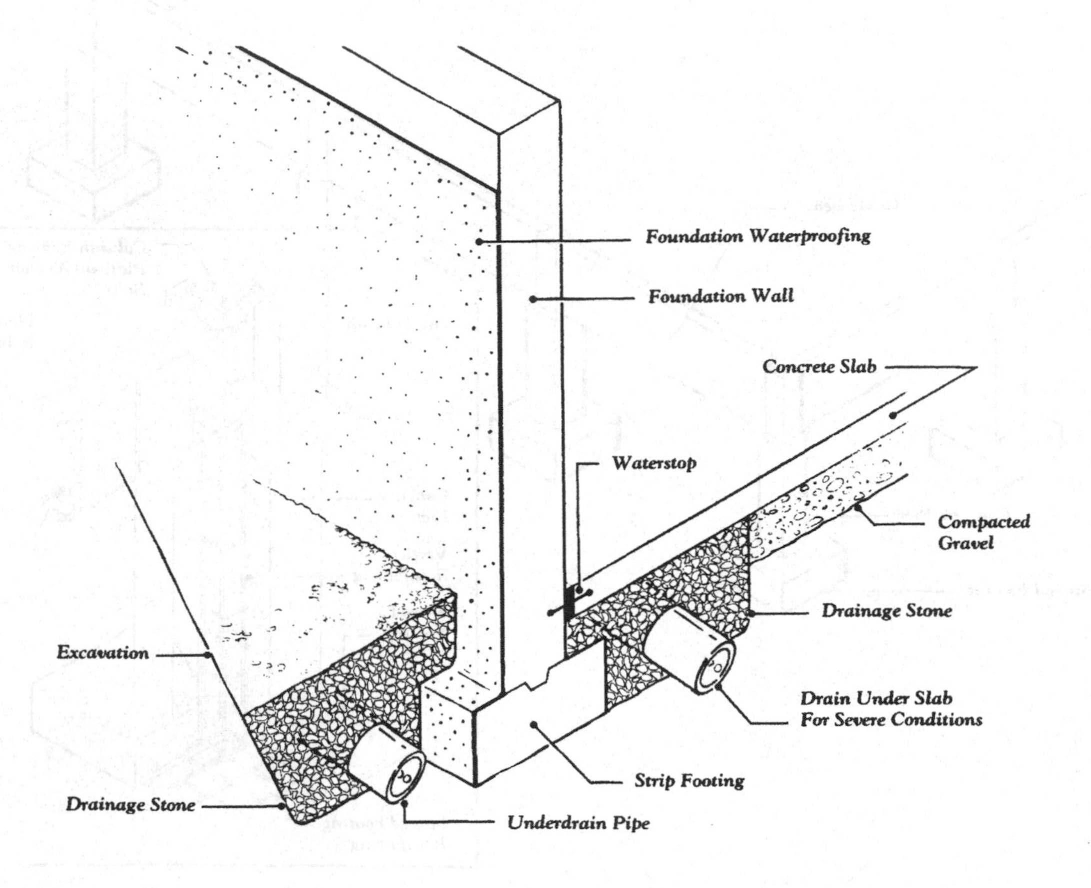

R.S. Means Co., Inc., *Fundamentals of the Construction Process*

Figure 3.24

There are two major types of piles: **end-bearing piles** and **friction piles**. End-bearing piles are driven through unstable soil to reach bedrock. Friction piles are driven far enough into cohesive soil to develop load support through "skin friction," between the soil and the outside surface of the pile.

Piles are driven into the ground by a pile driver. This is a crane with an attachment known as a pile hammer, that consists of weights lifted and then dropped or driven onto the pile head. The load capacity of a single pile is small compared to the loads delivered by the columns to the foundations, so piles are commonly installed in clusters. Pile clusters are capped with a concrete footing known as a **pile cap**, or slab, to evenly distribute loads to the individual piles as illustrated in Figure 3.25.

Piles are usually installed vertically, but occasionally they are installed at a slight angle. Piles installed at an angle are referred to as **battered**.

As previously stated, piles can be made from concrete (cast-in-place or precast), steel (H-pile or pipe pile), or timber. Each type is described in the following pages.

Concrete Piles: The first operation for a cast-in-place concrete pile is to drive a cylindrical steel shell with a **mandrel** until it reaches the specified capacity or location. A mandrel is a heavy metal tube inserted into hollow piles that allows them to be driven. The mandrel absorbs most of the impact and energy of the pile driver. The mandrel is then withdrawn, the shell inspected and, finally, is filled with concrete and reinforcing. Cast-in-place concrete piles may be treated for use in sea water, and can be easily altered in length.

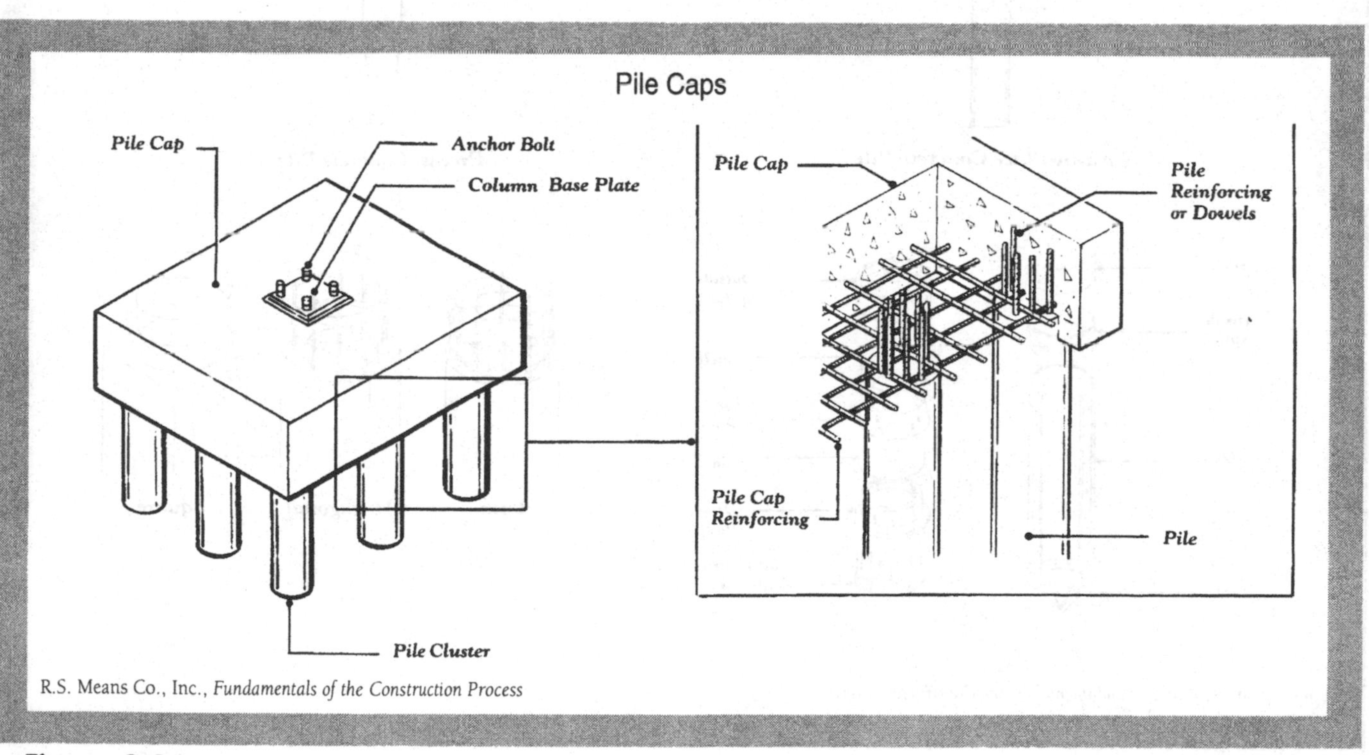

R.S. Means Co., Inc., *Fundamentals of the Construction Process*

Figure 3.25

Precast concrete piles are cast at a plant to a specified length and transported by trucks to the job site. They may be reinforced or post-tensioned. Post-tensioning is a method of increasing the strength of concrete by stressing tendons after the concrete has hardened. Like cast-in-place piles, they may be treated for use in sea water, but it is difficult and expensive to alter the lengths of such piles. In addition, precast concrete piles require heavy equipment for handling (unloading from trucks, lifting, etc.) and driving.

Examples of cast-in-place and precast concrete piles are shown in Figure 3.26.

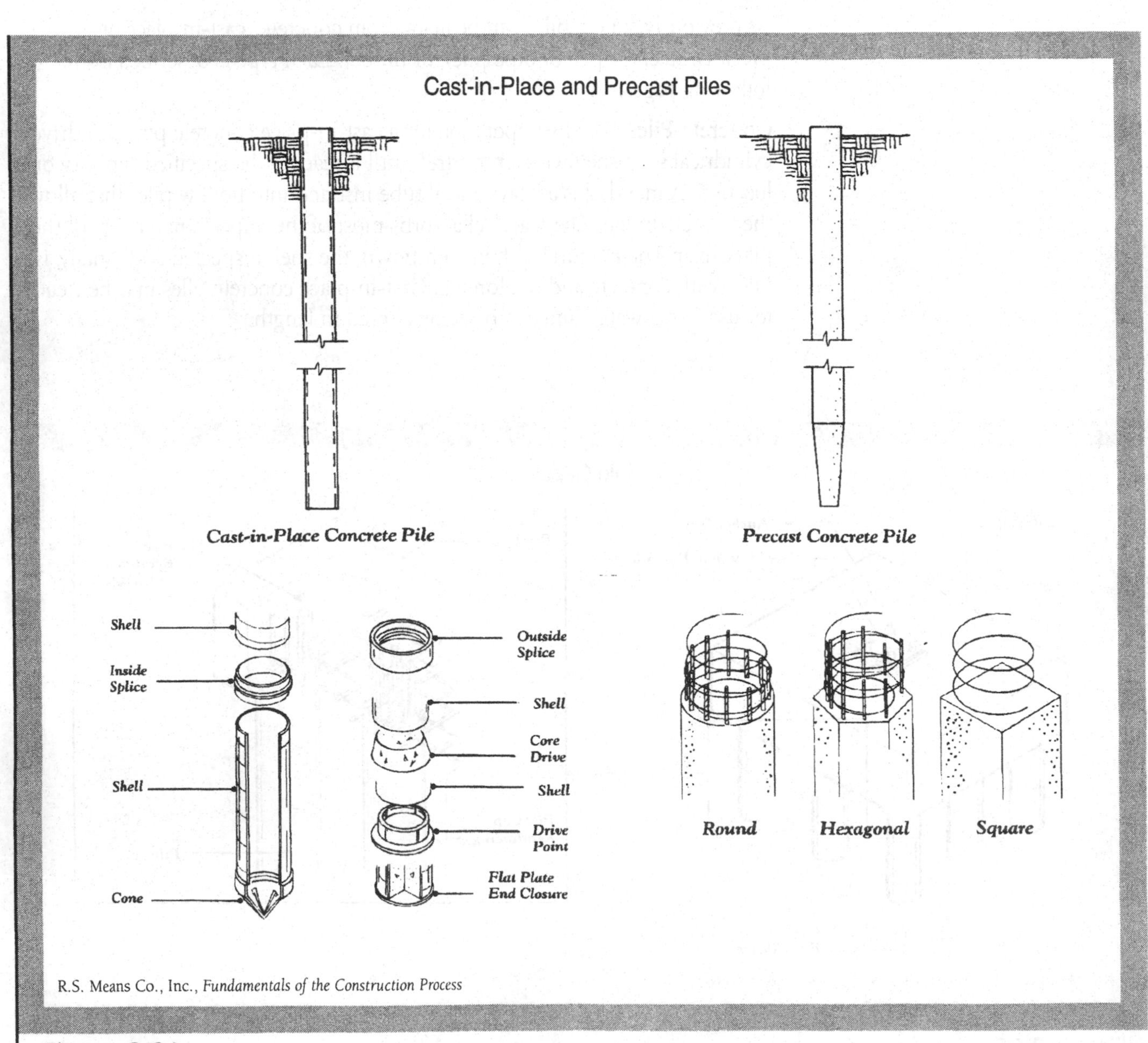

Figure 3.26

Steel Piles: Steel piles are strong and can withstand rough treatment. However, in some soil conditions, corrosion may be a problem. The two basic types of steel piles are **steel HP piles** and **steel pipe piles**.

Steel HP sections are hot-rolled wide flange members made for use in pile foundations. HP piles are seldom used as friction piles. Most often, they are used as end-bearing piles where long lengths of piles are required to be driven to refusal. HP piles can be brought to the site in convenient lengths and easily spliced as driving progresses, and withstand rough handling and driving conditions.

Steel pipe piles, also known as **composite piles**, are driven to a specified elevation and then filled with concrete. Steel pipe piles, like HP piles, are flexurally strong and can be easily cut or spliced. They can withstand rough handling and driving conditions and may be driven with a standard hammer without a mandrel. Examples of steel HP piles and pipe piles, with the necessary accessories, are shown in Figure 3.27.

Timber Piles: Timber piles are known for their relatively low cost and ease of handling. They are best suited under lightly loaded conditions, where driving is through soft strata. These piles can also be pressure-treated with creosote to prevent decay, although generally they are cut-off below the level of the water table to avoid the problem of decay.

Soil conditions are critical, however, as timber piles cannot withstand punishing driving through hard strata, rock, or boulders. Other driving limitations are that relatively small hammers must be used and timbers cannot be readily spliced. An illustration of a timber pile is shown in Figure 3.28.

Caissons:

Placing concrete into a deep drilled hole at least two feet in diameter creates a columnar foundation known as a caisson. This construction functions as a compression member and transfers building loads through the unstable soil to bedrock or stable hard stratum. The advantage of using a caisson over a pile is that there is no soil heaving, displacement vibration, or noise during installation, and the underlying bedrock can be visually inspected. Caissons are either reinforced or unreinforced, and either straight or belled out at the bearing level as illustrated in Figure 3.29.

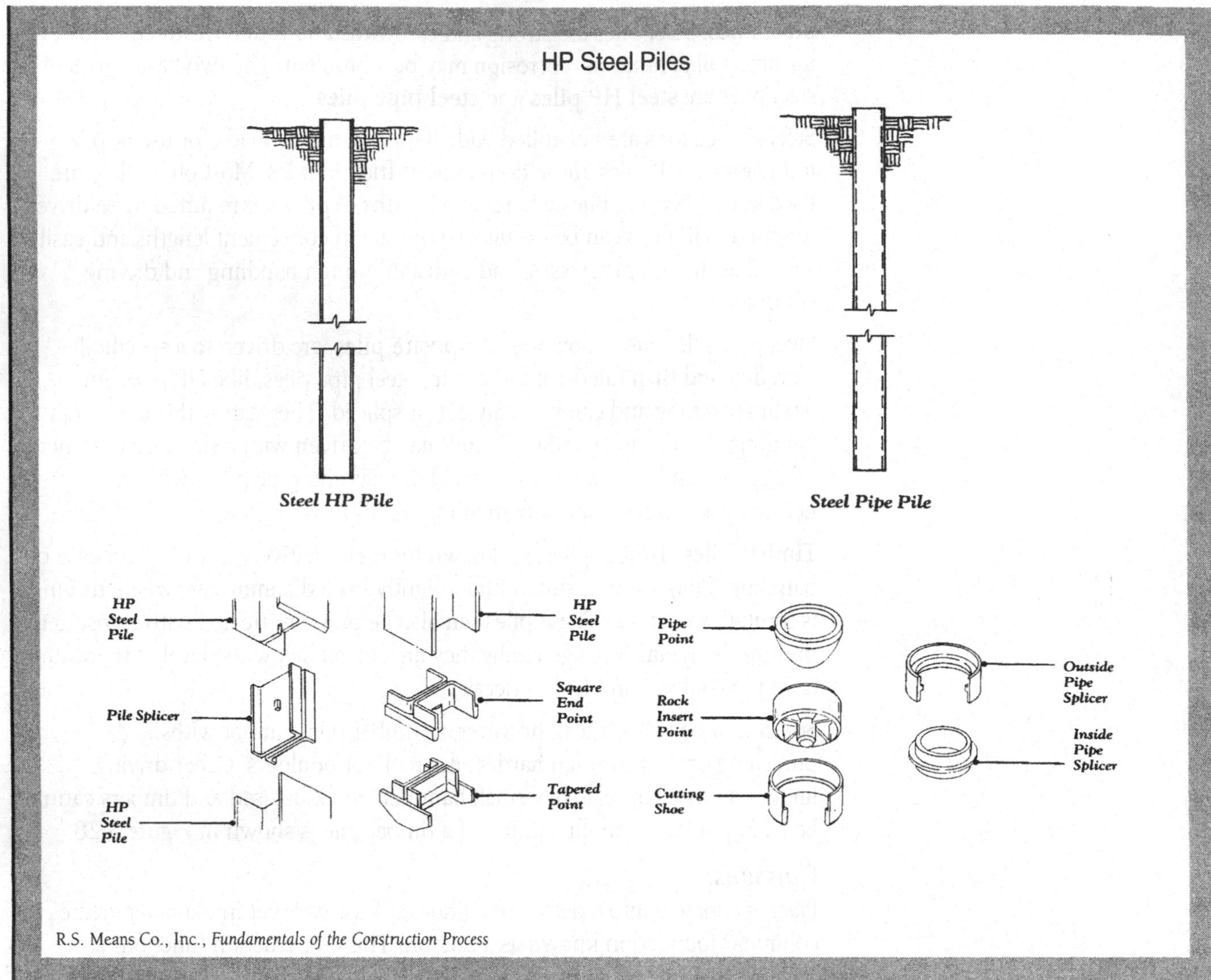

R.S. Means Co., Inc., *Fundamentals of the Construction Process*

Figure 3.27

There are three basic types of caissons: **belled caissons, straight caissons,** and **socketed** (or **keyed**) **caissons**.

Caissons are belled at the lower end to distribute the building load over a greater area and thus have a reduced bearing pressure on the soil. They should only be used in soils such as clays, hardpan, gravel, silt, and igneous rock, and are not recommended for shallow depths or poor soils. Steel casings line the drilled shaft for caissons used in weaker soils.

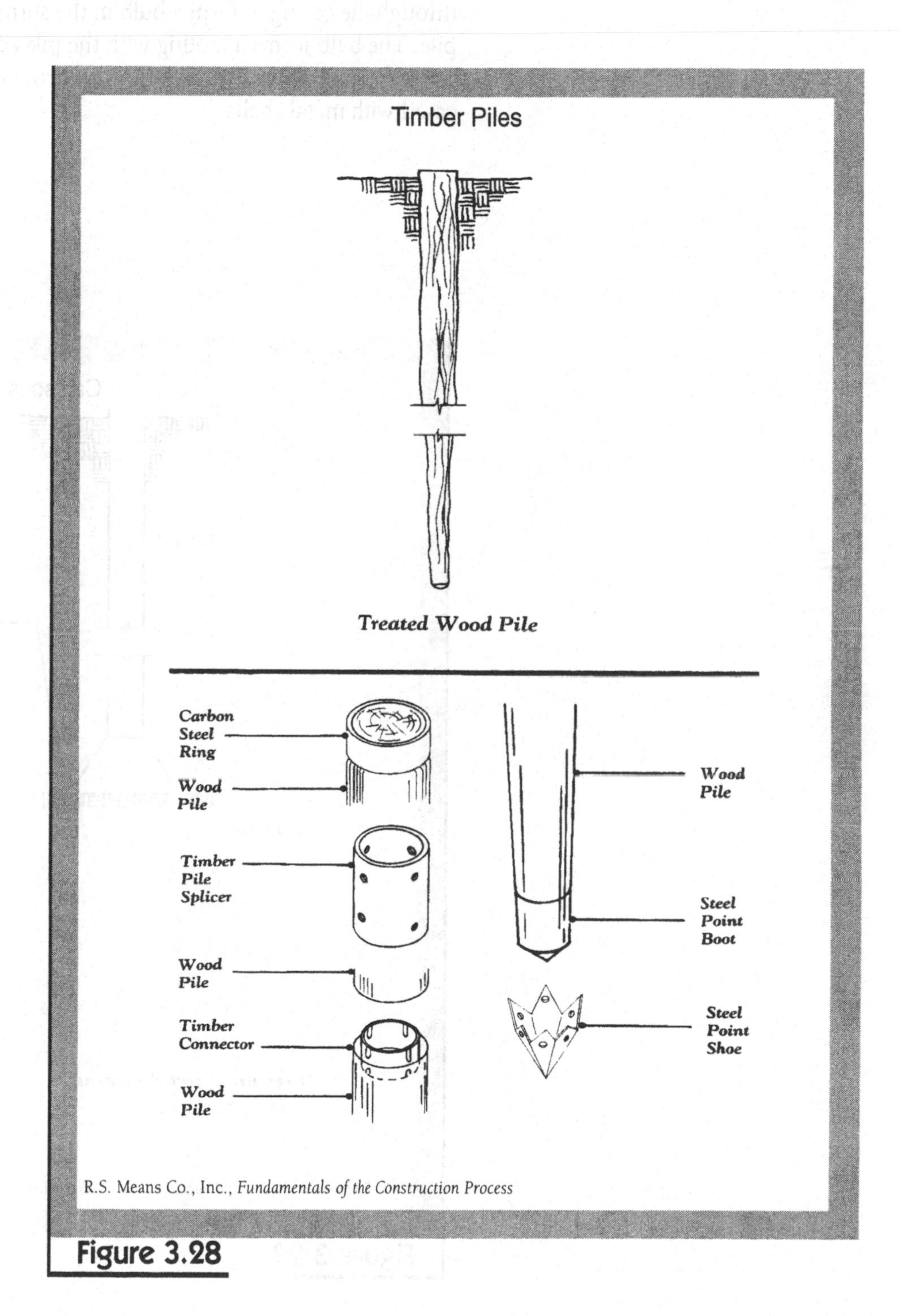

R.S. Means Co., Inc., *Fundamentals of the Construction Process*

Figure 3.28

Typical caissons may have a straight shaft when the supporting strata provides sufficient bearing for the end area.

Socketed caissons are used for extremely heavy loads. Installation of these piles involves sinking the shaft into rock to produce combined friction and bearing support action.

Pressure Injected Footings:

Pressure-injected footings or bulb end piles are placed in the same way as cast-in-place concrete piles, but differ in that a plug of uncured concrete is driven through the casing to form a bulb in the surrounding earth at the base of the pile. The bulb forms a footing with the pile acting as a column or pier. Piles of 25' or less may have the casing withdrawn, but those of 25' or more are usually cased with metal shells.

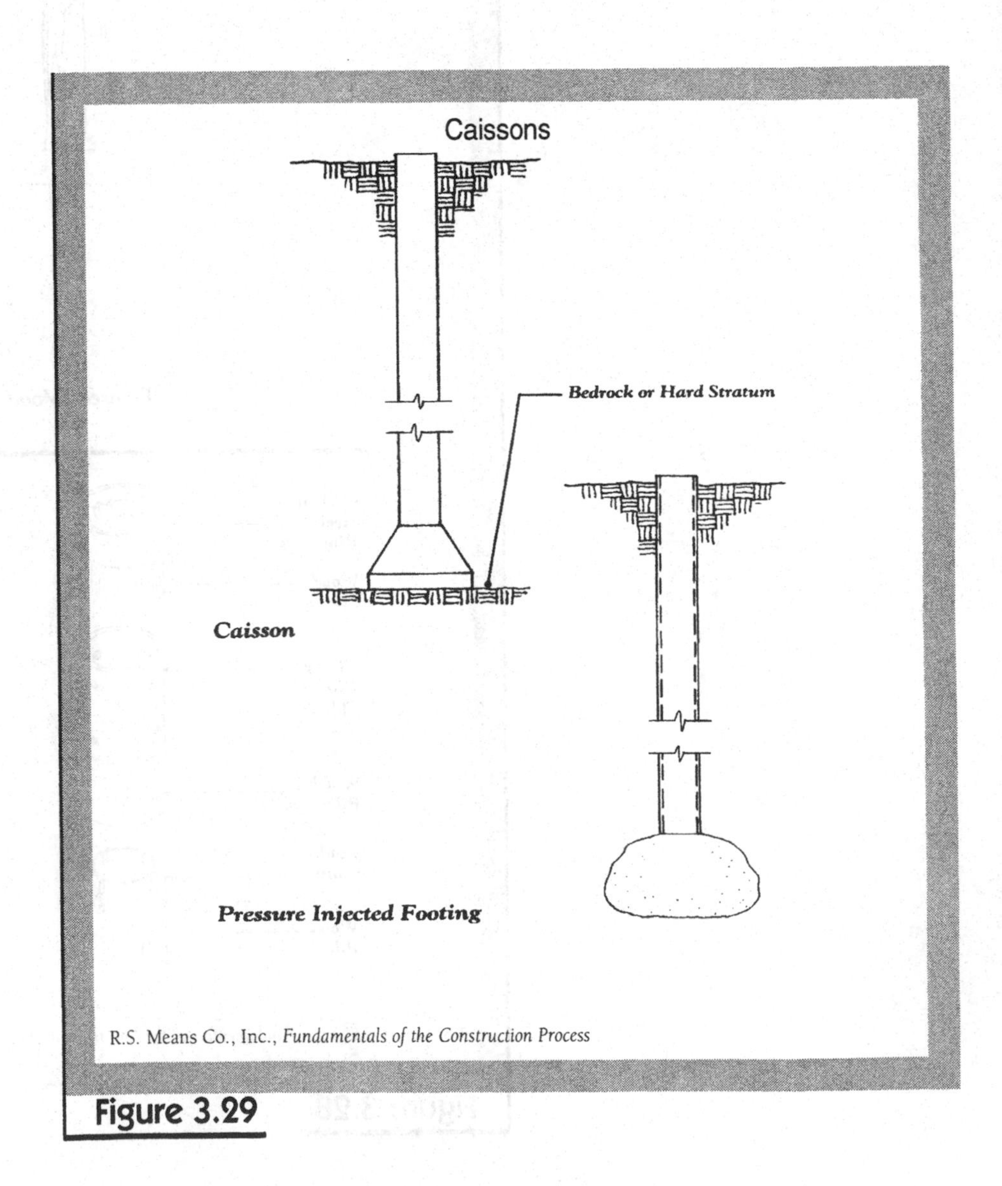

Figure 3.29

The following section is reprinted from *Basics for Builders: Plan Reading & Material Takeoff*, by Wayne J. DelPico, R.S. Means Co., Inc.

Structural Steel

Structural steel has the capacity to support large loads with a relatively compact size and shape. It can be used to support loads in both compression and tension, therefore making it an ideal material for flexural components. The variety of steel shapes and sizes available provides engineers with an economical solution to many structural design problems. Structural steel is also available in different strengths or grades for particular loading or stress conditions.

Structural Steel Shapes

Structural steel can best be defined as the members that make up the structural frame of the building that will transmit the load to the foundation. Figure 3.30 lists eight of the more common shapes.

Common Steel Sections

The upper portion of this table shows the name, shape, common designation and basic characteristics of commonly used steel sections. The lower portion explains how to read the designations used for the above illustrated common sections.

Shape & Designation	Name & Characteristics	Shape & Designation	Name & Characteristics
W	Wide Flange Parallel flange surfaces	M C	Miscellaneous Channel Infrequently rolled by some producers
S	American Standard Beam (I Beam) Sloped inner flange	L	Angle Equal or unequal legs, constant thickness
M	Miscellaneous Beams Cannot be classified as W, HP or S; infrequently rolled by some producers	T	Structural Tee Cut from W, M or S on center of web
C	American Standard Channel Sloped inner flange	H P	Bearing Pile Parallel flanges and equal flange and web thickness

R.S. Means Co., Inc., *Basics for Builders: Plan Reading & Material Takeoff*

Figure 3.30

Each shape is prefixed by a letter and numbers. These designations are more than simple identifications; they provide important information about the individual piece. The letter is the classification of the piece by shape. The first number refers to the actual or nominal depth in inches of the section. The second number refers to the weight per linear foot of the section, in pounds per LF.

The exceptions to this rule are the angle shapes. The first two numbers in an angle designation are the lengths of the "legs" of the particular angle, in inches. The third number is the thickness of each leg in inches. Figure 3.31 lists the designations and their corresponding depths and weights.

Structural steel is also available in different strengths, expressed as **yield stress.** Simply put, the yield stress, typically defined as **KSI** or **kips per square inch**, is the maximum allowable stress that can be exerted on the material before it fails. A kip is a unit of measure equal to 1,000 pounds. The different grades of steel are named according to the number of the test conducted by the

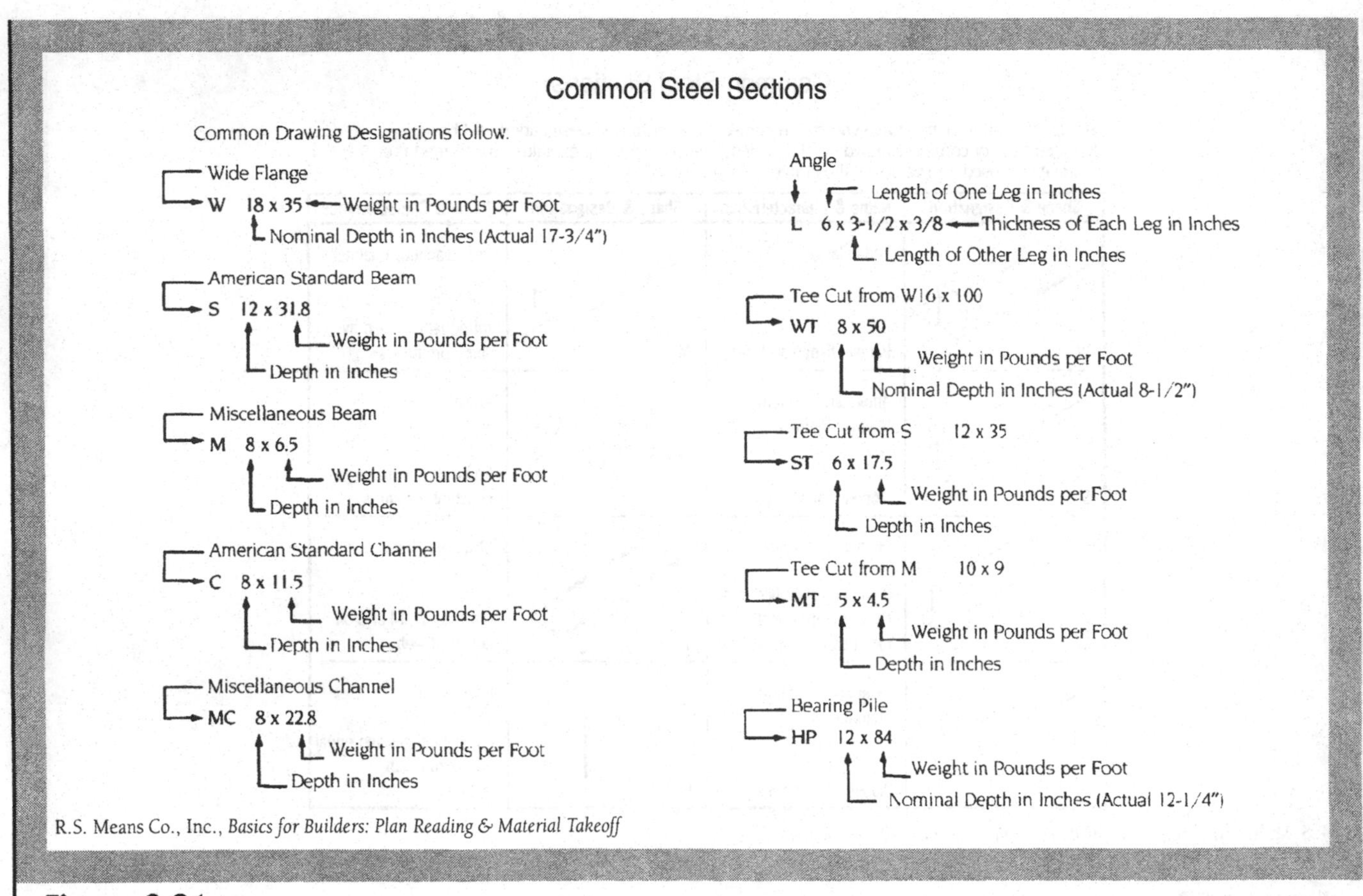

R.S. Means Co., Inc., *Basics for Builders: Plan Reading & Material Takeoff*

Figure 3.31

American Society of Testing and Materials (ASTM) to determine the characteristics of the species. Not all of the previously mentioned shapes are available in some of the more specialized grades of steel. Figure 3.32 lists some common structural steel grades.

Structural steel work is shown on structural drawings in plan views. Elevations, sections, and details are often added for clarity. The details show the connections to be used for the individual pieces. Figure 3.33 is a structural steel drawing shown in plan view, and the corresponding details.

Note: The term "DO" indicates duplication, similar to the expression "ditto," and is used to show more than one of the same designated beam, girder, joist, etc.

Many structural steel drawings provide the column lengths. The highest elevation on a column or beam is called the **top of steel** and is shown on the drawings as T.O.S. The proposed elevation for the top of the leveling plate is also given; the difference between the two is the length of the column.

Some shapes, such as angles, do not include a weight designation. Determining the weight of angles, plate steel of various thicknesses, and tube steel requires the use of a table in the *Manual of Steel Construction* published by the American Institute of Steel Construction.

Metal Joists

Steel joists are manufactured by welding hot-rolled or cold-formed sections to angle web or round bars to form a truss. Standard open-web and long-span steel joists were developed as an alternative to wood frame construction. The steel joist's capacity to carry loads spanning greater distances has made it popular for all types of light-occupancy construction. In addition, steel joists can be used in fire-rated construction. The joists' open webbing allows the passage of mechanical piping and electrical conduits without the drilling or coring of holes.

Common Structural Steel Specifications

ASTM Designation	Yield Stress in KSI	Description
A36	36	Most common carbon steel, all shape groups and plates and bars up to 8"
A529	42	All shape groups, as A36, but plates and bars up to 1/2" thick
A441	40-50	High-strength, low-alloy steel, shapes and plates, but in lesser variety
A572	42-65	High-strength, low-alloy steel, all shapes and plates
A242 & A588	42-50	High-strength, corrosion-resistant; drop in strength as sizes increase
A514	90-100	Quenched and tempered alloy; plates and bars only, with special care so as not to impair the heat treatment

R.S. Means Co., Inc., *Basics for Builders: Plan Reading & Material Takeoff*

Figure 3.32

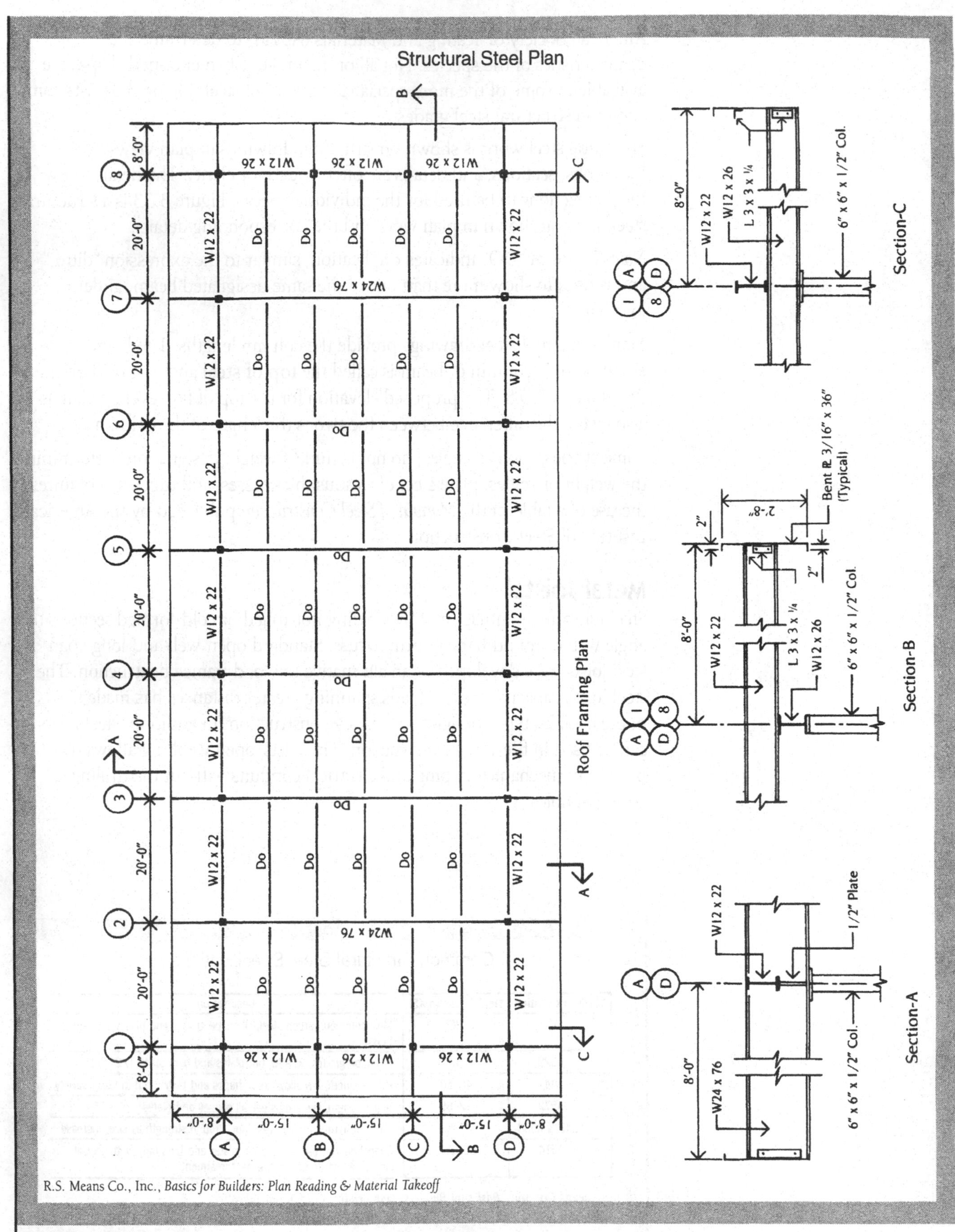

R.S. Means Co., Inc., *Basics for Builders: Plan Reading & Material Takeoff*

Figure 3.33

Types of Steel Joists: Steel joists can be classified according to one of the following three categories or series:

- **K-Series** are open-web, parallel-chord steel joists manufactured in standard depths of 8", 10", 12", 14", 16", 18", 20", 22", 24", 26", 28", and 30" and lengths up to 60'.
- **LH-Series** refers to long-span, open-web steel joists manufactured in depths of 18", 20", 24", 28", 32", 36", 40", 44" and 48", and lengths up to 96'.
- **DLH-Series** refers to deep long-span, open-web steel joists manufactured in depths of 52", 56", 60", 64", 68" and 72" and lengths up to 144'.

Figure 3.34 shows the typical details for both the K-Series and the LH- and DLH-Series open-web joists.

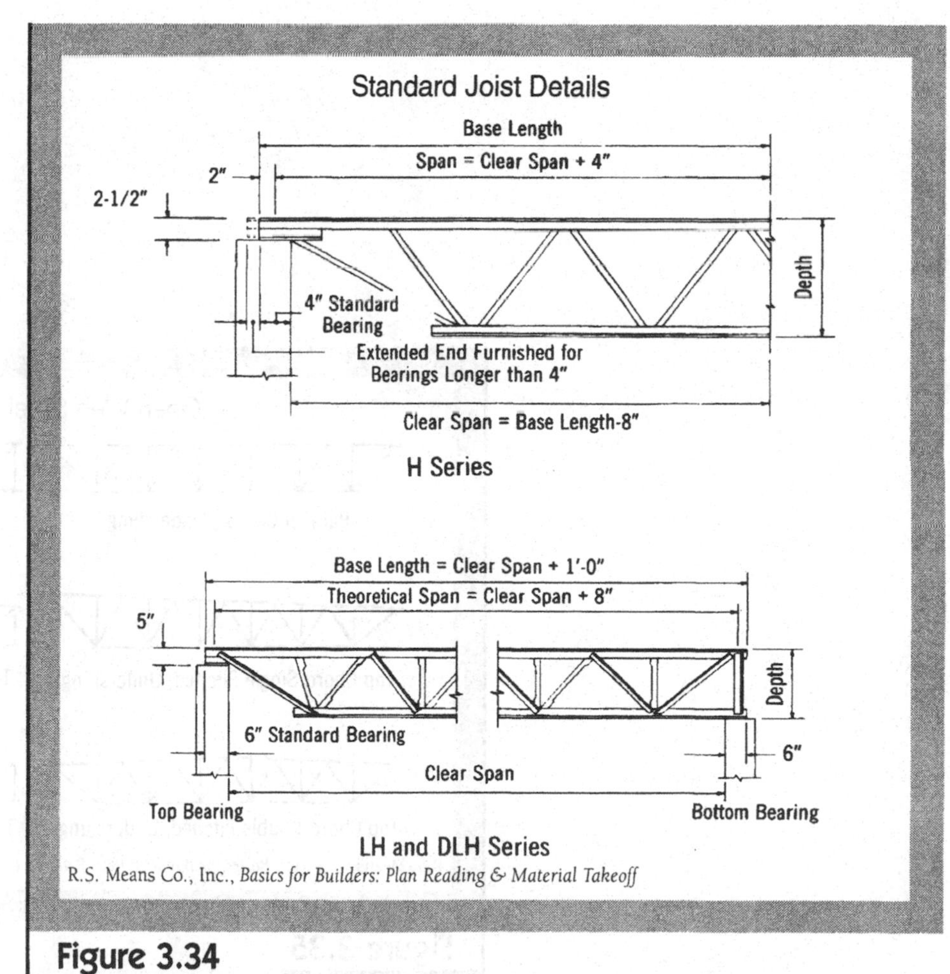

R.S. Means Co., Inc., *Basics for Builders: Plan Reading & Material Takeoff*

Figure 3.34

The LH- and DLH-Series are also available with top chords that are pitched or parallel to the bottom chords. The ends of the joists can be square ends or underslung. Figure 3.35 shows the different types of open-web steel joist designs available.

Joist designations are defined as follows:

24K10

The first number, **24**, is the depth in inches of the joist. The letter **K** indicates that it is a K-Series joist. The last number, **10**, indicates the load capacity/size of the chords.

In Figure 3.36, open-web steel joists are shown as they would appear in a typical structural drawing.

Bridging: Bridging is the lateral bracing of open-web steel joists to provide stiffness and stability and to prevent wracking of the joists. Bridging is most often accomplished with small lightweight steel angles bolted or welded perpendicular to the span of the joist. The locations of the courses of bridging are shown on the structural plans and are determined in accordance with the joist manufacturer's recommendation.

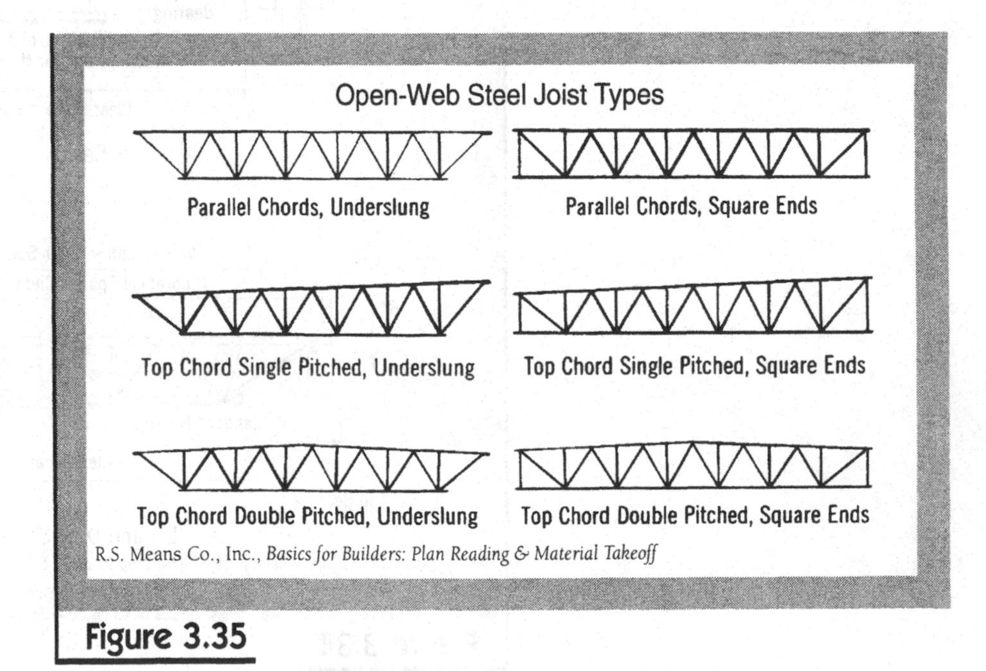

Figure 3.35

There are two main types of bridging:

- Horizontal bridging, which consists of two continuous steel members, one fastened to the top chord and one fastened to the bottom chord.
- Diagonal bridging, which consists of bracing that runs diagonally from the top of one chord to the bottom chord of the adjacent joist.

It is important that the terminations of bridging are securely fastened to the wall or beam, regardless of the bridging installation method.

Welding Connections

Welding is a method of joining steel elements by a process of fusion. It is used for buildings and bridges, and offers these advantages: less connection material than other methods, and a quieter process. Figure 3.37 provides some basic information on welding terminology and symbols.

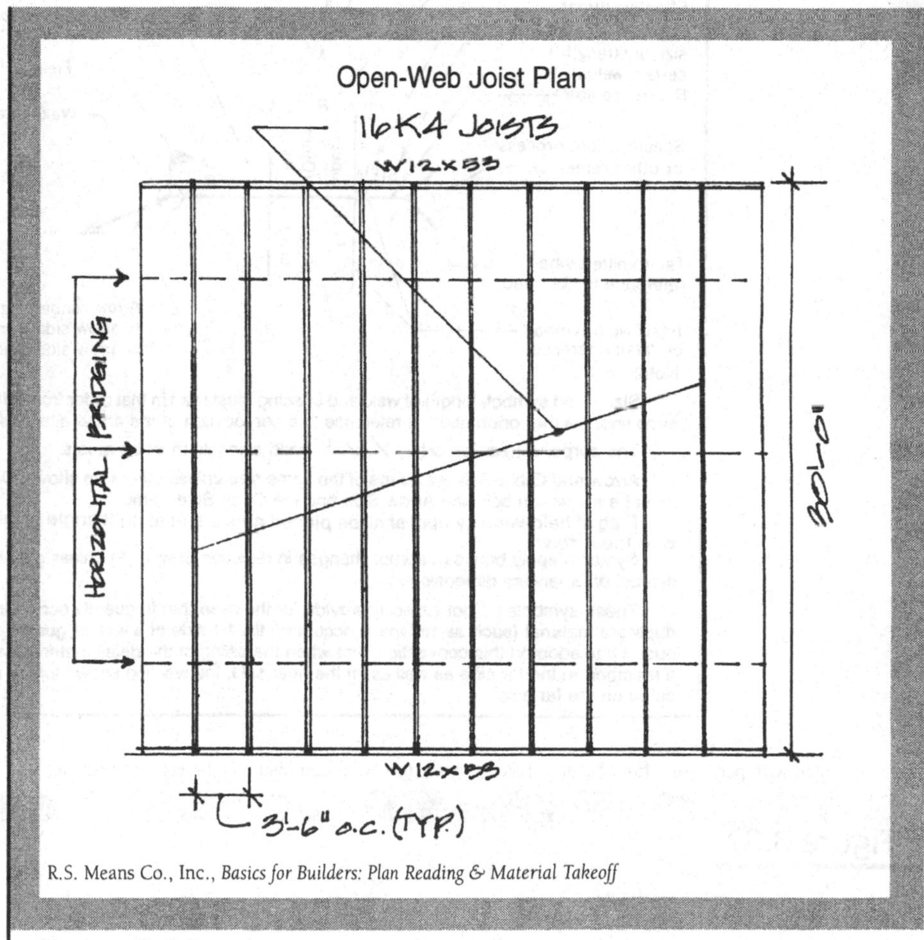

R.S. Means Co., Inc., *Basics for Builders: Plan Reading & Material Takeoff*

Figure 3.36

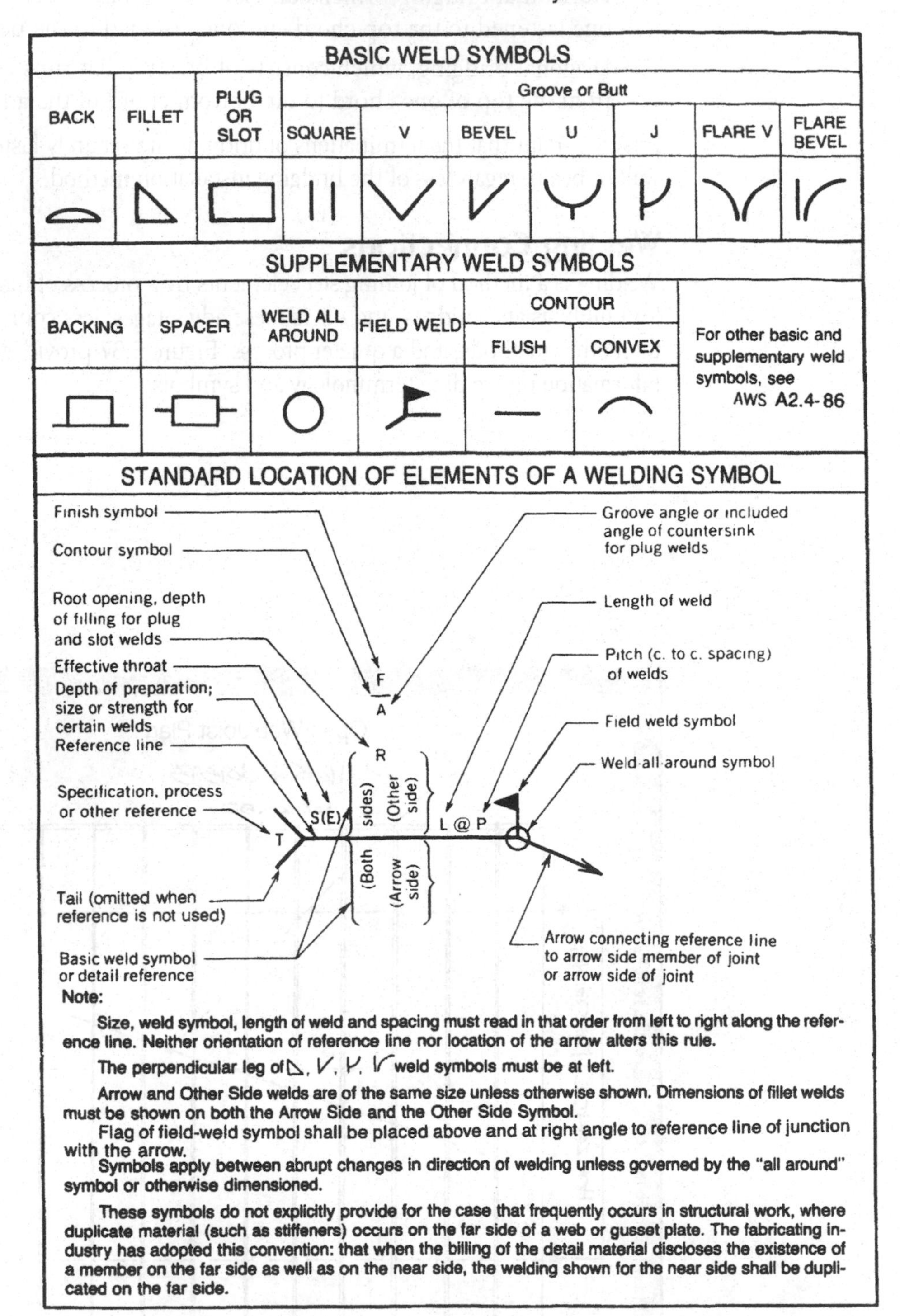

Reprinted with permission from *Manual of Steel Construction*, American Institute of Steel Construction, Inc.

Figure 3.37

The following section is reprinted with permission from *Design of Welded Structures*, by Omer W. Blodgett, The James F. Lincoln Arc Welding Foundation.

Determining Weld Size: Overwelding is one of the major factors of welding cost. Specifying the correct size of weld is the first step in obtaining low-cost welding. This demands a simple method to figure the proper amount of weld to provide adequate strength for all connections.

In strength connections, complete-penetration groove welds must be made all the way through the plate. Since a groove weld, properly made, has equal or better strength than the plate, there is no need for calculating the stress in the weld or attempting to determine its size. However, the size of a partial-penetration groove weld may sometimes be needed. When welding alloy steels, it is necessary to match the weld-metal strength to plate strength. This is primarily a matter of proper electrode selection and of welding procedures.

With fillet welds, it is possible to have too small a weld or too large a weld; therefore, it is necessary to determine the proper weld size.

Strength of Welds:

Many engineers are not aware of the great reserve strength that welds have. Figure 3.38 shows the recognized strength of various weld metals (by electrode designation) and of various structural steels.

Minimum Strengths Required of Weld Metals and Structural Steels

(AWS A5.1 & ASTM A233)
(as-welded condition)

	Material	Min. Yield Strength psi	Min. Tensile Strength psi
Weld Metals	E6010	50,000 psi	62,000 psi
	E6012	55,000	67,000
	E6024	58,000	62,000
	E6027	50,000	62,000
	E70XX	60,000	72,000
Steels	A7	33,000	60,000
	A373	32,000	58,000
	A36	36,000	58,000
	A441	42,000 46,000 50,000	63,000 67,000 70,000

Reprinted with permission from *Design of Welded Structures*, James F. Lincoln Arc Welding Foundation

Figure 3.38

Notice that the minimum yield strengths of the ordinary E60XX electrodes are over 50% higher than the corresponding minimum yield strengths of the A7, A373 and A36 structural steels for which they should be used.

Since many E60XX electrodes meet the specifications for E70XX classification, they have about 75% higher yield strength than the steel.

Submerged-Arc Welds:

AWS and AISC require that the bare electrode and flux combination used for submerged-arc welding shall be selected to produce weld metal having the tensile properties listed in Figure 3.39, when deposited in a multiple-pass weld.

Fillet Weld Size: The American Welding Society (AWS) has defined the effective throat area of a fillet weld to be equal to the effective length of the weld times the effective throat. The effective throat is defined as the shortest distance from the root of the diagrammatic weld to the face.

According to AWS, the leg size of a fillet weld is measured by the largest right triangle which can be inscribed within the weld. See Figure 3.40.

This definition would allow unequal-legged fillet welds, as in Figure 3.40(a). Another AWS definition stipulates the largest isosceles inscribed right triangle and would limit this to an equal-legged fillet weld, as shown in Figure 3.40(b).

Unequal-legged fillet welds are sometimes used to get additional throat area, hence strength, when the vertical leg of the weld cannot be increased. See Figure 3.41(a).

Minimum Properties Required of Automatic Submerged-Arc Welds
(AWS & AISC) (as-welded; multiple-pass)

Grade SAW—1	
tensile strength	62,000 to 80,000 psi
yield point, min.	45,000 psi
elongation in 2 inches, min.	25%
reduction in area, min.	40%

Grade SAW—2	
tensile strength	70,000 to 90,000 psi
yield point, min.	50,000 psi
elongation in 2 inches, min.	22%
reduction in area, min.	40%

Reprinted with permission from *Design of Welded Structures*, James F. Lincoln Arc Welding Foundation

Figure 3.39

Where space permits, a more efficient means of obtaining the same increase in throat area or strength is to increase both legs to maintain an equal-legged fillet weld with a smaller increase in weld metal. See Figure 3.41(b).

One example of this would be the welding of channel shear attachments to beam flanges. See Figure 3.42. Here the vertical leg of the fillet weld must be held to the thickness at the outer edge of the channel flange. Additional strength must be obtained by increasing the horizontal leg of the fillet.

The effective length of the weld is defined as the length of the weld having full throat. Further, the AWS requires that all craters shall be filled to the full cross-section of the weld.

In continuous fillet welds, this is no problem because the welder will strike an arc for the next electrode on the forward edge of the crater of the previous weld, then swing back into the crater to fill it, and then proceed forward for the remainder of the weld. In this manner no crater will be left unfilled.

In practically all cases of intermittent fillet welds, the required length of the weld is marked out on the plate and the welder starts welding at one mark and continues until the rim of the weld crater passes the other mark. In other words, the crater is beyond the required length of the intermittent fillet and is not counted.

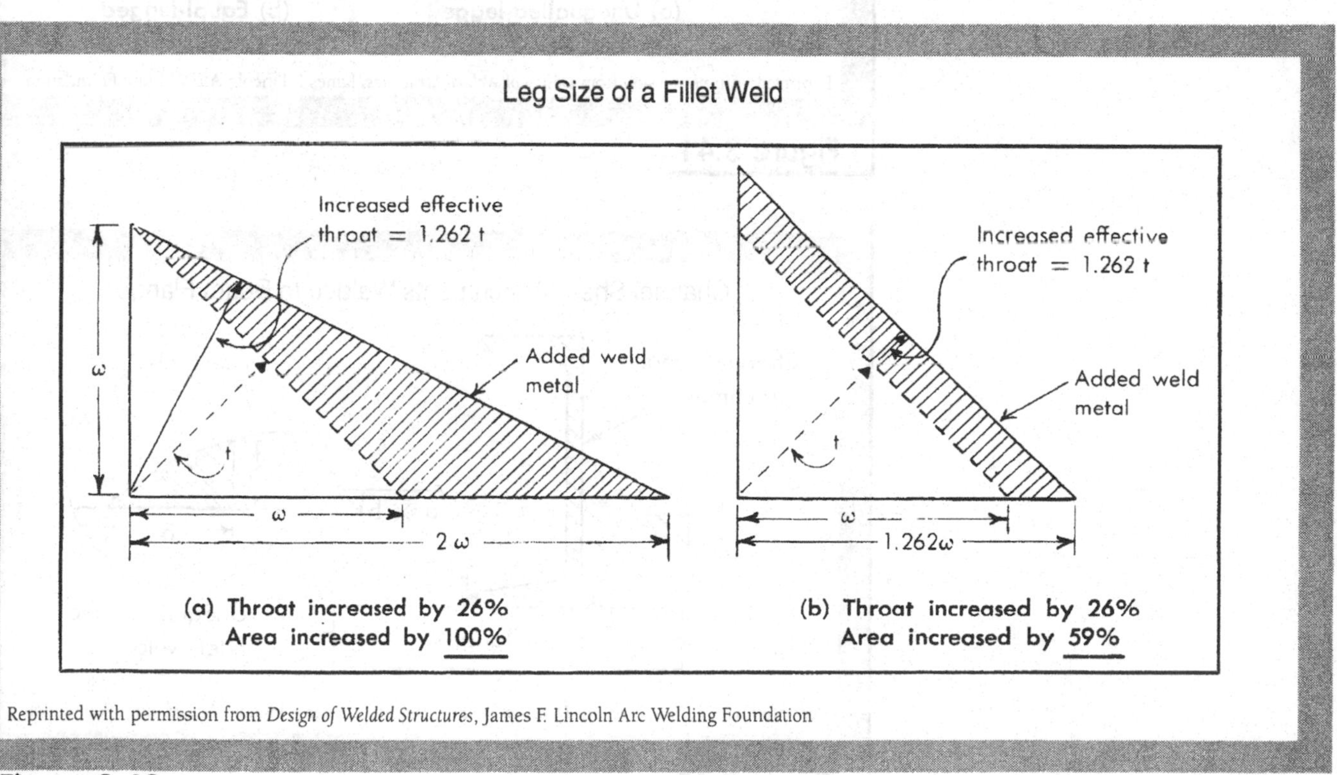

Reprinted with permission from *Design of Welded Structures*, James F. Lincoln Arc Welding Foundation

Figure 3.40

There may be some cases where the crater is filled and included in the weld length. This may be accomplished by filling the crater, or by using a method of welding part way in from one end, breaking the arc and welding in from the other end, and then overlapping in the central portion, thus eliminating any crater.

The effective throat is defined as the shortest distance between the root of the joint and the face of the diagrammatical weld. This would be a line from the root of the joint and normal to the flat face, as in Figure 3.43.

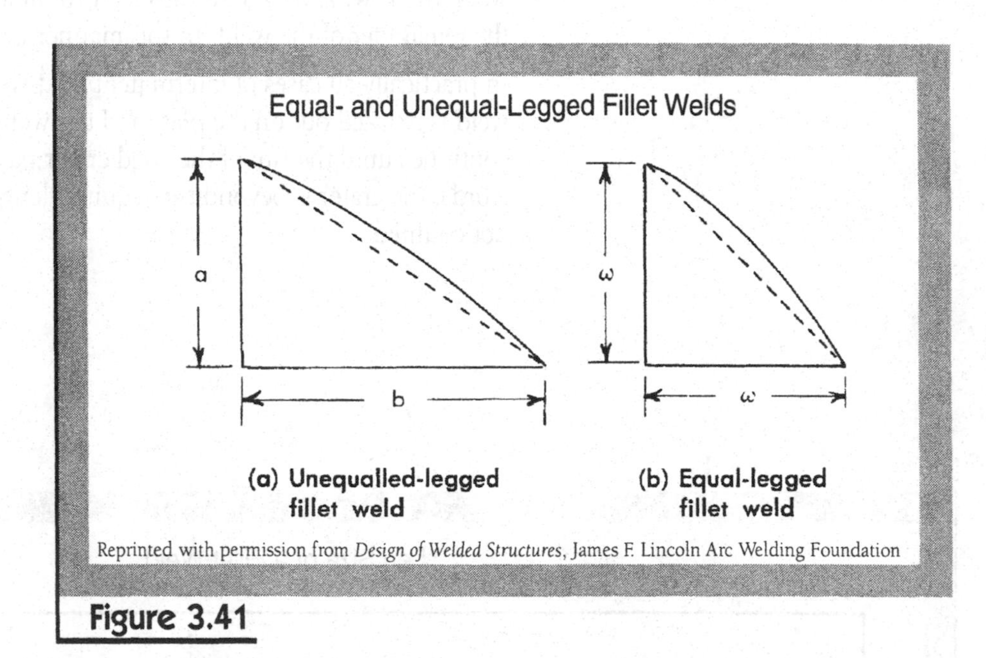

Reprinted with permission from *Design of Welded Structures*, James F. Lincoln Arc Welding Foundation

Figure 3.41

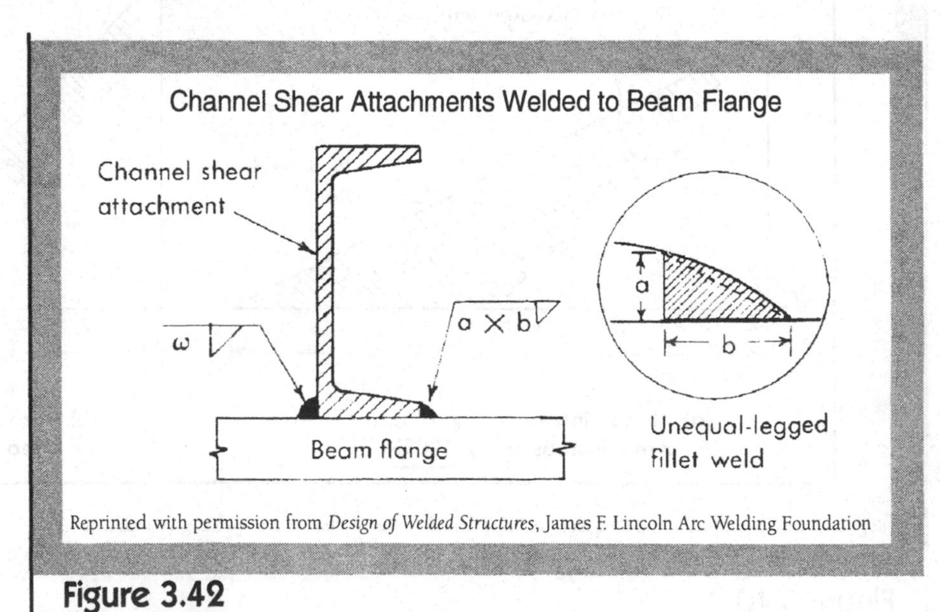

Reprinted with permission from *Design of Welded Structures*, James F. Lincoln Arc Welding Foundation

Figure 3.42

For an equal-legged fillet weld, the throat is equal to .707 times the leg size (w):

$t = .707\, w$

The allowable force on the fillet weld, 1" long is

$f = .707\, w\, t$

where:

f = allowable force on fillet weld, lbs per linear inch

w = leg size of a fillet weld, inches

t = allowable shear stress on throat of weld, psi

The AWS has set up several shear stress allowables for the throat of the fillet weld. These are shown in Figures 3.44 and 3.45 for the Building and Bridge fields.

Terminology: People who specify or are otherwise associated with welding often use the terms "joint" and "weld" rather loosely. For clarity in communication of instructions, it is desirable to keep in mind the basic difference of meaning between these two terms. This is illustrated in Figure 3.46.

The left-hand chart shows the five basic types of joints: butt, tee, corner, lap, and edge. Each is defined in a way that is descriptive of the relationship the plates being joined have to each other. Neither the geometry of the weld itself nor the method of edge preparation has any influence on the basic definition of the joint. For instance, the tee joint could be either fillet welded or groove welded.

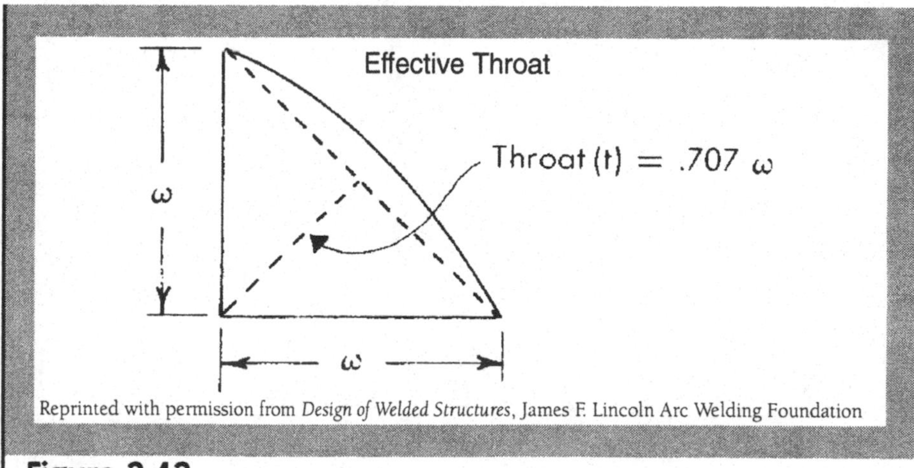

Reprinted with permission from *Design of Welded Structures*, James F. Lincoln Arc Welding Foundation

Figure 3.43

Allowables for Welds – Buildings

Type of Weld	Stress	Steel	Electrode	Allowable
Complete-Penetration Groove Welds	tension compression shear	A7, A36, A373	‡E60 or SAW-1	same as PL
		A441, A242*	E70 or SAW-2	
Partial-Penetration Groove Welds	tension transverse to axis of weld or shear on effective throat	A7, A36, A373	E60 or SAW-1	σ or τ = 13,600 psi
		A441, A242*	E60 low-hydrogen or SAW-1	
		A7, A373	E70 or SAW-2	
		A36	E70 or SAW-2	σ or τ = 15,800 psi
		A441, A242*	E70 low-hydrogen or SAW-2	
	tension parallel to axis of weld or compression on effective throat	A7, A36, A373	‡E60 or SAW-1	same as PL
		A441 or A242*	E70 or SAW-1	
Fillet Weld	shear on effective throat	A7, A36, A373	E60 or SAW-1	τ = 13,600 psi or f = 9600 ω lb/in
		A441, A242*	E60 low-hydrogen or SAW-2	
		A7, A373	E70 or SAW-2	
		A36	E70 or SAW-2	τ = 15,800 psi or f = 11,200 ω lb/in
		A441, A242*	E70 low-hydrogen or SAW-2	
Plug and Slot	shear on effective area	Same as for fillet weld		

* weldable A242

‡ E70 or SAW-2 could be used, but would not increase allowable

Reprinted with permission from *Design of Welded Structures*, James F. Lincoln Arc Welding Foundation

Figure 3.44

The right-hand chart shows the basic types of welds: fillet, square, bevel groove, V-groove, J-groove, and U-groove. The type of joint does not affect what we call the weld. Although the single bevel-groove weld is illustrated as a butt joint, it may be used in a butt, tee, or corner joint.

The complete definition of a welded joint must include description of both the joint and the weld.

Figures 3.47 and 3.48 provide formulas used to determine force and other properties of weld.

Allowables for Welds – Bridges

Type of Weld	Stress	Steel	Electrode	Allowable
Complete-Penetration Groove Welds	tension compression shear	A7, A373	‡E60 or SAW-1	Same as PL
Complete-Penetration Groove Welds	tension compression shear	A36 $\leqq$ 1" thick	‡E60 or SAW-1	Same as PL
Complete-Penetration Groove Welds	tension compression shear	A36 $>$ 1" thick	‡E60 low-hydrogen or SAW-1	Same as PL
Complete-Penetration Groove Welds	tension compression shear	A441, A242*	E70 low-hydrogen or SAW-2	Same as PL
Fillet Welds	shear on effective throat	A7, A373	‡E60 or SAW-1	τ = 12,400 psi or f = 8800 ω lb/in
Fillet Welds	shear on effective throat	A36 $\leqq$ 1" thick	‡E60 or SAW-1	τ = 12,400 psi or f = 8800 ω lb/in
Fillet Welds	shear on effective throat	A36 $>$ 1" thick	‡E60 low-hydrogen or SAW-1	τ = 12,400 psi or f = 8800 ω lb/in
Fillet Welds	shear on effective throat	A441, A242*	E70 low-hydrogen or SAW-2	τ = 14,700 psi or f = 10,400 ω lb/in
Plug and Slot	shear on effective area	A7, A373, A36 $\leqq$ 1" thick	‡E60 or SAW-1	12,400 psi
Plug and Slot	shear on effective area	A36 $>$ 1" thick A441, A242*	‡E60 low-hydrogen or SAW-1	12,400 psi

* weldable A242
‡ E70 or SAW-2 could be used, but would not increase allowable

Reprinted with permission from *Design of Welded Structures*, James F. Lincoln Arc Welding Foundation

Figure 3.45

Difference Between Welds and Joists

TYPES of JOINTS

Butt	(B)
Tee	(T)
Corner	(C)
Lap	(L)
Edge	(E)

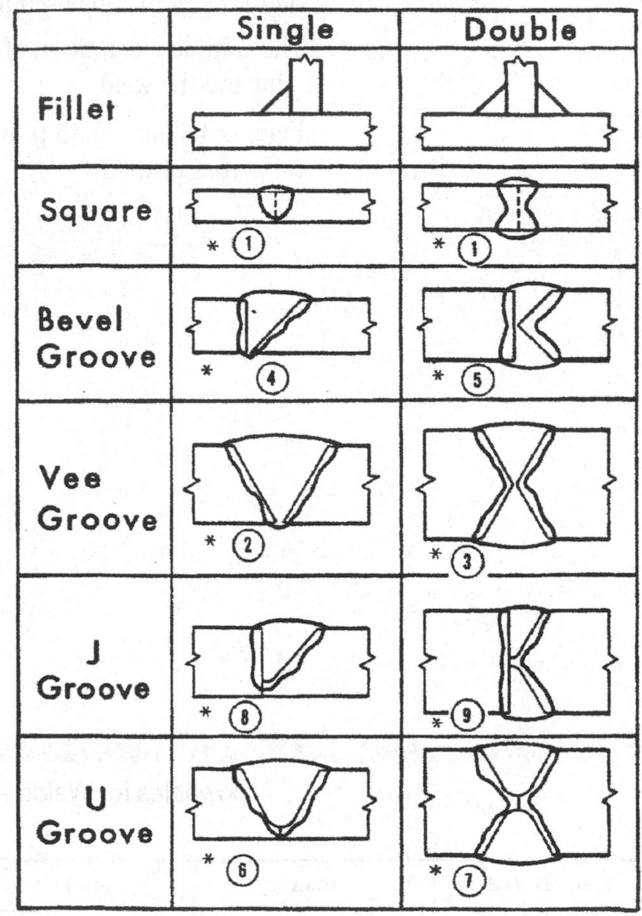

*Same Number Used on AWS Prequalified Joints

Reprinted with permission from *Design of Welded Structures*, James F. Lincoln Arc Welding Foundation

Figure 3.46

Finding Weld Size: The horizontal shear forces acting on the weld joining a flange to web, as in Figures 3.49 and 3.50, may be found from the following formula:

$$f = \frac{Vay}{In}$$

where:

f = force on weld, lbs./linear inches

V = total shear on section at a given position along beam, lbs.

a = area of flange held by weld, square inches

y = distance between the center of gravity of flange area and the neutral axis of whole section, inches

I = moment of inertia of whole section, inches

n = number of welds joining flange to web

Determining Force on Weld

Type of Loading		standard design formula stress lbs/in²	treating the weld as a line force lbs/in
	PRIMARY WELDS transmit entire load at this point		
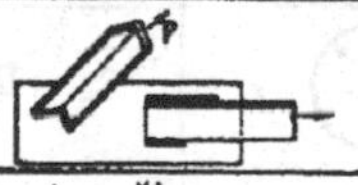	tension or compression	$\sigma = \frac{P}{A}$	$f = \frac{P}{A_w}$
	vertical shear	$\sigma = \frac{V}{A}$	$f = \frac{V}{A_w}$
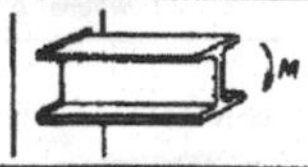	bending	$\sigma = \frac{M}{S}$	$f = \frac{M}{S_w}$
	twisting	$\sigma = \frac{TC}{J}$	$f = \frac{TC}{J_w}$
	SECONDARY WELDS hold section together - low stress		
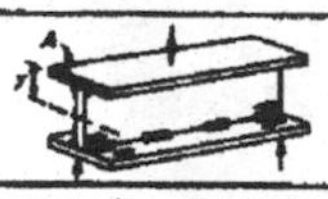	horizontal shear	$\tau = \frac{VAy}{It}$	$f = \frac{VAy}{In}$
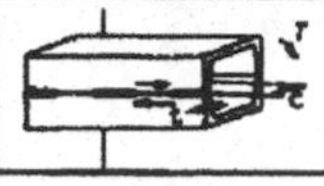	torsional horizontal shear*	$\tau = \frac{T}{2At}$	$f = \frac{T}{2A}$

A = area contained within median line.
(*) applies to closed tubular section only.

Reprinted with permission from *Design of Welded Structures*, James F. Lincoln Arc Welding Foundation

Figure 3.47

Properties of Weld Treated as Line

Outline of Welded Joint b=width d=depth	Bending (about horizontal axis x-x)	Twisting
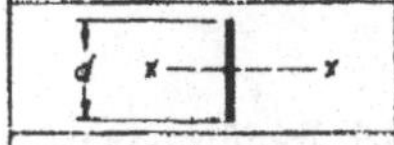	$S_w = \frac{d^2}{6}$ in.^{2}	$J_w = \frac{d^3}{12}$ in.^{3}
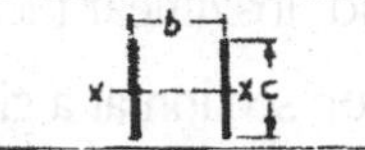	$S_w = \frac{d^2}{3}$	$J_w = \frac{d(3b^2 + d^2)}{6}$
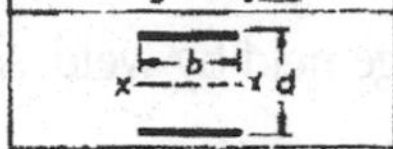	$S_w = bd$	$J_w = \frac{b^3 + 3bd^2}{6}$
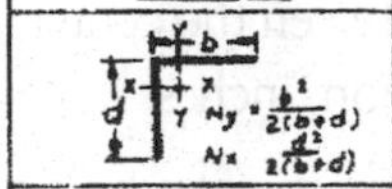 $N_y = \frac{b^2}{2(b+d)}$, $N_x = \frac{d^2}{2(b+d)}$	$S_w = \frac{4bd + d^2}{6} = \frac{d^2(4b + d)}{6(2b + d)}$ top bottom	$J_w = \frac{(b+d)^4 - 6b^2d^2}{12(b + d)}$
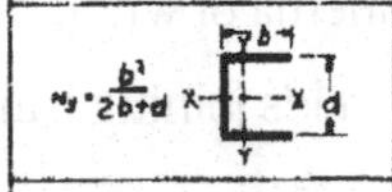 $N_y = \frac{b^2}{2b+d}$	$S_w = bd + \frac{d^2}{6}$	$J_w = \frac{(2b+d)^3}{12} - \frac{b^2(b + d)^2}{(2b+d)}$
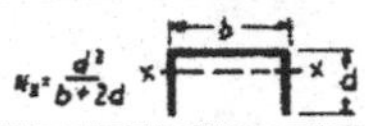 $N_x = \frac{d^2}{b+2d}$	$S_w = \frac{2bd + d^2}{3} = \frac{d^2(2b + d)}{3(b + d)}$ top bottom	$J_w = \frac{(b+2d)^3}{12} - \frac{d^2(b+d)^2}{(b + 2d)}$
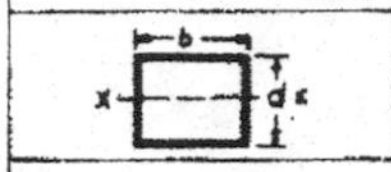	$S_w = bd + \frac{d^2}{3}$	$J_w = \frac{(b + d)^3}{6}$
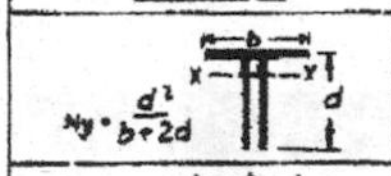 $N_y = \frac{d^2}{b+2d}$	$S_w = \frac{2bd+d^2}{3} = \frac{d^2(2b+d)}{3(b + d)}$ top bottom	$J_w = \frac{(b+2d)^3}{12} - \frac{d^2(b+d)^2}{(b + 2d)}$
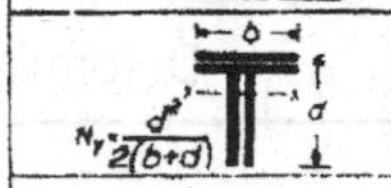 $N_y = \frac{d^2}{2(b+d)}$	$S_w = \frac{4bd+d^2}{3} = \frac{4bd^2 + d^3}{6b + 3d}$ top bottom	$J_w = \frac{d^3(4b + d)}{6(b + d)} + \frac{b^3}{6}$
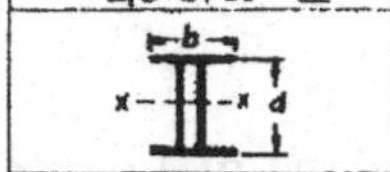	$S_w = bd + \frac{d^2}{3}$	$J_w = \frac{b^3 + 3bd^2 + d^3}{6}$
	$S_w = 2bd + \frac{d^2}{3}$	$J_w = \frac{2b^3 + 6bd^2 + d^3}{6}$
	$S_w = \frac{\pi d^2}{4}$	$J_w = \frac{\pi d^3}{4}$
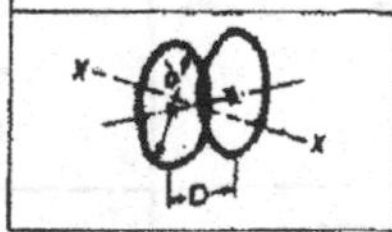	$I_w = \frac{\pi d}{2}\left(D^2 + \frac{d^2}{2}\right)$ $S_w = \frac{I_w}{c}$ where $c = \frac{\sqrt{D^2 + d^2}}{2}$	

Reprinted with permission from *Design of Welded Structures*, James F. Lincoln Arc Welding Foundation

Figure 3.48

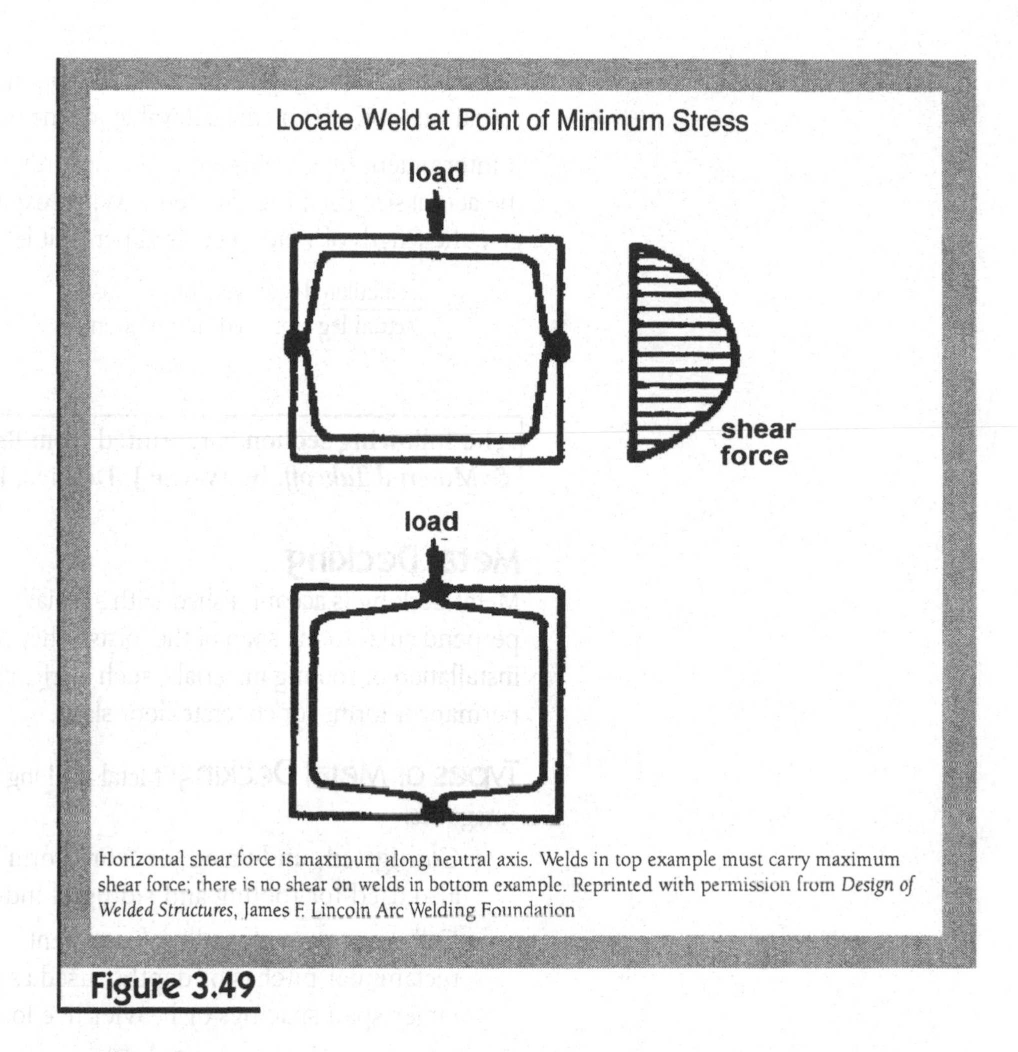

Horizontal shear force is maximum along neutral axis. Welds in top example must carry maximum shear force; there is no shear on welds in bottom example. Reprinted with permission from *Design of Welded Structures*, James F. Lincoln Arc Welding Foundation

Figure 3.49

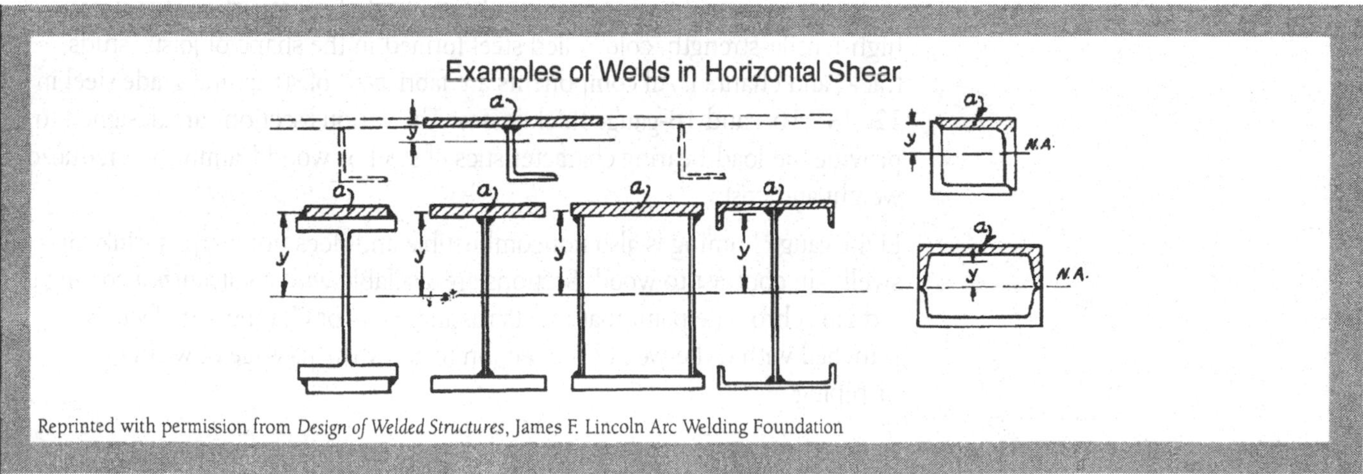

Reprinted with permission from *Design of Welded Structures*, James F. Lincoln Arc Welding Foundation

Figure 3.50

The leg size of the required fillet weld (continuous) is found by dividing this actual unit force (f) by the allowable for the type of weld metal used.

If intermittent fillet welds are to be used, divide this weld size (continuous) by the actual size used (intermittent). When expressed as a percentage, this will give the length of weld to be used per unit length.

$$\% = \frac{\text{calculated leg size (continuous)}}{\text{actual leg size used (intermittent)}}$$

The following section is reprinted from *Basics for Builders: Plan Reading & Material Takeoff*, by Wayne J. DelPico, R.S. Means Co., Inc.

Metal Decking

Metal decking is accomplished with specially formed sheets of steel applied perpendicular to the span of the joists. They serve as a substrate for the installation of roofing materials, such as rigid insulation and membrane, or as permanent forms for concrete floor slabs.

Types of Metal Decking: Metal decking can be classified in two main categories:

- Corrugated, undulated, or "corruform" concrete-fill permanent forms; also used for roofing and siding of industrial-type buildings.
- Cellular-type with well-defined bent contours in trapezoidal or rectangular pitches or depths; used as permanent concrete forms for larger-span spacings or heavier live load applications.

Metal decking is available in 18-, 20-, or 22-gauge thicknesses, in widths of 30" , and in standard lengths ranging from 14' to 31'. The depth of the decking section can vary with type and application, but is usually 1-1/2" to 2-1/2" deep. Special coatings and colors are available, but the most common finish is galvanized.

Light Gauge Metal Framing

Light-gauge metal framing refers to the method of construction that uses a high-tensile-strength, cold-rolled steel formed in the shape of joists, studs, track, and channel. All components are fabricated of structural grade steel in 12-,14-, 16-, and 18-gauge thicknesses. The various sections are designed to provide the load-bearing characteristics of steel or wood framing at a reduced weight and cost.

Light-gauge framing is also noncombustible and does not warp, shrink, or swell—in contrast to wood. Sections are available with a galvanized coating or red zinc chromate paint that resists rusting. Slots or channels are factory-punched within the web of the section to allow the passage of wiring or piping.

On-site cutting of the sections is done with a "chop" saw outfitted with a high speed metal-cutting blade. Layout and erection are similar to the procedures used for wood framing components.

Fastening of components can be done by bolting, screwing with self-tapping screws, or welding.

Joists: Joist sections are available in 6", 8", 10" and 12" depths, and flanges in 1-5/8" or 2-1/2" widths. The most common lengths are from 8' to 30' in 2' increments. Joists are used in floor and roof construction with the typical 12", 16" and 24" on-center spacings used in wood frame construction.

The following section is reprinted from *Fundamentals of the Construction Process*, by Kweku K. Bentil, R.S. Means Co., Inc.

Concrete

Concrete is a universal construction material; its component raw materials are inorganic, incombustible, highly versatile, and relatively low in cost in comparison to other materials.

Although the terms *concrete* and *cement* are sometimes used interchangeably outside of the construction industry, they have different meanings and are two distinct materials.

Concrete is made by mixing a paste of cement and water with sand and crushed stone, gravel, or other inert material. A rough idea of typical proportions of these components for concrete is shown in Figure 3.51. After mixing, usually accomplished for buildings in a transit truck, the plastic mixture is placed in forms and a chemical reaction called hydration takes place and the mass hardens.

Concrete is rated by its compressive strength after a 28-day curing period. The compressive strength most often specified for structures is 3000 pounds per square inch (psi). Concrete that is exposed to freezing temperatures such as curbs and sidewalks is usually 4000 psi. In some cases, concrete with much higher strength is required, from 6000 psi to as high as 19,000 psi. Specified strengths of concrete are produced by varying the proportions of cement, sand, or fine aggregate, coarse aggregate, and water.

Cast-in-place refers to concrete that is placed in a liquid state in the position and/or location it is to occupy permanently in a finished structure. Most cast-in-place concrete used on construction projects is ready-mixed, delivered to the construction site by special trucks, and then placed into forms by one of several methods: by a chute directly into formwork; into a hopper for distribution by wheelbarrows and mechanical buggies; into buckets to be hoisted by a crane; or into a concrete pump and pushed through a portable pipeline.

Once in the forms, the concrete is vibrated to eliminate air pockets, surround the reinforcing, and fill any voids in the framework. Slabs are "struck off" or screeded (leveled) after vibration but require additional finishing (floating and troweling) and curing.

Precast concrete is cast in the controlled environment of a casting plant where the quality of materials can be controlled to a much greater degree. The finished member is transported by truck to the jobsite and erected by crane.

Concrete may be reinforced, or *prestressed*, to partially counteract the stresses by the loads applied to a structure. The reinforcing can be stressed in the field, but is better controlled in a factory, such as a precast plant.

Figures 3.52 and 3.53 show desirable and undesirable concrete properties for various performance requirements.

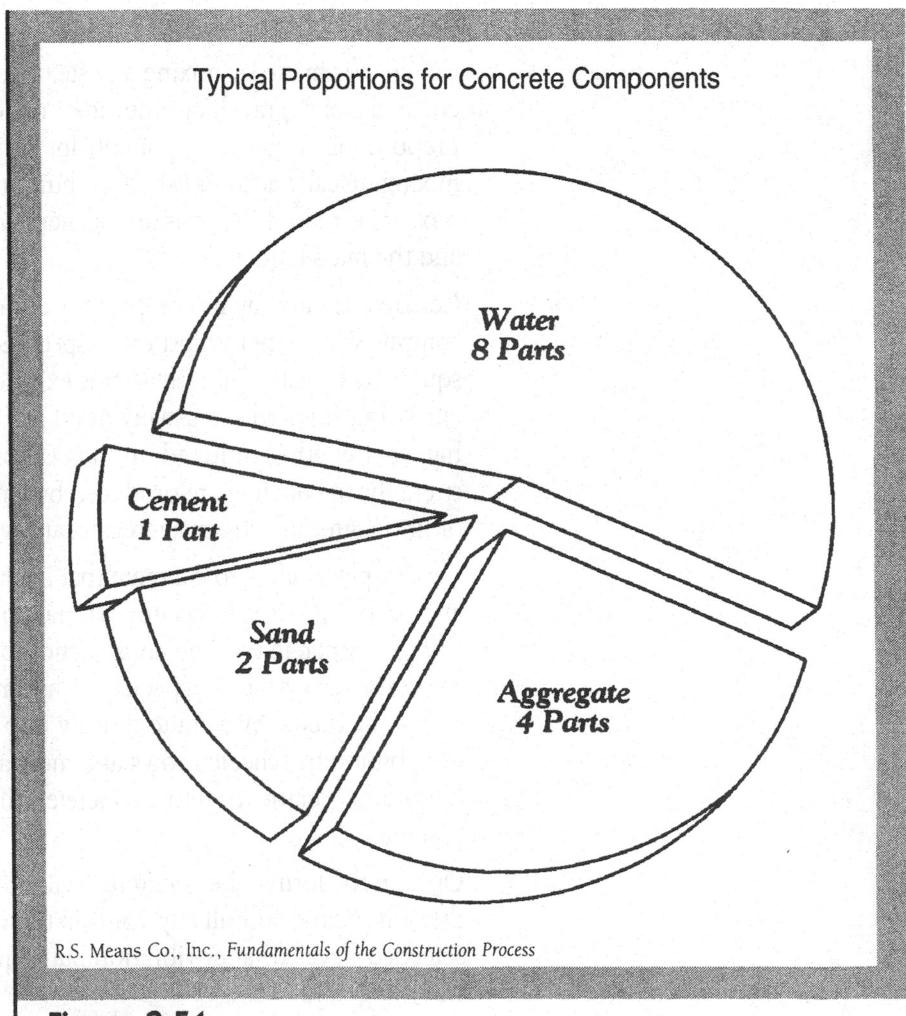

Figure 3.51

External Loads/Concrete Material Properties

Goal (performance requirements)	Results if the wrong material is selected (undesirable response)	Look for these properties	Avoid these!
Moving liquids	Erosion of surface	High density	Low density
		High compressive	Low compressive
		High tensile	Low tensile
Moving liquids and suspended solids	Erosion and abrasion of surfaces	High density	Low density
		High compressive	Low compressive
		High tensile	Low tensile
Vehicle wheels	Abrasion damage to surface	High density, high compressive strength	Low density, low compressive strength
	Edge spalling at joints	High compressive, tensile and bond strength, tensile anchorage into substrate	Low compressive, tensile and bond strength
Impact	Spalling	High tensile strength, internal tensile reinforcement	Low tensile strength
		High compressive strength	Low compressive strength
		Low modulus of elasticity	High modulus of elasticity
	Loss of bond	High bond strength, tensile anchorage into substrate	Low bond strength

R.S. Means Co., Inc., *Concrete Repair and Maintenance Illustrated*

Figure 3.52

Constructibility & Appearance Properties

	Goal (performance requirements)		Look for these properties	Avoid these!
Constructibility	Turn-around time 		Rapid strength gain	Slow strength gain
	Flowability 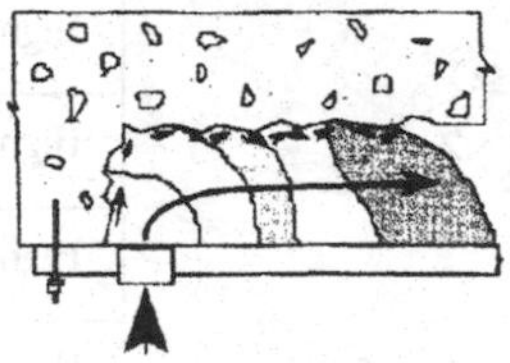		High slump	Low slump
			Small aggregate, fines, round shape	Large aggregate, angular shape, lack of fines
	Non sag 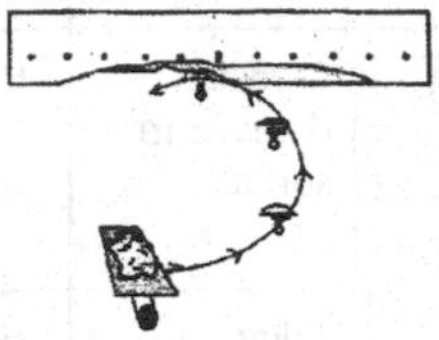		High internal cohesion, high adhesive grip	Low internal cohesion, low adhesive grip
	Forgiving "Murphy's Law"		Simple formulation, redundant	Complex formulation, dependent reactions

Goal (performance requirements)	Results if the wrong material is selected (undesirable response)		Look for these properties	Avoid these!
Appearance  CRACKS	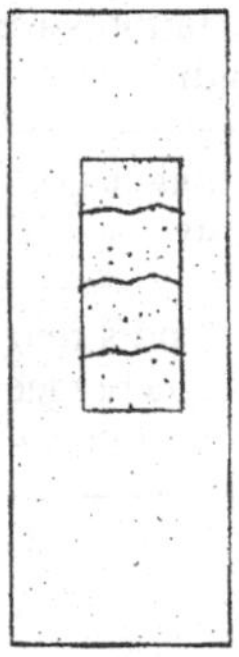	Cracking of surface from drying shrinkage*	Low drying shrinkage,* flexible surface membrane	High drying shrinkage*
		Cracking of surface in plastic stage	Low exotherm	High exotherm
			Low surface water loss during placement	High surface water loss during placement

*Refer to volume change affects included at the end of this section.

R.S. Means Co., Inc., *Concrete Repair and Maintenance Illustrated*

Figure 3.53

Strength of Concrete

Concrete strength requirements are prescribed by building code and the building design. A stronger concrete is not only structurally superior, but is more resistant to abrasion. Compressive and flexural strength are fundamental in the design of structures, slabs and pavements. Compressive strength is the measured maximum resistance of a concrete or mortar specimen to axial loading. It is expressed in pounds per square inch (psi) at 28 days after placement, and is designated by the symbol f'_c. Compressive strength is tested on specimens of mortar (2" cubes) or concrete (cylinders 6" diameter and 12" high). Most concrete for general use has a compressive strength between 3000 psi and 5000 psi. High-strength concrete is a minimum of 6000 psi, and can range as high as 20,000 psi.

Flexural strength indicates the material's ability to withstand bending. It is used in designing pavements and slabs on grade. Flexural strength can be determined based on compressive strength, if the empirical relationship between the two strengths is established for the size of the member and the materials. The flexural strength (or modulus of rupture) of normal-weight concrete is approximately 8–12% of the compressive strength and may be calculated as five to seven times the square root of the compressive strength. (See ACI 207.2R for formulas for calculating tensile strength.)

Strength is based on water-cement ratio and the curing process. Strengths increase as the water-cement ratios decrease. Concrete strength increases with time, as long as it is not permitted to dry too quickly, which slows down or stops the chemical reactions that develop strength. Concrete moisture must be controlled to achieve optimum strength. Unless specific admixtures are used to speed curing time, concrete placed in the field should be moist-cured continuously from the time it is placed until it has reached the desired characteristics.

Water-Cement Ratio

The quality of a concrete mix depends on several factors, including the ratio of water to cement, the aggregate, any admixtures and/or entrained air, and the time for curing, as well as moisture and temperature conditions. One of the most important factors affecting the strength of concrete is the water-to-cement ratio, expressed in pounds or gallons of water per sack of cement. While water is used to mix and place concrete in a plastic and workable state, too much thinning with water decreases the concrete's strength and makes it less weather-resistant. The correct proportions of water to cement are crucial to a quality mix.

The American Concrete Institute Building Code Requirements specify maximum permissible water-cement ratios for concrete when strength data from field experience or trial mixtures are not available. See Figures 3.54 and 3.54(b).

Proportioning by Water-Cement Ratio

4.4 – Proportioning by water-cement ratio

4.4.1 – If data required by Section 4.3 are not available, permission may be granted to base concrete proportions on water-cement ratio limits in Table 4.4.

TABLE 4.4 – MAXIMUM PERMISSIBLE WATER-CEMENT RATIOS FOR CONCRETE WHEN STRENGTH DATA FROM FIELD EXPERIENCE OR TRIAL MIXTURES ARE NOT AVAILABLE

Specified compressive strength, f'_c, psi*	Absolute water-cement ratio by weight	
	Non-air-entrained concrete	Air-entrained concrete
2500	0.67	0.54
3000	0.58	0.46
3500	0.51	0.40
4000	0.44	0.35
4500	0.38	†
5000	†	†

*28-day strength. With most materials, water-cement ratios shown will provide average strengths greater than indicated in Section 4.3.2 as being required.

†For strengths above 4500 psi (non-air-entrained concrete) and 4000 psi (air-entrained concrete), concrete proportions shall be established by methods of Section 4.3.

4.4.2 – Table 4.4 shall be used only for concrete to be made with cements meeting strength requirements for Types I, IA, II, IIA, III, IIIA, or V of "Specification for Portland Cement" (ASTM C 150), or Types IS, IS-A, IS(MS), IS-A(MS), I(SM), I(SM)-A, IP, IP-A, I(PM), I(PM)-A, IP(MS), IP-A(MS), or P of "Specification for Blended Hydraulic Cements" (ASTM C 595), and shall not be applied to concrete containing lightweight aggregates or admixtures other than those for entraining air.

4.4.3 – Concrete proportioned by water-cement ratio limits prescribed in Table 4.4 shall also conform to special exposure requirements of Section 4.5 and to compressive strength test criteria of Section 4.7.

4.5 – Special exposure requirements

4.5.1 – Normal weight and lightweight concrete exposed to freezing and thawing or deicer chemicals shall be air entrained with air content indicated in Table 4.5.1. Tolerance on air content as delivered shall be ±1.5 percent. For specified compressive strength f'_c greater than 5000 psi, air content indicated in Table 4.5.1 may be reduced 1 percent.

TABLE 4.5.1 – TOTAL AIR CONTENT FOR FROST-RESISTANT CONCRETE

Nominal maximum aggregate size, in.*	Air content, percent	
	Severe exposure	Moderate exposure
3/8	7-1/2	6
1/2	7	5-1/2
3/4	6	5
1	6	4-1/2
1-1/2	5-1/2	4-1/2
2†	5	4
3†	4-1/2	3-1/2

*See ASTM C 33 for tolerances on oversize for various nominal maximum size designations.

†These air contents apply to total mix, as for the preceding aggregate sizes. When testing these concretes, however, aggregate larger than 1-1/2 in. is removed by handpicking or sieving and air content is determined on the minus 1-1/2 in. fraction of mix. (Tolerance on air content as delivered applies to this value.) Air content of total mix is computed from value determined on the minus 1-1/2 in. fraction.

TABLE 4.5.2 – REQUIREMENTS FOR SPECIAL EXPOSURE CONDITIONS

Exposure condition	Maximum water-cement ratio, normal weight aggregate concrete	Minimum f'_c, lightweight aggregate concrete
Concrete intended to be watertight:		
(a) Concrete exposed to fresh water	0.50	3750
(b) Concrete exposed to brackish water or seawater	0.45	4250
Concrete exposed to freezing and thawing in a moist condition:		
(a) Curbs, gutters, guardrails or thin sections	0.45	4250
(b) Other elements	0.50	3750
(c) In presence of deicing chemicals	0.45	4250
For corrosion protection for reinforced concrete exposed to deicing salts, brackish water, seawater or spray from these sources	0.40*	4750*

*If minimum concrete cover required by Section 7.7 is increased by 0.5 in., water-cement ratio may be increased to 0.45 for normal weight concrete, or f'_c reduced to 4250 psi for lightweight concrete.

4.5.2 – Concrete that is intended to be watertight or concrete that will be subject to freezing and thawing in a moist condition shall conform to requirements of Table 4.5.2.

4.5.3 – Concrete to be exposed to sulfate-containing solutions shall conform to requirements of Table 4.5.3 or be made with a cement that provides sulfate resistance and used in concrete with maximum water-cement ratio or minimum compressive strength from Table 4.5.3.

4.5.3.1 – Calcium chloride as an admixture shall not be used in concrete to be exposed to severe or very severe sulfate-containing solutions, as defined in Table 4.5.3.

Figure 3.54

Cement

Cement, often referred to as portland cement, is a fine gray powder made from three main ingredients of silica, lime, and alumina. It often includes slag or flue dust from iron furnaces. The appropriate proportions of these materials are mixed by crushing, grinding, and blending. The resulting blend is then heated in a rotating kiln to very high temperatures to produce a vitrified product called clinker. The clinker is allowed to cool and then pulverized into powder form, usually with a small amount of gypsum to slow the curing process.

Various types of portland cement are used for different purposes. The American Society for Testing and Materials (ASTM) provides for five types of portland cement in ASTM C150.

Type I, CSA Normal, is general-purpose and appropriate for any use where special requirements (such as exposure to sulfates or unusual temperature increases) are not an issue. Type I is used for pavements, sidewalks, reinforced concrete buildings, culverts, water pipe, and masonry units.

Type II, CSA Moderate, is used in situations where moderate sulfate attack is possible, such as in drainage structures with higher than usual sulfate concentrations. Use of Type II minimizes temperature rise, a significant consideration when concrete is being placed in warm weather.

Type III, CSA High-Early-Strength, yields high strengths early after placement (usually one week or less). Type III is useful in situations where the forms must be removed quickly. Type III also reduces the controlled curing period in cold weather.

TABLE 4.5.3 – REQUIREMENTS FOR CONCRETE EXPOSED TO SULFATE-CONTAINING SOLUTIONS

Sulfate exposure	Water soluble sulfate (SO_4) in soil, percent by weight	Sulfate (SO_4) in water, ppm	Cement type	Normal weight aggregate concrete: Maximum water-cement ratio, by weight*	Lightweight aggregate concrete: Minimum compressive strength, f'_c psi*
Negligible	0.00–0.10	0–150	—	—	—
Moderate†	0.10–0.20	150–1500	II, IP(MS), IS(MS)	0.50	3750
Severe	0.20–2.00	1500–10,000	V	0.45	4250
Very severe	Over 2.00	Over 10,000	V plus pozzolan‡	0.45	4250

*A lower water-cement ratio or higher strength may be required for watertightness or for protection against corrosion of embedded items or freezing and thawing (Table 4.5.2).

†Seawater.

‡Pozzolan that has been determined by test or service record to improve sulfate resistance when used in concrete containing Type V cement.

Reprinted with permission from *Building Code Requirements for Reinforced Concrete*, American Concrete Institute

Figure 3.54b

Type IV, CSA Low Heat of Hydration, is used in situations where the heat generated by the installation must be kept to a minimum. Type IV has a slower curing rate than Type I, and is used for enormous concrete structures, such as large gravity dams.

Type V, CSA Sulfate-Resisting, is used where soils or groundwater are high in sulfate content. Like Type IV, Type V develops its strength at a slower rate than Type I.

Blended Hydraulic Cements are comprised of two or more types of the following byproduct materials: portland cement, ground granulated blast-furnace slag, fly ash and other pozzolans, hydrated lime, and pre-blended cement combinations of these materials. ASTM recognizes five classes of blended cements:

- *Type IS:* Portland blast-furnace slag cement, used for general concrete construction.
- *Type IP and P:* Portland-pozzolan cements. IP is used for general construction, and P is used for general construction where high early strengths are not necessary.
- *Type I (PM):* Pozzolan-modified portland cement, issued in general construction.
- *Type S:* Slag cement, used with portland cement as an ingredient in concrete or with lime in mortar, but not alone in structural concrete.

Expansive cements expand somewhat during the early hardening period after it has set. Expansive cement must meet ASTM C845 requirements. The advantage of expansive cement is its ability to control and reduce the incidence of drying shrinkage cracks.

Other types of cement include *air-entraining*, with improved resistance to freeze-thaw and scaling from, for example, chemicals used to melt ice and snow, and *white portland cement*, which is used for architectural/aesthetic considerations.

Ed. Note: Air-entrained concrete also provides greater workability than non-air-entrained concrete with the same water content.

For special circumstances, there are also waterproofed portland cement and plastic cement.

Concrete Admixtures

Admixtures are materials that are added to concrete to enhance certain properties for particular applications. Some types of admixtures include:

- fibrous reinforcing
- air-entraining
- water reducing
- plasticizing
- quick setting
- corrosion inhibiting

Fibrous Reinforcing: Fibrous reinforcing is a synthetic fiber that promotes resistance in conditions in which high impact and abrasion are anticipated. It provides excellent secondary reinforcing and crack control. Fibrous reinforcing can be used in thin sections, precast units, vaults, pipe, sidewalls, floor slabs, and similar applications. One and one-half pounds of fibrous reinforcing are used per cubic yard of concrete, and can be introduced with the aggregates or after all ingredients have been mixed.

Experience with filament glass fiber strands added to the concrete or plaster mix has shown, in a limited quantity of tests, that the fiber deteriorates from the portland cement unless the material was chopped from a polypropylene source. Otherwise, the material leaves a hole in the concrete or plaster after three to five years. There are not sufficient test results yet to determine the cause.

Air-Entraining Admixtures: Air-entrainment can be achieved by adding an admixture to concrete or by using air-entraining portland cement. Air-entraining admixtures introduce many air bubbles to the concrete, which results in better resistance to freezing and thawing in concrete slabs on grade. Air-entrainment also enhances watertightness.

Water Reducing Admixtures: This commonly used admixture is particularly appropriate when the mix design calls for a high-strength concrete.

Plasticizing: Plasticizers are used to create concrete that is extremely workable for high slump with the ability to flow for applications such as pumping. This admixture might be used for light commercial projects, for example, where concrete must be pumped to the second floor of a structure. Plasticizers are very useful for situations where a low water-cement ratio is desirable, along with a high degree of workability for ease of placement and consolidation. It is used for tremie concreting and other situations where high slumps are required.

Accelerators: This admixture promotes quick, high strength in concrete. It is available in chloride and non-chloride formula. (The non-chloride formula will not corrode reinforcing steel, metal decks, or other metal components.) Accelerators speed up the chemical reaction between portland cement and water, and accelerates the formation of gel—the binder that bonds concrete aggregates. Accelerated gel formation shortens the concrete's setting time, which in turn offsets the slow-setting effects of cold weather, while helping to increase the strength of the concrete. Use of accelerators can also reduce concrete curing time, enabling forms to be removed earlier.

Retarders: This substance is used to prolong the time it takes concrete to set. It can be used in hot weather to prevent concrete from stiffening before it is properly placed. Retarders are occasionally used as part of the concrete finishing process. Retarders are sometimes applied to forms to allow the washing away of a fine layer of concrete for an exposed aggregate finish.

Corrosion Inhibitors: Corrosion inhibitors added to the concrete during the batching process chemically inhibit the corrosive action of chlorides on reinforcing steel and other metals in concrete. The admixture is used in applications involving steel-reinforcement, and in post-tensioned, pre-stressed concrete that will come in contact with chlorides (such as near the ocean) or systems that are exposed to chemicals (such as parking garage decks and support structures).

Aggregates

Use of the correct aggregates is key to a successful concrete installation. Aggregates generally comprise approximately 60-75% of the concrete volume, and have a major effect on the concrete's properties and economy. Coarse aggregates usually consist of one or more gravels whose particles are between 3/8" and 1-1/2" in size. Fine aggregates consist of natural sand or crushed stone, with most particles smaller than 0.2". Aggregates should meet established standards. They should be clean, hard, strong and durable particles without any absorbed chemicals or clay coating that might interfere with the hydration and bond of the cement paste. Shale, soft or porous materials, and chert are to be avoided as they are not resistant to weathering and can contribute to surface defects ("pop-outs"). ASTM C33 specifies requirements for normal weight aggregates. ASTM also specifies tests that should be carried out to determine the quality of aggregates.

Placing & Finishing Concrete

The Portland Cement Association publishes *Design & Control of Concrete Mixtures*, an excellent reference for anyone involved in planning, designing or placing concrete. According to this publication, preparation for concrete placement involves "compacting, trimming, and moistening the subgrade; erecting the forms; and setting the reinforcing steel and other embedded items securely in place."

The moistening is done to keep the dry subgrade from absorbing water from the concrete, and to reduce the evaporation from the surface of the concrete. Moistening can be particularly important in hot, dry weather. Any snow or ice must be removed from the ground surface, and the subgrade cannot be frozen at the time of placement. All loose materials must be removed from hard surfaces such as rock or hardened concrete on which the new concrete will be placed. Slopes should be avoided on such existing surfaces; a horizontal or vertical face is ideal. Old hardened concrete is typically cleaned and roughened, and newly placed concrete roughened after it has hardened to allow a better bond with the next placement.

Concrete is generally placed in layers of uniform thickness, quickly enough to avoid seams, flow lines, and cold joints (planes of weakness) between the layers, but allowing enough time for consolidation of each layer. Vibration (internal or external) is used for consolidating concrete, causing it to settle in the forms. Layers are generally about 6-20" thick for reinforced members, and about 15-20" thick for mass work.

Subgrade preparation includes ensuring that the site is well-drained, of uniform bearing capacity, level or properly sloped, and free of sod, organic matter, and, as mentioned earlier, frost, snow, or ice. Issues affecting uniform support are discussed in more detail earlier in this chapter in "Soil Mechanics."

Testing

When concrete is put in place, it must be sampled using various testing methods, mostly set forth by ASTM and ACI. Concrete is tested in the field for water-to-cement ratio by means of a slump test, performed by rodding concrete from the transit mix chute into a conical steel form. When the form is removed, the amount of "drop," or "slump," is measured. The air content of concrete may also be tested. Compression test specimens are made and cured at specified intervals. Specimens are cured in the laboratory and a set may also be cured in the field, particularly when there is any question of field temperatures falling below 40 degrees or above 90 degrees. Tests are also made in molded concrete cylinders. The strength of test specimens must meet ACI requirements for working stress and ultimate strength. If the average strength of standard-cured cylinders falls below the required strength, the project architect or engineer may require changes in conditions or water-cement ratio, a cost adjustment is generally made according to the provisions of the contract.

Reinforcing Steel

To improve concrete's ability to resist tension and shear forces, deformed steel bars, called reinforcing, are placed in specified locations. The deformations allow the concrete to bond to the steel. The bars are held in position during placement of the concrete by tie wires or patented connectors. Concrete slabs are reinforced by welded wire fabric that consists of longitudinal and horizontal wires fused in position with a designated spacing.

The ACI Building Code provides details of reinforcement that address standard hooks, minimum bend diameters, bending, surface conditions, placement including tolerances and spacing, connections, lateral reinforcement for compression members and flexural members, and shrinkage and temperature reinforcement. The Concrete Reinforcing Steel Institute's *Manual of Standard Practice* provides material specifications for reinforcing bars and spirals, for welded wire fabric, and for bar supports. Figure 3.55 shows the identification marks used on ASTM standard bars, from the CRSI Manual. Figure 3.56 is a CRSI table showing the standards for reinforcing bars. Figure 3.57 lists common stock styles of welded wire fabric.

Building Code Requirements

The American Concrete Institute (ACI) has developed and published building code requirements for concrete construction, which have been adopted by the model and local building code authorities. ACI also gathers together its standards and committee reports in the annually revised ACI *Manual of Concrete Practice.* The ACI *Building Code Requirements for Reinforced Concrete* covers general requirements; standards for tests and materials; construction requirements, including concrete quality, mixing and placing concrete,

Identification Marks*—ASTM Standard Rebars

The ASTM specifications for billet-steel, rail-steel, axle-steel and low-alloy reinforcing bars (A615, A616, A617 and A706, respectively) require identification marks to be rolled into the surface of one side of the bar to denote the Producer's mill designation, bar size, type of steel, and minimum yield designation. Grade 60 bars show these marks in the following order.

1st—Producing Mill (usually a letter)
2nd—Bar Size Number (#3 through #11, #14, #18)
3rd—Type of Steel:

S for Billet (A615)
W for Low-Alloy (A706)
I for Rail (A616)
I R for Rail meeting Supplementary Requirements S1 (A616)
A for Axle (A617)

4th—Minimum Yield Designation

Minimum yield designation is used for Grade 60 and Grade 75 bars only. Grade 60 bars can either have one single longitudinal line (grade line) or the number 60 (grade mark). Grade 75 bars can either have two grade lines or the grade mark 75.

A grade line is smaller and is located between the two main ribs which are on opposite sides of all bars made in the United States. A grade line must be continued through at least 5 deformation spaces, and it may be placed on the same side of the bar as the other markings or on the opposite side. A grade mark is the 4th mark on the bar.

Grade 40 and 50 bars are required to have only the first three identification marks (no minimum yield designation).

VARIATIONS: Bar identification marks may also be oriented to read horizontally (at 90° to those illustrated). Grade mark numbers may be placed within separate consecutive deformation spaces to read vertically or horizontally.

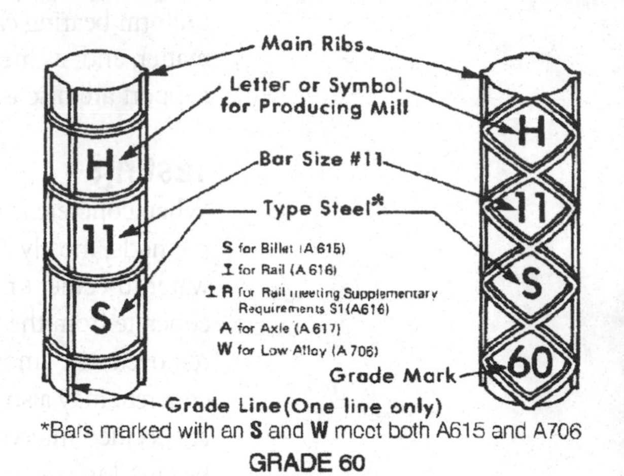

*Bars marked with an **S** and **W** meet both A615 and A706

GRADE 60

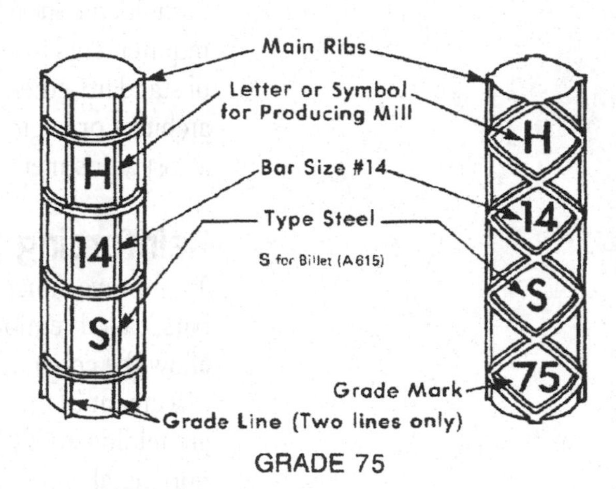

GRADE 75

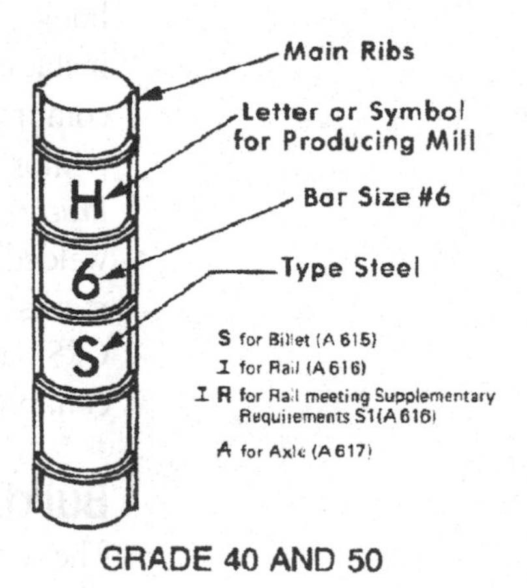

GRADE 40 AND 50

Reprinted with permission from *Manual of Standard Practice*, Concrete Reinforcing Steel Institute

Figure 3.55

ASTM STANDARD REINFORCING BARS

BAR SIZE DESIGNATION	AREA* SQ. INCHES	WEIGHT POUNDS PER FT.	DIAMETER* INCHES
#3	.11	.376	.375
#4	.20	.668	.500
#5	.31	1.043	.625
#6	.44	1.502	.750
#7	.60	2.044	.875
#8	.79	2.670	1.000
#9	1.00	3.400	1.128
#10	1.27	4.303	1.270
#11	1.56	5.313	1.410
#14	2.25	7.650	1.693
#18	4.00	13.600	2.257

Current ASTM Specifications cover bar sizes #14 and #18 in A615 Grade 60 and in A706 only.

***Nominal dimensions.**

Reprinted with permission from *Manual of Standard Practice*, Concrete Reinforcing Steel Institute

Figure 3.56

Common Stock Styles of Welded Wire Fabric

Certain styles of welded wire fabric as shown in Table 1 have been recommended by the Wire Reinforcement Institute as common styles. WWF manufacturers can meet specific steel area requirements when ordered for designated projects, or, in some localities, may be available from inventory.

TABLE 1—COMMON STYLES OF WELDED WIRE FABRIC*

Style Designation (W = Plain, D = Deformed)	Steel Area (in.²/ft)		Approximate Weight (lbs per 100 sq ft)
	Longitudinal	Transverse	
4 x 4-W1.4 x W1.4	0.042	0.042	31
4 x 4-W2.0 x W2.0	0.060	0.060	43
4 x 4-W2.9 x W2.9	0.087	0.087	62
4 x 4-W/D4 x W/D4	0.120	0.120	86
6 x 6-W1.4 x W1.4	0.028	0.028	21
6 x 6-W2.0 x W2.0	0.040	0.040	29
6 x 6-W2.9 x W2.9	0.058	0.058	42
6 x 6-W/D4 x W/D4	0.080	0.080	58
6 x 6-W/D4.7 x W/D4.7	0.094	0.094	68
6 x 6-W/D7.4 x W/D7.4	0.148	0.148	107
6 x 6-W/D7.5 x W/D7.5	0.150	0.150	109
6 x 6-W/D7.8 x W/D7.8	0.156	0.156	113
6 x 6-W/D8 x W/D8	0.160	0.160	116
6 x 6-W/D8.1 x W/D8.1	0.162	0.162	118
6 x 6-W/D8.3 x W/D8.3	0.166	0.166	120
12 x 12-W/D8.3 x W/D8.3	0.083	0.083	63
12 x 12-W/D8.8 x W/D8.8	0.088	0.088	67
12 x 12-W/D9.1 x W/D9.1	0.091	0.091	69
12 x 12-W/D9.4 x W/D9.4	0.094	0.094	71
12 x 12-W/D16 x W/D16	0.160	0.160	121
12 x 12-W/D16.6 x W/D16.6	0.166	0.166	126

*Many styles may be obtained in rolls.

Reprinted with permission from *Manual of Standard Practice*, Concrete Reinforcing Steel Institute

Figure 3.57

formwork, embedded pipes and construction joints, and reinforcement; analysis and design; strength and serviceability; flexure and axial loads; shear and torsion; slab, wall, and footing systems; precast and prestressed concrete; and special considerations, such as evaluation of existing structures and seismic design provisions.

Paving

Asphalt in bituminous paving is similar to portland cement in concrete in that both are used to bind the aggregate material. Unlike concrete, asphalt paving material is usually spread hot, except when cold mixes are used for patching and sub-bases. Asphalt paving material (bituminous asphaltic concrete) is created when asphalt cement is heated to a liquid form, then blended with an aggregate (crushed stone and sand). Asphalt pavement material consists of asphalt, aggregate, and air (voids). The strength of the asphalt paving is a function of the surface texture (fine aggregate) and its density, or compactness. Several types of asphalt are available. Among them are:

- *RC (Rapid Curing)*: contains naphtha or gasoline which evaporate quickly, leaving asphalt cement
- *MC (Medium Curing)*: contains a kerosene dilutant
- *SC (Slow Curing)*: contains an oil that is low in volatility
- *Emulsified asphalt*: a mixture of water and an emulsifying agent, and asphalt cement

Design Considerations

The major issues in pavement design are drainage and the size and amount of traffic that will be operating on the asphalt surface. If it is a parking lot with no anticipated traffic heavier than passenger vehicles, a basic design with 6" of base and 3–4" of asphalt will be sufficient. If heavy trucks will be operating on the pavement, then a base up to 12", with 6" of asphalt, may be needed. For loading docks, a concrete apron may be appropriate where the trailers will be parked. If a truck is parked for a considerable amount of time, it can depress the pavement underneath. The pavement surface must be strong enough to withstand distortion, resist wear from traffic, and provide both a smooth ride and skid-resistance. The pavement must also be well-bonded to the course below.

A well-designed pavement should be free of water after a rain shower. The site should drain out and away from the pavement. Large paved areas should have an elevation plan showing the proposed grades on a grid.

Ed. Note: See "Site Planning" and "Common Excavation Formulas and Calculations" earlier in this chapter for more on assessing and calculating site work requirements for a project.

The base and sub-base of the pavement system are important to its structural effectiveness. These two elements, with the asphalt surface above, distribute traffic wheel loads and must offer internal strength properties. Full-depth asphalt pavements have both tensile and compressive strength to resist internal stresses. *(See "Concrete Strength" in the "Concrete" section earlier in this chapter for definitions of tensile and compressive strength.)* Asphalt bases spread the wheel load over broader areas than untreated granular bases. Consequently, full-depth asphalt pavements require less total pavement structure in terms of thickness.

Aggregates for Asphalt Pavement

Aggregates are specified by grade for paving mixes. Gradation is determined by the material's ability to pass through a series of sieves, graduated in size. This process is called sieve analysis. Gradation specifications are established by industry associations, as well as federal, state and local agencies. Gradation may be described as percent passing, percent retained, and percent passing-retained. Coarse aggregate is material retained on the 2.36 mm (No. 8) sieve. Fine aggregate is all material that passes the 2.36 mm (No. 8) sieve. Mineral filler, fine aggregate, is material of which at least 70% passes the No. 200 sieve.

Two or more gradations of aggregates are often combined to create a blend that meets grading specifications for an asphalt mix. The following basic formula (from the Asphalt Institute's *Asphalt Handbook*) is used to determine the relative amounts of various aggregates needed to obtain a particular gradation.

$$P = Aa + Bb + Cc, \text{ etc.}$$

where

P = the percent of material that passes a particular sieve for the combined aggregates A, B, C, etc.

A, B, C, etc. = percentages of material passing a given sieve for aggregates A, B, C, etc.

a, b, c, etc. = proportions of aggregates A, B, C, etc., used in the combination where the total = 1.00.

Ed. Note: The Asphalt Handbook is highly recommended for its comprehensive coverage of asphalt properties, design, and placement information.

Placement Considerations

Contract documents should specify the aggregate mixture, gradation, quality, grade of the asphalt, and the heat ranges at which the aggregate is to be mixed and placed. An engineer designs the project and prepares the plans and specifications, then works with the contractors who will supervise the construction. The engineer inspects the work. The engineer and his or her inspector should discuss the operation before starting the work with the contractor's superintendents and foremen, and should plan the operation together. The *Asphalt Handbook* recommends discussing these important details:

1. Continuity and sequence of operations
2. Number of pavers needed for the project
3. Number and types of rollers needed
4. Number of trucks required

5. The chain of command for giving and receiving instructions
6. Reasons for possible rejection of the mix
7. Weather and temperature requirements
8. Traffic control

Asphalt concrete is usually placed on a base course. The base course should be well-graded material that is compacted to at least 98% optimum moisture (based on a modified Proctor test). The base course soils are classified in general terms by the predominant particle size or grading of particle sizes. These are usually gravel, coarse sand, medium sand, fine sand, silt, clay, and colloids. (See "Soil Mechanics" earlier in this chapter for more on soils analysis.) Well-graded soils contain a mixture of particle sizes and are generally free from organic material.

The base course should be checked for thickness, elevation and proper grading, and should be free of loose materials. Prior to placing the asphaltic concrete and before the binder course is installed, a tack coat of asphalt may be sprayed on. The binder course is placed first and contains larger stone sizes and sometimes stone chips. The wearing course, or top course, is made with coarse and fine sand. In cases such as the patching of a driveway or other use of only a small portion of paving material, only the wearing course is put down to save time. Asphaltic concrete is usually placed with a paving machine. The top course is usually hand-raked on the edges.

Truckloads of material should be inspected as they arrive, and the mix temperature checked on a regular basis. The paving inspector may reject loads that do not meet tolerance criteria. The paving crew must stay in close touch with the asphalt plant to communicate any changes needed in the mixture for subsequent truckloads. Records should be kept of loads placed, as well as any that are rejected.

The temperature of the mix is crucial. If too low, it will not compact properly. Too high and the curb may slough off. Temperatures generally range from 120-140 degrees Celsius (250-285 degrees Fahrenheit).

Compaction is accomplished with rolling equipment, as soon as possible after the hot asphalt mix has been spread. Rolling is usually done in three passes: first, the breakdown to compact the material; second, the intermediate rolling to create further density and seal the surface; and lastly, finish rolling to remove roller marks from previous passes.

Once rolling is complete, the surface should be checked to ensure there are no defects. Those that can be corrected by additional rolling should be addressed by placing fresh hot asphalt and compacting it before the surrounding area cools (below 85 degrees). Tolerances for smoothness must be ensured. Most specifications provide for a transverse variation no greater than 6 mm in 3 m, or 1/4" in 10'. Variations from the tolerance level in any layer should be corrected before placing the next layer.

Sampling and Testing

First the subgrade is tested for compaction. This can be done with a modified Proctor test or a nuclear density test. The modified Proctor test involves taking an in-place sample of the subgrade, then utilizing a lab test to determine the moisture content of the soil at which maximum compaction can be obtained. The nuclear density tests are more common now because results can be obtained instantly and they are nondestructive. This allows the contractor to correct any areas that are not adequately compacted.

Testing asphaltic concrete generally involves temperature monitoring during the installation and sampling of the material in place. Sampling can be done by core-drilling a sample and having a lab test it. Nuclear density testing can also be performed on in-place asphalt. The average of the densities from the samples obtained must meet targets based on percentages set by the lab (96%), by maximum possible (theoretical) density (92%), and by control strip density (99%).

Resurfacing Existing Roadways

The surface of an existing road should be examined before beginning any resurfacing work. Scarifying, recompacting, or repair of the old surface may be necessary before placing new material. If there are any soft spots or irregularities, these should be addressed. It may be necessary to patch or remove excess asphalt. Adequate time must be allowed for consolidation of patches before performing final surface preparations. Correction of drainage problems is extremely important.

These kinds of repairs often fall into the category of maintenance. In some cases, maintenance repairs may be addressed as temporary measures. Asphalt concrete and other types of hot mixes offer a more durable and lasting solution, and should be used when practical. Alternatives including cold-plant and road mix materials that contain medium-curing or emulsified asphalt can be used right away or stored for a short time. Materials with slow-curing asphalt or a solvent emulsion can be stored for a longer time for patching. Patching should be done before cracks lead to larger problem areas, the inevitable result of water entering the subgrade.

The following section is reprinted from *Fundamentals of the Construction Process*, by Kweku K. Bentil, R.S. Means Co., Inc.

Overview of the Construction Process

The construction process has become increasingly more complex in recent times as the industry attempts to meet the demands of a constantly growing and changing world. As a result, facilities, institutions, manufacturing companies, and other businesses are perpetually engaged in construction—creating new structures, adding to existing ones, or renovating, restoring, and preserving old buildings. For the professional who is not initiated into the field of construction, this many-faceted process may seem difficult to sort through and fully understand. To begin, the novice must be familiar with the key people involved in the construction process, the different phases of construction, the various types of construction, and ways to procure projects.

The Key Players in Construction

There are myriad individuals, professionals, and other entities involved in the construction of a building, but the key players in the industry are:

- the owner(s),
- the architect/engineer,
- general contractor,
- subcontractors,
- suppliers, and
- other consultants.

The relationships among these key parties are illustrated in Figure 3.58. Each party plays a distinct role in the construction process, described in the following paragraphs.

The Owner: Every construction project, no matter how large, small, complex, or simple, has an owner who recognizes the need for that project. The owner is the person, firm, organization, or agency that needs the construction work accomplished. The owner may be:

- an individual;
- a group of people;
- a firm (corporation, partnership, etc.);
- an organization (such as a church, civic club, etc.); or
- the government (such as federal agencies, states, counties or cities, and the military—the army, navy, air force, and marines).

Generally, the owner provides the financial resources to fund the project and the property on which the structure is to be built. The owner also makes binding decisions and approvals throughout the duration of a project. These decisions and approvals are, in some projects, delegated to a professional or consultant hired by the owner.

Traditionally, the owner initiates the construction process by his desire to invest in a structure or to meet his need for a structure. The owner then selects a construction professional (an architect, construction manager, or other) to assist him through the complex process, unless he has the required expertise himself, or within his own company. This professional, in serving his client, may secure the services of other professionals such as architects, engineers, real estate agencies, and mortgage and financing specialists.

Architect/Engineer: The architect/engineer transforms the dreams and ideas of the owner into plans and specifications. If hired by the owner in the early stages of the project, this professional may assist with some of the preliminary conceptual work, such as selecting a site and providing conceptual estimates.

During the design phase, the architect/engineer performs the design in several stages, each stage undergoing extensive review and approval by the owner and/or his representative. First, a **conceptual design** is prepared based on

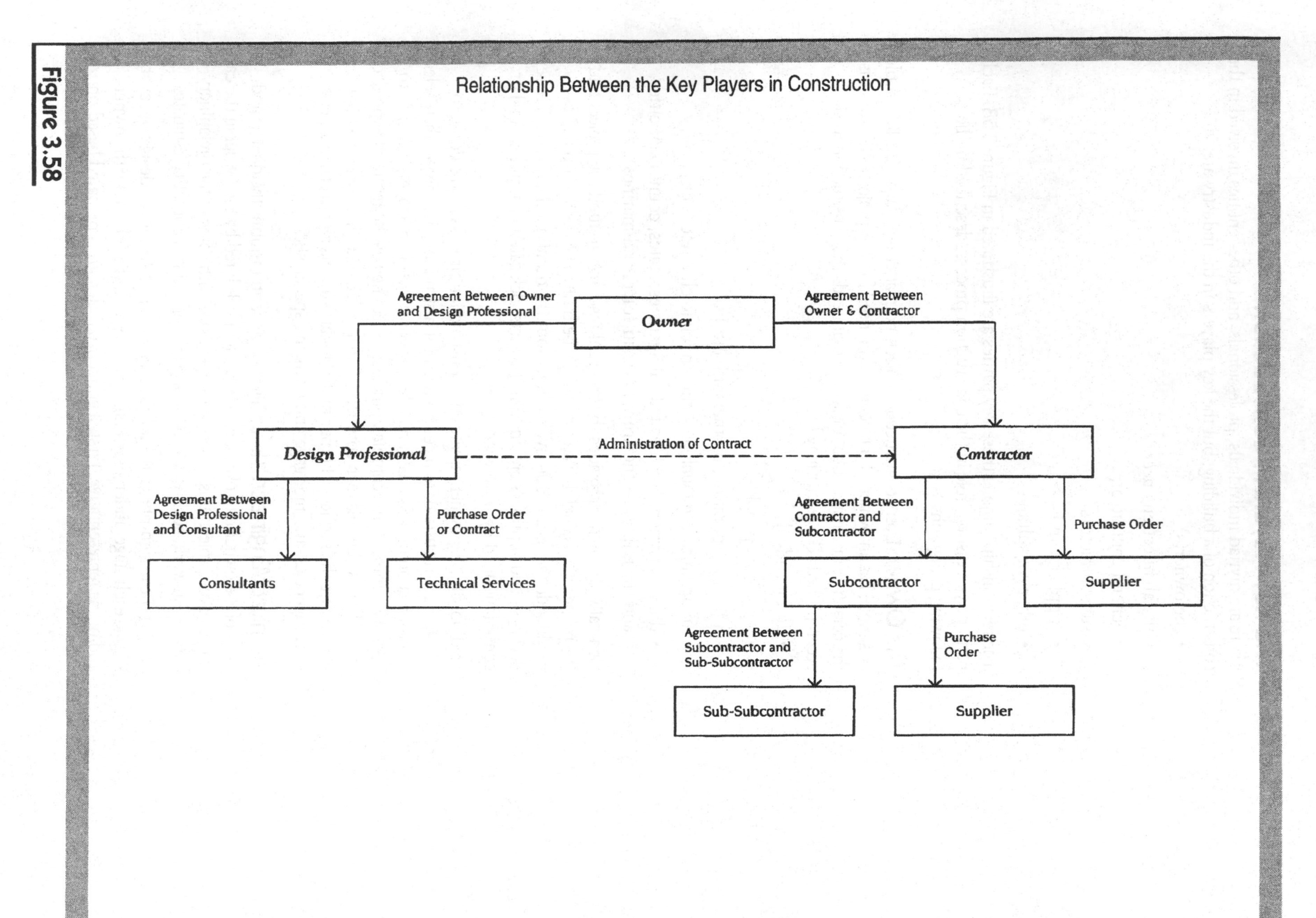
Relationship Between the Key Players in Construction
Agreement Between Owner and Design Professional
Owner
Agreement Between Owner & Contractor
Design Professional
Administration of Contract
Contractor
Agreement Between Design Professional and Consultant
Purchase Order or Contract
Agreement Between Contractor and Subcontractor
Purchase Order
Consultants
Technical Services
Subcontractor
Supplier
Agreement Between Subcontractor and Sub-Subcontractor
Purchase Order
Sub-Subcontractor
Supplier
R.S. Means Co., Inc., Plans, Specs & Contracts for Building Professionals

Figure 3.58

information provided by the owner. Then, a **preliminary design** is prepared. Once that is approved by the owner or his representative, a **semi-detailed design** is prepared. The next stage is the final detailed or **working drawings**, to be used for bidding or negotiating a construction contract.

The architect/engineer's responsibilities do not end here. He is responsible for the interpretation of the plans and specifications until the completion of the project and, depending on his contract with the owner, may be responsible for bid administration, inspections, and approval of progress payments.

General Contractor: The general (or prime) contractor transforms the plans and specifications prepared by the architect/engineer and owner into a physical structure. The contractor coordinates the work of the craftsmen (laborers, carpenters, plumbers, electricians), subcontractors, and material suppliers.

The prime contractor accepts the responsibility of completing the project for the agreed terms. Normally, the contractor signs an agreement with the owner to provide the labor, materials, equipment, and subcontracts required to complete the project in accordance with the plans and specifications, within a specified time period, for an agreed-upon price.

Subcontractors: Subcontractors play a major role in the construction phase of a project. The general contractor depends on subcontractors to perform portions of the work at appropriate intervals in order to maintain the project schedule. The subcontractor signs an agreement with the general contractor to provide all labor, equipment, and materials necessary to perform a specific portion of a given project for a designated price. These are portions of work the general contractor cannot or does not choose to perform with his own work force, such as masonry, finish carpentry, plumbing, and electrical work.

Suppliers: Another form of subcontractor is the supplier. Both contractors and subcontractors deal with suppliers. Suppliers are responsible for furnishing the materials and equipment to be incorporated into the project being constructed.

Other Consultants: On most construction projects, there is a need for consultants. These consultants may be hired by the contractor to provide expertise not available within the contractor's organization. Consultants on a construction project might include the following:

- **Specialists** (such as testing laboratories) perform the testing of soil and concrete, steel, asphalt, plumbing, and air conditioning work.
- **Schedulers** plan the progress of the project.
- **Claims Specialists** help prepare, negotiate, settle, or litigate claims that a contractor may have against an owner.
- **Certified Public Accountants** prepare independent and audited financial statements and offer tax planning advice.
- **Attorneys** offer legal advice.
- **Surveyors** locate the building properly on the site.

Phases of the Construction Process

While every construction project is unique, most follow a consistent pattern. The projects go through the following five major phases:

- Pre-bid phase
- Contract procurement phase
- Contract award phase
- Construction phase
- Operating and maintenance phase

These five phases are illustrated in Figure 3.59.

Pre-bid Phase: The pre-bid phase has two sub-phases: the conceptual phase and the design phase. The sub-phases are described in the following paragraphs.

Conceptual Phase:

The conceptual phase usually begins with an owner (or owners) recognizing the need for a structure, or having the desire to invest in a building. In this phase, all factors are considered and carefully examined to determine if the project is feasible. Factors such as the type of structure, size, possible locations, cost, availability of funds, source of the funding, market analysis, availability of utilities at the proposed site, and selection of the designer are weighed and parameters set. This early phase of a project is generally handled by the owner. However, an owner may consult other professionals during this phase to analyze the above factors.

Design Phase:

In the design phase, the needs, ideas, and/or dreams of the owner are transformed into plans and specifications. In this phase, an architect/engineer is selected and hired by the owner. Often several other design professionals are engaged by the architect/engineer. Specific design and other professionals consulted in addition to the architect/engineer vary depending on the type of construction project–residential, commercial, or industrial. Size, complexity, and the type of agreement signed between the owner and the architect/engineer also dictate the professionals involved. Some of these consultants are listed below.

- **Interior Designers** handle the design of the floor and wallcoverings, painting scheme, colors, and furnishings.
- **Landscape Architects** design the site layout of trees, parking lots, roadways, flowers, and grassed areas.
- **Structural Engineers** design the structural elements, such as slabs, columns, and beams, to withstand stresses from all loads during its useful life.

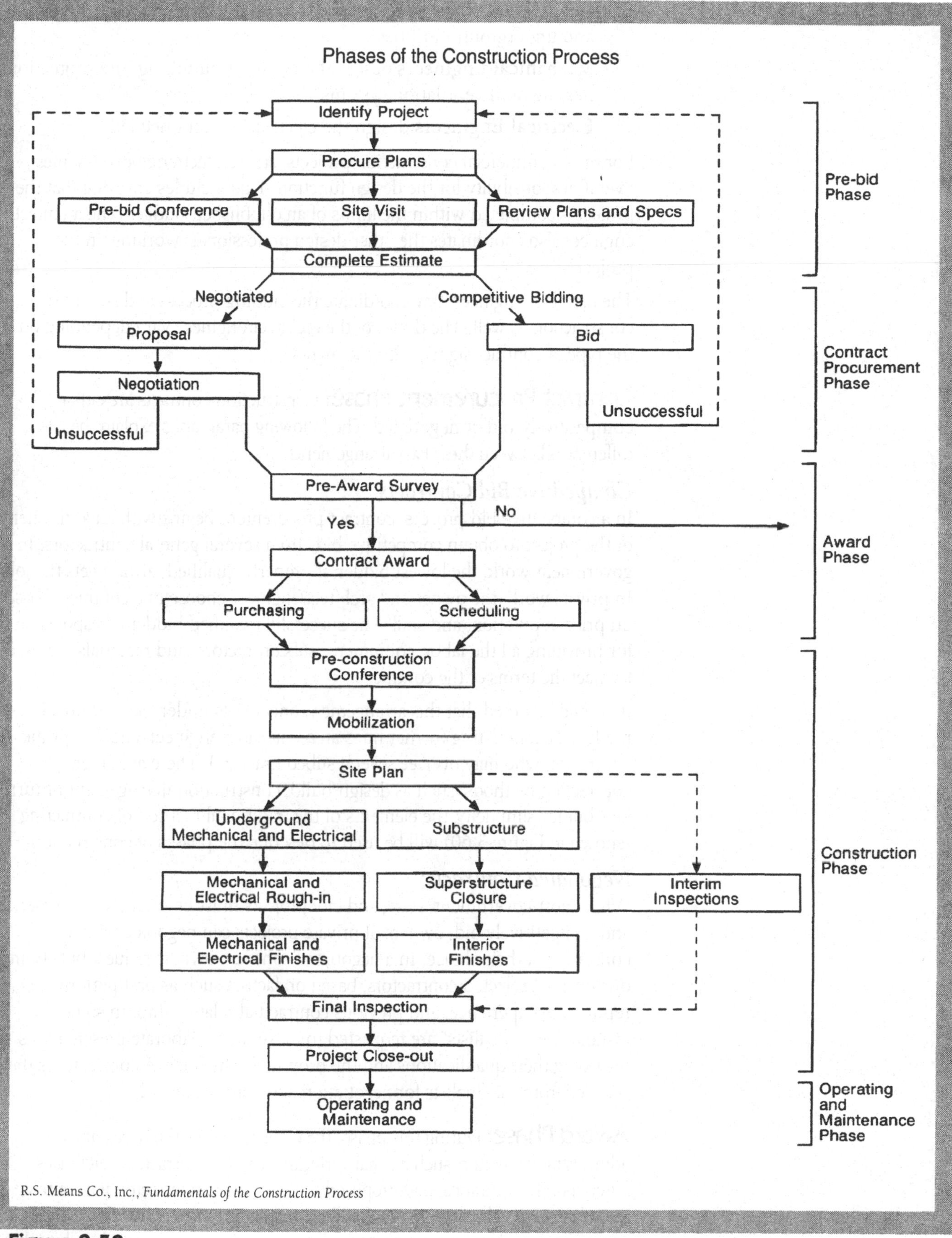

Phases of the Construction Process
Identify Project
Procure Plans
Pre-bid Conference
Site Visit
Review Plans and Specs
Complete Estimate
Negotiated
Competitive Bidding
Proposal
Bid
Negotiation
Unsuccessful
Unsuccessful
Pre-Award Survey
Yes
No
Contract Award
Purchasing
Scheduling
Pre-construction Conference
Mobilization
Site Plan
Underground Mechanical and Electrical
Substructure
Mechanical and Electrical Rough-in
Superstructure Closure
Interim Inspections
Mechanical and Electrical Finishes
Interior Finishes
Final Inspection
Project Close-out
Operating and Maintenance
Pre-bid Phase
Contract Procurement Phase
Award Phase
Construction Phase
Operating and Maintenance Phase
R.S. Means Co., Inc., *Fundamentals of the Construction Process*

Figure 3.59

- **Civil Engineers** design the required roads, curbs, drainage systems, and underground utilities.
- **Mechanical Engineers** design the required plumbing, fire protection, heating, and ventilation systems.
- **Electrical Engineers** design the electrical system needed.

For most commercial construction projects, an architect/engineer assumes overall responsibility for the design function. This includes ensuring that the project is completed within the limits of an established budget. The architect/engineer also coordinates the other design professionals working on the project.

The architect/engineer may coordinate the bidding process and supervise construction as well. The duties of the architect/engineer vary depending on the type of contract signed with the owner.

Contract Procurement Phase: Construction contracts are either **competitively bid** or **negotiated**. The following paragraphs explain the differences between these two arrangements.

Competitive Bid Contracts:

In a competitive bid process, contract procurement begins with advertisement of the project to obtain competitive bids from several general contractors. In government work, the lowest bidder, if properly qualified, always gets the job. In private work, the owner and architect/engineer choose one contractor based on price, reputation, and skill. The successful qualified bidder is responsible for providing all the labor, equipment, subcontractors, and materials required to meet the terms of the contract.

It should be noted that this arrangement only exists under the traditional method of contracting (owner hires an architect or engineer, and one prime contractor, who may oversee several subcontractors). There are other contracting methods, such as design/build, construction management or turn-key, but for simplicity, the elements of the traditional method of contracting (shown in Figure 3.60) will be used to describe the phases of construction.

Negotiated Contracts:

While most government, state, and city projects must be publicly advertised and competitively bid, owners of private projects can negotiate with a contractor of their choice. In a negotiated contract, owners request bids from one or more selected contractors, based on factors such as past performance, reputation, experience, and previous contractual relationship. In some instances, the "finalists" are requested to make more elaborate presentations regarding their qualifications and the proposed schedules. A contractor is then selected from the finalists for negotiation and contract award.

Award Phase: During this phase, the owner finalizes the necessary administrative matters such as final verification of the contractor's references, surety bonds, insurance, and proposed schedules. The contractor, if a contract award appears to be imminent, often begins to prepare the schedule of values, progress schedule, and assignment of the prospective site personnel such as the superintendent and the project manager.

Once the owner and the contractor sign a formal contract, most contractors immediately begin negotiating with prospective subcontractors and material suppliers. One must devote priority to subcontract work and materials needed at the beginning of the project, such as site preparation work, site utilities, reinforcing steel, door frames, and items to be embedded in the foundation and slab. Long lead items (items not manufactured or fabricated until an order is placed), or items that must be modified to meet project specifications and require several months advance notice in order to meet production and delivery schedules (such as mechanical and electrical equipment and elevators), should be ordered immediately.

Pre-Construction Conference

A pre-construction conference between the architect/engineer, the owner's representation, and the contractor takes place after the contract is awarded, before construction begins. This meeting is a forum to ask questions on technical and administrative aspects of the project.

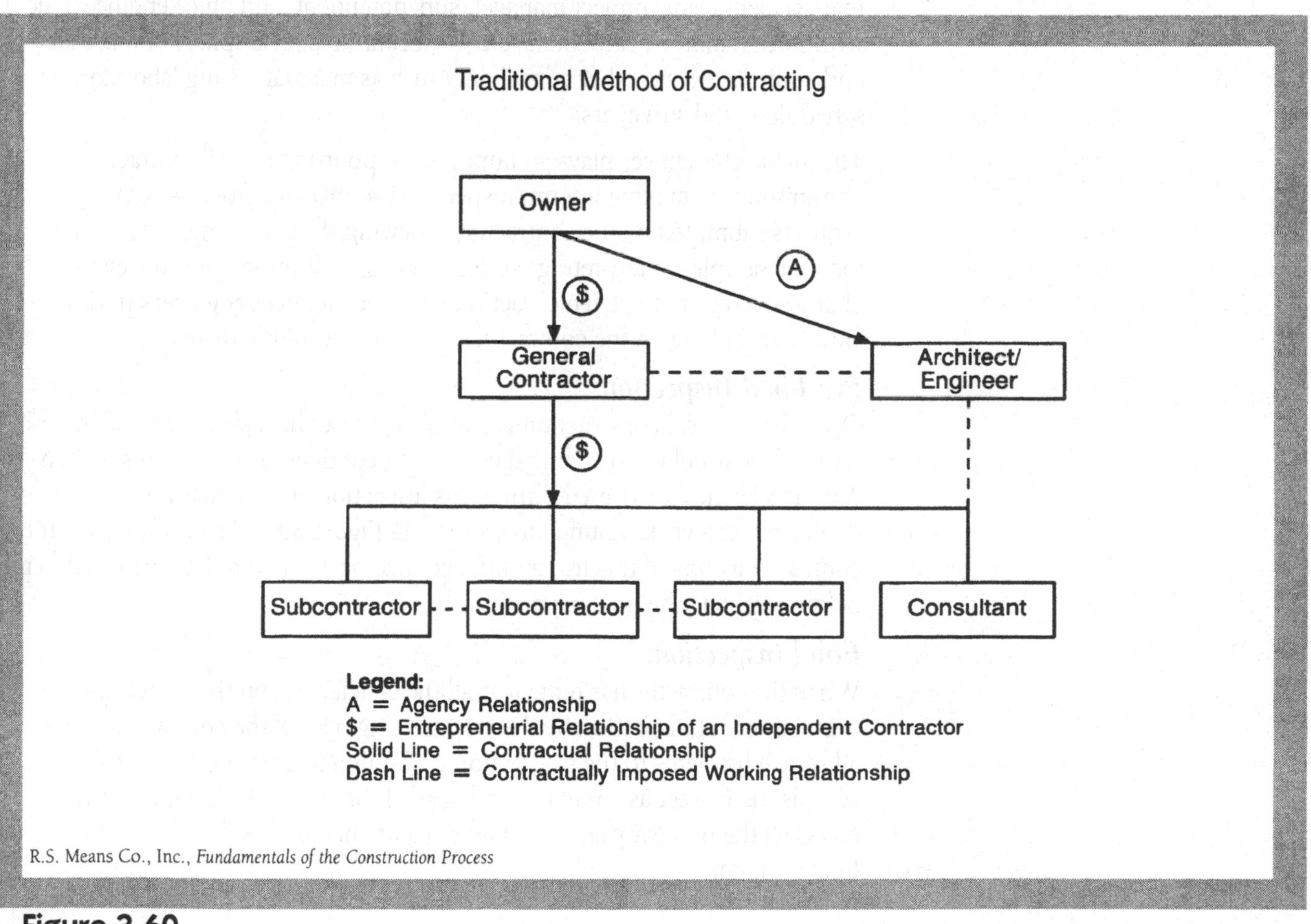

R.S. Means Co., Inc., *Fundamentals of the Construction Process*

Figure 3.60

Contractor's representatives attending this meeting include the following personnel:

- **Authorized Officer**, such as the president, vice-president, or construction manager.
- **Designated Project Manager**, usually based at the home office of the contractor who handles more than one contract at any given time.
- **Field Supervision Personnel**–any number of field supervision personnel, depending on the size, type, and complexity of a project; usually includes at least a superintendent, an office engineer, and an administrative employee.

In addition to the contractor's representatives, major subcontractors (mechanical, electrical, and site work), the architect/engineer, and the owner's site representative should attend the pre-construction meeting. At this meeting, all parties agree on responsibilities for portions of the project. Soon after the pre-construction conference—often within ten to fifteen days—the project moves into the construction phase.

Construction Phase: The construction phase transforms the needs, ideas, and dreams of the owner—plans and specifications—into a physical structure. The key players in this phase are the general contractor's project management team (project manager, superintendent, and office engineer), field workforce (laborers, carpenters, etc.), subcontractors, suppliers of materials and equipment, and other consultants such as material testing laboratories, schedulers, and surveyors.

The architect/engineer plays an important supporting role in contract administration, making interim inspections, approving progress payment requests submitted by the contractor, reviewing shop drawings, approving material samples, interpreting the plans and specifications, and reviewing changes to the contract. The exact scope of the architect/engineer's duties varies, depending on the contractual requirements of the owner.

Pre-Final Inspection:

Once the physical construction is complete, the architect/engineer, often with representatives of the owner and contractor, conducts a pre-final inspection. Any deficiencies uncovered during this inspection are summarized in a document known as a punch list, shown in Figure 3.61. The owner gives the contractor a copy of this list, and the contractor must complete the listed items before receiving the final payment.

Final Inspection:

When the contractor has remedied all the deficiencies on the punch list, he requests a final inspection by the architect/engineer. If the contractor corrects all punch list items to the satisfaction of the architect/engineer, he or she accepts the project as complete on behalf of the owner. When the contractor meets all the necessary administrative requirements, he submits a request for final payment.

Punch List

PROJECT: Admin. Bldg. & F.S.	**DATE OF WALKTHROUGH** ________
TIME START 9:15 AM	**TIME COMPLETE** 12:00 Noon
REPRESENTING SPONSOR:	**REPRESENTING CONTRACTOR:**
Chris Bell, Architect	Buck Buttress, Superintendent
Chuck Bloomfield, Engineer	U.N. Level, Foreman
N. Neilson, Inspector	S.O. Beit, Manager
	Doug Piper, Piper Piping Co.
	E.C. Hull, Sparks Electric

ITEMS REQUIRING CORRECTION	**CORRECTIVE ACTION**
1. M.H. cover too high at station 14+10	1. Piper Piping Co. will lower to proper elevation
2. Entrance door closer needs adjusting	2. Foreman will adjust
3. Sprinkler piping not painted	3. Painter was notified
4. Sheet metal flashing not straight	4. Sub was notified
5. Landscaping not com-pleted	5. Sub promises completion by Friday
6. Parking light doesn't turn on	6. Elec. sub is checking it
7. Rollup door needs adjusting	7. Sub was notified
8. Robe hook missing in men's room	8. Foreman will supply

R.S. Means Co., Inc., *Superintending for the General Contractor*

Figure 3.61

The administrative tasks necessary to finalize, or "close," a job are listed below.

- **Submit an affidavit of release of liens:** This document states that all the labor, materials, equipment, subcontractors, and consultants used on the project have been paid in full and that all applicable taxes (sales tax, payroll taxes, etc.) have been paid.
- **As-built drawings:** This is a set of the final drawings for the project, showing all changes made during construction. The changes are usually highlighted with red pen or pencil, or surrounded by an irregular line or "cloud."
- **Submit operating and maintenance manuals:** These are instructions for operating and maintaining all equipment and specialty items installed on the project. Manufacturers' warranties are included in this package.

Operating and Maintenance Phase: The operating and maintenance phase begins as soon as the owner accepts the completed structure. It involves periodic maintenance of the structure and the equipment installed within it. This phase may be simple or complicated, depending on the type of structure, its size, and the type of equipment housed within the structure.

Critical Path Method (CPM)

This method of scheduling has been found by many construction professionals to be better and more efficient than a bar chart in managing construction projects. CPM scheduling has the following major advantages:

- CPM identifies the dependence and interdependence of all activities in a schedule.
- From a CPM schedule, a project manager can immediately determine how a change in one or more activities impacts other activities in a schedule.
- CPM identifies activities that are critical to the timely completion of a project.
- This method enables a project manager to easily determine those activities that are not critical. Non-critical activities are tasks that may be delayed without affecting the overall project completion date.
- The CPM schedule shows the shortest time in which a project can be completed.

CPM is a method of developing a logic network of construction activities. Figure 3.62 shows how to graphically represent logic networks on a CPM schedule.

Representing Logic Networks

Examples of Arrow Diagrams and Interpretations

Diagram	Interpretation
I Activity J	The tail or the arrow ("I") represents the initiation or start of an activity. The arrowhead ("J") represents the completion of an activity.
L M	Activity "M" cannot start until activity "L" is completed, or the start of activity "M" is dependent upon the completion "L."
L M N	Activity "L" must be completed before activities "M" and "N" can start, i.e., activities "M" and "N" can be done concurrently, but cannot start until "L" is completed.
L N M O P Q R	Activities "L" and "M" can be done concurrently. Activity "N" cannot start until "L" and "M" are completed. Activities "O," "P," "Q" and "R" run concurrently, but all four cannot start until "N" is completed.
L M N	Activities "L" and "M" run concurrently, and have to be completed before "N" can start.
N L M P O	Activity "L" has to be completed before "M" can start. Activities "N" and "O" run concurrent under "L" and "M." Activity "P" cannot start until "L," "M," "N," and "O" are completed.
N Q L M P O	Activity "M" cannot start until "L" is completed. "N" and "O" run concurrent with "L" and "M." "P" and "Q" are concurrent, but cannot start until "L," "M," "N," and "O" are completed.

R.S. Means Co., Inc., *Fundamentals of the Construction Process*

Figure 3.62

The components of the logic network (shown in Figure 3.63) are explained below:

- The numbers 1, 2, 3, and 4 designate **Events**.
- The letters A, B, C, D, and E designate **Activities**.
- The dotted line (between events 3 & 4) is known as a **Dummy**.
- The line spanning between events 1 & 4 is called a **Hammock**.
- Activities "C" and "D" cannot start until event "2" is completed; the starts of activities "C" and "D" are **dependent** upon the completion of activity "B".

A CPM network may be prepared in two ways: by the **Arrow Diagram Method** or the **Precedence Diagram Method.**

The Arrow Diagram Method (ADM) is sometimes called the **I-J Method** or the **Activity-on-Arrow Method.** The Precedence Diagram Method (PDM) is also called the **Activity-on-Node Method**. The important aspects of these two methods are discussed in the following paragraphs.

The Arrow Diagram Method (ADM): In the arrow diagram method, arrows represent activities. The arrows are usually not drawn to scale and may go from left to right, right to left, up or down, or at an angle. However, in order to be consistent, avoid confusion, and for ease of use, it is usually recommended that the arrows go from left to right.

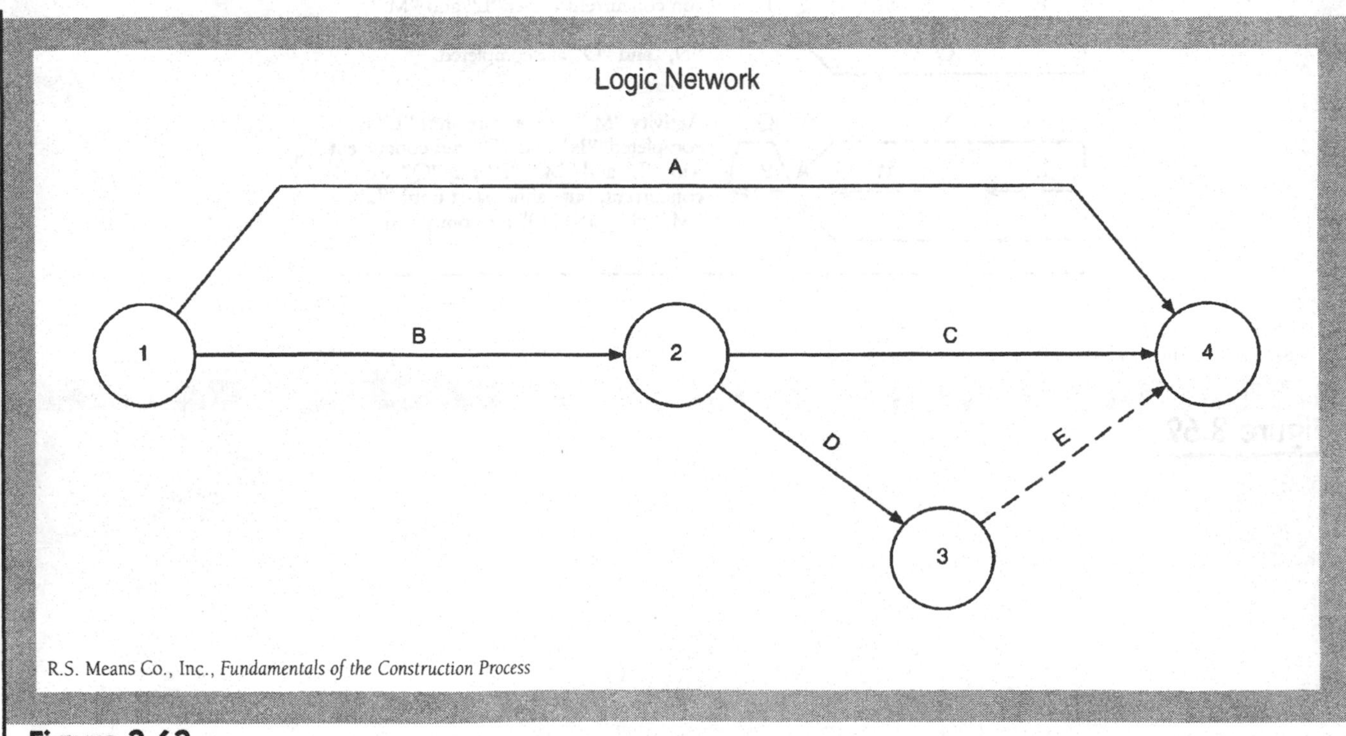

R.S. Means Co., Inc., *Fundamentals of the Construction Process*

Figure 3.63

For each activity (arrow) on the network, the person preparing the schedule must answer the following questions:

- When can this activity start and how long will it take to complete it?
- Which activities have to be completed before this activity can start?
- Which activities can be done concurrently with this activity?
- Which activities cannot begin until this activity is completed?

An example of an ADM network is illustrated in Figure 3.64.

The Precedence Diagram Method (PDM): Four main types of constraints are usually encountered on precedence diagrams. Whereas arrows represent activities in ADM, activity boxes or nodes represent activities in PDM. Figure 3.65 is a comparison of ADM and PDM constraints.

In using this method, the node (activity box) may be set up in several different ways. An example of a complete precedence network is illustrated in Figure 3.66.

Basic Scheduling Computation: Certain calculations are required during the preparation of a construction schedule. The following is a simplified guide on how to determine some of the basic values often needed in scheduling, based on the assumption that a project starts on day number one:

- Early Start (ES) plus for the first activity, or
  = EF of the preceding activity for other subsequent activities.
- Early Finish (EF) = ES plus the duration of the activity.
- Late Finish (LF) = LS of the subsequent activities, or
  = EF of the network (for the last activity or activities).
- Total Float (TF) = LS minus ES
  = LF minus EF
  = LF minus ES minus duration

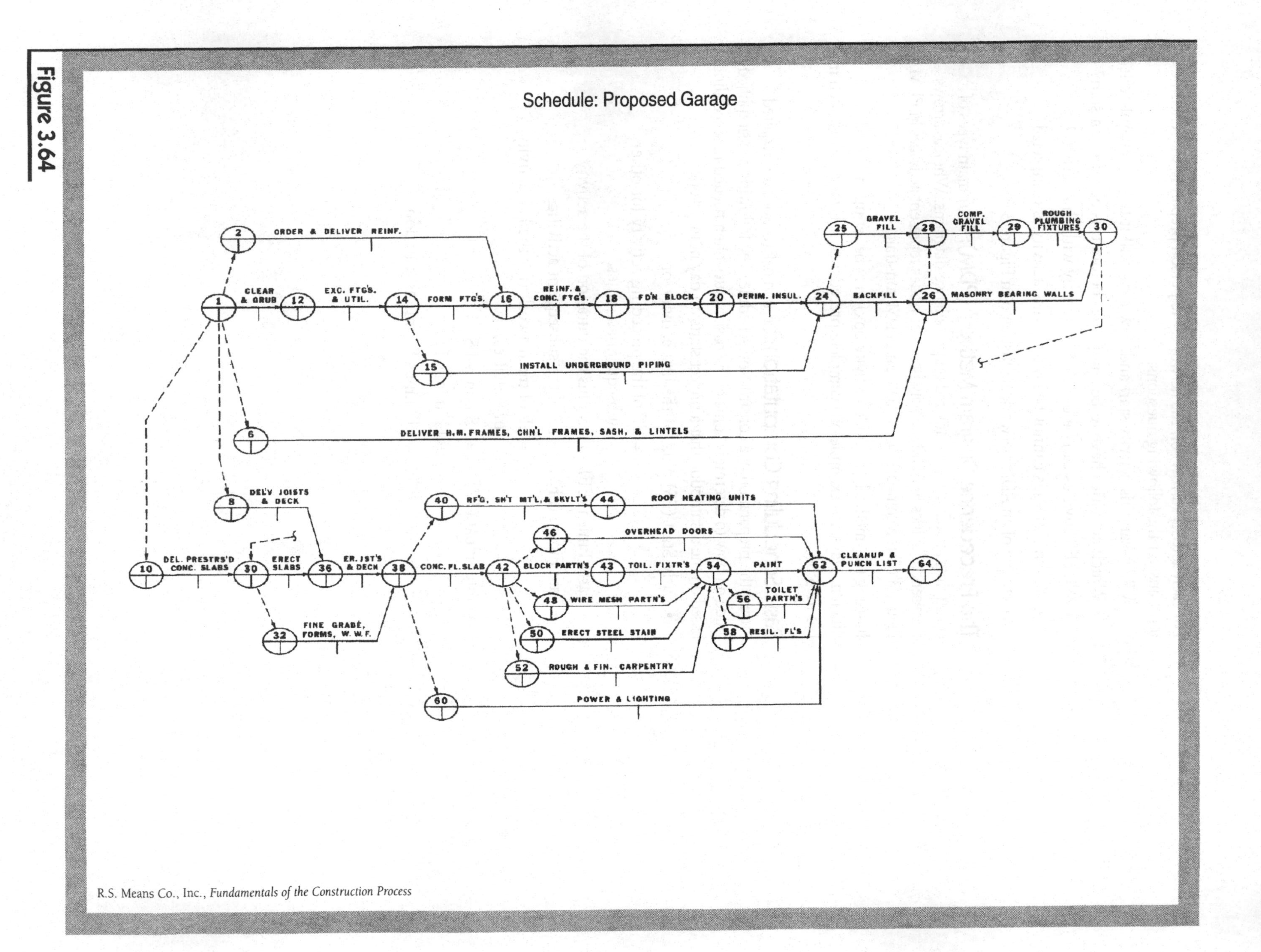

Schedule: Proposed Garage
ORDER & DELIVER REINF.
CLEAR & GRUB
EXC. FTGS. & UTIL.
FORM FTGS.
REINF. & CONC. FTGS.
FD'N BLOCK
PERIM. INSUL.
BACKFILL
MASONRY BEARING WALLS
GRAVEL FILL
COMP. GRAVEL FILL
ROUGH PLUMBING FIXTURES
INSTALL UNDERGROUND PIPING
DELIVER H.M. FRAMES, CHN'L FRAMES, SASH, & LINTELS
DEL'V JOISTS & DECK
DEL. PRESTRS'D CONC. SLABS
ERECT SLABS
ER. JST'S & DECK
FINE GRADE, FORMS, W. W. F.
CONC. FL. SLAB
RF'G, SH'T MT'L, & SKYLT'S
ROOF HEATING UNITS
OVERHEAD DOORS
BLOCK PARTN'S
TOIL. FIXTR'S
WIRE MESH PARTN'S
ERECT STEEL STAIR
ROUGH & FIN. CARPENTRY
POWER & LIGHTING
PAINT
TOILET PARTN'S
RESIL. FL'S
CLEANUP & PUNCH LIST
R.S. Means Co., Inc., Fundamentals of the Construction Process

Figure 3.64

Comparison of ADM and PDM Scheduling Methods

Precedence Scheduling can show various activity relationships to further develop network logic. Examples of relationships are shown below.

Connectors & Lag

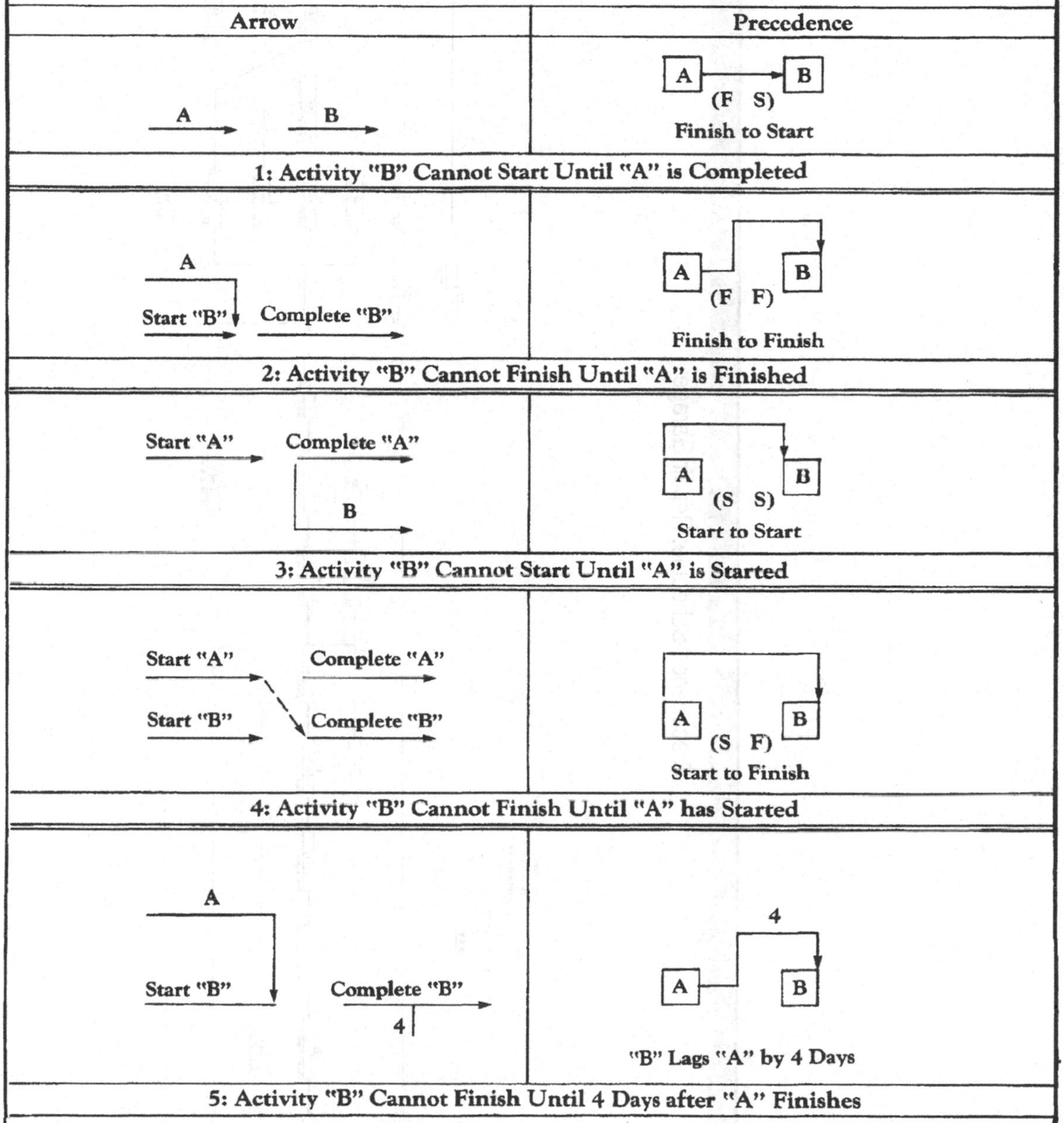

Arrow	Precedence
A → B →	A → B (F S) Finish to Start
1: Activity "B" Cannot Start Until "A" is Completed	
A; Start "B"; Complete "B"	A, B (F F) Finish to Finish
2: Activity "B" Cannot Finish Until "A" is Finished	
Start "A"; Complete "A"; B	A, B (S S) Start to Start
3: Activity "B" Cannot Start Until "A" is Started	
Start "A"; Complete "A"; Start "B"; Complete "B"	A, B (S F) Start to Finish
4: Activity "B" Cannot Finish Until "A" has Started	
A; Start "B"; Complete "B"; 4	4; A, B; "B" Lags "A" by 4 Days
5: Activity "B" Cannot Finish Until 4 Days after "A" Finishes	

If a time unit is added, the activity relationship would read as follows: Activity "A" must finish 4 days before activity "B" can finish. "A" leads "B" by 4 days or "B" lags "A" by 4 days.

R.S. Means Co., Inc., *Fundamentals of the Construction Process*

Figure 3.65

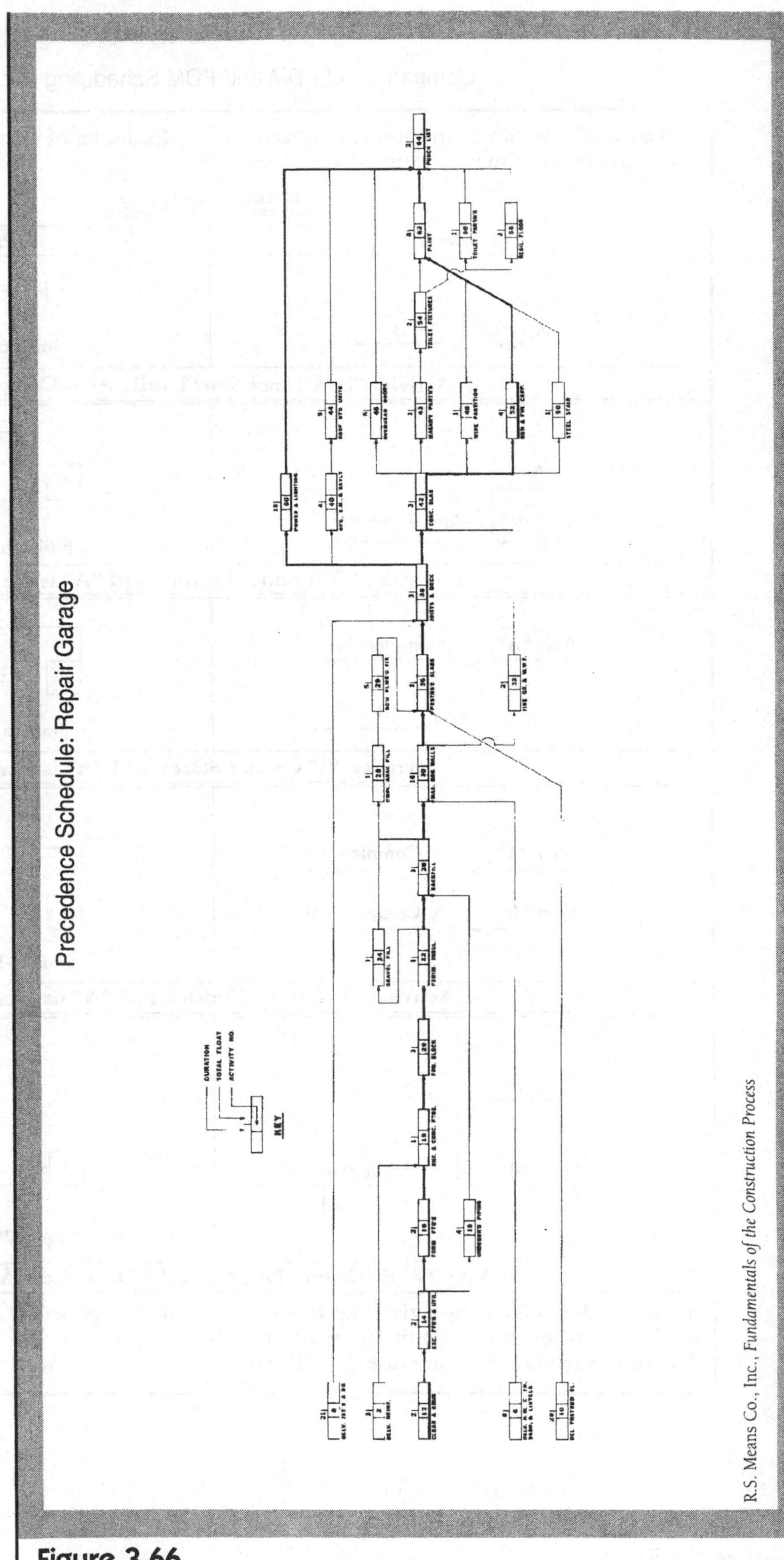

Precedence Schedule: Repair Garage

R.S. Means Co., Inc., *Fundamentals of the Construction Process*

Figure 3.66

Construction Contract Documents

The construction process is a concerted undertaking requiring the efforts of numerous professionals, the commitment of large sums of money, and considerable increments of construction professionals' time. In the past, this process was simple and there were relatively few people involved. Parties relied on trust, spoken agreements, and handshakes to uphold commitments. These handshake agreements have evolved into written, legally binding contracts, because of the complexity, magnitude, and the number of participants involved. Knowledge of the contents of construction documents simplifies the potentially overwhelming collection of agreements, plans, drawings, specifications, and other documents of the construction contract. This section describes the basic elements of construction contract documents.

Definition of a Contract

A contract is an agreement made in good faith between two or more capable parties to perform an act or acts. A contract may be invalid if signed under duress, under influence of another person, or if the objective is fraudulent. Contracts may be spoken or written, but are usually written because oral contracts are generally not recognized in court.

Elements of a Valid Contract: Every contract is unique, but certain elements are common to all valid contracts. These are listed below.

- The agreement must be mutual. (Both parties must agree to enter into the contract. In other words, there must be mutual consent between all parties to the contract.)
- There must be an offer and an acceptance.
- There must be some sort of consideration. (This is usually the promised value of the contract.)
- The purpose or objective of the contract must be clearly stated and must be lawful.
- There must be good faith or genuine intention on the part of the parties involved to fulfill the contents.
- The parties to the agreement must be capable of performing the designated tasks.

Types of Construction Contracts

There are many types of construction contracts. However, the following contracts are most commonly used in construction.

- Unit price contract
- Lump sum contract (also called the single fixed-price contract)
- Cost plus fixed fee contract
- Guaranteed maximum price contract

Unit Price Contract: Under the unit price contract, a contractor agrees to perform specific portions of work for predetermined unit prices. These unit prices usually include the contractor's direct and indirect costs, as well as the mark-up for job overhead, home office overhead, and profit. Since this type of contract is best used in situations where the quantities involved may vary or the scope may change, the total contract price is not firm, but depends on the actual quantities of units of work measured when the job is complete.

Unit price contracts may be used for heavy construction projects such as roads and highways. Contractors in these trades may be paid for the total number of cubic yards of base or tons of asphalt actually placed.

In this contractual arrangement, the contractor bids or negotiates unit prices on the project based on approximate quantities furnished by the owner.

Lump Sum Contract: Fixed lump sum contracts are used on most traditional, competitively bid projects. This contractual arrangement binds the contractor to perform all the work in accordance with the plans, specifications, terms, and general conditions of a contract for a fixed lump sum price, based on the materials and equipment described in the contract documents.

With this type of contract, most general contractors perform some portions of construction (such as masonry and concrete) with their own work forces. The remainder of the work is subcontracted to specialty contractors—painters, electricians, plumbers, air-conditioning mechanics, steel erectors, or roofers. The general contractor assumes the responsibility for all work done by his own forces and the subcontractors.

Cost Plus Fixed Fee Contract: This agreement is used primarily for negotiated contracts. With this type of contract, a contractor agrees to perform the work and the owner agrees to pay the contractor for the direct field costs, plus a fee to cover the contractor's home office costs and profit. The fee is usually based on the size and/or complexity of the work and may be expressed as a flat dollar amount or as a percentage of the total estimated cost of the project. One disadvantage with this type of contract, to an owner, is that the total cost of the project is unknown until the project is completed. Another problem is that the contractor does not have the incentive to minimize field costs. The advantages are that the owner can change design or materials during construction without complicated change orders or can start the project before the design is complete.

Guaranteed Maximum Price Contract: With this type of contract, the contractor agrees to perform all the work necessary to complete a construction project for a price that will not exceed a **pre-established maximum price**. Any costs above the guaranteed price are absorbed by the contractor. However, if the contractor completes the work for a price under the guaranteed maximum price, the savings is passed on to the owner unless there is an agreement (in the contract) to share any savings with the contractor. Contracts of this type may have a specific cost savings clause to motivate the contractor to perform the contract in the most economical manner, without sacrificing quality. Guaranteed maximum price contracts are used primarily for negotiated contracts.

Common Methods of Contracting

In the construction industry, it is not unusual to find many variations and/or combinations of the four main types of contracts being used. However, there are several established methods in which these four types are used. The most common methods are described in the following paragraphs.

The Traditional Method: Under this method, the owner selects an architect/engineer to design a project that can be built within a predetermined budget. When the design is completed and approved by the owner, one general contractor is selected to build the project. The contractor is selected through a competitive bid or negotiated process and reports directly to the owner or his representative.

The general contractor performs part of the work (usually between 15 and 40 percent) with his own work force and subcontracts the remainder to specialty contractors. The percentage performed by the general contractor could be more or less than the percentages quoted above. On most public contracts, a general contractor is required to perform at least 25 percent of the work with his own forces. For private or other contracts where no percentage is specified, general contractors perform only those portions of the work that do not require a specialist. The rest of the work is then subcontracted, since most subcontractors are specialists in their field and can buy materials and perform the work at a more productive and economical rate.

This method is used with any of the four types of contracts described in the previous section.

One of the main advantages of the traditional method is that the owner deals with only one contractor (the general contractor who coordinates and schedules all the work of his subcontractors). One of the main disadvantages of this method is that an adversarial situation could develop, with the general contractor on one side, the architect/engineer on the other, and the owner in the middle.

The Construction Management Method: This method, in its purest form, is run by a three-party team consisting of the owner, architect/engineer, and a construction manager. The owner contracts directly with several prime contractors (subcontractors who normally contract with the general contractor) instead of one general contractor. Unlike methods involving a general contractor, the owner assumes responsibility for work done by the subcontractors. The organization of the construction management method of contracting is illustrated in Figure 3.67. The major differences between the construction management method and the traditional method are presented in Figure 3.68.

The Turn-key Method: As the name turn-key implies, the owner accepts a project and pays for it when he can insert the key into the lock of the main entrance and take possession of the completed building.

In its purest form, the owner contracts with one organization, to whom he presents his building needs, an architectural program, a time frame for completion, and a budget. The turn-key organization then accepts overall responsibility for locating a site, completing the design, financing, and constructing a project. Often, this is accomplished by assembling a team of specialists that may include architects, engineers, bankers, contractors, and subcontractors to complete the project within a pre-established budget.

Design/Build: The design/build contract is similar to the turn-key contract in that the owner has only one contract for the design and construction of a project. The owner may provide the site, a schedule, and a budget. The sum and dates of progress payments are agreed upon by the contractor and the owner in advance.

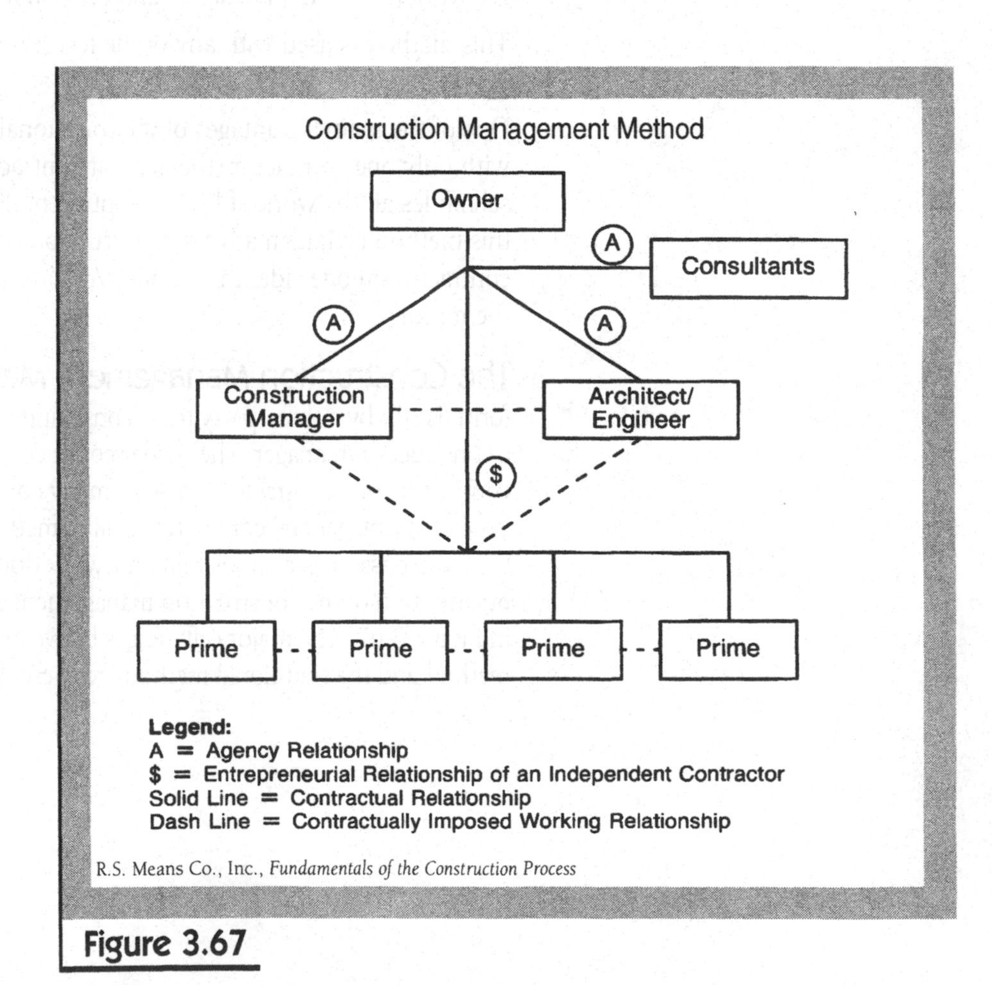

R.S. Means Co., Inc., *Fundamentals of the Construction Process*

Figure 3.67

Comparision of the Construction Management and Traditional Methods

Construction Management Method	Traditional Method
1. Several prime contractors contract directly with owner.	1. One general contractor contracts with owner.
2. Design and construction are handled as one single effort and has continuity.	2. Design and construction are two seperate efforts—no continuity.
3. Cohesive team effort between architect/engineer and construction manager.	3. Adversarial relationship between architect/engineer and general contractor.
4. Reduces layering of bonding.	4. Layering of bonding occurs with the general contractor furnishing bonds to the owner and subcontractors furnishing bonds to the general contractor.
5. Value analysis and cost control during design.	5. Usually less value analysis during the design phase.
6. No guaranteed costs at the onset of the project (unless it is a guaranteed maximum price contract).	6. Usually a guaranteed lump sum bid at the beginning of the project.
7. Integrated system but permits phased construction.	7. Phased construction coordinated by general contractor.
8. Construction manager has incentive to reduce costs to owner through value engineering.	8. General contractor has lump sum contract with owner.
9. Owner retains more control of the schedule and can plan cash flow to his or her advantage.	9. General contractor controls schedule.
10. Involvement of the construction manager during the planning and design phases provides the owner with source of independent information about probable costs and schedules.	10. Owner depends on the architect and general contractor for design, costs, and schedules.

R.S. Means Co., Inc., *Fundamentals of the Construction Process*

Figure 3.68

The Contract Package

The contract package for most construction projects consists of the following three elements.

- **The Contract**– owner-contractor agreement
- **The Specifications**– including the general conditions as well as addenda and any special provisions of the contract, where applicable
- **The Contract Drawings**– plans, with any revisions issued prior to the bid or negotiation date

An overview of these components is shown in Figure 3.69.

The Construction Contract: Every construction project involves a high cash flow and a high degree of risk. It is, therefore, important that written, legally binding, agreements are executed between all parties involved in the design and construction process. The main objectives of a construction contract are to define the practical and legal responsibilities and commitment of the parties involved; and to ensure that common-law risks, liabilities, and leverage are ascribed to the responsible party.

Construction contracts, like other contracts, are written in cumbersome legal language that reflects years of previous lawsuits and case law (judicial precedent). As such, the contents of a contract are often unclear to nonlegal people. An experienced construction attorney should be consulted during the development, review, and/or interpretation of a legal agreement.

A construction contract should, at least, include the following information.

- The date of the agreement
- The parties involved and their addresses
- A description or name of the project and its location
- The name of the architect/engineer, including an address
- A list of contract documents to be attached (drawings, specifications, addenda, etc.)
- A description of the scope of work to be performed
- The required starting date and completion date (including the number of calendar days)
- The amount of the contract
- Method and frequency of payment
- The rights and duties of all parties
- How changes in the work will be handled
- Provisions for termination by either party
- Other provisions as needed

While some firms often develop their own agreements, there are several standard forms available, developed by the American Institute of Architects (AIA) and the Associated General Contractors of America, Inc. (AGC).

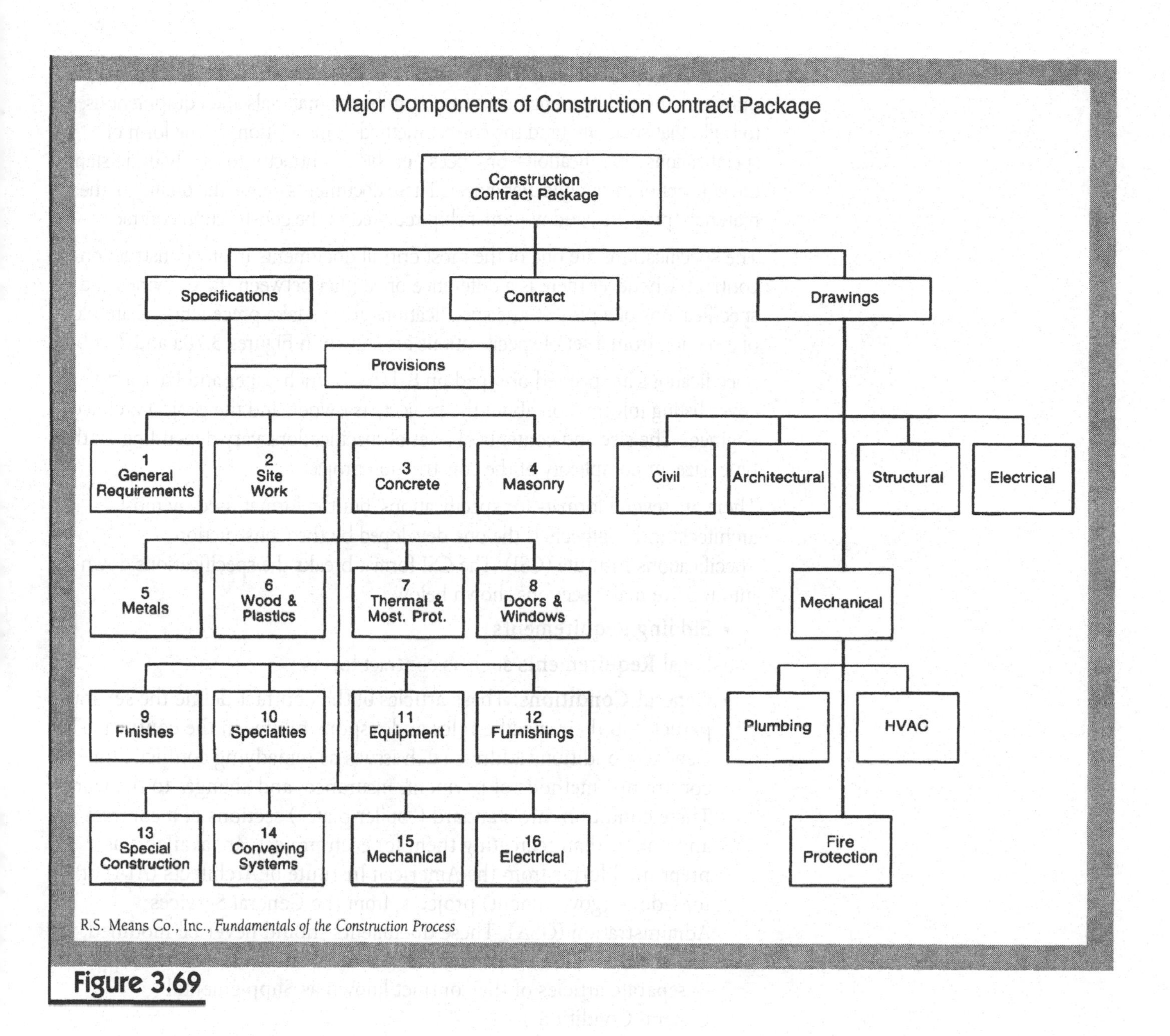
Major Components of Construction Contract Package
Construction Contract Package
Specifications
Contract
Drawings
Provisions
1 General Requirements
2 Site Work
3 Concrete
4 Masonry
5 Metals
6 Wood & Plastics
7 Thermal & Most. Prot.
8 Doors & Windows
9 Finishes
10 Specialties
11 Equipment
12 Furnishings
13 Special Construction
14 Conveying Systems
15 Mechanical
16 Electrical
Civil
Architectural
Structural
Electrical
Mechanical
Plumbing
HVAC
Fire Protection
R.S. Means Co., Inc., *Fundamentals of the Construction Process*

Figure 3.69

Contract Specifications: In order to build a structure from a set of drawings, the architect/engineer must describe the materials and equipment used to build that structure (and the correct method of installation) in the form of specifications. Specifications, or "specs," enable a contractor to clearly understand the concept of the architect/engineer. These documents define the quality of the materials, products, and workmanship required in the construction contract.

The specifications are one of the most critical documents in any construction contract; whenever there is a difference or conflict between the drawings and specifications of a project, the specifications always take precedence. Examples of a section from a set of specifications are shown in Figures 3.70a and 3.70b.

Specifications are printed or typed on 8-1/2 x 11 inch paper and have a stiff cover listing information about the project, its owner, and the project architect/engineer. The size and contents of a set of specifications vary depending on the type, size, or complexity of the construction project.

There are several formats for specifications, but the format used by most architects and engineers is the one developed by the Construction Specifications Institute (CSI). The CSI format breaks the specifications down into the five major sections shown below.

- **Bidding Requirements**
- **Legal Requirements** such as contract forms.
- **General Conditions:** These articles of the contract define the separate parties and their specific roles and responsibilities in the contract. The General Conditions address such issues as remedying conflicts, time constraints, methods of payment, insurance, and changes to the work. These conditions are standard ("boiler plate") sections of a contract and, rather than recreating them for each project, are available as preprinted forms from the American Institute of Architects (AIA) or, for federal (government) projects, from the General Services Administration (GSA). These documents should never be modified, and any differences from standard General Conditions should appear in separate articles of the contract known as Supplemental General Conditions.
- **Special Conditions:** These articles are unique conditions for the individual project and contain sections on shift work (premium time), site security, noise or dust control, and hazardous material handling.
- **Technical Specifications:** In this section, the architect/engineer provides the contractor with a comprehensive description of the required materials and equipment and their fabrication, quality, workmanship, and/or installation details. However, the architect/

Sample Specifications

SECTION 03300
CAST-IN-PLACE CONCRETE

PART 1 - GENERAL

1.01 REQUIREMENTS INCLUDED

Poured-in place concrete, foundations and other concrete items specified in other Sections.

1.02 RELATED REQUIREMENTS

A. Section 02800 - Site Improvements.

B. Section 03100 - Concrete Formwork.

C. Section 03200 - Concrete Reinforcement.

D. Section 03320 - Concrete Topping: Concrete topping and curbs over existing construction.

E. Section 03346 - Finishing Concrete Surfaces.

1.03 QUALITY ASSURANCE

A. Reference Standards: Comply with all applicable Federal, State and local codes, safety regulations, Portland Cement Assoc. Standards, Ready Mixed Concrete Assoc. Standards, Texas Aggregates Assoc. Standards and others referred to herein.

B. Tests and Submittals in accordance with Section 01410.

1. Mix Design and Tests: The mix design for all concrete be established by a testing laboratory under provisions of Section 01410. All tests shall be performed in accordance with standard procedures as follows:

   a. ASTM C 172 Standard Method of Sampling fresh concrete

   b. ASTM C 31 Standard Method of Making and Curing Concrete compressive and Flexural Strength. Test Specimens in the field.

   c. ASTM C 143 Standard Method of test for Slump of Portland Cement Concrete.

   d. ASTM C 39 Standard Method of test for Compressive Strength of Molded Concrete Cylinders.

2. Access: The Architect shall have access to all places where materials are stored, proportioned or mixed.

3. Proportions: The testing laboratory shall submit, prior to the start of concrete work, contemplated proportions and the results of preliminary 7 day compression test. Submit a separate set of proportions and test results for pumpcrete if used.

4. Slump test shall be made by the testing laboratory of concrete delivered to the site for each set of test cylinders.

5. Standard test cylinders of all concrete placed in the work shall be made by the testing laboratory. One (1) set of four (4) cylinders shall be taken for each 100 cubic yards or fraction there of poured on each day.

6. Two (2) cylinders of each set shall be tested at 7 days and two (2) cylinders to be tested at 28 days

7. Reports of above tests and field quality control tests: Provide copies of test reports:

   1 copy to Engineer

   2 copies to Architect

   2 copies to Contractor

8. Mill reports: The Contractor shall furnish mill reports of test of cement showing compliance with specifications.

9. All expenses for concrete design and testing shall be paid by the General Contractor

R.S. Means Co., Inc., *Plans, Specs & Contracts for Building Professionals*

Figure 3.70a

Sample Specifications (cont.)

C. Inspection: Inspection of Reinforcing Steel and Concrete Placement: Before any concrete is poured on any particular portion of project, reinforcing steel will be checked and approved by Architect or Engineer. Correct any errors or discrepancies before concrete is placed. Such checking and approval shall not relieve Contractor from his responsibility to comply with the Contract requirements.

1.04 REFERENCE STANDARDS

A. ASTM C33 - Concrete Aggregates.

B. ASTM C150 - Portland Cement

C. ACI 318 - Building Code Requirements for Reinforced Concrete

D. ASTM C494 - Chemical Admixtures for Concrete.

E. ASTM C94 - Ready-Mixed Concrete.

F. ACI 304 - Recommended Practice for Measuring, Mixing, Transporting and Placing Concrete.

G. ACI 305 - Recommended Practice for Hot Weather Concreting.

H. ACI 306 - Recommended Practice for Cold weather Concreting

I. ACI 301 - Specifications for Structural Concrete for Buildings

J. ACI 311 - Recommended Practice for Concrete Inspection.

1.05 SUBMITTALS

A. Submit product data in accordance with Section 01300.

B. Provide product data for specified products.

C. Submit manufactures' instructions in accordance with Section 01400.

D. Provide shop Drawings showing construction joints.

E. Provide schedule of pouring operations for approval before concreting operations begin.

F. Conform to Mix Design in accordance with 1.03-B.

1.06 PRODUCT DELIVERY, STORAGE AND HANDLING

Store materials delivered to the job and protect from foreign matter and exposure to any element which would reduce the properties of the material.

1.07 COORDINATION

A. Obtain information and instructions from other trades and suppliers in ample time to schedule and coordinate the installation of items furnished by them to be embedded in concrete so provisions for their work can be made without delaying the project.

B. Do any cutting and patching made necessary by failure or delay in complying with these requirements at no cost to Owner.

PART 2 - PRODUCTS

2.01 MATERIALS - STANDARD STRUCTURAL CONCRETE

A. Portland Cement: Type I and III shall conform to "Standard Specifications for Portland Cement" (ASTM C - 150) and shall be of domestic manufacture. Use only one brand of cement unless otherwise authorized by Architect.

B. Fine Aggregate: ASTM C33, natural bank sand or river sand, washed and screened so as to produce a minimum percentage of voids.

C. Normal Weight Coarse Aggregate: ASTM C33, gravel or crushed stone suitably processed, washed and screened, and shall consist of hard, durable particles without adherent coatings. Aggregate shall range from 1/4" to 1-1/4", well graded between the size limits.

Section 03300 Page 2

R.S. Means Co., Inc., *Plans, Specs & Contracts for Building Professionals*

Figure 3.70b

engineer assumes no responsibility for the adequacy of any procedure specified. The contractor is responsible for installing the final product by using his own skill, resources, experience, and ingenuity. This section is further divided into the sixteen divisions shown in Figure 3.69

Contract Drawings: It is impossible for an architect/engineer to communicate every thought, concept, and intention in the specifications. To supplement the specifications, architect/engineers prepare contract drawings, or **working drawings**. The number of sheets in a set of drawings depends on such factors as building size, complexity, and level of detail. Most working drawings include: site, architectural, structural, plumbing, mechanical, and electrical drawings, details, and schedules. These drawings are explained in the following paragraphs.

Site Drawings: These drawings provide information such as the geographic location of the project, a plot plan showing the general orientation of the project, contours (existing and proposed elevations), roads, parking lots, and the location of both existing and new site utilities. An example of a civil drawing is shown in Figure 3.71.

Architectural Drawings: All architectural information about a project is recorded here. Architectural drawings usually begin with floor plans starting from the lowest floor and progressing to the highest floor, followed by elevations, sections, schedules, and details. Examples of architectural drawings are provided in Figures 3.72a and 3.72b.

Structural Drawings: These are drawings of all structural supporting systems such as foundations, columns, floor systems, roof supporting systems, beams, and joists, showing their sizes, connection details, and location. Information on the type and strength of concrete and reinforcing may also be found here.

Plumbing Drawings: Plumbing drawings show water distribution piping and types of fixtures (sinks, toilet bowls, urinals, water fountains, etc.). They also show waste and vent systems, internal storm drainage, special piping, and fire protection standpipes.

Mechanical Drawings: These drawings provide information about the heating, ventilating, and air conditioning system such as ductwork, heating, and cooling equipment, including size, capacity, and arrangement.

Example of Civil Drawing

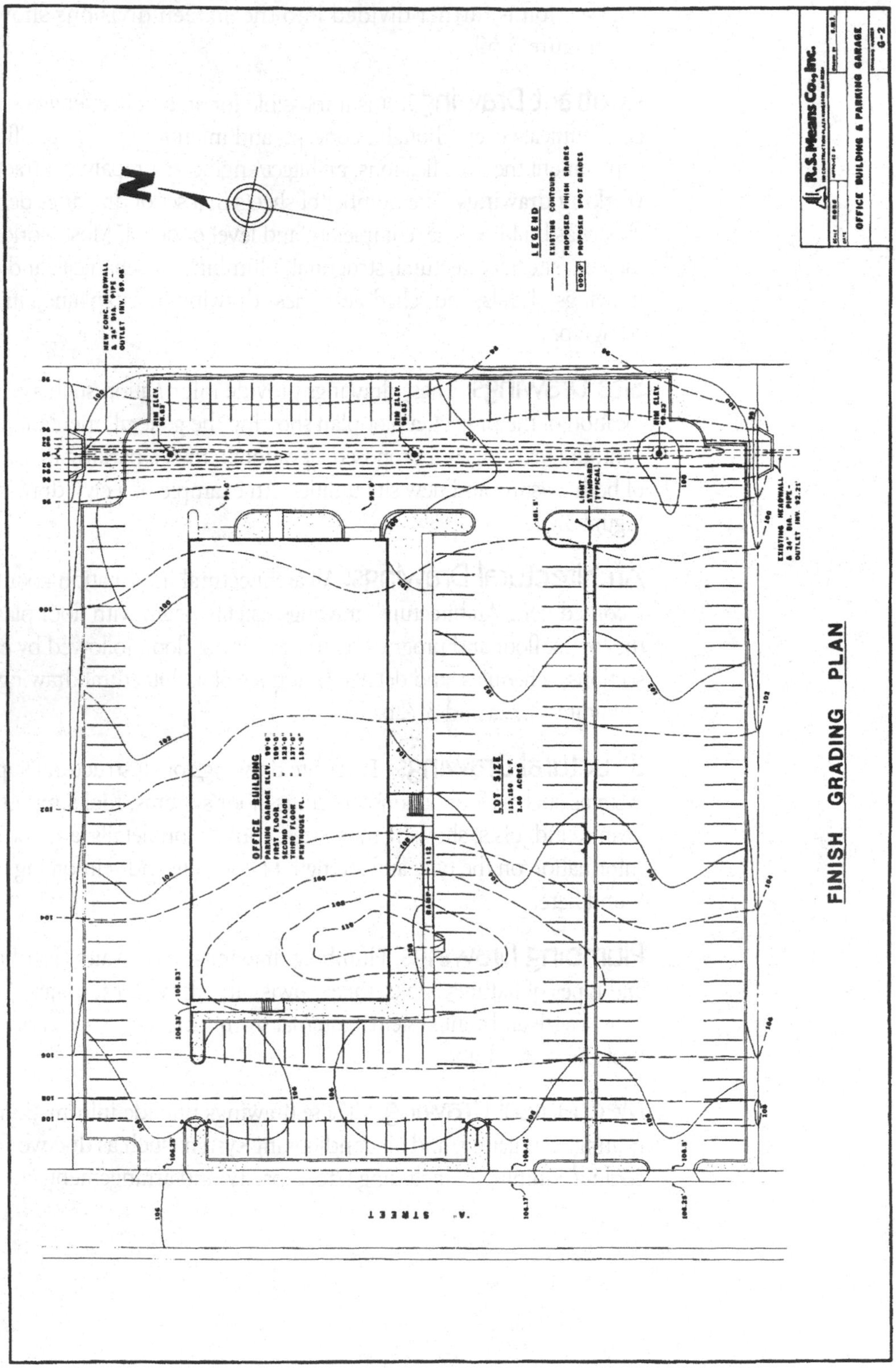

FINISH GRADING PLAN

R.S. Means Co., Inc., *Fundamentals of the Construction Process*

Figure 3.71

Figure 3.72a

Example of Architectural Drawing

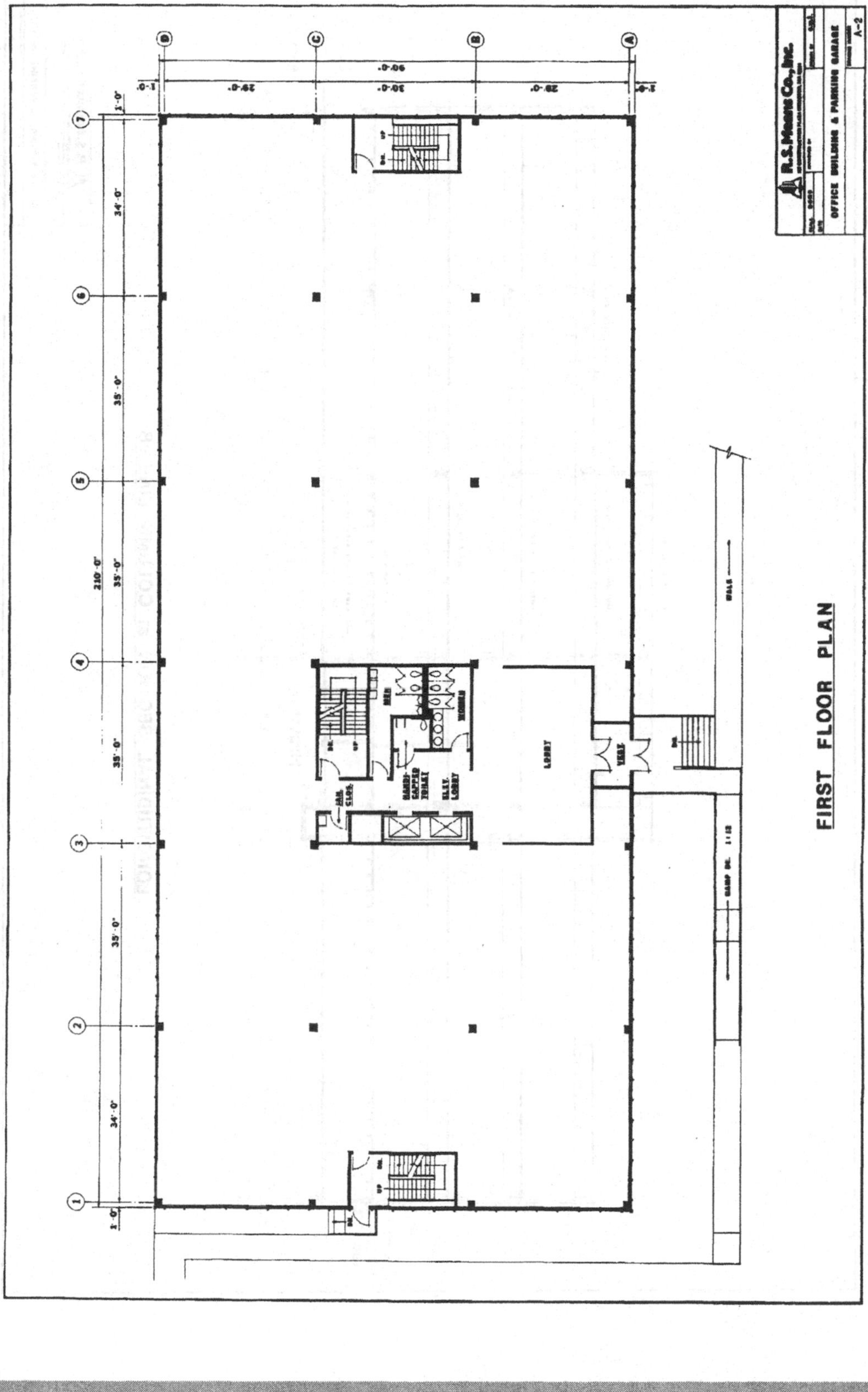

FIRST FLOOR PLAN

R.S. Means Co., Inc., *Fundamentals of the Construction Process*

Figure 3.72b

Example of Architectural Drawing (cont.)

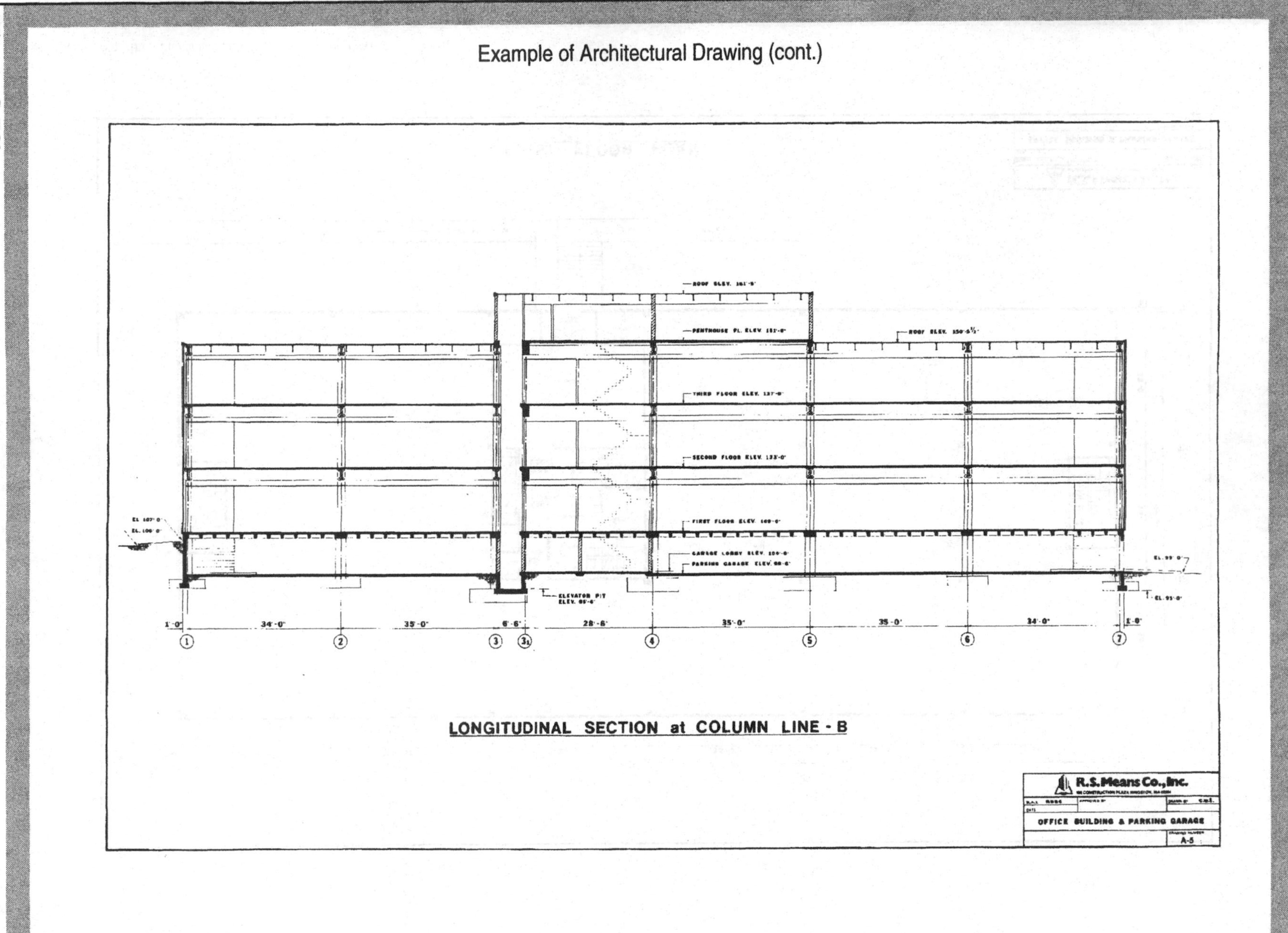

R.S. Means Co., Inc., *Fundamentals of the Construction Process*

Electrical Drawings: All information about electrical and electronic requirements of a building for fixtures, lighting, security, and other equipment is recorded here. The details of conduits and wiring are often prepared by the electrical subcontractor and recorded on the electrical shop drawings.

Details: These drawings provide expanded and additional information on items previously shown on the site, architectural, or structural drawings. In some cases, items that were previously drawn too small to read for lack of space, and/or to avoid overcrowding one sheet, are enlarged and illustrated with more detail here.

Schedules: Tables or schedules are usually provided to save time and avoid duplication of information throughout a set of drawings. Some examples of schedules commonly found in a set of working drawings include the following.

- **Door and window schedule:** This is a listing of all the doors and windows on a project by type, size, material, location, etc.
- **Finishes schedule:** This is a listing of all the interior finishes (floors, carpet, and tile; painting, wallpaper, and gypsum wallboard; ceiling tile, etc.) on the project with additional information such as location, color, and type.
- **Structural schedule:** This is a list of footings, columns, beams, and joists by size, location, material, and some form of identification.
- **Mechanical schedule:** This is a list of ducts (by size), fans, and pumps (by type, size, and capacity).
- **Electrical schedule:** This is a list of lighting fixtures, motors, generators, and transformers by size, type, capacity, and other relevant characteristics. Examples of two types of schedules (door schedules and room finish schedules) are illustrated in Figures 3.73 and 3.74.

The contract documents are the backbone of a building construction project. The various types of contracts—unit price, lump sum, cost plus fixed fee, and the guaranteed maximum price—all demand different payment arrangements and divisions of responsibilities. These types of contracts are used judiciously for the various contracting methods—the traditional, construction management, design/build, and turn-key methods. The contents of these documents are critical to the outcome of the project. The plans, specifications, drawings, and schedule must be clear and comprehensive to ensure that a project is built successfully to meet the owner's intents and needs.

Door and Frame Schedule

Means Forms

DOOR AND FRAME SCHEDULE

PROJECT

LOCATION

ARCHITECT

OWNER

PAGE OF

DATE

BY

DOOR NO.	SIZE			DOOR				FRAME					FIRE RATING		HARDWARE		REMARKS
	W	H	T	MAT.	TYPE	GLASS	LOUVER	MAT.	TYPE	DETAILS JAMB	HEAD	SILL	LAB	CON	SET NO.	KEYSIDE ROOM NO.	

R.S. Means Co., Inc., *Fundamentals of the Construction Process*

Figure 3.73

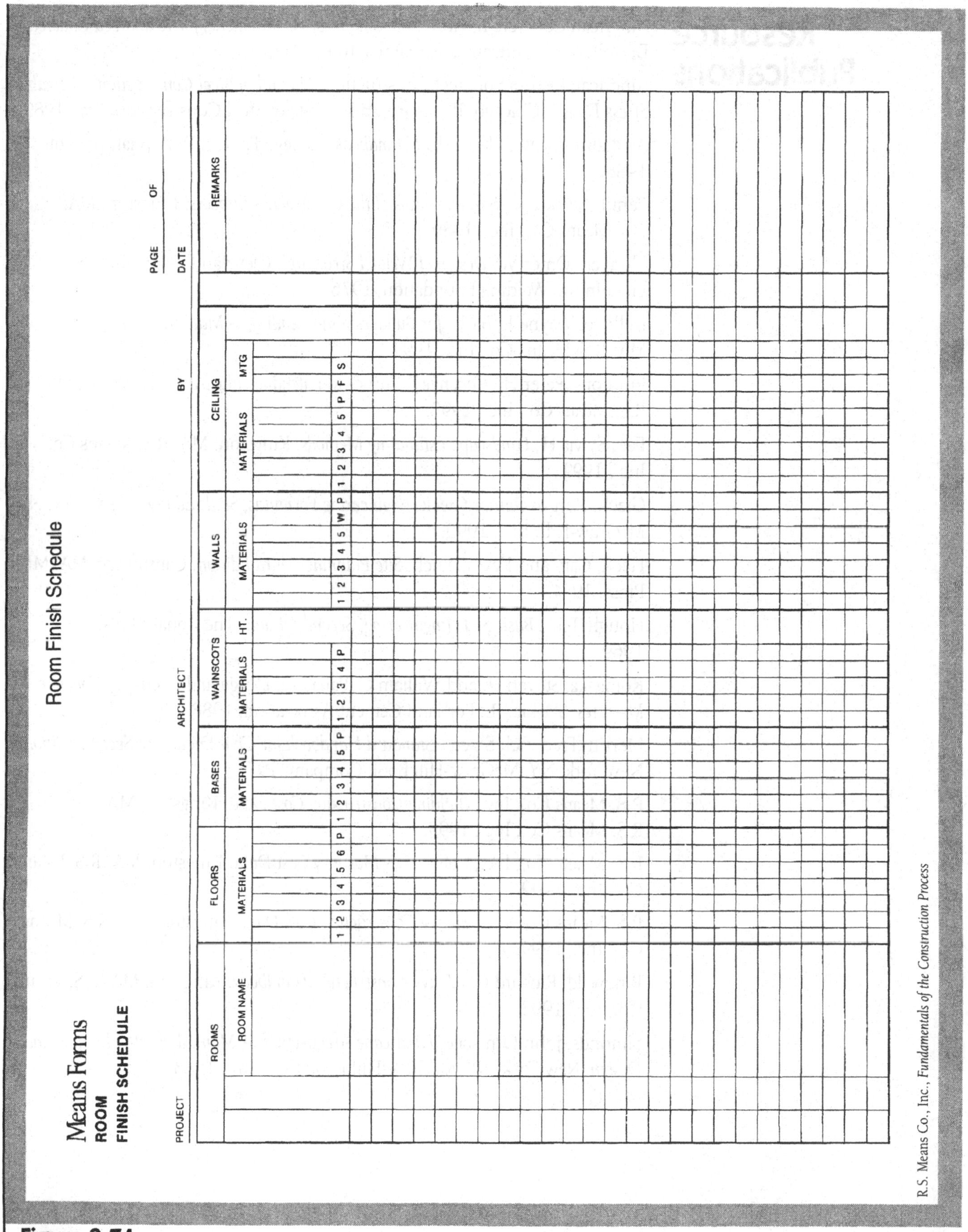

Figure 3.74

Room Finish Schedule

Means Forms
ROOM
FINISH SCHEDULE

PAGE ____ OF ____

PROJECT ____ ARCHITECT ____ BY ____ DATE ____

ROOMS			FLOORS							BASES						WAINSCOTS						WALLS							CEILING									REMARKS
		ROOM NAME	MATERIALS							MATERIALS						MATERIALS					HT.	MATERIALS							MATERIALS						MTG			
			1	2	3	4	5	6	P	1	2	3	4	5	P	1	2	3	4	P		1	2	3	4	5	W	P	1	2	3	4	5	P	F	S		

R.S. Means Co., Inc., *Fundamentals of the Construction Process*

Resource Publications

American Concrete Institute, *Building Code Requirements for Reinforced Concrete*. Detroit, MI: American Concrete Institute, 1984.

American Institute of Steel Construction, *Manual of Steel Construction: Allowable Stress Design*. Chicago, IL: American Institute of Steel Construction, Inc., 1989.

Asphalt Institute, *The Asphalt Handbook*. College Park, MD: Asphalt Institute, 1988.

Bentil, Kweku K., *Fundamentals of the Construction Process*. Kingston, MA: R.S. Means Co., Inc., 1989.

Blodgett, Omer W., *Design of Welded Structures*. Cleveland, OH: James F. Lincoln Arc Welding Foundation, 1976.

DelPico, Wayne J., *Basics for Builders: Plan Reading & Material Takeoff*. Kingston, MA: R.S. Means Co., Inc., 1994.

Emmons, Peter H., *Concrete Repair & Maintenance Illustrated*. Kingston, MA: R.S. Means Co., Inc., 1998.

Fee, Sylvia H, *Landscape Estimating Methods*. Kingston, MA: R.S. Means Co., Inc., 1999.

Gieck, Kurt & Reiner Gieck, *Engineering Formulas, Sixth Edition*. New York, NY: McGraw-Hill, Inc., 1990.

Hack, Gary and Kevin Lynch, *Site Planning, Third Edition*. Cambridge, MA: MIT Press, 1984.

Hough, B.K., *Basic Soils Engineering, Second Edition*, The Ronald Press Co., 1969.

Komaska, Steven H. and William C. Panarese, *Design and Control of Concrete Mixtures*. Skokie, IL: Portland Cement Association, 1988.

Merritt, Frederick S., ed., *Standard Handbook for Civil Engineers, Second Edition*. New York, NY: McGraw-Hill Book Company, 1976.

R.S. Means Co., Inc., *Building Construction Cost Data*. Kingston, MA: R.S. Means Co., Inc., 1999.

R.S. Means Co., Inc., *Concrete & Masonry Cost Data*. Kingston, MA: R.S. Means Co., Inc., 1999.

R.S. Means Co., Inc., *Heavy Construction Cost Data*. Kingston, MA: R.S. Means Co., Inc., 1999.

Ringwald, Richard C., *Heavy Construction Handbook*. Kingston, MA: R.S. Means Co., Inc., 1993.

Simonds, John Ormsbee, *Landscape Architecture: A Manual of Site Planning and Design*. New York, NY: McGraw-Hill Book Company, 1983.

For Additional Information

Steel

American Iron and Steel Institute
1101 17th Street, NW
Suite 1300
Washington, DC 20036
800-79-STEEL
www.steel.org

Concrete

American Concrete Institute (ACI)
38800 Country Club Drive
Farmington Hills, MI 48331
248-848-3700
www.aci-int.org

ACI offers many publications on the details of correct concrete design and construction. Among these publications are *Building Code Requirements for Reinforced Concrete, Building Movements & Joints,* and *Joints in Concrete Construction.*

Concrete Reinforcing Steel Institute
933 North Plum Grove Road
Schaumburg, IL 60195
847-517-1200
www.crsi.org

Publications include *Manual of Steel Construction* and *Manual of Standard Practice.*

National Concrete Masonry Association
2302 Horse Pen Road
Hemdon, VA 20171-3499
707-713-1910
www.ncma.org

NCMA offers publications that provide information on building code requirements for masonry structures; software; and educational programs for contractors, designers, and home owners.

Portland Cement Association
5420 Old Orchard Road
Skokie, IL 60077
847-966-6200
www.portcement.org

The Portland Cement Association publishes many books on mix design, placement, and testing.

Pavement

Asphalt Institute
P.O. Box 14055
Lexington, KY 40512-4053
606-288-4960
www.asphaltinstitute.org

Association of Asphalt Paving Technologists
1983 Sloan Place
St. Paul, MN 55117
612-293-9188

National Asphalt Pavement Association
5100 Forbes Blvd.
Lanham, MD 20706-4413
301-731-4748
www.hotmix.org

Chapter 4
Maintenance

Chapter 4 Maintenance

By Terry Wireman

In the past 20 years, management has become more production-oriented and at times has sacrificed long-term benefits for short-term profitability. Foreign competitors have taken advantage of this trend to develop strategic plans and build strong, complete organizations. One of the foremost areas of development for them is the maintenance function. Maintenance is extremely important to being competitive in the world market. If maintenance is to become a key factor in the survival and success of most U.S. corporations, management must change its view.

To survive in today's economy, industry must compete not only domestically, but internationally. All forms of production analysis, product reviews, and material reviews must be conducted and periodically checked. Statistical process control is only one of the new methods used to reduce operational costs.

Many industries are realizing that the maintenance function, while often viewed as a necessary evil, is also the last major area of cost reduction in both the public and private sectors. Facilities engineers and managers must understand the cost savings available and the impact that maintenance has on the company's ability to produce its product or provide its services.

Maintenance Goals & Objectives

A successful maintenance operation must be properly organized and structured. Determining the correct type of maintenance organization requires a clear understanding of its goals and objectives. If these are progressive and the maintenance organization is recognized as a contributor to the corporate bottom line, variations on some of the more conventional organizational structures can be used. **The typical goals and objectives for a maintenance organization include:**

1. **Provide maximum production or availability of facilities at the lowest cost and at the highest quality and safety standards.** This objective can be broken down into some smaller components including:

• *Maintaining existing equipment and facilities*

This is the primary reason for the existence of the maintenance organization. If a corporation's equipment or facilities are not operating, there is no advantage in having them. Thus, this first sub-objective is the keep-it-running charter of maintenance.

• *Equipment and facilities inspections and services*

These are generally referred to as *preventive/predictive maintenance programs*. This activity is designed to increase the availability of the equipment/facilities by reducing the number of unexpected breakdowns or service interruptions. It is one of the most critical parts of any program, designed to achieve the first main goal.

• *Equipment installations or alterations*

This is not the charge of maintenance in all organizations since it is performed by outside contract personnel. However, since maintenance must still maintain the equipment, they should be involved in any alterations or new equipment installations.

In viewing these three sub-goals, the maintenance organization will always attempt to maximize the company's resources, keeping the overall costs as low as possible, while ensuring the safety of personnel and the quality of the product/facilities.

2. Identify and implement cost reductions.

There are many ways to accomplish this goal. It is common to find adjustments made in:

- training
- repair procedures
- work planning
- tools

The labor or materials required to perform a specific job can often be reduced. In addition, any reduction in the time it takes to make repairs translates into reduced downtime (increased availability), a more significant cost factor than maintenance expenditures. When there are adjustments made to reduce costs, studies need to be conducted to show the before and after results. Quantifying the improvements is critical to maintaining management support for maintenance activities.

3. Provide accurate equipment maintenance records.

Maintenance records are generally collected as work orders and then must be compiled into reports showing meaningful information or trends. The problem is finding enough time to record valuable information on each individual work order.

Since most work is done in a reactionary mode, it is difficult to record events after the fact. For example, recording how many times a circuit breaker for a drive motor was reset in one week might seem somewhat insignificant to record on a work order. However, if the overload was due to an increased load on the motor by a bearing wearing inside the drive, it is worth noting that checking and discovering such wear beforehand could have prevented a catastrophic failure. Accurate record keeping is important if maintenance is really going to maintain equipment.

4. Collect Necessary Maintenance Cost Information.

This objective is related to the previous one. Cost information must, however, be more detailed than repair information. It is divided into these general categories:

- Labor costs
- Material costs
- Tool & equipment costs
- Contractor costs
- Lost production costs
- Miscellaneous costs

The importance of collecting this information is clear in maintenance budgeting. If accurate cost histories are not collected, how can the manager budget what next year's expenses will be, or should be? It is difficult to express to plant management, "We want to reduce maintenance manpower by 10% for next year," when you don't really know how the manpower resource was allocated for this year. Also if the labor costs are expressed only in dollar amounts, the differences in pay scales may make it difficult to determine how much manpower was used in total hours by craft. The information must be collected in dollars *and* in hours by craft. Where is the information collected? Collecting the cost information is again tied to work order control. Knowing the hours spent on the work order, times the labor rates of the individuals performing the work allows a calculation of the labor used for the work order. Adding up these charges over a given time period for all work orders gives you the total labor used. Adding up the work hours of each craft is another way to get the picture of necessary manpower resources, based on recent performance. Material costs are determined by tracking what parts were used on the job to each work order. The number of parts times their dollar value (obtained from stores or purchasing) allows the calculation of the total material dollars spent for a given time period.

Contractor and other cost information also must be collected at a work order level. Each work order form should have the necessary blanks for filling in this information. Only by tracking the information at the work order level can you roll up costs from equipment, to line, to department, to area, to total plant. Collecting the information at this level also identifies costs for equipment types, maintenance crafts, cost centers and so on. The data gathered through the work order can be used to perform detailed maintenance analysis. See "Standard Cost Justification" later in this chapter for more on collecting and analyzing maintenance cost information.

5. Optimize Maintenance Resources.

Making the most of the resources at hand is crucial. Few maintenance organizations have as many people, materials, or tools as they could use. Resources in short supply must be used carefully. For example, with good planning and scheduling practices, a proactive maintenance organization may double the productivity of their employees. While this may seem to be a bold statement, it has been accomplished countless times.

Maintenance supplies represent another area of savings. When good controls are put into place, organizations have seen as much as 15% reduction in inventory storage costs, while increasing the service level of the stores. These types of reductions, while improving service, are essential to optimizing the present resources.

Optimizing maintenance resources as highlighted in this section can only be achieved by good planning and scheduling practices.

6. Optimize Capital Equipment Life.
Any equipment, whether a complex industrial robot or your automobile, requires maintenance if it is to deliver its desired service life. The life cycle is optimized by an effective preventive maintenance program. Preventive Maintenance (PM) is a neglected discipline in many organizations throughout North America. Management's interest in short term gains has contributed to the neglect. Nevertheless, if capital equipment life is to be optimized, only effective preventive/predictive maintenance programs will achieve the goal.

7. Minimize Energy Usage.
While this objective may seem to be more production or operations-oriented, it is maintenance-related. How? Equipment and facilities that are properly maintained will require less energy to operate. For example, equipment with a poor maintenance schedule will have bearings without proper lubrication or adjustment, couplings not properly aligned, or gears misaligned, all of which contribute to poor performance and require more energy to operate. The key to achieving this objective is also a good preventive/predictive maintenance schedule.

Establishing the Services Maintenance Will Provide

Achieving the goals necessary to have a contributing maintenance organization will require some decisions concerning the maintenance organization and the type of maintenance service to be provided. Following are the types of decisions, beginning with the type of service required from the equipment.

Equipment Service Level

Equipment service level indicates the amount of time the equipment is available for its intended service. The amount of service required from the equipment and its resultant costs determine the type of maintenance philosophy a company will adopt. Six common philosophies are listed below.

1. *Breakdown maintenance* is just what it implies; the equipment is run until it breaks down. There is no preventive maintenance; the technicians work on only equipment that is malfunctioning. This is the most expensive way to perform maintenance. Equipment service is generally below acceptable levels, with product quality usually affected.

2. *Minor lube programs* are one step removed from breakdown maintenance programs. The equipment still is not overhauled until it breaks down, except that with the lube program, it takes longer for the equipment to break down. Unfortunately, many companies mistake the lube program for a preventive maintenance program. Equipment service level is still not satisfactory under this program.
3. *Preventive maintenance* includes the lubrication program (#2) plus routine inspections and adjustments. This allows many potential problems to be corrected before they occur. At this level of maintenance, equipment service levels will begin to enter the acceptable range for most operations.
4. *Predictive maintenance* is another type of preventive maintenance, utilizing technology that allows the forecasting of failures through monitoring and analysis of the condition of the equipment. The analysis is generally conducted through some form of trending of a parameter such as vibration, temperature, flow, and so forth. Predictive maintenance allows equipment to be repaired at times that do not interfere with production schedules. This removes one of the largest factors from the downtime cost. The equipment service level will be very high under this type of maintenance.
5. *Condition-based maintenance* is performed as needed, with the equipment monitored continually. Some plants will have PLCs (programmable logic controllers) wired directly to a computer to monitor the equipment condition in a real-time mode. Any deviation from the standard normal range of tolerances will cause an alarm (or in some cases a repair order) to be generated automatically. This real-time trending allows maintenance to be performed in the most cost-effective manner. This is the optimum maintenance cost versus equipment service level method available. However, the startup and installation cost can be high. Many companies are moving toward this type of maintenance.
6. *Zero failure maintenance* is used in any environment where the cost of a failure and the resulting production outage is very high. This type of maintenance combines several of the prior techniques to produce a maintenance environment where all critical points on equipment and processes are monitored in a real-time mode, with the data being charted and trended and projections made as to the service life remaining in each item. When the equipment or process becomes questionable, it is taken off-line, repairs are made, and it is returned to service. While this type of maintenance is sophisticated, it is also the most expensive. It is used only in processes where the cost can be justified.

Maintenance Staffing Options

Four methods are commonly utilized to staff the maintenance organization.

1. **Complete in-house staff** is the traditional approach in most U.S. companies. This is where the craft technicians performing the maintenance are direct employees of the company. All administrative functions (salary, benefits, and so forth) are the responsibility of the company.

2. **Combined in-house/contract staff** has become a more common approach to maintenance in the 1980s. In-house staff performs most of the maintenance, but outside contractors perform certain maintenance tasks, such as service on air-conditioners, equipment rebuilds, or insulation. This method can reduce the staff required for specific skill functions. If the contract personnel are not required full-time, this can contribute added savings.

3. **Contract maintenance staffs** use the company's own in-house supervisors, but contract outside employees. This method is common in Japan and is gaining popularity in the U.S. The contractor is responsible for providing the proper skilled individuals, which removes the burden of training and personnel administration from the company. The downside of the situation is not having the same employees all of the time. In this you do lose some familiarity with the equipment, but the interaction between the in-house supervision and the contract personnel can help to compensate.

4. **Complete contracting maintenance staff** includes all craftsmen, planners, and supervisors. The supervision generally reports to a plant engineer or plant manager, eliminating the need for any in-house maintenance personnel. While this program is not yet popular in the U.S., it can, when coupled with an operator-based PM program (explained in the PM section) prove to be cost-effective and a valid alternative to conventional maintenance organizations.

Geographical Maintenance Organization Structures

In addition to the types of maintenance organizations, there are also geographical breakdowns for the organization, regardless of how it is staffed.

Centralized maintenance organizations have all members of the organization reporting to a central location for assignment. All work requests are turned in to the central area for scheduling and dispatch. The utilization of the labor force is high, there is always work for them to do, since they are not assigned to a fixed geographical area. The drawback to this arrangement is that in plants with a large geographic area, response time to trouble calls can be lengthy, contributing to increased equipment downtime.

Area maintenance organizations have small maintenance shops spread throughout the plant. A certain number of employees are assigned to each area, with supervisors assigned to cover one or more areas, depending on the number of employees. This arrangement remedies the slow response time for breakdowns, since the employees are physically close to the equipment. The maintenance employees also develop some equipment ownership, since they are assigned to the equipment. The problem with this arrangement is the labor utilization. If the craft technicians have the equipment running and are up-to-date on the PM program, then they may have no work in their area. However, in another area, the craft technicians are overloaded, with perhaps several equipment items broken down.

The dilemma faced by supervision is whether to pull craft technicians from other areas to cover the breakdowns, hoping the equipment will not break down in the area from which the craft technicians are being drawn. The response time will be poor, and the utilization is not good. If all areas have the same workload, the area concept can work. It requires the effort of a dedicated staff, willing to be flexible if optimum utilization is to be achieved.

Combination organizations are a hybrid, incorporating the best of area and centralized. The concept is to station several small groups of employees near critical equipment, while keeping the main group in a centralized area. This allows for the emergency activities to be handled quickly, while the majority of the employees can be utilized for larger or scheduled repair jobs.

Growing Maintenance Organizations

Step One

Maintenance organizations tend to share the same growth pattern. When companies are small, they may only have one machine. The operator runs and maintains it, performing small repairs or services. If a large breakdown occurs, the operator disassembles the machine and sends the defective parts out for repair.

Step Two

As the company continues to grow, several machines are added. This necessitates adding several production workers. Since the production workers no longer depend on one machine, the first maintenance worker is added. This individual will be multi-skilled, capable of performing the variety of repairs that will be necessary.

Step Three

The third step in the growth pattern is adding more machines and production workers, and consequently, more maintenance workers. With this level of maintenance manpower, it is no longer convenient for the maintenance workers to report to the production supervisor, and a maintenance supervisor is put in place.

Step Four

The fourth step in the growth pattern is the specialization of maintenance personnel in their particular skill areas. It begins by having craft technicians become proficient at repairing a particular piece of equipment, or a particular type of repair. As the number of craft technicians continues to increase, the specialization continues.

Step Five

The fifth step in the growth pattern is the development of craft lines. This may be due to union influence, or the natural progression of step four. The lines can be strict or informal, but will become increasingly distinct.

Step Six

The sixth step in the growth pattern is the organization that becomes too large to manage from a central location. Several factors can contribute to the management problem. The internal geography of the plant is one factor. For example, if the plant covers several hundred acres, it may be physically impossible to manage it from one location, even with the help of radios, bicycles, or manned carts. This is when the organization is divided into the area concept described in the previous section. This allows small maintenance departments, such as paralleled by organizations 1-3. This step leads to interesting internal growth. Most organizations will develop two alternatives at this stage: *further internal growth or outside contracting.* Internal growth will develop central crafts or shops to support the area organizations. Thus, as in step one, when the repair is too large, complicated, or requires special equipment, it is sent to the central shops. As more work is required from the central shops, they tend to grow, while the area organizations tend to add employees only when new demands are made on their area (such as new equipment additions) or when attrition occurs.

Outside contracting occurs when the company either does not have the resources to implement central crafts or decides it is more cost-effective to contract with a local shop for machining, rewinding, or installation. The determining factors here are skill level of the contractor's workforce, response time, and synergism between the contractor and the company.

Step Seven

In the final growth step, the area organizations tend to revert back to the multi-craft concept, allowing for the maximum flexibility of the labor resources assigned to an area. To assist in the peak work periods, it is possible for the central organization to maintain a pool of qualified individuals, capable of being proficient in various areas of the plant. Some companies have as many as 30 or 40 area organizations within a single plant, coupled with central organizations and outside contracting. The area organizations report to a central maintenance management organization. This arrangement provides an optimum service/cost factor situation. Each company makes the policy decision several times before they find their optimum organization.

Organizational Charts

Once a company has decided on its growth pattern and organization arrangements, it must then decide on staffing assignments and develop the organizational chart. This means assigning responsibility and reporting lines. Most organizations use similar maintenance organizational charts, with modifications, depending on the size and development of the maintenance staff.

The operations, engineering, and maintenance managers should report to the same individual. This provides balanced input for plant management to make decisions. For example, if Maintenance needs access to the equipment for preventive maintenance or repairs, and Operations wants to continue running it instead of off-loading the production to other equipment, management can hear both sides and make the appropriate decision. If Maintenance reports to Operations, plant management may never hear how important the repairs are, and when the equipment does fail, the maintenance organization is blamed unfairly, when the situation was out of their control.

When Maintenance reports to Engineering, it is common to find project-type work given priority over preventive and repair maintenance tasks. This reduces the life of the capital equipment, with maintenance receiving the blame. Ideally, Maintenance reports to the plant manager directly.

Attitudes

The way in which management perceives maintenance, maintenance perceives its own role, and managers and craft technicians perceive one another's roles is also important. The attitude management shows toward the maintenance craft technicians helps to establish pride in workmanship. If an organization is in a "fire-fighting" mode, the workmanship suffers. Craft technicians will get into a habit of fixing it to "just get by." When actions are taken to change this approach and make it more proactive, the craft technicians may have difficulty adjusting. It is the same problem anyone has when trying to break habits.

Any in-fighting between Maintenance and Production/Operations/Facilities also must end. If Maintenance is to contribute to the overall profitability of the corporation, all parts of the organization must be given responsibility and accountability.

The Financial Impact of Maintenance

This discussion is broken down so that readers may apply the appropriate portions, and customize the cost justification to meet their own circumstances. Figure(s) 4.1-4.4 are a set of worksheets that can help a plant engineer calculate the possible savings a company may achieve by investing in improved maintenance policies and practices, including a Computerized Maintenance Management System (CMMS).

Standard Cost Justification

This portion of the cost justification is composed of four main parts:

- *Maintenance Labor Costs*
- *Maintenance Material Costs*
- *Project Cost Savings*
- *Downtime/Availability Costs*

Maintenance Labor Costs

Maintenance productivity in most American companies averages between 25%-35%. This translates into less than three hours per eight-hour shift of hands-on activities. Most of the lost productivity can be attributed to:

- Waiting for parts
- Waiting for information, drawings, instructions, etc.
- Waiting for the equipment to be shut down
- Waiting for rental equipment to arrive
- Waiting for other crafts to finish their part of the job
- Running from one emergency to another

While 100% productivity is an unrealistic goal for any maintenance organization, a more realistic percentage of 60% is achievable.

The productivity of maintenance technicians can be improved by concentrating on basic management techniques, such as:

- Planning jobs in advance
- Scheduling jobs and coordinating schedules with Operations
- Arranging for the parts to be ready
- Coordinating the tools, rental equipment, etc.
- Reducing the emergency work below the 50% level by PM

With computer assistance, planning time per job can be reduced, resulting in more jobs planned and coordinated, and more time for preventive maintenance activities. This helps reduce the number of emergency and breakdown activities. Fewer schedule changes help increase the productivity by reducing travel and waiting times. Companies that are successful in improving maintenance have increased productivity by as much as 28%.

Ed. Note: R.S. Means Co., Inc. publishes information for estimating maintenance labor costs. The annually updated Facilities Maintenance & Repair Cost Data *book provides labor hours and costs for maintenance and repair, preventive maintenance, and general (custodial) maintenance. This data can be used to develop a preventive maintenance program, and as the basis for a Zero-Base maintenance budget. See the Appendix of this book for sample pages.*

Maintenance Material Costs

Material costs are related to the frequency and size of the repairs made to the company's equipment. The sheer number of parts, in addition to the stores policies, purchasing policies, and overall inventory management practices contribute to the overall maintenance materials costs. Since little attention is paid to maintenance materials in some companies, inventories may be higher than necessary by some 20-30%. Excess inventory increases holding costs and makes materials unnecessarily expensive. The inability of the stores to service the maintenance department's needs results in "pirate" or "illegal" storage depots for just-in-case spares. This practice also drives up the cost of maintenance materials.

Good inventory controls enable companies to lower the value of the inventory and still maintain a service level of at least 95%. This enables the maintenance department to be responsive to the operations group, while increasing their

Maintenance Labor Costs

Calculation

1. Time wasted by personnel looking for spare equipment parts (Averages):

No inventory system	= 15-25%	________
Manual inventory system	= 10-20%	________
Work order system and inventory system	= 5-15%	________
Computerized inventory and manual work order system	= 0-5%	________

2. Time spent looking for information about a work order

Manual work order system	= 5-15%	________
No work order system	= 10-20%	________

3. Time wasted by starting wrong priority work order

Manual work order system	= 0-5%	________
No work order system	= 5-10%	________

4. Time wasted by equipment not being ready to work on (still in production)

Manual work order system	= 0-5%	________
No work order system	= 10-15%	________

5. **Total of all percentages of wasted time**
   1 + 2 + 3 + 4 ________
6. Total number of craftsmen ________
7. Multiply this figure (# 6) times 2080
   (normal hours worked by an employee in a year) ________
8. Multiply the percentage totals times the total number of hours for all craftsmen (5 x7) ________
9. Enter the average labor rate including benefits for a craftsman (sometimes called *burden rate*) ________
10. Multiply the potential savings in hours times the average labor rate (9 x 8) ________
11. Multiply the figure in line 10 times the percentage from the table below that best describes your facility

No work order or inventory system	=75-100%	
Manual work order system	= 50-75%	
Manual work order & inventory system	= 30-50%	
Computerized inventory & manual work order system	= 25-40%	________

12. **Total Savings** ________

This will represent the projected savings from labor productivity

Figure 4.1

Maintenance Material Costs

Calculation

1. Total dollar value of maintenance spares purchased per year ________
2. Percentage of time spares are already in stores when others are purchased ________
   - No inventory system = 25-30%
   - Manual inventory system = 10-20%
   - Computerized inventory system = 5-15%
3. Savings total (cost avoidance) ________
4. Additional savings (inventory overhead)
   Multiply line 3 times 30% ________
5. Estimated total inventory valuation ________
6. Estimated inventory reduction ________
   - No inventory system = 15-20%
   - Manual system = 5-10%
   (obsolete or unnecessary spares)
7. Estimated one time inventory reduction
   Multiply lines 5 & 6 ________
8. Estimated additional savings
   Multiply line 7 by 30% (holding cost reduction) ________
9. Number of stockouts causing downtime ________
10. Amount of downtime (in hours) ________
11. Cost of downtime (per hour) ________
12. Total cost of materials related downtime
    Multiply line 10 x line 11 ________
13. Percentage of savings obtainable ________
    - Current controls poor - 75%
    - Current controls fair - 50%
    - Current controls good - 25%
14. Savings in materials-related equipment downtime ________
    Multiply line 12 x line 13
15. **Total savings** ________
    (Add lines 3, 4, 7, 8, and 14)

Figure 4.2

own personal productivity. Successful companies have averaged 19% lower material costs and an overall 18% reduction in total inventory.

Project Cost Savings

In many companies, maintenance is involved in project, outage or refurbishing activities. These activities, if not properly controlled, can have a dramatic impact on the company's production capacity. The reason for this is that these activities are usually performed with the equipment in a "down" condition, with no production taking place during this time. Any time that can be eliminated from the project, outage, or refurbishing can be converted to production time.

Improved planning and coordination can be achieved with a disciplined use of a CMMS. This will often help to shorten the downtime, even if the company is currently using a project management system. Companies successful at project planning, coupled with good CMMS usage, have achieved an average 5% reduction in outage time.

Downtime/Availability Costs

These costs are the true savings for a company that is determined to improve maintenance policies and practices. Downtime cost for equipment may vary from several hundreds of dollars per hour to hundreds of thousands of dollars per hour. One company has several production lines in their plant, with the downtime on each being worth $1 million for 24 hours.

Project Cost Savings

Calculation

1. Number of major outages and overhauls per year ________
2. Average length (in days) of outage or overhaul ________
3. Cost of equipment downtime in lost sales ________
   Use hourly downtime rate times total hours of outages
4. Total estimated cost per year ________
   Multiply lines 1, 2 & 3
5. Estimated savings percentage ________
   No computerized work order system = 5-10%
   Project management system = 3-8%
   Project management system and inventory control system = 2-5%
6. **Total cost savings** ________
   Multiply lines 4 & 5

Figure 4.3

In some companies, levels of downtime can run as high as 30% or more. This results in lost sales opportunities and unnecessary expenditures for capital equipment, which puts the company in a weak competitive position. By dedicating the company to enforcing good maintenance policies and practices and utilizing the CMMS as a tracking tool, equipment downtime can be reduced dramatically. Successful companies average a 20% reduction in equipment downtime losses.

These worksheets were designed to give the plant engineer a guide to determine the impact maintenance has on the company financials. The guidelines are generic, but should help a plant engineer determine a return on investment for any maintenance improvement project that the company would approve.

Training

Training has been called one of the greatest weaknesses of the present maintenance structure in the U.S. It has been estimated that a company should spend approximately $1,200.00 per year for training of each maintenance supervisor, and $1,000.00 per year for each craft technician. In fact, if you don't provide some training for a craft technician in an 18-month time-period, their skills become dated.

In self-examination, when was the last formal training program for your craft technicians? For your supervisors? For your planners? The importance of training cannot be overstated. Without good quality training programs, a maintenance organization will never be cost-effective. We will examine the three areas just mentioned and explore some alternative methods of training.

Craft Training Programs

The first level, the apprentice training program, takes the "man (or woman) on the street" and gives the training necessary to become a skilled craft technician. The program must be a combination of on-the-job and classroom training. Most good programs will be three to four years in length, with hands-on lab sessions used with the classroom settings.

Some companies work with local vocational schools to fill entry-level craft positions. This allows the company to specify some of the material that must be covered in the program. The vocational school benefits because of the assistance in placing the students when they complete the program. Another option is university-level training, but this is generally used for more advanced training, later in a craft technician's career.

A second option for the craft technician entry level program is distance learning programs, which might be correspondence-type or on-line "cybercourses." This is completed by the craftperson at his or her own pace. This is one of the most economical methods used in maintenance training. Closely related are the "canned" programs which contain information related to each craft line. The disadvantage of the correspondence program is that the material is generic, dealing with theory more than real-world experience.

Downtime/Availability Costs

Calculation

1. Percentage of equipment downtime for year (If not known use estimate. Average for industry is 5% to 25%) __________
2. Total number of production hours for equipment for year __________
3. Total of all lost production hours for year (Multiply 1 & 2) __________
4. Multiply total lost production hours for year (from line 3) times your percentage from the table below. __________

   Presently using:

No work order system	=	25%
Work order system	=	20%
Work order and stores inventory system	=	10%

5. Total of downtime hours saved __________
6. Cost of downtime for 1 hour __________
7. TOTAL DOWNTIME COST SAVINGS __________
   (Multiply lines 5 and 6) Optional savings considerations:
8. Total direct labor wages and benefits times the total of all lost production hours __________
9. Lost sales for year (divide the total sales for year by the total number of yearly production hours.... Multiple this figure times the total downtime hours saved) __________
10. Increased production costs to make up production lost due to downtime. This would include the extra labor required on weekends or off shifts to operate the equipment, extra energy costs to operate the equipment, etc.

Total Projected Savings

1. Total from part A __________
2. Total from part B __________
3. Total from part C __________
4. Total from part D __________
5. Total savings possible from improvement program Total of lines 1 - 4) __________
6. Total projected price for improvement program __________
7. Payback (Divide line 6 by line 5) __________

Figure 4.4

While this theoretical approach is effective for engineering-level training, it is not satisfactory for craft-level training. A second problem is the lack of an available, interactive instructor to help the apprentice understand and apply the material being studied.

While the objection about the material being generic is true in most cases, there are at least two vendors who offer programs specific to the repair and troubleshooting of components. These programs have been successful and are described by craft technicians as beneficial. In fact, some of the journeymen, after they saw the quality of the material the apprentices were using, have gone out and bought their own set of materials for reference.

Another option is having training materials developed either internally or by an outside firm specializing in maintenance training materials. These materials are developed from a job needs analysis, a study that determines exactly what an individual needs to know to properly perform a specific job. The materials are then written and illustrated from this needs analysis. The materials are accepted readily, since they apply specifically to the craft technician's job. The largest obstacle to using this type of material is the cost. It is expensive to pay to have the needs analysis performed and the materials written, illustrated, printed, and bound. However, the advantage of having a job-specific training program that the apprentice can take at their own pace, can be worth the cost of this program.

The best option (also the most expensive) is to set up your own craft training center. It should have a classroom, lab, and real-world setting necessary to produce qualified, competent craft technicians. One such program started in the early '70s became a corporate model for one steelmaker. Of course, the hard times for steel in the mid '70s put an end to it, since training (especially for maintenance people) was viewed as an expendable expense. The program has since been restarted, but not on the scale it had in its prime. The program structure was designed for a "super-craft" environment, requiring proficiency in mechanical, electrical, and fluid power. The students were required to complete 1040 hours of classroom training, in addition to three years of on-the-job training.

The program took different forms depending on the economic situation. The apprentices attended class one day per week, one-week-per-month, or once every other month. Of these three formats, the optimum appeared to be the one-week-per-month format. This allowed the instructors to teach three weeks out of the month, with one week for preparation for the next cycle.

The five classrooms were set up with heavy tables (72"x30") where the students sat two to a table. (The metal top tables were necessary when the instructors would bring a display or lab project into the classroom.) There were five labs, three mechanical, one for fluid power, and two electrical (one D.C. control, one A.C. control). The curriculum consisted of the following:

- 260 hours D.C. electricity
- 260 hours A.C. electricity
- 260 hours mechanics
- 260 hours fluid power

The course outlines were not theory-intensive, concentrating instead on the maintenance and troubleshooting of the components and circuits. The apprentices received just enough design theory to understand why certain designs were used, but not enough to perform design (although at times the suggestion department wondered). For example, in the mechanics class dealing with V-belts, there was enough theory to understand speed differentials and forces generated by the rotation. The thrust of the discussion was how these forces create problems in maintaining the V-belts: why the tension requirements are critical, why alignment is important, why the bearing adjustment is important, why the condition of the sheaves is important, and so forth. These points made the material applicable to their jobs.

This brings us to the single most important factor in the program: the instructors, who spelled success or failure of the program. *If* the instructor could not relate the theory to the "real-world" needs of the apprentices, the course would be a waste of time. For this reason, the instructors were journeyman craft technicians with experience. They were required to have good presentation skills, to be able to develop logical, coherent outlines, and to develop tests for their materials. Without instructors who were respected for their job skills by their peers, the program would have been a failure.

Classroom or distance (correspondence or on-line) training must be conducted in conjunction with on-the-job training. This can be the most challenging part of the program; ensuring that if the apprentice is studying hydraulic circuits, they have a chance to work on some hydraulic circuits that month. Good communication with the apprentice's department heads and supervisors helps in coordination.

The reason for describing this program is not to paint a "pie-in-the-sky" picture, but rather to show the importance that some companies assign to craft training. If there is no effort to obtain a fully trained work force, do not expect to achieve a high level of skilled service from your maintenance work force.

Journeyman Training

Journeyman training is usually related to specific tasks or equipment maintenance procedures. Journeyman training courses can be conducted by in-house experts, vendor specialists, or outside consultants. The training may address a new technology that is being brought into the plant. For example, when vibration analysis was first being introduced into the maintenance environment, extensive training programs in the use of vibration analyzers were offered by the vendors and consultants. These programs were addressed to the journeyman-level craft technicians who would be involved in the programs.

When new equipment is purchased and installed in the plant, training programs are performed by the vendor on the care, maintenance, and troubleshooting of the equipment. Again, it is the journeyman-level craft technicians who are involved in the programs. Good journeyman craft training programs should be a part of any complete maintenance training program.

Cross-Training or "Pay for Knowledge"

This subject is included in the training section, since it is becoming increasingly common in progressive organizations. It is a sensitive subject, since it generally involves negotiations with the union representation. This type of program is essential if U.S. maintenance costs are to be brought in line with the costs incurred by overseas competition.

The cost savings for the company are found in planning and scheduling the maintenance activities. For example, consider a pump motor change-out. In a strict union environment, it would take:

- A pipe fitter to disconnect the piping
- An electrician to unwire the motor
- A millwright to remove the motor
- A utility person to move the motor to the repair area

The installation would go as follows:

- A millwright to install the motor
- A pipe fitter to connect the piping
- An electrician to wire the motor
- A machinist to align the motor

As can be seen, not only are many people involved, but the coordination to ensure that all crafts are available when needed without delay becomes extremely difficult. In a "multi-skilled" or "cross-trained" environment, there would be one or possibly two craft technicians sent to the job to complete all the job tasks. The cost and coordination advantages are obvious.

But what are the advantages for the employees? First, this approach helps to ensure maintenance is a profit center. This means being cost-effective, but also running the department as a business. For example, picture the following production scenario: Different people perform each of these tasks:

- Transport material to the job site
- Insert the part in the machine
- Drill the hole
- Finish the hole to specifications
- Machine the finish
- Take the part out of the machine
- Move the part to the next process.

This would not be a well managed, cost-effective operation, would it? Should maintenance be any different? Should not the goal be the same: to maximize the utilization of all assigned resources? Shouldn't the employees work to contribute to the profitability of the corporation? Our overseas competitors' employees do.

A second advantage is that multi-skilled employees have additional skills that ensure they are valuable contributors to the corporation's goals. This increases their self-esteem and value. Most of all, any cross-training effort must have financial rewards to the employees. Since multi-skilled employees are more valuable to the company, they should receive a higher rate of pay. For example:

- Apprentice – pay level 1
- Journeyman (one craft) – pay level 2
- Journeyman (1 craft, apprentice 1 craft) – pay level 3
- Journeyman (1 craft, apprentice 2 craft) – pay level 4
- Journeyman (2 craft, apprentice 1 craft) – pay level 5

And the list could go on, depending on the crafts involved. These types of programs are growing increasingly popular in the U.S., as they must. If we are to take advantage of potential cost savings, cross-training is one of the most important areas to investigate.

Planner Training

Maintenance planners should come from craft technicians who have good logistics aptitudes. However, planners need training beyond the skills required by craft technicians. They need programs teaching some of the following subject areas:

- Maintenance Priorities
- Maintenance Reporting
- Project Management
- Inventory Management
- Scheduling Techniques
- Computer Basics

If such training is not provided, it will be difficult to achieve the level of proficiency necessary to have a successful planning and scheduling program. Training is one of the most important factors in the development of a good maintenance planner. Where does one get the training materials necessary for a planner training program? The sources are similar to maintenance craft training materials. There are:

- Distance learning courses
- University-sponsored seminars
- Public seminars
- Maintenance consultants
- Maintenance software vendors

The content of the planner training program will vary depending on whether the organization is a facility, process industry, food service, vehicle fleet, etc. Each will have its own unique qualities, but many of the basic principles will be the same. Good, competent, skilled maintenance planners will pay dividends for the investment in their training many times over.

Supervisor Training

Front-line maintenance supervisor positions are filled 70% of the time from craft or planner promotions. This is so they will be familiar with the assignments that they will now be responsible to supervise. In a personal observation, it is rare when an engineer or another staff person can make the transition to becoming a front-line maintenance supervisor. The lack of craft knowledge or experience cannot be compensated for in actual job supervision.

One of the major mistakes that occurs when making a craft person into a front-line maintenance supervisor is that one week they are in the craft group, the next week they are the supervisor of that group. In many cases, there is no training given. It is almost as if management wants them to pick up their management skills by osmosis. Good supervisor training programs should be implemented before supervisory responsibilities are assumed. Some areas that should be addressed in these programs are:

- Time management
- Project management
- Maintenance management
- Management by objectives

The support and understanding of the front-line supervisor helps to determine the success or failure of many of the programs implemented by upper management. All staff people should receive the training necessary to ensure their careers are successful, so that the success carries on to the rest of the organization.

Work Order Systems

Once the organization is properly focused, staffed, and trained, it is necessary to focus on managing the maintenance function. To do so successfully, it is necessary to have the right data. The work order system is the mechanism for gathering this data. These documents are used to collect all necessary maintenance information. Work orders have been described in many different ways, but for the purpose of this text, we will use the following definition:

A request that has been screened by a planner, who has decided the work request is necessary and what resources are required to perform the work

Work orders should not be implemented by just the Maintenance Department, without regard for other parts of the organization, although Maintenance is the primary user of the work order. Maintenance requires information such as:

- Necessary equipment
- Required resources
- A description of the work
- Priority of the work
- Date the work must be completed

Other information may be required as well, depending on the type of facility or plant. The main point is that the maintenance organization must obtain the data needed to make sound management decisions. If reliable information cannot be obtained from the work order, it is unlikely to be available from another source.

Operations or Facilities also needs input into the work order process. They must be able to request work from Maintenance in an easy process. If they have to fill out fifteen forms in triplicate, chances are they will not participate in the use of the work order, which eliminates its effectiveness. Potential users of the work order system should need to fill in only brief information, such as:

- Equipment work is requested on
- Brief description of the request
- Date needed
- Requestor

This information can then be used by the planner to complete the work request, and convert it to a work order.

Engineering also needs input into the work order system, since they are usually charged with the effectiveness of the preventive/predictive maintenance (PM) programs. Of course, they will want to be able to request work for engineering services, but they also need access to historical information. The historical files, if accurate and properly maintained, can provide the information needed to operate a cost-effective preventive maintenance program. Without accurate information, the PM programs become guesswork. The engineering staff will need information such as:

- Mean time between failure
- Mean time to repair
- Cause of failure
- Repair type
- Corrective action taken
- Date of repair

The inventory and purchasing departments will also need information for the work order system, especially from the planned work backlog. If the work is planned properly, inventory and purchasing personnel will know what parts are needed and when they are needed. Good historical information on maintenance material usage will assist the inventory and purchasing personnel in setting maximum/minimum levels, order points, safety stock, and so forth on maintenance materials. The information required by inventory and purchasing includes:

- Part number
- Part description
- Quantity required
- Date required

Accounting needs information from the work order to properly charge the correct accounts for the labor and materials used to perform the maintenance task. This charge system is handled differently for many locations. The following are the most common types of accounting information gathered:

- Cost center
- Accounting number
- Charge account
- Departmental charge number

The use of this information on the work order is important if accounting is to cooperate in the use of the work order form.

Since upper management is interested in information that can be gathered from multiple work orders, the information must be easy to extract from the work order. The summary-type information should be compiled from completed work orders, work orders in process, and work orders awaiting scheduling. If the information is not easy to extract from the work order, managers can easily spend days gathering the information. The use of check-off boxes for key information fields can be invaluable. Computerized systems make this task easier, but only if they are properly designed.

Types of Work Orders

In any system, it is necessary to have several types of work orders. The most common are:

- Planned and scheduled
- Standing or blanket
- Emergency
- Shutdown or outage

Planned and scheduled work orders have already been described in brief. They are requested and screened by a planner. The resources are planned, the work is scheduled, information is entered into the system upon completion of the work, and then the work order is filed.

Standing work orders are written for the 5–30 minute quick jobs, such as resetting a circuit breaker or making a simple adjustment. If you were to write an individual work order for each of these small jobs, Maintenance would be buried in a mountain of detail which could not be compiled effectively into meaningful reports. It is for this reason that most quick-job work orders are written against an equipment number and charged to the work order number. The standing work order is not closed out. It remains open for a time period preset (by management); then it is closed and posted to history. A new standing work order number would be opened at that time.

One problem with standing work orders is that people feel they may be used like MasterCard® or Visa® cards for charging time for the craftsmen without accounting for it. This attitude occurs occasionally, but when the charges are closed out on the work order, offenders can be spotted. Computerized systems make it much easier, since they can quickly compile a list of all personnel who have charged time to a work order. Some of the more sophisticated systems can even display the percentage of time any craftwork charges to a type of work order. If offenders are suspected, it is easy to find them. Most employees will not abuse a standing work order system.

Emergency or breakdown work orders are generally written after the job is performed. Breakdowns require quick action, and there is not enough time to go through the usual planning and scheduling of the work order. In most cases, the craft technician, the supervisor, or the production supervisor will make out the emergency work order after the job is completed. The format of the emergency work order is similar to the work request, in that only brief, necessary information is required. When the work order is posted to the equipment history, it should be marked as an emergency work order, which allows for an analysis of the emergency work orders by:

- Equipment ID
- Equipment type
- Department
- Requestor

Analysis of the patterns of emergency work can be helpful in identifying certain trends, which can be invaluable when planning maintenance activities. The need for a central call-in point is to prevent overlap of assignments. If the calls are taken at various points, there may be several technicians dispatched to the same job.

When the technician or supervisor arrives at the job site, they may realize that the job is more involved than the call may have indicated. If the work required is going to be over a certain cost or time limit, the job is routed back to the planner for analysis. If it is going to be easier to coordinate and plan by scheduling, the planner takes control of the work order and schedules it as soon as the material and labor resources are available. This allows for a more cost-effective organization of labor productivity in deploying and performing maintenance tasks.

Shutdown or outage work orders are for work that is going to be performed as a project or during a time when the equipment or area is shut down for an extended period. These jobs are marked as outages or shutdowns, and should not appear in the regular craft backlog. This work is still planned, ensuring that the maintenance resource requirements for the shutdown/outage be known and ready before it begins. This approach prevents delays and maximizes the productivity of all employees involved.

In many cases, the work order information is entered into project management software in order to create a complete project schedule. Computerized Maintenance Management software typically does not include enough project management software features to make it an acceptable scheduling tool. Some vendors have included interfaces to project management systems, which tend to correct this problem.

Obstacles to Effective Work Order Systems

Preventive/Predictive Maintenance programs are keys to operating a work order system. If an organization is in a reactive (fire-fighting) mode, they have little or no time to operate a work order system. Clearly, it takes time to provide the information necessary to satisfy the work order system. When an organization runs from breakdown to breakdown, time to properly record the information simply is not available. If it is recorded, it is generally sketchy and inaccurate.

When companies are in a proactive mode, the work is planned on a regular schedule, with 25% or fewer emergency activities. In this situation, supervisors and planners have the time they need to properly utilize the work order reporting system. Without Preventive/Predictive Maintenance programs, it is impossible to utilize a work order information system properly.

A lack of control for the maintenance labor resource is a second factor that prevents optimum usage of a maintenance work order system. The following problems are common with labor resources:

- Insufficient personnel for one or all crafts
- Insufficient supervision of personnel
- Inadequate training of personnel
- Lack of accountability for work performed

Without controls in these areas, there may be inadequate or unacceptable resources available when the planner tries to schedule the work.

Ineffective stores controls can also reduce the work order system to total ineffectiveness when materials are required. If the planner does not have accurate, timely information concerning the materials in the maintenance stores, he or she cannot schedule the work. If the craft technician has a work order requiring certain materials, but the materials are not available when they are needed, the wasted time obtaining the materials lowers this individual's productivity. The planner, supervisors and craft technicians must have current information about the stock levels of maintenance inventory items. Most consultants feel that maintenance materials are the most essential part of a good planning program.

Poor planning disciplines affect the work order system because most of the information on the work order is not reliable. When this is the situation, work orders fall into disuse, resulting in discontinuance of the work order information flow. Job plans must be accurate and realistic if the work order system is to be successful. If companies do not have a work order planning system, they really do not have an effective work order system in place.

Lack of performance control is really a lack of follow-up on management's part. Once a job plan is produced or a work order is created, it should always be audited for compliance. This audit can highlight weaknesses in:

- Planning
- Scheduling
- Supervision
- Craft
- Skills

Any deficiencies can then be corrected. If performance controls are not used, the lack of accountability will create a disorganized effort, again allowing the work order to fall into disuse.

Inadequate or inaccurate equipment history hinders the work order system, since none of the information used to make management decisions will be reliable. The managers will not be able to base budget projections, equipment or facility repair forecasts, or labor needs on historical standards. This also highlights the fact that the work order system is not being accurately used, since the equipment/building history is built from the work order history file.

Unless care is taken to see that all posted data is accurate, the work order system will be unreliable and will not be used.

Planning and Scheduling

A recent survey polled maintenance managers on their top problems. Since there was more than one response per respondent, the totals were well over 100%. However, over 40% of the respondents felt scheduling was their biggest problem. Maintenance planning and scheduling is one of the most neglected disciplines today. We will explore some of the reasons for the lack of effective planning and scheduling and identify possible solutions.

Maintenance Planners and Supervisors

One of the major obstacles to maintenance planning and scheduling is management's reluctance to admit that planners are essential to the program. In fact, two-thirds of maintenance organizations in the United States do not even have planners. Maintenance organizations that do have planners often place responsibility on them for planning the activities of too many craft technicians. Planners should be responsible for 15 (an optimum) to 25 (absolute maximum) craft technicians. Supervisors should be responsible for overseeing the work of an average of ten craft technicians.

Why the difference between the two groups? This can best be answered by examining the job description(s) for the supervisor and planner.

Supervisor's Job Description

Responsibilities: While specific duties may vary from area to area, and from organization to organization, the following is an outline of the general responsibilities of a front-line maintenance supervisor:

Motivating the craft personnel. It is the supervisor's responsibility to see that each of the craft technicians they supervise is ready to perform his or her job each day. Using good management skills, the supervisor ensures that they are motivated to perform their assignments. Since the most effective way to lead is by example, the supervisor is ready to begin at starting time and is continually available to assist the craft technicians during the workday.

Determining the craft/skill/crew. This does not mean deciding how much craft manpower is required. This task has already been performed by the planner. Instead, the supervisor looks at the job and matches the skill levels of the craft technicians to the job. If it is a larger job, the supervisor matches the right skill level for several crafts, and assembles a team to accomplish the job. The supervisor is responsible for determining which individual employee will be working on each job.

Coordinating and following up. The supervisor is in the field with the craft technicians following up on the job, not sitting behind a desk, where it is impossible to see how the work is progressing.

Ensuring safety and quality. To accomplish this, the supervisor must be out with the craft technicians, and must have knowledge of job skills and techniques. If the supervisor is to be responsible for the hiring, firing, and pay reviews of the employees assigned to his or her supervision, he or she must have been properly trained to assess the craft technician's performance. A mistake often made is that the maintenance manager conducts the reviews, administers discipline, commendations, and so forth. Who better knows the work habits and skills of the employee? It is the direct supervisor, not the next level of management. Proper management training enables the supervisor to manage this responsibility effectively.

Technical competence. If a supervisor is to recommend improvements and cost reductions, he or she must understand the jobs his or her people are asked to perform and the processes they are asked to maintain, as well as the engineering principles involved in the design of the equipment being maintained.

Troubleshooting skills. Identifying the causes of failures or of repetitive breakdowns is an essential supervisory skill. Without it, technicians become mere parts changers. The supervisor's feedback on the cause of failure to the work order system can help make adjustments in the Preventive/Predictive Maintenance program, as well as give Engineering valuable feedback on design flaws.

Recommending the necessary skills and training programs for the craft technicians. Knowing what the technicians can and cannot do puts the supervisor in a position to recommend the training programs needed to improve job skills. Continued skill building enhances the knowledge and personal growth of the craft work force, improving the supervisor's ability to be an effective manager of the most important resource within the maintenance organization.

When examining the supervisor's job description, several points become clear. First, the supervisor must have in-depth job knowledge of the crafts he/she is supervising. This is paramount, for many job description tasks require knowledge of the work the craft technicians will be required to perform. Second, effective supervisors must be out on the floor interacting directly with the people they supervise on a daily basis. It is sad to say that in the majority of the maintenance organizations today, the job of the supervisor has become a glorified clerical position. This is wrong! A front-line maintenance supervisory position should be structured so that the supervisor spends no more than 25% of his or her time on paperwork, and the other 75% of their time with the technical people who report to them.

Cutbacks, with resulting elimination of clerical support help, have forced supervisors to assume duties not in their job description. In this climate of downsizing, often the worst mistake made by management is to not maintain dedicated planners to plan and schedule all maintenance activities. Maintenance departments must be properly staffed if supervisors are to fulfill their job assignments and maintain efficiency and cost-effectiveness within the organization.

The following section shows how the planner's role within the organization differs from that of the maintenance supervisor.

Planner's Job Description

Responsibilities: Reviewing the request. The planner's job starts when a work request is received. He/she reviews the request, ensuring it is not currently an active work order. The planner must also clearly understand what the requestor is asking for, so that the work plan will produce the desired results.

Visiting the job site. If the planner is unclear on what is requested, he or she will visit the job site. This serves two purposes: first, it ensures that the planner has a clear picture of what is requested; and second, it allows the opportunity to look for any safety hazards or other potential problems that may need to be addressed before the job is scheduled. If the planner is still unclear on what is being requested after visiting the job site, he or she must visit the requestor. This face-to-face discussion will help to ensure that the work is accurately understood before planning begins.

Estimating how many craft groups it will take to do the job and how long it should take them. This step is extremely important, for these estimates provide the foundation for scheduling accuracy.

Deciding what materials are needed. This is where accurate stores information is crucial. Without reliable information about on-hand quantities, maintenance planning will be inaccurate, resulting in unreliable schedules. The planner ensures that all materials are available in sufficient quantity before completing this step. The planner may not find the necessary parts to do the job stocked in the storeroom. It may become necessary to order them directly from the manufacturer. These are referred to as *no-stock items*, which should be ordered with the delivery date being the key to further processing of the work order.

Ensuring that all the required resources (labor, materials, tools, rental equipment, and contractors) are available before the work is scheduled. This will eliminate lost productivity, since everything is ready before the craft technicians begin to work on the job.

Maintaining a file of repetitive jobs. These jobs are performed the same way, using the same labor and materials each time. They are not done on a regularly scheduled basis, but with varying frequencies. The planner builds a file and statistically averages the actual labor hours and related costs each time the job is done so that this figure can become an estimate when the job is next scheduled. Aside from saving time in recalculating costs for similar repetitive jobs, this past data increases the accuracy of the estimates.

Another method the planner may use is to keep an historical file of work orders by equipment. When a job comes up that has been done before, the planner can go to the history file and pull out the previous work order. By copying the job steps, materials, and so forth from the previous work order, the job planning becomes easier. In addition, the planner should look for completion comments to ensure that the previous job plan did not overlook anything vital to the efficient performance of the task.

Maintaining control of the work order file and developing the craft backlog. The craft backlog is a total of all the labor requirements for work that is ready to schedule. This figure allows the planner to alert management to the need to add or decrease craft labor. The planner plots the backlog trends for a running six-month chart to show job-tracking trends.

Tracking the labor capacity for each week and taking enough work out of the backlog to make up the weekly schedule. This total will take into account vacations, sickness, and overtime. Tracking weekly labor capacity helps to ensure that an accurate schedule will be produced.

Producing a tentative weekly schedule and presenting it to management, who makes any needed changes and approves the schedule for the next week. The schedule is given to the maintenance supervisor at the end of the week, so he can prepare for the next week. It should be noted that the planner does not tell the supervisor when each job will be done or by whom; that is the supervisor's responsibility. The planner is responsible for weekly schedules; the supervisor for the daily schedules.

Receiving the work orders, noting any problems, and filing the completed work orders in the equipment building file. The work order file is kept in equipment sequence for easy access to equipment repair history. Each work order should include the following information:

- Date of repair
- Work order number
- Accumulated downtime
- Cause code
- Priority of work
- Actual labor
- Actual materials
- Total cost
- Year to date costs
- Life to date costs

This information can be compiled by management in databases and sorted into computerized reports for future decision-making.

Maintaining the equipment information, such as drawings, spare parts lists, and equipment manuals. These are used for reference by the entire maintenance organization, but are used particularly by the planner in performing their own work, since this information is a vital planning tool.

Working with Engineering to spot any excesses or deficiencies in the Preventive/Predictive Maintenance systems.

The planner's job is more paperwork-oriented than the supervisor's. The planner should expect to spend 75% of his or her time on paperwork or computer work, and the remaining 25% out on the floor, looking over equipment, parts, or spares. This is why planners and supervisors cannot be the same person; they both have full-time jobs, with highly specialized responsibilities and dissimilar functions within the maintenance organization.

Planner's Job Skills

First and foremost, the planner must have **good craft skills**. A planner who is effective in planning the job must be knowledgeable in how to do it. If job plans are not realistic and accurate, the planner's program will not have acceptance by the craft technicians, since poor planning increases their work and frustrations.

The planner must have **good communications skills**, since he or she is required to interact with multiple levels of management, operations/facilities, and engineering personnel. Poor communication skills can adversely affect the relationship Maintenance has with all or any of these groups.

The planner must have a good **aptitude for computer/paperwork**, since 75% of the planner's time is spent performing this type of activity. Some craft technicians cannot make the transition to planner for this reason; therefore, it is important to specify these requirements to them before they decide to become planners.

The planner must also have **the ability to clearly understand and convey instructions**. In communication with the requestors of work, instructions will be conveyed. If they are not understood, how can the planner transmit them to the craft technicians? Sketching ability may seem superfluous, but consider the following scenario. The planner is asked exactly what part needs to be changed. He or she picks up a pencil and piece of paper and begins to draw the part. It happens all the time. Sketching is an indispensable skill for a planner.

Most of all, the planner must be educated regarding the priorities and management philosophy of the organization. Without this understanding, it is difficult for the planner to function in a manner satisfactory to him- or herself, or to management. A clear understanding of the organization's goals can enhance the entire planning program's performance and can contribute to overall acceptance of the program.

Reasons & Solutions for Planning Program Failures

Examining typical failures in the past will help prevent future failures. One reason planning programs fail is the combination of unclear job descriptions and overlapping job responsibilities. This means that the first planner thinks the second planner is doing a task, while the second planner thinks the first planner is doing it. In such cases, no one plans the job, and the planning program loses credibility. Elimination of this problem involves strict planning lines and better communication within the planning organization structure. Whether the plans are by craft, crew, department, or supervisors, make the responsibilities clear, and then monitor them. This problem can be eliminated by establishing good management controls, strong communication, and a standardized methodology.

Unqualified planners may create unrealistic job plans that will destroy the credibility of the planning program. Proper job skills and training are crucial. The planner must then be given the opportunity to apply these skills and training. For the sake of the whole program, ineffective planners must be removed. If planners become careless, appropriate disciplinary procedures should be implemented. First check to make sure that management is not to blame.

Frequently, planners do not have enough time to properly plan: the ratio of planners to technicians is not correct. As discussed earlier, the ratio should be 1:15 at the optimum. A ratio of 1:20 could possibly be used if the working conditions and type of work planned are optimum. Anything above 1:25 spells certain failure for the program.

Consider the steps involved in planning a work order, as discussed previously. Could any one person plan for 25 work orders per day? How about 50 work orders per day? Consider how many work orders each craft technician completes each day. Multiply that times the number of technicians per planner, and you have the workload of the planner. Overloading the planner's workload is the most common problem; elimination of this problem helps to ensure a successful planning program.

Benefits of Planning

What kinds of benefits do we see from planning? First of all, there are the cost savings. Figure 4.5 shows documented figures of savings that companies have used in depicting planned maintenance costs versus breakdown or emergency maintenance costs.

These costs are based on identical jobs that are performed once in a breakdown mode, and once in a planned and scheduled mode. The cost savings from these three examples would pay for a comprehensive planning program for a considerable time period. In addition to the bottom line dollar savings, the increase in maintenance productivity also affects the morale of the work force. The national average for hands-on time is less than 50% for maintenance technicians. In some reactive organizations, it is below 25%. Why does this happen? Productivity losses result from reactive or unplanned work, and associated delays and losses are common.

In defense of the craft technicians, they are a skilled group. They, like any other craftsmen, want to do the best job possible. Lack of cooperation and coordination on management's part lessens their ability to do so.

Table I: A Comparison of Maintenance Costs

	Planned	Unplanned
Job 1	$30,000.00	$500,000.00
Job 2	$46,000.00	$118,000.00

Figure 4.5

Planning is an integral part of any successful maintenance organization. It affects everything from the bottom line to craft morale. If you have tried planning in the past and failed, consider some of these suggestions for success, and try them. If you are planning now, review this section and try to optimize these techniques.

Maintenance Scheduling

In its most simple terms, maintenance scheduling is the matching of maintenance labor and materials resources to the requests for those commodities. If it were that simple, maintenance scheduling would not be listed as one of the major problems of maintenance managers. The flow of scheduling starts with good job plans, prioritizing the work order, scheduling the work when resources are available, and completing the work when scheduled.

When planning the work order, the planner needs to track it through various stages. A planner would want to ensure that the work order had cleared all wait codes before the work was given the status "ready to schedule." For reasons previously discussed, scheduling work before it can be started decreases maintenance productivity.

The next step is to determine the available labor capacity for the scheduling period. The true labor capacity can be compared to a payroll check. It has a gross amount and a net amount. The gross is the hours worked times the pay rate. The net is what is left after taxes, social security, and other withholdings. Labor capacity also has a gross amount: total employees times the hours scheduled, plus overtime and/or contract workers. However, you can never expect that much work to be done, any more than you can expect to spend the gross amount on your paycheck. There are deductions from the labor capacity, such as unscheduled emergencies, absenteeism, allotments for PM work, or routine work.

Determining the Craft Backlog

Good scheduling also necessitates knowing the amount of work ahead of each craft. This is commonly called the *craft backlog*. The craft backlog can be accurately determined only if "real world" figures are used. It should be noted that accurate backlogs involve the open work orders that are ready to schedule, not work that is waiting for something to be resolved before it can be scheduled. Dividing this total hour figure by the craft's capacity for the week yields the craft's backlog in weeks.

Knowing the craft backlog in weeks helps to determine the staffing requirements for the craft group. A reasonable backlog is two to four weeks worth of work. Some companies will allow a two to eight week range, but if it goes beyond that time frame, requestors tend to bump up the priority and circumvent the scheduling process. A craft backlog of more than four weeks indicates a need for increased labor. This can be accomplished by:

- working overtime
- increasing contract labor
- transferring employees
- hiring employees

A craft backlog of less than two weeks indicates a need for reduced labor. The reduction may come in the following ways:

- limiting overtime
- decreasing contract labor
- transferring employees
- laying off employees
- cross-training employees to trades where help is needed

To properly manage the work force, it is necessary to trend the backlog over a time period. This helps to identify developing problems and to evaluate attempted solutions. A good graph should be for a rolling twelve-month time period.

Preparing the Schedule

With the above information, the planner is now ready to begin the schedule. It should be noted here that the planner is primarily concerned with a weekly schedule. This approach provides the maximum flexibility for handling unexpected delays caused by emergency/breakdown work, weather delays, production rush orders, and so forth.

Maintenance priorities are decided based on a variety of criteria, most significant of which is the measure of criticality or importance of the work. The simpler the priority system, the more widely accepted will be its use. The more complex it is, the fewer the people who will understand it or properly use it. Prioritization should be done with input from Maintenance and Operations, particularly regarding the importance of the equipment and the requested work. The higher the priority, the faster the work gets done. Some systems include an aging factor, which raises the priority so many points for each week the work order is in the backlog. This prevents "lifers," or work orders that never get done. Sometimes a problem is placed on the deferred maintenance list. This may be due to financial constraints, or simply a decision to address the problem during a future renovation.

The planner uses the status of the work order to begin the listing. All work already in process should be scheduled first, to eliminate jobs that are partially completed in the backlog. Jobs in progress are sorted within this status by priority. The next status level is those work orders previously scheduled, but not yet started. These also would be sorted by priority within the status. Next comes the work that is ready to schedule, again sorted by priority.

The planner deducts the hours required to address each work order from the available capacity for the craft group. When he or she runs out of hours of craft labor available, that is all the work the crew can be expected to accomplish for the next week. Any additional work orders on the list go into the backlog.

The planner then takes the schedule to a management meeting and presents the work scheduled for the next week. The maintenance manager, production/facilities manager, or the engineering manager may request that the planner defer some work orders in favor of getting some other work done. Once the schedule has been agreed to, the planner finalizes it and distributes copies to all parties involved, usually on the Friday of the preceding week. This ensures that there is agreement before the week starts.

The work order system is the key to managing maintenance. Not only do work orders provide the data necessary to properly manage maintenance, they also serve as the control documents for planning and scheduling maintenance work. Following the methods outlined in this section should enable the plant engineer to properly manage maintenance.

Inventory and Purchasing

Minimum Requirements

The inventory and purchasing staff has greater impact on maintenance productivity than any other support group. How does inventory and purchasing affect the maintenance organization? Previously, the point was made that maintenance work should be planned. Part of the job plan for maintenance is the detailing of all materials required to perform the work, ensuring they are in stock and available before the work is scheduled. Finding or transporting spare parts is a common source of delay, which can be eliminated if the job is properly planned.

On-line or real-time parts information is necessary to plan maintenance activities. The planner must know when selecting parts for a job whether they are in stock, out of stock, or in transit. The planner must have current information. If the work is planned based on information that is days, weeks or months old, when the craft technicians go to pick up the parts, they could experience delays. With current information, the planner will know what action to take. The minimum parts information the planner needs includes:

- Part number
- Part description
- Quantity on-hand
- Location of part
- Quantity reserved for other work
- Quantity on order
- Substitute part number

There is other information that the planner could use, but the above list will assist in preparing most jobs. However, if the above information is inaccurate or unreliable, the planner must physically check the store each time work is planned. This time-consuming activity lengthens the time it takes to properly plan a job, to the point where the planner will not be able to plan all of the work required.

Equipment "where used" listings are lists by equipment of all of the spare parts carried in the stores. These listings are important in several ways. First of all, they allow the planner quick access to parts information during the planning process. The planner will always know on what piece of equipment the work is being performed. This listing allows a quick look-up of parts information. If the part cannot be found on the list, it may need to be added to the list of spare parts by requesting that stores now carry it in stock. This list is important during a breakdown or emergency situation. When a part is needed, a quick scan of the spare parts list could save time looking for the part.

The planner must have accurate information on-hand. If the inventory system indicates that there are sufficient supplies of a part in the stores to perform a job, the planner may send a craft technician to get them. When the technician discovers the parts are not there, the inventory system loses credibility and usefulness to the maintenance department. If the planner or technician has to physically go to the store location and check each time a part is planned or requested, the maintenance department will experience a tremendous loss of productivity.

Projected delivery dates are important since no store will always have every part when it is requested. Knowing when the part will be delivered allows the planner to schedule the work based on that date. The vendors' delivery performance must be reliable. There would be another loss of maintenance productivity if a job was scheduled for a certain week, only to find that when the job was started, the parts never were delivered as promised.

Improving Inventory Management to Enhance Productivity

The points described so far in this section focus on the minimum requirements for a maintenance inventory system. To raise the level of the inventory system so it can enhance the productivity of the maintenance department would include tracking balances for issues, reserves, and returns. Returns in a production inventory system indicate how many items have been returned to the vendor. In a maintenance inventory system, returns indicate how many parts have been planned for a job, issued to a work order, but were not needed, and so were returned to the stores for credit. This indicator becomes a measure of the planner's performance. This is important since, if a planner always planned too many parts for each job, then the inventory stock level would be higher than required. This ties up unnecessary capital in spares that the company could put to use elsewhere in the business.

In many companies, *asset tracking*, or movement of rebuildable spares, is important. As with any other part of the organization, stores and purchasing should be monitored for performance. The indicators mentioned are useful to track performance levels for the stores and purchasing groups. Poor performance by these two groups will have a dramatic impact on the maintenance organization. It is good practice to copy maintenance managers on inventory and purchasing reports. The maintenance manager can use this information for comparisons regarding maintenance performance. Any conflicts between the two groups can then be discussed and remedied.

Types of Maintenance Spares

Maintenance has many different types of spares that need to be tracked through the inventory function. Examining the types of spares will help maintenance managers ensure that correct controls are placed on the more important items.

Bin stock items are materials that have little individual value, but high-volume usage. Examples include small bolts, nuts, washers, and cotter pins. These items are usually placed in an open issue area. Their usage is not tracked to individual work orders like larger items. The best way to maintain the free issue items is the two-bin method. The items are kept in an open carousel bin where the craft technicians can get what they need when they need it. When the bin becomes empty, the store clerk puts the new box in the bin, and orders two more boxes. By the time the bin is emptied, the boxes are delivered and the cycle starts over again.

Bin stock controlled issue items are similar to the free issue items, except their access is limited. The stores clerk will hand the items out, while still not requiring a requisition or work order number for the item. The stock levels should be maintained similarly to the free-issue stock levels using the two-bin method.

Critical or insurance spares are those items that may not have much usage, but due to order, manufacture, and delivery times, must be kept in stock in case they are needed. Stocking decisions for these items should take into account the cost of lost production or the amount of downtime that will be incurred if the part is not stocked. If this cost is high, it will be better to stock the item than to risk the cost of a breakdown. Since the cost of these items is usually high on a per-unit basis, it is important that they receive proper care while in storage. This means a heated, dry, weather-proof storage area. If the spare remains in storage for six months, a year, or longer, good storage conditions will prevent its deterioration.

Rebuildable spares include items like pumps, motors, and gearcases, for which the repair cost (materials and labor) is less than the cost to rebuild it. Depending on the size of the organization, the spare may be repaired by maintenance technicians or departmental shop personnel, or sent outside the company to a repair shop. These items are also generally high-dollar spares and must be kept in good environmental conditions. Their usage, similar to that of the critical spares, must be closely monitored and tracked. Lost spares of this type can result in considerable financial loss.

Consumables are items that are taken from the stores and used up or thrown away after a certain time period. These items might include flashlight batteries, soap, oils, and greases. Their usage is tracked and charged to a work order number or accounting code. Historical records may be studied and charted to determine the correct levels of stock to carry for each item. If problems develop with the stock level, the inventory level can be adjusted on a periodic basis.

In some companies, *tools and equipment* are kept in the storeroom or in a tool crib and issued like inventory items. The difference is that the tools are returned when the job is completed. The tool tracking system will track the tool's location, who has it, what job it is being used on, and the date returned. This type of system is used only to track tools with a specifically high value or where there are only a relative few in the entire company. This system should not be used to track ordinary hand tools.

Scrap Materials and Parts

When Maintenance is involved in construction, or outside contractors are doing construction work in the plant, there are generally surplus or residual materials left over. Since there is no place else to put them, they end up in the maintenance stores. These residual or surplus items can become a problem in the stores. If the parts are not going to be used again in the short term (1–6 months), they should be returned to the vendor for credit. If they are going to be used, or are critical spares, they should be assigned a stock number and properly stored. A word of caution: storing these items to have them "just in case" is expensive. If the storeroom becomes a junkyard, it is costing the company money that most employees do not realize.

Over a period of time, all stores accumulate scrap or other useless spare parts. At least once per year, the stocking policies should be reviewed. If there are scrap items, get rid of them. One manager had an interesting method he used to clean out a stores location that was turned over to him. Together with a supervisor, a planner and a craft technician, he went through the stores and identified every item they could. The rest of the items were piled outside the storeroom with a sign saying: "If you recognize any of these parts, put an identification tag on them." After two weeks of people identifying parts, anything left over was scrapped. Whether managers realize it or not, it is costly to keep spares.

One additional note on maintenance storerooms: There is a philosophy that all maintenance stores should be open. This philosophy is incorrect. It is important to have accurate and timely inventory information and there must be controls placed on the movement of certain maintenance spares. Open stores, or failing to monitor the individuals having access to the stores, eliminates any controls. Parts can be used without anyone knowing where they went. Someone may move them within the stores and no one else will know where they are. This type of system is expensive and will not allow a maintenance organization to use their materials effectively. Closed stores are critical to improving maintenance stores inventory control.

The costs of maintenance inventories have been mentioned throughout this section, but consider some of the common hidden costs. The total cost for carrying an item in stores may be as high as 30–40% of the value of the item per year. For someone with an inventory of ten million dollars, to think that three or four million dollars are required each year to maintain that inventory is staggering. This is the critical reason why it is so important to carry only as much of each item as is really required. Anything over that amount is waste that is deducted directly from the corporate bottom line. Inventory control is critical and should not be overlooked in any effort to improve a maintenance organization.

Cost Savings Considerations

Since the costs of inventories are so high, what other efforts can be made to curb or control these costs? Standardization of equipment, supplies and suppliers has proven to be a major source of savings for organizations. For example, standardizing equipment can help reduce inventory. Imagine a plant with 15 presses. If each press was made by a different manufacturer, few, if any, of the parts would be interchangeable. What does this do to the inventory? There would have to be 15 sets of spares for each of the presses. Imagine the total cost for the inventory for such an arrangement. But what if the 15 presses were from the same manufacturer? Many of the spare parts would be interchangeable. What does this do to the inventory? Instead of 15 sets of spares, there may only be five sets. The odds of more than five of the presses needing the same part at the same time are very small. Think of the savings in carrying costs alone for ten sets of spares. Multiply this number times the number of different types of multiple equipment items in the plant, and it can quickly add up. Similar cost savings can be applied to the standardization of building heating and cooling systems or door hardware.

What about maintenance supplies or suppliers? Studies have shown that consolidating supplies and suppliers has saved large percentages of total inventory costs. This is one area where we can learn from the Japanese. They keep the number of suppliers low and receive better prices and service, since each of the suppliers receives more business from them. Simplifying these relationships helps all involved.

Reduction of obsolete, spoiled, or vanished parts is accomplished through better inventory control and closed storerooms, as mentioned earlier. These points cannot be overemphasized. Substantial savings can be generated through inventory control which will, at the same time, improve the service Maintenance receives from the inventory and purchasing functions.

Maintenance Controls

Unfortunately, there are many organizations where Maintenance and Stores/ Purchasing do not cooperate. In fact, only 50% of the organizations polled in a survey allowed Maintenance any controls over their inventory. This is alarming, since Maintenance is responsible for budgeting for repair materials. It is being held responsible for something it cannot control.

In organizations that are controlled by internal politics, Maintenance usually loses. Inventory and Purchasing often influence upper management to a point that negates the effectiveness of the maintenance organization. When proper organizational support is pushed aside or overlooked, the entire organization suffers, and many times the maintenance organization gets the blame, when it does not have control. If the organization allows Maintenance to control its own resources (including spare parts), it can become a profit center, enhancing corporate profitability and productivity.

Preventive Maintenance

Preventive maintenance has become a term with broad and varied definitions. In this chapter we will provide a generic definition, discuss the types of preventive maintenance, and offer guidelines for implementing a good preventive maintenance program.

Preventive maintenance is any planned maintenance activity that is designed to improve equipment life and avoid any unplanned maintenance activity. In its simplest form it can be compared to the service schedule for an automobile. Certain tasks are scheduled at varying frequencies, all designed to keep the automobile from experiencing any unexpected breakdowns. Preventive maintenance for equipment is no different.

Why is preventive maintenance important? One reason is increased automation in industry. The more automated the equipment, the more components that could fail and cause the entire piece of equipment to be taken out of service. Routine services and adjustments can keep the automated equipment in the proper condition to provide uninterrupted service.

Just-In-Time (JIT) manufacturing is becoming more common in the U.S. JIT requires that the materials being turned into finished goods arrive at each step of the process just in time to be processed. JIT eliminates unwanted and unnecessary inventory. The downside to JIT is that it also requires a high degree of equipment availability. This means the equipment must be ready to operate when a production demand is made, and it must not break down during the operating cycle. Without the buffer inventories (and high costs) traditionally found in U.S. processes, preventive maintenance is necessary to prevent equipment downtime.

If equipment does fail during an operational cycle, there will be delays in making the product and delivering it to the customer. In these days of intense competition, delays in delivery can result in a lost customer. Preventive maintenance needs equipment reliable enough to produce a production schedule dependable enough to give a customer firm delivery dates.

In many cases, when equipment is not reliable enough to schedule to capacity, companies purchase another identical piece of equipment. In case the first one breaks down on a critical order, they now have a back-up or a spare. With the price of equipment today, this can be an expensive solution to a common problem, particularly when unexpected equipment failures can be reduced, if not almost eliminated, by a good preventive maintenance program. With equipment availability at its highest possible level, redundant equipment is not required.

Reducing Insurance Inventories

Maintenance carries many spare parts just in case the equipment breaks down. Operations carries additional in-process inventory for the same reason. Good preventive maintenance programs allow the maintenance department to know the condition of the equipment and thereby prevent breakdowns. The savings from reduction (in some cases, elimination) of insurance inventories can finance the entire preventive maintenance program.

Increasing Equipment "Uptime"

In manufacturing and process operations, each production process is dependent on the previous process. In many manufacturing companies, the processes are divided into *cells*. Each cell is viewed as a separate process or operation, and each is dependent on the previous cell for the necessary materials to process. An uptime of 97% might be acceptable for a stand-alone cell; however, if ten cells, each with a 97% uptime are tied together to form a manufacturing process, the total uptime for the process would be 71%. This is an unacceptable rate in any process. Preventive maintenance must be used to raise the uptime to even higher levels.

Longer equipment life is a result of performing the needed services on the equipment when required. Using the example of an automobile again, if it is serviced at prescribed intervals, it will deliver a long and useful life. However, if it is neglected, and the oil never changed, it will have a shorter useful life. Since some industrial equipment is even more complex than the newer computerized automobiles, service requirements may be extensive and critical.

Energy Savings

Preventive maintenance reduces the energy consumption for the equipment to its lowest possible level. Well-serviced equipment requires less energy to operate, since all bearings, mechanical drives, and shaft alignment will receive timely attention. By reducing these drains on the energy used by a piece of equipment, overall energy usage in a plant could amount to a 5% reduction, yet another economic incentive to help justify a good preventive maintenance program.

Higher Quality Products

Higher product quality is a direct result of a good preventive maintenance program. Poor, out-of-tolerance equipment never produced a quality product. World-class manufacturing experts realize and emphasize that rigid, disciplined preventive maintenance programs produce high-quality products. To achieve the quality required to compete in the world markets today, preventive maintenance programs will be required.

Change of Attitude

If operations or facilities were organized and operated the same way as most maintenance organizations, we would never get any products or services when we needed them. An attitude change is necessary to give maintenance the priority it needs. This may also involve a management viewpoint change. Current management philosophy tends to sacrifice long-term planning for short-term returns. This attitude is what causes problems for maintenance organizations, leading to reactive maintenance with little or no controls. When maintenance is given its due attention, it can become a profit center, producing positive, bottom-line improvements to the company.

One important point should be made before an organization attempts to discuss and select types of PM. No preventive maintenance program will be truly successful without strong support from upper management of the plant facility. There are enough demands made by plant management to get in the way of allowing time to perform maintenance on the equipment instead of running it wide open, that without commitment to the program, PM will never be performed. Or it will be too little, too late. This support is the cornerstone for any PM program.

Types of Preventive Maintenance

Routine maintenance, such as lubrication, cleaning, and inspections is the first step in beginning a preventive maintenance program. These service steps take care of small problems before they cause equipment outages. The inspections may reveal deterioration which can be repaired through the normal planned and scheduled work order system. One problem develops in companies that have this type of program: They stop here, thinking this constitutes a preventive maintenance program. However, this is only a start; there is more a company can and should do.

Proactive replacements means replacing of deteriorating or defective components before they can fail. This scheduling of the repairs eliminates the high costs related to a breakdown. These components are usually found during inspection or routine service. One caution should be noted: Replacement should be for only those components that are in danger of failure. Excessive replacement of components "thought" to be defective can inflate the cost of the preventive maintenance program. Only known defective or "soon to fail" components should be changed.

Scheduled refurbishings are generally found in utility companies, continuous process-type industries or in cyclic facilities usage, such as colleges or school systems. During the shutdown or outage, all known or suspected defective components are changed out. The equipment or facility is restored to a condition where it should operate relatively trouble-free until the next outage. These projects are scheduled using a project management type of software, allowing the company to have a time line for starting and completing the entire project. All resource needs are known in advance, with the entire project being planned.

Predictive maintenance is a more advanced form of the inspections described earlier in this section. Using presently available technology, inspections can be performed that detail the condition of virtually any component of a piece of equipment. Some of the technologies include:

- vibration analysis
- spectrographic
- lubrication oil analysis
- infrared scanning
- SPA

Condition Monitoring takes predictive maintenance one step further, by performing the inspections in a "real-time" mode. This is done by taking the signals from sensors installed on the equipment and feeding them into the computer. The computer monitors and trends the information, allowing maintenance to be scheduled when it is needed. This approach eliminates error on the part of the technicians making the readings out in the field. The trending is useful for scheduling the repairs at times when production is not using the equipment.

Reliability engineering is the final step in preventive maintenance. If problems with equipment failures still persist after using the aforementioned tools and techniques, engineering should begin to analyze other parts or elements that have been previously overlooked. If not, then a design engineering study should be undertaken to consider possible modifications to the equipment to correct the problem.

Incorporating all of the above techniques into a comprehensive preventive maintenance program will enable a plant or facility to optimize the resources dedicated to the PM program. Neglecting any of the above areas can result in a PM program that is not cost-effective.

Preventive Maintenance Program Indicators

Each of these indicators can be used as an argument for improvements in an existing program or as justification for starting a program. For example, if equipment utilization is below 90%, it is clear that the equipment is not being serviced correctly. If there is a PM program in place, it needs rapid adjustment, before management decides it is of no value and does away with it.

If there is high wait time for the machine operators when the equipment does fail, this indicates a failure. Major failures should be detected by a good PM program before they occur. If there are numerous major failures, the PM program must be changed to address the problems before support is lost for the program.

If the breakdowns can be traced to lack of lubrication or adjustments, the PM program is to blame again. It is necessary again to adjust the program quickly to address the problems. A good PM program should remedy all lubrication and service-related failures.

The following section is reprinted from *Cost Planning and Estimating for Facilities Maintenance*, R.S. Means Co., Inc.

Predictive Maintenance or Condition Monitoring

Condition Monitoring, or *Predictive Maintenance*, is the philosophy that maintenance should be performed based on the condition of the system or equipment, rather than on a time (or interval) basis. Condition Monitoring involves continuous or periodic monitoring and diagnosis of equipment and components in order to forecast failures. In some cases, it is the most effective maintenance that can be performed.

Condition Monitoring is analogous to a physician monitoring and diagnosing a patient's condition. While the physician cannot predict when you will die, he or she can monitor your condition and provide you with a plan to maintain and improve your health which will, hopefully, extend your life. *Predictive* implies *positive* control of machinery condition. Condition Monitoring will improve the overall health of machines, including extending life and diagnosing most impending failures. Other names for the approach are used in various industries or agencies; for example, NASA uses the term *Predictive Testing and Inspection* (PT&I).

Condition Monitoring has been emerging as an important maintenance tool as more people have become aware of its benefits, and as the costs associated with monitoring the condition of systems and equipment have decreased. Prior to the 1950s, there was little discussion regarding the maintenance of systems and equipment. Most maintenance was intuitive or breakdown (reactive), and there was little examination of the relationship between failures and maintenance. In the 1960s, the airline industry set out to improve the effectiveness of maintenance in order to reduce costs and increase reliability without sacrificing safety. *Reliability-Centered Maintenance*, a book by Stanley Nowland and Howard Heap, was the first detailed discussion on the subject and is the basis for modern Reliability-Centered Maintenance (RCM) programs. A key element of RCM is the understanding that time-based maintenance is sometimes not the most effective maintenance method. Time-based maintenance may introduce problems into otherwise healthy machines or, in extreme cases, result in premature overhaul or replacement. Based on this understanding, it became apparent that, when possible, a time-based inspection of systems and equipment would result in more effective utilization of maintenance resources. In the eighties, advances in technology and proven results in the aerospace, military, utilities, and process industries raised the awareness of time-based inspections, which have become known as Condition Monitoring or Predictive Maintenance.

Condition Monitoring is a subset of periodic maintenance and is forecast-oriented. Inspection techniques that are nonintrusive/nonrestrictive are always preferred, in order to avoid introducing problems. In addition, Condition Monitoring generally involves data collection devices, data analysis, and computer databases to store and trend information.

Proven Condition Monitoring Techniques

Vibration Analysis:

When people think of Condition Monitoring, they often think of vibration analysis. The technology and techniques have been developing for over 30 years, and over 78% of all manufacturing or processing plants use vibration analysis. Vibration analysis of rotating machines such as motors, pumps, fans, and gears is widely accepted as a viable technique to identify changing conditions. Reduced costs of test equipment and data management (primarily computers), availability of training, and development of computer-based expert systems are all contributing to this acceptance.

The technique measures machinery movement (vibration), typically through the use of an accelerometer, and examines the vibration spectrum to identify and trend frequencies of interest. Some frequencies are associated with a machine's design, regardless of its condition. For example, a healthy fan or rotary compressor may have a frequency that is equal to the machine speed times the number of fan blades. The vibration analysts may monitor this frequency to note changes in the amplitude that indicate a degrading condition. Other frequencies, for example, those associated with rolling element bearings, may be a sign of bearing damage and will alert the analysts to the start of bearing failure. It is common for electric motor problems, such as broken rotor bars or stator eccentricity, to be seen in vibration associated with electrical line frequency.

The vibration data may be collected with a portable device for periodic monitoring, or a continuous monitoring system may be installed for costly or critical systems. Analysis of the vibration data requires a detailed understanding of machinery operations and of vibration analysis techniques.

Infrared Thermography:

Infrared Thermography (IRT) is the application of infrared detection instruments to identify pictures of temperature differences (thermogram). The test instruments used are noncontact, line-of-sight, thermal measurement and imaging systems. Because IRT is a noncontact technique, it is especially attractive for identifying hot/cold spots in energized electrical equipment, large surface areas such as boilers, roofs, and building walls, and other areas where "stand off" temperature measurement is necessary.

IRT inspections are identified as either *qualitative* or *quantitative*. The quantitative inspection is concerned with the accurate measurement of the temperature of the item of interest. The qualitative inspection concerns relative differences, hot and cold spots, and deviations from normal or expected temperature ranges. Qualitative inspections are significantly less time-consuming than quantitative because the thermographer is not concerned with highly accurate temperature measurement. What the thermographer does obtain is highly accurate temperature differences (T) between like components. For example, a typical motor control center will supply three-phase power, through a circuit breaker and controller, to a motor. Current flow through the three-phase circuit should be uniform, which means that the components within the circuit should have similar temperatures, one to the other. Any uneven heating, perhaps due to dirty or loose connections, would quickly be identified with IRT imaging systems.

IRT can be utilized to identify degrading conditions in electrical systems such as transformers, motor control centers, switchgear, switchyards, or power lines. In mechanical systems, IRT can identify blocked flow conditions in heat exchangers, condensers, transformers, cooling radiators, and pipes. It can also be used to verify fluid level in large containers such as fuel storage tanks, and degraded refractory in boilers and furnaces.

Lubricating Oil Analysis:

Lubricating oil analysis is performed on in-service machines to monitor and trend emerging conditions, confirm problems identified through other means, such as vibration, and to troubleshoot known problems. Tests have been developed to address indicators of the machine mechanical wear condition, the lubricant condition, and to determine if the lubricant has become contaminated. Contaminated oil can contain water, dirt, or oil that was meant for another application. If not corrected, these contaminants can quickly lead to machine damage and failure. Lube condition trending, such as depletion of additives, can identify when the oil should be changed. Material, such as metal or seal particles, can identify machine damage before catastrophic failure, allowing for less costly repair.

Other techniques include Ultrasonic Noise Testing (identifies arcing in electrical equipment and leaks in gas systems), Ultrasonic Thickness Measurement (used on pipe wall), Flow Measurement, Valve Operation Analysis, Corrosion Monitoring, Process Parameter Analysis, and Insulation Resistance Trending of motors and circuits.

Analysis Techniques

Pattern Recognition:

Often machines exhibit recognizable operation patterns. Deviations from the pattern or norm are indications of changes that may identify the onset of failure. For example, the infrared thermography inspections discussed earlier are seeking to identify unexpected thermal patterns.

Test Against Limits or Ranges:

This should be done for parameters or conditions that do not follow continuous trends or repeatable patterns. This is useful in instrument calibration.

Relative Comparison of Data:

Look for change as related to earlier data or from another baseline (such as similar equipment). This requires stable plant/equipment conditions.

Statistical Process Analysis (also called "Parameter Control Monitoring"):

This type of analysis generally uses process or maintenance data that already exists or is collected. It applies statistical techniques to process or maintenance data in order to identify deviation from the norm.

Correlation Analysis:

The most powerful technique is the one that uses data from multiple sources, related technologies, or different analysts. This cross-reference chart in Figure 4.6 illustrates how several of the techniques discussed earlier can be related to confirm the condition diagnosis.

Many companies are benefiting from Condition Monitoring. A few of the success stories:

- Allied-Signal's Chesterfield Plant reduced unscheduled repairs from 33% to 4% through use of vibration analysis.
  AIPE Facilities, March/April 1991
- Companies save $6 to $10 for each dollar spent on Condition Monitoring. One DuPont plant saved $22 for each dollar spent. Another DuPont plant increased up-time from 50% to 86% through

the use of Condition Monitoring. In a study by Mobil Oil, the company documented a 62% reduction in rotating equipment failures through Condition Monitoring.
Maintenance Technology, February 1994

- Houston Lighting & Power, using infrared thermography inspection in a predictive maintenance program to reduce maintenance costs, ". . . realized a potential savings totaling $13 million by averting forced outages and increasing plant efficiency."
Power Engineering, June 1994
- Finding and repairing 324 air leaks saves company $52,000/year.
Maintenance Technology, January 1994
- Extending oil drain intervals enables company to save on oil, machine downtime, and labor hours.
Industrial Maintenance & Plant Operation, January 1996

Predictive Maintenance and Condition Monitoring Definitions

Condition Monitoring: The continuous or periodic monitoring and diagnosis of equipment and components in order to forecast failures.

Maintenance: Work performed to retain facility, system, or equipment capacity/availability in a safe and reliable condition to retain function.

Predictive Maintenance: See Condition Monitoring.

Preventive Maintenance: Work performed at set intervals (calendar, run time, cycles, etc.) to prevent failure. Interval-based preventive maintenance is often the appropriate approach to maintenance for systems that have defined, age-related wear patterns. Condition monitoring is a subset of preventive maintenance because the inspection is interval-based.

Combined Use of Various Techniques to Determine Equipment Condition

	Ultrasonic Noise	Infrared Thermography	Motor Circuit Evaluation	Wear Particles	Lubricant Condition	Temperature	Vibration
Temperature							•
Lubricant Condition							
Wear Particles					•	•	•
Motor Circuit Evaluation							•
Infrared Thermography			•			•	•
Ultrasonic Noise		•	•	•		•	•
Visual		•		•		•	

Figure 4.6

Proactive Maintenance: Those work items that are performed to avoid functional degradation. In the operation and maintenance portion of a system's life cycle, this could include establishing refurbishment standards, analyzing failures to identify trends or root causes, replacing obsolete components prior to failure, and other actions that seek out failure modes and identify solutions to prevent them.

Repair: The work needed to restore function. Repair can range from minor adjustment to major overhaul.

Run-to-Failure: Often called *reactive maintenance* or *breakdown maintenance*, work is performed only when deterioration in a machine's condition causes a functional failure. Run-to-Failure is typically associated with a high percentage of the total work being unplanned maintenance, high replacement part inventories, and the inefficient use of maintenance personnel. However, Run-to-Failure is acceptable for noncritical, low-cost equipment.

Management Reporting and Analysis

It has been said that:

- To manage, you must have controls
- To have controls, you must have measurement
- To have measurement, you must have information
- To have information, you must collect data

Reports are time-consuming to compile manually. Computer-generated output can vary from a simple database or spreadsheet, sorted and compiled into a report to a sophisticated CMMS. Companies using a CMMS have an advantage, since many of these reports are included in the system. If they are not, then there is usually a report writer that can be used to construct these reports.

Computerized reporting has another advantage, especially with the relational databases; they can also produce meaningful graphic representations of the information. These graphs can be more helpful in describing trends and patterns than columns of figures and can also be included in reports to upper management. It would be beneficial to graph as many of these reports as possible.

Maintenance Reports

The following reports are organized into categories with their needed review by the maintenance staff. The last general category is for reports that should be produced on an "as needed" basis.

I. Daily Reports

These are reports that should be produced daily for review by the appropriate maintenance personnel and managers.

Work Summary Report

This report lists each work order that is either currently in progress or that has been closed out in the past day. This report allows a quick look at the prior day's activity, and shows estimated versus actual totals for the following categories:

- Maintenance labor
- Maintenance materials
- Equipment downtime

The report should be divided into two sections: *work orders completed* and *work orders still in process.* Each of these sections should be sorted by priority. This should range from emergency, to planned and scheduled, to preventive maintenance, to routine. This summary of work performed allows the manager to answer quickly any questions that operations may ask during a daily review meeting.

Schedule Progress Report

This report shows a listing of only the work scheduled for the week and the present status of each of the work orders. It should show the actual versus estimated figures, similar to the work order summary report. The difference is that only the work orders appearing on the schedule will be listed. This report should conclude with a summary showing the number of work orders scheduled to be closed out during the week versus those that have been closed out. A percentage could even be included as *percent of scheduled work completed.* This allows the manager to monitor the progress of the schedule and make any adjustments necessary during the week to ensure schedule completion.

Emergency Report

This report lists all of the emergency work requested in the past day. It should be in two parts. The first should be a line summary showing labor hours and materials dollars used. The second part of the report should be more detailed, showing what craft technicians worked on the job, what parts were used, and any detail or completion notes. This allows the manager a quick look at the breakdowns/emergency work for the last day. If a particular breakdown/emergency needs clarification, the second part of the report provides the required details.

Reorder Report

This is a list of all inventory items that have reached their reorder point during the past day. Depending on the maintenance and purchasing relationship, this list is used to generate purchase requisitions or purchase orders. If maintenance is not involved in the ordering, it may be an information-only report for their purposes. If the organization is multi-warehouse, this report should be divided by warehouse. This allows the option of transferring between warehouses or shows the need for the order. The report should show on-hand, reserved, and minimum quantities for each item listed.

II. Weekly Reports

In addition to the daily reports, each week there is additional information required to properly manage maintenance. This section provides examples of these reports.

Schedule Compliance Report

This report compares the results from the previous week's activity to the schedule for the same week. It should begin with *all work completed, all work not completed*, and finally, *all work not started*. This allows for detailed analysis of the week's activity. There should also be a comparison of time allotted for emergency activities versus the time actually used for emergency activities. This will give an indication of why more or less work was completed than scheduled. There could also be a comparison of work scheduled to be worked on versus total work completed. This summary figure could be used as the efficiency percentage. Tracking this percentage over a 6-12 month time period provides a complete picture on scheduling efficiency.

PM Compliance Report

This report lists the PM tasks scheduled to be completed for the previous week and their present status. This would be broken down by:

- Tasks completed as scheduled
- Tasks started as scheduled, but not completed
- Tasks scheduled, but not started

This report gives a quick overview of the status of the preventive maintenance program for the previous week. A detailed section of the report may also be used to show who performed the work, what parts were used, completion comments, and any related work orders that were written.

PM Due/Overdue Report

This report lists all overdue PM tasks. It should sort from the oldest to the newest tasks. If possible, the report could start with a specified time period, *such as over eight weeks overdue,* and sort down to the ones that became overdue last week. This allows for a quick look at the numbers critically overdue, down to those that are just overdue. This report is most beneficial if a summary line of the numbers and percentage of work contained in each overdue category is listed. This report should be a good indicator of the condition of the PM program.

Work Order Status Report

This is a listing of all work orders in the backlog. It should be available in several formats, listed below with a brief explanation of their need:

- *By Department*

This report is distributed to the departmental supervisors so they can see the status of all of the work requested for their department. This allows them to see the status of their department (including "pet" projects) without continually calling Maintenance.

• *By Equipment*
This report allows quick access to all work presently requested for all equipment. This serves as a quick reference for anyone requesting work to prevent duplicate work orders in the system. They can review if a similar job is requested for the equipment.

• *By Planner/Supervisor*
This report gives some indication of the workload for both the planner and supervisor. Balance can be achieved in the organization by offloading work from one planner or supervisor to others during peak work times.

• *Past Due Work Order Report*
This report is a listing of all work orders in the current backlog sorted by the "date needed" field. This report should be sorted from the most to the least overdue. Work orders without any date-needed information should not be included on this report. Status of all work orders should be included on this report (except completed). The reason is that if a work order is being held up for materials unavailable (for example), then if it is overdue by a considerable amount, the planner may be able to find some manner of expediting or substituting the materials to complete the work order. This report can also provide some good feedback information to engineering, stores, and operations.

• *Backlog By Craft/Crew/Department*
This report is used to track the amount of work for each craft or crew group. In some organizations, the work may even be backlogged by department. This should be a summary type report, showing the totals for each craft group in total hours and then in number of weeks work. A second option should be a trending report showing the weekly backlog for each craft group for the last year. This will highlight trends, seasonal peaks, and so forth, allowing management to make sound, justifiable staffing decisions.

III. Monthly Reports

In addition to looking at the daily and weekly reports, other information is more meaningful to review on a monthly basis.

Completed Work Orders Report

This report is in two parts (summary and detail), listing all work orders completed during the month. The summary report shows total work orders closed; actuals versus estimates for the man-hours with labor costs; material costs; and contractor costs. The summary should be available by several sorts:

- By requesting department
- By equipment type
- By equipment ID

This style of summary report assists in spotting trouble areas. The detail report is a work-order-by-work-order listing . The sorts used for the summary report should also be used for the detail reports. This allows the manager the ability to investigate any discrepancies in detail.

Planning Effectiveness Report

This report is used to highlight the effectiveness of the planners. The report should show all the work orders for each planner, listing the actual costs compared to the planned costs. In some computerized systems it may be helpful to specify a tolerance percentage to reduce the amount of information the manager has to review. A summary by planner is beneficial, with a listing of the most effective to least effective planner at the conclusion of the report. Any major discrepancies should be reviewed at a detailed level.

Supervisory Effectiveness Report

This report is identical to the planners report except that it shows the effectiveness of the supervisors. The report should show all the work orders for each supervisor, listing the actual costs compared to the planned costs. Again computerized systems allow use of a tolerance percentage to reduce the amount of information the manager has to review. A summary by supervisor can list the most effective to the least effective supervisor, with major discrepancies evaluated in detail.

Downtime Report

This report compares the actual versus estimated downtime for all closed work orders for the month. This is a critical report since it includes not just the planned and scheduled work, but also the emergency/breakdown work. Rather than being a measure of planning and supervising effectiveness, this report is a measure of the effectiveness of the entire organization. As has been mentioned in previous material, PM programs, planning, supervising, stores, purchasing, and organizational coordination all play a part in controlling downtime. A specified tolerance could be beneficial in keeping the report shorter. Any major discrepancies should be investigated, with appropriate correctional measures taken.

Budget Variance Report

This report should compare the actual figures for all maintenance expenses to the budgeted figures for each category. Depending on the plant or facility budgeting procedures, they may be broken into labor, material, contractor, tools, and so forth. The report should provide an opportunity for the manager to correct potential problems before it is too late.

Conclusion

In many organizations, the facilities or plant engineer has responsibility for the maintenance function. As shown by the material in this chapter, properly managing the maintenance function is time-consuming and requires attention to detail and focus. Yet, with all of the facility or plant manager's other responsibilities, the maintenance function may wind up being delegated to secondary status. This should not be allowed to happen. The maintenance function in the organization can be a *profit generator*. First, reducing expenses by controlling maintenance costs is an expense avoidance. The money not spent translates directly to the company's bottom line. In addition to the cost avoidance savings, there are also the savings generated by the plant assets performing at their designed rate and function. This helps the company avoid investing in unnecessary assets, which keeps the return on net assets high. The result is shareholders confident in the company's ability to manage its business.

Resource Publications

Applied Management Engineering, PC, and Harvey H. Kaiser. *Maintenance Management Audit*. R.S. Means Company, Inc., 1991.

Facilities Maintenance & Repair Cost Data 1999. Published annually by R.S. Means Company, Inc. (An excerpt from this book appears in the Appendix.)

Liska, Roger W., PE, AIC. *Means Facilities Maintenance Standards*. R.S. Means Company, Inc., 1988.

Magee, Gregory H., PE. *Facilities Maintenance Management*. R.S. Means Company, Inc., 1988.

Matulionis, Raymond C., and Joan C. Freitag, eds. *Preventive Maintenance of Buildings*. Van Nostrand Reinhold, 1991.

Cost Planning & Estimating for Facilities Maintenance. R.S. Means Company, Inc., 1996.

Westerkamp, Thomas A. *Maintenance Manager's Standard Manual*. Prentice Hall, 1993.

Wireman, Terry. *Computerized Maintenance Management Systems, 2nd ed.* New York Industrial Press, 1994.

Wireman, Terry. *Industrial Maintenance*. Reston Publishing Company, 1983.

Wireman, Terry. *Inspection and Training for T.P.M.* New York Industrial Press, 1992.

Wireman, Terry. *Preventive Maintenance*. Reston Publishing Company, 1984.

Wireman, Terry. *Total Productive Maintenance: An American Approach*. New York Industrial Press, 1991.

Wireman, Terry. *World Class Maintenance Management*. New York Industrial Press, 1990.

Chapter 5

Energy Efficiencies

Chapter 5 Energy Efficiencies

Energy in all its varied forms—electricity, steam, natural gas, etc.,—is a significant overhead cost for all facilities. Facilities engineers are responsible for controlling operating costs, and using energy appropriately will save your facility money.

Flexibility in choosing energy types is one method of reducing costs of operations. Some forms of energy, for example, vary in cost depending on the time of year, time of day or government policy. To take advantage of these and other cost-saving opportunities, you need to know how and why your facility uses energy. Then, utilize that knowledge to choose the type of energy that will help you maintain an efficient operation at the best cost possible.

A first step in moving your facility towards optimum energy efficiency is to begin with an energy audit. Whether you choose to ultimately hire an energy management firm, or use your own staff to make energy-saving changes, it is important to understand your facility's energy use.

The following section is reprinted with permission from *Total Productive Facilities Management*, by Richard W. Sievert, Jr., R.S. Means Co., Inc.

Energy Costs

Are you paying more for electricity and fuel than is necessary? Are your facilities energy-efficient? As energy is used more efficiently, product cost can be reduced and profits improved. An energy audit is necessary to identify energy consumers within your building, understand specific energy usage patterns within the building, and identify opportunities for reducing energy consumption.

Measurable goals should be set for reducing facility energy consumption. An orderly accounting of energy used in a building should be compared against some standard of performance or budget; for example, kilowatts of energy consumed, or cost per given time period. Opportunities for energy conservation can typically be found in HVAC equipment, air distribution systems, temperature control systems, electrical distribution and lighting systems, and production equipment. Heat losses and gains through floors, walls, ceilings, roofs, doors, and windows, and occupancy levels and activities also affect energy use.

Significant energy savings often may be realized with relatively minor modifications or investment. Changes in operation or maintenance procedures may be all that is necessary to achieve significant energy savings. Recalibration of controls, regular filter changes, and coil cleaning are examples of simple procedures that save energy. The energy conservation measure may also be as simple as shutting off lights and setting back temperatures during unoccupied periods. Reduced lighting levels, and using energy-saving motors and luminaries or task lighting can reduce energy consumption as well. If a pump or fan is running improperly, adjustments will save money. Sometimes, larger energy-consuming air handling equipment and systems are operated during evenings or weekend unoccupied periods when a smaller, more efficient system could be used instead to treat a specific area, such as a computer room or break room.

Figure 5.1 is an outline of procedures for conducting a facilities energy audit.

Some heating and cooling systems may have been oversized back when energy costs were relatively inexpensive. In other situations, in order to reduce front-end engineering costs and minimize the risk of under-sizing, engineers may have been inclined to oversize systems to avoid the analysis necessary for optimum system sizing and efficient operation. Boilers, pumps, furnaces, chillers, fans, lighting systems and motors may all have been oversized. Heating and cooling loads should be calculated to determine how much heating and cooling must actually be produced to meet the space loads and to condition ventilation air. Heat generated from lights, people, and equipment is a primary source of indoor cooling loads. Duct distribution systems should be examined for proper sizing, layout, and air balancing. Duct leakage should be minimal. Leaking water, steam, or inert gas may seem quite small as it escapes into the air, but over time it can represent a sizable amount of energy. Based on past experience, the scheduling of chillers is one improvement that saves large amounts of energy. Figure 5.2 shows energy performance ratios useful in conducting an audit and analyzing the results.

Today there are a variety of sophisticated energy cost-controlling devices. These include optimum start/stop of equipment, variable frequency drives which reduce fan speeds when demand decreases, outdoor economizers (for free cooling), electrical demand charge limiting, computerized energy management systems, duty cycling, night setbacks, chiller optimization, and process heat recovery systems.

You will want to set goals for reduction of energy usage based on the audit and select the energy conservation measures with the highest payback. Once an energy conservation program is established, building systems will need a routine examination to monitor energy usage and ensure efficiency.

Energy-Operational Audit (Typical Procedures)

I. Initial Review

A. Building record drawings
   1. Mechanical
   2. Controls
   3. Electrical
   4. Architectural

B. Utility records and operating logs
   1. Gas
   2. Electric
   3. Oil
   4. Operation logs

Note: These records should be reviewed for at least the two previous years.

C. Utilization
   1. Present usage
   2. Operating schedule
   3. Environmental requirements for space conditioning

Note: For some facilities it may be necessary to electronically monitor the electrical power usage and demand characteristics.

D. Maintenance and repair records for all major pieces of equipment

II. System Review–Office Phase

A. Review records and documents
   1. Building record drawings
   2. Utility records
   3. Maintenance records

B. Determine the type and kind of mechanical equipment and controls intended to serve specific areas, as well as the characteristics and limitations of the equipment and controls.

C. Compare original equipment design intent to equipment capability and operating requirements.

D. Analyze HVAC systems with regard to comfort or process conditioning criteria and load handling capability
   1. Part load performance
   2. Full load performance

E. Evaluate controls, operating sequences and control parameters

F. Interface with utility companies
   1. Discuss systems revisions to achieve more favorable rate structures
   2. Review economic feasibility of converting from one energy source to another

III. System Review–Field Phase

A. Survey property for comformance to plans
   1. Mechanical and equipment schedule
   2. Electrical
   3. Architectural

B. Interview management and operating personnel to determine inadequacies of present systems and controls

C. Perform an in-depth site survey of each operating system

D. Conduct operational tests to validate the performance of controls, equipment and systems as required

E. Perform selected field measurements with engineering test equipment

IV. Analysis and Report Phase

A. Perform preliminary load calculations to compare system capabilities with present demand

B. Interview management and operating personnel to determine inadequacies of present systems and controls
   1. Automated controls
   2. Mechanical equipment status
   3. Energy utilization
   4. Useful economic life vs. present age and condition

C. Identify alternate control sequences for mechanical systems
   1. Intermittent operation based on time clock and/or temperature parameters

D. Mechanical services
   1. Preventive maintenance
   2. Revision or remedial work on systems and equipment

R.S. Means Co., Inc., *Total Productive Facilities Management*

Figure 5.1

The Effects of Utility Deregulation on Energy Costs

The deregulation of utilities will be an increasingly important development for facility managers as competition is introduced into these businesses. In the old structure, electric utility companies owned their own power plants, transmission systems, and distribution networks, and exclusively covered a certain geographic region. In the new system, the three components are separated, transmission and distribution continue to be regulated, and generation (deregulated) is available from more than one producer or broker. The other option is continuing to purchase power from the same local utility, with the three components bundled together.

The natural gas industry has already been similarly deregulated in some areas, where customers purchase from a selected provider. The gas is transported from a distribution point by the local utility.

It is expected that deregulation will bring about significant savings to energy purchasers due to better efficiencies and an ability to select not only providers, but specific services. Legislation is in development for the setting of deregulation guidelines.

Energy Performance Ratios

$$\frac{\text{Kilowatt–Hours}}{\text{Square Foot}}$$

A month-by-month record of a building's electric energy use pinpoints the times when energy use is largest and therefore energy saving opportunities are greatest.

$$\frac{\text{Square Foot}}{\text{Cooling (Tons of Air Conditioning)}}$$

Data needed to determine cooling capacity requirements include: production equipment heat rejection rates, occupant levels and activities, quantity of outside air for exhaust make-up and ventilation, humidity control, indoor temperature requirements and heat gain through the building envelope. Cooling expressed in tons of air conditioning represents a substantial percentage of overall energy cost. According to IFMA's *Benchmarks II Report*, for commercial offices, the average cooling requirement is 47 S.F. per cooling ton. The average monthly utility consumption is 2.2 kilowatt-hours per S.F. (Actual figures vary significantly by geographical region, type of building construction, and occupant density.)

$$\frac{\text{Electric Cost}}{\text{Square Foot}}$$

For example purposes, energy costs may range from $3 to over $6 per square foot. Energy cost may represent from 1% to over 4% of sales revenues for a printer.

R.S. Means Co., Inc., *Total Productive Facilities Management*

Figure 5.2

U.S. Utility Deregulation Legislation

Similar to the telecommunications industry the utilities industry is undergoing changes as a result of federal legislation designed to create competition and in theory lower costs to the consumer. Two major acts of U.S. federal legislation are changing the way natural gas and electrical utilities operate today and in the future.

Natural Gas: The Federal Energy Regulatory Commission enacted "FERC 686" on April 8, 1992 to require competition over time within the natural gas pipeline industry. FERC 686 requires the natural gas pipeline companies to separate the sales of natural gas from the transportation costs of the gas. This has drastically changed the way local gas companies purchase and receive natural gas. This also creates competition in the wholesale natural gas industry.

More and more states are now allowing for the retail commodity sale of natural gas from the "Well-Head" (or source of generation) plus cost of regulated pipeline services to the "City Gate" (or point of connection to the interstate pipeline companies) and the "Local Distribution Company" (or local gas distributor).

Electricity: The legislation for the electric transmission industry "FERC 888" was enacted on April 24, 1996. FERC 888 requires the electric transmission companies to allow any qualified company to utilize their transmission wires for a set fee. This allows for many new forms of generation and transmission of electricity thus creating competition in the electric wholesale power industry. The industry has started to move the competition from the wholesale side to the retail side.

Also more and more states are now allowing for the retail commodity sale of electricity from the "Generator" (or source of generation) plus cost of transmission services to the "Pool Transmission Facility (PTF)" (or point of connection to the transmission companies) and the "Local Distribution Company" (or local electric distributor). There is more information available on FERC's web page at www.ferc.fed.us.

Purchasing Energy: New Choices

There are many trends in the operation of a facility or plant that affect the type and/or amount of energy required. The way you purchase energy is also a matter of cost vs. production requirements. Since the deregulation of the natural gas industry, many facilities now use a combination of combustion fuels for the operation of boilers.

Energy Service Companies

Deregulation means facilities will be able to buy their power from their traditional utilities or from other utilities, or brokers who purchase power for resale. It also means a company can now produce its own power and sell the extra to a utility company. Countless scenarios are possible. It is a complex question that is not easily answered without outside advice.

Outsourcing is a major trend in the facilities industry and the main reason is the high level of technical expertise required for each trade or responsibility. Energy analysis and procurement is one area that is being contracted. The requirements of energy procurement and energy management have become very complicated and are continuously changing. Choosing an Energy Service Company is a difficult task and an important one.

Most *Energy Service Companies,* or ESCOs, have developed due to the mandated Demand Side Management (DSM) programs from the electric industry. Some of the companies even started as part of a local distribution company and were spun off to become their own company.

Before choosing an ESCO you need to:

- Assemble copies of two years of your energy bills including fuel oil, natural gas or propane, and electricity.
- Look at your consumption trends including peak periods or any penalties you may be paying.
- Get a copy of the tariffs from your local distribution companies (electric and natural gas).
- Also get a copy of the current contract with your oil supplier and/or propane supplier.
- If your contracts and/or tariffs are tied to an index for the cost of the commodity, look for the index. Many indexes can be found in one of the following web pages: www.nymex.com or www.powermarketers.com or www.energysource.com.

Once all of your historical data is assembled you need to decide which one of the many types of energy service companies your facility requires. They include:

Arbitrage Specialists:

Arbitrage specialists are consultants that analyze a facility's energy use and develop systems for the facility to change fuel based upon cost of operation. These consultants have a good financial background and must have an understanding of your facility because they tell the plant engineer which fuel to use. In some cases they act as your agent for the procurement of fuels.

Fuel Arbitrage is a method of operation that takes advantage of a facility's flexibility to reduce the cost of operations. This allows the plant engineer to change fuels based upon cost of procurement and cost of operation. Often the plant engineer must take into consideration many other factors that may affect the total cost of operation including but not limited to:

- Energy required for preheating of fuels,
- Pumping requirements, and
- Emission requirements of the facility.

In the purchasing of fuels, there are of course external forces that can affect the cost of operations including: weather, fuel markets and foreign policies. Depending upon the method of fuel procurement, many outside forces can be mitigated.

Basically your business's ability to bear risk is directly proportional to the potential savings that can be found in the fuels markets. For example, "futures" are a way of financing the pumping and development of wells. If a company decides to buy "futures" for its procurement of energy it is counting on the fact that a particular well will produce the amount of fuel specified in the contract. You must also secure a method of transporting that fuel from the well to the company's facility. Intrastate oil and natural gas pipelines, ship or truck are all possible methods of transportation. All of these methods have contractual obligations from both parties. Depending upon the size and quantity of fuel consumed you may want to contract with a fuel broker or marketer to meet the needs of your facility while passing on to them some of the risks associated with fuel procurement.

Energy Marketers: These companies are the commodity brokers of the deregulated energy business. They secure large volumes of natural gas, heating oil, propane and electricity from the source and arrange for transportation to your facility. By purchasing large quantities they benefit by economies of scale. Typically they have many pipeline contracts to use as well as storage in some of the local tank farms or LNG facilities to handle peak loads. Depending upon the size, some companies hedge futures by purchasing options for future contracts of delivery. By doing this they commit to buy a set volume in a future month without committing the total cost of the commodity. This method reduces some of the risk of a volatile fuel market.

Demand Side Management Companies:

These companies specialize in reducing baseline energy consumption (typically electricity). They will analyze "time of use" data and your historical electric bills to look for unusual spikes in the demand curve. Typically demand is measured in kVA, which is a reflection of your facility's reactive power. Therefore many demand side management companies can save you money quickly by calculating the required power factor correction and installing the proper capacitors.

Electric demand is typically the peak consumption of electricity during any 15-minute period of time within one month. The focus of this process is to determine how to modify your equipment or its operation to minimize the peaks. Peak power consumption also reduces the life of electrical distribution equipment. Minimizing the peak periods reduces your electric costs and increases the life of electric distribution equipment.

Ratchet Clauses: Some utilities have a *ratchet clause* on the demand portion of the tariff. A ratchet is typically used to supplant the capacity of the service portion of your electric costs whereby the utility measures the peak electric consumption in a 15-minute period and uses that peak for an extended period such as six months or until a higher peak occurs, whichever comes first.

A Demand-Side Energy Service Company (ESCO) usually offers:

- Design and development services,
- Financing for the project,
- Installing and maintaining new equipment,
- Monitoring of energy consumption to verify savings, and
- Taking responsibility for the project generating the energy savings predicted.

ESCOs address numerous issues related to energy use. They now look at the available energy sources or the supply-side of energy management and provide numerous ancillary services—such as energy load shaping, power quality assistance, energy brokering, and even hazardous waste disposal and water and sewage management services that go well beyond the original energy conservation (or demand-side) issues. Many ESCOs still focus on either supply or demand side energy management and it may be best to have two ESCOs and get the best of each type of expertise.

Whether it is better energy load management, more efficient lighting and HVAC systems, new motors on the production floor or a new centralized power control system, installation of more energy-efficient equipment and controls is still at the heart of much of the work performed by ESCOs. The increased energy-efficiency of the customer is essential because ESCOs work is performance-based. The amount the energy service company earns is dependent on the energy savings of the customer.

Typically a company's contract with an ESCO runs from 7 to 10 years and maintenance of the new equipment during that time is usually included as part of the contract price. Training of the company's personnel in future maintenance is often included to assure a smooth transition at the end of the contract.

Fuel Blind Approach Energy Service Company:

This is the most comprehensive approach to energy service. However, the cost tends to be much higher. These companies analyze all of the above options as well as how your company operates to look at each operation or process from a cost of operation approach. Many of the options require life cycle costing analysis to calculate the true costs of equipment installation, operating costs, the cost of the service and payback through energy savings.

Submetering is one of the truly analytical ways to learn about electricity use at your facility and to calculate savings after energy-saving improvements are made. Submetering has actually become a necessity in the operating of any plant and is explained in detail in "Analyzing Your Energy Profile" later in this chapter. Many different departments and operations become isolated profit centers for a business, and the energy use of each department or profit center must be determined. By disaggregating the various loads throughout a facility you become inherently aware of the individual load profiles for each subprocess. If you compare the load profiles for each operation to the cost of energy consumed, it may be cost-efficient to put an operation on a different schedule, a second shift for example.

Many other fuel blind approach methods look at combining systems when they use similar processes. If the operation of one load center requires a small process and another load center utilizes a similar process, combining the two may well produce a lower cost of operation.

An example of this is a hospital where boilers are operated year round for many reasons, including domestic water production, moisture reduction in the summer and heat in the winter. In this same facility electric autoclaves are used to sterilize surgical equipment. The plant engineer wants to operate the autoclaves with the boilers but finds that the steam pressure required is inadequately supplied by the boilers. However, by installing a steam compressor to the existing steam distribution you reduce the electric demand by 80% and increase the load to the boilers. This increase in load of the boilers brings the firing range of the burners to a much more efficient level. Therefore the net energy in to energy out is vastly improved. This is one example. However, you must also take into consideration that the reduction in resistive load may impact the power factor of the facility.

Now that you have an understanding of the different approaches, you should develop a relationship with an energy service provider that meets your facility's requirements. Look for a provider that has performed well for similar operations. Also, look for corporate guarantees that have the proper backing to stand up if the energy conservation measures do not deliver the savings promised.

An important issue for a plant manager considering hiring an Energy Service Company is to assess the competency and experience of the firm. Recent publicity of the importance of energy-savings to a company's bottom line and all the new developments in energy-saving equipment have encouraged many individuals to jump into energy-saving consulting and call themselves an ESCO. Of course, not all of these firms offer the same quality and types of services. It is important to thoroughly research the abilities of the ESCO and to understand what the firm can and can not offer.

- Is it a demand-side or supply-side focused firm you are looking at?
- What is their previous experience with the equipment and operations your company uses?
- Are they accredited by NAESCO?

NAESCO is the National Association of Energy Service Companies and thoroughly examines an ESCO's competency and business practices before it grants accreditation. At publication, there were 18 accredited ESCOs and one accredited lighting service company. NAESCO may be contacted at 1615 M Street, NW, Suite 800, Washington, DC 20036. Telephone: (202) 822-0950. Website: www.naesco.org.

Energy and the Environment

While the government is lessening regulations on utilities, industrial processes are coming under greater scrutiny. During the rapid industrial growth over the past hundred years, cost of operation was the key to profitability. Many industrial processes created byproducts that have since been found to be potentially hazardous, and governments have responded by developing regulations that control hazardous conditions. Industry has been challenged to keep up with the regulations while still maintaining profitability. New technology has been developed to address these regulations and keep industry efficient while still staying profitable.

"Greenhouse" Emissions

International Cooperation: As the regulations have been developed, research has also found that the pollution or wastes can migrate between countries. One of the recent developments in waste mitigation is the "Kyoto Protocol", which is an agreement between all of the global industrial nations to create a "Positive Global Climate Change". US Secretary of Energy Bill Richardson said, "Significant change toward solving the challenge of climate change will come directly from retail competition in electricity markets. As competition begins to flourish in the industry, utilities will respond by increasing power plant efficiencies to reduce costs. New business opportunities will present themselves, as companies bundle together electricity sales and energy-solving technologies. The Energy Department estimates that the Clinton Administration's Competition Plan will reduce greenhouse gas emissions by up to 40 million metric tons of carbon equivalent by 2010." Facilities professionals will be charged with reducing greenhouse gasses by cutting the amount of energy used within their company's facilities.

New Opportunities: The old paradigm in industry was that significant action on reducing greenhouse gas emissions runs counter to the bottom line. The new paradigm—one embraced and pushed forward by the Department of Energy—is that clean business is good business. New technologies mean we can expand the economy at the same time we improve the environment. The DOE is an incubator for many of the developing technologies that will positively impact greenhouse gas emissions and our nation's economy. There are many research papers and information regarding energy on the DOE's web page: www.doe.gov.

In many states the building codes have been changing to reflect maximum parameters for energy consumption. One example is Massachusetts, which takes into consideration the energy use of a proposed building. Baseboard electric heat as a sole source of heating, for example, would not meet the revised energy code. For more information or examples see the Massachusetts Safety Department Energy Forum on the web: www.magnet.state.ma.us.

Deregulation of utilities in the United States offers plants and other large energy consumers an opportunity to benefit from competition between the traditional utility and new energy providers. Utilities generating power will no longer have control of transmission and/or distribution. It will be possible to buy power from generating facilities at great distances, perhaps realizing a considerable savings. Transmission (the movement in bulk of electricity) and distribution (the delivery of electricity to the end consumer) costs will be an additional cost. The transmission and distribution utilities remain under regulation and will likely retain exclusive territories. The opportunity therefore exists only in the generating of electricity—it makes it possible for more companies to enter the business. The electricity will still be carried as it was, through a regulated distribution and transmission system. The changes presented by regulation certainly center attention on the cost of electricity. It also highlights the importance of controlling energy consumption to save money.

Analyzing Your Energy Profile

Before a company can make informed decisions on which provider(s) to use, a thorough analysis of the demand and consumption of energy must be done. It can not be done with the utility bill alone, but it is a place to start. Almost all utility bills for commercial and industrial consumers include two sections:

- Usage (or Consumption) which is measured in kilowatt hours (kWh).
- Demand which is measured in kilowatts (kW).

Usage and demand both play a role in determining the facility's electricity rate. The rate is determined by the utility by examining electrical usage throughout the day and charging by the highest rate of consumption. The power company should be able to show you exactly how to read and understand your bill. Refer to Chapter 9, "Electrical Engineering" for further information on reading your utility bill.

Beyond your current power charges, you need to understand exactly how and when your major equipment is demanding and using power. Your current utility may offer energy consulting services that can assist you in improving your energy efficiency profile by evaluating current needs and suggesting energy-saving projects before you invest in capital equipment or change your operations. When working with utility energy consultants, you should find out if they are Certified Energy Managers by the Association of Energy Engineers. This association may be contacted at Association of Energy Engineers, 4025 Pleasantdale Rd., Suite 420, Atlanta, GA 30340, www.aeecenter.org.

Some experts say that to get the best possible price on your facility's energy needs, you need to know your utility use in 15-minute increments for at least a year. This allows assessments by time of day, week, season and year. The lowest electricity rates go to the companies with the flattest load profile because their use is predictable. Bringing down the peaks in your facility's load profile will help cut your energy costs. Demand charges are many times greater than consumption charges, so using submetering to reduce peak demand and flatten load profile is an important step in reducing energy costs.

Submetering

To achieve the best possible understanding of power usage and demand at your facility, *submetering* is a necessity. Submeters are metering devices that are able to monitor usage. Installed after the main meter, a submeter is able to monitor the energy demand and usage of a particular location, circuit or user. There are three types of submeters: socket or current transformer type, electromechanical and electronic. The main advantage of submetering is being able to pinpoint energy usage. Whether the submeters are installed for departments, processes, or tenant monitoring, they provide valuable energy data that can help reduce demand and use.

Electronic submeters are usually the current choice because they are:

- Compact, less expensive, and easier to install because they require no feedthrough wiring or subpanels.
- Precise—currently certified to ANSI C12.1 and C12.16 accuracy.
- Provide a range of functions that allow evaluation of individual machines and processes—information that can be used for demand-side management such as demand shedding.
- Can be hooked up to a PC to provide remote real-time monitoring and control.

Analyzing electrical demand and then re-thinking and rescheduling processes will help lower energy costs, and consequently, operating costs. Energy management systems that monitor electrical use are helpful in tracking and then lowering peak demand. Once target levels are determined, the monitoring system can trigger an alarm that will allow plant staff to switch off loads. Switching on and off in cycles equipment, such as freezers, air handlers and chillers that do not need to be in continuous operation, can help lower peak demand.

The following section is reprinted with permission from *Understanding Building Automation Systems*, by Reinhold A. Carlson, PE and Robert A. Di Giandomenico, R.S. Means Co., Inc.

HVAC Energy Management

A major source of energy savings, in most buildings, is through reduced run time of HVAC systems. Electricity is saved by reducing fan and/or pump operations. Additional savings are achieved through reduced outdoor air intake, thereby relieving a burden on the heating and cooling equipment. Some energy management programs use outdoor air for cooling. This reduces the load on the mechanical cooling equipment, resulting in reduced energy consumption. Reset programs can maintain the temperature of air delivered to the spaces at optimum levels to satisfy comfort conditions. This eliminates the energy waste that results from overheating and subcooling.

Electrical power companies are promoting programs to reduce the demand for electricity and relieve pressure on their limited generating capacity. Many companies offer rebate incentives to encourage customers to install energy management systems that will reduce building electrical demand. For example, the replacement of inefficient light fixtures and bulbs can result in savings on electrical costs. Additional savings can be realized by reducing lighting levels, and turning off lights during unoccupied periods.

All energy management programs should be approached with a degree of caution. If not properly applied, these programs could damage expensive equipment, compromise comfort conditions, and have a negative impact on air quality. What follows is a discussion of some of the energy management programs used today.

Duty Cycle Program

The duty cycle program reduces electrical energy consumed by the fan in an HVAC system by cycling the fan on and off. The off periods are a function of space temperature, as sensed by a space temperature sensor. When space temperature is at the midpoint of a preprogrammed comfort range, the fan can be turned off for an extended period of time without affecting comfort. As space temperature approaches the end of the comfort range, the off periods are decreased (Figure 5.3).

The operator programs the duty cycle period, the comfort range in degrees Fahrenheit (for example, 68°F–78°F), minimum on and off times, and the maximum off time. These times are set to avoid the equipment damage that can result from too rapid cycling. When the space temperature is at the midpoint of the comfort range, as sensed by a space temperature sensor, loads such as fans are turned off and outdoor air dampers are closed for the maximum off time in the period. When space temperature deviates from the midpoint, the off periods shorten proportionally. At the extreme ends of the comfort range, which usually occur at outdoor design conditions, the duty cycle program cancels and the load is energized continuously.

Multiple space sensors may be used for space temperature compensated duty cycling programs. In this case, the program selects the highest space temperature sensed for cooling and the lowest temperature sensed for heating. When multiple electrical loads are duty cycled, the program automatically adjusts the cycle periods so that they do not occur simultaneously, thereby causing a power surge. This is shown in Figure 5.4.

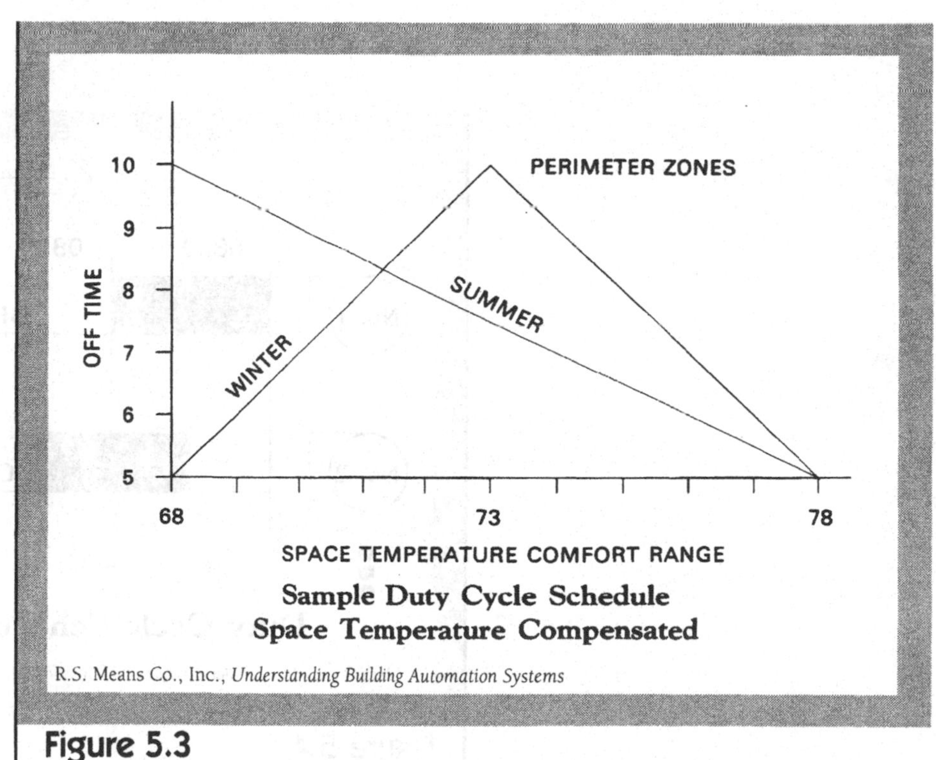

Sample Duty Cycle Schedule
Space Temperature Compensated

R.S. Means Co., Inc., *Understanding Building Automation Systems*

Figure 5.3

There are certain cautions that should be exercised in the application of the duty cycle program. First, the best candidates for this program are heating, ventilating, and air conditioning unit fans (under 100 horsepower only), exhaust fans, and hot water pumps. HVAC units and exhaust fans serving critical areas such as computer rooms, clean rooms, laboratories, operating rooms, and areas where toxic materials or plastics of any kind are stored, should not be duty cycled.

Second, many questions have been raised about the effect of duty cycling on electric motors. Research has shown that the only serious problem that may arise is overheating of the motor during startup. The two important factors to consider when duty cycling electric motors are the minimum time between starts (minimum interval) and the minimum off time. Figure 5.5 depicts guidelines for cycling induction motors.

The minimum off time is determined by the cooling requirements of the motor after running at its normal operating temperature. For small motors, three minutes is adequate cooling time before restarting. Notice in Figure 5.5 that the interval and the off time are extended as the motor size increases. As a general rule, motors larger than 100 horsepower should not be duty cycled. If HVAC fan motors of this size are installed, alternatives should be considered, such as cycling inlet vanes or mixing dampers, to achieve energy savings. However, belts and magnetic starters, if properly sized and installed, will be minimally affected by a properly applied duty cycle program.

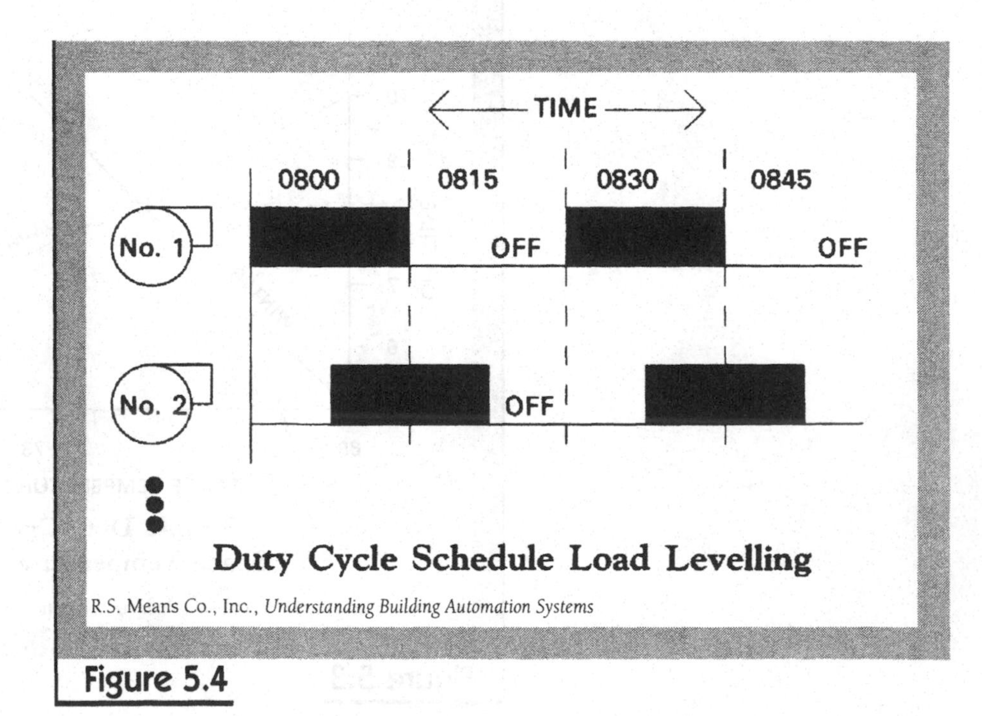

Figure 5.4

Power Demand Limiting Program

Electrical utility companies charge for energy consumption over an established billing period (usually one month). These charges cover fuel costs, operating costs, and the utility's investment in generating equipment.

The charges are calculated based on readings taken from the building's electric meter. Every commercial building has a kilowatt-hour meter. This meter records, on a continuous basis, the kilowatt-hours of electricity consumed by the building. At the end of the billing period, the utility company reads the meter and bills the customer for the energy consumed. The charges are frequently based on a sliding scale, so that the more energy consumed, the lower the unit cost. Added to the cost per kilowatt-hour is a fuel adjustment charge to cover the fluctuating cost of fuel used to produce electricity during that billing period. This is a fixed charge per kilowatt-hour regardless of the number of kilowatt-hours consumed.

Since the demand for electricity by commercial customers is heavier at certain times of the day than at others, the utilities must have equipment and transmission lines capable of handling these peaks. Although these peak demand periods occur only occasionally and for relatively short periods of time, the generating capability must nonetheless be available. *Demand charges* were established to pay for the installation and maintenance of this excess generating capability. Figure 5.6 shows a typical building load profile.

Motor Cycling Guidelines

Motor Cycling Guidelines			NEMA Design Types A–F	
HP	**1/4–10**	**10–20**	**20–50**	**50–100***
Min Interval (Minutes)	10	20	30	40
Min Off Time (Minutes)	3	5	7	7
Suggested Max Off Time	5	7	10	10
Min % kW Savings	30%	25%	23%	17%

*With reduced voltage start only

R.S. Means Co., Inc., *Understanding Building Automation Systems*

Figure 5.5

Demand charges are based on the average kilowatts consumed during a demand interval. A *demand interval* is a specific period of time established by the utility company. This period of time varies depending on the utility company, but in most cases is 15 or 30 minutes. Demand charges are based on the maximum demand (demand peak) during any demand interval in the billing period. Many utility companies have a ratchet clause which states that the customer must pay for this maximum demand for eleven months. If, during the eleven-month period, a new, higher demand peak occurs, this will establish the basis for the demand charge over the next eleven months.

The power demand limiting program monitors electrical consumption during each and every demand interval, and sheds (turns off) assigned loads as required to reduce demand. Before such a program can be implemented, however, the building owner or engineer must determine which loads in the building are "sheddable" for brief periods of time without creating serious environmental problems. For instance, shedding air-conditioning for a computer room or an operating room would not be acceptable, nor would shutting down a ventilating unit for hospital patient areas or intensive care units. Air-conditioning units and exhaust fans supplying general office areas, on the other hand, can be shed for brief periods, as can units supplying conference rooms, lobbies, coffee shops, and cafeterias. Also, the set point of large, electrically driven chillers can be set up a few degrees as a temporary demand limiting strategy.

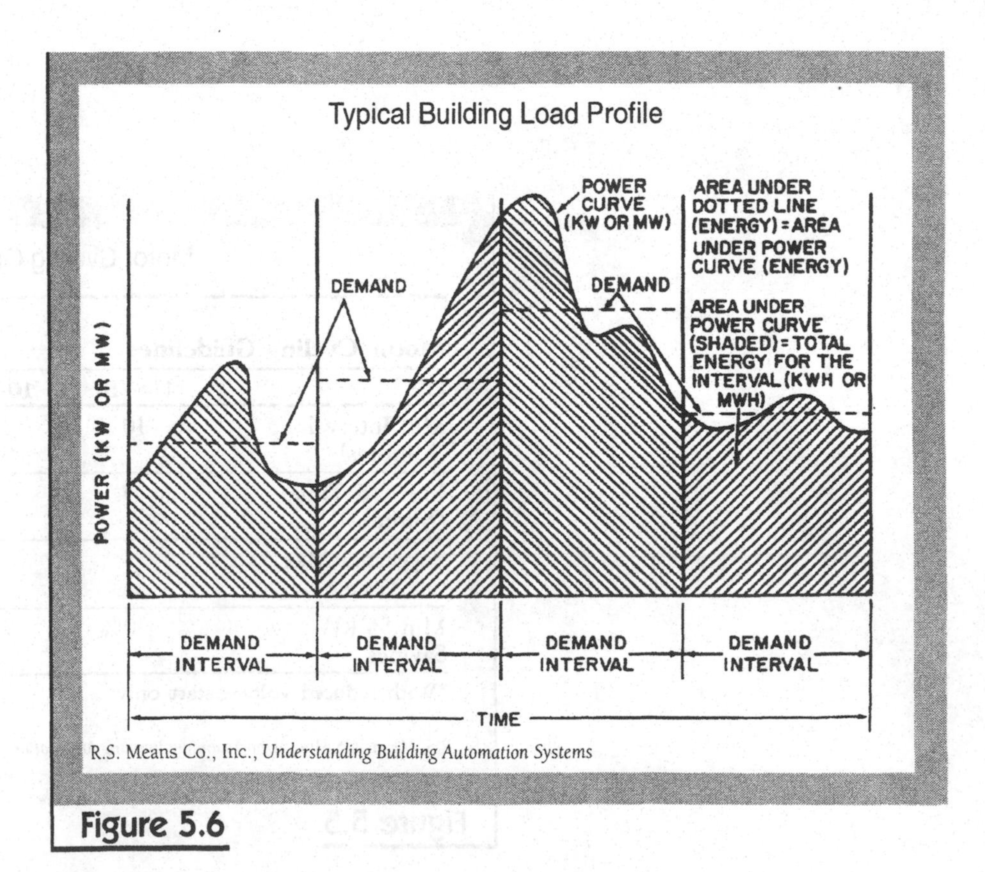

R.S. Means Co., Inc., *Understanding Building Automation Systems*

Figure 5.6

Loads are prioritized according to their importance to environmental and comfort conditions in the building. Low-priority loads are shed first and restored last. High-priority are the last to be shed and the first to be restored. If multi-speed motors are assigned to the load shed program, they can be shed in multiple steps.

The power demand limiting program, monitors power consumption from the data input of a pulse transmitter located on the building's electric meter. However, not all building meters are equipped with a pulse transmitter. The building owner should check with the power company before applying this program to determine whether or not his building has one. If the transmitter does not exist, there will be a charge by the power company for the installation.

The power demand limiting program, during the 15- or 30-minute demand interval, computes the energy consumed and forecasts the power demand at the end of the interval. After completing the forecast, the program adds or sheds loads to maintain the demand level within programmed limits.

If all assigned loads have been shed and electrical consumption continues to increase, the program automatically shifts the demand limit upward. When seasonal variations in electrical demand occur, the program compensates rather than maintain the original limits. Figure 5.7 is a sketch showing how the basic demand limiting program works.

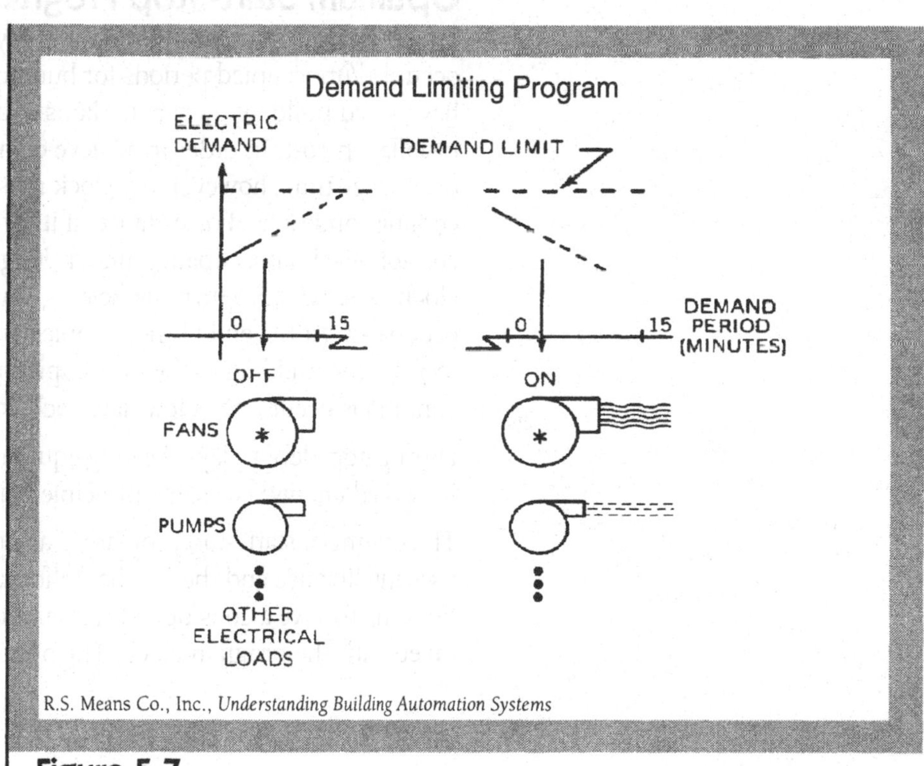

R.S. Means Co., Inc., *Understanding Building Automation Systems*

Figure 5.7

Unoccupied Period Program

With the exception of residential-type buildings, most buildings have occupied and unoccupied periods. Office buildings are typically occupied 40 to 48 hours per week, which means that they are unoccupied approximately 72% of the time. Hospitals have certain areas, such as administrative offices, gift shops, cafeterias, x-ray departments, and operating rooms which are not occupied full-time. If unoccupied buildings or building zones are heated, cooled, or ventilated to the same levels as when occupied, energy is being wasted.

Considerable energy savings can be achieved by setting temperatures back during unoccupied periods in the heating season, and up during unoccupied periods in the cooling season. Savings on hot water radiation systems can be achieved by decreasing water temperatures during the unoccupied periods. Air-handling systems, which are cycled to maintain reduced temperatures during unoccupied periods, should have their fresh air dampers closed.

The unoccupied period program, or night cycle program, is primarily a heating season function. It can, however, maintain a high space temperature limit during the cooling season, if desired. A space temperature sensor maintains space temperature at preset levels during unoccupied periods by cycling the heating or cooling source. If multiple sensors are used, the program will control from the lowest temperature in the heating season and the highest temperature in the cooling season. If humidity is of critical importance in a building or building zone, a relative humidity sensor can be installed to override the program and cycle the air-handling system to maintain a high or low space relative humidity level.

Optimum Start-Stop Program

Program clocks have been used for many years to automatically determine the occupied/unoccupied periods for buildings or building zones. These devices have saved building owners millions of dollars in energy costs at a minimal installation cost. In order to achieve comfortable building temperature at occupancy time, however, the clock must be set for outdoor design heating or cooling conditions. For instance, if it takes two hours to bring a building up to comfort levels at occupancy, under design heating conditions, the program clock must be set to turn the heating equipment on two hours prior to occupancy. At the end of the occupied period, the program clock is usually set to place the building in the unoccupied mode. However, such design conditions occur only a few days each winter.

During non-design days, heating equipment that starts two hours early results in wasted energy. The same principle applies to the cooling season.

The optimum start-stop program is an adaptive energy-saving program that uses intelligence and the flywheel effect (energy retention capacity) of a building to save a considerable amount of energy beyond that which can be saved with the program clock. The optimum start program monitors space and outdoor air temperatures several hours prior to the programmed occupancy time of the building or zone. If the space temperature is within comfort limits, the heating or cooling equipment will be started exactly at building or zone occupancy time. If space temperature is not within comfort limits, the program will calculate the correct start-up time necessary to achieve comfortable temperatures at occupancy time. After weekends or holiday periods, a building may require longer warm-up or cool-down periods. The optimum start program automatically compensates by extending the run time of the heating or cooling equipment prior to occupancy.

Figure 5.8 is a graphic representation of the optimum start program for both the summer and winter modes.

Figure 5.9 shows a typical optimum stop graphic representation for the heating mode.

The *optimum stop program* occurs at the end of the building or zone occupancy time. It calculates the earliest time during the occupancy period that the heating or cooling equipment can be shut down, allowing the flywheel effect of the building to stabilize temperature levels until the end of the period.

Both optimum start and optimum stop programs can function with multiple space sensors. The program will be sensitive to the area or zone sensor which is exposed to the most extreme temperatures.

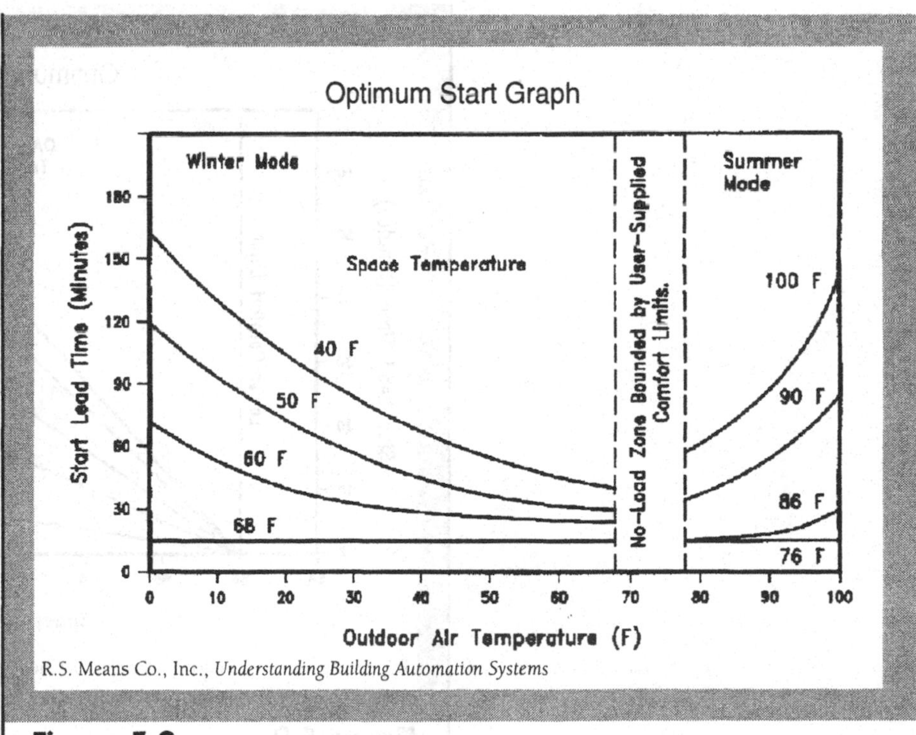

R.S. Means Co., Inc., *Understanding Building Automation Systems*

Figure 5.8

Unoccupied Night Purge Program

During the summer cooling season it is not unusual for the outdoor air temperature to drop considerably at night. Frequently, during the early morning hours prior to building occupancy time, the outdoor air temperature is below building space temperatures. This cool outdoor air can be utilized to cool the building, thereby eliminating the need for mechanical cooling during early morning occupancy hours. This free cooling will generate energy savings and also save wear on the mechanical cooling equipment.

At a preprogrammed time in the early morning hours, the program begins to monitor space and outdoor temperature and humidity. If space conditions indicate a need for cooling, and if outdoor air conditions are suitable, the *night purge program* is initiated. The program starts the HVAC supply fan and associated exhaust fan, and opens the outdoor air damper 100%. Warm air from the building continues to be purged until the space temperature and relative humidity indoors reach the same levels as the outdoor air conditions, or until the morning start-up program begins. The outdoor air temperature must be above a preselected minimum to ensure that the program is operable only during the cooling season. Figure 5.10 shows a flow chart of the night purge program operation.

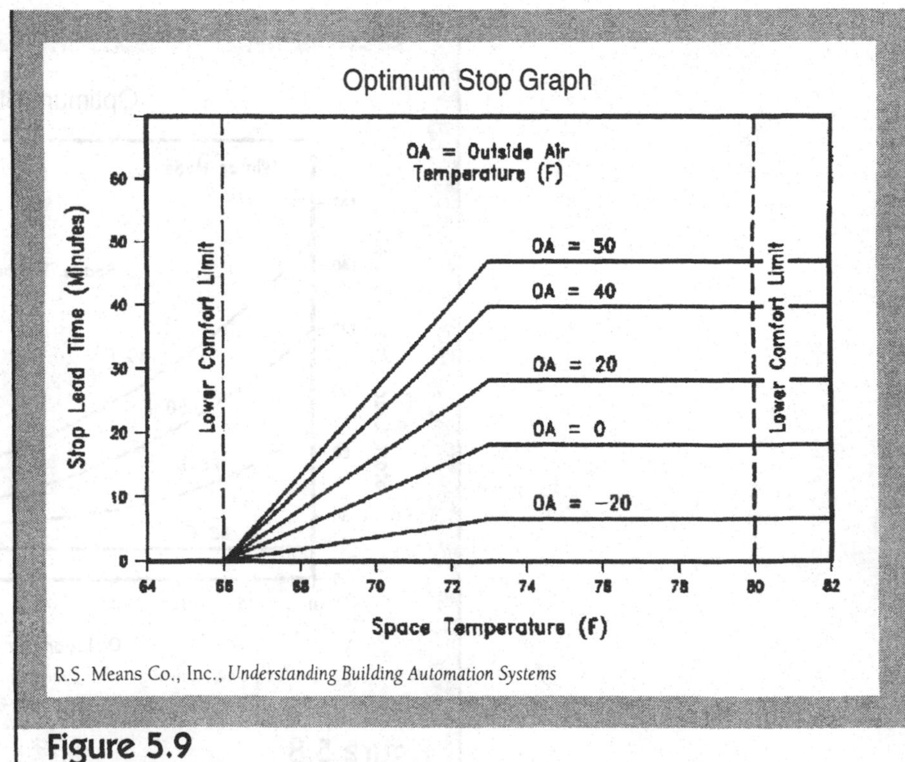

R.S. Means Co., Inc., *Understanding Building Automation Systems*

Figure 5.9

The night purge program can be applied to most HVAC systems that are capable of using 100% outdoor air. Some package-type HVAC units and rooftop units are limited mechanically to admit 10 or 20% outdoor air, and therefore do not qualify.

Enthalpy Program

The traditional mixed air economizer control system saves energy by using outdoor air for cooling rather than mechanical cooling equipment. This control loop functions by sensing dry bulb temperatures. When the outdoor air dry bulb temperature reaches a level at which it is no longer suitable for cooling, the fresh air damper closes to the minimum position and the mechanical refrigeration equipment becomes operable. It is at this point that the *enthalpy control system* becomes important.

The economizer system, since it measures dry bulb temperature only, keeps the system on return air throughout the cooling season. The enthalpy control system, however, selects the air source (return air or outdoor air) which has the lowest total heat (*enthalpy*), and therefore requires the least amount of heat removed by the cooling coil.

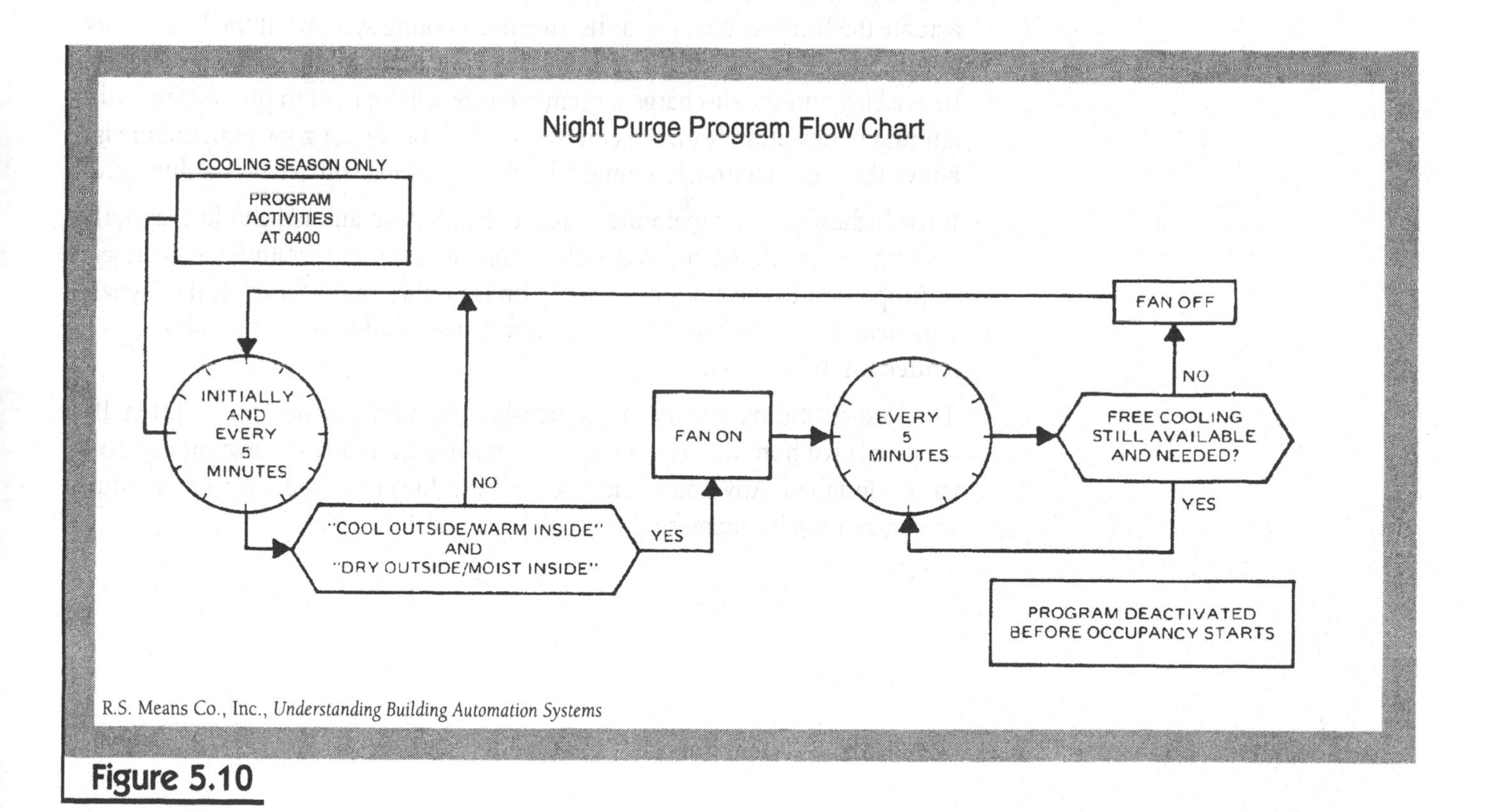

R.S. Means Co., Inc., *Understanding Building Automation Systems*

Figure 5.10

The enthalpy program monitors the temperature and relative humidity or dewpoint of the outdoor and return air. During occupied periods when cooling equipment is in operation, the program calculates the total heat, or enthalpy, of the outdoor air and return air. It then positions the outdoor air and return air dampers to use the air source with the lowest total heat or least enthalpy.

This program will work with HVAC systems that have sprayed or unsprayed cooling coils. The system must have the capability of operating on 100% outdoor air, or 100% return air.

Load Reset Program

HVAC systems are sized for design conditions, or the peak load during the heating and cooling seasons. These peak load periods are of short duration and occur infrequently, which means that most systems operate under partial load conditions most of the year. This results in inefficient performance. If the systems are not properly controlled, energy is wasted.

The *load reset program* controls heating and/or cooling to maintain comfort conditions in the building while consuming a minimum amount of energy. This is accomplished by resetting the heating and cooling discharge air temperatures to satisfy the zone with the greatest heating or cooling load. When no heating or cooling is required, the program de-energizes both sources.

If an HVAC system supplies multiple zones with different exposures such as North, South, or locations (interior, exterior), the heating and cooling requirements could vary widely. The load reset program will accept and evaluate inputs from multiple zone sensors and issue signals, if necessary, to actuate the heating, cooling, or heating and cooling systems. If the lowest zone temperature is below the preprogrammed comfort level, the heating source will be enabled and the discharge air temperature will be reset in proportion to the amount of deviation from the comfort level. If the lowest zone temperature is above the preprogrammed comfort level, the heating source will be shut off.

If the highest zone temperature is above the preprogrammed comfort level, the cooling source will be enabled and the discharge air temperature will be reset in proportion to the amount of deviation from the comfort level. If the highest zone temperature is below the preprogrammed comfort level, the cooling source will be shut off.

The load reset program also has a high-limit humidity or dewpoint option. If space relative humidity is critical, space relative humidity or dewpoint sensors can be installed. Any one of these sensors can lower the cooling reset schedule to compensate for increasing space relative humidity levels.

In a terminal reheat system, the HVAC unit discharge air temperature will be reset to the point at which the zone sensor with the greatest demand for cooling is satisfied. In dual duct and multi-zone units, the cold deck is reset according to the zone sensor with the greatest demand for cooling, and the hot deck is reset according to the sensor with the greatest demand for heating. When all space sensors are satisfied at the preprogrammed comfort levels, both the heating and cooling sources will be shut off.

Figure 5.11 shows how discharge air temperature in a hot and cold deck system is reset as a function of the sensor with the greatest demand for heating or cooling.

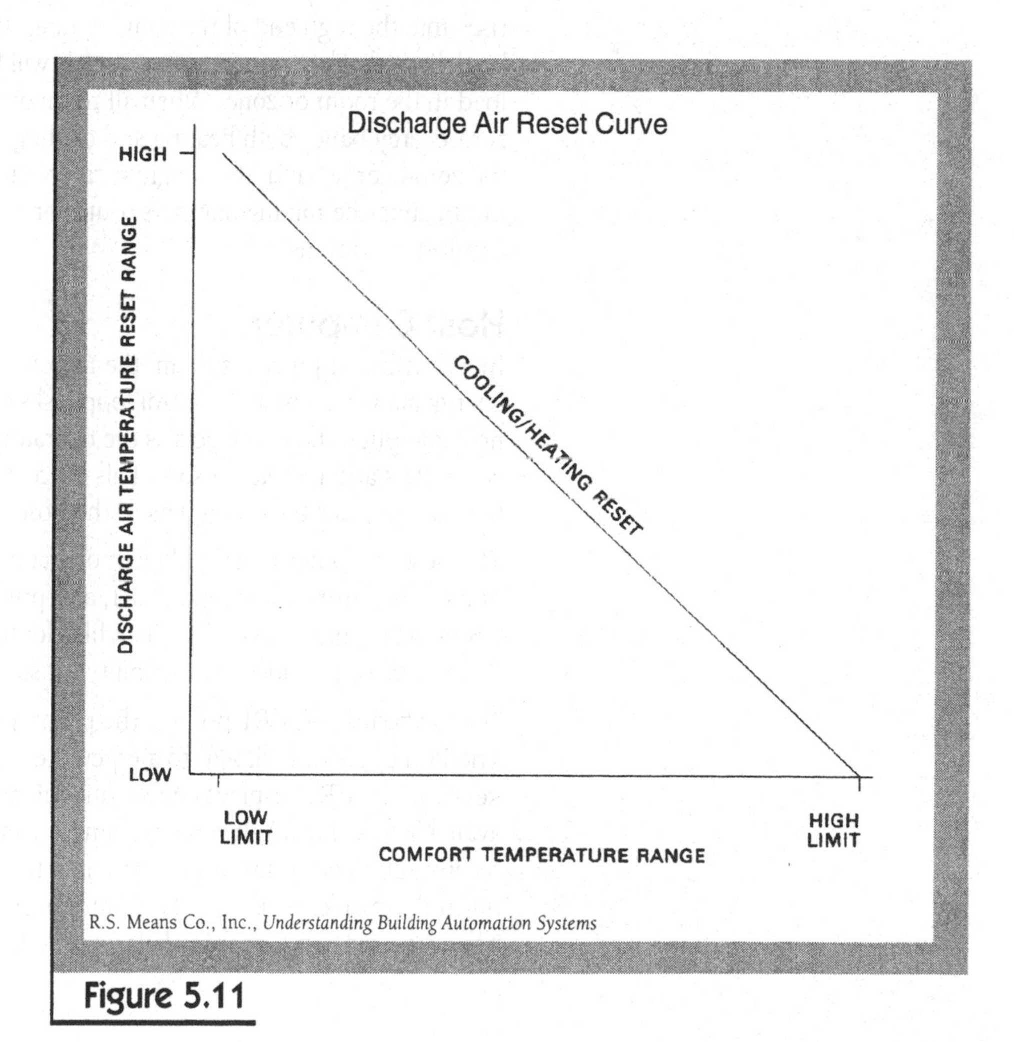

Figure 5.11

Zero-Energy Band Program

This program finds its application in HVAC systems that have heating and cooling capability. The program saves energy by avoiding simultaneous heating and cooling of air delivered to spaces. The space comfort range is divided into three sections: heating, cooling, and zero-energy band. In the zero-energy band portion of the comfort range, both heating and cooling sources are disabled. When space temperatures drop to the lower end of the comfort range, the heating source is energized and the discharge air temperature is reset upward. When space temperature enters the upper end of the comfort range, the cooling source is energized and the discharge air temperature is reset downward. The reset action that takes place in the heating and cooling sections of the comfort range is a preprogrammed function of the temperature range of the heating or cooling section. In other words, a load reset program operates in both the heating and cooling sections of the comfort range (See Figure 5.12).

When an HVAC system serves different zones, or rooms, the heating and cooling requirements will vary widely. Therefore, when the temperature in the coldest room or zone falls into the low end of the comfort range (the heating section), heat will be delivered. The temperature of the air, however, will be reset according to the degree of penetration into the heating section. Cooling follows a similar process. When the temperature in the warmest room or zone rises into the high end of the comfort range (the cooling section), cool air will be delivered. The temperature of the air will be reset according to the cooling load in the room or zone. When all room or zone temperatures are within the zero-energy band, both heating and cooling sources will be disabled. Within the zero-energy band, the program can reset discharge air temperature by modulating the mixing dampers (outdoor air and return air) to maintain space comfort conditions.

Host Computer

In a distributed process system, the remote DDC controllers perform their energy management and control loop tasks on a stand-alone basis. The central host computer, however, acts as the operator-machine interface to the controllers and remote sensors. This is accomplished over a communications bus, which links the controllers to the host.

The host computer is an intelligent operator-machine interface which consists of hard disk drives, keyboard, CRT, and printer. The hard disk holds all operating programs as well as data files for the PC host and controllers. The diskette drive provides the capability for storage of historical data and logs.

The keyboard and CRT provide the primary operator interface with the system. The PC keyboard is similar to a typewriter keyboard and includes a numeric keypad. The CRT displays conversational menus which guide the operator to available functions. When a programming task is to be performed, the operator receives guidance on what data to enter, through conversational prompting and fill-in-the-blanks requests. The printer operates in a receive-only mode and provides hard copy records of alarms and logs. All logs, such as all points,

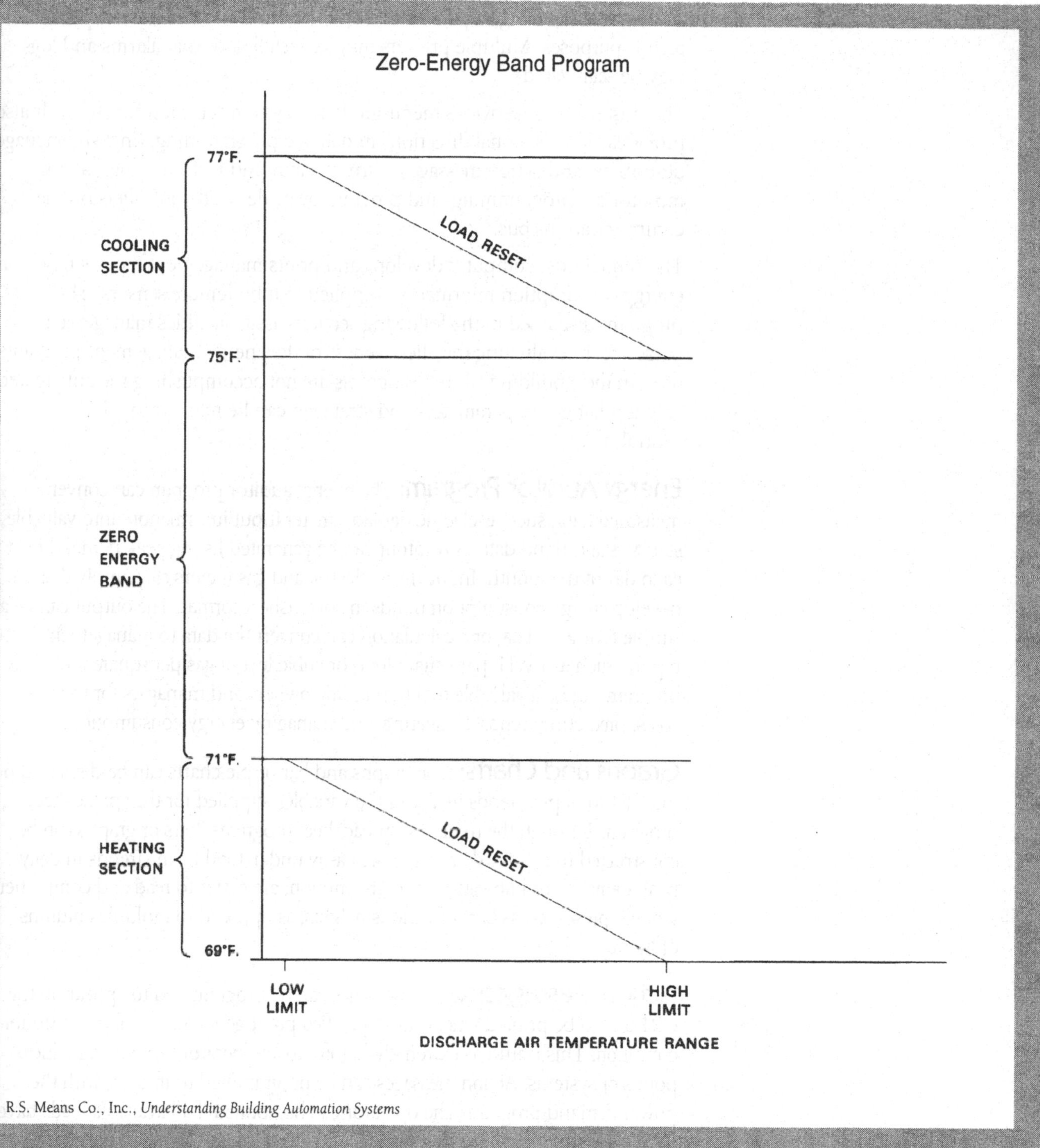

R.S. Means Co., Inc., *Understanding Building Automation Systems*

Figure 5.12

alarms, trends, and groups, can be requested from a log menu and printed for record purposes. Multiple printers may be installed so that alarms and logs may be segregated.

The host software provides menu-guided or system selection functions. It also provides conversational directions in data file programming, English language descriptors and action messages, current status and historical logs, alarm monitoring, programming, and program override of all controllers on the communications bus.

The central host computer develops and prints management reports based on energy consumption information supplied to it by remote sensors. The programs discussed in the following sections assist facilities management personnel in evaluating the effectiveness of the energy management programs used in their buildings. If the programs are not accomplishing the anticipated results, changes in parameters and strategies can be made through the central host.

Energy Auditor Program: The energy auditor program can convert measured data, such as electric demand meter input information, into valuable statistics and trend data. A printout can be generated listing peak demand for each day of the month. Inputs from electric and gas meters can supply data to develop energy consumption trends in spreadsheet format. The output can be a simple listing of data, or a calculation can convert the data to management-level reports such as kWH per square foot, or cubic feet of gas per square foot. This program can be a valuable tool to building owners and managers for tracking costs, predicting trends, budgeting, and managing energy consumption.

Graphs and Charts: Line graphs and bar or pie charts can be displayed on the CRT to depict trends in any of the variables supplied for the spreadsheet program. Through the trend log spreadsheet programs, bars or graphs can be constructed to display data in a more easily understood form. Trends in daily peak demand, or kilowatt-hour consumption, are easier to read and comprehend when represented as bars of various heights, as opposed to multiple columns of figures.

Action Messages: Action messages can be programmed to appear on the CRT and to be printed whenever a specified point generates an alarm or trouble condition. This feature is extremely helpful to an operator monitoring critical points or systems. Action messages can be programmed to interact with the power demand program. The operator can be notified if all available loads have been shed and a new demand peak is imminent. The CRT and printer can display corrective action instructions.

All of the energy management programs outlined in this section are included in the DDC controller software. As energy costs escalate and energy shortages occur, these programs offer an answer to the control and management of building operating costs.

Assigned personnel can receive alarms at authorized telephone numbers during predefined periods of the day or night. When an off-normal condition occurs on one or more specified points, the software automatically dials the preprogrammed telephone number or its alternate and issues the appropriate voice message. The program then issues a request for commands.

An authorized operator can dial into the host computer from an off-site location and request the status of a point or issue commands. Specific status points and command points can be assigned and programmed for access through the dial-in program. This program enables building operations personnel to remain in contact with their building after regular business hours.

Monitoring Utility Use

Utilities Metering Program

It is frequently the responsibility of the building manager, director of facilities, or a member of the building owner's staff to periodically review and keep ongoing records of energy consumed by the facility. These records include the amount of electricity or fuel (natural gas or oil), purchased steam, or any other heating and/or cooling medium purchased from an outside source. In many cases, the building owner or his operating staff must rely on the accuracy of the vendor's meter for their energy consumption records. However, many times a single meter records energy consumption for an entire facility with multiple tenants. The question then is how to fairly allocate costs to each tenant. Energy costs are sometimes allocated on a square foot basis or on the basis of occupancy hours. Unfortunately, there is no accurate way of distributing the costs equitably unless individual meters are installed for each tenant.

A Building Automation System (BAS) provides the means to dynamically monitor and record a facility's energy consumption on a real-time basis. The strategic installation of utility meters by individual tenant, department, floor, zone, or mechanical system provides a method of fairly assessing a tenant's energy consumption. In addition, the BAS can monitor the outdoor air as well as space temperature and relative humidity. This information, combined with the building energy consumption data, can be used in historical trend graphs to determine the impact on energy consumption by these measured variables. The wealth of building operating cost information that can be provided by a BAS can help a building owner or operating staff to accurately define energy consumption areas, and make more informed decisions on how to control the energy consumption of HVAC and lighting systems.

Tenant Energy Monitoring Program

Air-conditioning in most commercial buildings today is provided by constant volume single-zone air handling systems, heat pumps, or variable air volume systems. Most of the automatic temperature control companies manufacture direct digital control (DDC) controllers which regulate zone level terminal units supplied by these systems. These terminal units may be reheat coils, variable air volume boxes, or zone heat pumps.

The DDC zone controllers, linked by a communications bus, can send air flow and temperature information to a central host personal computer. The computer, through energy algorithms, can calculate heating or cooling energy supplied to a tenant zone or floor. This provides an excellent means of determining how much energy a tenant consumed during overtime occupancy periods. The BAS can then automatically invoice tenants on a monthly basis for energy consumed in their areas. The invoice will document the invoice date, time period, and energy consumption.

Figure 5.13 shows how a variable air volume box can be monitored for energy consumption. The temperature of the air entering the box (called *primary air*), space temperature, and air flow through the box are monitored by the DDC zone controller over a specified period of time. The central host computer, using these variables, computes the energy consumed per hour and applies the number of hours of operation to determine the total amount of energy consumed during that time period. The computer then applies to this figure the cost per energy unit to determine total cost of energy used in that time period.

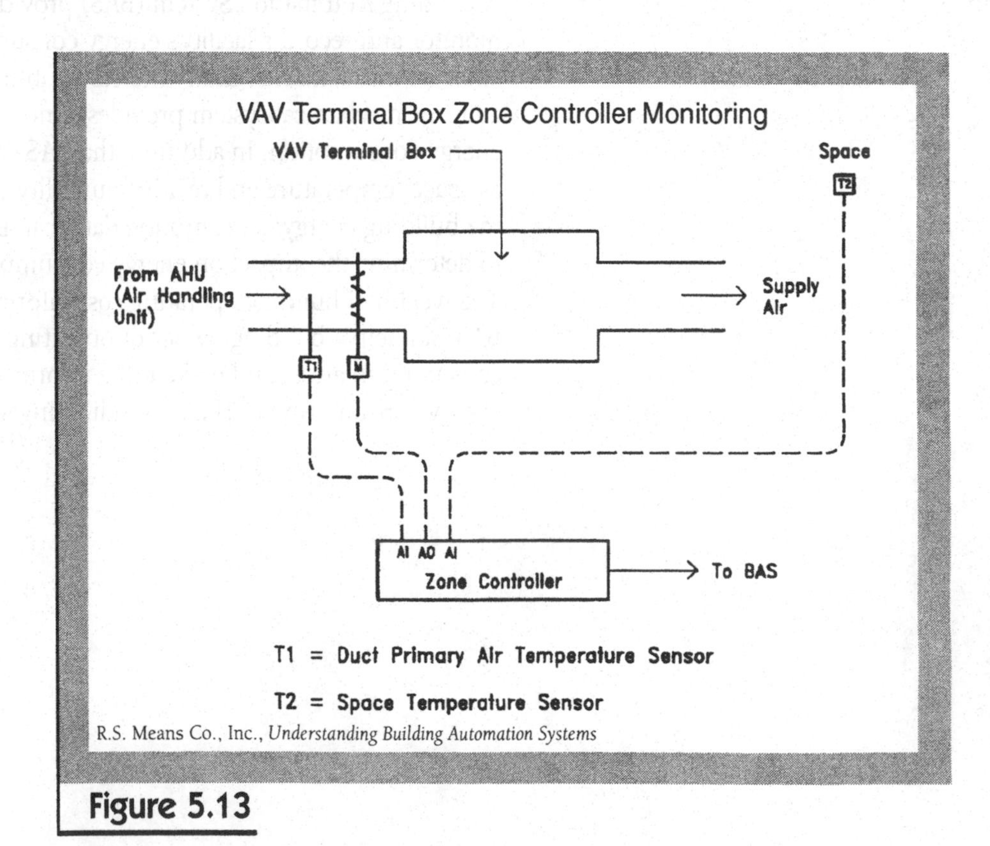

R.S. Means Co., Inc., *Understanding Building Automation Systems*

Figure 5.13

Heating/Cooling Plant Efficiency Program

One of the primary goals of a building or facilities manager is to provide a comfortable, clean, and safe environment in the buildings he or she manages, and to do so in the most cost-effective manner. The rising cost of energy, and the changes to ventilation codes to improve indoor air quality, make this task extremely difficult. As previously stated, a sound preventive maintenance program on all heating and cooling equipment is the first step toward achieving this goal.

The next logical step is to continuously monitor the efficiency of the central heating and cooling plants (boilers and chillers). These primary heating and cooling sources frequently represent the largest portion of the energy bills for the physical plant. A small decrease in operating efficiency of these large central systems can result in a significant increase in energy consumption and its associated costs. Therefore, it is important that plant operating personnel be aware of any downward trend in plant operating efficiency so that corrective action can be taken immediately.

The same sensors used to monitor and control energy consumption can be used to monitor the efficiency of the heating or cooling plant. The simple equation for efficiency is:

$$\% \text{ Efficiency} = \frac{\text{output}}{\text{input}}$$

These sensors are strategically located in a central heating or cooling plant. Through a DDC controller, they constantly transmit critical information to a central host computer. This information can be used by the computer to develop energy consumption spreadsheets, graphs, or charts. The building manager or owner can then use this valuable information to monitor energy consumption trends and plan budgets.

Fossil Fuel Steam Heating Plant

Figure 5.14 depicts a method of monitoring the efficiency of an oil/gas-fired steam heating plant. Oil or gas input to the dual burner is measured to complete the input side of the equation. A system flow meter measures output from the boiler. Figure 5.15 illustrates a method used by the BAS to calculate percent efficiency.

Fossil Fuel Hot Water Heating Plant

Monitoring the efficiency of a hot water boiler is similar to that shown for the fossil fuel steam heating plant. The flow of oil and natural gas is measured to provide the input side of the equation. The boiler output, however, is determined by measuring water flow through the boiler and the temperature pickup of the water as it passes through the boiler. Figure 5.16 shows a method of monitoring the efficiency of an oil/gas-fired hot water heating plant. Figure 5.17 illustrates a method used by the BAS to calculate percent efficiency.

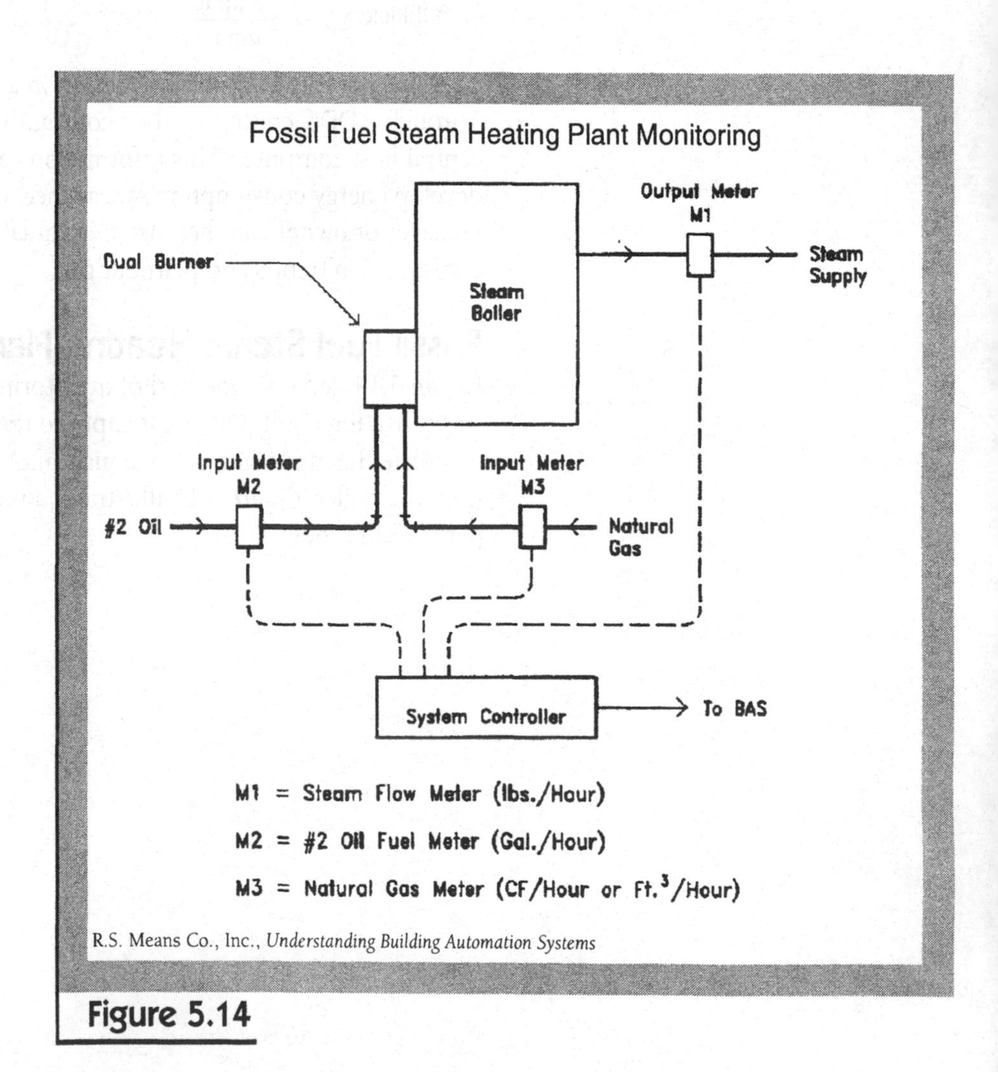

R.S. Means Co., Inc., *Understanding Building Automation Systems*

Figure 5.14

Fossil Fuel Steam Heating Plant Percent Efficiency Calculation

Step 1: Determine Consumed Input Energy (BTU's)

#2 Oil:

$$\underbrace{\frac{(\text{Gal})}{(\text{Hour})}}_{\substack{\text{Meter M2} \\ \text{Fig. 5.14}}} \times \underbrace{\frac{139{,}600\ (\text{BTU})}{(\text{Gal})}}_{\text{Conversion Factor}} \times \underbrace{(\text{Hours})}_{\text{User-defined Time Period}} = \#\,2\ \text{Oil-BTU's}$$

Natural Gas:

$$\overbrace{\frac{(\text{ft}^3)}{(\text{Hour})}}^{\text{Meter M3}} \times \overbrace{\frac{950\ \text{to}\ 1{,}150(\text{BTU})}{(\text{ft}^3)}}^{} \times \overbrace{(\text{Hours})}^{} = \text{Nat. Gas-BTU's}$$

Step 2: Determine Used Output Energy (BTU's)

Steam:

$$\underbrace{\frac{(\#)}{(\text{Hour})}}_{\substack{\text{Meter M1} \\ \text{Fig. 5.14}}} \times \underbrace{\frac{1{,}000\ (\text{BTU})}{(\#)}}_{\text{Conversion Factor}} \times \underbrace{(\text{Hours})}_{\text{User-defined Time Period}} = \text{Steam-BTU's}$$

Step 3: Determine Percent Efficiency (% Eff.)

$$\%\ \text{Eff.} = \frac{\text{Output}}{\text{Input}} \times 100\% = \frac{\#\,2\ \text{Oil or Nat. Gas (BTU's)}}{\text{Steam (BTU's)}} \times 100\%$$

R.S. Means Co., Inc., *Understanding Building Automation Systems*

Figure 5.15

Electric Chiller Plant

Monitoring the efficiency of an electrically driven chiller plant is similar to that shown for the fossil fuel hot water plant. Electrical power consumed by the chiller is measured for the input part of the equation. The output is determined by measuring water flow through the evaporator, and the temperature difference of the water as it passes through the evaporator. Figure 5.18 shows a method of monitoring the efficiency of an electrically driven chiller. Figure 5.19 shows a method used by the BAS to calculate percent efficiency.

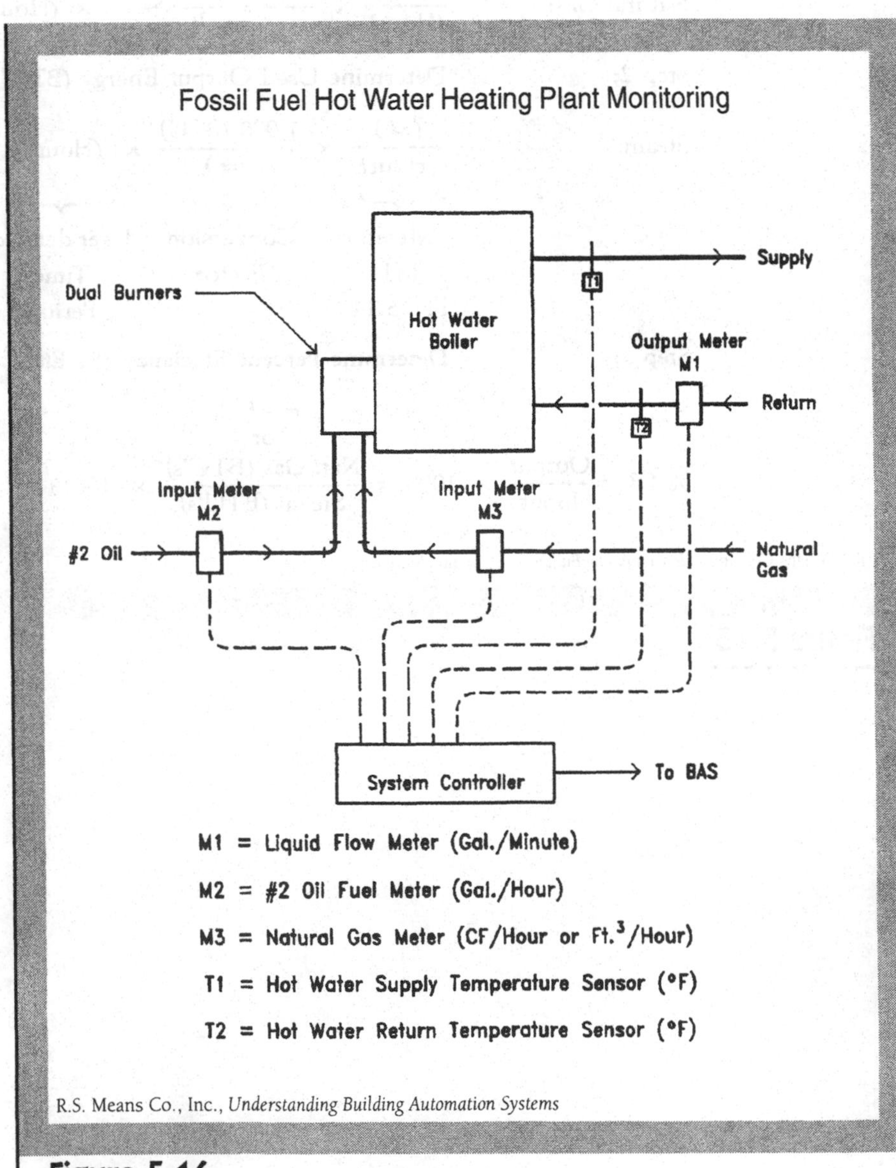

R.S. Means Co., Inc., *Understanding Building Automation Systems*

Figure 5.16

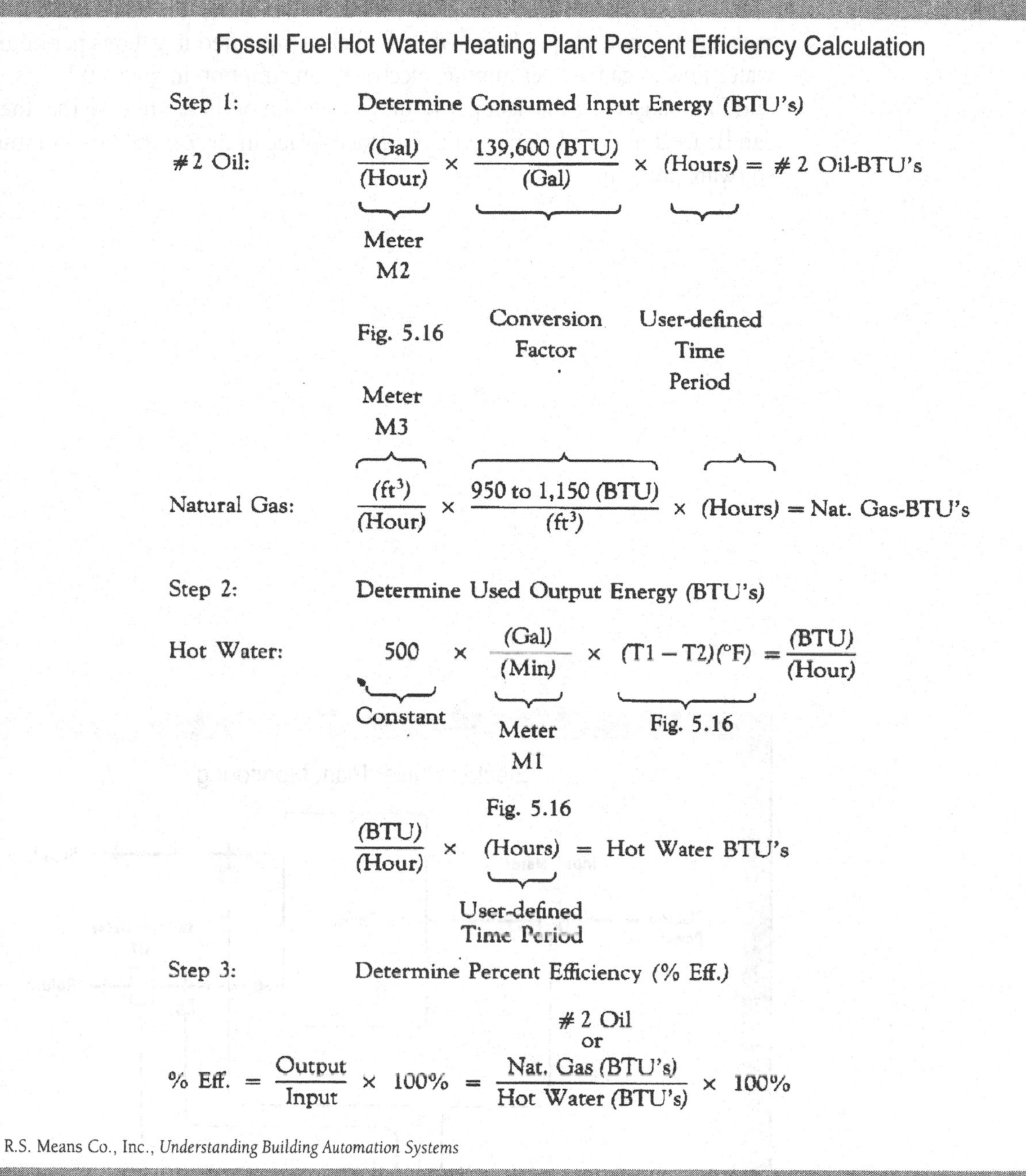
Fossil Fuel Hot Water Heating Plant Percent Efficiency Calculation
Step 1:
Determine Consumed Input Energy (BTU's)
#2 Oil:
(Gal) / (Hour) × 139,600 (BTU) / (Gal) × (Hours) = #2 Oil-BTU's
Meter
M2
Fig. 5.16
Conversion
Factor
User-defined
Time
Period
Meter
M3
Natural Gas:
(ft³) / (Hour) × 950 to 1,150 (BTU) / (ft³) × (Hours) = Nat. Gas-BTU's
Step 2:
Determine Used Output Energy (BTU's)
Hot Water:
500 × (Gal) / (Min) × (T1 – T2)(°F) = (BTU) / (Hour)
Constant
Meter
M1
Fig. 5.16
Fig. 5.16
(BTU) / (Hour) × (Hours) = Hot Water BTU's
User-defined
Time Period
Step 3:
Determine Percent Efficiency (% Eff.)
% Eff. = Output / Input × 100% = #2 Oil or Nat. Gas (BTU's) / Hot Water (BTU's) × 100%
R.S. Means Co., Inc., Understanding Building Automation Systems

Figure 5.17

The engineering units for the variables are clearly different. Steam flow is measured in pounds per hour, while fuel oil is measured in gallons per hour, water flow in gallons per minute, electrical consumption in kilowatt hours, and natural gas in cubic feet per hour. Conversion of these units, so that they can be used in the efficiency equation, takes place in the central host computer software program.

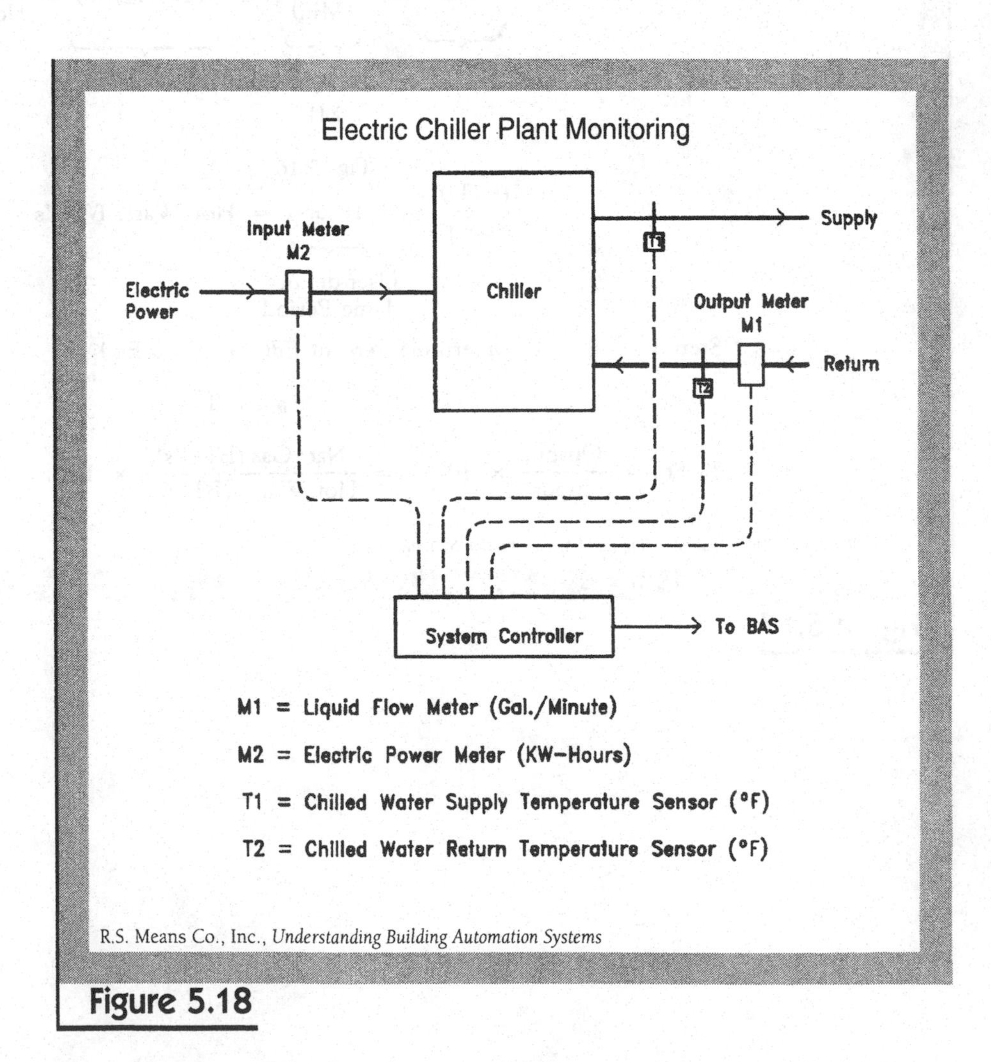

R.S. Means Co., Inc., *Understanding Building Automation Systems*

Figure 5.18

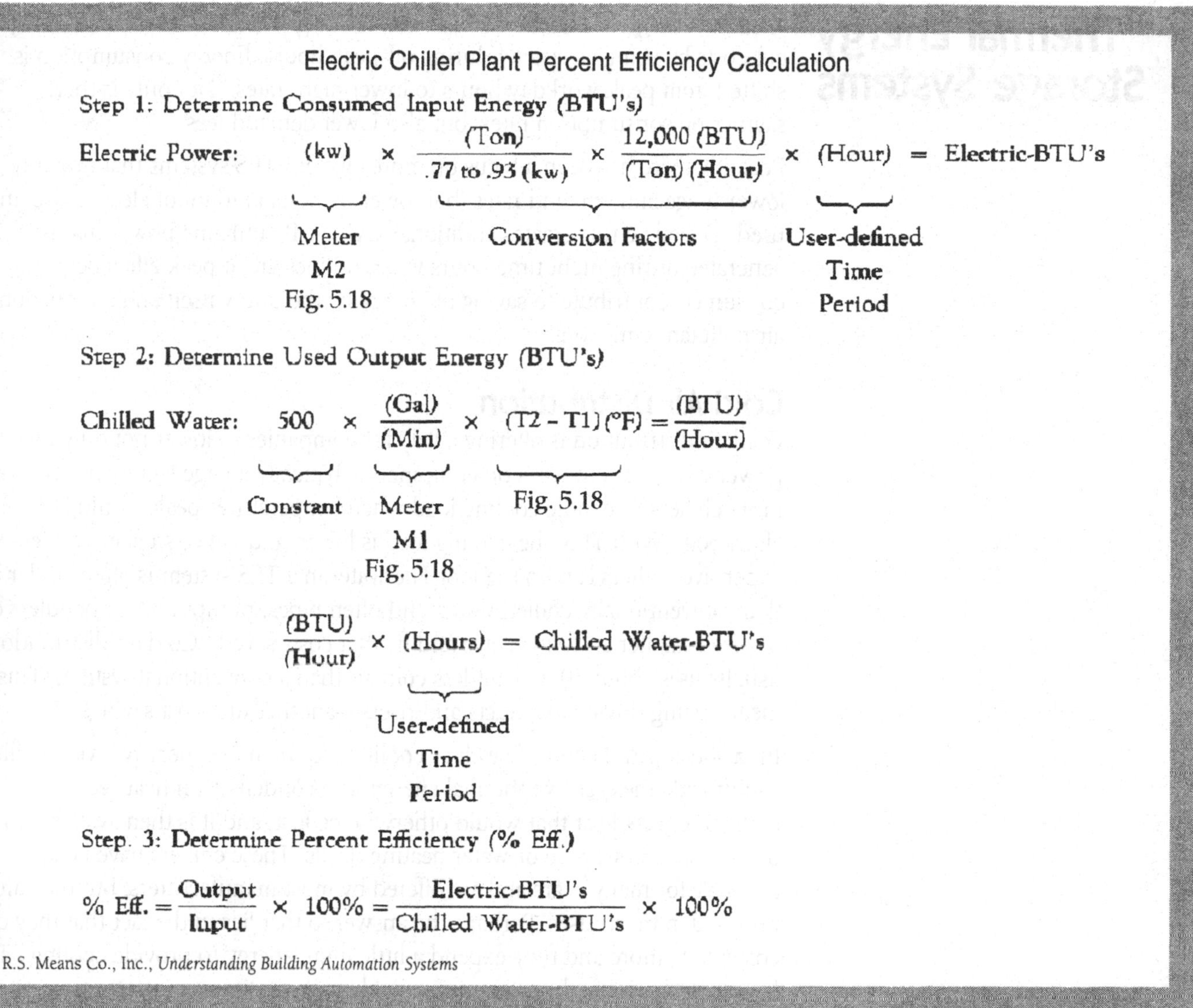
Electric Chiller Plant Percent Efficiency Calculation
Step 1: Determine Consumed Input Energy (BTU's)
Electric Power: (kw) × (Ton) / .77 to .93 (kw) × 12,000 (BTU) / (Ton) (Hour) × (Hour) = Electric-BTU's
Meter M2 Fig. 5.18
Conversion Factors
User-defined Time Period
Step 2: Determine Used Output Energy (BTU's)
Chilled Water: 500 × (Gal) / (Min) × (T2 – T1) (°F) = (BTU) / (Hour)
Constant
Meter M1 Fig. 5.18
Fig. 5.18
(BTU) / (Hour) × (Hours) = Chilled Water-BTU's
User-defined Time Period
Step. 3: Determine Percent Efficiency (% Eff.)
% Eff. = Output / Input × 100% = Electric-BTU's / Chilled Water-BTU's × 100%
R.S. Means Co., Inc., Understanding Building Automation Systems

Figure 5.19

Thermal Energy Storage Systems

Thermal energy storage (TES) systems offer large electrical consumers a way to cut costs by using electricity during off-peak hours. Energy consumption is shifted from peak workday hours to lower night rates. Not only is there savings on consumption rates, but also lower demand fees.

Recently there has been a focus on more efficient TES systems that not only lower usage and demand rates, but lower the total amount of electricity being used. Of course this creates additional savings. By utilizing power that is generated during night time hours when utilities are at peak efficiency, consumers contribute to saving of energy by the utility itself and a reduction in air pollutant emissions.

Cold Air Distribution

Cold air distribution is offering a way for companies to lower not only their power costs, but cut costs on equipment. Typically a large facility has two or more chillers to handle cooling load when temperatures peak in mid-afternoon. With TES, the cooling load is lower, and hence smaller and less expensive chillers can do the job. The water in a TES system is often colder than conventionally chilled water, and often pipes, pumps and air handlers can be sized smaller at design stage and further costs saved. Cold air distribution usually uses about 30 percent less cold air than a conventional system. This means sizing down fans, ducts and risers—another first-cost savings.

In facilities with heating as well as cooling requirements, heat-recovery chillers can provide inexpensive thermal energy. The condenser on heat-recovery chillers collects heat that would otherwise be lost, and it is then available to be used for process, space or water heating needs. These chillers have been available for many years and are offered by most manufacturers, but they are not used in most cases. Perhaps the answer to that lies in the fact that they do cost a little more and they expend a little more energy to provide cooling. The heat they recapture, however, can quickly make an impact on energy costs.

For further information on thermal energy storage systems and heat recovery chillers, check with the Air-Conditioning and Refrigeration Institute (ARI) at 4301 North Fairfax Drive, Suite 425, Arlington, VA 22203. Telephone is: (703) 524-8800; Fax: (703) 528-3816; and website: www.ari.org. The Electric Power Research Institute (EPRI) is at 3412 Hillview Avenue, Palo Alto, CA 94304-1395. Telephone is (650) 855-2000; and website:www.epri.com.

The following section is reprinted with permission from *HVAC Systems Evaluation*, by Harold R. Colen, PE, R.S. Means Co., Inc.

Meeting the Demand for Power

The dramatic increase in the cost of energy, especially electricity, has prompted many changes in the HVAC industry. Variable air volume and the microprocessor are now the foundation of standard HVAC design. The latest advancements have been inspired by the electric power shortage, as electric utilities struggle to meet peak demands for power.

Low Temperature Supply Air

Thermal Storage: One method of easing the burden of generating sufficient power to meet the peak demand is to produce and store chilled water or ice during the off-peak hours of electric demand, for use during the peak hours of electric energy usage. Low temperature supply air is a cousin to this type of cooling storage (ice bank). Generating and storing ice makes it possible to create air temperatures as low as 40°F.

For most air-conditioning systems, the fan energy required to transport the supply and return air represents about 40% of the cooling energy. Figure 5.20 is an example that demonstrates these relationships.

Comparison of Fan Energy to Total Cooling Energy

Example: A 100-ton chiller in use 10 hours per day, 20 days per month, over a 6-month cooling season, at 1,000 full load ton-hours (flth) of cooling per year, 12 month fan operation for ventilation and intermediate season cooling, $0.10 per kwh.

Chiller:	100 kw x 1,000 flth =	100,000 kwh
Chw pump:	26.2 kw x 10 h/d x 20 d/m x 6 m/y	31,440 kwh
Cwp:	8.6 kw	
Ct fan:	4.0	
Total	12.6 kw x 1,000 flth =	12,600 kwh
Evaporator fans:	40 kw x 10 h/d x 20 d/m x 12 m/y =	96,000
	Total	240,040 kwh

Total yearly cooling and ventilation energy use: 240,040 kwh
Evaporator fan percentage of total cooling and ventilation yearly energy consumption is 40%. (96,000/240,040).

R.S. Means Co., Inc., *HVAC Systems Evaluation*

Figure 5.20

Reduction in Fan Energy

If the quantity of supply and return air circulated to conditioned spaces is reduced, the fan energy necessary to move that air is also reduced. The volume of air needed to cool a space depends on the temperature difference between the supply air and the space temperature.

The sensible heat gain of the space to be cooled warms the cold air supply to the space to the desired room temperature. The latent (moisture) heat generated inside the conditioned space adds moisture to the supply air. Most comfort air-conditioning systems are designed for a 55°F supply air temperature. Heat from the supply fan and from the conditioned space warms the supply air up to 75°F.

The following calculations determine the supply air quantities for a 216,000 btuh internal sensible heat load.

$$\text{cfm} = \frac{\text{btuh (internal sensible heat)}}{1.08 \times \text{td (temperature difference between room air and supply air temperatures)}}$$

$$\text{For 55°F supply air: cfm} = \frac{216{,}000}{1.08 \times (75\text{–}55)} = 10{,}000 \text{ cfm}$$

$$\text{For 45°F supply air: cfm} = \frac{216{,}000}{1.08 \times (75\text{–}45)} = 6{,}667 \text{ cfm}$$

If 45°F rather than 55°F supply air is delivered to a space, a one-third reduction of the 55°F air volume will produce the same cooling benefit. The 45°F supply air reduces the relative humidity of the space, so some additional latent heat is removed at the cooling coil. The 45°F supply air system actually requires less than two-thirds of the 55°F air, because less supply fan heat is added to the supply air.

The fan brake horsepower adds heat to the fan supply air. The fan brake horsepower (bhp) is equal to:

$$\frac{\text{cfm} \times \text{sp}}{6356 \times \text{efficiency}}$$

Assuming a duct sp of 1" and a fan efficiency of 75%, the fan bhp due to the fan moving 10,000 cfm through the 1" static pressure duct system is:

$$\frac{10{,}000 \text{ cfm} \times 1 \text{ sp}}{6356 \times 0.75} = 2.1 \text{ bhp}$$

The fan bhp due to the fan moving 6667 cfm through the 1" static pressure duct systems is:

$$\frac{6667 \text{ cfm} \times 1 \text{sp}}{6356 \times 0.75} = 1.39 \text{ bhp}$$

Two-thirds of the fan energy is required for the 45°F supply air system, as compared to a 55°F supply air system.

$$\frac{1.39}{2.1} = 0.66$$

The amount of fan heat that is added to the supply air is determined by the fan brake horsepower.

$$1 \text{ bhp} = 2545 \text{ btuh}$$

$$\text{Temperature difference (td)} = \frac{\text{btuh}}{\text{cfm} \times 1.08}$$

1.08 is the constant to convert cfm to lbs./hr. x specific heat.

$$\text{ft}^3 = 0.24 \times 60 \text{ min./hr.}/13.34 \text{ ft}^3/\text{lb.} = 1.08$$

$$\text{td} = \frac{1.39 \text{ bhp} \times 2545 \text{ btuh}}{6667 \text{ cfm} \times 1.08} = 0.49°\text{F}$$

Fan heat raises the supply air temperature 0.49°F for every inch of static pressure that the fan operates against. The total air supply moved by the supply fan is the sum of the supply air required to remove the sensible heat from the conditioned space, plus the supply air required to remove the fan heat, plus the supply air necessary to remove the duct heat gain. The psychrometric chart in Figure 5.21 diagrams the air condition leaving the cooling coil until it warms up to the room temperature.

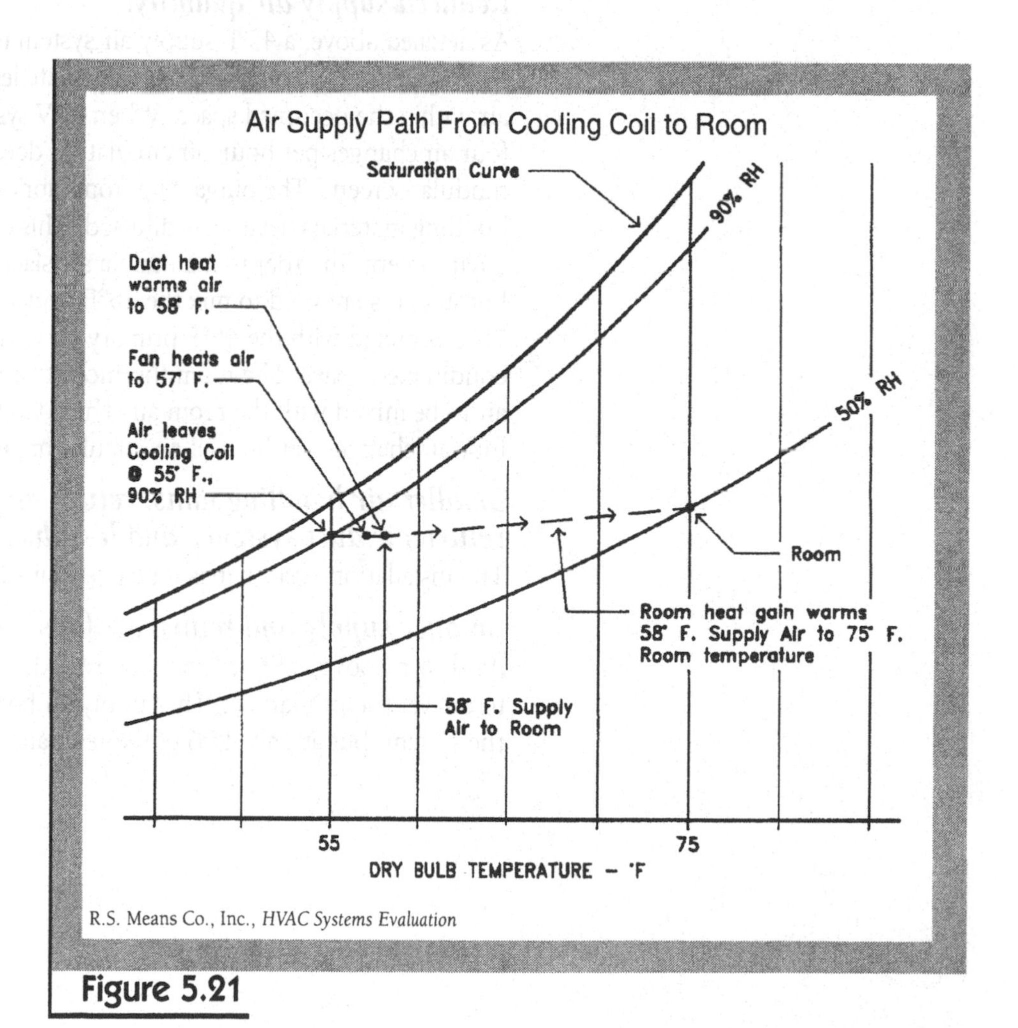

R.S. Means Co., Inc., *HVAC Systems Evaluation*

Figure 5.21

The supply air quantity for the 55°F system =

$$\frac{216{,}000 \text{ btuh} + (2.1 \text{ bhp} \times 2545 \text{ btuh per bhp})}{1.08 \times (75-55)} = 10{,}247$$

The air supply volume must be increased from 10,000 cfm to 10,247 cfm, to compensate for the heat added by the supply fan.

$$\frac{10{,}274}{10{,}247} = 1.025$$

2-1/2% more supply air is needed to offset the supply fan heat.

The supply air quantity for the 45°F supply air system =

$$\frac{216{,}000 \text{ btuh} + (1.39 \text{ bhp} \times 2545)}{1.08 \times (75-45)} = 6775 \text{ cfm}$$

$$\frac{6775}{6667} = 1.016$$

Only 1.6% more supply air is needed to offset the 45°F supply air fan heat. The 45°F supply air system requires only 66% of the supply air of a 55°F supply air system.

$$\frac{6667 \text{ cfm}}{10{,}247 \text{ cfm}} = 0.66$$

Advantages:

Reduced supply air quantity:

As detailed above, a 45°F supply air system uses 34% less primary air than a 55°F system. It is not desirable to circulate less than four air changes of supply air within the occupied space. When VAV systems reduce the airflow below four air changes per hour, air circulation does not take place below the office modular screens. The outgassing from fabrics, floor coverings, furniture and building materials never gets diffused. This creates an uncomfortable environment. In order to maintain a satisfactory air change rate, supply air booster fans are used to mix the 45°F primary air with return air. Room air at 75°F is mixed with the 45°F primary air to deliver 55°F supply air to the conditioned space. The room thermostat controls the amount of 45°F primary air to be mixed with the room air. The conditioned space always experiences four air changes per hour of air circulation, regardless of the internal load.

Smaller air handling units, return air fans, primary supply air and return air duct systems, and less duct insulation:

The installation cost savings for these items could amount to $1.80/sf.

Smaller supply and return air fans:

As shown above, 34% less air is delivered, and 34% less fan energy is required to deliver the primary air. The supply air booster fans add fan horsepower to the system, but at only 25% of the resistance of the primary air system.

Less chilled water gpm, piping, insulation, and smaller chilled water pumps:

The water quantity is determined by the following:

$$8.3 \text{ lb./gal.} \times 60 \text{ min./h} = 500$$

$$\text{gpm} = \frac{\text{btuh}}{500 \times \text{td}}$$

$$\text{gpm (45°F water)} = \frac{750{,}000 \text{ btuh}}{500 \times (60\text{–}15)} = 100 \text{ gpm}$$

$$\text{gpm (40°F water)} = \frac{750{,}000 \text{ btuh}}{500 \times (60\text{–}40)} = 75 \text{ gpm}$$

$$\text{gpm (35°F water)} = \frac{750{,}000 \text{ btuh}}{500 \times (60\text{–}35)} = 60 \text{ gpm}$$

Lower chilled water temperatures result in 25% to 40% less gpm and pump energy, as well as a savings of 15–25% of the installation cost of chilled water pumps, piping, and insulation.

Building Envelope

The building envelope is the physical enclosure of a building—its foundation, walls and roof. It is the physical and thermal barrier between the outside elements and interior spaces, and as such is central to determining the interior environment: the size, shape and appearance of the physical space, the energy required to heat, cool and light the area, and the comfort of the people working within.

Building Heat Losses

The roof will have a significant impact on a building's heat transmission. Heat transmission through a roof can be reduced by:

1. Use of shiny or light-colored roofing materials for heat reflection.
2. Use of optimum insulation quantities.
3. Increased ventilation in attic spaces.
4. Design of roof pitches and overhangs to increase the lag time of heat flow.
5. Use of building materials that reduce heat transmission.

The following section is reprinted with permission from *Roofing: Design Criteria, Options, Selection*, by R.D. Herbert III, R.S. Means Co., Inc.

Roof insulation can serve a dual purpose: to reduce heat transfer, and to act as a stable, uniform substrate for the roof membrane. In most cases, roof insulation is installed *between the structural deck and the roof membrane.* One exception, the Inverted Roof Membrane Assembly (IRMA), is used in severely cold climates. In this arrangement, the insulation rests above the waterproofing membrane. This system is also known as Protected Membrane Roofing (PMR).

There are many types and thicknesses of material available for roof insulation. The most common are described in the following paragraphs.

Ed. Note: Figure 5.22 is a chart comparing the characteristics of various roof insulation materials.

Note: Composite board is a combination of two of the following boards, with a facing of a different material. The two boards are laminated in the factory to produce a product with the advantages of both materials.

Polyisocyanurate Foam Board: Polyisocyanurate board is a dimensionally stable, closed-cell foam which normally has a glass fiber facing bonded to it in order to receive hot moppings or adhesive. It has a high thermal efficiency (R = 7.2 in.) and relatively low cost.

There have been some problems with polyisocyanurate board. The most publicized is thermal drift of the "R" value as the insulation ages. There have also been incidences of facer separation. It is best to use materials from the same supplier whenever practical, and to ensure that the board manufacturer also cross-references, or approves, the bitumen and adhesive(s) used.

Polyurethane Foam Board: Polyurethane and urethane foam boards have been in common use for a longer period than polyisocyanurate, and were the efficiency "leaders" before the advent of phenolic and polyisocyanurate, but overall, it is still an efficient insulation board.

Polystyrene Foam Board: Expanded polystyrene (EPS) is a polymer (plastic) impregnated with a foaming agent which, when exposed to heat, creates a uniform structure resistant to moisture penetration. Polystyrene has an "R" value of 3.85–4.76 per inch and a compressive strength of 21–27 psi (pounds per square inch) at 1.5 pounds per cubic foot of density. As a comparison, the compressive strength of mineral board (perlite) is 35 psi.

Most expanded polystyrene is used under loose-laid, ballasted single-ply systems. The type of expanded polystyrene often used in these cases is a material with a density of one pound per cubic foot, with an average "R" value of 4.2 per inch. This type of polystyrene is similar to that used in the manufacture of inexpensive picnic coolers.

Characteristics of Various Roof Insulation Materials

Characteristics	TYPE OF INSULATING BOARD								
	Polyisocyanurate Foam	Polyurethane Foam	Extruded Polystyrene	Molded Polystyrene	Cellular Glass	Mineral Fiber	Phenolic Foam	Wood Fiber	Glass Fiber
Impact resistant	G	G	G	F	G	E	G	E	F
Moisture resistant	E	G	E	G	E	G	E		
Fire Resistant	E				E	E	E		E
Compatible with bitumens	E	G	F	F	E	E	E	E	G
Durable	E	E	E	E	E	G	F	E	E
Stable "k" value			E	E	E	E		E	E
Dimensionally stable	E	E	E	E	E	E	E	E	G
High thermal resistance	E	E	E	G	F	F	E	F	G
Available tapered slabs	Y	Y	Y	Y	Y	Y	Y	Y	Y
"R" value per in. thickness*	7.20	6.25	4.76	3.85-4.35	2.86	2.78	8.30	1.75-2.00	4.00
Thicknesses available	1"-3"	1"-4"	1"-3½"	½"-24"	1½"-4"	¾"-3"	1"-4"	1"-3"	¾"-2½"
Density (lb./ft.³)	2.0	1.5	1.8-3.5	1.0-2.0	8.5	16-17	1.5	22-27	49
Remarks	Prone to "thermal drift"	Prone to "thermal drift" Note "A": Should be overlaid with a thin layer of wood fiber, glass fiber or perlite board, with staggered joints.	Somewhat sensitive to hot bitumen & adhesive vapors Note "A": Should be overlaid with a thin layer of wood fiber, glass fiber or perlite board, with staggered joints.	Somewhat sensitive to hot bitumen & adhesive vapors Note "A": Should be overlaid with a thin layer of wood fiber, glass fiber or perlite board, with staggered joints.			Prone to "Thermal drift" relatively new & untested.	Expands with moisture —holds moisture.	Prone to damage from moisture infiltration.

E = EXCELLENT
G = GOOD
F = FAIR

R.S. Means Co., Inc., *Roofing: Design Criteria, Options, Selection*

Figure 5.22

Using mechanized gravel buggies for ballast installation tends to damage this lighter density EPS. The lighter material is also more likely to "shuffle" or crack under the membrane. The industry favors using a board with a minimum density of 1.5 pounds of the extruded material, which is far more durable.

Extruded polystyrene: Extruded polystyrene is a closed-cell foam with a low capacity for water absorption (0.6% by volume), which makes it ideal for use in the insulated roof membrane assembly (IRMA). Its "R" value is approximately 4.8 per inch.

Installation: Most building codes require the installation of a 1/2 inch fire-rated gypsum board underlayment when polystyrene is used over a steel deck. The NRCA also recommends that, when using polystyrene board, a thin (minimum 1/4 inch) layer of perlite or fiberboard "recoverboard" be overlaid, with joints staggered from the insulation board joints below. Polystyrene board must also be protected from ultraviolet light.

Cellular Glass Board: Cellular glass board is no longer used as often as it once was for roof insulation. Its lower "R" value per inch and relatively high density (thus, weight) has rendered it less popular than the newer materials. It is, however, a moisture-resistant, stable, and durable material.

Mineral Board: Mineral board is composed of expanded perlite, cellulose binders, and waterproofing agents. Low water absorption, stability, superior fire-resistance, and the best compressive resistance of all the insulations currently used, make this material extremely popular. The biggest drawback of mineral board is its relatively low "R" value (2.78 per inch). It becomes impractical and expensive to use 4 inches of this material to achieve an "R" value of 10 when the same value can be attained using only 1-1/2 inches of foam insulation.

When mineral board is used as the base layer of a two-layer insulation system, it must be mechanically attached to the deck. A vapor retarder should then be installed or mopped over the mineral board, followed by a high performance insulation. This system ensures secure attachment to the deck and positive control of vapor, which would otherwise tend to be driven into the insulation during the winter months. By installing two separate layers of insulation with staggered joints, thermal bridging of the fasteners is eliminated, as is bridging at joints. Placing the vapor retarder above the fasteners and maintaining a temperature at the fasteners above the dew point reduces corrosion of the fasteners and the deck.

Phenolic Foam Board: Phenolic foam is popular with specifiers because of its "R" value of 8.3 per inch and its competitive pricing. It is a fire-resistant, dimensionally stable, foamed plastic. It is friable (subject to crumbling), and will fracture if abused. There have been instances where phenolic foam "dished" (raised at the corners) under singly-ply membranes. If soaked, phenolic foam will absorb moisture and lose much of its thermal efficiency. Although not really tested, it is reasonable to assume that phenolic foam is affected by thermal drift, as are the other foams that use fluorocarbon blowing agents. (Thermal drift is the reduction in "R" value that occurs when the fluorocarbon used as the foaming agent in manufacture vacates the material over time, and is replaced with atmospheric air, which has a lower "R" value.)

Wood Fiber Board: Wood fiber board is used as a combination decking and insulation board for many roof systems. It is stronger and more durable than the other board insulation products, and makes a far better attachment substrate. However, these attributes are offset by its relatively low thermal resistance. Many specifiers use a composite board or a combination of wood fiber board and one of the high efficient foams to create a solid, efficient system upon which to overlay a roof membrane.

Glass Fiber Board: Glass fiber insulation board is comprised of glass fibers, bound by a resinous binder and rolled into rigid board. A top surface *(facing)* of asphalt-adhered kraft paper or foil is applied as protection. This type of insulation is very efficient (R = 4 per inch) and has been tested by both Factory Mutual and Underwriters' Laboratory. This board is relatively soft and prone to moisture, which attacks the binder and causes the material to collapse.

Refer to Figure 5.22 for a comparison of characteristics of various roof insulation materials.

Heat Loss Through Walls

Figure 5.23 provides factors for determining heat loss from industrial buildings and warehouses, and Figure 5.24 contains correction factors for outside design temperature. Figures 5.25, 5.26 and 5.27 provide further information on heat transmission, and "U" and "R" values.

Exterior wall insulation is often installed when the exterior finish is remodeled. Rigid board insulation in foundation and basement applications is installed between the existing wall and new siding material.

R155-050 Factor for Determining Heat Loss for Various Types of Buildings

General: While the most accurate estimates of heating requirements would naturally be based on detailed information about the building being considered, it is possible to arrive at a reasonable approximation using the following procedure.

1. Calculate the cubic volume of the room or building.
2. Select the appropriate factor from R155-050. Note that the factors apply only to inside temperatures listed in the first column and to 0°F outside temperature.
3. If the building has bad north and west exposures, multiply the heat loss factor by 1.1.
4. If the outside design temperature is other than 0°F, multiply the factor from R155-050 by the factor from R155-060.
5. Multiply the cubic volume by the factor selected from R155-050. This will give the estimated BTUH heat loss which must be made up to maintain inside temperature.

Building Type	Conditions	Qualifications	Loss Factor*
Factories & Industrial Plants General Office Areas 70°F	One Story	Skylight in Roof	6.2
		No Skylight in Roof	5.7
	Multiple Story	Two Story	4.6
		Three Story	4.3
		Four Story	4.1
		Five Story	3.9
		Six Story	3.6
	All Walls Exposed	Flat Roof	6.9
		Heated Space Above	5.2
	One Long Warm Common Wall	Flat Roof	6.3
		Heated Space Above	4.7
	Warm Common Walls on Both Long Sides	Flat Roof	5.8
		Heated Space Above	4.1
Warehouses 60°F	All Walls Exposed	Skylights in Roof	5.5
		No Skylights in Roof	5.1
		Heated Space Above	4.0
	One Long Warm Common Wall	Skylight in Roof	5.0
		No Skylight in Roof	4.9
		Heated Space Above	3.4
	Warm Common Walls On Both Long Sides	Skylight In Roof	4.7
		No Skylight in Roof	4.4
		Heated Space Above	3.0

Note: This table tends to be conservative particularly for new buildings designed for minimum energy consumption.

R.S. Means Co., Inc., *Means Mechanical Cost Data 1999*

Figure 5.23

R155-060 Outside Design Temperature Correction Factor (for Degrees Fahrenheit)

Outside Design Temperature	50	40	30	20	10	0	-10	-20	-30
Correction Factor	0.29	0.43	0.57	0.72	0.86	1.00	1.14	1.28	1.43

R155-070 and R155-080 provide a way to calculate heat transmission of various construction materials from their U values and the TD (Temperature Difference).

1. From the Exterior Enclosure Division or elsewhere, determine U values for the construction desired.
2. Determine the coldest design temperature. The difference between this temperature and the desired interior temperature is the TD (temperature difference).
3. Enter R155-070 or R155-080 at correct U Value. Cross horizontally to the intersection with appropriate TD. Read transmission per square foot from bottom of figure.
4. Multiply this value of BTU per hour transmission per square foot of area by the total surface area of that type of construction.

R.S. Means Co., Inc., *Means Mechanical Cost Data 1999*

Figure 5.24

R155-070 Transmission of Heat

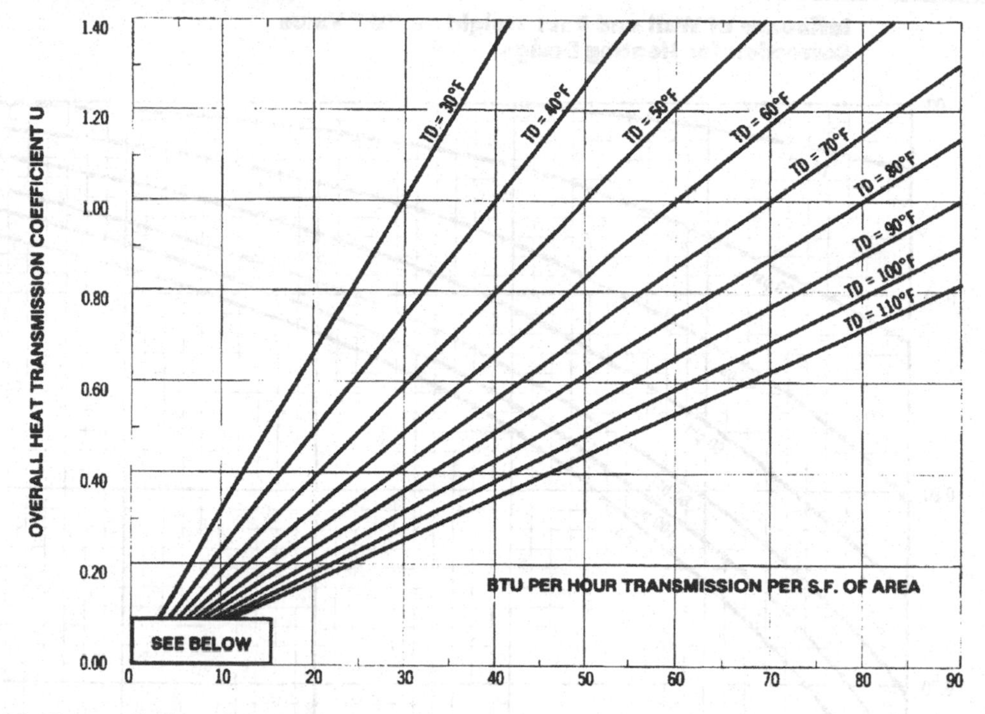

R155-080 Transmission of Heat (Low Rate)

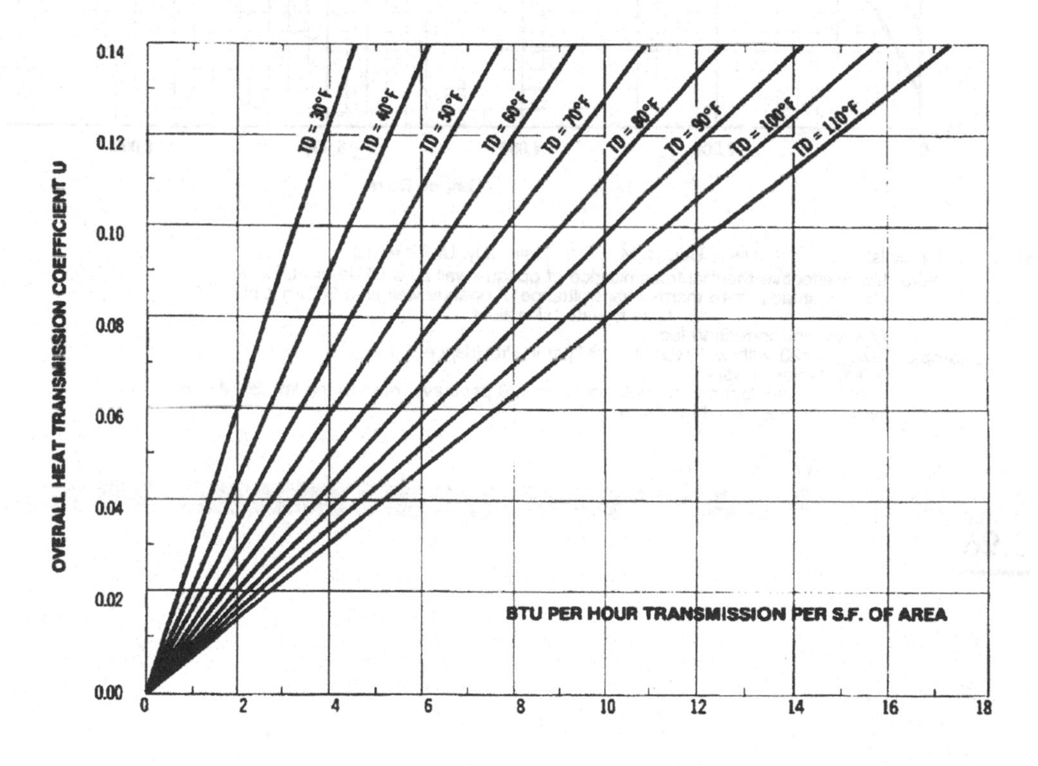

R.S. Means Co., Inc., *Means Mechanical Cost Data 1999*

Figure 5.25

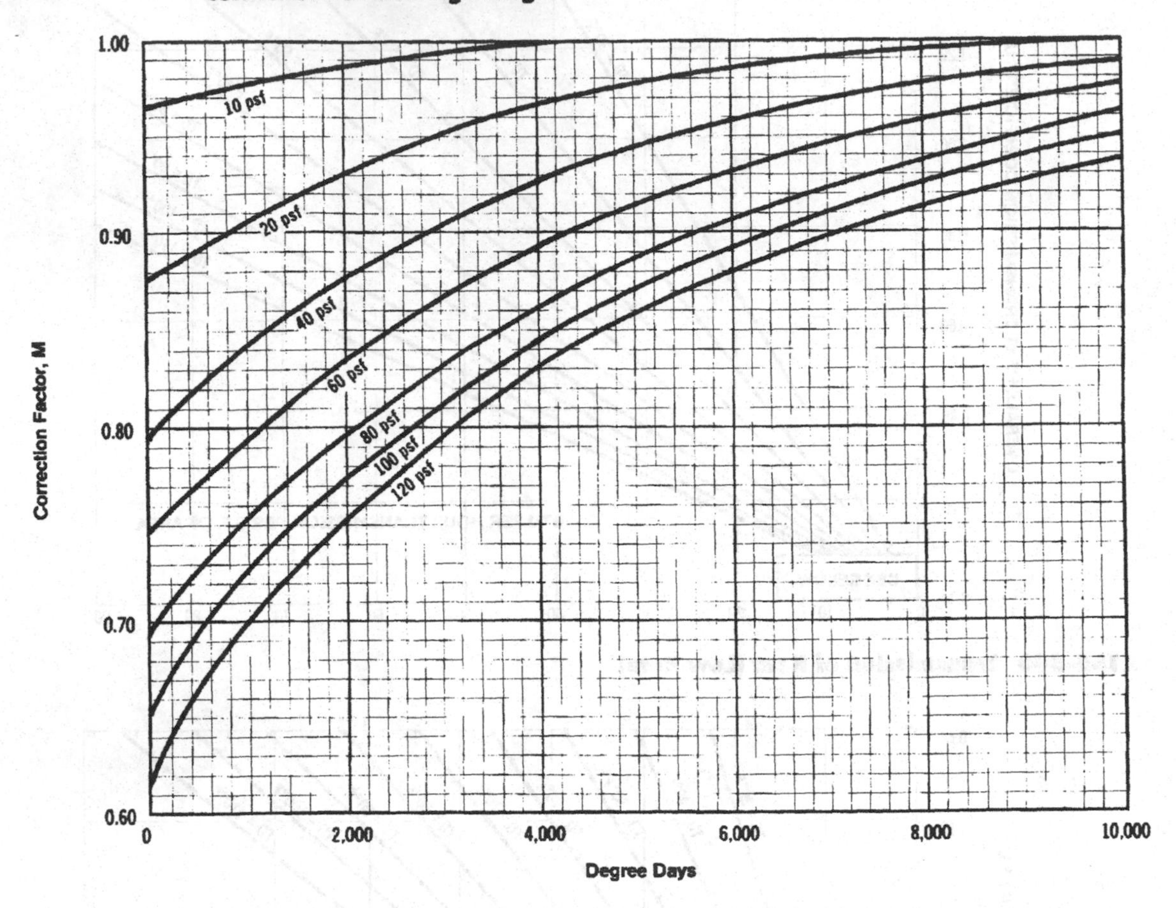

Effective "U" for walls: $Uw = Uw_{ss} \times M$ (similarly, Ur for roofs)

where Uw = effective thermal transmittance of opaque wall area BTU/h x Ft.^{2} x °F

Uw_{ss} = steady state thermal transmittance of opaque wall area BTU/h x Ft.^{2} x °F (steady state "U" value)

M = weight correction factor

Example: Uw_{ss} = 0.20 with wall weight = 120 psf in Providence, R.I. (6000 degree days)

Enter chart on bottom at 6000, go up to 120 psf curve, read to the left .88 M = .88 and Uw = 0.20 x 0.88 = 0.176

R.S. Means Co., Inc., *Means Assemblies Cost Data 1999*

Figure 5.26

Reference Aids | **R15.3-400** | **Thermal Properties of Materials**

Resistances ("R") of Building and Insulating Materials

Material	Wt./Lbs. per C.F.	R per Inch	R Listed Size
Air Spaces and Surfaces			
Enclosed non-reflective spaces, E=0.82			
50°F mean temp., 30°/10°F diff.			
.5"			.90/.91
.75"			.94/1.01
1.50"			.90/1.02
3.50"			.91/1.01
Inside vert. surface (still air)			0.68
Outside vert. surface (15 mph wind)			0.17
Building Boards			
Asbestos cement, 0.25" thick	120		0.06
Gypsum or plaster, 0.5" thick	50		0.45
Hardboard regular	50	1.37	
Tempered	63	1.00	
Laminated paper	30	2.00	
Particle board	37	1.85	
	50	1.06	
	63	0.85	
Plywood (Douglas Fir), 0.5" thick	34		0.62
Shingle backer, .375" thick	18		0.94
Sound deadening board, 0.5" thick	15		1.35
Tile and lay-in panels, plain or			
acoustical, 0.5" thick	18		1.25
Vegetable fiber, 0.5" thick	18		1.32
	25		1.14
Wood, hardwoods	48	0.91	
Softwoods	32	1.25	
Flooring Carpet with fibrous pad			2.08
With rubber pad			1.23
Cork tile, 1/8" thick			0.28
Terrazzo			0.08
Tile, resilient			0.05
Wood, hardwood, 0.75" thick			0.68
Subfloor, 0.75" thick			0.94
Glass			
Insulation, 0.50" air space			2.04
Single glass			0.91
Insulation Blanket or Batt, mineral, glass			
or rock fiber, approximate thickness			
3.0" to 3.5" thick			11
3.5" to 4.0" thick			13
6.0" to 6.5" thick			19
6.5" to 7.0" thick			22
8.5" to 9.0" thick			30
Boards			
Cellular glass	8.5	2.63	
Fiberboard, wet felted			
Acoustical tile	21	2.70	
Roof insulation	17	2.94	
Fiberboard, wet molded			
Acoustical tile	23	2.38	
Mineral fiber with resin binder	15	3.45	
Polystyrene, extruded,			
cut cell surface	1.8	4.00	
smooth skin surface	2.2	5.00	
	3.5	5.26	
Bead boards	1.0	3.57	
Polyurethane	1.5	6.25	
Wood or cane fiberboard, 0.5" thick			1.25

Material	Wt./Lbs. per C.F.	R per Inch	R Listed Size
Insulation Loose Fill			
Cellulose	2.3	3.13	
	3.2	3.70	
Mineral fiber, 3.75" to 5" thick	2-5		11
6.5" to 8.75" thick			19
7.5" to 10" thick			22
10.25" to 13.75" thick			30
Perlite	5-8	2.70	
Vermiculite	4-6	2.27	
Wood fiber	2-3.5	3.33	
Masonry Brick, Common	120	0.20	
Face	130	0.11	
Cement mortar	116	0.20	
Clay tile, hollow			
1 cell wide, 3" width			0.80
4" width			1.11
2 cells wide, 6" width			1.52
8" width			1.85
10" width			2.22
3 cells wide, 12" width			2.50
Concrete, gypsum fiber	51	0.60	
Lightweight	120	0.19	
	80	0.40	
	40	0.86	
Perlite	40	1.08	
Sand and gravel or stone	140	0.08	
Concrete block, lightweight			
3 cell units, 4"-15 lbs. ea.			1.68
6"-23 lbs. ea.			1.83
8"-28 lbs. ea.			2.12
12"-40 lbs. ea.			2.62
Sand and gravel aggregates,			
4"-20 lbs. ea.			1.17
6"-33 lbs. ea.			1.29
8"-38 lbs. ea.			1.46
12"-56 lbs. ea.			1.81
Plastering Cement Plaster,			
Sand aggregate	116	0.20	
Gypsum plaster, Perlite aggregate	45	0.67	
Sand aggregate	105	0.18	
Vermiculite aggregate	45	0.59	
Roofing			
Asphalt, felt, 15 lb.			0.06
Rolled roofing	70		0.15
Shingles	70		0.44
Built-up roofing .375" thick	70		0.33
Cement shingles	120		0.21
Vapor-permeable felt			0.06
Vapor seal, 2 layers of			
mopped 15 lb. felt			0.12
Wood, shingles 16"-7.5" exposure			0.87
Siding			
Aluminum or steel (hollow backed)			
oversheathing			0.61
With .375" insulating backer board			1.82
Foil backed			2.96
Wood siding, beveled, ½" x 8"			0.81

R.S. Means Co., Inc., *Means Assemblies Cost Data 1999*

Figure 5.27

The following information is from the United States Department of Energy's Energy Efficiency and Renewable Energy Network (EREN). It is currently available on the web at www.eren.doe.gov/buildings/purchs_windows.html.

Purchasing for Energy Efficiency: Windows and Doors

The building envelope is the material that separates the inside from the outside of a building. The envelope defines how the building interacts with the external environment, particularly with regard to energy. The building envelope refers to the roof, walls, foundation, insulation, seals, doors, and fenestration (design and placement of windows). The concerns of a building envelope are infiltration, exfiltration, solar gains through the windows, heat loss through the windows, internal heat transfer, and heat gains and losses through the ceiling, walls, and floor.

There is no standard equation for the best envelope. Before deciding on the materials to use, a builder needs to look at the orientation (placement, i.e., southerly exposure) and the geographic area of the building, the number of sunny days, temperature, and humidity. Using envelope materials that are best suited for specific climate conditions will improve the energy efficiency of the building. Some general information on the components of the envelope are presented below. To learn more about building envelope design considerations, see Oak Ridge National Laboratory's Building Envelope Research website at www.ornl.gov.

Fenestration

Fenestration, or window sizing and glazing (the glass and coatings used in the window), should be part of the integrated building design, since it strongly affects daylighting, passive solar heating, and natural ventilation.

In terms of the building envelope, the window's U-value (a measure of its insulating ability—low U-value means better insulating ability) is the critical factor. The U-value of the window as a whole should be used, because efficient glazings can be compromised with poor frame designs. Aluminum frames, for instance, conduct heat well and should not be used in cold climates. See the National Fenestration Rating Council for a Certified Products Directory that lists U-values for more than 20,000 certified products.

Window size and location are determined by the daylight, passive solar heating, and natural ventilation needs. In cold climates, large south-facing windows allow significant solar energy into the house and also provide daylighting; properly sized overhangs can prevent overheating in the summer. In hot climates, north-facing windows can provide daylighting without heating the house.

East- and west-facing windows generally cause excessive heat gains in the summer and heat losses in the winter, and are usually sized small. Although overhangs are impractical for east- and west-facing windows, vertical shading can be used, or trees and shrubs can be strategically located to shade the windows. Landscaping has other benefits, including natural cooling and protection from the wind. See the EREC fact sheet, "Landscaping for Energy Efficiency."

In climates where natural ventilation makes sense, windows need to be located on both the windward and leeward sides of the building. Roughly equivalent-sized openings on both sides of the building will maximize the air flow through the building. The windows should be offset so that the air will circulate through the building rather than blowing directly through it.

Doors

For solid doors, insulated metal or fiberglass doors are the best choice. Since many commercial buildings use glass doors, the glazing choice for the door should be evaluated in terms of the door's orientation and the insulating, solar heating (or shading), and daylighting roles that the door may help fulfill. When evaluating doors, consider the U-value (heat transfer coefficient) for the door as a whole, as some frames will conduct heat readily.

For heavily trafficked buildings, air infiltration through doors is an important energy consideration. Revolving doors and double-doored entryways should be considered in this case.

Glazings

Glazings in both windows and glass doors can have dramatic effects on the energy performance of a building. A wide variety of coatings and configurations are now available to achieve a range of design goals. In addition to a glazing's U-value, the other important factors are the daylight transmittance (the amount of visible light the window lets in) and the shading coefficient (the amount of heat the glazing lets in—glazings with low shading coefficients allow less heat in).

Glazings are now available with high daylight transmittances and low shading coefficients, allowing daylighting without heat gain. These windows are a good choice when solar heating is not the goal. For windows intended for solar heating, obviously a high shading coefficient is needed.

"Low-e" (low-emittance) coatings are a recent innovation for glazings. Low-e coatings reduce the heat emitted from the warm pane to the cool pane and can significantly lower the U-value of glazings. In cold climates, low-e coatings are put on the interior of the outside pane; in hot climates, on the exterior of the inside pane. Note that low-e coatings also reduce the glazing's shading coefficient, and may not be the best choice when solar heating is desired.

For more information about glazings, see the EREC fact sheet, "Advances in Glazing Materials for Windows."

Lighting

Lighting, which can account for as much as 40% of a facility's energy costs, offers relatively simple strategies to get started saving energy. New fixture choices such as a louvered parabolic reflector, four-bulb luminaire with electronic ballast offers several advantages. It's easy to re-lamp, provides as much illumination as an old eight-bulb fluorescent troffer and uses about 18 percent less energy. It also gets rid of old fluorescent fixtures that qualify as hazardous wastes since they have slight amounts of mercury in the bulbs and PCBs in the ballasts.

The following information is from the United States Department of Energy's Energy Efficiency and Renewable Energy Network (EREN). It is currently available on the web at www.eren.doe.gov/buildings/purchs_elec.html.

Purchasing for Energy Efficiency

Lighting includes both electric lighting and natural lighting from the sun, or daylighting. The whole-buildings approach to design often includes daylighting because lighting is one of the major energy uses in a building. Combining daylighting with energy-efficient electric lighting will achieve the best lighting with the lowest energy use.

Buildings that incorporate these features not only use less electricity, but also have lower cooling loads because less heat is being generated by the lights. By accounting for this, a whole building design can balance the extra cost of daylighting strategies with the lower cost of smaller air-conditioning equipment. Although the heating system may also need to be sized larger, it is much more efficient to heat a building with an efficient central heater than with light bulbs. Light bulbs also create uneven heat loads in buildings, which makes constant and even temperature control harder to achieve.

Electric Lighting

Lighting accounts for 31% of the energy use in commercial buildings. Significant reductions in energy use can be achieved by installing energy-efficient lights, fixtures, and controls. In recent years, new technologies such as electronic ballasts, dimmable fluorescents, and others have revolutionized the lighting industry. Retrofits to install new technologies are often cost effective, paying back within a few years.

Compact fluorescent light bulbs can replace incandescent light bulbs in most commercial applications, and they are now widely available. To demonstrate the high efficiency of these light bulbs, compare the light output of a 60-watt incandescent bulb, at 775 lumens, to the light output of a 32-watt T-8 compact fluorescent bulb, at 3,050 lumens. That is roughly four times the light at half the energy use.

Other options available for commercial buildings include improved reflectors for standard fluorescent fixtures, allowing the number of bulbs per fixture to be reduced, and electronic ballasts, which can save up to 25% of energy use compared to older magnetic ballasts. A new technology that holds great promise for commercial buildings is the Sulfur lamp. These lamps produce a very bright, near-sunlight quality light at high efficiencies. An installation in the U.S. Department of Energy's Forrestal Building reduced energy use by 60%. Because the lamps are extremely bright, "light pipes" typically carry the light from one central sulfur lamp to evenly light a large floor space. Lawrence Berkeley National Laboratory has developed an alternative approach that uses uplighters, which direct the light upward and bounce it off the ceiling.

Sulfur lamps are also an excellent choice for energy-efficient outdoor lighting. Another energy-efficient choice is high-intensity discharge lamps, such as mercury, metal halide, and high-pressure sodium lamps.

Ed. Note: For further information on lighting options and selection, please refer to Chapter 9, "Electrical Engineering."

Water Usage

Water supply and demand are issues that have been getting much more press over the past few years. We have seen the drought in California in the late 1980s result in strict water conservation rules. States experiencing rapid growth such as Arizona, Nevada, Texas, and Florida are already taking measures to conserve or will very soon. Expanded demand for domestic use, irrigation and industrial use is far outstripping local supply. The quality of the water supply is as much an issue as quantity. Higher levels of pollution found in recent years translate into higher costs for water treatment and an increasing recycling of water within industrial facilities.

Water Recycling and Reuse

Two methods to conserve water are water recycling and water reuse. Water recycling is the reuse of water again for the same purpose. This may involve treating the water before it is suitable for reuse. The processes take place within the plant. Before initiating a program of recycling there are several issues to consider:

- Where can recycled water be used?
- How is the water quality changed after use?
- What is the minimum quality required?
- What treatment, if any, is required for recycling?

Water reuse is the utilization of wastewater or reclaimed water from a facility such as a municipal water treatment facility. Local water reuse regulations will govern how the water may be used. Most often it is used for purposes such as irrigation, fire protection, watering plants and lawns, and some industrial uses. Questions that need to be answered include:

- Where can reused water be utilized?
- What is the water quality needed?
- What are local sources of suitable wastewater, and how may it be transported?

Reuse of Cooling Water

Industry use of water for cooling is one of the largest uses of water in the U.S. Water is most often utilized for cooling equipment that generates heat or in processes that require condensing of gas. Utilization of water for once-through cooling—cooling a heat source by surrounding it with water and then discharging it—accounts for very large water usage.

Reusing this water by capturing it, cooling it and utilizing it again, and maybe several times for the same process or another process, would mean a great amount of water savings.

Recent Regulations and Innovations

(EPACT) Motor Mandate: Motor efficiency became the law when the U.S. Energy Policy Act (EPACT) mandate was enacted in late 1997. It applies to general purpose, industrial motors either imported or manufactured in the U.S. that are three-phase 230/460 V AC motors from 1 to 200 H.P., T-frame with three digit frame numbers, foot-mounted, NEMA Design A or B, rated for continuous duty.

The law only applies to general purpose motors. Special purpose or definite-use motors that cannot be used for general purpose applications are exempt. Four-pole and six-pole motors with Design C torque are not covered by the Act.

The energy efficiency of the new standard motors is better than old standard-efficiency motors, but usually no higher than premium-efficiency models previously produced.

The Act specified a cutoff manufacturing date for the old motors of October 24, 1997, so it is likely these motors may still be available and will surely be still in use. There is no law against purchasing these old motors nor is there any requirement to replace an existing motor. However, in the interests of energy-efficiency, choosing a new efficient motor is likely to be a wise choice. Your facility's specific needs have to be considered. The higher rpm frequently found on premium-efficiency motors may actually cancel out any energy savings on some direct drive fans or pumps.

For further information on motors and the Energy Policy Act, contact the Department of Energy at www.motor.doe.gov.

IEEE 841 Motor: In severe service applications the IEEE 841 motor is likely the motor of choice. The IEEE wrote new specifications for the 841 motor in 1995, adding new attributes and made it a standard instead of a recommendation. The IEEE 841-1994 standard applies to all severe duty, Totally Enclosed Fan Cooled (TEFC) squirrelcage induction motors 600 V and below, plus 2,300 and 4,000 V motors 500 H.P. and above.

Its higher price is quickly offset by savings through its excellent energy efficiency and long-term reliability. Developed to power rotating equipment, such as compressors, pumps and blowers, it was a project of the Petroleum and Chemical Industry Committee of the IEEE.

For more information on motors, see Chapter 9, "Electrical Engineering."

Compressed Air Challenge: The Department of Energy launched a new energy-saving initiative, The Compressed Air Challenge, in early 1998. A private/public initiative, it targets air compressors and air systems with the purpose of lowering air emissions, reducing system inefficiencies, and capturing energy savings. The reasoning is that although compressed systems may have been efficient when first installed, many have been modified, enlarged and then ignored over the years, resulting in a less-than-efficient current system.

One of the tenets behind the challenge is a belief that compressed air systems are generally not well understood. The threat to production if a system is scheduled for an upgrade or maintenance often prevents necessary improvements.

The program's initial goal is to increase efficiency 10 percent by 2010. It is a modest improvement, but supporters believe once there is a better understanding of the issue, improvements of 25–30 percent or more will be seen.

The 10 percent improvement is expected to produce a total industry savings of $150 million per year and reduce the amount of carbon contributing to the greenhouse effect by 700,000 tons. The Department of Energy says this is equivalent to taking 130,000 automobiles off the road. See the D.O.E. website at www.knowpressure.org for more information.

Steam Challenge: The latest energy-saving program out of the Department of Energy is The Steam Challenge. It is a voluntary program developed in cooperation with the Alliance to Save Energy that seeks to:

- Raise efficiency by 20 percent within 12 years
- Increase use of efficient industrial steam systems
- Help industry utilize a systems approach in the design and use of boilers, distribution systems, and steam applications

To accomplish these goals, the program will provide steam plant operators with technical assistance to improve efficiency and achieve greater productivity with lower production costs. There is also an education component that promotes the energy-saving and environmental benefits of efficient steam systems. They will be qualifying plant personnel in best technology practices, have five showcase steam systems, and offer software and other information. The program's website is www.oit.doe.gov/steam.

References

The following charts provide some basic formulas and conversion factors that can be used in completing basic energy conservation questions. Further information on mechanical, HVAC, and electrical subjects can be found in the corresponding chapters of this book.

Units of Measure and Their Equivalents

Length
1 in. = 25.4 mm.
1 mm. = .03937 in.
1 ft. = 30.48 cm.
1 meter = 3.28083 ft.
1 micron = .001 mm.

Area
1 sq. in. = 6.4516 sq. cm.
1 sq. ft. = 929.03 sq. cm.
1 sq. cm. = 0.155 sq. in.
1 sq. cm. = 0.0010764 sq. ft.

Volume
1 cu. in. = 16.387 cu. cm.
1 cu. ft. = 1728 cu. in.
1 cu. ft. = 7.4805 U.S. gal.
1 cu. ft. = 6.229 British gal.
1 cu. ft. = 28.317 liters
1 U.S. gal. = 0.1337 cu. ft.
1 U.S. gal. = 231 cu. in.
1 U.S. gal. = 3.785 liters
1 British gal. = 1.20094 U.S. gal.
1 British gal. = 277.3 cu. in.
1 British gal. = 4.546 liters
1 liter = 61.023 cu. in.
1 liter = 0.03531 cu. ft.
1 liter = 0.2642 U.S. gal.

Weight
1 ounce av. = 28.35 g.
1 lb. av. = 453.59 g.
1 gram = 0.03527 oz. av.
1 kg. = 2.205 lb. av.
1 cu. ft. of water = 62.425 lb.
1 U.S. gal. of water = 8.33 lb.
1 cu. in. of water = 0.0361 lb.
1 British gal. of water = 10.04 lb.
1 cu. ft. of air at 32° F. & 1 atm = 0.080728 lb.

Velocity
1 ft. per sec. = 30.48 cm. per sec.
1 cm. per sec. = .032808 ft. per sec.

Flow
1 cu. ft. per sec. = 448.83 gal. per min.
1 cu. ft. per sec. = 1699.3 liters per min.
1 U.S. gal. per min. = 0.002228 cu. ft. per sec.
1 U.S. gal. per min. = 0.06308 liters per sec.
1 cu. cm. per sec. = 0.0021186 cu. ft. per min.

Density
1 lb. per cu. ft. = 16.018 kg. per cu. meter
1 lb. per cu. ft. = .0005787 lb. per cu. in.
1 kg. per cu. meter = 0.06243 lb. per cu. ft.
1 g. per cu. cm. = 0.03613 lb. per cu. in.

Viscosity
1 Centipoise = .000672 lb. per ft. sec.
1 Centistoke = .00001076 sq. ft. per sec.

Pressure
1 in. of water = 0.03613 lb. per sq. in.
1 in. of water = 0.07355 in. of Hg.
1 ft. of water = 0.4335 lb. per sq. in.
1 ft. of water = 0.88265 in. of Hg.
1 in. of mercury = 0.49116 lb. per sq. in.
1 in. of mercury = 13.596 in. of water
1 in. of mercury = 1.13299 ft. of water
1 atmosphere = 14.696 lb. per sq. in.
1 atmosphere = 760 mm. of Hg.
1 atmosphere = 29.921 in. of Hg.
1 atmosphere = 33.899 ft. of water
1 lb. per sq. in. = 27.70 in. of water
1 lb. per sq. in. = 2.036 in. of Hg.
1 lb. per sq. in. = .0703066 kg. per sq. cm.
1 kg. per sq. cm. = 14.223 lb. per sq. in.
1 dyne per sq. cm. = .0000145 lb. per sq. in.
1 micron = .00001943 lb. per sq. in.

Energy
1 B.T.U.* = 777.97 ft. lbs.
1 erg = 9.4805 x 10^{-11} B.T.U.
1 erg = 7.3756 x 10^{-8} ft. lbs.
1 kilowatt hour = 2.655 x 10^{6} ft. lbs.
1 kilowatt hour = 1.3410 h.p. hr.
1 kg. calorie = 3.968 B.T.U.

Power
1 horsepower = 33,000 ft. lb. per min.
1 horsepower = 550 ft. lb. per sec.
1 horsepower = 2,546.5 B.T.U. per hr.
1 horsepower = 745.7 watts
1 watt = 0.00134 horsepower
1 watt = 44.26 ft. lbs. per min.

Temperature
Temperature Fahrenheit (F) = 9/5 Centigrade (C) + 32 = 9/4 R + 32
Temperature Centigrade (C) = 5/9 Fahrenheit (F) − 32 = 5/4 R
Temperature Reaumur (R) = 4/9 Fahrenheit (F) − 32 = 4/5 C
Absolute Temperature Centigrade or Kelvin (K) = Degrees C + 273.16
Absolute Temperature Fahrenheit or Rankine (R) = Degrees F + 459.69

Heat Transfer
1 B.T.U. per sq. ft. = .2712g. cal. per sq. cm.
1 g. calorie per sq. cm. = 3.687 B.T.U. per sq. ft.
1 B.T.U. per hr. per sq. ft. per °F = 4.88 kg. cal. per hr. per sq. m. per °C:
1 Kg. cal. per hr. per sq. m. per °C = .205 B.T.U. per hr. per sq. ft. per °F:
1 Boiler Horsepower = 33,479 B.T.U. per hr.

*B.T.U. Formula (energy required to heat any substance):
B.T.U./Hr. = Matl. Wt. (lbs.) x Temp. Rise (°F) x Spec. Heat.

R.S. Means Co., Inc., *Means Estimating Handbook*

Figure 5.28

Conversion Factors and Equivalents

Conversion Factors and Equivalents		
Multiply	**By**	**To Obtain**
Acres	43.560	Square Feet
B.T.U.'s	0.2530	Kilogram-calories
B.T.U.'s	778.2	Foot-pounds
B.T.U.'s	.0002928	Kilowatt-hours
B.T.U.'s per min.	12.97	Foot-pounds/sec.
B.T.U.'s per min.	0.02356	Horsepower
B.T.U.'s per min.	0.01757	Kilowatts
B.T.U.'s per min.	17.57	Watts
Centimeters	0.3937	Inches
Cubic Feet	7.481	Gallons
Cubic feet/min.	0.1247	Gallons/sec.
Cubic feet/min.	62.43	Lbs. of water/min.
Cubic feet water	62.43	Lbs. of water
EDR	240	B.T.U.'s
Ft. head	2.31	Lbs.
Ft. of Water	62.43	Lbs./Sq. Ft.
Ft. of Water	0.8826	Inches of mercury
Ft. of Water	0.4335	Lbs./Sq. In.
Ft./min.	0.01136	Miles/Hr.
Foot-pounds	.001285	B.T.U.'s
Gallons	231	Cubic inches
Gallons/min.	.002228	Cu. Ft./sec.
Gallons water	8.345	Lbs. of water
Grains (troy)	0.06480	Grams
Grams	0.03527	Ounces
Horsepower	42.44	B.T.U.'s/min.
Horsepower	33,000	Ft.-Lbs./min.
Horsepower	550	Ft.-Lbs./sec.
Horsepower	745.7	Watts
Horsepower (Boiler)	33,479	B.T.U.'s/Hr.
Horsepower (Boiler)	9.804	Kilowatts
Horsepower-hours	2547	B.T.U.'s
Inches	2.540	Centimeters
Inches of mercury	1.133	Ft. of water
Inches of mercury	0.4912	Lbs./Sq. In.
Inches of water	0.03613	Lbs./Sq. In.
Kilograms	2.2046	Pounds
Kilowatts	56.92	B.T.U.'s/min.
Kilowatts	1.341	Horsepower
Kilowatt-hours	3415	B.T.U.'s
Liters	0.2642	Gallons
Miles	5280	Feet
Ounces	437.5	Grains
Ounces	28.35	Grams
Pounds	7000	Grains
Pounds	453.6	Grams
Pounds of water	27.68	Cubic inches
Pounds of water	0.1198	Gallons
Pounds of water	7000	Grains
Pounds of water/min.	0.2669	Cu. Ft./sec.
Square feet	144	Square inches
Square inches	1.273×10^{-6}	Circular mils
Square miles	640	Acres
Temp. (degs. C) + 273	1	Abs. temp. (degs. K)
Temp. (degs. C) + 17.8	1.8	Temp. (degs. Fahr.)
Temp. (degs. F) −32	5/9	Temp. (degs. Cent.)
Ton (Refrigeration)	200	B.T.U.'s/min.
Watts	.001341	Horsepower
Watt-hours	3.415	B.T.U.'s

R.S. Means Co., Inc., *Means Estimating Handbook*

Figure 5.29

Resource Publications

Carlson, Reinhold A., PE, and Robert A. Di Giandomenico. *Understanding Building Automation Systems.* R.S. Means Company, Inc.

Colen, Harold R., PE. *HVAC Systems Evaluation.* R.S. Means Company, Inc.

Herbert, R.D., III. *Roofing: Design Criteria, Options, Selection.* R.S. Means Company Inc.

Sievert, Richard W., Jr. *Total Productive Facilities Management.* R.S. Means Company, Inc.

U.S. Department of Energy, Energy Efficiency and Renewable Energy Network (EREN). *Purchasing for Energy Efficiency: Windows and Doors.* www.eren.doe.gov/buildings/purchs_windows.html

U.S. Department of Energy, (EREN). *Electric Lighting.* www.eren.doe.gov/buildings/purchs_elec.html

For Additional Information

Air Conditioning and Refrigeration Institute (ARI)
4301 North Fairfax Drive, Suite 425
Arlington, VA 22203
www.ari.org

Electric Power Institute (EPRI)
3412 Hillview Avenue
Palo Alto, CA 94304
www.epri.com

Association of Energy Engineers
4025 Pleasantdale Road, Suite 420
Atlanta, GA 30340
www.aeecenter.org

The Energy Foundation, a non-profit grant-making organization promoting energy efficiency and renewable energy, has a good list of energy-related web sites. www.ef.org.

Environmental Protection Agency (EPA) Office of Water. www.epa.gov/OW

Department of Energy (DOE)
The primary field office for DOE energy efficiency programs is:
U.S. Department of Energy
Golden Field Office
1617 Cole Boulevard
Golden, CO 80401-3393
(303) 272-4742

There are numerous articles that can be found on U.S. Department of Energy sites on the web:

- DOE Energy Efficiency and Renewable Energy Network (EREN) offers many articles on increasing energy efficiency. www.eren.doe.gov
- DOE Office of Industrial Technologies has extensive information on increasing industrial energy efficiency. www.oit.doe.gov.
- Funded by DOE, The Energy Efficiency and Renewable Energy (EE\RE) Program at Oak Ridge National Laboratory researches and develops energy efficient and renewable energy technology. It offers information on developing technology for building systems, industrial processes, utilities, and transportation. www.ornl.gov

National Association of Energy Service Companies
1615 M Street, NW, Suite 800
Washington, DC 20036
(202) 822-0950
www.naesco.org

Solstice at http://solstice.crest.org/efficiency/index.shtml has an extensive listing of energy-related web sites including information on building, government programs, case studies, and references.

Chapter 6

Heating, Ventilation & Air Conditioning

Chapter 6 Heating, Ventilation & Air Conditioning

The importance of maintaining and optimizing a facility's heating, ventilation, and air conditioning (HVAC) system cannot be overstated. While all associated building systems carry a certain level of criticality, there is not one that is more continuous, more encompassing, and, when operated properly, more invisible to the occupants than the HVAC system that maintains building comfort.

The unfortunate aspect of HVAC systems operation is that if the system and facilities staff are successful in doing their job, their efforts go unnoticed, unheralded, and, perhaps, unappreciated. Should, however, building comfort decline (for whatever reason) the response is certain, swift, and, at times, unforgiving. It is important, therefore, for facilities personnel to have an understanding or working knowledge of this subject. This knowledge base should include basic HVAC theory; system selection and design; and system operation and maintenance.

This chapter is meant to provide facilities managers and personnel with a practical overview of HVAC, including some basic principles of design and operation. Most of the text is from *HVAC: Design Criteria, Options, Selection, Second Edition* and *HVAC Systems Evaluation*, both published by R.S. Means Co., Inc. The former focuses on design/selection, while the latter covers operations and maintenance issues. Both are highly recommended for their comprehensive, clear treatment of HVAC issues.

The following text is reprinted with permission from *HVAC: Design Criteria, Options, Selection, Second Edition* by William H. Rowe, III, PE, AIA, R.S. Means Co., Inc.

HVAC Heating and Cooling Loads

The size of heating and cooling equipment is determined by the "load" it must carry. For a boiler, the load is the amount of heat that must be pushed into the building per hour to keep it warm. For cooling, the load is the amount of heat per hour that must be taken away from the building to keep it cool. Methods for determining both heating and cooling loads are discussed in this section.

Heating Loads

Heat is measured in British thermal units (Btu), calories (c), or joules (J). One Btu equals the amount of energy needed to heat one pound of water 1°F (See Figure 6.1). One calorie equals the amount of energy needed to heat one gram of water 1 °C. One calorie equals 4.2 joules. The joule is the more common measure of heat content.

Heat rates are expressed in the following units:
British thermal unit per hour – Btu/hr.
Joules per second – J/s
(One J/s equals 1 watt – W)

Because smaller units are easier to work with, 1,000 Btu/hr. is often written as 1 MBH. 1000 J/s is written as 1 kW.

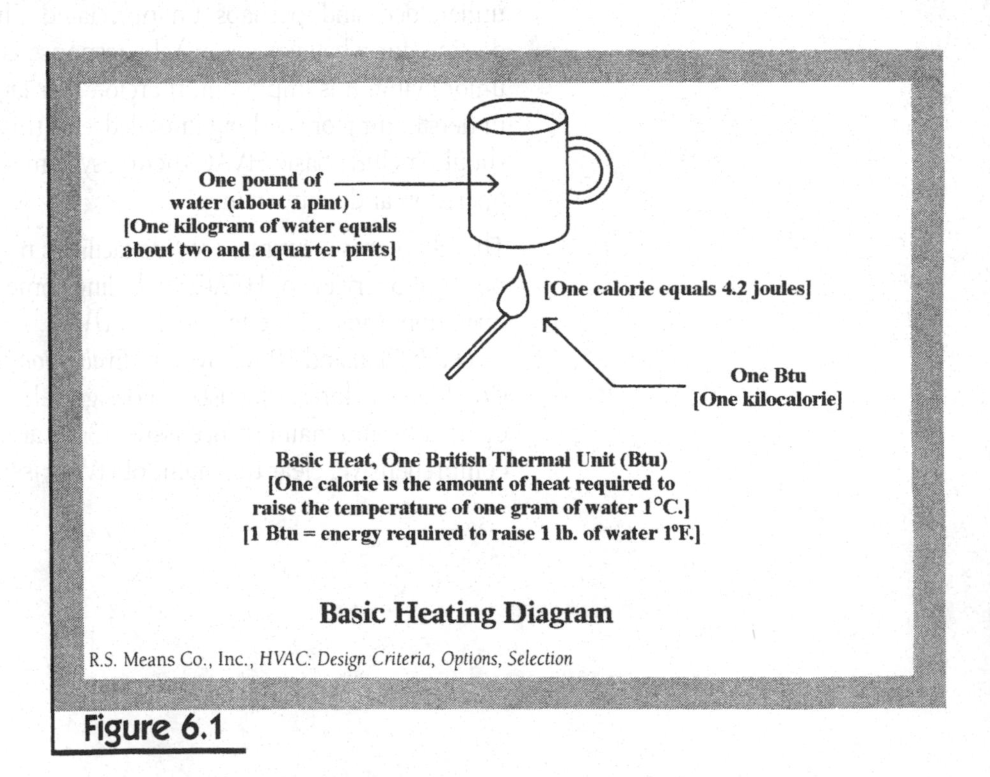

One pound of
water (about a pint)
[One kilogram of water equals
about two and a quarter pints]
[One calorie equals 4.2 joules]
One Btu
[One kilocalorie]
Basic Heat, One British Thermal Unit (Btu)
[One calorie is the amount of heat required to
raise the temperature of one gram of water 1°C.]
[1 Btu = energy required to raise 1 lb. of water 1°F.]
Basic Heating Diagram
R.S. Means Co., Inc., HVAC: Design Criteria, Options, Selection

Figure 6.1

Water: The properties of water are basically constant in the 32–212°F [0–100 °C]* range. Water is easy to work with because it is a liquid with nearly constant physical characteristics over the range of temperatures normally used in heating, and is readily available and chemically stable. For these reasons, water is the most commonly used heating fluid.

* *SI metric equivalents are provided for most imperial units. The SI units appear in brackets [ ].*

Heating and Sensible Heat: Heating systems are often designed without humidification; all of the heat energy is used to warm the air in a room. When heating systems are designed without humidification, all of the added heat is termed *sensible heat,* because every Btu added to the air is devoted only to raising the air temperature. This heat can be felt or "sensed." When humidification is incorporated into heating systems the heat load (number of Btus necessary) increases by approximately 30 percent.

Heat Loss: The heating load for a building is determined by calculating the amount of heat loss from the building to the outside during the winter design condition. (The common winter design condition is the cold winter night in a particular climate that is surpassed 97-1/2% of the time.)

There are three ways in which heat is lost from a building: *conduction*, *convection*, and *radiation*. The total MBH of these three types of heat loss equals the heat load, or the amount of heat that must be put into the building by the heating system to offset these losses and maintain the proposed comfort zone.

Conduction transfers heat in a chain-like manner, from higher-temperature to lower-temperature molecules. As one molecule heats up, it transfers heat to the molecule next to it. With this type of heat transference, the molecules are stationary and the heat moves, or is conducted, through them. For example, if you hold one end of a steel poker and use the other end to stir a fire, your hand gets warmer as the heat from the fire moves, or is conducted, along the shaft to your hand. Similarly, a building's heat is conducted to the outdoors through the solid surfaces of walls, roofs, floor slabs, glass doors, and windows. The rate of heat depends on the type of material and the temperature difference between adjacent materials. Good conductors, such as metals, transfer more heat than good insulators.

Convection transfers heat as warm molecules actually move from one place to another. If a current of air was passed over the heated poker noted above, some of the heat of the rod would be transferred to the passing air, or convected. In buildings, heat is convected from the interior to the outdoors by air that leaks, or infiltrates, through cracks around windows and doors, and by the exhaust and ventilation air that moves between the interior and the exterior of the building. Most HVAC distribution systems work by convection. The hot water that moves heat from the boiler through the pipes to the fan coil units and the warm air furnace that distributes heat to the room via ductwork are moving heat around the building by convection, either by force or by gravity.

Radiation transfers heat by electromagnetic waves. Sunlight is the most common, and most spectacular, example. Heat from the sun reaches the earth by radiating through 93 million miles of space. In the steel poker example, heat is radiated when it gets "red hot," at which time the heat can be felt several feet away. Heat is radiated onto buildings by the sun during the day. Solar radiation is primarily considered as a factor in calculating the cooling load, where it must be overcome. It is not a significant factor in calculating the heating load.

Figure 6.2 illustrates the basic processes of conduction, convection, and radiation. The heat losses that commonly occur in buildings are illustrated in Figure 6.3. Each type of heat loss is described in the following sections.

There are several methods used to calculate heating and cooling loads. (See Figure 6.4.) They are comparable. This book uses the ASHRAE method.

Cooling

Cooling is almost universally accomplished by blowing cool air into a room or building. For example, by supplying the proper quantity of air at 60°F [15 °C] to the room temperature of a space, the air will mix and the temperature will be maintained at the design temperature (plus or minus 78°F [25 °C]). The cool air absorbs the heat gained from transmission through walls, windows, and doors; infiltration of warm air from around doors and windows; solar radiant energy; and internal heat gains from lights, people, and equipment.

The common unit of measure for cooling is the ton, which is derived from an earlier period when ice was used for cooling. One ton of cooling is equal to 12,000 Btus [3.5 kW] per hour. It takes 144 Btus [152 kJ] to melt one pound of ice at 32°F [0 °C]; when one ton of ice melts in one 24-hour period, it has absorbed 12,000 Btus [3.5 kW] per hour.

Just as boilers and furnaces produce heating, a refrigeration system produces cooling. Heating equipment adds heat to a building; cooling equipment, in contrast, subtracts heat, or more simply, pushes heat away from the building. Basic cooling systems are shown on the cooling ladder in Figure 6.5.

Cool air is delivered by the air conditioning unit, neutralizes the warm-air buildup (heat gain), and maintains the desired room temperature. Some of the return air from the space is exhausted to the outdoors. The balance of the return air is mixed with the outdoor air, is cooled, and is redistributed to the space. Figure 6.6 illustrates a basic cooling layout.

There is a significant difference between *making* heat (heating) and *transferring* heat (cooling). The principle of heating is easy to understand. It is just like boiling a cup of water. In heating a building, fuel is burned in the combustion chamber of a boiler or furnace, and the heat energy from the combustion warms the water, steam, or air for the system. Each unit of energy from combustion is added to the system, starting at a high temperature and stepping down along the distribution path.

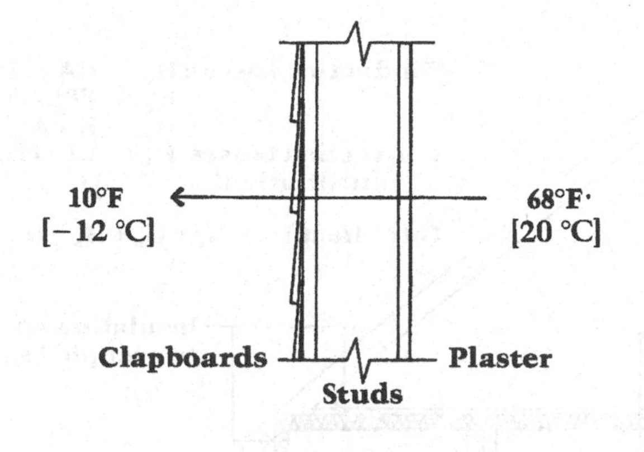

Conduction

Heat moves from hot to cold — molecules transfer heat from one to another as heat moves through them.

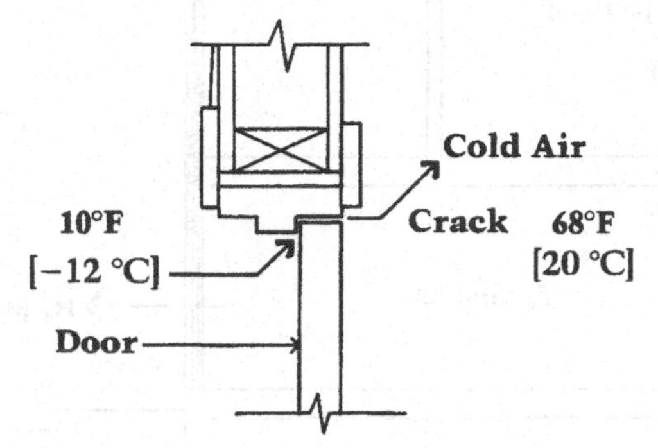

Convection

Cold air infiltrates into a space by physically moving from outdoors to indoors.

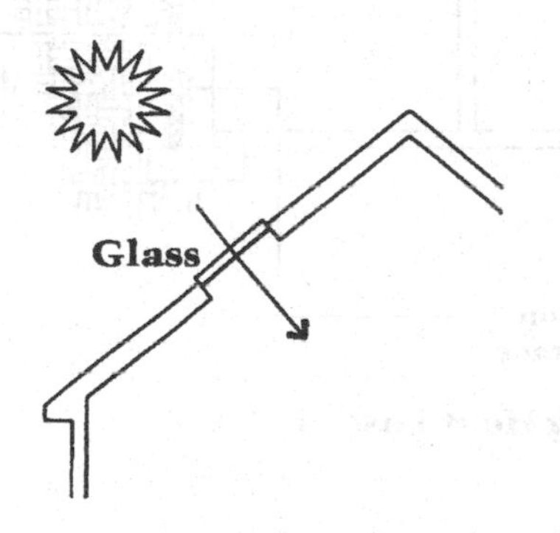

Radiation

Solar energy penetrates the building wall by passing through transparent glass surfaces.

Conduction, Convection, and Radiation

R.S. Means Co., Inc., *HVAC: Design Criteria, Options, Selection*

Figure 6.2

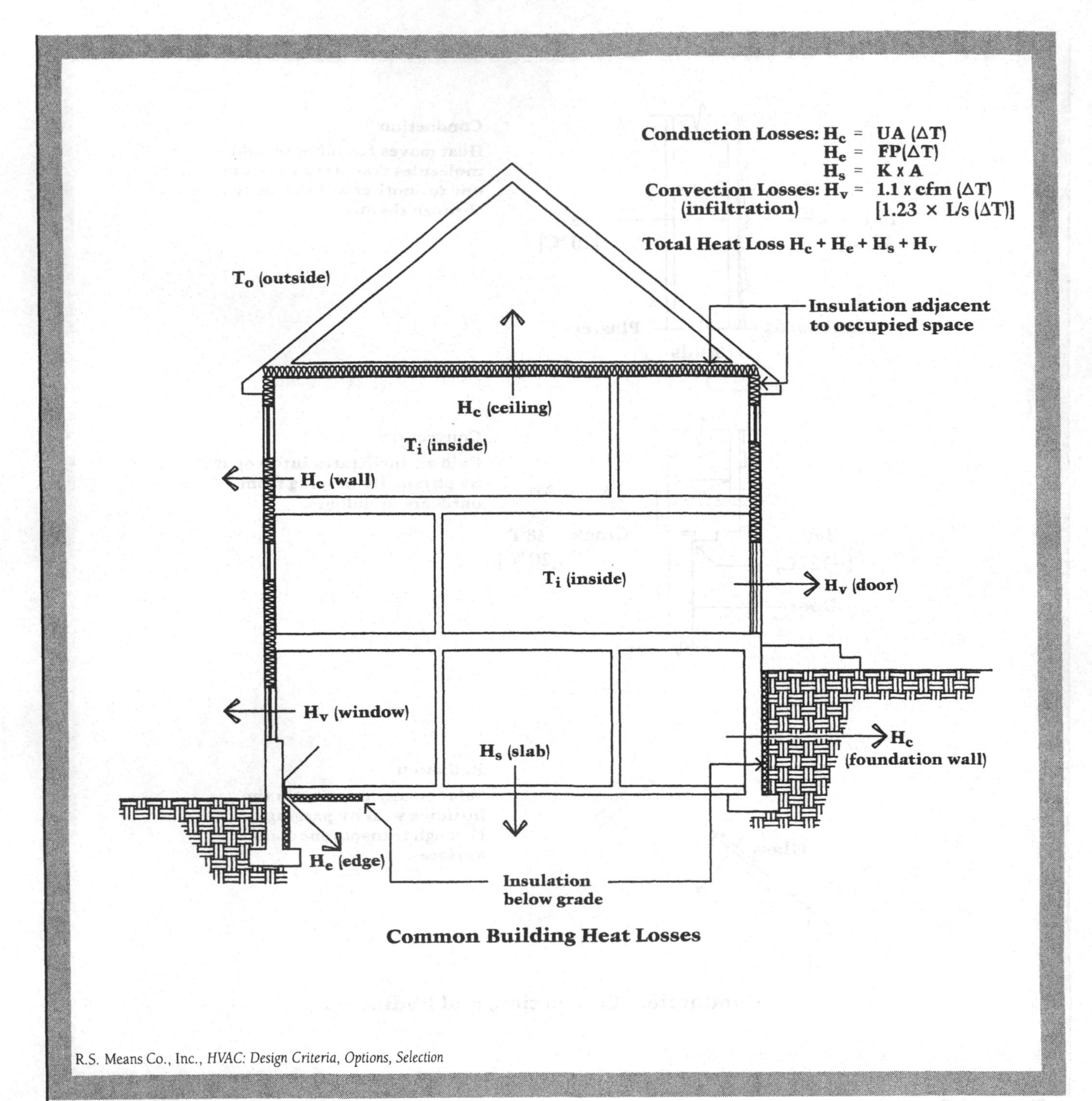

Common Building Heat Losses

R.S. Means Co., Inc., *HVAC: Design Criteria, Options, Selection*

Figure 6.3

Cooling is more complicated. It is obviously not a problem to cool a space in the winter—simply opening the windows will cool off a space, as warm air flows "downhill" (outside to colder temperatures) to be replaced by the cooler outside air. Opening windows in the summer, however, will not cool a space. The heat inside must be pushed "uphill" to the outdoors. In the summer, the supply air temperature of 60°F [15 °C] is made by taking the heat from the return air mixture and rejecting it to the warmer outdoor air.

Heat Loss Formulas

Conduction Heat Losses

Losses Through Structure (Walls, Roofs, Doors, Windows)

$H_C = UA(\Delta T)$

H_C = Conduction loss (BTU/hr.) [W]

U = Overall heat transfer coefficient (Btu/hr. × ft.^{2} × °F) [W/m^2.K]

A = Surface area (s.f.) [m^2]

ΔT = Temperature difference (°F) [K]

Losses Through Edges of Slab

$H_E = FP(\Delta T)$

H_E = Heat loss through edge (Btu/hr.) [W]

F = Edge factor (Btu/hr. ft. °F) [W/m^2.K/m]

P = Perimeter of slab (ft.) [m]

ΔT = Indoor temperature minus outdoor temperature

Losses Through Slabs and Basement Walls

H_S = K × A *or*

= 2 [0.6] × Area for floors and basement walls 4 feet below grade *or*

= 4 [1.2] × Area for floors and walls from 0 to 4 feet below grade

H_S = Heat loss through slab

K = Constant (Btu/hr. s.f.) [W/m^2]

A = Area (s.f.) [m^2]

Convection Heat Losses

H_V = 1.1 cfm (ΔT) [1.23 × L/s × ΔT]

H_V = Convection loss (Btu/hr.) [W]

1.1 = Constant

cfm = Air entering building (cubic feet per minute) [m^3/s]

ΔT = Temperature difference (°F) [°C or K]

Total Heat Loss (Btu/hr.)

$H_T = H_C + H_E + H_V + H_S$

H_C = Conduction losses (walls, windows, floor, roof, attics, crawl spaces, garages, basements)

H_E = Edge losses, when applicable

H_S = Slab losses (grades, foundations, walls)

H_V = Infiltration and ventilation

All numbers shown in brackets are metric equivalents

R.S. Means Co., Inc., *HVAC: Design Criteria, Options, Selection*

Figure 6.4

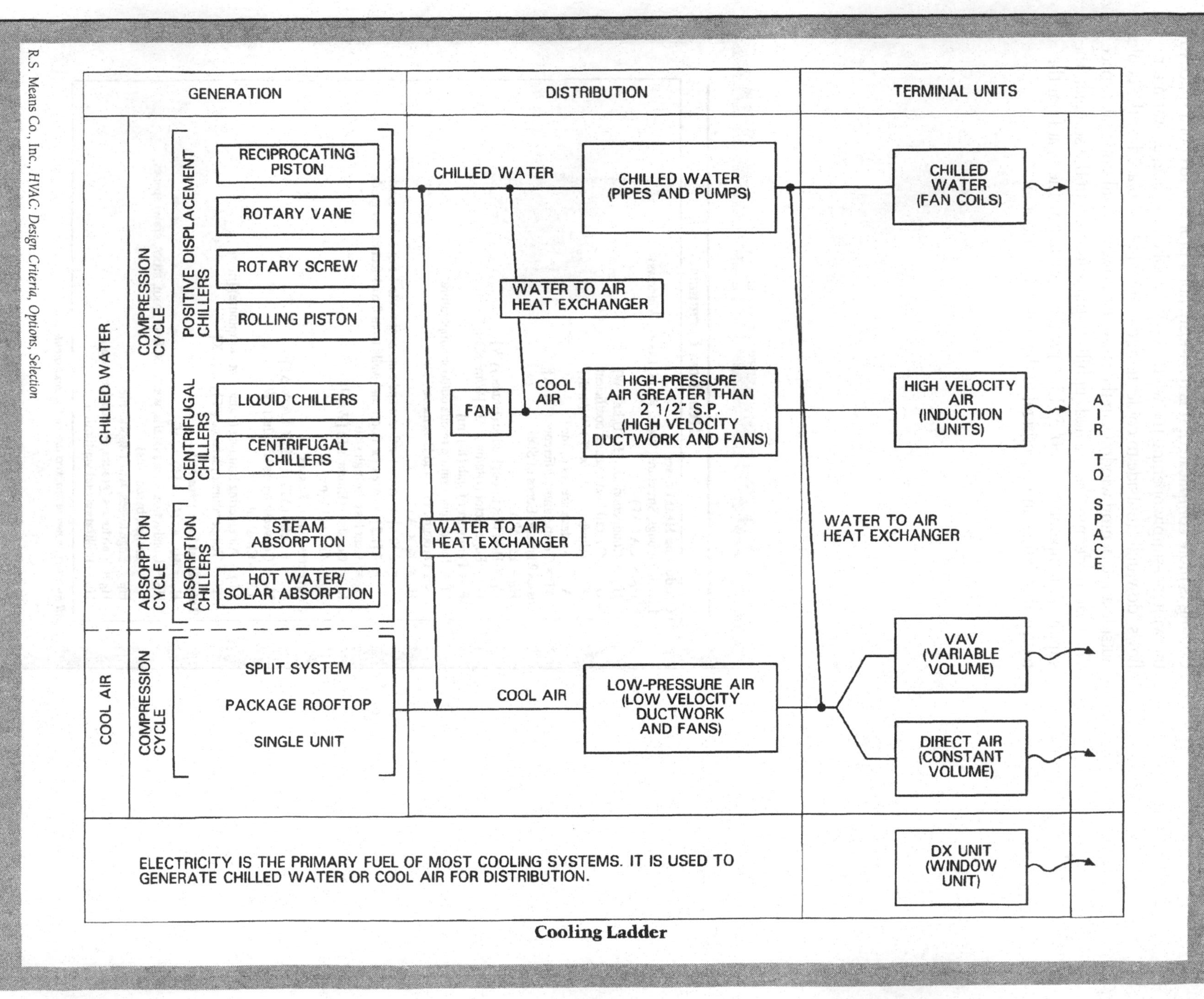

Cooling Ladder

R.S. Means Co., Inc., *HVAC: Design Criteria, Options, Selection*

Figure 6.5

This is very similar to what happens with a household refrigerator. The inside of a refrigerator is kept cooler than the surrounding kitchen. To maintain cooling, the refrigeration cycle takes the heat to be removed from the inside of the refrigerator and pumps it outside. The process of taking the heat from inside the space and putting it into the refrigerant warms the refrigerant gas. The added energy of the compressor raises the temperature and pressure of the refrigerant gas further, to about 110°F [43 °C]. The exposed condenser coils on the back of the refrigerator are cooled by the room air temperature, which turns the refrigerant gas back into a liquid because of the heat reduction. This process keeps the inside of the refrigerator at about 40°F [4 °C]. (The compressor is the only major device that consumes energy in the refrigerator.) The temperatures and layouts for basic heating and cooling systems are shown in Figure 6.7.

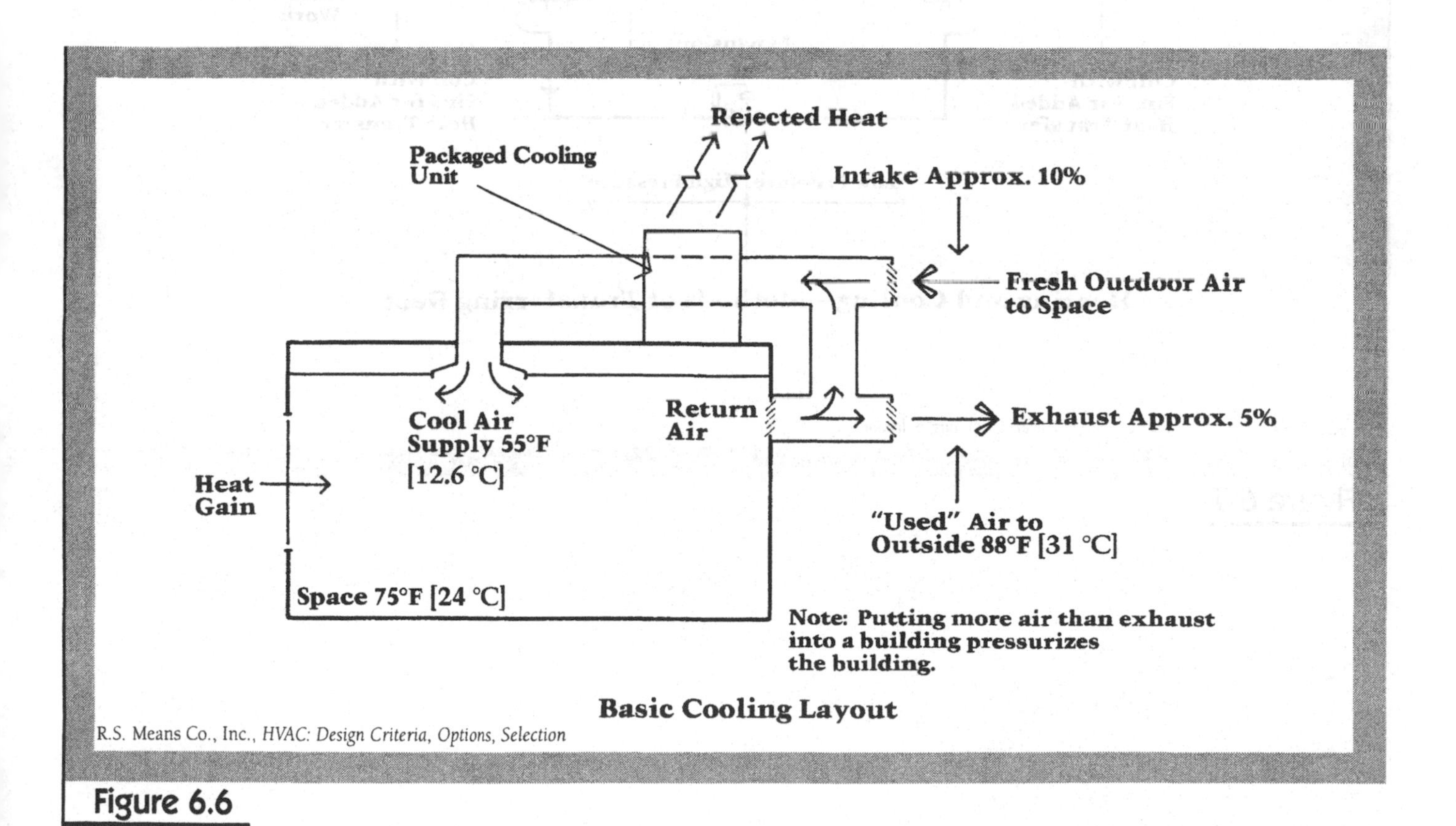

Basic Cooling Layout

R.S. Means Co., Inc., *HVAC: Design Criteria, Options, Selection*

Figure 6.6

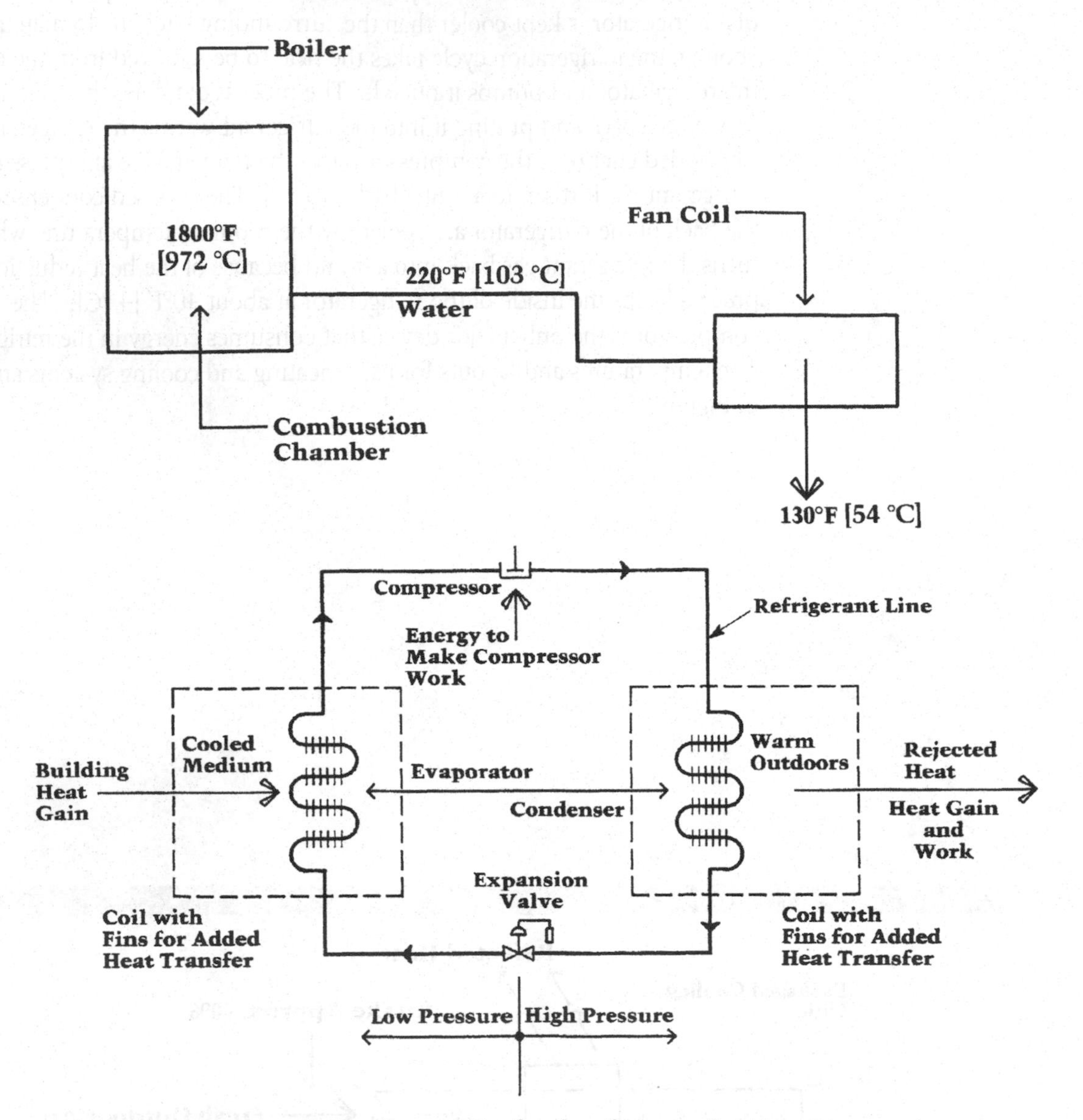

Heating and Cooling—Methods of Transferring Heat

R.S. Means Co., Inc., *HVAC: Design Criteria, Options, Selection*

Figure 6.7

In heating, each Btu [Joule] of heat burned produces nearly one Btu [Joule] of building heat. In cooling, each Btu [Joule] to be rejected requires only about one-third to one-fifth of a Btu [Joule] of compressor energy. The effectiveness of cooling is measured by the coefficient of performance. The **coefficient of performance** (COP) equals the amount of cooling energy produced, or heat gain removed (in Btus/hour), divided by the energy consumed to produce the work of the compressor cooling (in Btus/hour).

$$\text{COP} = \frac{\text{amount of cooling energy produced}}{\text{energy consumed to produce the cooling}}$$

The coefficient of performance generally ranges from 3.5 to 5 Btus for most applications.

Cooling Loads

The size of a cooling system is based on the load it must carry. In cooling, the load is expressed in Btus/hour or tons [watts or kilowatts]. One ton of cooling equals 12,000 Btus/hour, or 12 MBH [one kilowatt of cooling equals 1000 watts]. The cooling load represents the amount of heat that must be removed from a building, usually referred to as heat gain. In addition to the heat gain from conduction and convection (computed similarly to heat losses), cooling loads must take into account radiation and internal heat gains from lights, appliances, power, and people. All of these factors represent additional loads to a cooling system.

Cooling loads are based on the statistics for a hot summer day that is surpassed only 2-1/2% of the time in a particular climate. The critical time and date (which varies with each building) is determined by designers, because cooling load calculations are more complicated than heating load calculations. The primary reason for this complexity is the solar energy heat gain factor, for which computations are elaborate. The solar orientation of the building and the amount of glass receiving radiant energy varies not only from building-to-building, but also day-by-day, and even hour-by-hour. Furthermore, the heat "stored" in a building depends on the overall mass of the building, the hours of operation, and the length of time that lights are on in the building each day. The color of the building and type of sunscreen are also important factors. Light colors reflect more sun and lower the building heat load. Awnings, shades, and venetian blinds also cut out solar radiation and lower solar heat gain.

The actual cooling load calculation can involve many iterations or tests during the cooling season. The resulting design conditions may include many variations. For example, one room facing east may have its heaviest heat gain at 10:00 a.m. on June 19, while another room facing west any have its design load (heaviest heat gain) at 5:00 p.m. on August 21; the maximum overall load for the building as a whole may occur at 3:00 p.m. on July 20. The interior temperature may swing with the varied loads as the cooling equipment responds.

In addition to sensible heat (a change in the air temperature that is felt or "sensed"), cooling systems must also account for *latent heat gains*. Latent heat is the energy that is required to change a solid to a liquid or a liquid to a gas; no temperature change occurs during this process. In cooling, latent heat is the energy resulting from condensation of moisture in the air. Condensing water vapor is the reverse of boiling. It takes approximately 1,000 Btus, or 1 MBH, to boil or condense a pound of water. [It takes approximately 2300 joules to boil or condense 1 kg of water.] Boiling normally occurs at 212°F [100 °C]. However, it is possible to vaporize 50°F [10 °C] water. It is also possible to pass humid air at 80°F [27 °C] over a cooling coil and produce 50°F [10 °C] air with a much lower humidity.

The energy required to vaporize the 50°F [10 °C] water and dehumidify the 80° F [27 °C] air is the latent heat of vaporization. Latent heat for cooling occurs when outdoor air (which contains moisture) is introduced into a building. People and cooking equipment also add moisture (latent heat) to a space. Sensible and latent heat gains are illustrated in Figure 6.8. Figure 6.9 illustrates the five categories of common building heat gains, which are listed below. Figure 6.10 provides formulas for calculating heat gain generated by these sources.

- Conduction heat gain
- Convection heat gain
- Solar heat gain
- Internal heat gain (lights, motors, cooking, etc.)
- People

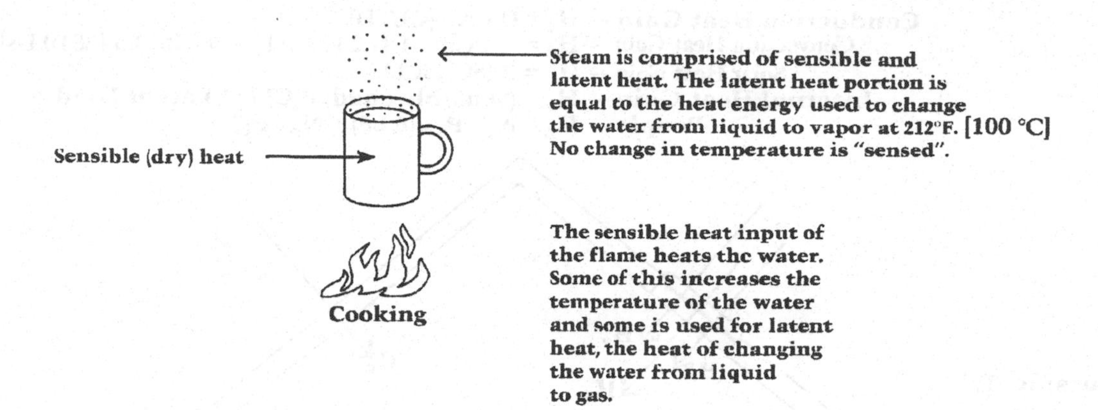

Steam is comprised of sensible and latent heat. The latent heat portion is equal to the heat energy used to change the water from liquid to vapor at 212°F. [100 °C] No change in temperature is "sensed".
Sensible (dry) heat
Cooking
The sensible heat input of the flame heats the water. Some of this increases the temperature of the water and some is used for latent heat, the heat of changing the water from liquid to gas.

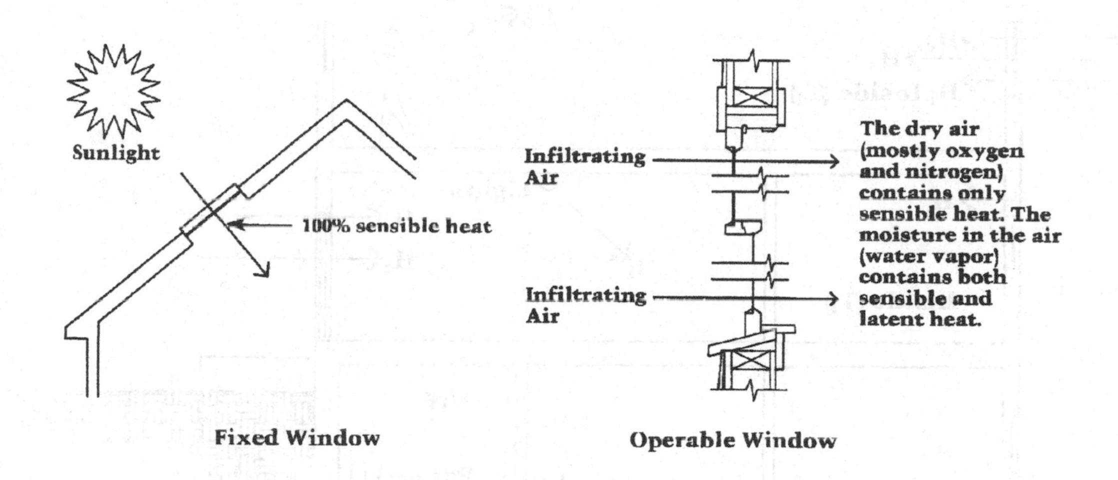

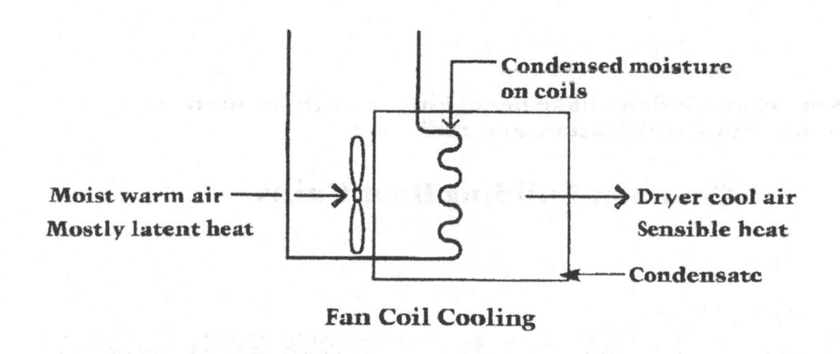

Sensible and Latent Heat Gains

R.S. Means Co., Inc., *HVAC: Design Criteria, Options, Selection*

Figure 6.8

Conduction Heat Gain — $H_c = U \cdot A \cdot (CLTD)$
Convection Heat Gain — $H_v = 1.1 \cdot cfm \cdot \Delta T\ [1.23 \cdot L/s \cdot \Delta T] + 4{,}840\ cfm\ [2810\ L/s]\ (\Delta W)$
Solar Heat Gain — $H_s = A \cdot (SC) \cdot (SCL)$
Internal Heat Gain — H_i = (Sensible Load) • (CLF) + Latent Load
People — $H_p = N_o \cdot P_s \cdot (CLF) + N_o \cdot P_L$

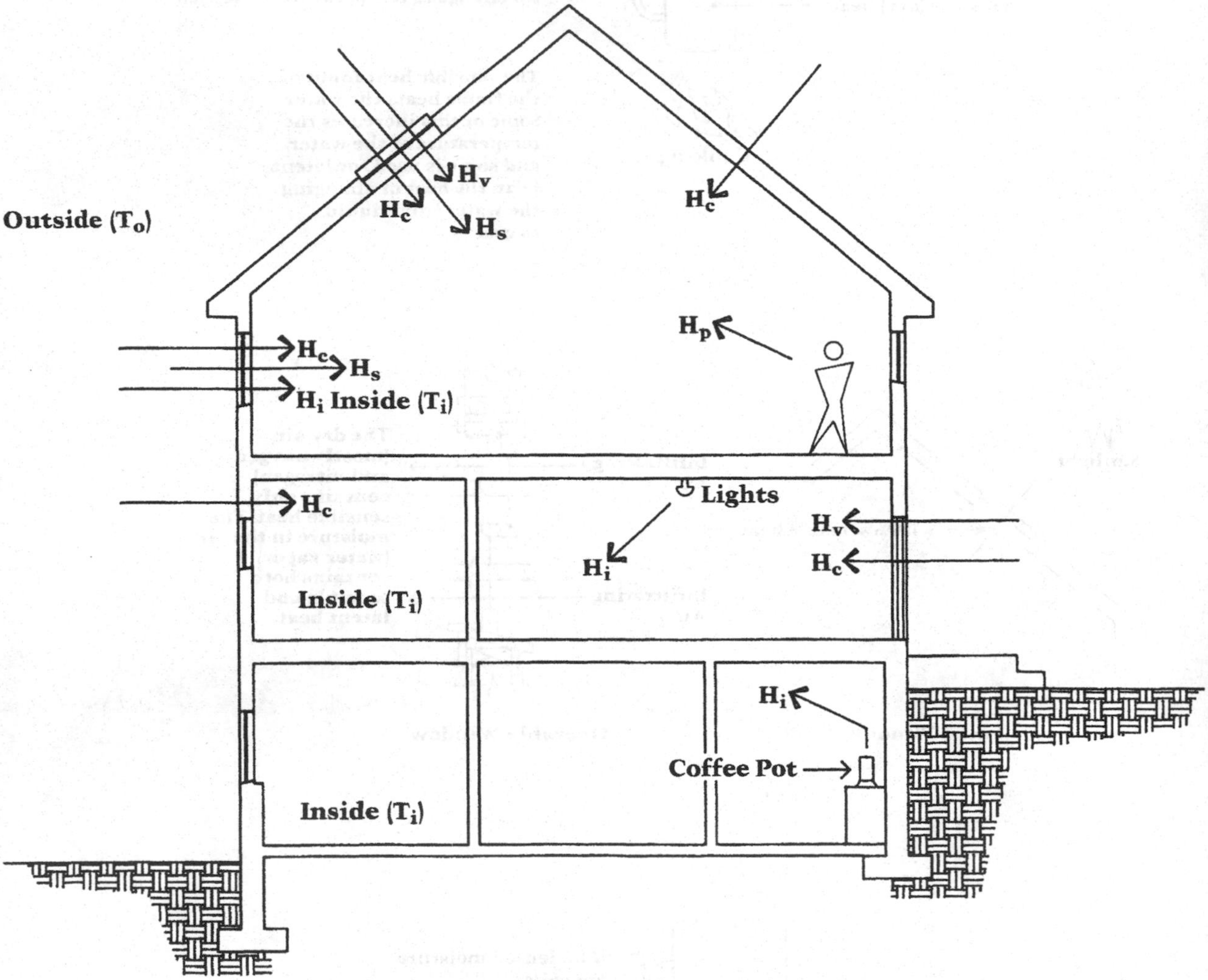

Outside (T_o)
H_v H_c H_s H_c
H_p
H_c H_s
H_i Inside (T_i)
H_c
Lights
H_i
H_v
H_c
Inside (T_i)
H_i
Coffee Pot
Inside (T_i)

Note that windows have heat gains from three sources: conduction, infiltration, and radiation.

Common Building Heat Gains

R.S. Means Co., Inc., *HVAC: Design Criteria, Options, Selection*

Figure 6.9

Heat Gain Formulas

Conduction Heat Gain

H_C = UA (CLTD)

H_C = Conduction heat gain (Btu/hr.) [W]

U = Overall heat transfer coefficient (Btu/hr. s.f. °F) [W.m²/K]

A = Surface area (s.f.) [m²]

CLTD = Cooling load temperature difference (°F) [°C]

Convection Heat Gain (Infiltration and Mechanical Ventilation)

H_V = |1.1 cfm [1.23 L/s](ΔT)| + |4,840 cfm [2810 L/s](ΔW)|

(Sensible heat gain) (Latent heat gain)

ΔT = Temperature difference between space and outside air (°F)

ΔW = Moisture content difference between inside and outside air (lbs. water/lbs. dry air)

Solar Radiation Heat Gain

H_S = A (SC) (SCL)

H_S = Radiant heat gain (Btu/hr.) [W]

A = Glass area (s.f.) [m²]

SC = Shading coefficient

SCL = Solar cooling load factor

Internal Heat Gain

H_i = (sensible load) CLF + (latent load)

H_i = Internal heat gain (Btu/hr.)

CLF = Cooling load factor

People Heat Gain

$H_P = N_O P_S CLF + N_O P_L$

H_P = People heat gain (Btu/hr.) [W]

N_O = Number of occupants

P_S = Sensible heat gain per person (Btu/hr.) [W]

CLF = Cooling load factor, usually 1.0

P_L = Latent heat gain per person (Btu/hr.) [W]

Total Heat Gain

$H_T = H_C + H_V + H_S + H_I + H_P$

All numbers shown in brackets are metric equivalents

R.S. Means Co., Inc., *HVAC: Design Criteria, Options, Selection*

Figure 6.10

The following section is reprinted with permission from *HVAC: Design Criteria, Options, Selection, Second Edition* by William H. Rowe, III, PE, AIA, R.S. Means Co., Inc.

Psychrometrics

Psychrometrics is the analysis of the properties of air as it passes through an HVAC process such as cooling, heating, humidification, dehumidification, or any combination thereof. The psychrometric chart (Figure 6.11) is a convenient and concise summary of the properties of air.

The Psychrometric Chart

Air in any condition can be represented by one point on the psychrometric chart; from this point, seven properties of the air can be read. The seven properties that can be read from any point are as follows (See Figure 6.11):

- dry bulb temperature (°F) [°C] (line #1)
- wet bulb temperature (°F) [°C] (line #2)
- moisture content or specific humidity (lbs. of moisture per lb. of dry air or grains of moisture per lb. of dry air) [g of moisture per g of dry air] (line #3)
- relative humidity (%) (line #4)
- specific volume (cu.ft./lb.) [m^2/kg] of air (line #5)
- enthalpy (Btu/lb. of air) [kJ/kg of dry air] (line #2)
- dew point (°F) [°C] (line #3)

If any two of these properties are known, the other five can be found by plotting the point on the psychrometric chart.

Dry bulb temperature (°F) [°C] is the temperature of the air as read by a normal thermometer.

Wet bulb temperature (°F) [°C] is the temperature registered on a thermometer that has a wet gauze wrapped around its bulb and air blowing across it at 50 fpm [0.25 m/s]. Water evaporates from the gauze as it absorbs heat from the fluid in the bulb. Consequently, the wet bulb temperature is lower than the dry bulb termperature unless the air is saturated (100% humidity), in which case no evaporation can take place, and the dry bulb and wet bulb temperatures are equal. At a given dry bulb temperature, the wet bulb temperature is an indication of the relative humidity of the air.

Moisture content or specific humidity (lbs. water/lb. air or grains water/lb. air) [g water/g dry air] is a measure of the amount of water in the air.

Relative humidity (%) is an indication of the degree of saturation of the air.

Specific volume (cu.ft./lb. of dry air) [m^3/kg dry air] of the air is the volume that one pound of air occupies.

Enthalpy (Btu/lb.) [kJ/kg] is a measure of the quantity of energy in the air.

Dew point (°F) [°C] is the temperature at which the air is saturated. At the dew point, the dry bulb temperature is equal to the wet bulb temperature.

Figure 6.11

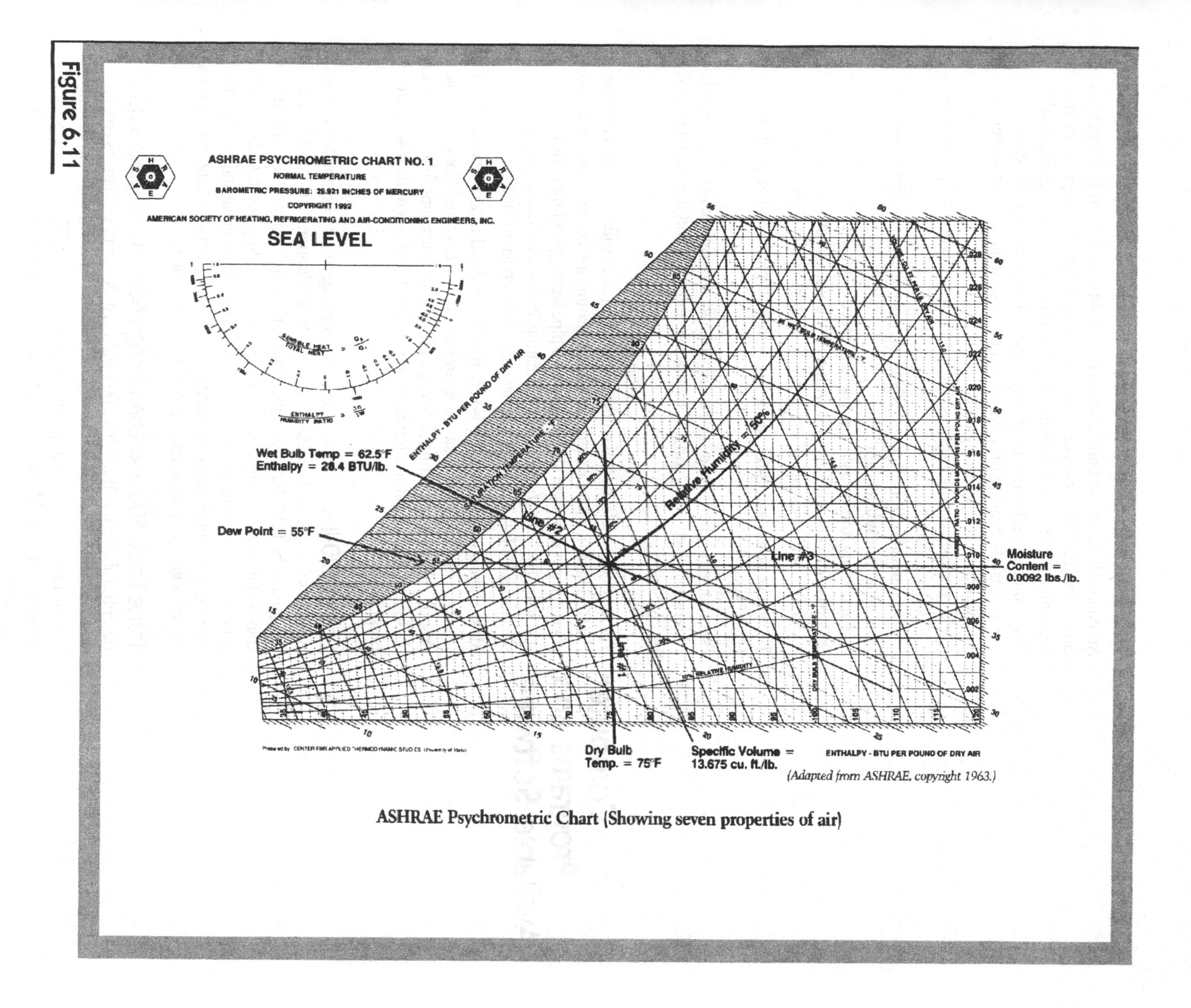

ASHRAE Psychrometric Chart (Showing seven properties of air)

The Psychrometric Process

Air enters HVAC equipment at one set of conditions and leaves at a new set of conditions. Each of these conditions can be represented by a point on the psychrometric chart; a line drawn between these two points represents the HVAC process performed by the equipment. The processes include:

- a sensible heating process, which may occur as air passes over a hot water or steam heating coil;
- a sensible cooling process, which may occur when air passes over a dry cooling coil;
- a humidification process, which may occur when air passes over a steam humidifier; and
- a cooling and moisture reduction process, which would occur as air passes over a cooling coil with a dew point lower than the dew point of the air (also called a sensible and latent cooling process).

The above processes represent the functions of typical equipment in an HVAC system.

Psychrometric charts for a wide variety of purposes can be obtained from the American Society of Heating, Refrigerating, and Air-Conditioning Engineers, Inc. (ASHRAE).

Computer Programs and Available Software

There is a wide selection of computer programs and software targeted to the HVAC designer. These programs include heating and cooling load calculations, energy analysis, system sizing, duct and pipe sizing, psychrometrics, specialized system analysis, cost estimating, electronic catalogs, equipment selections, facility and plant management, and service management.

Most of the software available is very user-friendly and is accompanied by good reference manuals. The many programs are so varied it would be impossible to describe all of them, but brief descriptions of a few programs follow. (Contact these manufacturers for information on the latest versions of their software programs.)

Carrier E20-II: The Carrier E20-II software program is a comprehensive set of programs that covers the wide range of HVAC applications. The most critical program is the HAP (Hourly Analysis Program), which utilizes the ASHRAE transfer function method for detailed heating and cooling load calculations. Other programs include engineering economic analysis, commercial and residential load estimating and operating cost analysis, duct design, refrigerant and water piping design, and an electronic catalog for Carrier equipment.

Trane Trace 600 and Load Design 600: Like the Carrier package described above, the Trane Trace 600 and Load Design 600 packages cover a wide range of HVAC design aids and include programs ranging from heating and cooling load analysis to equipment sizing and selection. The Trane CDS selection programs are available to designers to help in sizing, selecting, and specifying equipment.

Customized Software: Customized software is available when an energy management, building management, or direct digital control (DDC) system is installed in a building or facility. These systems are designed to coordinate with existing HVAC systems in the building or facility. They offer programs that aid in preventative maintenance schedules, energy management programs, load shedding, energy-saving control systems, and systems analysis. Generally, these software packages are available only as part of a new DDC- or microprocessor-based control system. The software can be obtained from controls companies such as Johnson Controls, Honeywell, Landys and Gyr Powers, and Andover Controls.

Other HVAC Packages: Other packages are available from local, national, and specialized software companies throughout the country. These packages cover the full range of HVAC system requirements and analysis, from simple U-value calculations to energy consumption estimates and troubleshooting systems in operation.

Manufacturers' Software: With the increased use of CADD (computer aided drafting and design), most equipment manufacturers have developed or are developing software or electronic products that aid the designer in equipment selection and specification. Many of these software packages provide details and drawings compatible with common CADD programs, so information can be imported onto working drawings with relative ease.

The following section is reprinted with permission from *HVAC: Design Criteria, Options, Selection, Second Edition* by William H. Rowe III, PE, AIA, R.S. Means Co., Inc.

Fan Design Characteristics

Fans are used most often in open systems. The air from the fan is transmitted through ducts and discharged into or exhausted from the building spaces. When air is recirculated within a system, it is still not a tight enough system to be called closed. Leakage, infiltration, exhaust air, makeup air, and room losses mean that the air is supplied and returned by separate open systems.

Size and Location

Fans must be of an appropriate size to overcome the friction losses in the ducts and fittings. Because air is so light relative to water (0.075 pounds per cubic foot [1.2 kg/m^3] for air versus 62.5 pounds per cubic foot [1000 kg/m^3] for water), the pressure head to lift the air from one height to another is small and usually ignored. The weight of the water cannot be overlooked in pumping systems where the density of water is so much more than that of air. As with pumps, fans are sized to provide the proper flow (cfm) [L/s] and pressure (in. wg) [Pa] for the particular system. There are performance curves for fans and fan laws to regulate their use, just as there are for pumps.

Ed. Note: System pressure loss is calculated based on duct design. Fan static pressure is the sum of all losses in the ductwork, including terminal (diffusers) operating pressure, static pressure, and total airflow. All must be quantified in order to size a fan properly.

Fans may be exposed in the area served, as in a bathroom ceiling exhaust, or remotely located, either on a roof or in a mechanical room. Fans may be direct drive or belt driven. A direct drive fan is less expensive, but might send objectionable noise directly into the ductwork. Belt drives contain a belt that wraps around two pulleys, or sheaves—one on the motor drive and the other around the fan drive. Belt drives offer more flexibility in performance since the diameter of the sheaves can be changed to regulate the fan speed.

Fan Types

There are two broad categories of fans commonly used in driving systems: centrifugal fans and axial flow fans (See Figure 6.12). Centrifugal fans take in air at the side (circular opening) and discharge it radially into the supply stream (rectangular opening). Axial flow fans are mounted in the air stream and boost the pressure of the air. Figure 6.12 may be used to determine the basic fan type. Fans supply air from as low as 50 cfm [25 L/s] to more than 500,000 cfm [236,000 L/s] and at pressures from 0.1" wg [30 Pa] to 6" wg [1775 Pa]. Fan speeds can be adjusted by using two-speed motors, variable-pitch sheaves or blades, or variable-frequency motors. Noise control is a particular concern in fan design; noise ratings are taken into account in fan selection. When fans are near combustible fumes, nonsparking materials are usually specified for blades and housing.

Fan Curves

Each type of fan has a characteristic performance curve. This information is supplied by the manufacturer. As with pumps, the fan curves plot pressure versus flow and indicate other characteristics, such as efficiencies and horsepower over the range of operation. Typical fan performance characteristics are shown in Figure 6.13 for forward, backward, and radial pitched centrifugal fans.

As with pumps, it is important to select a fan large enough to handle the pressure and flow required for the system. Oversizing slightly (by 10–15 percent) is common practice. Comparisons between various manufacturers and models should be made to select for overall durability, ease of maintenance, cost, and efficiency.

Fan Laws

As with pumps, fan laws may be used to predict fan performance under conditions different from the conditions listed in a fan curve. The fan laws, shown in Figure 6.14, are used to determine the proper fan speed for a system. Slowing the fan to a lower speed is always less expensive than running it at high speed and partially closing dampers (throttling).

Fan Arrangement

Fans may be placed in series or parallel in a manner similar to pumps. In practice, however, this approach is rarely taken because fans come in a wide range of capacities suitable for most installation conditions. Also, the compressibility of air (as compared to water) makes it more difficult to properly pair fans in series or parallel.

When design conditions in a system vary, it is more common to select a two-stage fan or, in larger installations, to utilize variable pitched blades. Advancements have recently been made in varying the frequency of power to motors from the normal 60 hertz. Further developments are likely in this area.

Selection Guide for Fan Types

Fan Type	Type and Impeller Design	Use	Remarks
Centrifugal	**Airfoil:** Airfoil-contour blades curve away from direction of rotation.	General HVAC. Usually for larger systems, where cost is justified by energy savings.	Highest efficiency of any centrifugal type. Power reaches maximum near peak efficiency.
	Backward-Curved: Blades curve away from direction of rotation.	Same as airfoil; used in corrosive environments that might damage airfoil blades. Also used in high-speed applications.	Similar to airfoil; slightly less efficient.
	Radial: Blades are not curved.	Industrial materials handling. Medium speed applications.	High pressure characteristics. Easy to repair. Not common in HVAC. Least efficient of centrifugal types.
	Forward-Curved: Blades curve toward direction of rotation.	Low-pressure HVAC applications. Small residential and packaged units. Low-speed applications.	Fans in series.
Axial	**Propeller:** 2 – 5 blades on a small hub.	Large volume at low pressure. Ventilation and make-up air applications.	Low efficiency, noisy. Best when not ducted.
	Tubeaxial: 4 – 8 blades on a medium-sized hub. Blades are either airfoil or single thickness.	Heavy duty propeller fans used where space is tight. Industrial exhaust. Low and medium pressure HVAC applications.	More efficient than propeller type, and capable of developing more static.
	Vaneaxial: Many blades, large hub. Blades can be adjustable.	General HVAC and industrial applications. Used where space is tight and costs permit.	Most efficient propeller type, with high-pressure capability. Medium flow rate. Good downstream air distribution, more compact than centrifugal fans.

R.S. Means Co., Inc., *HVAC: Design Criteria, Options, Selection*

Figure 6.12

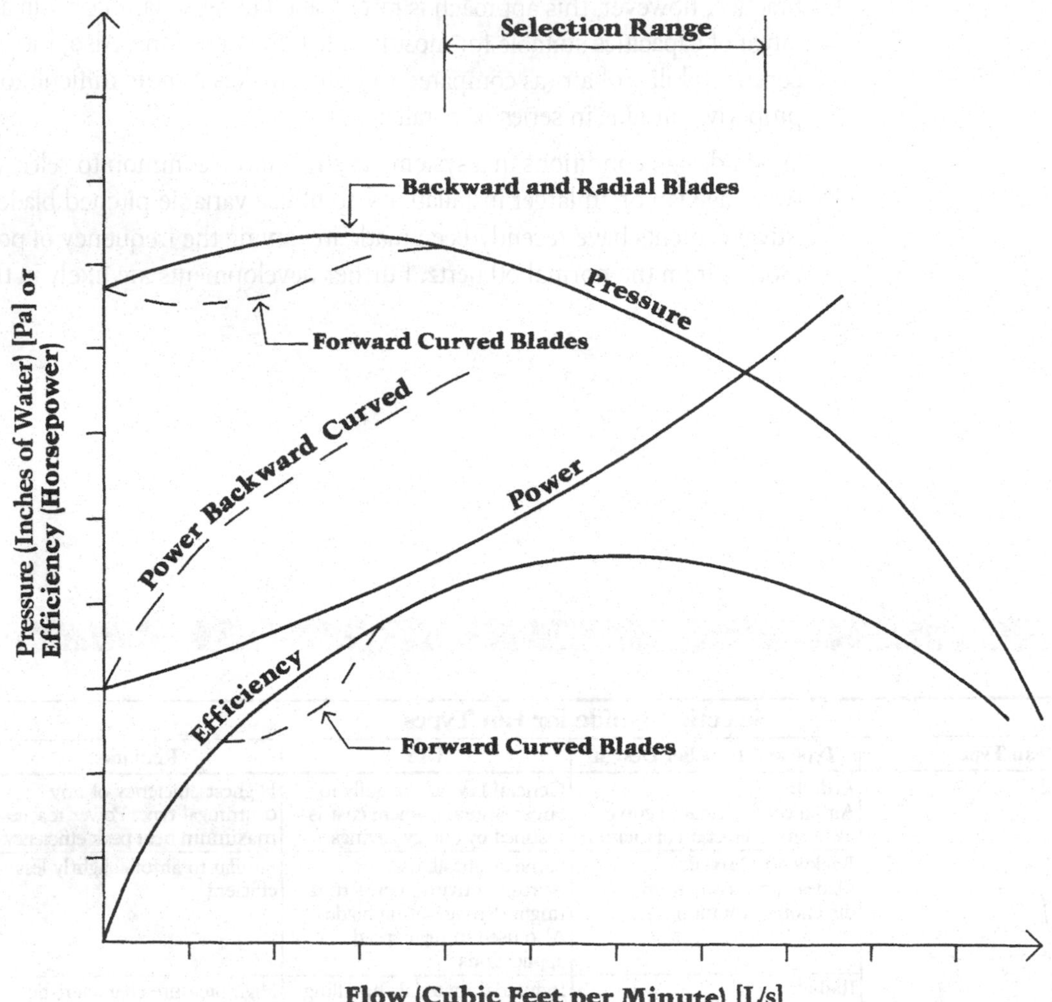

Typical conditions for pressure, power, and efficiency are shown as solid lines. The exceptions are indicated by dashed lines. Backward curved blades have a different power line. Forward curved blades have a characteristic "dip" in their pressure and efficiency curves at low speeds. Always select above this range to avoid "hunting".

Typical Fan Performance Curves

R.S. Means Co., Inc., *HVAC: Design Criteria, Options, Selection*

Figure 6.13

LAWS OF FAN PERFORMANCE

Fan laws are used to predict fan performance under changing operating conditions or fan size. They are applicable to all types of fans.

The fan laws are stated in *Table 5.* The symbols used in the formulas represent the following quantities:

Q — Volume rate of flow thru the fan.

N — Rotational speed of the impeller.

P — Pressure developed by the fan, either static or total.

Hp — Horsepower input to the fan.

D — Fan wheel diameter. The fan size number may be used if it is proportional to the wheel diameter.

W — Air density, varying directly as the barometric pressure and inversely as the absolute temperature.

FAN LAWS

VARIABLE	CONSTANT	NO.	LAW	FORMULA
SPEED	Air Density, Fan Size, Distribution System	1	Capacity varies as the Speed.	$\frac{Q_1}{Q_2} = \frac{N_1}{N_2}$
		2	Pressure varies as the square of the Speed.	$\frac{P_1}{P_2} = \left(\frac{N_1}{N_2}\right)^2$
		3	Horsepower varies as the cube of the Speed.	$\frac{Hp_1}{Hp_2} = \left(\frac{N_1}{N_2}\right)^3$
FAN SIZE	Air Density, Tip Speed	4	Capacity and Horsepower vary as the square of the Fan Size.	$\frac{Q_1}{Q_2} = \frac{Hp_1}{Hp_2} = \left(\frac{D_1}{D_2}\right)^2$
		5	Speed varies inversely as the Fan Size.	$\frac{N_1}{N_2} = \frac{D_2}{D_1}$
		6	Pressure remains constant.	$P_1 = P_2$
	Air Density, Speed	7	Capacity varies as the cube of the Size.	$\frac{Q_1}{Q_2} = \left(\frac{D_1}{D_2}\right)^3$
		8	Pressure varies as the square of the Size.	$\frac{P_1}{P_2} = \left(\frac{D_1}{D_2}\right)^2$
		9	Horsepower varies as the fifth power of the Size.	$\frac{Hp_1}{Hp_2} = \left(\frac{D_1}{D_2}\right)^5$
AIR DENSITY	Pressure, Fan Size, Distribution System	10	Speed, Capacity and Horsepower vary inversely as the square root of Density.	$\frac{N_1}{N_2} = \frac{Q_1}{Q_2} = \frac{Hp_1}{Hp_2} = \left(\frac{W_2}{W_1}\right)^{1/2}$
	Capacity, Fan Size, Distribution System	11	Pressure and Horsepower vary as the Density.	$\frac{P_1}{P_2} = \frac{Hp_1}{Hp_2} = \frac{W_1}{W_2}$
		12	Speed remains constant.	$N_1 = N_2$

(Courtesy Carrier Corporation, McGraw-Hill Book Company)

Footnote: Cross-references in table footnotes pertain to the source from which tables were taken; they do not refer to chapters or pages in this text.

Figure 6.14

Equipment Data Sheets

Figures 6.15a and 6.15b contain equipment data sheets for fans. These sheets describe the efficiency, useful life, typical uses, capacity range, special considerations, accessories, and additional equipment needed for driving system components.

These sheets can be used in the selection of each piece of equipment. A Means line number, where applicable, is included. Costs can be determined by looking up the line numbers in the current edition of *Means Mechanical Cost Data.*

The following section is reprinted with permission from *HVAC Systems Evaluation* by Harold R. Colen, R.S. Means Co., Inc.

Ductwork Design Characteristics

Ductwork represents approximately one-third of the installation cost of an air-conditioning system and one-half of the heartache endured by owners, engineers, and mechanical contractors. It should be noted that duct design influences the construction cost, fan operating cost, and noise generation of the duct system. The importance of properly reviewing design considerations cannot be overemphasized.

Construction Cost Factors

The following considerations influence the installed cost of ductwork. Some are expressed as "rules of thumb," or guidelines that should be followed in designing a ductwork system.

Minimize the Number of Fittings: The cost of a vaned elbow is approximately ten times the cost of five feet of straight duct. Try to eliminate as many fittings as possible. Rather than installing transitions to make a small reduction in the duct size after branch takeoffs, continue the same duct size.

Keep the Aspect Ratio Low: Aspect ratio is the ratio of the width of a duct to its height. A 48" x 12" duct has an aspect ratio of 4. As Figure 6.16 demonstrates, the installed cost of a duct with an aspect ratio of 4 is almost 150% that of a square duct.

Use as Much Round Duct as Possible: As Figure 6.16 points out, the installed cost of a round duct is almost 80% less than that of a duct with an aspect ratio of 4.

Try to Keep Duct Pressure Low: Figure 6.17 shows that the installed cost of a 3"-4" wg duct class is 45% more than a duct with a 1/2"-1" wg pressure class.

Seal Ductwork: The leakage due to unsealed ductwork means that ductwork and air handling equipment must be increased to compensate for the air loss.

Driving Equipment Data Sheets for Fans

Centrifugal Ceiling or Cabinet Exhaust Fan	**Right angle ceiling or cabinet exhaust fan**	**Means No.: 157-290**
	Typical use: Exhaust	**cfm Range: 50 – 2,500 [25 – 1200 L/s]**
	Comments: Usually direct drive. Larger sizes are used for conference rooms and offices. Smaller models are used in toilets, and often switched with lights. Check for compatible voltages. Typically low pressure, short duct run to exterior or plenum.	
Centrifugal Sidewall Exhaust Fan	**Centrifugal exhaust fan**	**Means No.: 157-290**
	Typical use: Exhaust	**cfm Range: 100 – 7,000 [47 – 3300 L/s]**
	Comments: Direct or belt drive. Mounted on the exterior, where interior mounting space is limited, and higher flow rate is required.	
Centrifugal In-line Fan	**Centrifugal in-line fan**	**Means No.: 157-290**
	Typical use: Supply or exhaust	**cfm Range: 100 – 1,200 [47 – 5665 L/s]**
	Comments: Direct or belt drive. Greater performance than cabinet exhausters, usually concealed above ceiling. Not very attractive.	
Centrifugal Utility Fan	**Centrifugal utility fan**	**Means No.: 157-290**
	Typical use: Exhaust	**cfm Range: 300 – 180,000 [140 – 85 000 L/s]**
	Comments: Direct or belt drive. Roof mounted, very efficient. Typically for exhaust in low- to medium-pressure applications. Typically uses airfoil blades.	
Centrifugal Tubular Fan	**Centrifugal tubular fan**	**Means No.: 157-290**
	Typical use: Exhaust	**cfm Range: 1,600 – 120,000 [750 – 56 650 L/s]**
	Comments: Similar to the centrifugal utility fan, but designed for in-line type of mounting.	
Centrifugal Roof Fan	**Centrifugal roof fan**	**Means No.: 157-290**
	Typical use: Exhaust	**cfm Range: 1,100 – 37,000 [520 – 17 500 L/s]**
	Comments: Direct or belt drive. Typically for exhaust, but can be used for supply. Commonly used at the top of toilet, kitchen or dryer stacks in commercial buildings.	
Centrifugal Fans – Variable Pitch Blade	**Variable pitch blade**	**Means No.: 157-290**
	Typical use: VAV systems	
	Comments: Activated on pressure. Efficient.	
Centrifugal Fan – Variable Speed	**Variable speed**	**Means No.: 157-290**
	Typical use: VAV systems	
	Comments: Frequency of current to motor is varied from 60 Hz (cycles/second) to adjust speed of blades. Economical.	

All numbers shown in brackets are metric equivalents.

R.S. Means Co., Inc., *HVAC: Design Criteria, Options, Selection*

Figure 6.15a

Driving Equipment Data Sheets for Fans (continued)

Propeller Paddle Blade Air Circulator	**Propeller air circulator**	**Means No.: 157-290**
	Typical use: Air circulation at ceiling of space.	**cfm Range: 100–7,000 [47–3 300 L/s]**
	Comments: Usually direct drive. Weak pressure, good airflow. Typically used in high spaces to prevent stratification. Rotation can be reversed for heating or cooling applications.	
Propeller Sidewall	**Propeller sidewall fan**	**Means No.: 157-290**
	Typical use: Exhaust	**cfm Range: 200–90,000 [95–42 500 L/s]**
	Comments: Direct or belt drive. Typically not ducted. Used for cooling and ventilation in the summer. Often used to keep mechanical and electrical room from overheating. In smaller, residential applications, this can be used to draw cooler basement air to the upper floors.	
Propeller Tube Axial	**Propeller tube axial fan**	**Means No.: 157-290**
	Typical use: Supply or exhaust	**cfm Range: 4,000–100,000 [1900–47 500 L/s]**
	Comments: Designed for the industrial environment, such as high temperatures and exhausting gases and particulate matter. Typically roof mounted.	
Propeller Vane Axial Supply or Exhaust Fan	**Propeller vane axial**	**Means No.: 157-290**
	Typical use: Supply or exhaust	**cfm Range: 300–72,000 [140–34 000 L/s]**
	Comments: The best flow and pressure characteristics for any propeller fan.	
Propeller Roof Exhaust or Supply Fan	**Propeller roof fan**	**Means No.: 157-290**
	Typical use: Supply or exhaust	**cfm Range: 300–72,000 [140–34 000 L/s]**
	Comments: Similar to the centrifugal roof exhaust fan.	

All numbers shown in brackets are metric equivalents.

R.S. Means Co., Inc., *HVAC: Design Criteria, Options, Selection*

Figure 6.15b

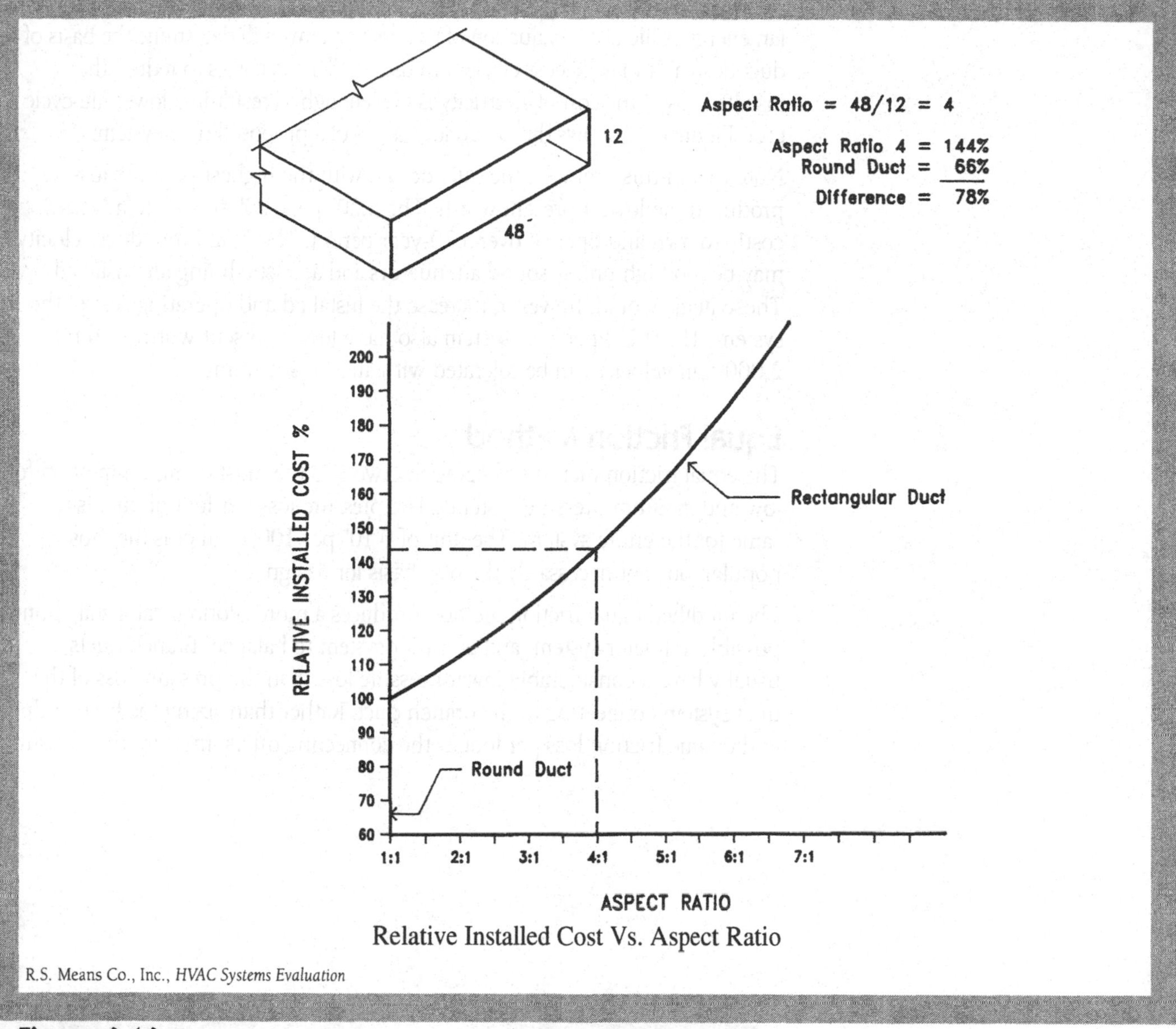

Relative Installed Cost Vs. Aspect Ratio

R.S. Means Co., Inc., *HVAC Systems Evaluation*

Figure 6.16

Analyze Operating Costs: The static pressure loss of a duct system affects fan energy. A life cycle evaluation of the duct systems will determine the basis of duct design. It may be cost-efficient to use smaller duct sizes to reduce the installed cost if the cost of electricity is low enough to result in a lower life-cycle cost. Figure 6.18 shows the life cycle analysis of a proposed duct system.

Note that in this example, the duct design with the highest pressure loss produced the lowest present worth. The 0.20" per 100' duct system is less costly to own and operate over a 20-year period. The 2,325 fpm duct velocity may be too high unless sound attenuators and acoustic lining are installed. Those items would, however, increase the installed and operating cost of the system. The 0.15" per 100' system also has a lower present worth, and the 2,000 fpm velocity can be tolerated with an acoustic lining.

Equal Friction Method

The equal friction method of sizing ductwork is the most common method for low and medium pressure systems. The pressure loss per foot of duct is the same for the entire system. The unit of 0.10" per 100' of duct is the most popular, but not necessarily the best basis for design.

The modified equal friction method produces a more economical installation, possibly a quieter system, and an easier system to balance. Branch ducts usually have a considerably lower pressure loss than the pressure loss of the duct system connecting to the branch duct. Rather than sizing the branch duct at the same friction loss per foot as the connecting ducts, increase the pressure

Relative Duct System Fabrication & Installation Costs for the Same Size Duct

Duct Pressure Class	Cost Ratio
0" – 1/2"	1.0
1/2" – 1"	1.05
1" – 2"	1.15
2" – 3"	1.40
3" – 4"	1.50
4" – 6"	1.60
6" – 10"	1.80

R.S. Means Co., Inc., *HVAC Systems Evaluation*

Figure 6.17

loss in the branch ducts so that the pressure loss in the branch duct will be equal to the pressure loss of the connecting duct run. The branch duct sizes will be reduced to "burn up" the excess pressure rather than using only a volume damper for this purpose. The duct size in the branch should not be reduced to a point where the duct velocity will exceed the maximum recommended velocity. The example in Figure 6.19 uses the modified equal friction method.

Life Cycle Duct System Evaluation

12,000 sq. ft. building, 20,000 cfm, system operates 3,000 hrs. per yr., energy cost $0.10/kwh, installed cost of ductwork $4/lb. 12% interest, 20-year period, capital recovery factor 0.22526

Assume the duct system at 0.10"/100 ft. friction loss is 1.5 lb. per sq. ft. of building area. Duct system installed cost is 1.5 lb./sq. ft. x 12,000 sq. ft. x $4/lb. = $72,000

Duct friction in/100 ft.	Total duct s.p.	bhp	kw	Energy cost $/yr.	Inst. cost $	Main duct size	Main duct vel. fpm	Main duct perim. lin. ft.
.10	1"	4.5	3.4	1,020	72,000	72" x 24"	1,660	16
.15	1.5"	6.8	5.1	1,530	63,000	60" x 24"	2,000	14
.20	1.96"	8.8	6.6	1,980	56,250	52" x 24"	2,325	12.5

bhp = cfm x s.p./6356 x eff

$$\left(\frac{V'}{V''}\right)^2 = \frac{P'}{P''}$$

bhp = 20,000 x 1/6356 x .7 = 4.5 bhp

$$\left(\frac{2000}{1660}\right)^2 = \frac{P'}{1}, P = 1.5''$$

bhp = 20,000 x 1.5/6356 x .7 = 6.9 bhp
20,000 x 1.96/6356 x .7 = 8.8 bhp

$$\left(\frac{2325}{1660}\right)^2 = \frac{P'}{1}, P = 1.96''$$

Duct Friction	.10/100 ft.	.15/100 ft.	.20/100 ft.
Initial Cost	$ 72,000	$ 63,000	$ 56,250
Operating Cost			
1,020 x 1/0.22526	4,528		
1,530 x 1/0.22526		6,792	
1,980 x 1/0.22526			8,789
Present Worth	$ 76,528	$ 69,792	$ 65,039

R.S. Means Co., Inc., *HVAC Systems Evaluation*

Figure 6.18

Using Figure 6.19, the duct design procedure for the 15,000 cfm system for a bank would be as follows:

- Establish the cfm that each section of duct is to carry.
  Determine the maximum allowable main duct velocity from Figure 6.20. Note that when noise generation is the controlling factor, a maximum main duct velocity of 1,500 fpm is recommended for banks.
- Starting with the cfm at the fan, determine the area of the duct at that location, using the formula:

$$\text{area (S.F.)} = \frac{\text{cfm}}{\text{velocity (fpm)}}$$

$$\text{area} = \frac{15{,}000 \text{ cfm}}{1{,}500 \text{ fpm}} = 10 \text{ S.F.}$$

cfm = cubic feet/min.
S.F. = square feet
fpm = feet per minute

- When the duct area is established, determine the duct dimensions to suit the available space. Assume that 24" is the maximum possible duct height.

  10 S.F. x 144 (Sq. inches per S.F.)/24" = 60",
  Therefore, select 60" x 24" supply duct.

Figure 6.21 is the chart for finding the equivalent rectangular duct size of a round duct that has the same friction loss as the rectangular duct.

Use the chart in Figure 6.21 to determine the equivalent round duct size for the 60" x 24" duct. The length of adjacent sides is at the top horizontal line (6-30) and at the left vertical column (6-96). The equivalent round duct is found at the intersection of the two adjacent sides. The length of the 24" side at the top line is matched with the 60" adjacent side at the left vertical column. The 40.4 equivalent round duct is located at the intersection of the two lines. (Over 27% more metal is required for rectangular duct than for an equivalent friction loss round duct.)

Figure 6.22 gives the friction of air in straight round ducts for a particular cfm and velocity.

Locate 15,000 cfm at the air quantity line at the left vertical line of Figure 6.22. Follow the 15,000 cfm line horizontally until it intersects the 41" duct diameter diagonal line. Draw a vertical line from the intersection point to the left, and read 0.065" per 100' on the left horizontal line. The friction loss for equal friction design for this system is 0.065" per 100'. All of the duct in the longest run will be sized at a friction loss of 0.065" per 100'. The 1,600 fpm velocity for the 41" round duct handling is 15,000 cfm. The velocity would be different for any other duct size of equal friction loss. The velocity of any other equal friction loss rectangular duct will always be less.

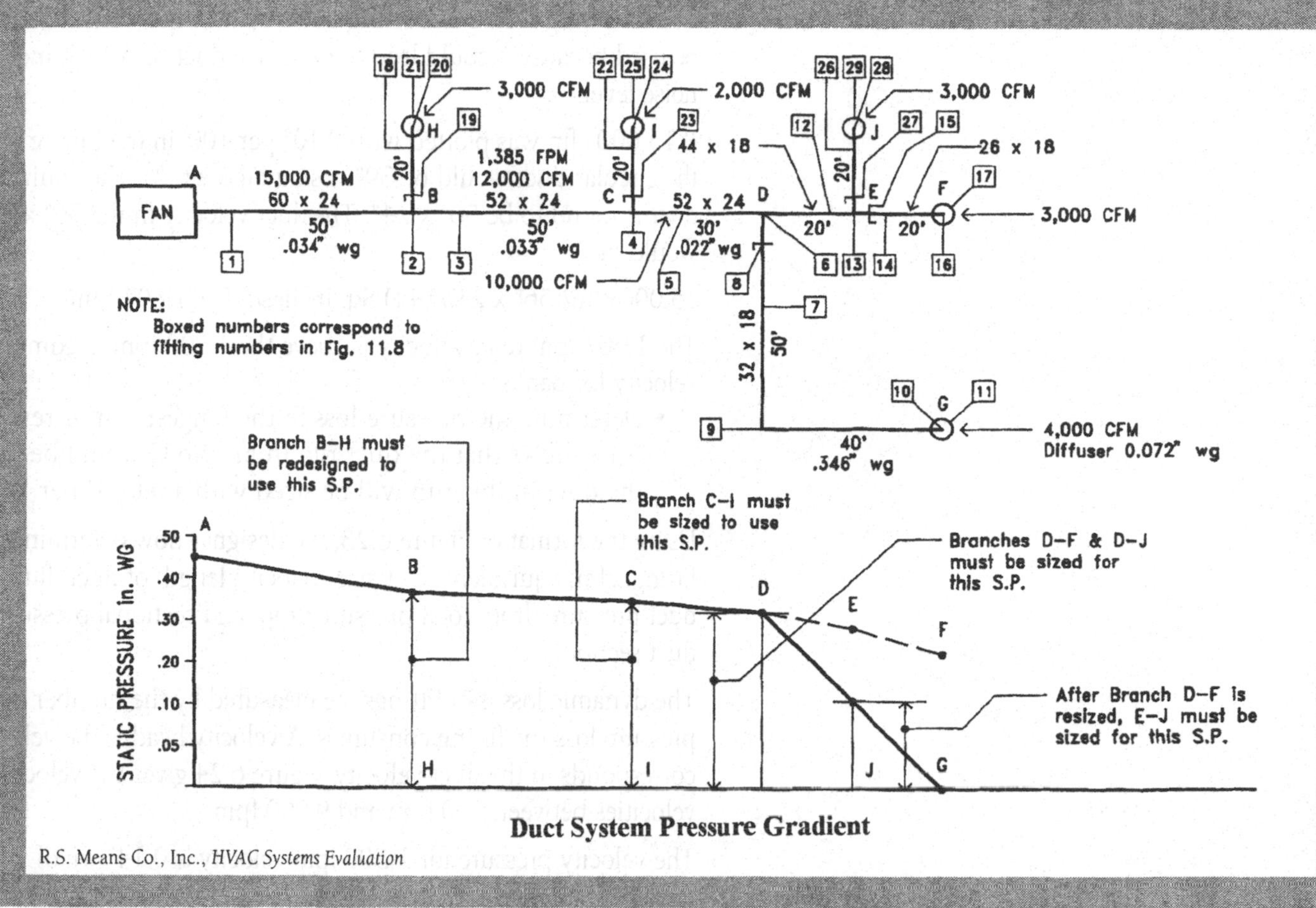

R.S. Means Co., Inc., *HVAC Systems Evaluation*

Figure 6.19

Recommended Maximum Duct Velocities for Low Velocity Systems (FPM)

Application	Controlling Factor Noise Generation Main Ducts	Controlling Factor—Duct Friction: Main Ducts Supply	Main Ducts Return	Branch Ducts Supply	Branch Ducts Return
Residences	600	1000	800	600	600
Apartments Hotel Bedrooms Hospital Bedrooms	1000	1500	1300	1200	1000
Private Offices Directors Rooms Libraries	1200	2000	1500	1600	1200
Theatres Auditoriums	800	1300	1100	1000	800
General Offices High Class Restaurants High Class Stores Banks	1500	2000	1500	1600	1200
Average Stores Cafeterias	1800	2000	1500	1600	1200
Industrial	2500	3000	1800	2200	1500

Courtesy Carrier Air Conditioning Company

Figure 6.20

The following method of sizing ducts by the equal friction method should n
be used because it could lead to excessive duct velocities and the resulting hi
noise level.

If 15,000 cfm was plotted with 0.10" per 100' in the Figure 6.22 friction cha
the circular duct would be 39". Using Figure 6.21, the equivalent rectangula
duct is found to be 56" x 24". The duct velocity in the 56" x 24" duct
would be:

15,000 cfm/(56" x 24"/144) Sq. Inches/S.F. = 1,607 fpm

The 1,607 fpm duct velocity is above the 1,500 fpm maximum allowable velocity for banks.

- Determine the pressure loss in the longest run. A review of Figure 6.19 shows that the duct run from A to G would be the longest. All the duct in this run will be sized with a 0.065" per 100' friction loss

Using the format of Figure 6.23, the designer now determines the size, type
fitting, cfm, equivalent duct size, velocity, length of duct, fitting coefficient, duct pressure drop, total pressure drop, and sectional pressure drop for each duct section.

The dynamic losses of fittings are measured by the number of velocity heads pressure loss the fitting consumes. A velocity head is the velocity pressure th
corresponds to the duct velocity. Figure 6.24 gives the velocity pressure for t
velocities between 300 fpm and 9,000 fpm.

The velocity pressure for 1,500 fpm velocity is 0.14" wg of water. The numb
of velocity pressures that a fitting expends in pressure loss is called *Coefficient C.*

Recommended Duct Design Procedures

Figures 6.25a–h show ASHRAE's recommended HVAC duct design procedur
as presented in the *1997 ASHRAE Fundamentals Handbook.*

Ed. Note: The text within this excerpt uses the figure numbering system of the ASHRAE Fundamentals Handbook. Note that these ASHRAE figure numbers appe in parentheses before heading.

Circular Equivalents of Rectangular Duct for Equal Friction and Capacity[a]

Lgth Adj[b]	Length of One Side of Rectangular Duct (a), in. 4.0	4.5	5.0	5.5	6.0	6.5	7.0	7.5	8.0	9.0	10.0	11.0	12.0	13.0	14.0	15.0	16.0
3.0	3.8	4.0	4.2	4.4	4.6	4.7	4.9	5.1	5.2	5.5	5.7	6.0	6.2	6.4	6.6	6.8	7.0
3.5	4.1	4.2	4.6	4.8	5.0	5.2	5.3	5.5	5.7	6.0	6.3	6.5	6.8	7.0	7.2	7.5	7.7
4.0	4.4	4.6	4.9	5.1	5.3	5.5	5.7	5.9	6.1	6.4	6.7	7.0	7.3	7.6	7.8	8.0	8.3
4.5	4.6	4.9	5.2	5.4	5.7	5.9	6.1	6.3	6.5	6.9	7.2	7.5	7.8	8.1	8.4	8.6	8.8
5.0	4.9	5.2	5.5	5.7	6.0	6.2	6.4	6.7	6.9	7.3	7.6	8.0	8.3	8.6	8.9	9.1	9.4
5.5	5.1	5.4	5.7	6.0	6.3	6.5	6.8	7.0	7.2	7.6	8.0	8.4	8.7	9.0	9.3	9.6	9.9

Lgth Adj.[b]	Length of One Side of Rectangular Duct (a), in. 6	7	8	9	10	11	12	13	14	15	16	17	18	19	20	22	24	26	28	30	Lgth Adj.[b]
6	6.6																				6
7	7.1	7.7																			7
8	7.6	8.2	8.7																		8
9	8.0	8.7	9.3	9.8																	9
10	8.4	9.1	9.8	10.4	10.9																10
11	8.8	9.5	10.2	10.9	11.5	12.0															11
12	9.1	9.9	10.7	11.3	12.0	12.6	13.1														12
13	9.5	10.3	11.1	11.8	12.4	13.1	13.7	14.2													13
14	9.8	10.8	11.4	12.2	12.9	13.5	14.2	14.7	15.3												14
15	10.1	11.0	11.8	12.6	13.3	14.0	14.6	15.3	15.8	16.4											15
16	10.4	11.3	12.2	13.0	13.7	14.4	15.1	15.7	16.4	16.9	17.5										16
17	10.7	11.6	12.5	13.4	14.1	14.9	15.6	16.2	16.8	17.4	18.0	18.6									17
18	11.0	11.9	12.9	13.7	14.5	15.3	16.0	16.7	17.3	17.9	18.5	19.1	19.7								18
19	11.2	12.2	13.2	14.1	14.9	15.7	16.4	17.1	17.8	18.4	19.0	19.6	20.2	20.8							19
20	11.5	12.6	13.5	14.4	15.2	16.0	16.8	17.5	18.2	18.9	19.5	20.1	20.7	21.3	21.9						20
22	12.0	13.0	14.1	15.0	15.9	16.8	17.6	18.3	19.1	19.8	20.4	21.1	21.7	22.3	22.9	24.0					22
24	12.4	13.5	14.6	15.6	16.5	17.4	18.3	19.1	19.9	20.6	21.3	22.0	22.7	23.3	23.9	25.1	26.2				24
26	12.8	14.0	15.1	16.2	17.1	18.1	19.0	19.8	20.6	21.4	22.1	22.9	23.5	24.2	24.9	26.1	27.3	28.4			26
28	13.2	14.5	15.6	16.7	17.7	18.7	19.6	20.5	21.3	22.1	22.9	23.7	24.4	25.1	25.8	27.1	28.3	29.5	30.6		28
30	13.6	14.9	16.1	17.2	18.3	19.3	20.2	21.1	22.0	22.9	23.7	24.4	25.2	25.9	26.6	28.0	29.3	30.5	31.7	32.8	30
32	14.0	15.3	16.5	17.7	18.8	19.8	20.8	21.8	22.7	23.5	24.4	25.2	26.0	26.7	27.5	28.9	30.2	31.5	32.7	33.9	32
34	14.4	15.7	17.0	18.2	19.3	20.4	21.4	22.4	23.3	24.2	25.1	25.9	26.7	27.5	28.3	29.7	31.0	32.4	33.7	34.9	34
36	14.7	16.1	17.4	18.6	19.8	20.9	21.9	22.9	23.9	24.8	25.7	26.6	27.4	28.2	29.0	30.5	32.0	33.3	34.6	35.9	36
38	15.0	16.5	17.8	19.0	20.2	21.4	22.4	23.5	24.5	25.4	26.4	27.2	28.1	28.9	29.8	31.3	32.8	34.2	35.6	36.8	38
40	15.3	16.8	18.2	19.5	20.7	21.8	22.9	24.0	25.0	26.0	27.0	27.9	28.8	29.6	30.5	32.1	33.6	35.1	36.4	37.8	40
42	15.6	17.1	18.5	19.9	21.1	22.3	23.4	24.5	25.6	26.6	27.6	28.5	29.4	30.3	31.2	32.8	34.4	35.9	37.3	38.7	42
44	15.9	17.5	18.9	20.3	31.5	22.7	23.9	25.0	26.1	27.1	28.1	29.1	30.0	30.9	31.8	33.5	35.1	36.7	38.1	39.5	44
46	16.2	17.8	19.3	20.6	21.9	23.2	24.4	25.5	26.6	27.7	28.7	29.7	30.6	31.6	32.5	34.2	35.9	37.4	38.9	40.4	46
48	16.5	18.1	19.6	21.0	22.3	23.6	24.8	26.0	27.1	28.2	29.2	30.2	31.2	32.2	33.1	34.9	36.6	38.2	39.7	41.2	48
50	16.8	18.4	19.9	21.4	22.7	24.0	25.2	26.4	27.6	28.7	29.8	30.8	31.8	32.8	33.7	35.5	37.2	38.9	40.5	42.0	50
52	17.1	18.7	20.2	21.7	23.1	24.4	25.7	26.9	28.0	29.2	30.3	31.3	32.3	33.3	34.3	36.2	37.9	39.6	41.2	42.8	52
54	17.3	19.0	20.6	22.0	23.5	24.8	26.1	27.3	28.5	29.7	30.8	31.8	32.9	33.9	34.9	36.8	38.6	40.3	41.9	43.5	54
56	17.6	19.3	20.9	22.4	23.8	25.2	26.5	27.7	28.9	30.1	31.2	32.3	33.4	34.4	35.4	37.4	39.2	41.0	42.7	44.3	56
58	17.8	19.5	21.2	22.7	24.2	25.5	26.9	28.2	29.4	30.6	31.7	32.8	33.9	35.0	36.0	38.0	39.8	41.6	43.3	45.0	58
60	18.1	19.8	21.5	23.0	24.5	25.9	27.3	28.6	29.8	31.0	32.2	33.3	34.4	35.5	36.5	38.5	40.4	42.3	44.0	45.7	60
62		20.1	21.7	23.3	24.8	26.3	27.6	28.9	30.2	31.5	32.6	33.8	34.9	36.0	37.1	39.1	41.0	42.9	44.7	46.4	62
64		20.3	22.0	23.6	25.1	26.6	28.0	29.3	30.6	31.9	33.1	34.3	35.4	36.5	37.6	39.6	41.6	43.5	45.3	47.1	64
66		20.6	22.3	23.9	25.5	26.9	28.4	29.7	31.0	32.3	33.5	34.7	35.9	37.0	38.1	40.2	42.2	44.1	46.0	47.7	66
68		20.8	22.6	24.2	25.8	27.3	28.7	30.1	31.4	32.7	33.9	35.2	36.3	37.5	38.6	40.7	42.8	44.7	46.6	48.4	68
70		21.1	22.8	24.5	26.1	27.6	29.1	30.4	31.8	33.1	34.4	35.6	36.8	37.9	39.1	41.2	43.3	45.3	47.2	49.0	70
72			23.1	24.8	26.4	27.9	29.4	30.8	32.2	33.5	34.8	36.0	37.2	38.4	39.5	41.7	43.8	45.8	47.8	49.6	72
74			23.3	25.1	26.7	28.2	29.7	31.2	32.5	33.9	35.2	36.4	37.7	38.8	40.0	42.2	44.4	46.4	48.4	50.3	74
76			23.6	25.3	27.0	28.5	30.0	31.5	32.9	34.3	35.6	36.8	38.1	39.3	40.5	42.7	44.9	47.0	48.9	50.9	76
78			23.8	25.6	27.3	28.8	30.4	31.8	33.3	34.6	36.0	37.2	38.5	39.7	40.9	43.2	45.4	47.5	49.5	51.4	78
80			24.1	25.8	27.5	29.1	30.7	32.2	33.6	35.0	36.3	37.6	38.9	40.2	41.4	43.7	45.9	48.0	50.1	52.0	80
82				26.1	27.8	29.4	31.0	32.5	34.0	35.4	36.7	38.0	39.3	40.6	41.8	44.1	46.4	48.5	50.6	52.6	82
84				26.4	28.1	29.7	31.3	32.8	34.3	35.7	37.1	38.4	39.7	41.0	42.2	44.6	46.9	49.0	51.1	53.2	84
86				26.6	28.3	30.0	31.6	33.1	34.6	36.1	37.4	38.8	40.1	41.4	42.6	45.0	47.3	49.6	51.7	53.7	86
88				26.9	28.6	30.3	31.9	33.4	34.9	36.4	37.8	39.2	40.5	41.8	43.1	45.5	47.8	50.0	52.2	54.3	88
90				27.1	28.9	30.6	32.2	33.8	35.3	36.7	38.2	39.5	40.9	42.2	43.5	45.9	48.3	50.5	52.7	54.8	90
92					29.1	30.8	32.5	34.1	35.6	37.1	38.5	39.9	41.3	42.6	43.9	46.4	48.7	51.0	53.2	55.3	92
96					29.6	31.4	33.0	34.7	36.2	37.7	39.2	40.6	42.0	43.3	44.7	47.2	49.6	52.0	54.2	56.4	96

Reprinted with permission from *1989 ASHRAE Handbook—Fundamentals*

Figure 6.21

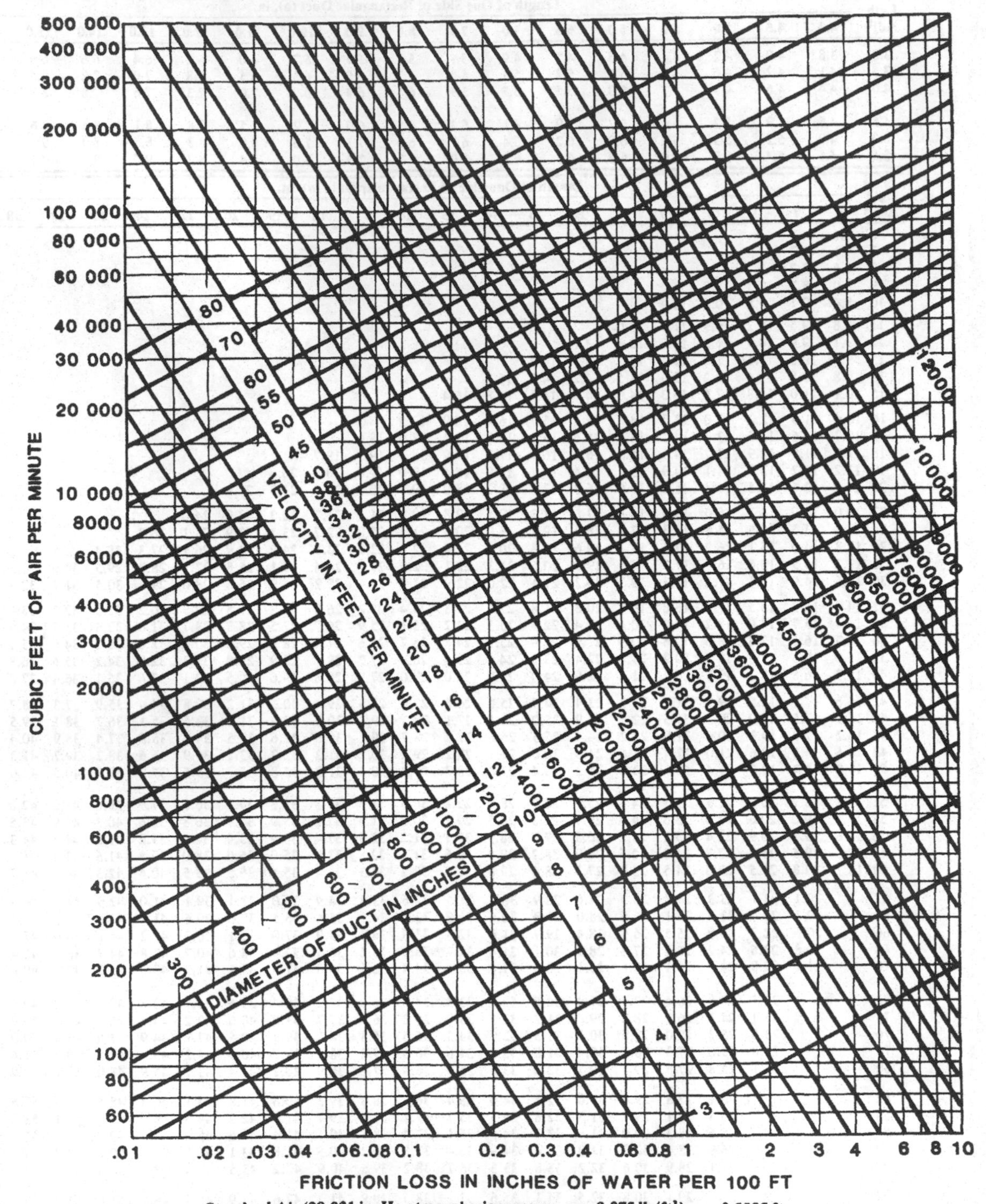

Standard Air (29.921 in. Hg atmospheric pressure, $\rho = 0.075$ lb/ft^3), $\varepsilon = 0.0005$ ft

Fig. C-1 Friction Chart for Average Ducts (See Table 1)

Reprinted with permission from 1989 *ASHRAE Handbook—Fundamentals*

Figure 6.22

Node	No.	Fitting No.	Type of Fitting	Air-flow cfm	Duct Size	Vel fpm	Vel Pres	Duct L'th ft	Fitting Loss Coef	Duct P.D. ft/100	Tot P.D. in	Sect P.D. in
Duct Calculation Form												
A-B	1	–	Duct	15,000	60x24	1500	.14	50	–	.065	.033	
	2	5-23	T,main	15,000	–	1500	.14	–	.01	–	.001	.034
B-C	3	–	Duct	12,000	52x24	1385	.12	50	–	.065	.033	
	4	5-23	T,main	12,000	–	1385	.12	–	0	–	0	.033
C-D	5	–	Duct	10,000	52x24	1300	.11	30	–	.065	.020	
	6	5-23	T,main	10,000	–	1300	.11	–	.02	–	.002	.022
D-G	7	–	Duct	4,000	32x18	1000	.06	90	–	.065	.059	
	6	5-27	T,br	4,000	32x18	1000	.06	–	.66	–	.040	
	8	6-2	Dpr	4,000	32x18	1000	.06	–	.04	–	.002	
	9	3-8	Elbow	4,000	32x18	1000	.06	–	.38	–	.023	
	10	3-6	Tap	4,000	24x24	1000	.06	–	1.20	–	.072	
	11	mfg	Diff	4,000	24x24	–	–	–	–	–	.150	.346
D-E	12	–	Duct	6,000	44x18	1090	.08	40	–	.065	.025	
	13	5-23	T,main	6,000	–	1090	.08	–	.02		.002	.027
E-F	14	–	Duct	3,000	26x18	923	.06	20	–	.065	.013	
	15	6-2	Dpr	3,000	26x18	923	.06	–	.04	–	.002	
	16	3-6	Tap	3,000	24x24	923	.06	–	1.20	–	.072	
	17	mfg	Diff	3,000	24x24	–	–	–	–	–	.015	.102
B-H	2	5-27	T,br	3,000	26x18	923	.06	–	.72	–	.043	
	18	–	Duct	3,000	26x18	923	.06	20	–	.065	.013	
	19	6-2	Dpr	3,000	26x18	923	.06	–	.04	–	.002	
	20	3-6	Tap	3,000	24x24	923	.06	–	1.20	–	.072	
	21	mfg	Diff	3,000	24x24	923	.06	–	–	–	.015	.146
C-I	4	5-27	T,br	2,000	24x14	860	.05	–	.75	–	.0389	
	22	–	Duct	2,000	24x14	860	.05	20	–	.065	.013	
	23	6-2	Dpr	2,000	24x14	860	.05	–	.04	–	.002	
	24	3-6	Tap	2,000	24x24	860	.05	–	1.20	–	.060	
	25	mfg	Diff	2,000	24x24	860	–	–	–	–	.015	.128
E-J	13	5-27	T,br	3,000	26x18	923	.06	–	.72	–	.043	
	26	–	Duct	3,000	26x18	923	.06	20	–	.065	.013	
	27	6-2	Dpr	3,000	26x18	923	.06	–	.04	–	.002	
	28	3-6	Tap	3,000	24x24	923	.06	–	1.20	–	.072	
	29	mfg	Diff	3,000	24x24	923	.06	–	–	–	.015	.145

System Pressure Loss = (A-B) .022 + (B-C) .033 + (C-D) .022 + (D-G) .346 = .435"

R.S. Means Co., Inc., *HVAC Systems Evaluation*

Figure 6.23

Velocity, fpm	Velocity Pressure, in. H_2O	Velocity, fpm	Velocity Pressure, in. H_2O	Velocity, fpm	Velocity Pressure, in. H_2O	Velocity, fpm	Velocity Pressure, in. H_2O	Velocity, fpm	Velocity Pressure, in. H_2O
300	0.01	2050	0.26	3800	0.90	5550	1.92	7300	3.32
350	0.01	2100	0.27	3850	0.92	5600	1.95	7350	3.37
400	0.01	2150	0.29	3900	0.95	5650	1.99	7400	3.41
450	0.01	2200	0.30	3950	0.97	5700	2.02	7450	3.46
500	0.02	2250	0.32	4000	1.00	5750	2.06	7500	3.51
550	0.02	2300	0.33	4050	1.02	5800	2.10	7550	3.55
600	0.02	2350	0.34	4100	1.05	5850	2.13	7600	3.60
650	0.03	2400	0.36	4150	1.07	5900	2.17	7650	3.65
700	0.03	2450	0.37	4200	1.10	5950	2.21	7700	3.70
750	0.04	2500	0.39	4250	1.13	6000	2.24	7750	3.74
800	0.04	2550	0.41	4300	1.15	6050	2.28	7800	3.79
850	0.05	2600	0.42	4350	1.18	6100	2.32	7850	3.84
900	0.05	2650	0.44	4400	1.21	6150	2.36	7900	3.89
950	0.06	2700	0.45	4450	1.23	6200	2.40	7950	3.94
1000	0.06	2750	0.47	4500	1.26	6250	2.43	8000	3.99
1050	0.07	2800	0.49	4550	1.29	6300	2.47	8050	4.04
1100	0.08	2850	0.51	4600	1.32	6350	2.51	8100	4.09
1150	0.08	2900	0.52	4650	1.35	6400	2.55	8150	4.14
1200	0.09	2950	0.54	4700	1.38	6450	2.59	8200	4.19
1250	0.10	3000	0.56	4750	1.41	6500	2.63	8250	4.24
1300	0.11	3050	0.58	4800	1.44	6550	2.67	8300	4.29
1350	0.11	3100	0.60	4850	1.47	6600	2.71	8350	4.35
1400	0.12	3150	0.62	4900	1.50	6650	2.76	8400	4.40
1450	0.13	3200	0.64	4950	1.53	6700	2.80	8450	4.45
1500	0.14	3250	0.66	5000	1.56	6750	2.84	8500	4.50
1550	0.15	3300	0.68	5050	1.59	6800	2.88	8550	4.56
1600	0.16	3350	0.70	5100	1.62	6850	2.92	8600	4.61
1650	0.17	3400	0.72	5150	1.65	6900	2.97	8650	4.66
1700	0.18	3450	0.74	5200	1.69	6950	3.01	8700	4.72
1750	0.19	3500	0.76	5250	1.72	7000	3.05	8750	4.77
1800	0.20	3550	0.79	5300	1.75	7050	3.10	8800	4.83
1850	0.21	3600	0.81	5350	1.78	7100	3.14	8850	4.88
1900	0.22	3650	0.83	5400	1.82	7150	3.19	8900	4.94
1950	0.24	3700	0.85	5450	1.85	7200	3.23	8950	4.99
2000	0.25	3750	0.88	5500	1.89	7250	3.28	9000	5.05

This table is based on the following equation:

$$V = 1096.5 \sqrt{\frac{VP}{\text{Air Density}}}$$

Where: $1096.5 = 60\sqrt{2 \times 322 \times \frac{62.3}{12}} = 60 \times 18.3 = 1097.1$

Velocities vs. Velocity Pressures

Reprinted with permission from *1989 ASHRAE Handbook—Fundamentals*

Figure 6.24

HVAC DUCT DESIGN PROCEDURES

The general procedure for HVAC system duct design is as follows:

1. Study the building plans, and arrange the supply and return outlets to provide proper distribution of air within each space. Adjust calculated air quantities for duct heat gains or losses and duct leakage. Also, adjust the supply, return, and/or exhaust air quantities to meet space pressurization requirements.
2. Select outlet sizes from manufacturers' data (see Chapter 31).
3. Sketch the duct system, connecting supply outlets and return intakes with the air-handling units/air conditioners. Space allocated for supply and return ducts often dictates system layout and ductwork shape. Use round ducts whenever feasible.
4. Divide the system into sections and number each section. A duct system should be divided at all points where flow, size, or shape changes. Assign fittings to the section toward the supply and return (or exhaust) terminals. The following examples are for the fittings identified for Example 6 (Figure 16), and system section numbers assigned (Figure 17). For converging flow fitting 3, assign the straight-through flow to section 1 (toward terminal 1), and the branch to section 2 (toward terminal 4). For diverging flow fitting 24, assign the straight-through flow to section 13 (toward terminals 26 and 29) and the branch to section 10 (toward terminals 43 and 44). For transition fitting 11, assign the fitting to upstream section 4 [toward terminal 9 (intake louver)]. For fitting 20, assign the unequal area elbow to downstream section 9 (toward diffusers 43 and 44). The fan outlet diffuser, fitting 42, is assigned to section 19 (again, toward the supply duct terminals).
5. Size ducts by the selected design method. Calculate system total pressure loss; then select the fan (refer to Chapter 18 of the 1996 *ASHRAE Handbook—Systems and Equipment*).
6. Lay out the system in detail. If duct routing and fittings vary significantly from the original design, recalculate the pressure losses. Reselect the fan if necessary.
7. Resize duct sections to approximately balance pressures at each junction.
8. Analyze the design for objectionable noise levels, and specify sound attenuators as necessary. Refer to the section on System and Duct Noise.

Example 8. For the system illustrated by Figures 16 and 17, size the ductwork by the equal friction method, and pressure balance the system by changing duct sizes (use 1 in. increments). Determine the system resistance and total pressure unbalance at the junctions. The airflow quantities are actual values adjusted for heat gains or losses, and ductwork is sealed (assume no leakage), galvanized steel ducts with transverse joints on 4 ft centers (ε = 0.0003 ft). Air is at standard conditions (0.075 lb_m/ft^3 density).

Because the primary purpose of Figure 16 is to illustrate calculation procedures, its duct layout is not typical of any real duct system. The layout includes fittings from the local loss coefficient tables, with emphasis on converging and diverging tees and various types of entries and discharges. The supply system is constructed of rectangular ductwork; the return system, round ductwork.

Solution: See Figure 17 for section numbers assigned to the system. The duct sections are sized within the suggested range of friction rate shown on the friction chart (Figure 9). Tables 11 and 12 give the total pressure loss calculations and the supporting summary of loss coefficients by sections. The straight duct friction factor and pressure loss were calculated by Equations (19) and (20). The fitting loss coefficients are from the *Duct Fitting Database* (ASHRAE 1994). Loss coefficients were calculated automatically by the database program (not by manual interpolation). The pressure loss values in Table 11 for the diffusers (fittings 43 and 44), the louver (fitting 9), and the air-measuring station (fitting 46) are manufacturers' data.

The pressure unbalance at the junctions may be noted by referring to Figure 18, the total pressure grade line for the system. The system resistance P_t is 2.63 in. of water. Noise levels and the need for duct silencers were not evaluated. To calculate the fan static pressure, use Equation (18):

$$P_s = 2.63 - 0.50 = 2.1 \text{ in. of water}$$

where 0.50 in. of water is the fan outlet velocity pressure.

Reprinted with permission from *1997 ASHRAE Handbook—Fundamentals*

Figure 6.25a

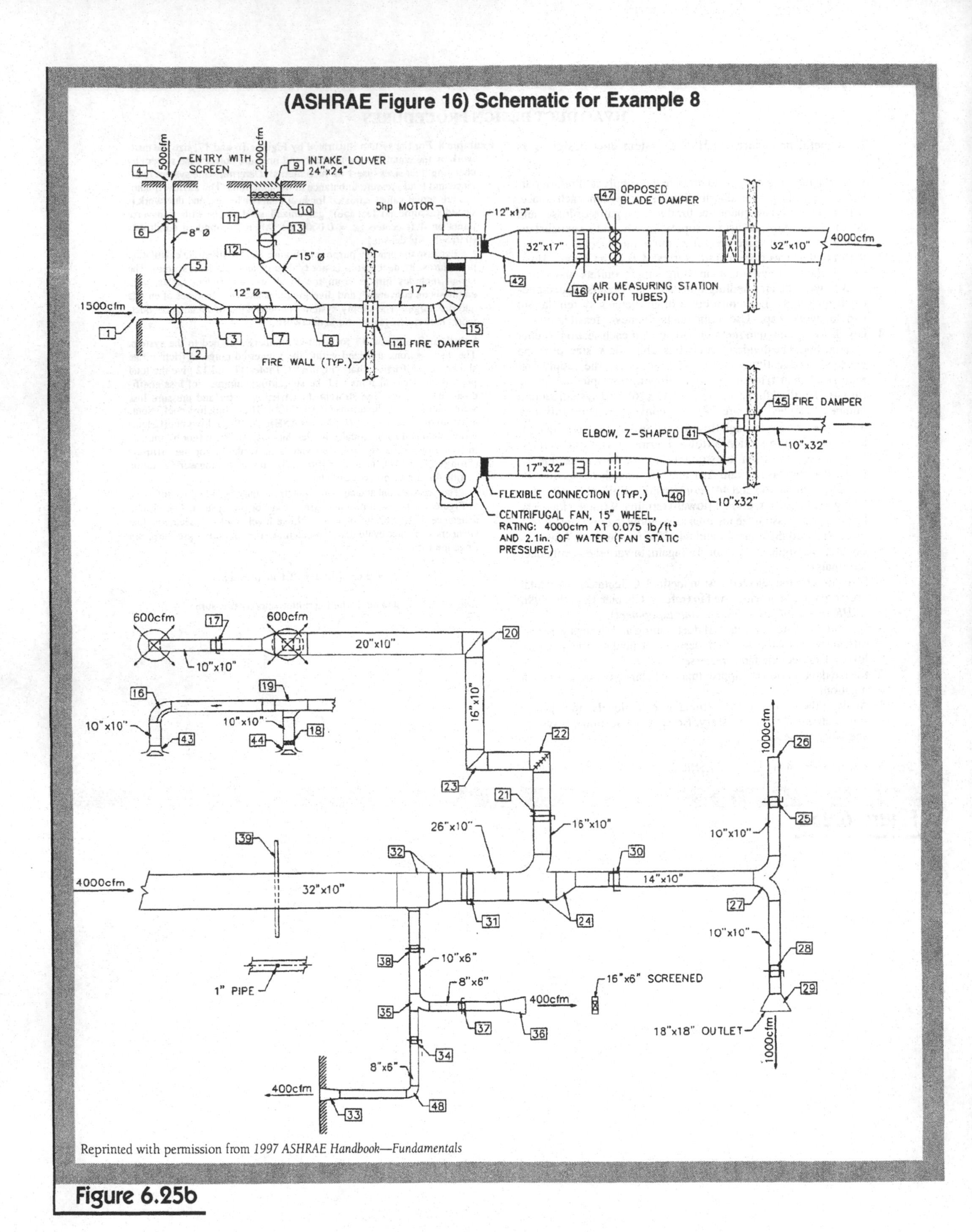
(ASHRAE Figure 16) Schematic for Example 8
ENTRY WITH SCREEN
500cfm
2000cfm
INTAKE LOUVER 24"x24"
5hp MOTOR
12"x17"
OPPOSED BLADE DAMPER
32"x17"
32"x10"
4000cfm
AIR MEASURING STATION (PITOT TUBES)
8" Ø
15" Ø
12" Ø
17"
1500cfm
FIRE DAMPER
FIRE WALL (TYP.)
FIRE DAMPER
ELBOW, Z-SHAPED
10"x32"
17"x32"
FLEXIBLE CONNECTION (TYP.)
CENTRIFUGAL FAN, 15" WHEEL, RATING: 4000cfm AT 0.075 lb/ft³ AND 2.1in. OF WATER (FAN STATIC PRESSURE)
600cfm
600cfm
20"x10"
10"x10"
16"x10"
26"x10"
16"x10"
14"x10"
1000cfm
4000cfm
32"x10"
1" PIPE
10"x6"
8"x6"
400cfm
16"x6" SCREENED
18"x18" OUTLET
400cfm
Reprinted with permission from 1997 ASHRAE Handbook—Fundamentals

Figure 6.25b

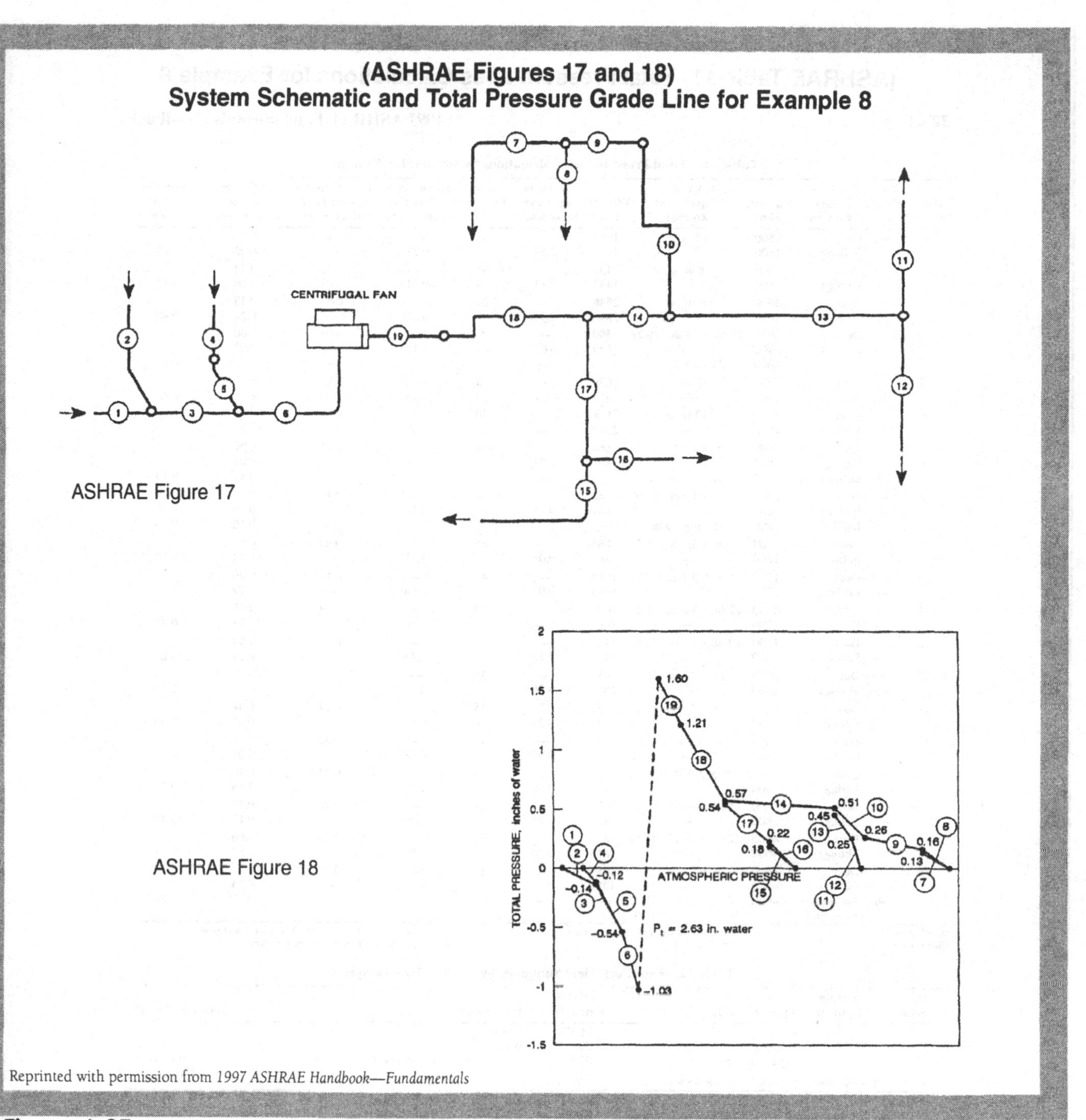
(ASHRAE Figures 17 and 18)
System Schematic and Total Pressure Grade Line for Example 8
CENTRIFUGAL FAN
ASHRAE Figure 17
ASHRAE Figure 18
TOTAL PRESSURE, inches of water
ATMOSPHERIC PRESSURE
P_t = 2.63 in. water
1.60
1.21
0.57
0.54
0.51
0.45
0.26
0.25
0.22
0.18
0.16
0.13
-0.12
-0.14
-0.54
-1.03
Reprinted with permission from 1997 ASHRAE Handbook—Fundamentals

Figure 6.25c

(ASHRAE Table 11) Total Pressure Loss Calculations for Example 8

Table 11 Total Pressure Loss Calculations by Sections for Example 8

Duct Section[a]	Fitting No.[b]	Duct Element	Airflow, cfm	Duct Size (Equivalent Round)	Velocity, fpm	Velocity Pressure, in. of water	Duct Length,[c] ft	Summary of Fitting Loss Coefficients[d]	Duct Pressure Loss/100 ft,[e] in. of water	Total Pressure Loss, in. of water	Section Pressure Loss, in. of water
1	—	Duct	1500	12 in. φ	1910	—	15	—	0.40	0.06	
	—	Fittings	1500	—	1910	0.23	—	0.33	—	0.08	0.14
2	—	Duct	500	8 in. φ	1432	—	60	—	0.39	0.23	
	—	Fittings	500	—	1432	0.13	—	−0.71	—	−0.09	0.14
3	—	Duct	2000	12 in. φ	2546	—	20	—	0.67	0.13	
	—	Fittings	2000	—	2546	0.40	—	0.67	—	0.27	0.40
4	—	Duct	2000	24 in. × 24 in. (26.2)	500	—	5	—	0.01	0.00	
	—	Fittings	2000	—	500	0.02	—	1.11	—	0.02	
	9	Louver	2000	24 in. × 24 in.	—	—	—	—	—	0.10[f]	0.12
5	—	Duct	2000	15 in. φ	1630	—	55	—	0.23	0.13	
	—	Fittings	2000	—	1630	0.17	—	1.73	—	0.29	0.42
6	—	Duct	4000	17 in. φ	2538	—	30		0.45	0.14	
		Fittings	4000	—	2538	0.40	—	0.87	—	0.35	0.49
7	—	Duct	600	10 in. × 10 in. (10.9)	864	—	14	—	0.11	0.02	
	—	Fittings	600	—	864	0.05		0.26	—	0.01	
	43	Diffuser	600	10 in. × 10 in.	—	—	—	—	—	0.10[f]	0.13
8	—	Duct	600	10 in. × 10 in. (10.9)	864	—	4	—	0.11	0.00	
	—	Fittings	600	—	864	0.05	—	1.25	—	0.06	
	44	Diffuser	600	10 in. × 10 in.	—	—	—	—	—	0.10[f]	0.16
9	—	Duct	1200	20 in. × 10 in. (15.2)	864	—	25	—	0.08	0.02	
	—	Fittings	1200	—	864	0.05	—	1.67	—	0.08	0.10
10	—	Duct	1200	16 in. × 10 in. (13.7)	1080	—	45	—	0.13	0.06	
	—	Fittings	1200	—	1080	0.07	—	2.66	—	0.19	0.25
11	—	Duct	1000	10 in. × 10 in. (10.9)	1440	—	10	—	0.29	0.03	
	—	Fittings	1000	—	1440	0.13	—	1.68	—	0.22	0.25
12	—	Duct	1000	10 in. × 10 in. (10.9)	1440	—	22	—	0.29	0.06	
	—	Fittings	1000	—	1440	0.13	—	1.45	—	0.19	0.25
13	—	Duct	2000	14 in. × 10 in. (12.9)	2057	—	35	—	0.47	0.16	
	—	Fittings	2000	—	2057	0.26	—	0.16	—	0.04	0.20
14	—	Duct	3200	26 in. × 10 in. (17.1)	1772	—	15	—	0.27	0.04	
	—	Fittings	3200		1772	0.20	—	0.12	—	0.02	0.06
15	—	Duct	400	8 in. × 6 in. (7.6)	1200	—	40	—	0.32	0.13	
	—	Fittings	400	—	1200	0.09		0.58		0.05	0.18
16	—	Duct	400	8 in. × 6 in. (7.6)	1200	—	20	—	0.32	0.06	
	—	Fittings	400	—	1200	0.09	—	1.74	—	0.16	0.22
17	—	Duct	800	10 in. × 6 in. (8.4)	1920	—	22	—	0.70	0.15	
	—	Fittings	800	—	1920	0.23	—	0.76	—	0.17	0.32
18	—	Duct	4000	32 in. × 10 in. (18.8)	1800	—	23	—	0.25	0.06	
	—	Fittings	4000	—	1800	0.20		2.91	—	0.58	0.64
19	—	Duct	4000	32 in. × 17 in. (25.2)	1059	—	12	—	0.06	0.01	
	—	Fittings	4000	—	1059	0.07	—	4.71	—	0.33	
	46	Air-measuring station	4000	—	—	—	—	—	—	0.05[f]	0.39

[a]See Figure 17.
[b]See Figure 16.
[c]Duct lengths are to fitting centerlines.
[d]See Table 12.
[e]Duct pressure based on a 0.0003 ft absolute roughness factor.
[f]Pressure drop based on manufacturers' data.

Table 12 Loss Coefficient Summary by Sections for Example 8

Duct Section	Fitting Number	Type of Fitting	ASHRAE Fitting No.[a]	Parameters	Loss Coefficient
1	1	Entry	ED1-3	$r/D = 0.2$	0.03
	2	Damper	CD9-1	$\theta = 0°$	0.19
	3	Wye (30°), main	ED5-1	$A_s/A_c = 1.0$, $A_b/A_c = 0.444$, $Q_s/Q_c = 0.75$	0.11 (C_s)
Summation of Section 1 loss coefficients					0.33
2	4	Entry	ED1-1	$L = 0$, $t = 0.064$ in. (16 gage)	0.50
	4	Screen	CD6-1	$n = 0.70$, $A_1/A_o = 1$	0.58
	5	Elbow	CD3-6	60°, $r/D = 1.5$, pleated	0.27
	6	Damper	CD9-1	$\theta = 0°$	0.19
	3	Wye (30°), branch	ED5-1	$A_s/A_c = 1.0$, $A_b/A_c = 0.444$, $Q_b/Q_c = 0.25$	−2.25 (C_b)
Summation of Section 2 loss coefficients					−0.71
3	7	Damper	CD9-1	$\theta = 0°$	0.19
	8	Wye (45°), main	ED5-2	$A_s/A_c = 0.498$, $A_b/A_c = 0.779$, $Q_s/Q_c = 0.5$	0.48 (C_s)
Summation of Section 3 loss coefficients					0.67

[a]*Duct Fitting Database* (ASHRAE 1991) data for fittings reprinted in the section on Fitting Loss Coefficients.

Figure 6.25d

(ASHRAE Table 12) Loss Coefficient Summary for Example 8

Duct Design **32.25**

Table 12 Loss Coefficient Summary by Sections for Example 8 (*Concluded*)

Duct Section	Fitting Number	Type of Fitting	ASHRAE Fitting No.[a]	Parameters	Loss Coefficient
4	10	Damper	CR9-4	$\theta = 0°$, 5 blades (opposed), $L/R = 1.25$	0.52
	11	Transition	ER4-3	$L = 30$ in., $A_o/A_1 = 3.26$, $\theta = 17°$	0.59
Summation of Section 4 loss coefficients					1.11
5	12	Elbow	CD3-17	45°, mitered	0.34
	13	Damper	CD9-1	$\theta = 0°$	0.19
	8	Wye (45°), branch	ED5-2	$Q_b/Q_c = 0.5$, $A_s/A_c = 0.498$, $A_b/A_c = 0.779$	1.20 (C_b)
Summation of Section 5 loss coefficients					1.73
6	14	Fire damper	CD9-3	Curtain type, Type C	0.12
	15	Elbow	CD3 9	90°, 5 gore, $r/D = 1.5$	0.15
	—	Fan and system interaction	ED7-2	90° elbow, 5 gore, $r/D = 1.5$, $L = 34$ in.	0.60
Summation of Section 6 loss coefficients					0.87
7	16	Elbow	CR3-3	90°, $r/W = 0.70$, 1 splitter vane	0.14
	17	Damper	CR9-1	$\theta = 0°$, $H/W = 1.0$	0.08
	19	Tee, main	SR5-13	$Q_s/Q_c = 0.5$, $A_s/A_c = 0.50$	0.04 (C_s)
Summation of Section 7 loss coefficients					0.26
8	19	Tee, branch	SR5-13	$Q_b/Q_c = 0.5$, $A_b/A_c = 0.50$	0.73 (C_b)
	18	Damper	CR9-4	$\theta = 0°$, 3 blades (opposed), $L/R = 0.75$	0.52
Summation of Section 8 loss coefficients					1.25
9	20	Elbow	SR3-1	90°, mitered, $H/W_1 = 0.625$, $W_o/W_1 = 1.25$	1.67
Summation of Section 9 loss coefficients					1.67
10	21	Damper	CR9-1	$\theta = 0°$, $H/W = 0.625$	0.08
	22	Elbow	CR3-10	90°, single-thickness vanes, design 2	0.12
	23	Elbow	CR3-6	$\theta = 90°$, mitered, $H/W = 0.625$	1.25
	24	Tee, branch	SR5-1	$r/W_b = 1.0$, $Q_b/Q_c = 0.375$, $A_s/A_c = 0.538$, $A_b/A_c = 0.615$	1.21 (C_b)
Summation of Section 10 loss coefficients					2.66
11	25	Damper	CR9-1	$\theta = 0°$, $H/W = 1.0$	0.08
	26	Exit	SR2-1	$H/W = 1.0$, Re = 122,500	1.00
	27	Wye, dovetail	SR5-14	$r/W_c = 1.5$, $Q_{b1}/Q_c = 0.5$, $A_{b1}/A_c = 0.714$	0.60 (C_b)
Summation of Section 11 loss coefficients					1.68
12	28	Damper	CR9-1	$\theta = 0°$, $H/W = 1.0$	0.08
	29	Exit	SR2-5	$\theta = 19°$, $A_1/A_o = 3.24$, Re = 130,000	0.77
	27	Wye, dovetail	SR5-14	$r/W_c = 1.5$, $Q_{b2}/Q_c = 0.5$, $A_{b2}/A_c = 0.714$	0.60 (C_b)
Summation of Section 12 loss coefficients					1.45
13	30	Damper	CR9-1	$\theta = 0°$, $H/W = 0.71$	0.08
	24	Tee, main	SR5-1	$r/W_b = 1.0$, $Q_s/Q_c = 0.625$, $A_s/A_c = 0.538$, $A_b/A_c = 0.615$	0.08 (C_s)
Summation of Section 13 loss coefficients					0.16
14	31	Damper	CR9-1	$\theta = 0°$, $H/W = 0.38$	0.08
	32	Tee, main	SR5-13	$Q_s/Q_c = 0.8$, $A_s/A_c = 0.813$	0.04 (C_s)
Summation of Section 14 loss coefficients					0.12
15	48	Elbow	CR3-1	$\theta = 90°$, $r/W = 1.5$, $H/W = 0.75$	0.19
	33	Exit	SR2-6	$L = 18$ in., $D_h = 6.86$	0.28
	34	Damper	CR9-1	$\theta = 0°$, $H/W = 0.75$	0.08
	35	Tee, main	SR5-1	$r/W_b = 1.0$, $Q_s/Q_c = 0.5$, $A_s/A_c = 0.80$, $A_b/A_c = 0.80$	0.03 (C_s)
Summation of Section 15 loss coefficients					0.58
16	36	Exit	SR2-3	$\theta = 20°$, $A_1/A_o = 2.0$, Re = 70,000	0.63
	36	Screen	CR6-1	$n = 0.8$, $A_1/A_o = 2.0$	0.08
	37	Damper	CR9-1	$\theta = 0°$, $H/W = 0.75$	0.08
	35	Tee, branch	SR5-1	$r/W_b = 1.0$, $Q_b/Q_c = 0.5$, $A_s/A_c = 0.80$, $A_b/A_c = 0.80$	0.95 (C_b)
Summation of Section 16 loss coefficients					1.74
17	38	Damper	CR9-1	$\theta = 0°$, $H/W = 0.6$	0.08
	32	Tee, branch	SR5-13	$Q_b/Q_c = 0.2$, $A_b/A_c = 0.187$	0.68 (C_b)
Summation of Section 17 loss coefficients					0.76
18	39	Obstruction, pipe	CR6-4	Re = 15,000, $y = 0$, $d = 1$ in., $S_m/A_o = 0.1$, $y/H = 0$	0.17
	40	Transition	SR4-1	$\theta = 22°$, $A_o/A_1 = 0.588$, $L = 18$ in.	0.04
	41	Elbows, Z-shaped	CR3-17	$L = 42$ in., $L/W = 4.2$, $H/W = 3.2$, Re = 240,000	2.51
	45	Fire damper	CR9-6	Curtain type, Type B	0.19
Summation of Section 18 loss coefficients					2.91
19	42	Diffuser, fan	SR7-17	$\theta_1 = 28°$, $L = 40$ in., $A_o/A_1 = 2.67$, $C_1 = 0.59$	4.19 (C_o)
	47	Damper	CR9-4	$\theta = 0°$, 8 blades (opposed), $L/R = 1.39$	0.52
Summation of Section 19 loss coefficients					4.71

[a]*Duct Fitting Database* (ASHRAE 1994) data for fittings reprinted in the section on Fitting Loss Coefficients.

Figure 6.25e

INDUSTRIAL EXHAUST SYSTEM DUCT DESIGN

Chapter 26 of the 1995 *ASHRAE Handbook—Applications* discusses design criteria, including hood design, for industrial exhaust systems. Exhaust systems conveying vapors, gases, and smoke can be designed by equal friction, static regain, or T-method. Systems conveying particulates are designed by the constant velocity method at duct velocities adequate to convey particles to the system air cleaner. For contaminant transport velocities, see Table 2 in Chapter 26 of the 1995 *ASHRAE Handbook—Applications*.

Two pressure-balancing methods can be considered when designing industrial exhaust systems. One method uses balancing devices (e.g., dampers, blast gates) to obtain design airflow through each hood. The other approach balances systems by adding resistance to ductwork sections (i.e., changing duct size, selecting different fittings, and increasing airflow). This self-balancing method is preferred, especially for systems conveying abrasive materials. Where potentially explosive or radioactive materials are conveyed, the prebalanced system is mandatory because contaminants could accumulate at the balancing devices. To balance systems by increasing airflow, use Equation (51), which assumes that all ductwork has the same diameter and that fitting loss coefficients, including main and branch tee coefficients, are constant.

$$Q_c = Q_d(P_h/P_l)^{0.5} \quad (51)$$

where

Q_c = airflow rate required to increase P_l to P_h, cfm
Q_d = total airflow rate through low-resistance duct run, cfm
P_h = absolute value of pressure loss in high-resistance ductwork section(s), in. of water
P_l = absolute value of pressure loss in low-resistance ductwork section(s), in. of water

For systems conveying particulates, use elbows with a large centerline radius-to-diameter ratio (r/D), greater than 1.5 whenever possible. If r/D is 1.5 or less, abrasion in dust-handling systems can reduce the life of elbows. Elbows are often made of seven or more gores, especially in large diameters. For converging flow fittings, a 30° entry angle is recommended to minimize energy losses and abrasion in dust-handling systems. For the entry loss coefficients of hoods and equipment for specific operations, refer to Chapter 26 of the 1995 *ASHRAE Handbook—Applications* and to ACGIH (1995).

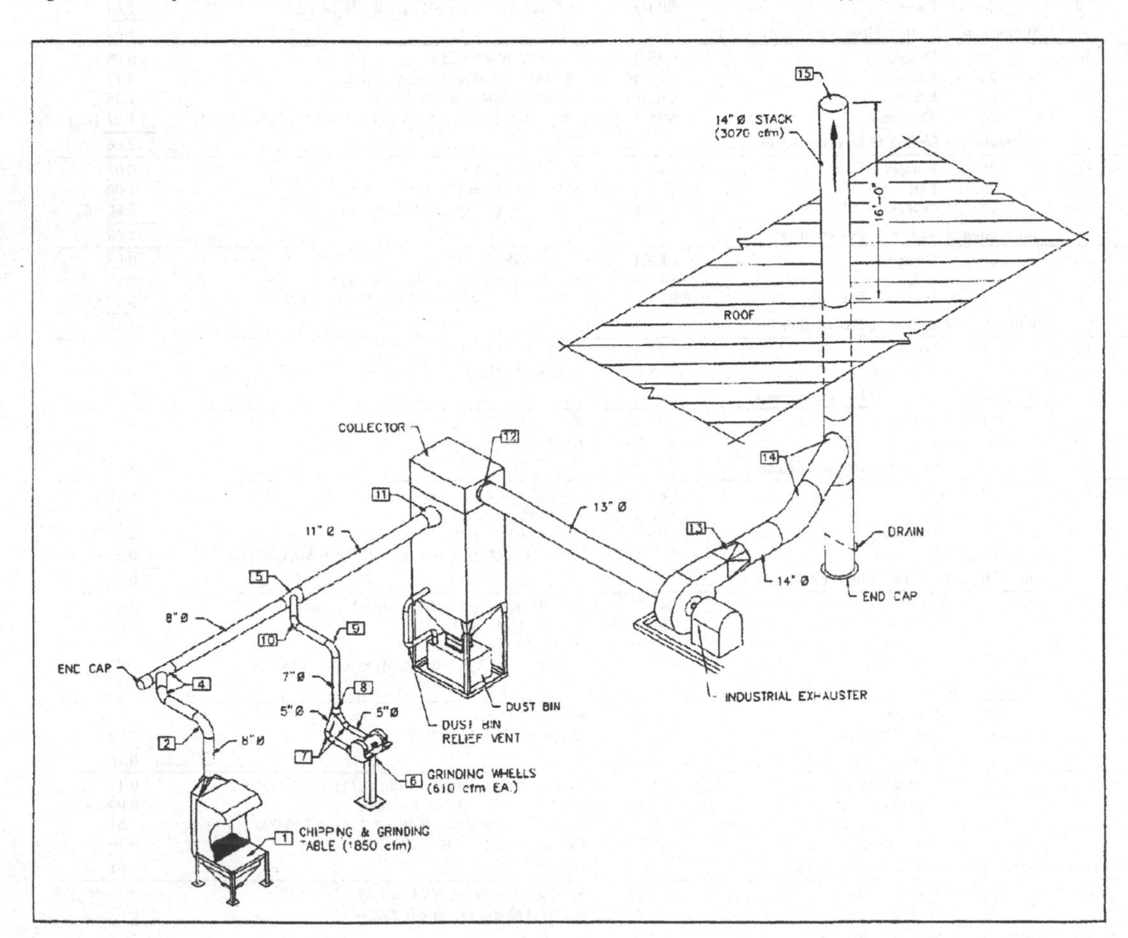

Figure 6.25f

Duct Design (Cont.)

Example 9. For the metalworking exhaust system in Figures 19 and 20, size the ductwork and calculate the fan static pressure requirement for an industrial exhaust designed to convey granular materials. Pressure balance the system by changing duct sizes and adjusting airflow rates. The minimum particulate transport velocity for the chipping and grinding table ducts (sections 1 and 5, Figure 20) is 4000 fpm. For the ducts associated with the grinder wheels (sections 2, 3, 4, and 5), the minimum duct velocity is 4500 fpm. Ductwork is galvanized steel, with the absolute roughness being 0.0003 ft. Assume that air is standard and that duct and fittings are available in the following sizes: 3 in. through 9.5 in. diameters in 0.5 in. increments, 10 in. through 37 in. diameters in 1 in. diameter increments, and 38 in. through 90 in. diameters in 2 in. diameter increments.

The building is one story, and the design wind velocity is 20 mph. For the stack, use Design J shown in Figure 13 in Chapter 15 for complete rain protection. The stack height, determined by calculations from Chapter 15, is 16 ft above the roof. This stack height is based on minimized stack downwash; therefore, the stack discharge velocity must exceed 1.5 times the design wind velocity.

Solution: For the contaminated ducts upstream of the collector, initial duct sizes and transport velocities are summarized below. The 4474 fpm velocity in sections 2 and 3 is acceptable because the transport velocity is not significantly lower than 4500 fpm. For the next available duct size (4.5 in. diameter), the duct velocity is 5523 fpm, significantly higher than 4500 fpm.

Duct Section	Design Airflow, cfm	Transport Velocity, fpm	Duct Diameter, in.	Duct Velocity, fpm
1	1800	4000	9	4074
2,3	610 each	4500	5	4474
4	1220	4500	7	4565
5	3020	4500	11	4576

The following tabulation summarizes design calculations up through the junction after sections 1 and 4.

Design No.	D_1, in.	Δp_1, in. of water	Δp_{2+4}, in. of water	Imbalance, $\Delta p_1 - \Delta p_{2+4}$
1	9	1.46	3.09	−1.63
2	8.5	2.00	3.08	−1.08
3	8	2.79	3.00	−0.21
4	7.5	3.92	2.88	+1.04

$Q_1 = 1800$ cfm
$Q_2 = 610$ cfm; $D_2 = 5$ in. dia.
$Q_3 = 610$ cfm; $D_3 = 5$ in. dia.
$Q_4 = 1220$ cfm; $D_4 = 7$ in. dia.

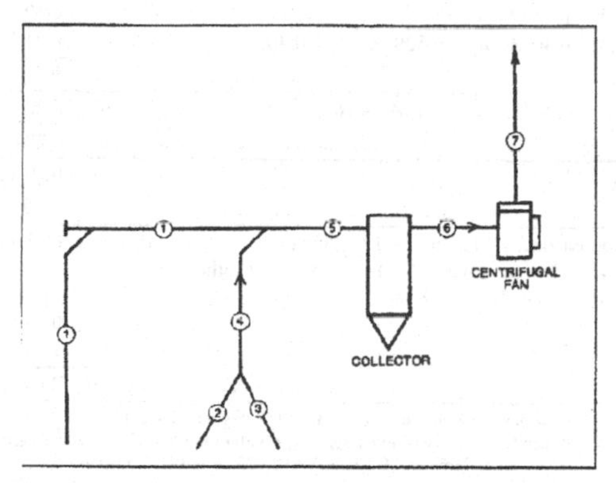

Fig. 20 System Schematic with Section Numbers for Example 9

For the initial design, Design 1, the imbalance between section 1 and section 2 (or 3) is 1.63 in. of water, with section 1 requiring additional resistance. Decreasing section 1 duct diameter by 0.5 in. increments results in the least imbalance, 0.21 in. of water, when the duct diameter is 8 in. (Design 3). Because section 1 requires additional resistance, estimate the new airflow rate using Equation (51):

$$Q_{c,1} = (1800)(3.00/2.79)^{0.5} = 1870 \text{ cfm}$$

At 1870 cfm flow in section 1, 0.13 in. of water imbalance remains at the junction of sections 1 and 4. By trial-and-error solution, balance is attained when the flow in section 1 is 1850 cfm. The duct between the collector and the fan inlet is 13 in. round to match the fan inlet (12.75 in. diameter). To minimize downwash, the stack discharge velocity must exceed 2640 fpm, 1.5 times the design wind velocity (20 mph) as stated in the problem definition. Therefore, the stack is 14 in. round, and the stack discharge velocity is 2872 fpm.

Table 13 summarizes the system losses by sections. The straight duct friction factor and pressure loss were calculated by Equations (19) and (20). Table 14 lists fitting loss coefficients and input parameters necessary to determine the loss coefficients. The fitting loss coefficients are from the *Duct Fitting Database* (ASHRAE 1994). The fitting loss coefficient tables are included in the section on Fitting Loss Coefficients for illustration but can not be obtained exactly by manual interpolation since the coefficients were calculated by the duct fitting database algorithms (more significant figures). For a pressure grade line of the system, see Figure 21. The fan total pressure, calculated by Equation (16), is 7.89 in. of water. To calculate the fan static pressure, use Equation (18):

$$P_s = 7.89 - 0.81 = 7.1 \text{ in. of water}$$

where 0.81 in. of water is the fan outlet velocity pressure. The fan airflow rate is 3070 cfm, and its outlet area is 0.853 ft² (10.125 in. by 12.125 in.). Therefore, the fan outlet velocity is 3600 fpm.

The hood suction for the chipping and grinding table hood is 2.2 in. of water, calculated by Equation (19) from Chapter 26 of the 1995 *ASHRAE Handbook—Applications* [$HS = (1 + 0.25)(1.74) = 2.2$ in. of water, where 0.25 is the hood entry loss coefficient C_o, and 1.74 is the duct velocity pressure P_v a few diameters downstream from the hood]. Similarly, the hood suction for each of the grinder wheels is 1.7 in. of water:

$$HS_{2,3} = (1 + 0.4)(1.24) = 1.7 \text{ in. of water}$$

where 0.4 is the hood entry loss coefficient, and 1.24 in. of water is the duct velocity pressure.

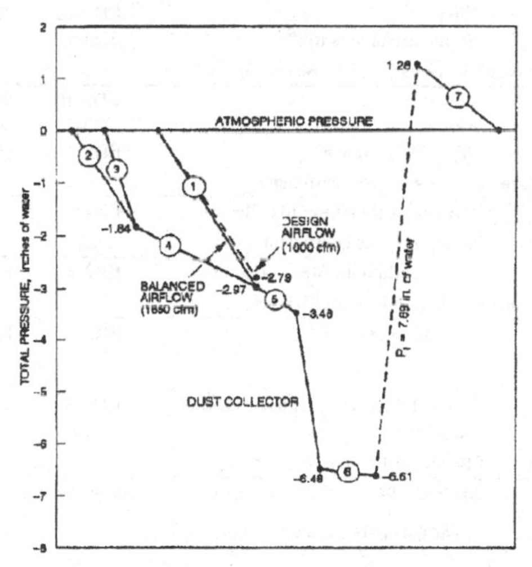

Fig. 21 Total Pressure Grade Line for Example 9

Figure 6.25g

Total Pressure Loss and Loss Coefficient for Example 9

Table 13 Total Pressure Loss Calculations by Sections for Example 9

Duct Section[a]	Duct Element	Airflow, cfm	Duct Size	Velocity, fpm	Velocity Pressure, in. of water	Duct Length,[b] ft	Summary of Fitting Loss Coefficients[c]	Duct Pressure Loss/100 ft, in. of water[d]	Total Pressure Loss, in. of water	Section Pressure Loss, in. of water
1	Duct	1850	8 in. ϕ	5300	—	22.5	—	4.63	1.04	
	Fittings	1850		5300	1.75	—	1.10	—	1.93	2.97
2,3	Duct	610	5 in. ϕ	4474	—	9	—	5.94	0.53	
	Fittings	610		4474	1.24	—	1.06	—	1.31	1.84
4	Duct	1220	7 in. ϕ	4565	—	11.5	—	4.08	0.47	
	Fittings	1220		4565	1.30	—	0.51	—	0.66	1.13
5	Duct	3070	11 in. ϕ	4652	—	8.5	—	2.44	0.21	
	Fittings	3070		4652	1.35	—	0.22	—	0.30	0.51
—	Collector,[e] fabric	3070	—	—	—	—	—	—	3.0	3.0
6	Duct	3070	13 in. ϕ	3331	—	12	—	1.05	0.13	
	Fittings	3070		3331	0.69	—	0.00	—	0.00	0.13
7	Duct	3070	14 in. ϕ	2872	—	29	—	0.72	0.21	
	Fittings	3070		2872	0.51	—	2.09	—	1.07	1.28

[a] See Figure 20.
[b] Duct lengths are to fitting centerlines.
[c] See Table 14.
[d] Duct pressure based on a 0.0003 ft absolute roughness factor.
[e] Collector manufacturers set the fabric bag cleaning mechanism to actuate at a pressure difference of 3.0 in. of water between the inlet and outlet plenums. The pressure difference across the clean media is approximately 1.5 in. of water.

Table 14 Loss Coefficient Summary by Sections for Example 9

Duct Section	Fitting Number	Type of Fitting	ASHRAE Fitting No.[a]	Parameters	Loss Coefficient
1	1	Hood[b]	—	Hood face area: 3 ft by 4 ft	0.25
	2	Elbow	CD3-10	90°, 7 gore, $r/D = 2.5$	0.11
	4	Capped wye (45°), with 45° elbow	ED5-6	$A_b/A_c = 1$	0.64 (C_b)
	5	Wye (30°), main	ED5-1	$Q_s/Q_c = 0.60$, $A_s/A_c = 0.529$, $A_b/A_c = 0.405$	0.10 (C_s)
Summation of Section 1 loss coefficients					1.10
2,3	6	Hood[c]	—	Type hood: For double wheels, dia. = 22 in. each, wheel width = 4 in. each; type takeoff: tapered	0.40
	7	Elbow	CD3-12	90°, 3 gore, $r/D = 1.5$	0.34
	8	Symmetrical wye (60°)	ED5-9	$Q_b/Q_c = 0.5$, $A_b/A_c = 0.51$	0.32 (C_b)
Summation of Sections 2 and 3 loss coefficients					1.06
4	9	Elbow	CD3-10	90°, 7 gore, $r/D = 2.5$	0.11
	10	Elbow	CD3-13	60°, 3 gore, $r/D = 1.5$	0.19
	5	Wye (30°), branch	ED5-1	$Q_b/Q_c = 0.40$, $A_s/A_c = 0.529$, $A_b/A_c = 0.405$	0.21 (C_b)
Summation of Section 4 loss coefficients					0.51
5	11	Exit, conical diffuser to collector	ED2-1	$L = 24$ in., $L/D_o = 2.18$, $A_1/A_o = 16$	0.22
Summation of Section 5 loss coefficients					0.22
6	12	Entry, bellmouth from collector	ER2-1	$r/D_1 = 0.20$	0.00 (C_1)
Summation of Section 6 loss coefficients					0.00
7	13	Diffuser, fan outlet[d]	SR7-17	Fan outlet size: 10.125 in. by 12.125 in., $A_o/A_1 = 1.596$ (assume 14 in. by 14 in. outlet rather than 16 in. round), $L = 18$ in.	0.45 (C_o)
	14	Capped wye (45°), with 45° elbow	ED5-6	$A_b/A_c = 1$	0.64 (C_b)
	15	Stackhead	SD2-6	$D_e/D = 1$	1.0
Summation of Section 7 loss coefficients					2.09

[a] *Duct Fitting Database* (ASHRAE 1994) data for fittings reprinted in the section on Fitting Loss Coefficients.
[b] From *Industrial Ventilation* (ACGIH 1995, Figure VS-80-19).
[c] From *Industrial Ventilation* (ACGIH 1995, Figure VS-80-11).
[d] Fan specified: Industrial exhauster for granular materials: 21 in. wheel diameter, 12.75 in. inlet diameter, 10.125 in. by 12.125 in. outlet, 7.5 hp motor.

Reprinted with permission from *1997 ASHRAE Handbook—Fundamentals*

Figure 6.25h

Exhaust Ventilation System Design

by D. Jeff Burton, PE, CIH

Before exhaust ventilation system design begins, the objectives of the project must be well defined (usually as part of the commissioning process). Figure 6.26 shows a simple local exhaust system and a list of the types of information normally gathered.

In the specialty of exhaust ventilation, four design approaches have been used over the years: the blast gate method, the plenum method, the equivalent foot method and the velocity pressure (VP) method. The most widely used approach today is the VP method because it results in a "balanced design," and input factors are readily available.

The VP method estimates system losses as fractions of velocity pressure:

$SP_{loss} = K_{loss} \bullet VP \bullet d$ where

SP_{loss} = conversion of system static pressure to heat/vibration/noise, inches wg (U.S. units)

K_{loss} = the loss factor (obtained from literature and manufacturer sources) for the fitting, unitless; for example, a typical elbow has a loss factor of $K_{loss} = 0.40$

VP = the average velocity pressure in the duct, inches wg (U.S. units)

d = the density correction factor, unitless

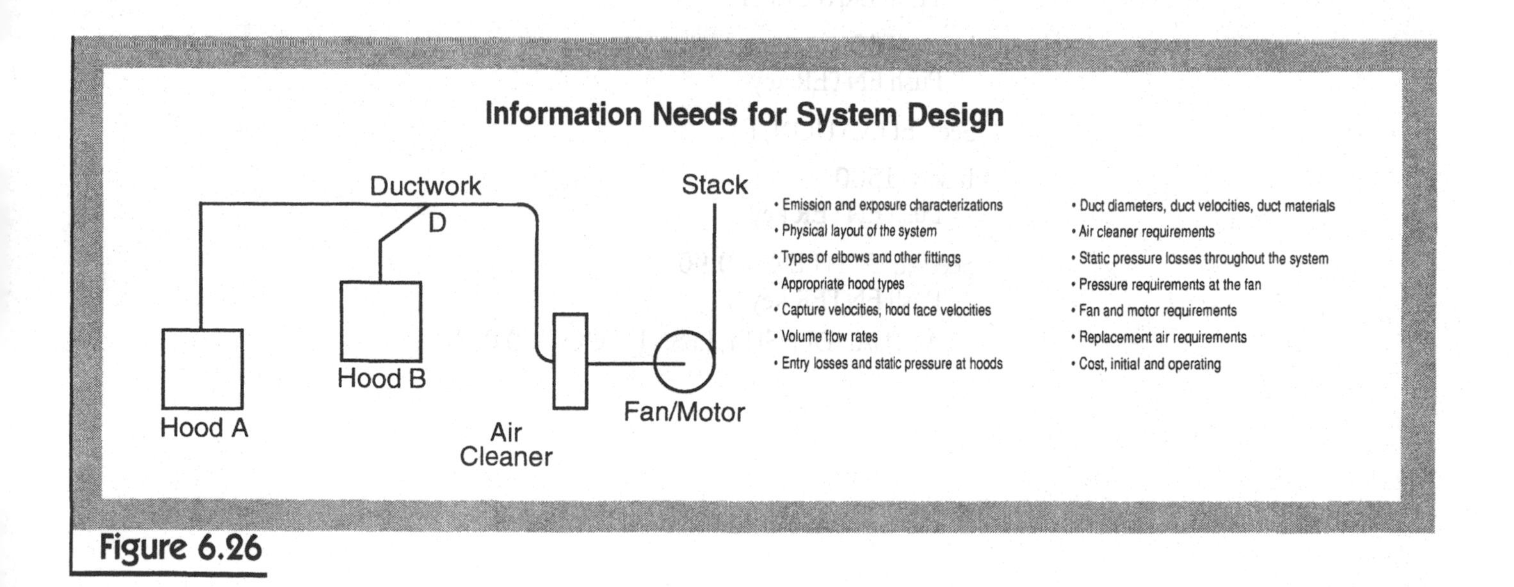

Information Needs for System Design
Ductwork
D
Stack
Hood A
Hood B
Air Cleaner
Fan/Motor
• Emission and exposure characterizations
• Physical layout of the system
• Types of elbows and other fittings
• Appropriate hood types
• Capture velocities, hood face velocities
• Volume flow rates
• Entry losses and static pressure at hoods
• Duct diameters, duct velocities, duct materials
• Air cleaner requirements
• Static pressure losses throughout the system
• Pressure requirements at the fan
• Fan and motor requirements
• Replacement air requirements
• Cost, initial and operating

Figure 6.26

The following examples show how calculations can be made using a pre-programmed calculator system. (This example uses IVE's VentCalc.) For more information on this product, see "Resources" at the end of this chapter.

Sample Calculation: Estimating Actual Air Density

Program 0 (zero) uses the Ideal Gas Law to estimate a Density Correction Factor, d. for use in calculations for nonstandard air.

The calculator assumes STP at: BP = 29.92" Hg, T = 70°F, RH = 0%.

(This is the ACGIH definition of STP and is very close to ASHRAE's.)

The calculator asks which measure you will use for atmospheric pressure: barometric pressure (BP) or elevations? You will input the temperature in Fahrenheit and the BP in inches of Hg or the elevation in feet. The calculator then shows "d = (some number)" and the actual air density in lbs/cubic foot. The SI (metric) program is similar.

All subsequent calculations are automatically corrected for air density.

Example

What is the air density correction factor (d)?
Elevation = 1,500 feet above sea level;
Air temp = 100°F; run Program 0 (zero).

Title: DENSITY CORRECTION FACTOR, d
Push **ENTER** key

See: USING FOR PRESSURE:
BAROM PRES = 1
ELEVATION = 2
WHICH?

Insert: For elevation, insert "2"
Push **ENTER** key

See: INPUT DATA:
TEMPERATURE, F?

Insert: **100**
Push **ENTER** key

See: ELEVATION, FT?

Insert: **1500**
Push **ENTER** key

See: d, UNITLESS = **0.90**
Push **ENTER** key
ACTUAL DENSITY, LBS/CU FOOT = **0.067**

Sample Calculation: Local Exhaust Duct Design

Programs 1–5 estimate airflow parameters in two duct runs—No. 1 (the main) and No. 2 (the branch or stack). In our sample calculation there is a main and a stack. See Figure 6.26. The approach is simple and accurate, as demonstrated below:

1. Prepare preliminary design plans as usual (e.g., conduct emission studies, lay out duct funs; estimate desired Q and transport velocities; choose duct materials, fittings, hoods, elbow types, branch entry types; estimate loss factors for hoods and fittings; and so forth.) The workbook that accompanies the preprogrammed calculator provides more information.
2. Fill in IVE VentCalc Design Worksheet Rows 1–18 from your plans. (See Figure 6.27.)
3. Input information from the worksheet to Program 1 and to Program 2 for the stack.
4. After you have input all the data, the calculator will give you the Ideal Duct Diameter. Write it in Row 19.
5. Choose an appropriate whole-number trial diameter and write it in Row 20.
6. Push the **ENTER** key. The calculator asks for a trial diameter. After inputting an appropriate whole-size trial diameter, the calculator will output each of the items in Rows 21–34, as appropriate.

In the sample problem, the system will move Q = 1,200 acfm against a fan total pressure of four inches wg (See Row 32.) We will balance the system—using the blast gate or fan speed—to achieve a hood static pressure of SP_h = 0.94 inches wg. It also suggests a rated two-horsepower motor.

When balancing a junction (two ducts merging), you would run IVE Program 3. Program 4 estimates an appropriate duct diameter to achieve balance, if necessary. Program 5 estimates the Fan Total pressure and provides rough estimates for motor sizing.

The column titled "Source" refers to this kind of data found in charts provided in the *Industrial Ventilation Workbook*. This kind of data may be obtained from other sources as well.

Final Note: It is important to work through a number of sample calculations prior to actually attempting design of real ventilation systems. Unless you are trained and appropriately licensed, you should obtain the review and design services of a registered professional engineer (PE) before actually constructing any ventilation system designed using the VentCalc approach.

IVE VentCalc Design Worksheet

For use with TI-82

		Project: Lab Fume Hood System Design Date:				Initial Design		Resize/Redesign	
		Run ID: Main and Stack				Run 1	Run 2	Run 1	Run 2
Prog.	Info	Row Item	Source	Units (US/SI)	Cue	Prog.	Prog.	Prog.	Prog.
	I	1. Duct ID	Plans	" FROM" --> "TO"	-	A-B	B-C	NA	NA
	N	2. Desired flow rate	Chart 11 (or VentCalc Prog 6)		Q	1200	1200		
	P	3. Duct Length	Plans	feet	L	35	12		
	U	4. Target duct velocity	Chart 9, 11	fpm	VT	2500	2500		
	T	5. Duct Roughness	Chart 10	unitless	R	0.85	0.85		
	S	6. Total slot length	Plans (no slot = 0)	feet	Ls	12	0		
		7. Desired slot velocity	Plans (no slot=0)	fpm	Vs	1200	0		
	F	8. Acceleration factor	(Hood=1)	unitless	-	1	0		
1	A	9. Duct entry K from Hood	Chart 11	unitless	-	0.50	0		
	C	10. Elbow K	Chart 13	unitless	-	0.60	0		
or	T	11. Branch entry K	Chart 14	unitless	-	0	0		
	O	12. System effect K	Chart 16	unitless	-	0.40	0		
2	R	13. Other K (Gate housing)		unitless	-	0.10	0		
	S	14. Sum of factors (Sum Rows 8-13)		unitless	Sum	2.6	0.10		
		15. SP @ "FROM" (See previous Row 29)		inch wg	-	0	0		
	S	16. Other SP user ( )		inch wg	-	0	0		
	P	17. Other SP user (Air Cleaner)		inch wg	-	2.0	0		
		18. Sum of SP	(Sum rows 15-17)	inch wg	Sum	2.0	0		
	O	19. Ideal Duct diameter		inch	D1	9.38	9.38		
	U	20. Trial Duct diameter, D1		inch	D1	9	9		
	T	21. Actual duct velocity		fpm	V	2716	2716		
	P	22. Duct velocity pressure		inch wg	VP	0.46	0.46		
	U	23. Duct friction loss		inch wg	-	0.37	0.13		
	T	24a Junction VP increase (if "FROM" is a juncton)		inch	-	0	0		
	S	24b Plenum loss (when using a slot hood)		inch wg	-	0.25	0		
		25. Total SP at "TO" location		inch wg	-	3.82	0.17		
	If	26. SPh (when coming FROM a hood)		inch wg	SPh	0.94	-		
		27. Slot width		inch	Ws	1.0	-		
	B	28. Ratio of SP's	(OK = 0.8 to 1.2)	unitless	Ratio	-			
3	A	29. Governing SP		inch wg	Gov	-			
	L	30. Corected flow rate		cfm	Qact	-	-		
4	D1	31. New Trial Diameter		Yes or No size	-	-	-		
	F	32. Fan Total Pressure		inch wg	FTP	4.0			
5	A	33. air horsepower		hp	ahp	1.44			
	N	34. rated power motor		hp	rhp	>1.9			

Figure 6.27

The following section is reprinted with permission from *HVAC: Design Criteria, Options, Selection, Second Edition* by William H. Rowe, PE, AIA, R.S. Means Co., Inc.

Refrigeration

All cooling systems use refrigeration equipment to produce cooling. There are six refrigeration systems:

- Compression cycle
- Absorption cycle
- Steam jet cycle
- Air cycle
- Thermoelectric cycle
- Solar cycle

Since compression and absorption refrigeration systems account for over 98 percent of all installations, this discussion is limited to these two areas. The steam jet, air, thermoelectric, and solar cycles are rarely installed.

The type of refrigeration cycle selected depends on cost and capacity considerations. Figures 6.28 and 6.29 are selection guides for refrigeration equipment and all types of cooling systems.

Compression Cycle

The most common refrigeration system is the **compression cycle,** which is used in most installations. The compression cycle is the most effective means of heat removal per pound of refrigerant. Figure 6.30 illustrates the compression refrigeration cycle with its four basic pieces of equipment: evaporator, compressor, condenser, and expansion valve. A refrigerant flows through the four pieces of equipment, forming a loop in which the refrigerant changes from a liquid to a gas and then back to a liquid. This refrigerant has a low boiling point. Nearly all refrigerants boil below 0°F [–17.8 °C]. Beyond the expansion valve, the boiling (evaporation) takes place. In expanding from the liquid to the gaseous state, the refrigerant absorbs heat from the evaporator and the space surrounding the evaporator. This changes the refrigerant from a liquid to a gas as it absorbs heat from the air surrounding the evaporator (room temperature), causing it to "boil." This phase-change heat transfer is the reverse of steam heating; the medium in cooling changes from a liquid to a gas, whereas in heating, the medium changes from a gas to a liquid as heat is transferred.

The gas temperature and pressure must then be raised significantly above the outdoor temperature by the compressor to allow the outdoor temperature to cool the refrigerant, condense it back to liquid form, and return it for recycling around the loop. The refrigerant has absorbed heat twice: once in the evaporator, where the building heat gain is removed; and again from the compressor, which has added energy to raise the gas temperature above the outdoor air temperature. The total heat absorbed by the refrigerant, from the heat of evaporation and the heat of compression, is finally rejected to the outside by the condenser, and the refrigerant repeats the loop.

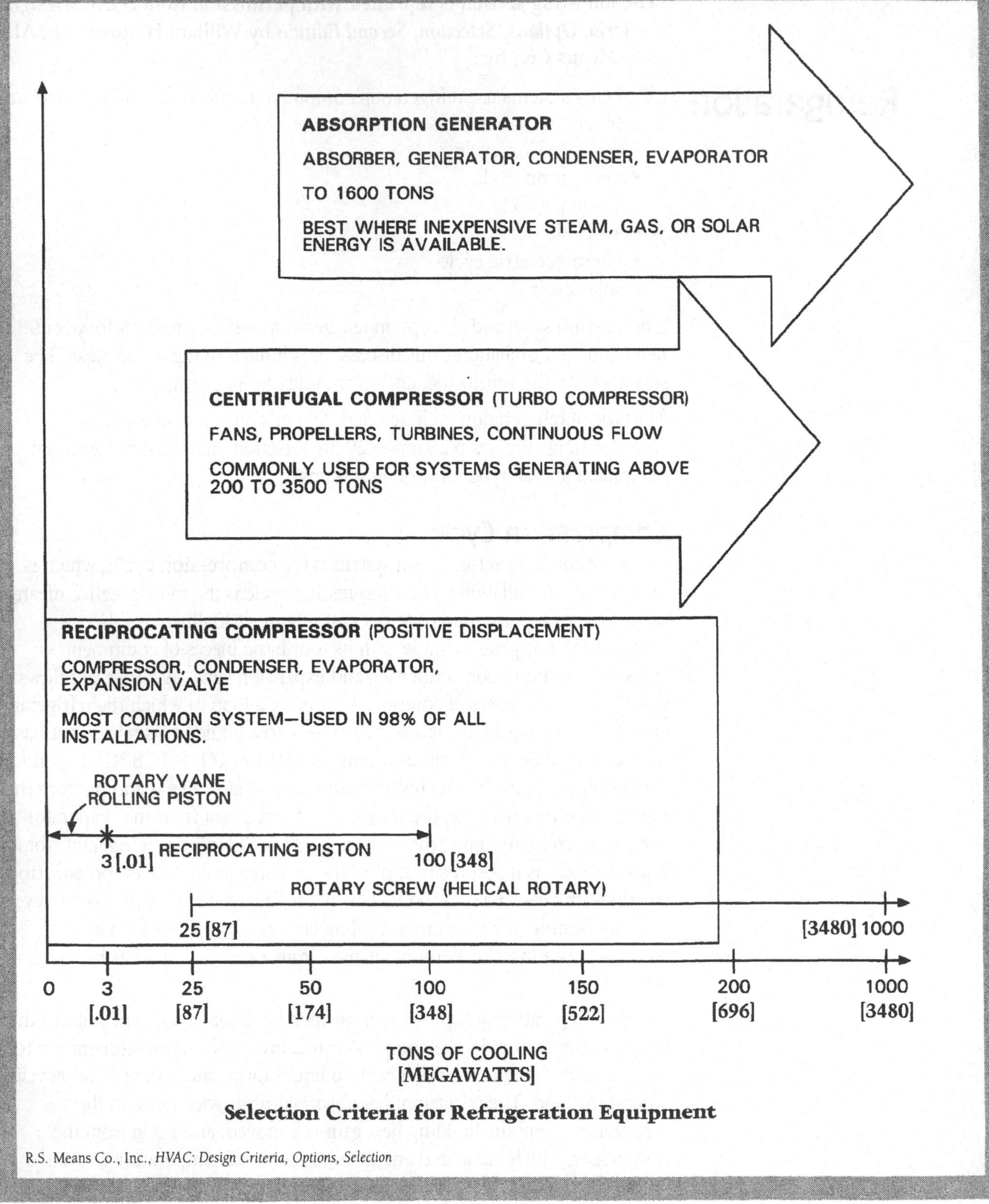

Selection Criteria for Refrigeration Equipment

R.S. Means Co., Inc., *HVAC: Design Criteria, Options, Selection*

Figure 6.28

Guidelines for Cooling System Selection

System	Subsystem		Range of Tonnage [megawatts]	Common Applications	Remarks
Compression Cycle	Positive Displacement	Reciprocating piston	3 – 100 [.01 – 348]	All buildings	Noise isolation recommended.
		Rotary vane	0 – 3 [0 – .01]	Refrigerators Window AC units	
		Rotary screw	25 – 1,000 [87 – 3480]	Commercial Institutional	Variable speed possible.
		Rolling piston	0 – 3 [0 – .01]	Refrigerators Window AC units	
		Centrifugal compressors (turbocompressors)	50 – 1,000 [174 - 3480]	All buildings	Produce chilled water or brine for circulation to coils and terminal units.
Absorption Cycle			3 – 3,500 [.01 – 12 180]	All buildings	Inexpensive steam, gas, solar most common requirement.
Steam Jet Cycle			10 – 1,000 [34.8 – 3480]	Foods and chemicals Freeze drying	Special applications.
Air Cycle			10 – 150 [34.8 – 522]	Aircraft	Special applications.
Thermoelectric Cycle			5 – 50 [17.4 – 174]	Low-rise buildings	Thermoelectric solar panels generate electricity. Good in remote sunny locations.
Solar Cycle			50 – 100 [174 – 348]	Low-rise buildings	Hot water solar panels supply absorption units.

All numbers shown in brackets are metric equivalents.

R.S. Means Co., Inc., *HVAC: Design Criteria, Options, Selection*

Figure 6.29

COMPRESSOR
With each stroke of the piston, the pressure of the refrigerant gas is increased as it passes to the condenser. This causes the temperature of the refrigerant to be simultaneously raised and set significantly above the outdoor temperature where it can reject heat.

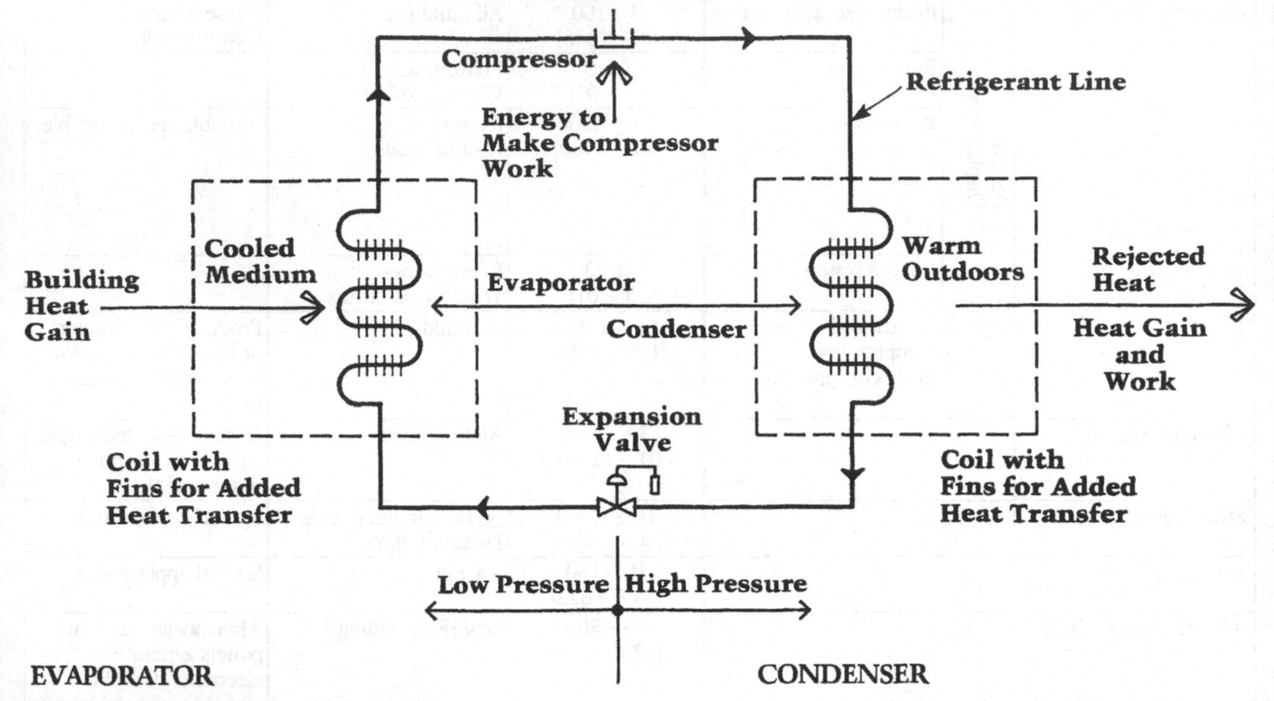

EVAPORATOR

In the evaporator the refrigerant, which has a low boiling point, changes from a liquid to a gas (evaporates, boils). This change of phase from liquid to gas causes it to absorb heat from the medium to be cooled. The gas then carries the absorbed heat away from the medium and to the compressor.

CONDENSER

The condenser receives the hot gases from the compressor and cools them with the cooler outdoor air. This causes the gases to liquify or condense, some heat to be absorbed from the medium, and the compressor to be rejected to the outdoors.

EXPANSION VALVE
The hot liquid refrigerant from the condenser, which is at a high pressure level, flows through the orifice of the expansion valve. The orifice causes some of the hot liquid refrigerant to turn to gas at the lower pressure of the evaporator side. This "misting", or boiling, of some of the refrigerant lowers the temperature and pressure of the remaining liquid.

The Compression Refrigeration Cycle

R.S. Means Co., Inc. *HVAC: Design Criteria, Options, Selection*

Figure 6.30

The building heat gain is absorbed by the evaporator at no operating cost other than that used by the fan evaporator, which, while it may be considerable, is still less than that used by the compressor. It is the compressor that uses the most significant amount of energy and costs the most money to run. As previously explained, the effectiveness of a cooling system is measured by the coefficient of performance. Another term used in cooling is the **energy efficiency ratio** (E.E.R.). The energy efficiency ratio equals the heat gain in Btu/hour removed by the equipment, divided by the watts of energy consumed to cool the space.

$$E.E.R. = \frac{\text{heat gain removed by equipment (Btu/hr.)}}{\text{watts of energy consumed to cool the space}}$$

Since 1 watt equals 3.41 Btu/hour, the energy efficient ratio equals 3.41 times the value of the coefficient of performance.

1 watt = 3.421 Btu/hr., therefore

E.E.R. = 3.41 x C.O.P.

Energy efficient ratio values of 9 to 12 are typical, and a value of over 10.5 is considered good.

The type of compression system selected depends on the overall cooling load required, the type of facility, the cost of the system, and the way in which the requirements of each system can be accommodated. Compression cycle systems are either reciprocating compressors (positive displacement) or centrifugal compressors (turbo).

Because there are many criteria that establish system selection, more than one type of compressor may be adequate for a given application. When this occurs, the final selection is based on either the initial cost or the life cycle cost of the system.

Reciprocating compressors, or positive displacement compressors, are the most common type of cooling equipment, and are used almost exclusively for systems with up to 50 tons [180 kW] capacity. Positive displacement compressors "squeeze" the refrigerant by using a piston, vane, or screw to compress the gas.

The four different types of reciprocating compressors are listed below. Figure 6.31 illustrates examples of these types.

- Reciprocating piston
- Rotary vane
- Rotary screw
- Rolling piston

Because they are small, reciprocating compressor units are usually single packaged units, which are air-cooled and located on the roof or through a wall.

Chillers produce chilled water for distribution to terminal cooling units. They often contain multiple compressors and are able to provide multi-stage cooling.

Some reciprocating compressors are used in split systems, which means that they are divided into two parts. An evaporator, an expansion valve, and a supply air fan make up one-half of the unit, which is located in or near the space to be cooled. The compressor and air-cooled condenser (condensing unit) make up the other half of the unit, which is usually located outdoors away from the space, typically on the roof. Reciprocating compressors are best suited to low-rise office and residential buildings with many tenants, since each tenant can be held responsible for the individual split system serving the space. The practical distance between the evaporator/valve/fan and the compressor/condenser unit is approximately 50 feet [15.25 m], limiting pressure and temperature changes in the refrigerant lines that connect the two halves of the system.

Centrifugal compressors are used in larger installations, typically above 200 tons [720 kW], although some manufacturers have models for loads as low as 50 tons [180 kW]. The centrifugal compressors continually "squeeze" the refrigerant gases by forcing a continual stream of the gas through a fan-like device. This forces the gas into a high-pressure line, which simultaneously raises its temperature. Figure 6.32 illustrates a typical centrifugal chiller.

Type	Description	Comments
Reciprocating Piston Compressor	A cylinder, piston-type compressor with crankshaft and inlet and outlet valves utilizing the two-cycle method of intake and compression.	May be electric drive or powered by an internal combustion engine or a steam turbine. This is the most common compressor used in the 3 to 50 ton [.01 – 174 megawatt] range.
Rotating Vane Rotary Compressor	An off-center roller with two oscillating vanes attached rotates within the housing, compressing the gas and discharging through exhaust vane-type valves.	High volumetric efficiency. Electric drive. Most common use is 3 tons [.01 megawatts] and smaller.
Rotary Screw (Helical Rotary) Compressor	The single main helical rotor acts against and in conjunction with the two-gate rotors to compress the gas and discharges through vane-type valves.	Wide load range possible. Common in 25 to 1,000 ton [87 – 3480 megawatt] capacities. May be electric drive or powered by an internal combustion engine.
Rolling Piston-Type Rotary Compressor	The eccentric shaft and the rotating roller combine with a single oscillating vane (attached to the housing) in compressing the gas and discharging through vane-type valves.	High volumetric efficiency. Most common usage in 3 tons [.01 megawatts] and smaller. Electric drive.

All numbers shown in brackets are metric equivalents.

Compressors may be open-type or hermetic. In an open-type, the shaft connects the motor and compressor. The integrity is maintained through a shaft seal. In hermetic design, the motor and compressor are self-contained and are sealed within an air-tight, gas-cooled housing.

Rotary compressors maintain a continuous suction pressure, negating the need for inlet or suction valves.

Halocarbons and ammonia are the two common types of refrigerant used in these compressors.

Types of Positive Displacement Refrigeration Compressors

R.S. Means Co., Inc., *HVAC: Design Criteria, Options, Selection*

Figure 6.31

Absorption Cycle

The absorption cycle is typically used in buildings where there is both a high cooling load (over 200 tons [720 kW]) and a heat source that is inexpensive in the summer, such as solar or steam. Absorption cycle is the second most popular cooling method after compression. Because absorption machines use water and salt solutions and do not use refrigerants (specifically CFCs), they are environmentally more attractive than compressors.

The absorption cycle involves the principle that a salt or salt solution has an affinity for water or water vapor. A salt shaker that clogs in the summer when the moisture content of the air is high is a simple example of this. Thus, as a highly concentrated salt solution, such as anhydrous lithium bromide, is sprayed over a pool of water, some of the water is induced to vaporize and mix with the salt spray. This happens because chemically, the salt attracts water, and the concentrated salt spray displaces some of the normally occurring water vapor, creating a partial vacuum of water vapor above the pool of water. These factors combine to draw the water from a liquid to a vapor, which mixes with the salt spray. Whenever water changes from a liquid to a gas, approximately 1,000 Btus [300 W] of heat are required to vaporize each pound of water. As the water vaporizes, it takes the 1,000 Btus/pound [650 W/kg] from the liquid pool and energizes the water into vapor. The heat leaves the water pool, chilling it considerably to below 50°F [9 °C].

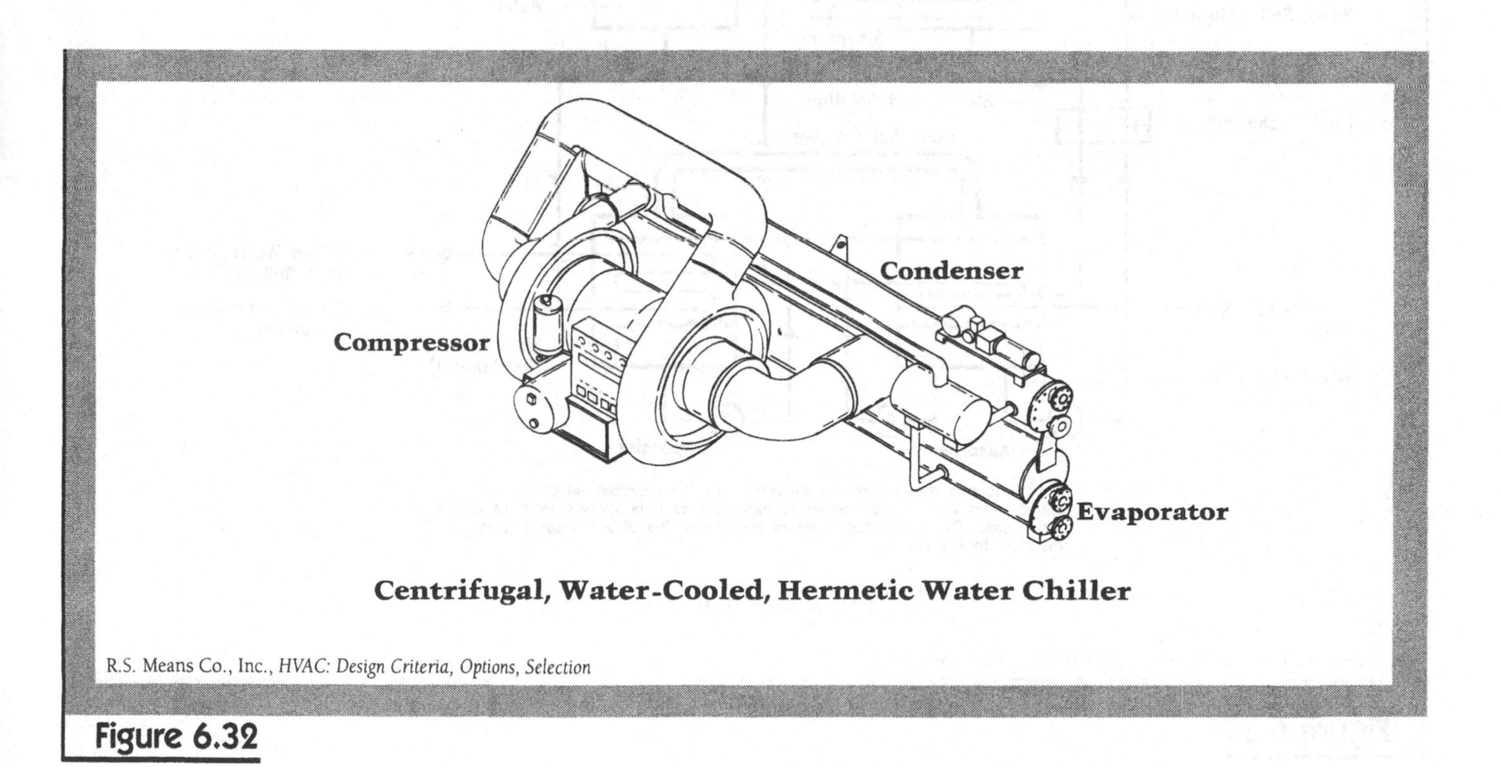

Centrifugal, Water-Cooled, Hermetic Water Chiller

R.S. Means Co., Inc., *HVAC: Design Criteria, Options, Selection*

Figure 6.32

Absorption systems produce chilled water by a method different than compression refrigeration machines, but within the same overall tonnage as centrifugal chillers. Depending on relative installation costs, efficiencies, structural loads, and operating costs, compression or absorption systems can be interchanged while other piping and HVAC systems are left intact.

Refer to the schematic of an absorption chiller in Figure 6.33. The salt solution or absorbent (lithium bromide) is sprayed in one chamber—the absorber—and the refrigerant (water) is sprayed in another chamber—the evaporator. Both elements are sprayed to increase their surface area, thereby increasing the potential for attraction. As the refrigerant is attracted to the absorbent chamber (via a connecting tube), the refrigerant changes from a liquid to a vapor and mixes with the absorbent, diluting it. This vaporization achieves two things.

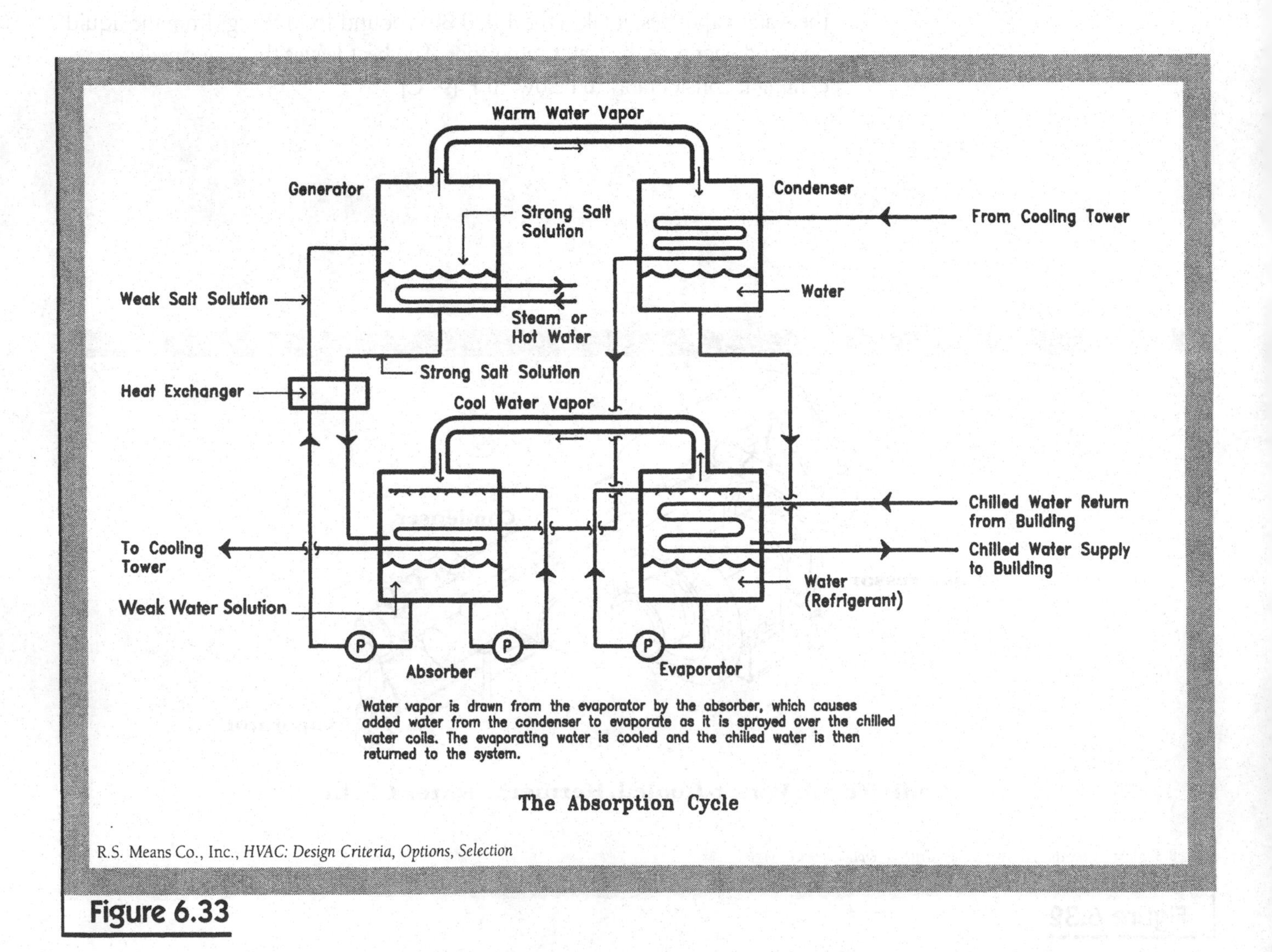

The Absorption Cycle

R.S. Means Co., Inc., *HVAC: Design Criteria, Options, Selection*

Figure 6.33

First, by drawing some refrigerant away from the refrigerant chamber, a partial vacuum is created, which draws more refrigerant into the vapor stage by lowering its boiling point. Second, because energy is required to convert a liquid to a vapor, and the vapor carries this potential energy, heat is drawn from the refrigerant that remains, and it is chilled considerably.

Water, for example, requires approximately 1,000 Btus [650 W/kg] per pound to convert from a liquid to a vapor. The chilled water return from the building passes through the evaporator chamber to be cooled, and supplies the building with chilled water. In the absorber, the water vapor (refrigerant) releases the heat of vaporization as it is converted back into a liquid and absorbed by the weak salt solution. This heat tends to warm up the absorber chamber but is carried away by the cooling tower water, which dumps this heat in the rooftop cooling tower. The weak salt solution in the absorber is pumped up by the generator, which "regenerates" the strong salt solution by separating the water and salt for recirculation. Within the generator, the weak salt solution is heated in order to separate the water (volatile refrigerant) from the salt (absorbent). The concentrated absorbent returns to the absorber, where it may be reused and may capture more water vapor from the evaporator. The heat exchanger between the absorber and the generator recaptures some of the heat for the weak absorbent solution, and removes some heat from the strong absorbent on its way back to the absorber. The warm water vapor travels from the generator to the condenser, where it is condensed. The heat is removed via the cooling tower water. The condensed refrigerant then returns to the evaporator to start the cycle again.

Absorption equipment is large, heavy, and requires a heat source to provide the energy for the system. It is most advantageous when the summer cost of the heat source is low. Absorption systems have a natural potential for use with solar energy, because the heat from the sun increases as the need for cooling and the availability of heat to drive the absorption system increase.

There are several elements that must be carefully considered in the design and construction of an absorption chiller.

- The absorbent should have a strong affinity to the refrigerant.
- The refrigerant should be sufficiently more volatile than the absorbent so that a minimal amount of heat can separate them in the generator.
- Corrosive substances should be matched by materials that are corrosion-resistant.
- Because CFCs are not used in absorbent chillers, the negative environmental impact of these machines is limited. The designer should note, however, that some absorbent-refrigerant pairs are not permitted in residential applications because of the toxic or volatile nature of the chemicals used.
- Since there is no solid phase in the absorption cycle, some absorbents have the potential to crystallize, clogging the machine or upsetting the delicate balance between the chemicals involved. Many machines on the market have integral safety features that can both detect and

inhibit the formation of crystals in the cycle; the designer should be careful to include such features to prevent blockages in the cycle.
- Absorption equipment is large and heavy, requiring special space and structural considerations.
- The primary power requirement for absorption equipment, aside from electricity to run the pumps, is a heat source. The heat sources that are most commonly used are steam, gas, and solar; each potential source should be evaluated.

The selection of the refrigeration cycle usually establishes the principal component parts. The parts are normally furnished by the equipment manufacturer, either as a package unit or in separate, but compatible, component parts.

Condensers

After the type of refrigeration cycle and compressor (where applicable) are chosen, the condenser must be selected. There are three principal types of condensers: air-cooled condensers, water-cooled condensers, and evaporative condensers (see Figure 6.34).

Air-cooled condensers are typically used for systems where refrigerant is the medium in the equipment. The refrigerant coils are placed directly in the outdoor air. Mounted outdoors on a roof or through a window, air-cooled condensers are suitable for loads up to 50 tons [180 kW]. Air around the coils is required for circulation to reject the heat. Fans are usually provided to increase performance. Household refrigerators, common window air conditioners, simple package rooftop units, and split systems all use air-cooled condensers.

Closed-circuit condenser water coolers are similar in appearance and operation to air-cooled condensers. These units cool condenser water or a water glycol solution rather than a refrigerant gas. This type of water cooler is used for winter operation in climates subject to below-freezing temperatures.

Water-cooled condensers cool the refrigerant coils with water. By using a shell and tube heat exchanger, water from sources such as cooling towers (described below), or river, lake, spray pond, or ground water can be used.

Evaporative condensers utilize water that is sprayed across refrigerant coils to boost the cooling effect by evaporative cooling. They are similar to air-cooled condensers in that a fan is used to remove the heat and vapor from the surface of the coils, thus cooling the refrigerant.

A special type of condenser arrangement that is very common in large applications involves a **cooling tower** for condenser heat removal. The condenser water at the water-cooled condenser is sent to the cooling tower, where it is cooled and returned to the condenser. Cooling towers operate similarly to evaporative condensers, in that sprayed water and a fan are used to remove heat.

Cooling towers operate with either a closed or an open loop. In an open loop, the water that cools the condenser is sprayed in a large chamber (open to the atmosphere), where air is used to remove the heat. In a closed loop, the water

Selection Criteria for Condensers

Type	Tonnage [megawatt] Range	Common Uses	Location	Special Requirements
Air-Cooled Condenser 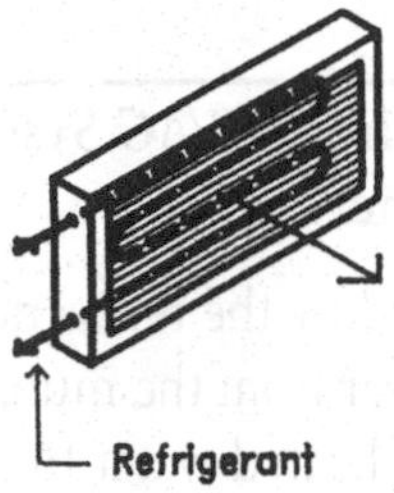 Refrigerant	0–50 [0–174]	Package units using refrigerant only Split system	Rooftop Through wall	Limit run and pipe for split systems.
Water-Cooled Condenser 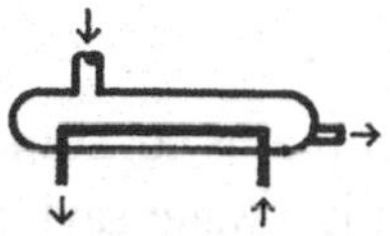	3–150 [.01–522]	Small water cooled units using domestic water (where permitted) or water from ponds, lakes, rivers, or groundwater	Rooftop Ground mount	Cooling tower or other water source is required in urban setting.
Evaporative Condenser 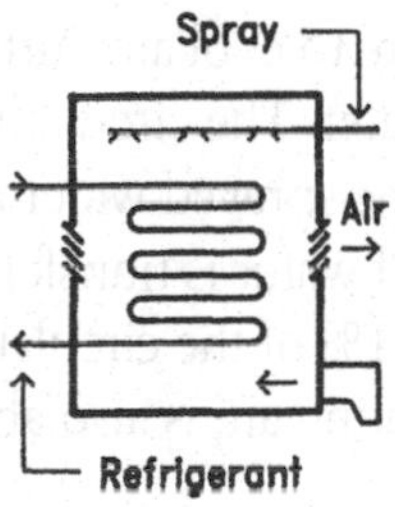 Spray Air Refrigerant	10-250 [34.8–870]	Intermediate facilities	Rooftop Ground mount	Freeze protection.
Cooling Tower 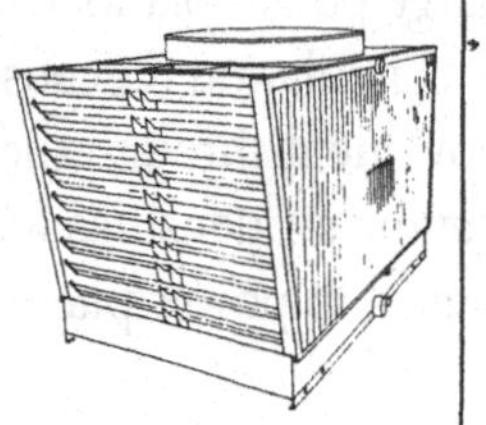	50–1,000 [174–3430]	High-rise buildings	Rooftop Ground mount	Cools water for water-cooled condenser.

R.S. Means Co., Inc., *HVAC: Design Criteria, Options, Selection*

Figure 6.34

from the condenser is enclosed in a coil and is sprayed with water from a separate loop. Cooling towers are very efficient, and can bring the temperature of the water to within 5–10°F [2.75–5.5 °C] of the ambient dry bulb temperature. Common operating temperatures for condenser water are 95°F [35 °C] to the cooling tower and 85°F [29 °C] to the condenser.

The following section is reprinted by permission from *HVAC Systems Evaluation* by Harold R. Colen, R.S. Means Co., Inc.

Refrigeration Condensing Systems

In the early days of air-conditioning, the heat generated by the refrigeration cycle was removed from the condenser by passing water from the municipal water system over the condenser and discharging that heated water to the sewer system as waste. Today, this method of water cooling is either prohibitive costwise or restricted by local codes or ordinances in order to conserve water. Indeed, the demand for potable water has exceeded supply in many localities, to the point that limitations on its availability have curtailed new construction.

Cooling Towers

Cooling towers reduce the amount of the potable water that is consumed to about 5% of the total circulated water. A cooling tower is a water saving device that uses heat and mass transfer to cool water. Figure 6.35 shows that the water to be cooled is distributed in the cooling tower by spray nozzles, to expose a very large water surface to the atmospheric air brought into contact with the spray water.

A fan circulates relatively cool and dry atmospheric air to the water. Some of the latent heat is taken from the water by this exposure to cool air. Part of the water is evaporated, thereby cooling the remaining water. The circulating air accepts the transfer of the heat of vaporization from the sprayed water and discharges it into the atmosphere. Some of the mass of water is transferred to the air stream. The evaporation rate is approximately 1% of the circulating water rate. Drift, the water mass transfer to the circulating air, is also about 1% of the water flow.

Cooling Tower Types

Natural Draft

Natural draft cooling towers are used primarily in large power plants where the savings of fan energy and the elimination of plume recirculation (because of the tower height) justify the high initial cost. Air movement through the tower is propelled by the difference in the entering and leaving air densities (chimney effect). Hyperbolic natural draft cooling towers are used in power plants.

Mechanical Draft

More mechanical draft cooling towers are in use than any other type. Forced draft towers push the air through the tower, whereas induced draft towers draw the air through the tower. Propeller fans are used in induced draft towers when the air pressure loss through the tower is low. Figure 6.36 shows a forced draft and an induced draft cooling tower.

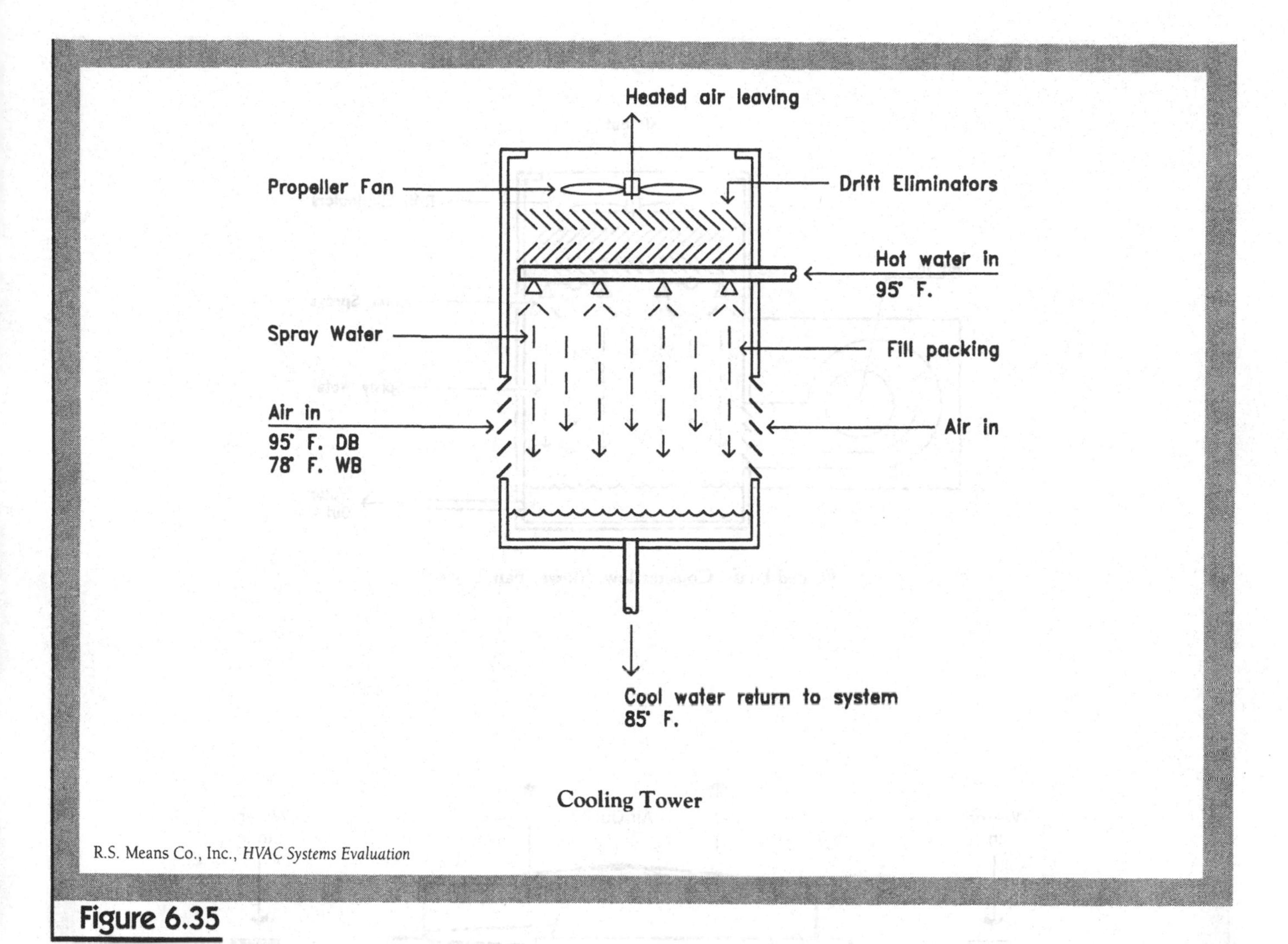

Cooling Tower

R.S. Means Co., Inc., *HVAC Systems Evaluation*

Figure 6.35

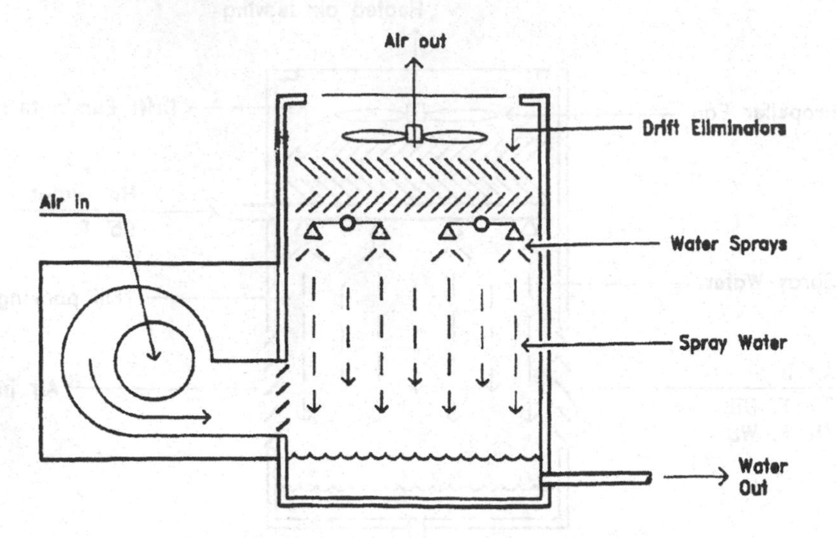

Forced Draft, Counterflow, Blower Fan Tower

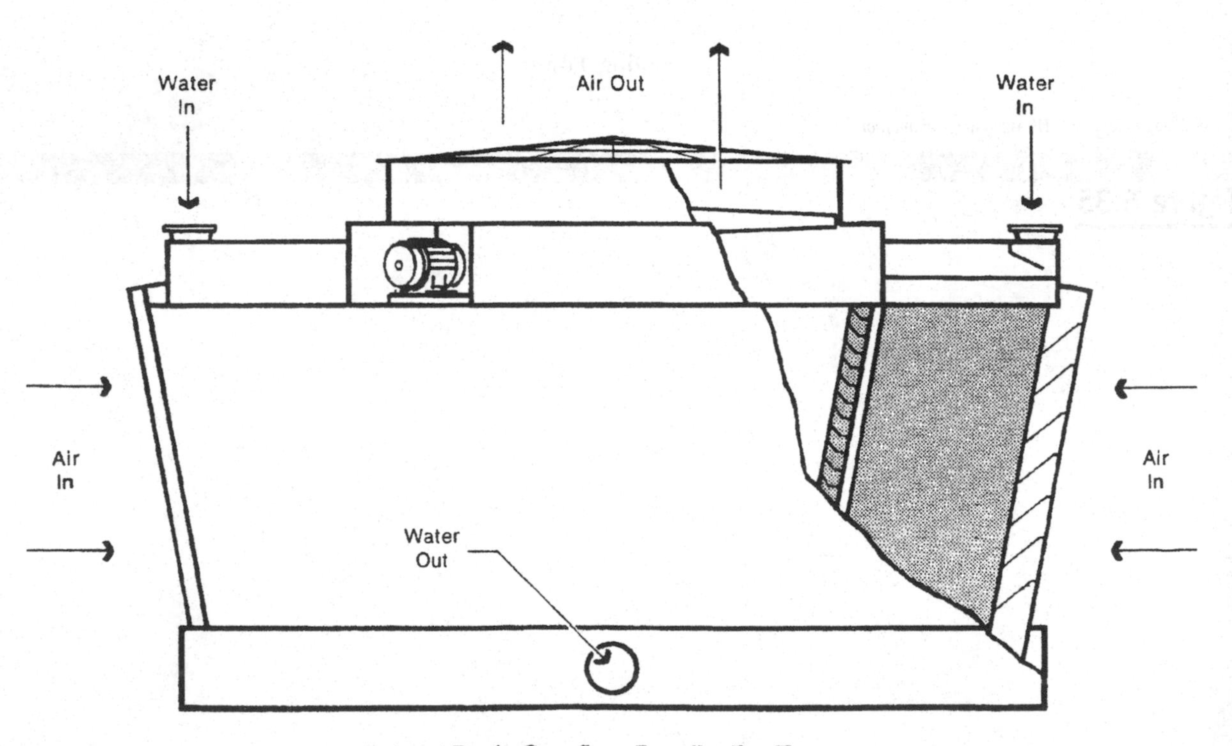

Induced Draft, Crossflow, Propeller Fan Tower

Courtesy of the Marley Cooling Tower Company

Figure 6.36

Range and Approach: The cooling tower "range" is the temperature difference between the entering and leaving water (ECWT and LCWT). Many designs are based on water entering the tower at 95°F and leaving at 85°F, for a 10°F range. The "approach" is how close the leaving water temperature is to the entering air wet bulb temperature. This is dependent on the design wet bulb temperature for the region where the tower is to be installed. If the leaving water is 85°F and the entering wet bulb is 78°F, the approach will be 7°F. Figure 6.37 is a diagram of range and approach.

Economic Cooling Tower Selection

Figure 6.38 shows that decreasing the approach from 15°F to 7.5°F doubles the size of the tower. It may be possible to use a slightly higher cooling tower cold water temperature, during peak load conditions, to reduce the cooling tower size.

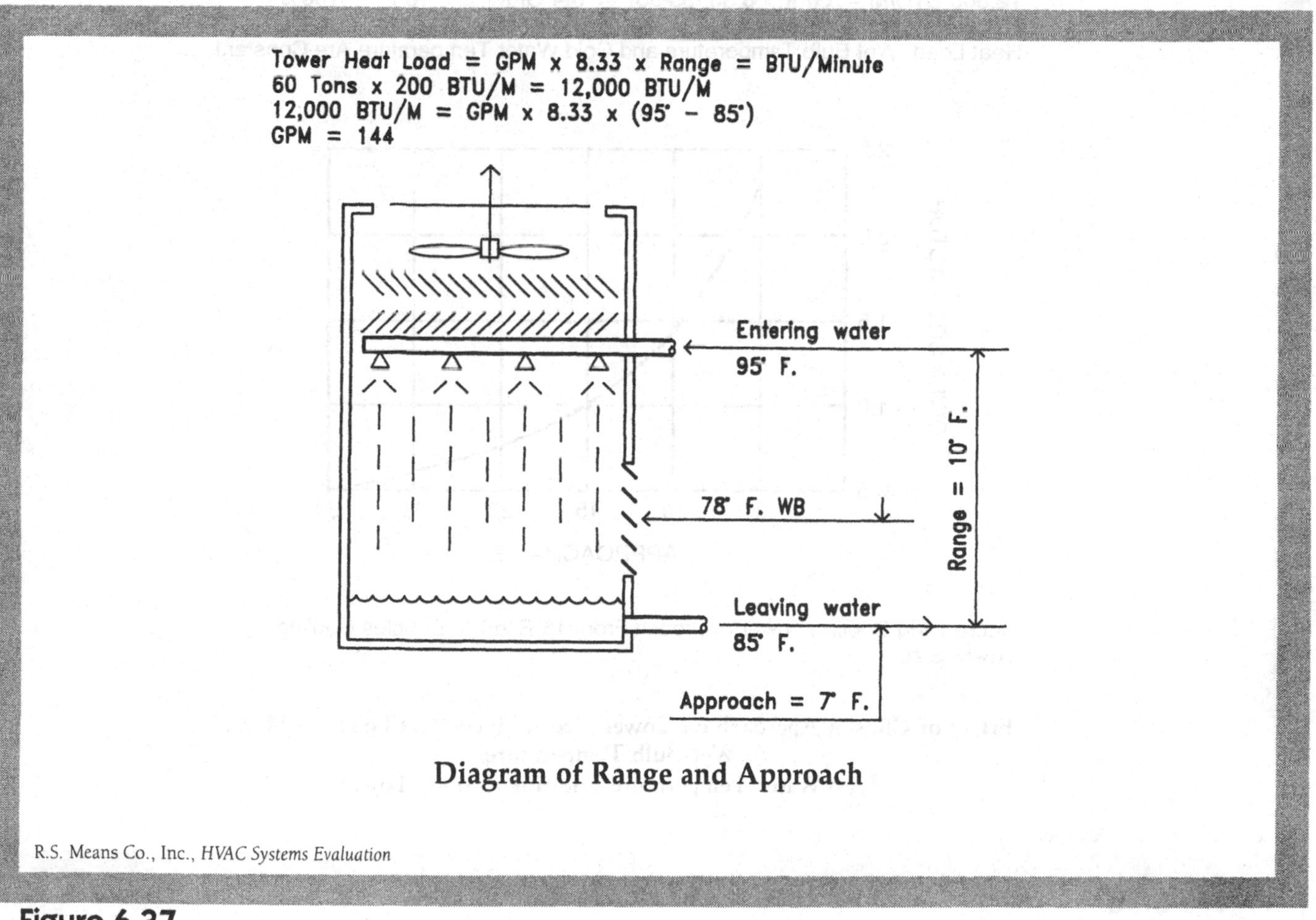

Diagram of Range and Approach

R.S. Means Co., Inc., *HVAC Systems Evaluation*

Figure 6.37

Courtesy of the Marley Cooling Tower Company

Before making a cooling tower selection, evaluate the cooling tower performance with a lower gpm than would be used in a "normal" selection. Cooling towers for refrigeration chillers are usually selected for a water flow rate of 3 gpm/ton. Figure 6.39 shows that for the same heat transfer (capacity), wet bulb temperature, and tower leaving water temperature, tower size can be reduced 10% by:

- decreasing the gpm from 144 to 118, or
- increasing the range by 22%.

Decreasing the gpm from 144 to 118 and increasing the range by 50% will reduce the required tower size by 20%. If 85°F water leaves the tower, circulating less water to the tower will raise the tower entering water temperature to 100°F instead of 95°F. Reducing the (design) gpm through the tower will reduce the tower size to 80% of the 95°F inlet water tower size. The

BTU/M = 6 PM x 8.33 x Range — Range

12,000 BTU/M = GPM x 8.33 (95-85) = 144 GPM — 10°F — 100%

12,000 BTU/M = 118 x 8.33 x (T_O-85) = T_O = 97.2° F — 12.2°F — 122%

Heat Load, Wet Bulb Temperature and Cold Water Temperature Are Constant

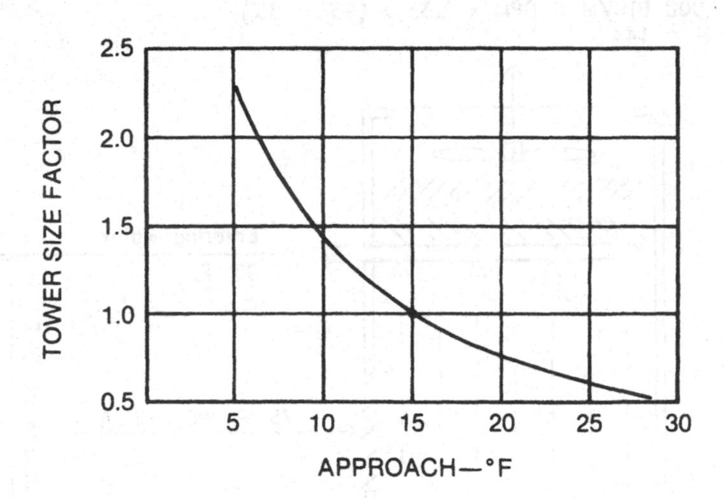

Decreasing Cooling Tower Approach From 15°F to 7°F Doubles Cooling Tower Size

Effect of Chosen Approach on Tower Size at Fixed Heat Load, GPM, and Wet-Bulb Temperature

T_O = Water Temperature Entering Cooling Tower

Courtesy of the Marley Cooling Tower Company

Figure 6.38

installation cost is further reduced with the use of 3" pipe size instead of 4" pipe size. The pump size is one-third smaller, and the pump energy required will be one-third less, forever. The smaller cooling tower also means a permanently lower requirement for cooling tower fan energy.

If the efficiency of the towers is the same, the size requirement of a cooling tower depends on the entering wet bulb temperature for the approach, the leaving water temperature, and the range. The range is determined by the refrigeration machine selection. The cooling capacity and efficiency of the refrigeration machines vary inversely with the head pressure. As the head pressure increases, the cooling capacity decreases and the kWs increase.

Figure 6.40 shows that the condenser performance is dependent on the gpm, the difference between the condensing temperature and the entering condenser water temperature (ECWT), and the condenser size (heat exchange surface).

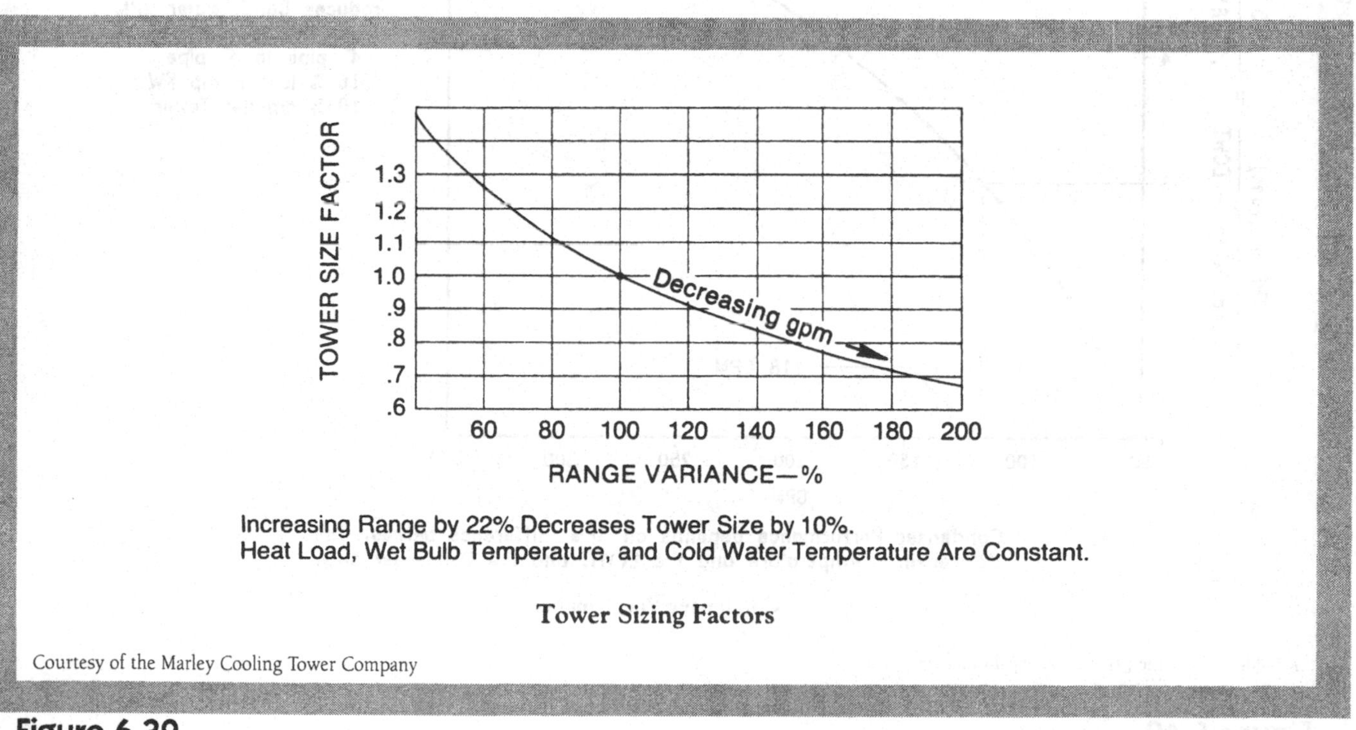

Tower Sizing Factors

Courtesy of the Marley Cooling Tower Company

Figure 6.39

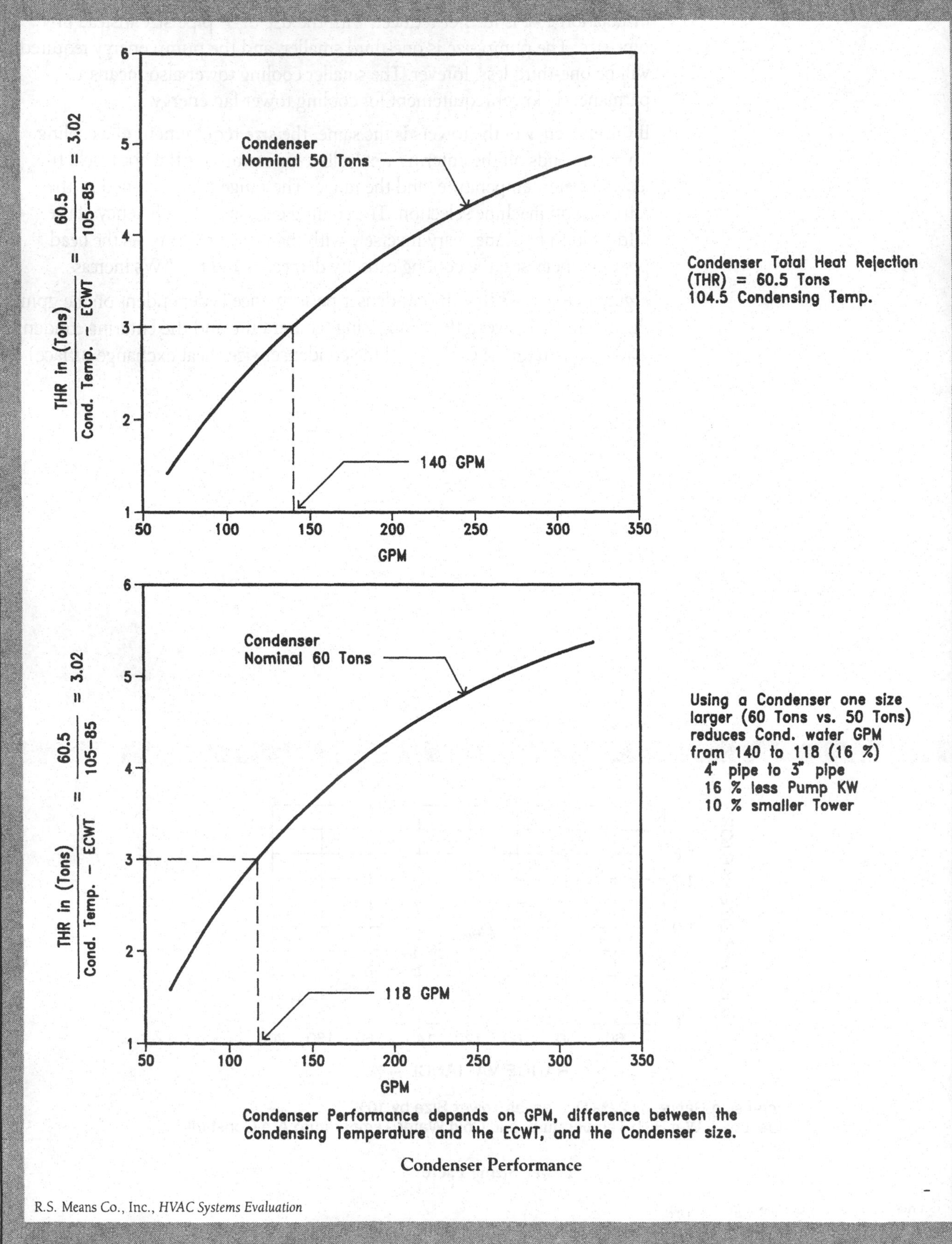

Condenser Performance

R.S. Means Co., Inc., *HVAC Systems Evaluation*

Figure 6.40

Figure 6.40 indicates that 60.5 tons of heat rejection, with 85°F entering condenser water, can be handled with 145 gpm through a nominal 50-ton condenser, and 118 gpm through a nominal 60-ton condenser. The total heat rejection (THR) remains the same. Only the gpm and the water temperature rise (range) change.

btu/minute = gpm x 8.33 lbs./gal. x range

12,000 btu/m = 144 gpm x 8.33 lbs./gal. x 10°F

12,000 btu/m = 118 gpm x 8.33 lbs./gal. x (T_o-85°F)

T_o = 97.2°F

By reducing the condenser water circulation from 144°F to 118°F, the water temperature entering the tower is increased from 95°F to 97.2°F. Under these conditions, the cooling tower size can be reduced by 10%. The reduced gpm will lower the operating cost, and could result in an additional installed cost savings, if pipe and condenser water pump sizes can be reduced.

Interference

Interference (illustrated in Figure 6.41) occurs when a portion of the saturated tower effluent contaminates the air intake of the downwind tower. Cooling towers should be installed so that the prevailing winds will carry the effluent away from the towers. If physical restraints cannot arrange cross-wind airflow, the tower selection entering air wet bulb temperature should be elevated by as much as three degrees.

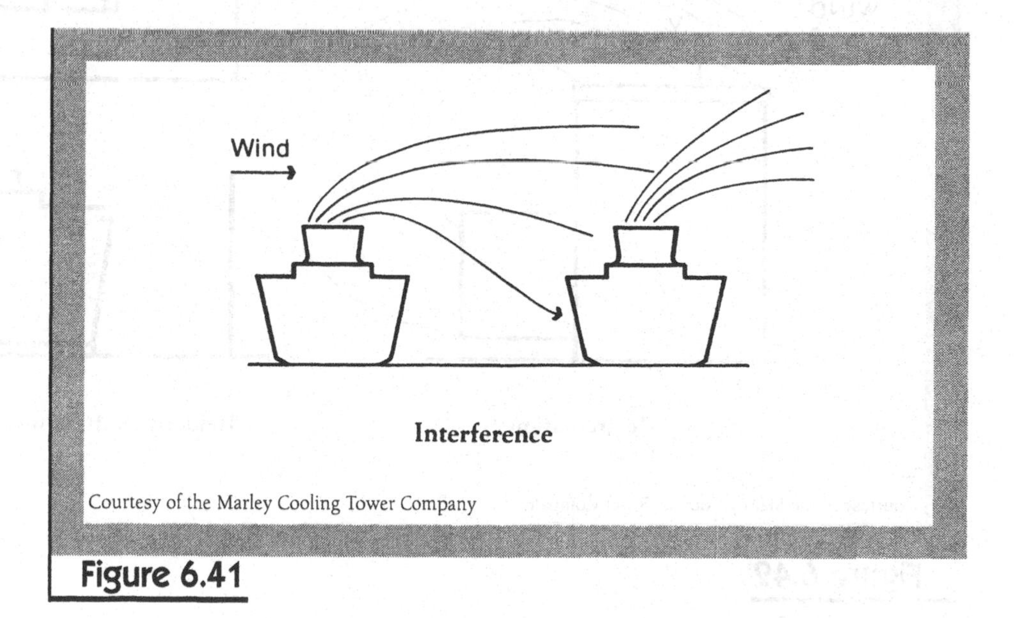

Figure 6.41

Walls or Enclosures

Restrictions to the air entering the tower can impede airflow through the tower and reduce the tower's performance. The distance from the wall to the induced draft tower (Figure 6.42) should be no less than the height of the tower. Induced draft tower fans cannot operate against much static pressure, so the velocity entering the tower should remain low.

Forced draft towers have centrifugal fans to overcome the tower-high internal static pressure. A slightly higher static pressure imposed by a restricted air intake is not as significant as it is with an induced draft tower. The lower air discharge velocity of the forced draft tower makes recirculation of effluent air more likely than the high velocity of the induced draft cooling tower. The wall proximity to a forced draft tower should be no less than twice the tower width, to minimize recirculation. This arrangement is shown in Figure 6.43.

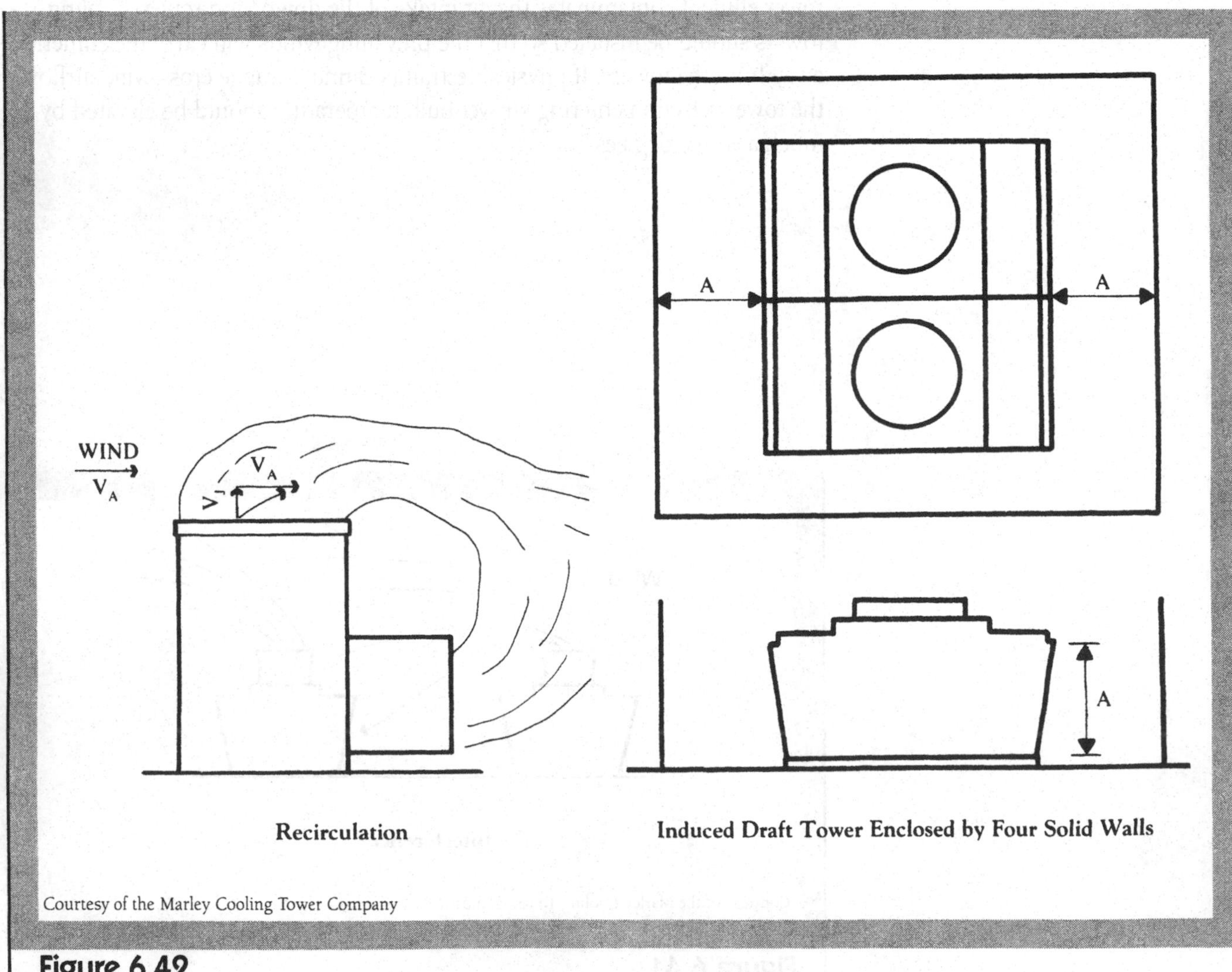

Figure 6.42

Where critical recirculation cannot be avoided, the design wet bulb temperature for induced draft towers should be increased 1°F, and 2°F for forced draft towers.

Free Cooling

"Free cooling" with cooling towers is accomplished by using the cold condenser water produced by the cooling tower during low ambient wet bulb conditions, to produce chilled water. When the wet bulb temperature is sufficiently reduced, the cold water leaving the cooling tower can reduce or temporarily eliminate the use of mechanical refrigeration, to produce the

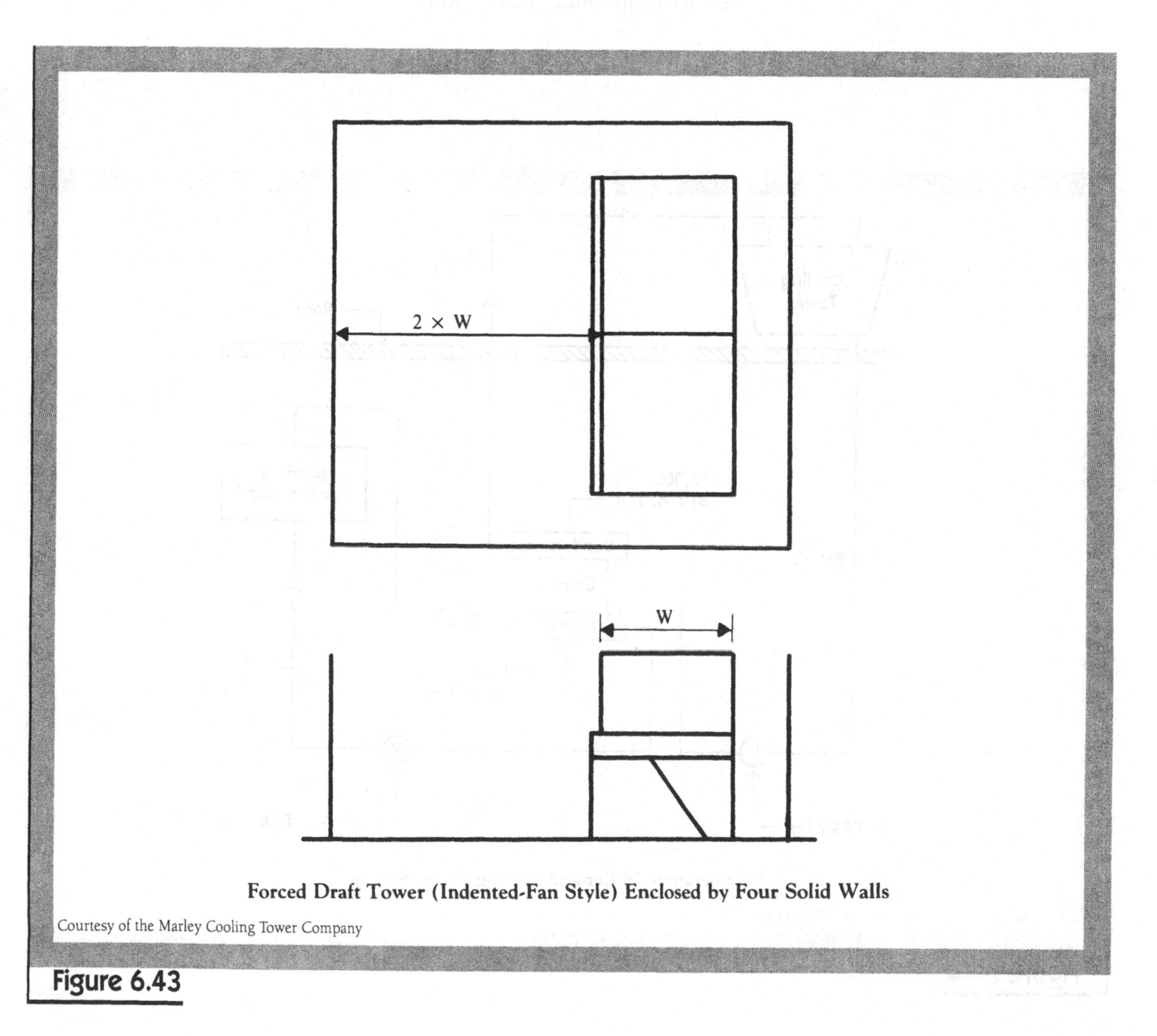

Forced Draft Tower (Indented-Fan Style) Enclosed by Four Solid Walls

Courtesy of the Marley Cooling Tower Company

Figure 6.43

chilled water. In northern latitudes, free cooling hours could be available 75% of the total yearly operating hours, but less than 20% of southern climate operating hours. Figure 6.44 shows a typical refrigeration chiller and cooling tower system.

Figure 6.45 shows the possible condenser water temperatures that can be produced by cooling towers at various wet bulb temperatures. The cooling load is usually lower when wet bulb temperatures are depressed. Process loads, such as computer rooms, are an exception. At 50% of peak cooling loads, 45°F condenser water can be produced at a 35°F wet bulb temperature. This will permit chilled water cooling without using the refrigeration chiller. At 50°F web bulb temperature, the tower can produce 57°F at half load. A 57°F chilled water temperature can usually handle the cooling needs under part load conditions. For example, there are over 4,700 hours in Kansas City when the wet bulb temperature is below 50°F.

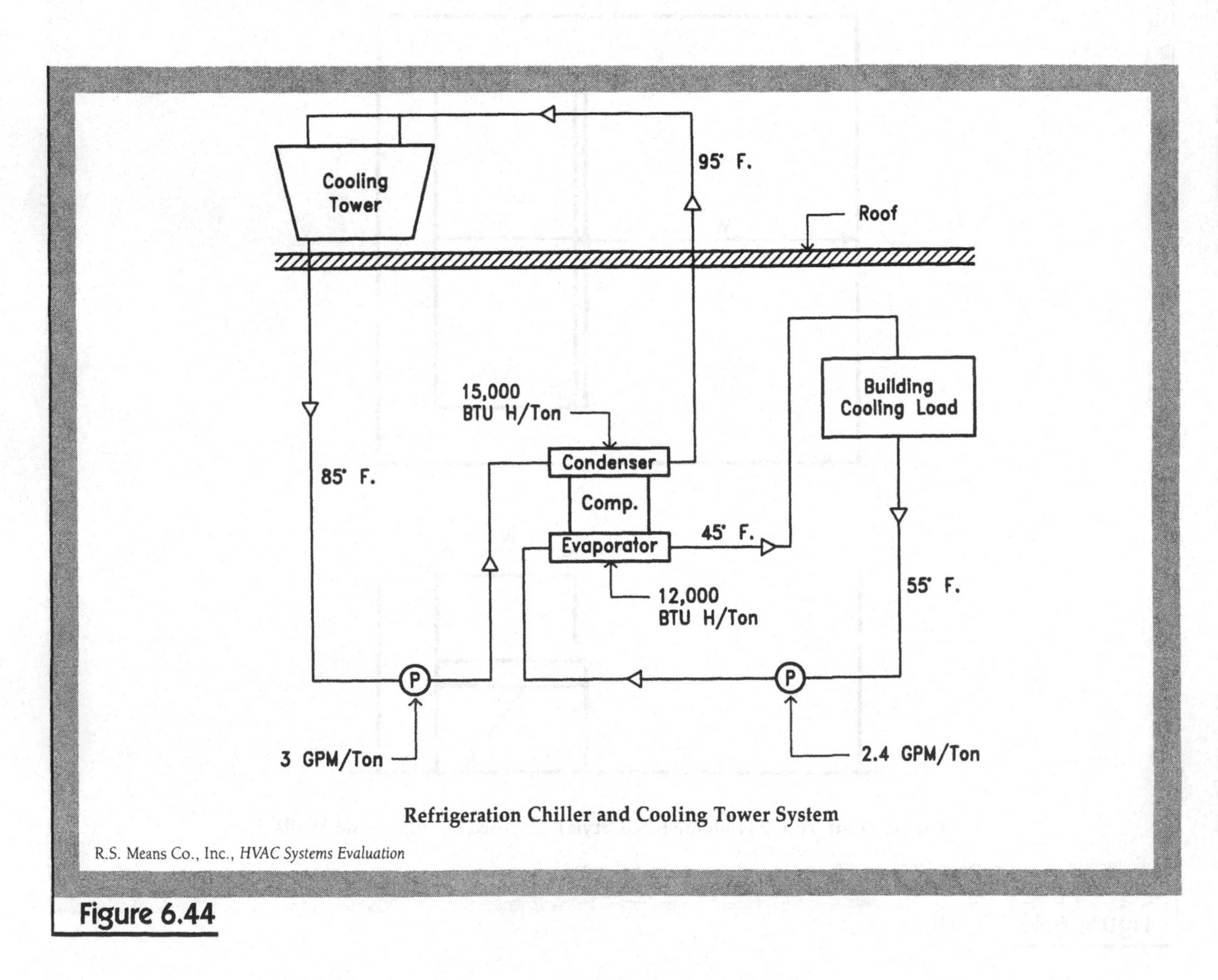

Refrigeration Chiller and Cooling Tower System

R.S. Means Co., Inc., *HVAC Systems Evaluation*

Figure 6.44

Figure 6.46 shows a direct cooling system in which cold condenser water is filtered and introduced directly into the chilled water system.

Direct systems are efficient because a heat exchange process is eliminated with cold condenser water circulating through the chiller. Cooling tower water is dirty, however, and the water strainer must be relied upon to remove contaminants that would otherwise interfere with the chiller heat exchanger. Chillers and cooling coil performance are based on a 0.0005 fouling factor. Any increase in the fouling factor could wreak havoc on the cooling performance, and create tremendous problems for operating personnel.

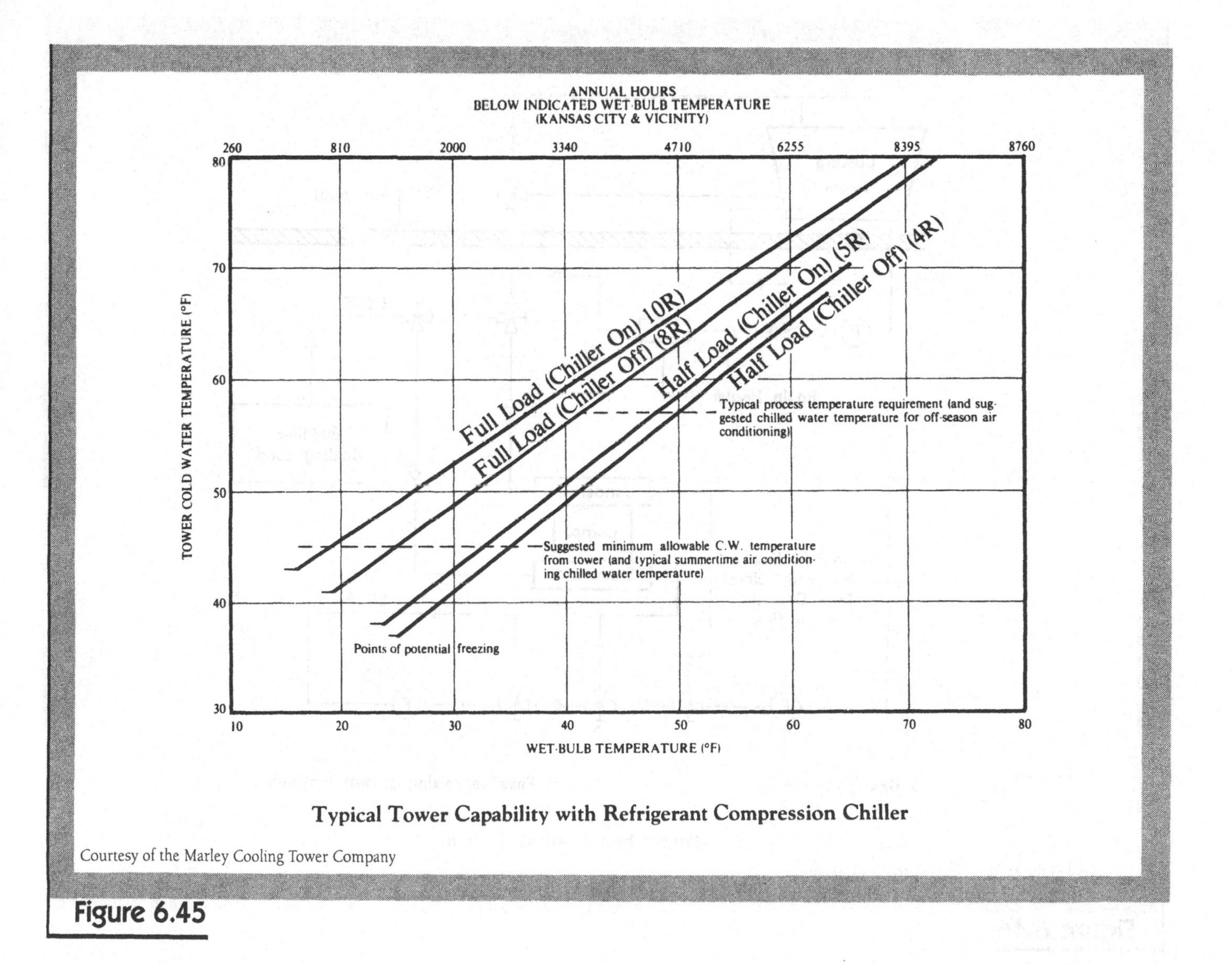

Typical Tower Capability with Refrigerant Compression Chiller

Courtesy of the Marley Cooling Tower Company

Figure 6.45

Indirect Free Cooling System

Contamination of the chilled water system can be eliminated by installing a heat exchanger piped as per Figure 6.47, to separate the condenser and chilled water circuits. Heat exchangers are available that will produce a 2°F to 4°F temperature differential across the heat exchanger. When the compressor is not operating, the cooling tower does not have to remove the compressor heat (heat of compression), so it is able to lower the range from 10°F to 8°F. The 8°F rise in condenser water temperature enables the heat exchanger to produce a substantial 10°F temperature rise on the cooling coil side.

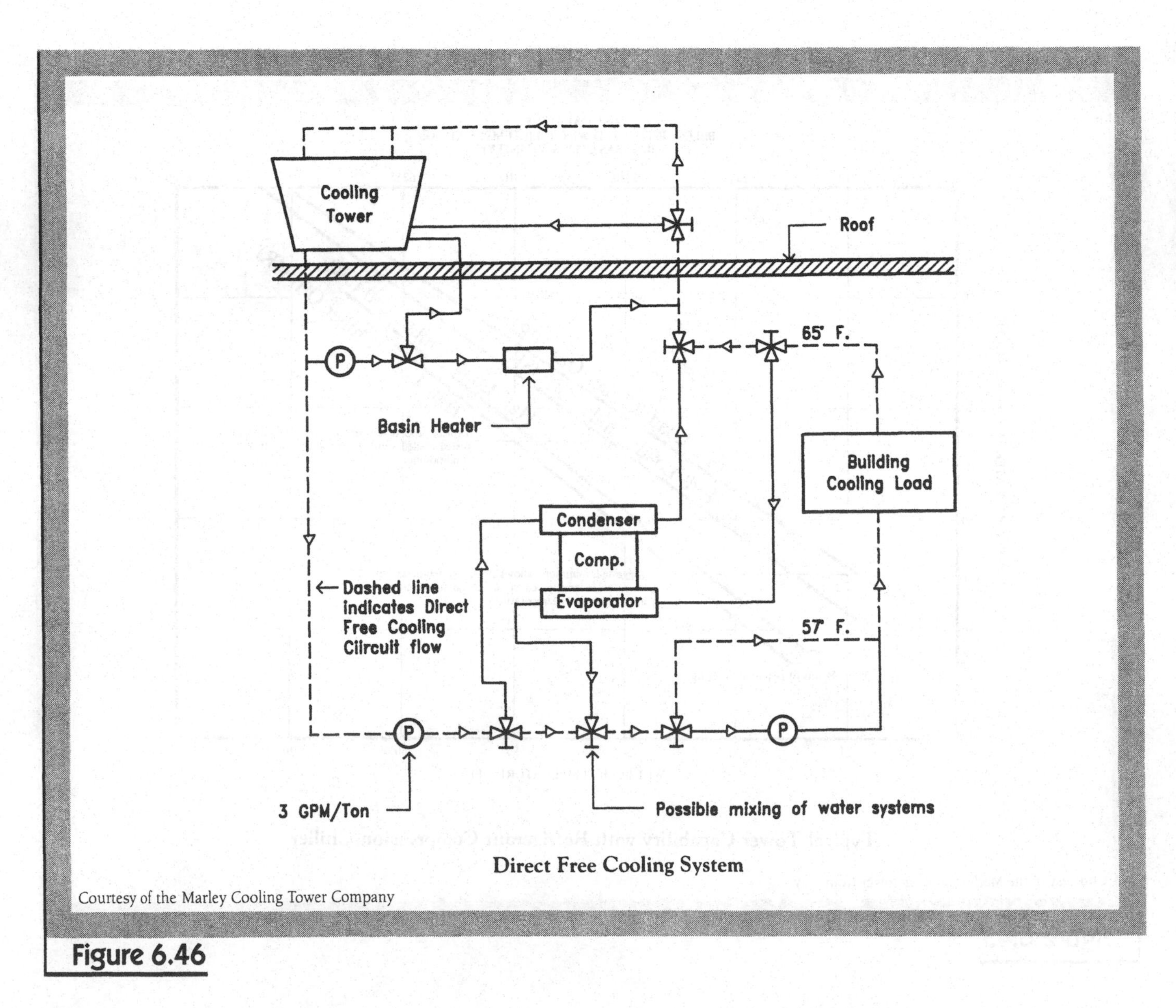

Direct Free Cooling System

Courtesy of the Marley Cooling Tower Company

Figure 6.46

Advantages

- **Lower condenser water supply temperature** — Condenser water supply temperatures within 5°F of the wet bulb temperature are possible. Condenser water supply temperatures of 85°F are standard with 78°F wet bulb. A refrigeration condensing temperature of 105°F is common with 85°F tower water, resulting in 0.886 kW/ton performance. Air-cooled systems operate at a 120°F condensing temperature with 95°F air on the condenser, and consume 1.10 kW/ton.
- **Usually lowest installed cost** — Air-cooled systems under 40 tons with remote air-cooled condensers usually cost about 30% more to install than water-cooled systems with cooling towers. The installed cost of air-cooled systems over 40 tons with remote air-cooled condensers is approximately 40% higher than water-cooled systems with cooling towers. The installed cost of packaged air-cooled chillers is only very slightly (1–2%) higher than water-cooled systems with cooling towers.

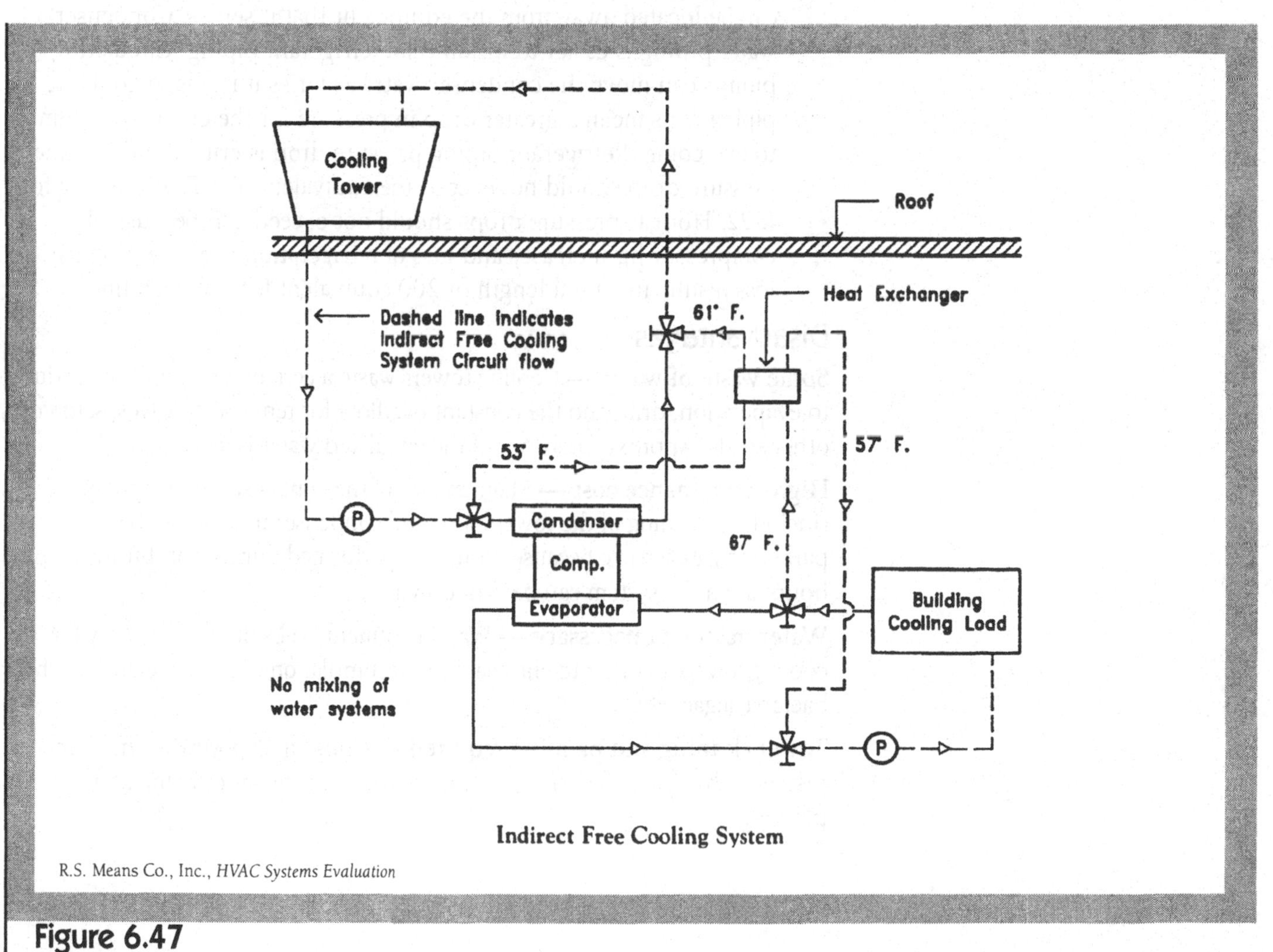

Indirect Free Cooling System

R.S. Means Co., Inc., *HVAC Systems Evaluation*

Figure 6.47

- **Possible lowest operating cost** — Water-cooled reciprocating chillers are 25% more efficient than air-cooled chillers (0.886 to 1.1 kW/ton). Water-cooled centrifugal chillers are 69% more efficient than reciprocating air-cooled chillers at peak load. In southern climates, even with condenser water system treatment costs, the operating cost of water-cooled systems is still less than that of air-cooled systems. The operating cost of air-cooled systems in northern climates can be less than water-cooled systems.
- **Availability of free cooling** — Free cooling, described earlier in this chapter, is relatively simple to implement with cooling tower systems and closed circuit coolers. Air-cooled systems cannot provide the benefit of free cooling when outside air for cooling is not utilized.
- **Capable of handling very large chillers** — Cooling towers can service all sizes of refrigeration systems. They are available to handle flow rates up to half a million gpm or more. Packaged air-cooled reciprocating chillers are made up to 200 tons. Although air-cooled centrifugal chillers are available, they are rarely used because of the high initial and operating costs.
- **Heat rejection equipment can be a considerable distance from the chiller** — There is almost no limit to the distance that a cooling tower can be located away from the equipment that it serves. Condenser water piping is easier to install than refrigerant piping. Circulating pumps can move the condenser water as far as it needs to go. Long piping runs mean a greater drop in pressure for the circulating pump to overcome. Refrigerant piping pressure drop is critical. Liquid line pressure drop should not exceed the equivalent of 2°F, or 5.8 psig for R-22. Hot gas pressure drops should not exceed 2°F, because of compressor hp increases and loss of refrigeration capacity. The 2°F loss results in a total length of 200 equivalent feet for each line.

Disadvantages

Some waste of water — Cooling towers waste a certain amount of water due to evaporation, drift, and the constant overflow for removal of leaves, scum or other solids. Approximately 5% of the circulated water is lost.

High maintenance cost — Maintenance of the condenser water system (including cleaning of the tower basin and condenser tubing) can be particularly expensive because it must be performed during non-business hours when the system can be shut down.

Water treatment necessary — Water treatment is absolutely necessary for cooling towers in order to eliminate the accumulation of contaminants, such as bacteria, algae, etc.

Tower cleaning and painting required — Constant exposure to the elements results in the threat of corrosion, and the need for regular cleaning and painting.

Must be winterized or drained to prevent freezing

Refilling required — The basin and piping must be refilled with water when cooling is needed again in spring (after they have been drained for winter weather to avoid freezing).

Regular inspection and maintenance required — Fill, eliminators, float valves, fans, basin headers, spray trees, etc., all require regular inspection and repair (maintenance).

Moisture carryover (drift) — Water carryover from the tower air discharge can be damaging to property, such as automobiles, windows, and building materials that are contacted by the water.

Circulating pump required

Greater weight and structural supports required

Exhaust plume during cool weather (fog)

The following section is reprinted with permission from *HVAC Systems Evaluation* by Harold R. Colen, P.E.

Heat Recovery Systems

Heat recovery is the reclamation and use of heat energy that otherwise would be wasted. The waste heat is substituted for a portion of the heating or cooling energy that would normally be required for heating the building, outside ventilation, makeup air, or domestic hot water. Cooling energy can also be saved by transferring a portion of the warm outside ventilation air to the cool building exhaust air. It is important to be sure that the energy expended in the reclamation process does not exceed the value of the reclaimed energy over a period of time.

This section describes and compares four different types of heat recovery systems:

- Air-to-Air Runaround System
- Heat Pipe System
- Fixed Plate Heat Exchanger
- Waste Heat Boiler

Air-to-Air Runaround System

The air-to-air runaround system is a means of extracting a portion of the heat from the warm exhaust air and transferring that heat to the outside ventilation air for the building. The exhaust air heat source may be toilet exhaust air, general exhaust air, or heat from the boiler and incinerator flue gases entering the chimney. Figure 6.48 is an example of an air-to-air runaround system that reclaims part of an animal facility's building exhaust air, and transfers that heat to the outside ventilation air.

In a runaround heat recovery system, a pump circulates the 40% glycol solution through the piping system that connects the coils. The glycol temperature leaving the outside air coil could be below 30°F. The cold glycol could condense and freeze the moisture on the exhaust air coil. A freeze control thermostat must sense the glycol temperature leaving the outdoor air coil and divert the warm glycol leaving the exhaust air coil to mix with the cold glycol, to keep the glycol entering the exhaust air coil above 30°F.

The quality of the exhaust air must be evaluated to determine if filters should be installed to protect the coils. If filters are not installed, frequent coil cleaning will be required. Coil selections should not exceed six rows and eight fins per inch to facilitate coil cleaning.

The fan energy expended by the added resistance of the filters and coils, plus the glycol pump energy, must be calculated for an entire year's operation. This is an added operating expense that must be compared to the heat recovery energy savings. Figure 6.49 is an example of a heat recovery energy study. Runaround systems are about 50% efficient. The installed cost of a runaround system is about the same as a fixed plate heat exchanger (discussed later in this chapter) and half the installed cost of a heat pipe.

Note that the additional fan and pump energy costs of the example in Figure 6.49 exceed the preheat steam cost savings. In this case, the installation of an air-to-air runaround system is not justified.

Heat Pipe

The heat pipe is an air-to-air heat reclamation device. As Figure 6.50 shows, the refrigeration tube is continuous from the exhaust duct to the outside air duct. Each tube contains liquid refrigerant that evaporates when it absorbs heat from the warm exhaust air. The refrigerant migrates as a gas to the cold end of the tube, where it condenses and releases heat to the cold outside air. The

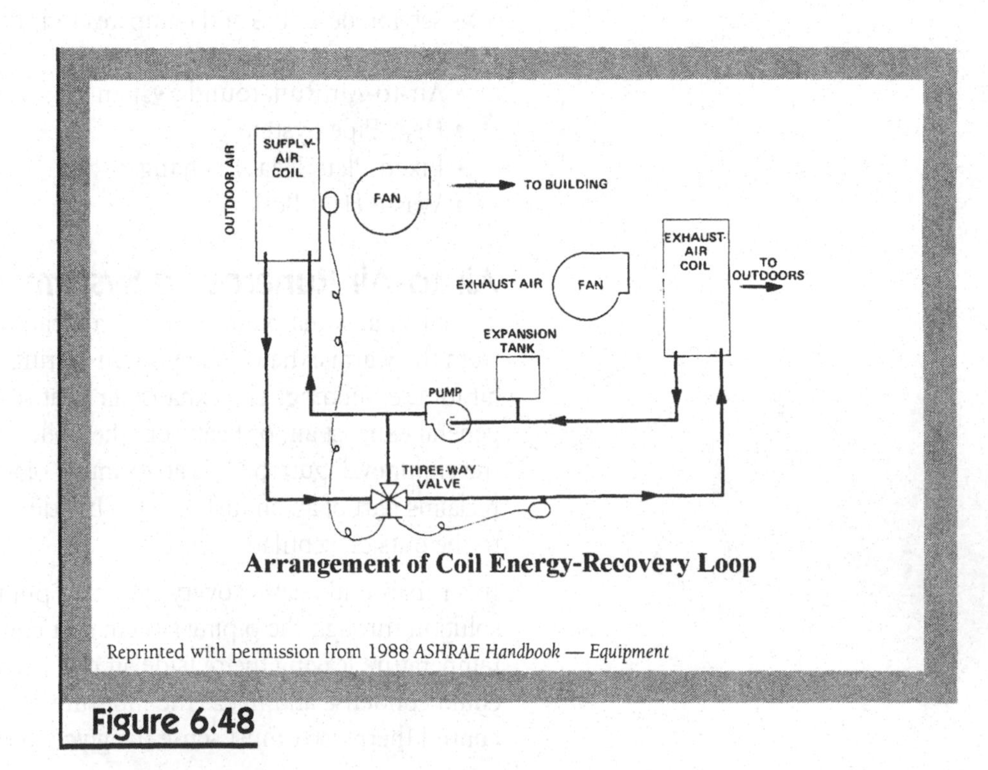

Arrangement of Coil Energy-Recovery Loop

Reprinted with permission from 1988 *ASHRAE Handbook — Equipment*

Figure 6.48

condensed liquid then runs back down the pipe to the warm end of the tube to start another cycle. Tilt controls are available to control the heat exchange by regulating the refrigerant flow (by tilting the tubes to slow or speed up refrigerant flow).

Heat pipes are about 60% efficient, but cost about twice as much to install as runaround coils. Since the continuous heat pipe tubes must be situated in both outside air and the exhaust ducts, the two air streams must be brought close to each other at the heat pipe. This is not usually easy to accomplish, since outside air and exhaust louvers are not normally close to each other.

Heat Recovery Energy Study
New York City, using an electric rate of $0.15/kwh, steam cost $10 per 1,000 #. 80,000 cfm of outside air at 0°F, 104,000 cfm exhaust air at 75°F, 8,760 hours per year. Glycol pump operates 4,000 hours per year.
Present Cost of Heating 80,000 cfm x 1.08 x 117,811 degree hours to 65°F = 10,178,870 MBH 10,178,870 # steam x $10/1,000 # = 100,179 per year.
Energy saved by heat recovery is 50% of present usage. Cost of energy saved is $100,179 per year x 0.50 = $50,090 per year.
Energy penalty of runaround system. Outside air coil static pressure, 1.0 in wg.
80,000 cfm x 1.0 in wg x 0.75 kw/hp / 6356 x 0.7 eff = 13.5 kw
Exhaust air coil static pressure, 1.72 in wg Exhaust air coil filter static pressure, 0.80 in wg Total additional exhaust static pressure, 2.52 in wg
104,000 cfm x 2.52 in wg x 0.75 kw/hp / 6356 x 0.7 eff. = 44 kw
Glycol pump 340 gpm, 60 ft. head, 40% glycol factor of 2.
340 gpm x 60 ft. x 2 x 0.75 kw/hp / 3969 x 0.7 eff. = 11 kw.
Total fan kw penalty = 13.5 + 44 = 57.5 kw Cost of additional fan energy = 57.5 kw x 8,760 hrs./yr. x $0.15/kwh = $75,555 per year
Cost of pump energy 11 kw x 4,000 hrs./yr. x $0.15/kwh = $6,600/yr.
Total electric penalty cost = $75,555 + $6,600 = $82,155/year
Energy penalty $82,155 - energy saved $50,000 = $32,155 additional cost every year
Cost of installing runaround system is $200,000.
Save $200,000 capital outlay and $32,155 per year and do nothing.
Heat recovery involving additional coil and filter pressure drops 8,760 hours per year. In high electric rate areas, may not be advisable.

R.S. Means Co., Inc., *HVAC Systems Evaluation*

Figure 6.49

Fixed Plate Heat Exchangers

Fixed plate heat exchangers are static devices that are configured in a cross-flow arrangement. The primary heat exchange plate separates the air streams. This essentially eliminates leakage between the air streams. There is no secondary heat exchange fluid involved, so that a broad temperature range is possible. Figure 6.51 shows four types of basic plate heat exchangers.

The efficiency of the fixed plate heat exchanger is about 60%. The installed cost of the plate heat exchanger is approximately the same as the runaround coils and one-half that of a heat pipe. The duct connections for a fixed plate heat exchanger are similar to the heat pipe arrangement. The exhaust and outside air ducts must be brought to the common heat exchange device.

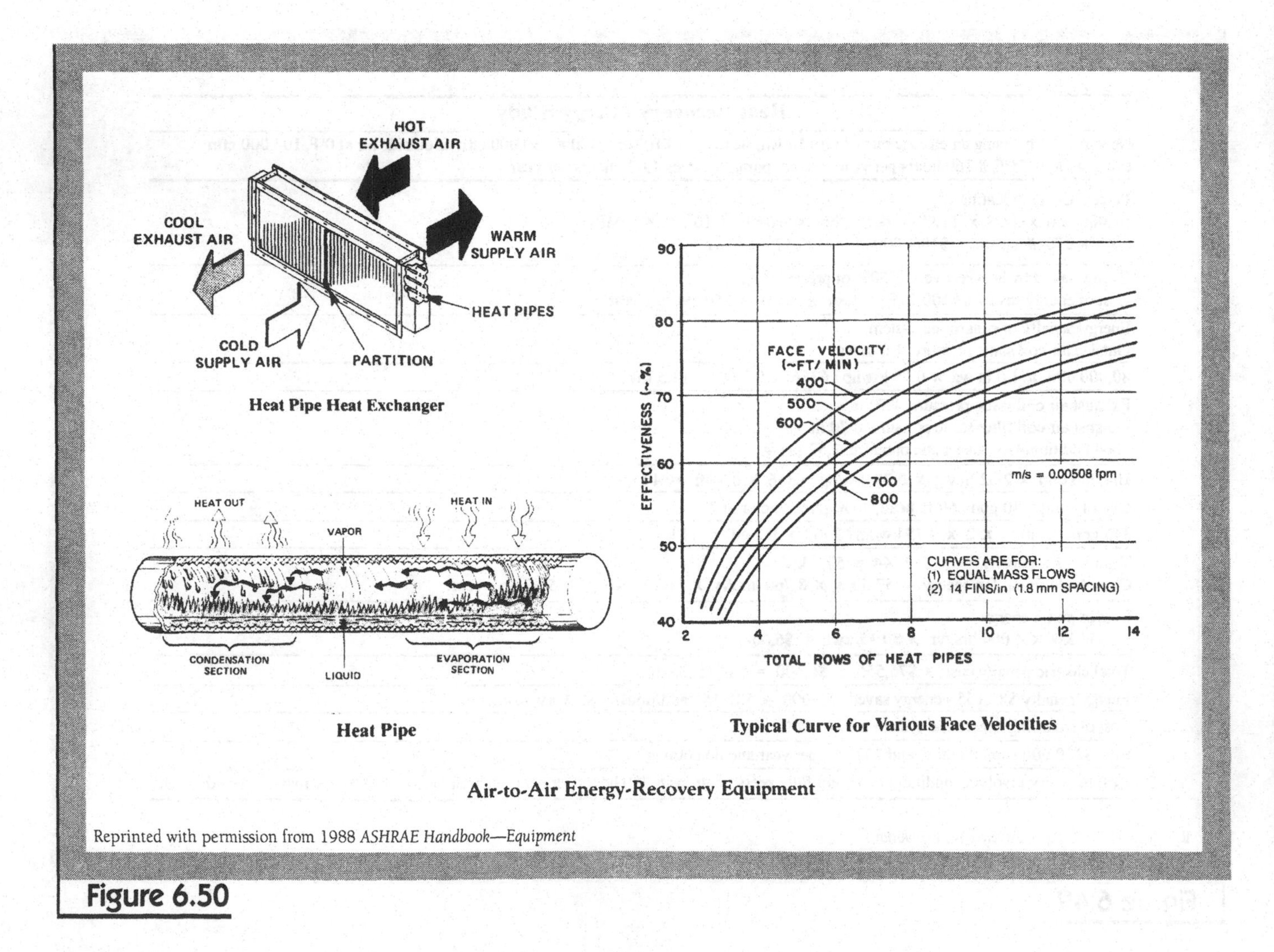

Air-to-Air Energy-Recovery Equipment

Reprinted with permission from 1988 *ASHRAE Handbook—Equipment*

Figure 6.50

Waste Heat Boilers

Exhaust combustion gases are passed through waste heat boilers so that the 800°F turbine or 1,800°F incinerator flue gas can convert the water in the waste heat boiler heat exchanger to high pressure steam. A 2,800 kW gas turbine can deliver 17,500 lbs. of 150 psi steam per hour by passing 800°F exhaust gases through the waste heat boiler. To prevent condensation, the flue gas temperature must not be allowed to drop below 300°F. Figure 6.52 shows a waste heat boiler and turbine generator arrangement.

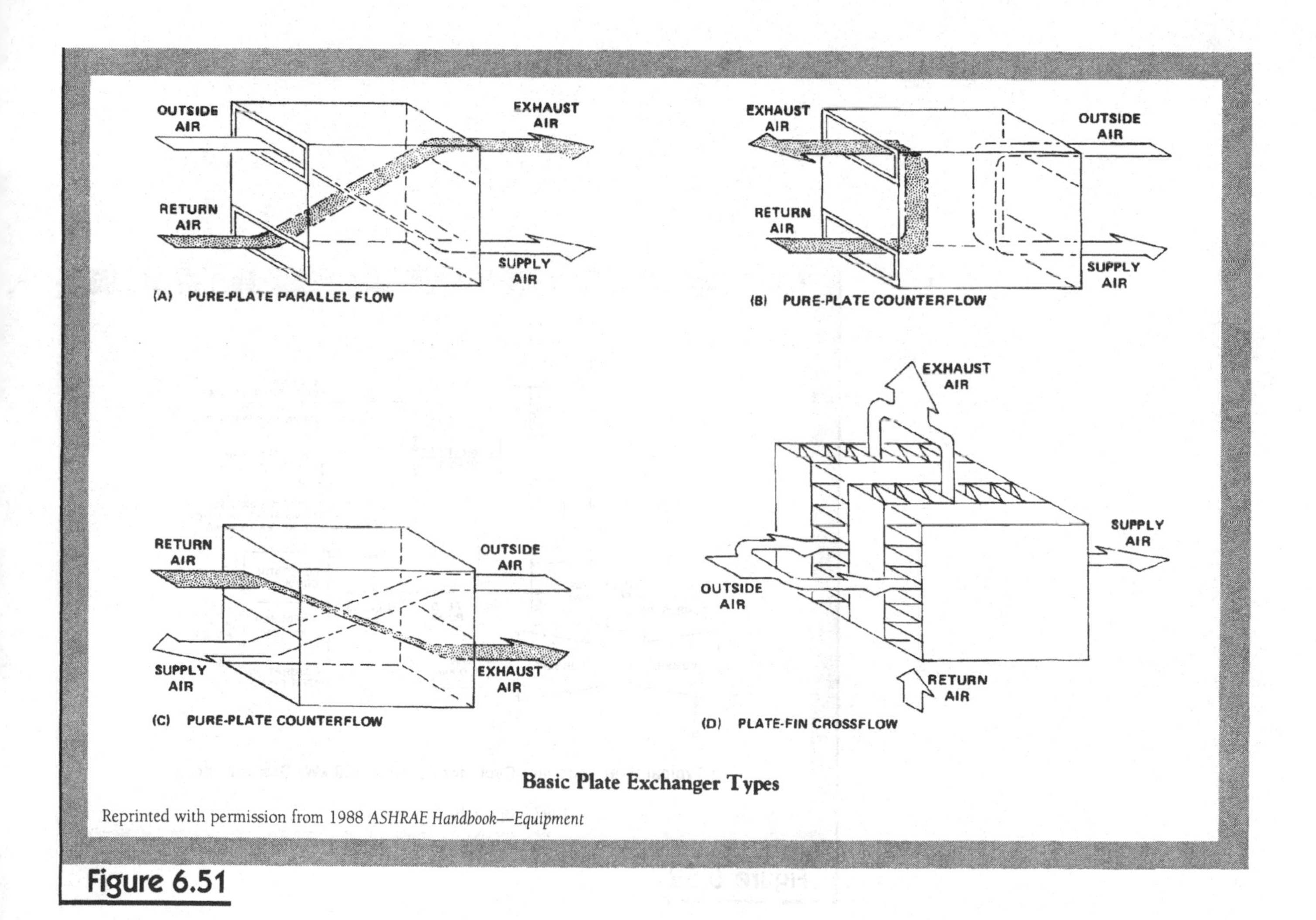

Basic Plate Exchanger Types

Reprinted with permission from 1988 *ASHRAE Handbook—Equipment*

Figure 6.51

The following section is reprinted with permission from the 1996 *ASHRAE Systems and Equipment Handbook*

Air Filtration — Types & Performance

Panel Filters

Viscous Impingement Filters: These are panel filters made up of coarse fibers with a high porosity. The filter media are coated with a viscous substance, such as oil (also known as adhesive), which causes particles that impinge on the fibers to stick to them. Design air velocity through the media is usually in the range of 200 to 800 fpm. These filters are characterized by a low pressure drop, low cost, and good efficiency on lint, but exhibit low efficiency on normal atmospheric dust. They are commonly made 1/2 to 4 in. thick. Unit panels are available in standard and special sizes up to about 24 in. by 24 in. This type of filter is commonly used in residential furnaces and air conditioning, and is often used as a prefilter for higher-efficiency filters.

A number of different materials are used as the filtering medium, including coarse (15 to 60 mm diameter) glass fibers, coated animal hair, vegetable fibers, synthetic fibers, metallic wools, expanded metals and foils, crimped screens, random-matted wire, and synthetic open-cell foams. The arangement of the medium in this type of filter involves three basic configurations:

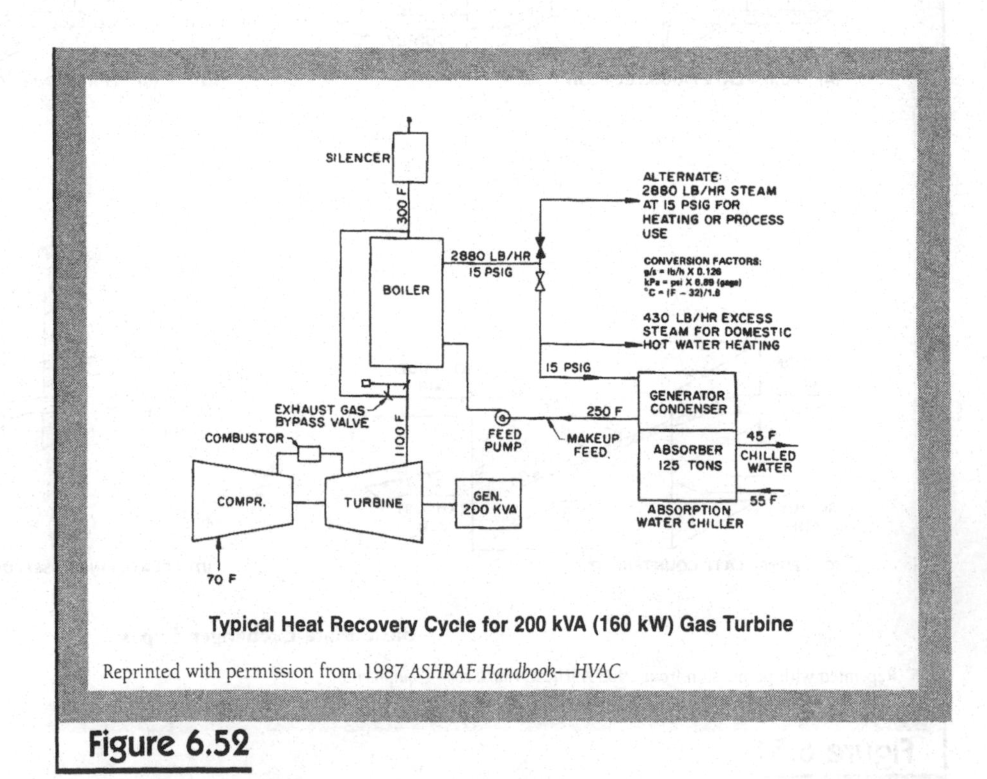

Typical Heat Recovery Cycle for 200 kVA (160 kW) Gas Turbine

Reprinted with permission from 1987 *ASHRAE Handbook—HVAC*

Figure 6.52

Sinuous media. The filtering medium (consisting of corrugated metal or screen strips) is held more or less parallel to the airflow. The direction of airflow is forced to change rapidly in passing through the filter, thus giving inertial impingement of dust on the metal elements.

Formed-screen media. Here, the filter media (screens or expanded metal) are crimped to produce high porosity media that avoid collapsing. Air flows through the media elements and dust impinges on the wires. The relatively open structure allows the filter to store substantial quantities of dust and lint without plugging.

Random fiber media. Fibers, either with or without bonding material, are formed into mats of high porosity. Media of this type are often designed with fibers packed more densely on the leaving air side than on the entering air side. This arrangement permits both the accumulation of larger particles and lint near the air-entering face of the filter and the filtration of finer particles on the more closely packed air-leaving face. Fiber diameters may also be graded from coarse at the air-entry face to fine at the exit face.

Although viscous impingement filters usually operate in the range of 300 to 600 fpm, they may be operated at higher velocities. The limiting factor, other than increased flow resistance, is the danger of blowing off agglomerates of collected dust and the viscous coating on the filter.

The loading rate of a filter depends on the type and concentration of the dirt in the air being handled and the operating cycle of the system. Manometers, static pressure gages, or pressure transducers are often installed to measure the pressure drop across the filter bank. From the pressure drop, it can be determined when the filter requires servicing. The final allowable pressure drop may vary from one installation to another; but, in general, unit filters are serviced when their operating resistance reaches 0.5 in. of water. The decline in filter efficiency (which is caused by the absorption of the viscous coating by dust, rather than by the increased resistance because of dust load) may be the limiting factor in operating life.

The manner of servicing unit filters depends on their construction and use. Disposable viscous impingement, panel-type filters are constructed of inexpensive materials and are discarded after one period of use. The cell sides of this design are usually a combination of cardboard and metal stiffeners. Permanent unit filters are generally constructed of metal to withstand repeated handling. Various cleaning methods have been recommended for permanent filters; the most widely used involves washing the filter with steam or water (frequently with detergent) and then recoating it with its recommended adhesive by dipping or spraying. Unit viscous filters are also sometimes arranged for in-place washing and recoating.

The adhesive used on a viscous impingement filter requires careful engineering. Filter efficiency and dust-holding capacity depend on the specific type and quantity of adhesive used; this information is an essential part of test data and filter specifications. Desirable adhesive characteristics, in addition to efficiency and dust-holding capacity, are (1) a low percentage of volatiles to prevent excessive evaporation; (2) a viscosity that varies only slightly within the service temperature range; (3) the ability to inhibit growth of bacteria and mold spores; (4) a high capillarity or the ability to wet and retain the dust particles; (5) a high flash point and fire point; and (6) freedom from odorants or irritants.

Typical performance of viscous impingement unit filters operating within typical resistance limits is shown as Group I in Figure 6.53.

Dry-Type Extended-Surface Filters: The media in dry-type air filters are random fiber mats or blankets of varying thicknesses, fiber sizes, and densities. Bonded glass fiber, cellulose fibers, wool felt, synthetics, and other materials have been used commercially. The media in filters of this class are frequently supported by a wire frame in the form of pockets, or V-shaped or radial pleats. In other designs, the media may be self-supporting because of inherent rigidity or because airflow inflates it into extended form such as with bag filters. Pleating of the media provides a high ratio of media area to face area, thus allowing reasonable pressure drop.

In some designs, the filter media is replaceable and is held in position in permanent wire baskets. In most designs, the entire cell is discarded after it has accumulated its maximum dust load.

The efficiency of dry-type air filters is usually higher than that of panel filters, and the variety of media available makes it possible to furnish almost any degree of cleaning efficiency desired. The dust-holding capacities of modern dry-type filter media and filter configurations are generally higher than those of panel filters.

The placement of coarse prefilters ahead of extended-surface filters is sometimes justified economically by the longer life of the main filters. Economic considerations should include the prefilter material cost, changeout labor, and increased fan system power use. Generally, prefilters should be considered only if they can substantially reduce the part of the dust that may plug the protected filter. A prefilter usually has an atmospheric dust-spot efficiency of 70% or more. Temporary prefilters are worthwhile during building construction to capture heavy loads of coarse dust. HEPA-type filters of 95% DOP efficiency and greater should always be protected by prefilters of 80% or greater ASHRAE atmospheric dust-spot efficiency. A single filter gage may be installed when a panel prefilter is placed adjacent to a final filter. Because the prefilter is normally changed on a schedule, the final filter pressure drop can be read without the prefilter in place every time the prefilter is changed.

Typical performance of some types of filters in this group, when they are operated within typical rated resistance limits and over the life of the filters, is shown as Groups II and III in Figure 6.53.

The initial resistance of an extended-surface filter varies with the choice of media and the filter geometry. Commercial designs typically have an initial resistance from 0.1 to 1.0 in. of water. It is customary to replace the media when the final resistance of 0.5 in. of water is reached for low resistance units, and 2.0 in. of water for the highest resistance units. Dry media providing higher orders of cleaning efficiency have a higher resistance to airflow. The operating resistance of the fully dust-loaded filter must be considered in the system design, because that is the maximum resistance against which the fan will operate. Variable air volume systems and systems with constant air volume controls prevent abnormally high airflows or possible fan motor overloading from occurring when filters are clean.

Flat panel filters with media velocity equal to duct velocity are possible only in the lowest efficiency units of the dry type (open-cell foams and textile-denier nonwoven media). Initial resistance of this group, at rated airflow, is mainly between 0.05 and 0.25 in. of water. They are usually operated to a final resistance of 0.50 and 0.70 in. of water.

In extended-surface filters of the intermediate efficiency ranges, the filter media area is much greater than the face area of the filter; hence, velocity through the filter media is substantially lower than the velocity approaching the filter face. Media velocities range from 6 to 90 fpm, although the approach velocities run to 750 fpm. Depth in direction of airflow varies from 2 to 36 in.

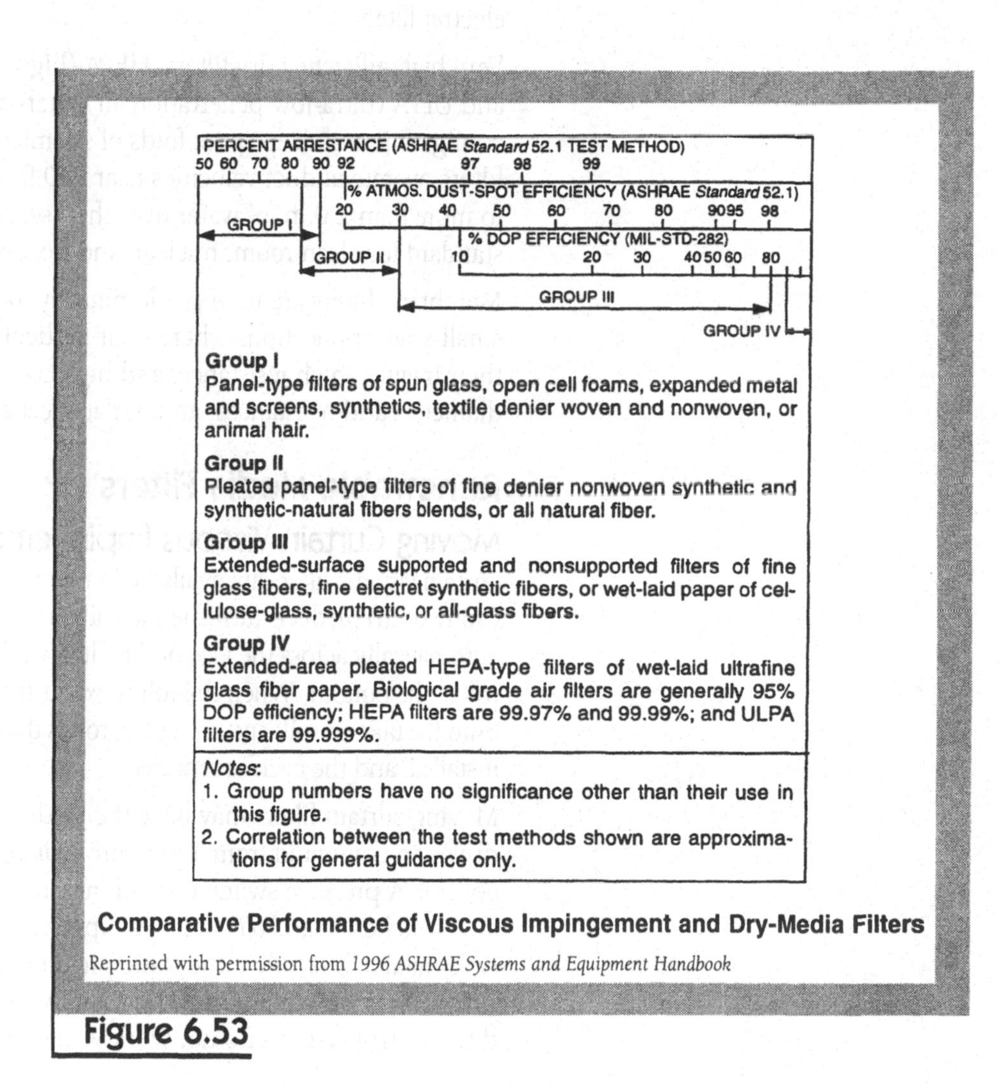

Comparative Performance of Viscous Impingement and Dry-Media Filters

Reprinted with permission from *1996 ASHRAE Systems and Equipment Handbook*

Figure 6.53

Filter media used in the intermediate efficiency range include those of (1) fine glass or synthetic fibers, 0.7 to 10 mm in diameter, in mat form, up to 1/2 in. thick; (2) thin nonwoven mats of fine glass fibers, cellulose, or cotton wadding; and (3) nonwoven mats of comparatively large diameter fibers (more than 30 mm) in greater thicknesses (up to 2 in.).

Electret filters are composed of electrostatically charged fibers. The charges on the fibers augment the collection of smaller particles by Brownian diffusion with Coulombic forces caused by the charges on the fibers. There are three types of these filters: resin wool, electret, and an electrostatically sprayed polymer. The charge on the resin wool fibers is produced by friction during the carding process. During production of the electret, a corona discharge injects positive charges on one side of a thin polypropylene film and negative charges on the other side. These thin sheets are then shredded into fibers of rectangular cross-section. The third process spins a liquid polymer into fibers in the presence of a strong electric field, which produces the charge separation. The efficiency of the charged-fiber filters is due to both the normal collection mechanisms of a media filter and the strong local electrostatic effects. The effects induce efficient preliminary loading of the filter to enhance the caking process. However, dust collected on the media can reduce the efficiency of electret filters.

Very high-efficiency dry filters, HEPA (high-efficiency particulate air) filters, and ULPA (ultra-low penetration air) filters are made in an extended-surface configuration of deep space folds of submicrometre glass fiber paper. Such filters operate at duct velocities near 250 fpm, with resistance rising from 0.5 to more than 2.0 in. of water over their service life. These filters are the standard for clean room, nuclear, and toxic-particulate applications.

Membrane filters are used predominately for air sampling and specialized small-scale applications where their particular characteristics compensate for their fragility, high resistance, and high cost. They are available in many pore diameters and resistances, and in flat sheet and pleated forms.

Renewable Media Filters

Moving-Curtain Viscous Impingement Filters: Automatic moving-curtain viscous filters are available in two main types. In one type, random-fiber (nonwoven) media is furnished in roll form. Fresh media is fed manually or automatically across the face of the filter, while the dirty media is rewound onto a roll at the bottom. When the roll is exhausted, the tail of the media is wound onto the takeup roll, and the entire roll is thrown away. A new roll is then installed, and the cycle is repeated.

Moving-curtain filters may have the media automatically advanced by motor drives on command from a pressure switch, timer, or media light-transmission control. A pressure switch control measures the pressure drop across the media and switches *on* and *off* at chosen upper and lower set points. This control saves media, but only if the static pressure probes are located properly and unaffected by modulating outside air and return air dampers. Most pressure-drop control systems do not work well in practice. Timers and media light-transmission controls help to avoid these problems; their duty cycles can usually be adjusted to provide satisfactory operation with acceptable media consumption.

Filters of this replaceable roll design generally have a signal indicating when the roll of media is nearly exhausted. At the same time, the drive motor is de-energized so that the filter cannot run out of media. The normal service requirements involve insertion of a clean roll of media at the top of the filter and disposal of the loaded dirty roll. Automatic filters of this design are not, however, limited in application to the vertical position. Horizontal arrangements are available for use with makeup air units and air-conditioning units. Adhesives must have qualities similar to those for panel-type viscous impingement filters, and they must withstand media compression and endure long storage.

The second type of automatic viscous impingement filter consists of linked metal mesh media panels installed on a traveling curtain that intermittently passes through an adhesive reservoir. In the reservoir, the panels give up their dust load and, at the same time, take on a new coating of adhesive. The panels thus form a continuous curtain that moves up one face and down the other face. The media curtain, continually cleaned and renewed with fresh adhesive, lasts the life of the filter mechanism. The precipitated dirt must be removed periodically from the adhesive reservoir.

The resistance of both types of viscous impingement automatically renewable filters remains approximately constant as long as proper operation is maintained. A resistance of 0.40 to 0.50 in. of water at a face velocity of 500 fpm is typical of this class.

Moving-Curtain Dry-Media Filters: Random-fiber (nonwoven) dry media of relatively high porosity are also used in moving-curtain (roll) filters for general ventilation service. Operating duct velocities near 200 fpm are generally lower than those of viscous impingement filters.

Special automatic dry filters are also available, which are designed for the removal of lint in textile mills and dry cleaning establishments and the collection of lint and ink mist in pressrooms. The medium used is extremely thin and serves only as a base for the buildup of lint, which then acts as a filter medium. The dirt-laden media is discarded when the supply roll is used up.

Another form of filter designed specifically for dry lint removal consists of a moving curtain of wire screen, which is vacuum-cleaned automatically at a position out of the airstream. Recovery of the collected lint is sometimes possible with such a device.

Performance of Renewable Media Filters: ASHRAE arrestance, efficiency, and dust-holding capacities for typical viscous impingement and dry renewable media filters are listed in Table 1 in Figure 6.54.

Electronic Air Cleaners

Electronic air cleaners can be highly efficient filters using electrostatic precipitation to remove and collect particulate contaminants such as dust, smoke, and pollen. The designation *electronic air cleaner* denotes a precipitator for HVAC air filtration. The filter consists of an ionization section and a collecting plate section.

In the ionization section, small-diameter wires with a positive direct current potential between 6 and 25 kV are suspended equidistant between grounded plates. The high voltage on the wires creates an ionizing field for charging particles. The positive ions created in the field flow across the airstream and strike and adhere to the particles, thus imparting a charge to them. The charged particles then pass into the collecting plate section.

The collecting plate section consists of a series of parallel plates equally spaced with a positive direct current voltage of 4 to 10 kV applied to alternate plates. Plates that are not charged are at ground potential. As the particles pass into this section, they are forced to the plates by the electric field on the charges

Table 1 Performance of Renewable Media Filters (Steady-State Values)

Description	Type of Media	ASHRAE Weight Arrestance, %	ASHRAE Atmospheric Dust-Spot Efficiency, %	ASHRAE Dust-Holding Capacity, g/ft²	Approach Velocity, fpm
20 to 40 μm glass and synthetic fibers, 2 to 2 1/2 in. thick	Viscous impingement	70 to 82	<20	60 to 180	500
Permanent metal media cells or overlapping elements	Viscous impingement	70 to 80	<20	NA (permanent media)	500
Coarse textile denier nonwoven mat, 1/2 to 1 in. thick	Dry	60 to 80	<20	15 to 70	500
Fine textile denier nonwoven mat, 1/2 to 1 in. thick	Dry	80 to 90	<20	10 to 50	200

Reprinted with permission from *1996 ASHRAE Systems and Equipment Handbook*

Figure 6.54

they carry; thus, they are removed from the airstream and collected by the plates. Particulate retention is a combination of electrical and intermolecular adhesion forces and may be augmented by special oils or adhesives on the plates. Figure 6.55 shows a typical electronic air cleaner cell.

In lieu of positive direct current, a negative potential also functions on the same principle, but more ozone is generated.

With voltages of 4 to 25 kV dc, safety measures are required. A typical arrangement makes the air cleaner inoperative when the doors are removed for cleaning the cells or servicing the power pack.

HVAC Equations

The following section (Figure 6.56) contains HVAC equations in Imperial and Metric units.

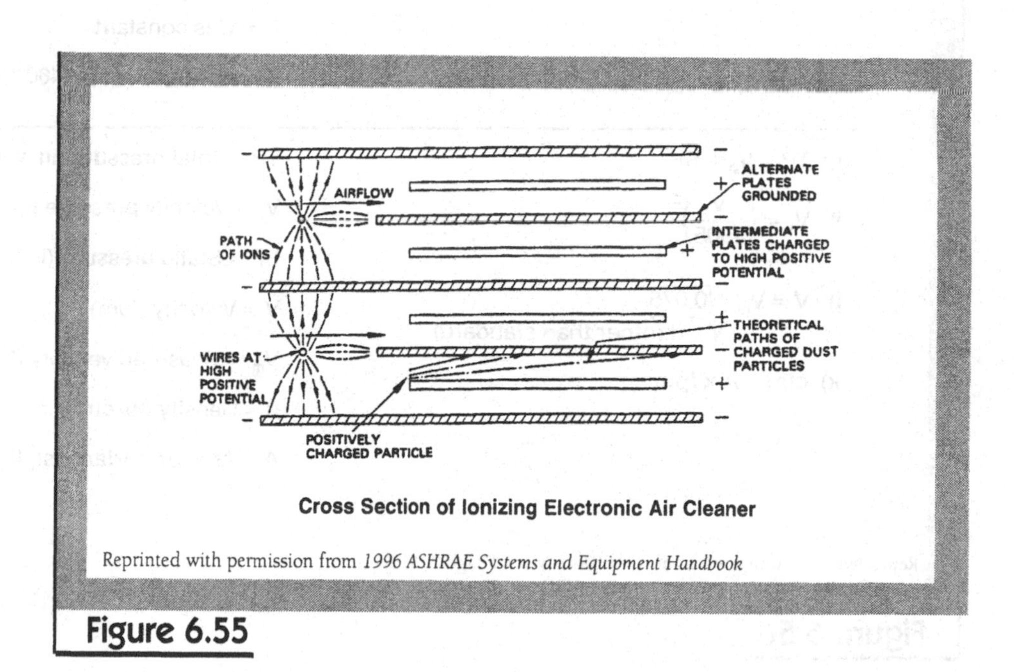

Cross Section of Ionizing Electronic Air Cleaner

Reprinted with permission from *1996 ASHRAE Systems and Equipment Handbook*

Figure 6.55

HVAC EQUATIONS IN U.S. UNITS (BRITISH SYSTEM)

1. AIR EQUATIONS (10)

a) $V = 1096 \sqrt{\frac{V_p}{d}}$

or for standard air:

$V = 4005 \sqrt{V_p}$

$\left(d = 1.325 \frac{P_b}{T}\right)$

V = Velocity (fpm)

V_p = Velocity pressure (in w.g.)

d = Density (lb/cu ft)

P_b = Absolute static pressure (in. Hg) (Barometric pressure + static pressure)

T = Absolute Temp. (460° + °F)

---

b) $Q = 60 \times C_p \times d \times cfm \times \Delta t$

or for standard air:

$Q\ (sens.) = 1.08 \times cfm \times \Delta t$

c) $Q\ (Lat.) = 4840 \times cfm \times \Delta W$

d) $Q\ (Total) = 4.5 \times cfm \times \Delta h$

e) $Q = A \times U \times \Delta t$

f) $R = \frac{1}{U}$

Q = Heat Flow (Btu/hr)

C_p = Specific Heat (Btu/lb/°F)

Δt = Temperature Difference (°F)

ΔW = Humidity Ratio (lb H_2O/lb dry air)

Δh = Enthalpy Diff. (Btu/lb dry air)

A = Area or Surface (sq ft)

U = Heat transfer coefficient

R = Thermal resistance

g) $\frac{P_1V_1}{T_1} = \frac{P_2V_2}{T_2} = R$

P = Absolute pressure (lb/sq ft)

V = Total Volume (cu ft)

R = Gas constant

T = Absolute temp. (460° + °F)

---

h) $TP = V_p + SP$

i) $V_p = \left(\frac{V}{4005}\right)^2$

j) $V = V_m \sqrt{\frac{0.075}{d}}$ (other than standard)

k) $cfm = A \times fpm$

TP = Total pressure (in w.g.)

V_p = Velocity pressure (in w.g.)

SP = Static pressure (in w.g.)

V = Velocity (fpm)

V_m = Measured velocity (fpm)

d = Density (lb/cu ft)

A = Area or Surface (sq ft)

Reprinted with permission from *HVAC Duct System Design, SMACNA*

Figure 6.56

HVAC EQUATIONS IN METRIC UNITS

1. AIR EQUATIONS (10)

a) $V = 1.288\sqrt{V_p\left(\frac{1.2}{d}\right)}$

or for standard air:

($d = 1.2$ kg/m³)

$V = 1.3\sqrt{V_p}$

V = Velocity (m/s)

V_p = Velocity Pressure (pascals)

d = Density (kg/m³)

---

b) $Q = 3.6 \times C_p \times d \times l/s \times \Delta t$

or for standard air:

($c_p = 0.28$ W/kg/°C)

Q (sens.) $= 1.21 \times l/s \times \Delta t$

c) Q (Lat.) $= 3.0 \times l/s \times \Delta W$

d) Q (Total Heat) $= 1.27 \times l/s \times \Delta h$

e) $Q = A \times U \times \Delta t$

f) $R = \frac{1}{U}$

Q = Heat Flow (watts)

C_p = Specific Heat (W/kg/°C)

Δt = Temperature Difference (°C)

ΔW = Humidity Ratio (g H_2O/kg dry air)

Δh = Enthalpy Diff. (kJ/kg dry air)

A = Area of Surface (m²)

U = Heat transfer coefficient

R = Thermal Resistance

g) $\frac{P_1V_1}{T_1} = \frac{P_2V_2}{T_2} = R$

P = Absolute Pressure (pascals)

V = Total Volume (m³)

R = Gas Constant

T = Absolute Temperature (°K)

---

h) $TP = V_p + SP$

i) $V_p = \left(\frac{V}{1.3}\right)^2 = \frac{d}{2} \times V^2$

j) $V = 1.4\sqrt{\frac{V_p}{d}}$

k) $m^3/s = A \times m/s$

TP = Total Pressure (pascals)

V_p = Velocity Pressure (Pa)

SP = Static Pressure (Pa)

V = Velocity (m/s)

d = Density (kg/m³)

A = Area (m²)

Reprinted with permission from *HVAC Duct System Design*, SMACNA

Figure 6.56 cont.

U.S. UNITS

2. FAN EQUATIONS (10)

a) $\frac{cfm_2}{cfm_1} = \frac{rpm_2}{rpm_1}$

b) $\frac{SP_2}{SP_1} = \left(\frac{rpm_2}{rpm_1}\right)^2$

c) $\frac{bhp_2}{bhp_1} = \left(\frac{rpm_2}{rpm_1}\right)^3$

d) $\frac{\text{rpm (fan)}}{\text{rpm (motor)}} = \frac{\text{Pitch diam. motor pulley}}{\text{Pitch diam. fan pulley}}$

cfm = cu ft/min

rpm = revolutions/min

SP = Static pressure (in w.g.)

bhp = brake horsepower

3. HYDRONIC EQUATIONS (10)

a) $Q = 500 \times gpm \times \Delta t$

b) $\frac{\Delta P_2}{\Delta P_1} = \left(\frac{gpm_2}{gpm_1}\right)^2$

c) $\Delta P = \left(\frac{gpm}{C_v}\right)^2$

d) $whp = \frac{gpm \times H \times Sp.\ Gr.}{3960}$

e) $bph = \frac{gpm \times H \times Sp.\ Gr.}{3960 \times E_p}$

f) $NPSHA = P_a \pm P_s + \frac{V^2}{2_g} - P_{vp}$

g) $h = \frac{fLv^2}{2gD}$

gpm = Gallons per minute

Q = Heat Flow (Btu/hr)

Δt = Temperature diff. (°F)

ΔP = Pressure diff. (psi)

C_v = Valve constant

whp = Water horsepower

bhp = Brake horsepower

H = Head (ft w.g.)

Sp. Gr. = Specific gravity (use 1.0 for water)

E_p = Efficiency of pump (%/100)

NPSHA = Net positive suction head available

P_a = Atm. press. (use 34 ft w.g.)

P_s = Pressure at pump centerline (ft w.g.)

V = Velocity head at point P_s (fpm)

P_{vp} = Absolute vapor pressure (ft w.g.)

g = Gravity acceleration (32.2 ft/sec^2)

h = Head loss (ft)

f = Friction factor (Moody)

L = Length of pipe (ft)

D = Internal diameter (ft)

v = Velocity (ft/sec)

Reprinted with permission from *HVAC Duct System Design*, SMACNA

Figure 6.56 cont.

METRIC UNITS

2. FAN EQUATIONS (10)

a) $\frac{m^3/s_2}{m^3/s_1} = \frac{rad/s_2}{rad/s_1}$

b) $\frac{SP_2}{SP_1} = \left(\frac{rad/s_2}{rad/s_1}\right)^2$

c) $\frac{kW_2}{kW_1} = \left(\frac{rad/s_2}{rad/s_1}\right)^3$

d) $\frac{\text{rad/s (fan)}}{\text{rad/s (motor)}} = \frac{\text{Pitch diam. (mm) motor pulley}}{\text{Pitch diam. (mm) fan pulley}}$

m^3/s = cubic metres/second
rad/s = radians/second

SP = Static pressure (Pa)

kW = kilowatts

3. HYDRONIC EQUATIONS (10)

a) $Q = 4190 \times m^3/s \times \Delta t$

b) $\frac{\Delta P_2}{\Delta P_1} = \left(\frac{m^3{}_2}{m^3{}_1}\right)^2$

c) $\Delta P = \left(\frac{m^3}{C_v}\right)^2$

d) $WP = 0.1020 \times m^3/s \times H \times d$

e) $BP = \frac{WP}{E_p}$

f) $NPSHA = P_a \pm P_s + \frac{V^2}{2g} - P_{vp}$

g) $h = \frac{fLv^2}{2gD}$

Q = Heat Flow (kilowatts)

Δt = Temperature Difference (°C)

m^3 (used for large volumes)

ΔP = Pressure Diff. (Pa or kPa)

C_v = Valve constant

WP = Water power (W or kW)

H = Head (k P_a)

d = Density (kg/m^3)

BP = Brake power (W or kW)

E_p = Efficiency of Pump (%/100)

NPSHA = Net positive suction head available

P_a = Atm. press. (pascals)
(Std. Atm. press. = 101 325Pa)

P_s = Pressure at pump centerline (Pa)

$\frac{V^2}{2g}$ = Velocity pressure at point P_s (Pa)

P_{vp} = Absolute vapor pressure (Pa)

g = Gravity acceleration (9.807 m/s^2)

h = Head loss (m)

f = Friction factor (Moody, metric)

L = Length of pipe (m)

D = Internal diameter (m)

v = Velocity (m/s)

Reprinted with permission from *HVAC Duct System Design*

Figure 6.56 cont.

U.S. UNITS

4. ELECTRIC EQUATIONS (10)

a) $\text{Bhp} = \dfrac{\text{I x E x P.F. x Eff.}}{746}$ (Single Phase)

b) $\text{Bhp} = \dfrac{\text{I x E x P.F. x Eff. x 1.73}}{746}$ (Three Phase)

c) $E = IR$

d) $P = EI$

e) $\dfrac{\text{F.L. Amps* x Voltage*}}{\text{Actual Voltage}} = \text{Actual F.L. Amps}$

*Nameplate ratings

I = Amps (A)
E = Volts (V)
P.F. = Power factor
R = ohms (Ω)
P. = watts (W)

Table 6-15 AIR DENSITY CORRECTION FACTORS (U.S. Units) (10)

Altitude (ft)		–	Sea Level	1000	2000	3000	4000	5000	6000	7000	8000	9000	10,000
Barometer	in Hg	–	29.92	28.86	27.82	26.82	25.84	24.90	23.98	23.09	22.22	21.39	20.58
	in w.g.	–	407.5	392.8	378.6	365.0	351.7	338.9	326.4	314.3	302.1	291.1	280.1
Air Temp.		–40°	1.26	1.22	1.17	1.13	1.09	1.05	1.01	0.97	0.93	0.90	0.87
°F		0°	1.15	1.11	1.07	1.03	0.99	0.95	0.91	0.89	0.85	0.82	0.79
		40°	1.06	1.02	0.99	0.95	0.92	0.88	0.85	0.82	0.79	0.76	0.73
		70°	1.00	0.96	0.93	0.89	0.86	0.83	0.80	0.77	0.74	0.71	0.69
		100°	0.95	0.92	0.88	0.85	0.81	0.78	0.75	0.73	0.70	0.68	0.65
		150°	0.87	0.84	0.81	0.78	0.75	0.72	0.69	0.67	0.65	0.62	0.60
		200°	0.80	0.77	0.74	0.71	0.69	0.66	0.64	0.62	0.60	0.57	0.55
		250°	0.75	0.72	0.70	0.67	0.64	0.62	0.60	0.58	0.56	0.58	0.51
		300°	0.70	0.67	0.65	0.62	0.60	0.58	0.56	0.54	0.52	0.50	0.48
		350°	0.65	0.62	0.60	0.58	0.56	0.54	0.52	0.51	0.49	0.47	0.45
		400°	0.62	0.60	0.57	0.55	0.53	0.51	0.49	0.48	0.46	0.44	0.42
		450°	0.58	0.56	0.54	0.52	0.50	0.48	0.46	0.45	0.43	0.42	0.40
		500°	0.55	0.53	0.51	0.49	0.47	0.45	0.44	0.43	0.41	0.39	0.38
		550°	0.53	0.51	0.49	0.47	0.45	0.44	0.42	0.41	0.39	0.38	0.36
		600°	0.50	0.48	0.46	0.45	0.43	0.41	0.40	0.39	0.37	0.35	0.34
		700°	0.46	0.44	0.43	0.41	0.39	0.38	0.37	0.35	0.34	0.33	0.32
		800°	0.42	0.40	0.39	0.37	0.36	0.35	0.33	0.32	0.31	0.30	0.29
		900°	0.39	0.37	0.36	0.35	0.33	0.32	0.31	0.30	0.29	0.28	0.27
		1000°	0.36	0.35	0.33	0.32	0.31	0.30	0.29	0.28	0.27	0.26	0.25

Standard Air Density, Sea Level, 70°F = 0.075 lb/cu ft

Reprinted with permission from *HVAC Duct System Design*

Figure 6.56 cont.

METRIC UNITS

4. ELECTRIC EQUATIONS (10)

a) $\text{Bhp} = \dfrac{\text{I x E x P.F. x Eff.}}{746}$ (Single Phase)

b) $\text{Bhp} = \dfrac{\text{I x E x P.F. x Eff. x 1.73}}{746}$ (Three Phase)

c) $E = IR$

d) $P = EI$

e) $\dfrac{\text{F.L. Amps* x Voltage*}}{\text{Actual Voltage}} = \text{Actual F.L. Amps}$

*Nameplate ratings

I = Amps (A)
E = Volts (V)
P.F. = Power factor
R = ohms (Ω)
P. = watts (W)

Table 6-16 AIR DENSITY CORRECTION FACTORS (Metric Units) (10)

Altitude (m)	Sea Level	250	500	750	1000	1250	1500	1750	2000	2500	3000
Barometer (kPa)	101.3	98.3	96.3	93.2	90.2	88.2	85.1	83.1	80.0	76.0	71.9
Air Temp. °C 0°	1.08	1.05	1.02	0.99	0.96	0.93	0.91	0.88	0.86	0.81	0.76
21°	1.00	0.97	0.95	0.92	0.89	0.87	0.84	0.82	0.79	0.75	0.71
50°	0.91	0.89	0.86	0.84	0.81	0.79	0.77	0.75	0.72	0.68	0.64
75°	0.85	0.82	0.80	0.78	0.75	0.73	0.71	0.69	0.67	0.63	0.60
100°	0.79	0.77	0.75	0.72	0.70	0.68	0.66	0.65	0.63	0.59	0.56
125°	0.74	0.72	0.70	0.68	0.66	0.64	0.62	0.60	0.59	0.55	0.52
150°	0.70	0.68	0.66	0.64	0.62	0.60	0.59	0.57	0.55	0.52	0.49
175°	0.66	0.64	0.62	0.62	0.59	0.57	0.55	0.54	0.52	0.49	0.46
200°	0.62	0.61	0.59	0.57	0.56	0.54	0.52	0.51	0.49	0.47	0.44
225°	0.59	0.58	0.56	0.54	0.53	0.51	0.50	0.48	0.47	0.44	0.42
250°	0.56	0.55	0.53	0.52	0.50	0.49	0.47	0.46	0.45	0.42	0.40
275°	0.54	0.52	0.51	0.49	0.48	0.47	0.45	0.44	0.43	0.40	0.38
300°	0.51	0.50	0.49	0.47	0.46	0.45	0.43	0.42	0.41	0.38	0.36
325°	0.49	0.48	0.47	0.45	0.44	0.43	0.41	0.40	0.39	0.37	0.35
350°	0.47	0.46	0.45	0.43	0.42	0.41	0.40	0.39	0.38	0.35	0.33
375°	0.46	0.44	0.43	0.42	0.41	0.39	0.38	0.37	0.36	0.34	0.32
400°	0.44	0.43	0.41	0.40	0.39	0.38	0.37	0.36	0.35	0.33	0.31
425°	0.42	0.41	0.40	0.39	0.38	0.37	0.35	0.34	0.33	0.32	0.30
450°	0.41	0.40	0.38	0.37	0.36	0.35	0.34	0.33	0.32	0.31	0.29
475°	0.39	0.38	0.37	0.36	0.35	0.34	0.33	0.32	0.31	0.29	0.28
500°	0.38	0.37	0.36	0.35	0.34	0.33	0.32	0.31	0.30	0.28	0.27
525°	0.37	0.36	0.35	0.34	0.33	0.32	0.31	0.30	0.29	0.27	0.26

Dry air weight at Sea Level, 21°C = 1.205 kg/m³

Reprinted with permission from *HVAC Duct System Design*, SMACNA

Figure 6.56 cont.

Resource Publications

ASHRAE Fundamentals Handbook. American Society of Heating, Refrigerating, and Air Conditioning Engineers, Inc., Atlanta GA.

ASHRAE Handbook: HVAC Applications. American Society of Heating, Refrigerating, and Air Conditioning Engineers, Inc., Atlanta GA.

ASHRAE Handbook: HVAC Systems and Equipment. American Society of Heating, Refrigerating, and Air Conditioning Engineers, Inc., Atlanta GA.

HVAC: Design Criteria, Options, Selection, by William H. Rowe, PE, AIA. R.S. Means Co., Inc., Kingston, MA, 1994.

HVAC Duct System Design, Sheet Metal and Air Conditioning Contractors National Association, Inc., Vienna, VA, 1977.

HVAC Systems Evaluation, by Harold R. Colen, PE. R.S. Means Co., Inc., Kingston, MA, 1990.

Trane Air Conditioning Manual. The Trane Company, La Crosse, WI, 1996.

Industrial Ventilation Workbook, 4th ed., by D. Jeff Burton, PE, CIH. IVE Press, Inc., Bountiful, UT, 1999.

VentCalc Calculator-Aided System Design Software Package. IVE, Inc., Bountiful, UT. Programs include duct design, airflow in duct work, dilution-related equations, and other ventilation and O H & S calculations.

Chapter 7

Mechanical Engineering

Chapter 7 Mechanical Engineering

Mechanical engineers are concerned with the design and operation of mechanical devices or systems, particularly those that consume or produce energy. Examples of mechanical systems are everywhere: pump and piping systems, heating and air-conditioning systems, and steam generators, to name a few.

Mechanical engineering encompasses several basic engineering sciences, including physics, materials science, solid mechanics, fluid mechanics, and thermodynamics. These sciences form the theoretical foundation of the field. Although facilities engineers and managers are rarely called upon to perform mechanical engineering design, a basic comprehension of these engineering sciences is useful and sometimes necessary in understanding the many complex mechanical systems found in modern facilities.

Engineers use mathematics to apply these mechanical engineering principles to practical situations in facilities engineering and management—to create new designs, predict behavior, minimize cost, and optimize operations. Several mathematical formulas are listed in this chapter; others can be found in the *Fundamentals of Engineering Discipline Specific Reference Handbook*, published by the National Council of Examiners for Engineering and Surveying.

The first section of this chapter provides brief overviews of three major areas of mechanical engineering:

- Statics & Dynamics, including principles such as moments, conditions of equilibrium, levers, and motion and acceleration;
- Fluid Mechanics, including hydrostatic pressure, tank discharge, and an introduction to pumps;
- Thermodynamics, including energy, work and power, laws of energy (conservation, potential, kinetics, spring, and efficiency); and temperature and vapors.

The second section, entitled "Additional Information," provides more detail on some of the topics addressed in the first section and delves into some additional areas of mechanical engineering, including steam, pulleys, free body diagrams, moment of inertia, and strain and stress. The second section may be of particular interest to participants in the AFE CPE Review Program.

Chapter 7 concludes with an overview of pumps, hydraulics, and air compressors.

Ed. Note: Fire protection is not addressed in this chapter because of the excellent resources readily available to the facility engineer. Foremost among these is the National Fire Protection Agency, which offers a variety of publications and other materials on fire safety regulations. The NFPA is also an excellent source for information on protection products like detection and extinguishing equipment. Finally, the Uniform Building Code delineates requirements for specific usages in Section 904: Fire Extinguishing Systems.

The following sections are reprinted with permission from the *Engineer-In-Training Reference Manual*, by Michael R. Lindeburg, Professional Publications, Inc.

INTERNAL AND EXTERNAL FORCES

An *external force* is a force on a rigid body caused by other bodies. The applied force can be due to physical contact (i.e., pushing) or close proximity (e.g., gravitational, magnetic, or electrostatic forces). If unbalanced, an external force will cause motion of the body.

An *internal force* is one which holds parts of the rigid body together. Internal forces are the tensile and compressive forces within parts of the body, as found from the product of stress and area. Although internal forces can cause deformation of a body, motion is never caused by internal forces.

MOMENTS

Moment is the name given to the tendency of a force to rotate, turn, or twist a rigid body about an actual or assumed pivot point. (Another name for moment is *torque*, although torque is used mainly with shafts and other power-transmitting machines.) When acted upon by a moment, unrestrained bodies rotate. However, rotation is not required for the moment to exist. When a restrained body is acted upon by a moment, there is no rotation.

An object experiences a moment whenever a force is applied to it.[5] Only when the line of action of the force passes through the center of rotation (i.e., the actual or assumed pivot point) will the moment be zero.

Moments have primary dimensions of length × force. Typical units are foot-pounds, inch-pounds, and newton-meters.[6,7]

COUPLES

Any pair of equal, opposite, and parallel forces constitute a *couple*. A couple is equivalent to a single moment vector. Since the two forces are opposite in sign, the x-, y-, and z-components of the forces cancel out. Therefore, a body is induced to rotate without translation. A couple can be counteracted only by another couple. A couple can be moved to any location without affecting the equilibrium requirements.

[5]The moment may be zero, as when the moment arm length is zero, but there is a (trivial) moment nevertheless.

[6]Units of kilogram-force-meter have also been used in metric countries.

[7]Foot-pounds and newton-meters are also the units of energy. To distinguish between moment and energy, some authors reverse the order of the units. Therefore, pound-feet and meter-newtons become the units of moment. This convention is unnecessary and not universal, since the context is adequate to distinguish between the two.

DISTRIBUTED LOADS

If an object is continuously loaded over a portion of its length, it is subject to a *distributed load*. Distributed loads result from *dead load* (i.e., self-weight), hydrostatic pressure, and materials distributed over the object.

If the load per unit length at some point x is $w(x)$, the statically equivalent concentrated load, F_R, can be found from Eq. 32.27. The equivalent load is the area under the loading curve.

$$F_R = \int_{x=0}^{x=L} w(x)\,dx \qquad 32.27$$

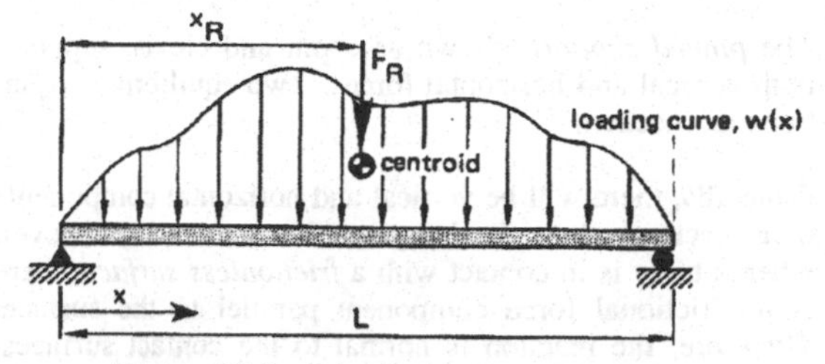

Figure 32.6 Distributed Loads on a Beam

The location, x_R, of the equivalent load coincides with the centroid of the area under the loading curve and is referred to in some problems as the *center of pressure*.

CONDITIONS OF EQUILIBRIUM

An object is static when it is stationary. To be stationary, all of the forces on the object must be in equilibrium.[11] For an object to be in equilibrium, the resultant force and moment vectors must both be zero.

$$\mathbf{F}_R = \sum \mathbf{F} = 0 \qquad 32.34$$

$$F_R = \sqrt{F_{R,x}^2 + F_{R,y}^2 + F_{R,z}^2} = 0 \qquad 32.35$$

$$\mathbf{M}_R = \sum \mathbf{M} = 0 \qquad 32.36$$

$$M_R = \sqrt{M_{R,x}^2 + M_{R,y}^2 + M_{R,z}^2} = 0 \qquad 32.37$$

[11]Thus the term *static equilibrium*, though widely used, is redundant.

REACTIONS

The first step in solving most statics problems is to determine the reaction forces (i.e., the *reactions*) supporting the body. The manner in which a body is supported determines the type, location, and direction of the reactions.

For beams, the two most common types of supports are the roller support and the pinned support. The *roller support*, shown as a cylinder supporting the beam, supports vertical forces only. Rather than support a horizontal force, a roller support simply rolls into a new equilibrium position. Only one equilibrium equation (i.e., the sum of vertical forces) is needed at a roller support. Generally, the terms *simple support* and *simply-supported* refer to a roller support.

The *pinned support*, shown as a pin and clevis, supports both vertical and horizontal forces. Two equilibrium equations are needed.

Generally, there will be vertical and horizontal components of a reaction when one body touches another. However, when a body is in contact with a *frictionless surface*, there is no frictional force component parallel to the surface. Therefore, the reaction is normal to the contact surfaces. The assumption of frictionless contact is particularly useful when dealing with systems of spheres and cylinders in contact with rigid supports. Frictionless contact is also assumed for roller and rocker supports.[12]

LEVERS

A *lever* is a simple mechanical machine with the ability to increase an applied force. The ratio of the load-bearing force to applied force (i.e., the *effort*) is known as the *mechanical advantage* or *force amplification*. As Fig. 32.12 shows, the mechanical advantage is equal to the ratio of lever arms.

$$\text{mechanical advantage} = \frac{F_{\text{load}}}{F_{\text{applied}}} = \frac{\text{applied force lever arm}}{\text{load lever arm}} = \frac{\text{distance moved by applied force}}{\text{distance moved by load}} \qquad 32.45$$

[12]Frictionless surface contact, which requires only one equilibrium equation, should not be confused with a frictionless pin connection, which requires two equilibrium equations. A pin connection with friction introduces a moment at the connection, increasing the number of required equilibrium equations to three.

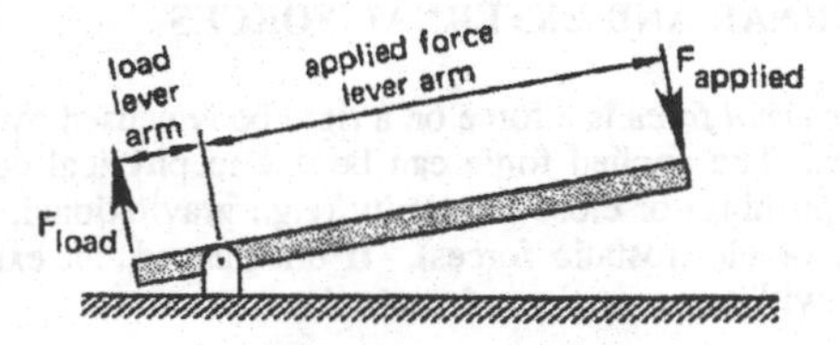

Figure 32.12 A Lever

CENTROID OF AN AREA

The *centroid* of an area is analogous to the center of gravity of a homogeneous body.[1] The centroid is often described as the point at which a thin homogeneous plate would balance. This definition, however, combines the definitions of centroid and center of gravity, and implies gravity is required to identify the centroid, which is not true.

The location of the centroid of an area bounded by the x- and y-axes and the mathematical function $y = f(x)$ can be found by the *integration method* by using Eqs. 39.1 through 39.4. The centroidal location depends only on the geometry of the area and is identified by the coordinates (x_c, y_c). Some references place a bar over the coordinates of the centroid to indicate an average point, such as $(\overline{x}, \overline{y})$.

$$x_c = \frac{\int x\,dA}{A} \qquad 39.1$$

$$y_c = \frac{\int y\,dA}{A} \qquad 39.2$$

$$A = \int f(x)\,dx \qquad 39.3$$

$$dA = f(x)\,dx = g(y)\,dy \qquad 39.4$$

The locations of the centroids of *basic shapes*, such as triangles and rectangles, are well known. The most common basic shapes have been included in the enclosed figure. There should be no need to derive centroidal locations for these shapes by the integration method.

[1]The analogy has been simplified. A three-dimensional body also has a centroid. However, the centroid and center of gravity will not coincide unless the body is homogeneous.

Centroids and
Moments of Inertia of
Common Geometric Shapes

Rectangle

$\bar{I}_{x'} = \frac{1}{12}bh^3$

$\bar{I}_{y'} = \frac{1}{12}b^3h$

$I_x = \frac{1}{3}bh^3$

$I_y = \frac{1}{3}b^3h$

$J_O = \frac{1}{12}bh(b^2 + h^2)$

$\bar{x} = b/2$

$y = h/2$

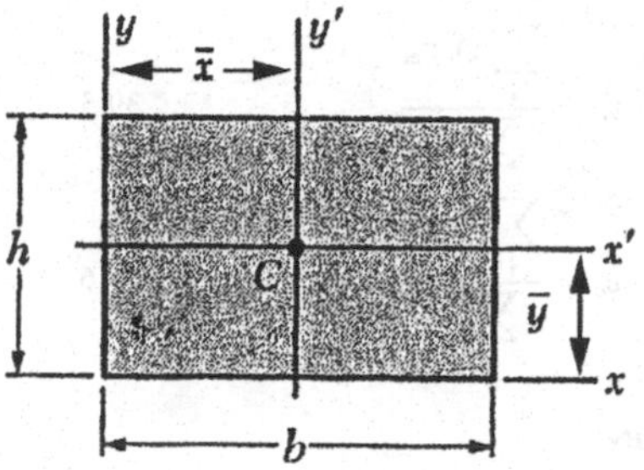

Triangle

$\bar{I}_{x'} = \frac{1}{36}bh^3$

$I_x = \frac{1}{12}bh^3$

$\bar{y} = \frac{h}{3}$

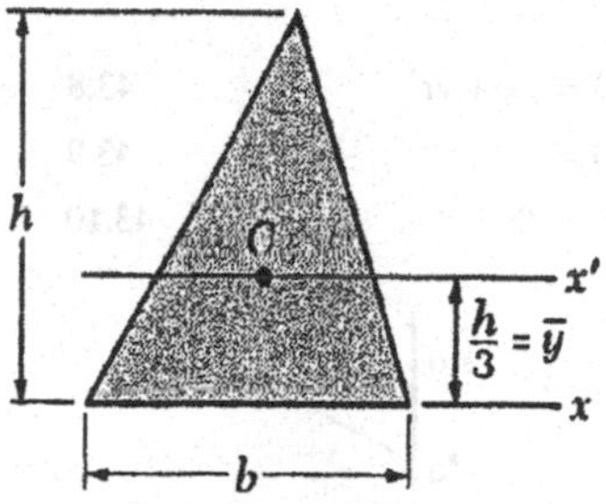

Circle

$I_x = I_y = \frac{1}{4}\pi r^4$

$J_O = \frac{1}{2}\pi r^4$

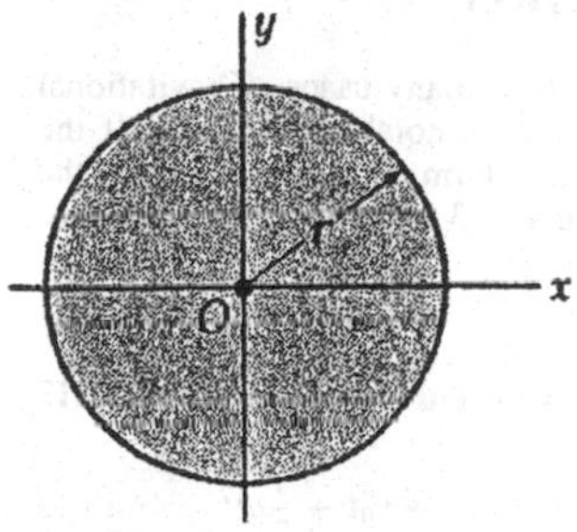

Semicircle

$I_x = I_y = \frac{1}{8}\pi r^4$

$J_O = \frac{1}{4}$

$\bar{y} = \frac{4r}{3\pi}$

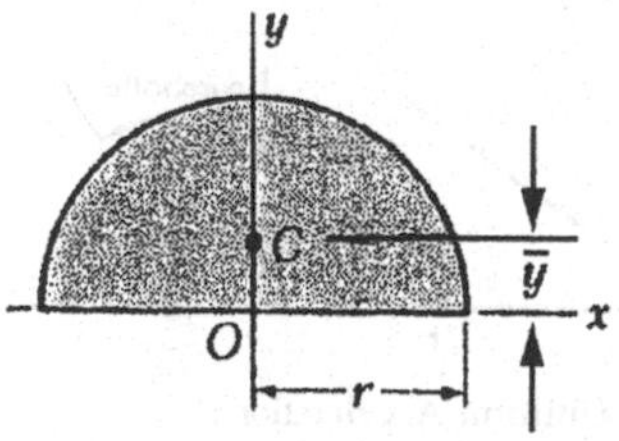

The centroid of a complex area can be found from Eqs. 39.5 and 39.6 if the area can be divided into the basic shapes in enclosed figure.

$$x_c = \frac{\sum_i A_i x_{ci}}{\sum A_i} \qquad 39.5$$

$$y_c = \frac{\sum_i A_i y_{ci}}{\sum A_i} \qquad 39.6$$

UNIFORM MOTION

The term *uniform motion* means uniform velocity. The velocity is constant and the acceleration is zero. For a constant velocity system, the position function varies linearly with time.

$$s(t) = s_0 + \mathrm{v}t \qquad 43.8$$

$$\mathrm{v}(t) = \mathrm{v} \qquad 43.9$$

$$a(t) = 0 \qquad 43.10$$

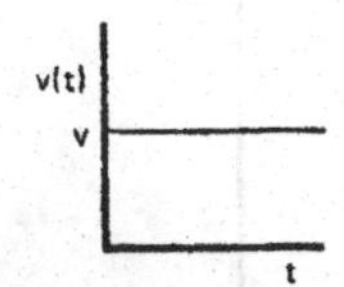

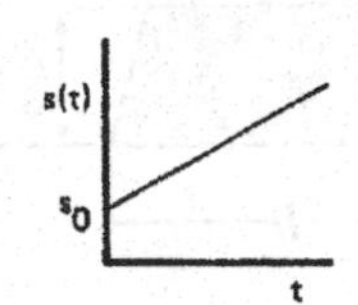

Figure 43.2 Constant Velocity System

UNIFORM ACCELERATION

The acceleration is constant in many cases. (Gravitational acceleration, where $a = g$, is a notable example.) If the acceleration is constant, the a term can be taken out of the integrals in Eqs. 43.12 and 43.13.

$$a(t) = a \qquad 43.11$$

$$\mathrm{v}(t) = a\int dt = \mathrm{v}_0 + at \qquad 43.12$$

$$s(t) = a\iint dt^2 = s_0 + \mathrm{v}_0 t + \frac{1}{2}at^2 \qquad 43.13$$

Table 43.1 summarizes the equations required to solve most uniform acceleration problems.

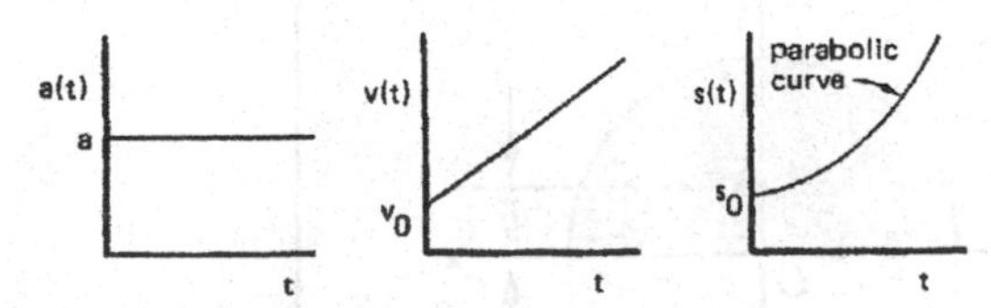

Figure 43.3 Uniform Acceleration

Table 43.1
Uniform Acceleration Formulas[a]

to find	given these	use this equation
a	$t, \mathrm{v}_0, \mathrm{v}$	$a = \frac{\mathrm{v} - \mathrm{v}_0}{t}$
a	t, v_0, s	$a = \frac{2s - 2\mathrm{v}_0 t}{t^2}$
a	$\mathrm{v}_0, \mathrm{v}, s$	$a = \frac{\mathrm{v}^2 - \mathrm{v}_0^2}{2s}$
s	t, a, v_0	$s = \mathrm{v}_0 t + \frac{1}{2}at^2$
s	$a, \mathrm{v}_0, \mathrm{v}$	$s = \frac{\mathrm{v}^2 - \mathrm{v}_0^2}{2a}$
s	$t, \mathrm{v}_0, \mathrm{v}$	$s = \frac{1}{2}t(\mathrm{v}_0 + \mathrm{v})$
t	$a, \mathrm{v}_0, \mathrm{v}$	$t = \frac{\mathrm{v} - \mathrm{v}_0}{a}$
t	a, v_0, s	$t = \frac{\sqrt{\mathrm{v}_0^2 + 2as} - \mathrm{v}_0}{a}$
t	$\mathrm{v}_0, \mathrm{v}, s$	$t = \frac{2s}{\mathrm{v}_0 + \mathrm{v}}$
v_0	t, a, v	$\mathrm{v}_0 = \mathrm{v} - at$
v_0	t, a, s	$\mathrm{v}_0 = \frac{s}{t} - \frac{1}{2}at$
v_0	a, v, s	$\mathrm{v}_0 = \sqrt{\mathrm{v}^2 - 2as}$
v	t, a, v_0	$\mathrm{v} = \mathrm{v}_0 + at$
v	a, v_0, s	$\mathrm{v} = \sqrt{\mathrm{v}_0^2 + 2as}$

[a]The table can be used for rotational problems by substituting α, ω, and θ for a, v, and s, respectively.

DYNAMIC EQUILIBRIUM

An accelerating body is not in static equilibrium. Accordingly, the familiar equations of statics ($\sum F = 0$ and $\sum M = 0$) do not apply. However, if the so-called *inertial force*, ma, is included in the static equilibrium equation, the body is said to be in *dynamic equilibrium*.[4,5] This is known as *D'Alembert's principle*. Since the inertial force acts to oppose changes in motion, it is negative in the summation.

$$\sum \mathbf{F} - m\mathbf{a} = 0 \quad \text{[SI]} \qquad 44.17(a)$$

$$\sum \mathbf{F} - \frac{m\mathbf{a}}{g_c} = 0 \quad \text{[U.S.]} \qquad 44.17(b)$$

It should be clear that D'Alembert's principle is just a different form of Newton's second law, with the ma term transposed to the left-hand side.

The analogous rotational form of the dynamic equilibrium principle is

$$\sum \mathbf{T} - I\alpha = 0 \quad \text{[SI]} \qquad 44.18(a)$$

$$\sum \mathbf{T} - \frac{I\alpha}{g_c} = 0 \quad \text{[U.S.]} \qquad 44.18(b)$$

FLAT FRICTION

Friction is a force that always resists motion or impending motion. It always acts parallel to the contacting surfaces. The frictional force, F_f, exerted on a stationary body is known as *static friction*, *Coulomb friction*, and *fluid friction*. If the body is moving, the friction is known as *dynamic friction* and is less than the static friction.

The actual magnitude of the frictional force depends on the *normal force*, N, and the *coefficient of friction*, f, between the body and the surface.[6] For a body resting on a horizontal surface, the normal force is the weight of the body.

$$N = mg \quad \text{[SI]} \qquad 44.19(a)$$

$$N = \frac{mg}{g_c} \quad \text{[U.S.]} \qquad 44.19(b)$$

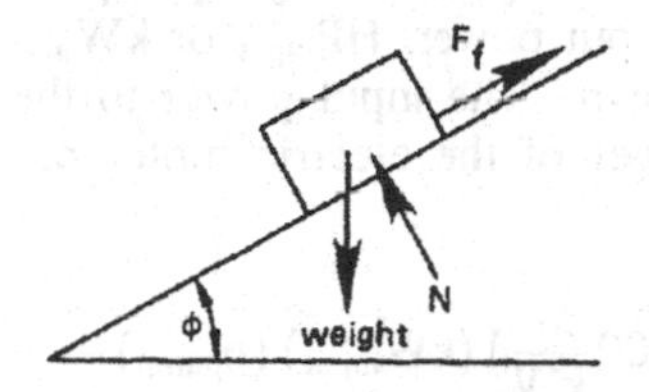

Figure 44.6 Frictional and Normal Forces

If the body rests on an inclined surface, the normal force is calculated from the weight.

$$N = mg\cos\phi \quad \text{[SI]} \qquad 44.20(a)$$

[4]Other names for the *inertial force* are *inertia vector* (when written as ma), *dynamic reaction*, and *reversed effective force*. The term $\sum \mathbf{F}$ is known as the *effective force*.

[5]"Dynamic" and "equilibrium" are contradictory terms. A better term is simulated equilibrium, but this form has not caught on.

[6]The symbol μ is also widely used by engineers to represent the coefficient of friction.

The maximum static frictional force, F_f, is the product of the coefficient of friction, f, and the normal force, N. [The subscripts s and k are used to distinguish between the static and dynamic (kinetic) coefficients of friction.]

$$F_{f,\max} = f_s N \qquad 44.21$$

The frictional force acts only in response to a disturbing force. If a small disturbing force (i.e., a force less than $F_{f,\max}$) acts on a body, then the frictional force will equal the disturbing force, and the maximum frictional force will not develop. This is known as the *equilibrium phase*. The *motion impending phase* is where the disturbing force equals the maximum frictional force, $F_{f,\max}$. Once motion begins, however, the coefficient of friction drops slightly, and a lower frictional force opposes movement. These cases are illustrated in Fig. 44.7.

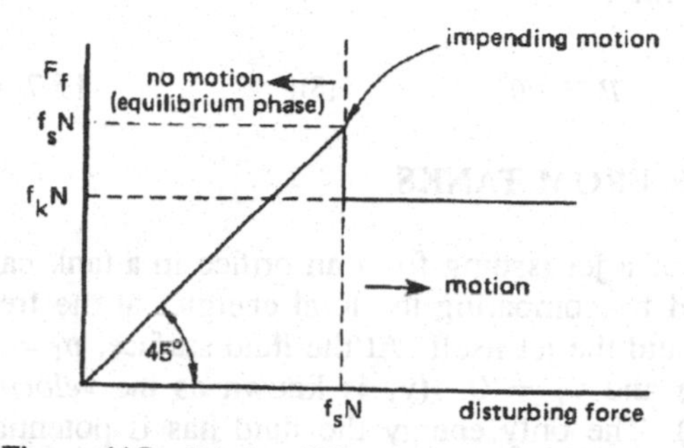

Figure 44.7 Frictional Force versus Disturbing Force

A body on an inclined plane will not begin to slip down the plane until the component of weight parallel to the plane exceeds the frictional force. If the plane's inclination angle can be varied, the body will not slip until the angle reaches a critical angle known as the *angle of repose* or *angle of static friction*, ϕ. Equation 44.22 relates this angle to the coefficient of static friction.

$$\tan\phi = f_s \qquad 44.22$$

HYDROSTATIC PRESSURE

Hydrostatic pressure is the pressure a fluid exerts on an immersed object or container walls.[4] Pressure is equal to the force per unit area of surface.

$$p = \frac{F}{A} \qquad 15.6$$

FLUID HEIGHT EQUIVALENT TO PRESSURE

Pressure varies linearly with depth. The relationship between pressure and depth for an incompressible fluid is given by Eq. 15.7.

$$p = \rho g h \quad \text{[SI]} \qquad 15.7(a)$$

DISCHARGE FROM TANKS

The velocity of a jet issuing from an orifice in a tank can be determined by comparing the total energies at the free fluid surface and the jet itself. At the fluid surface, $p_1 = 0$ (atmospheric) and $v_1 = 0$. (v_1 is known as the *velocity of approach.*) The only energy the fluid has is potential energy. At the jet, $p_2 = 0$. All of the potential energy difference $(z_1 - z_2)$ has been converted to kinetic energy. The theoretical velocity of the jet can be derived from the Bernoulli equation. Equation 17.61 is known as the equation for *Torricelli's speed of efflux*.

$$v_t = \sqrt{2gh} \qquad 17.61$$

$$h = z_1 - z_2 \qquad 17.62$$

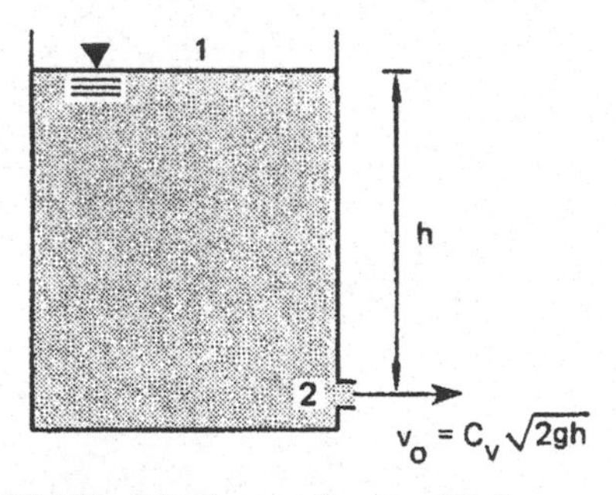

Figure 17.10 Discharge from a Tank

[4]The term *hydrostatic* is used with all fluids, not just water.

INTRODUCTION TO PUMPS AND TURBINES

A *pump* adds energy to the fluid flowing through it. The amount of energy that a pump puts into the fluid stream can be determined by the difference between the total energy on either side of the pump. In most situations, a pump will add primarily pressure energy. The specific energy added (a positive number) on a per-unit mass basis (i.e., J/kg or ft-lbf/lbm) is given by Eq. 17.53.

$$E_A = E_{t,2} - E_{t,1} \qquad 17.53$$

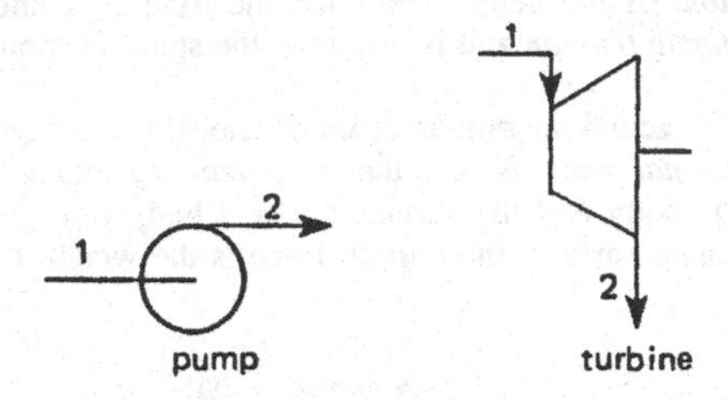

Figure 17.6 Pump and Turbine Representation

The specific energy added by a pump can also be calculated from the input power, HP_{input} or kW_{input}, if the mass flow rate is known. The input power to the pump will be the output power of the electric motor or engine driving the pump.

$$E_A = \frac{\left(1000 \frac{\text{W}}{\text{kW}}\right)(\text{kW}_{\text{input}})(\eta_{\text{pump}})}{\dot{m}} \quad \text{[SI]} \qquad 17.55(a)$$

$$E_A = \frac{\left(550 \frac{\text{ft-lbf}}{\text{sec-hp}}\right)(\text{HP}_{\text{input}})(\eta_{\text{pump}})}{\dot{m}} \quad \text{[U.S.]} \qquad 17.55(b)$$

The *water horsepower*, WHP, (also known as the *hydraulic horsepower* and *theoretical horsepower*), is the amount of power actually entering the fluid.

$$\text{WHP} = (\text{HP}_{\text{input}})\,\eta_{\text{pump}} \qquad 17.56$$

A *turbine* extracts energy from the fluid flowing through it. As with a pump, the energy extraction can be obtained by evaluating the Bernoulli equation on both sides of the turbine and taking the difference. The energy extracted (a positive number) on a per-unit mass basis is given by Eq. 17.57.

$$E_E = E_{t,1} - E_{t,2} \qquad 17.57$$

EXTENDED BERNOULLI EQUATION

The original Bernoulli equation assumes frictionless flow and does not consider the effects of pumps and turbines. When friction is present, and when there are minor losses such as fittings and other energy-related devices in a pipeline, the energy balance is affected. The *extended Bernoulli equation* takes these additional factors into account.

$$(E_p + E_v + E_z)_1 + E_A = (E_p + E_v + E_z)_2 + E_E + E_f + E_m \quad 17.58$$

$$\frac{p_1}{\rho} + \frac{v_1^2}{2} + z_1 g + E_A = \frac{p_2}{\rho} + \frac{v_2^2}{2} + z_2 g + E_E + E_f + E_m \quad \text{[SI]} \quad 17.59(a)$$

As defined, E_A, E_E, and E_f are all positive terms. None of the terms in Eq. 17.58 is negative.

The concepts of sources and sinks can be used to decide whether the friction, pump, and turbine terms appear on the left or right side of the Bernoulli equation. An *energy source* puts energy into the system. The incoming fluid and a pump contribute energy to the system. An *energy sink* removes energy from the system. The leaving fluid, friction, and a turbine remove energy from the system. In an energy balance, all energy must be accounted for, and the energy sources just equal the energy sinks.

$$\sum E_{\text{sources}} = \sum E_{\text{sinks}} \quad 17.60$$

Therefore, the energy added by a pump always appears on the left-hand side of the Bernoulli equation. Similarly, the frictional energy loss always appears on the right-hand side.

PUMPING POWER

The energy (head) added by a pump can be determined from the difference in total energy on either side of the pump. Writing the Bernoulli equation for the discharge and suction conditions produces Eq. 18.5.

$$h_A = h_{t(d)} - h_{t(s)} \quad 18.5$$

$$h_A = \frac{p_d - p_s}{\rho g} + \frac{v_d^2 - v_s^2}{2g} + (z_d - z_s) \quad \text{[SI]} \quad 18.6(a)$$

The pumping power depends on the head added (h_A) and the mass flow rate. For example, the product $\dot{m} h_A$ has the units of foot-pounds per second (in customary U.S. units), which can be easily converted to horsepower. Pump output power is known as *hydraulic power* or *water power*. Hydraulic power is the net power actually transferred to the fluid.

Horsepower is the unit of power in the United States and other non-SI countries, which gives rise to the terms *hydraulic horsepower* and *water horsepower* (WHP). Various relationships for finding the hydraulic horsepower are given in Table 18.2.

Table 18.2
Hydraulic Horsepower Equations[(a)]

	Q in gpm	$\dot{m}$ lbm/sec	$\dot{V}$ in cfs
h_A in feet	$\frac{h_A Q(\text{SG})}{3956}$	$\frac{h_A \dot{m}}{550} \times \frac{g}{g_c}$	$\frac{h_A \dot{V}(\text{SG})}{8.814}$
Δp in psi	$\frac{\Delta p Q}{1714}$	$\frac{\Delta p \dot{m}}{(238.3)(\text{SG})} \times \frac{g}{g_c}$	$\frac{\Delta p \dot{V}}{3.819}$
Δp in psf	$\frac{\Delta p Q}{2.468 \times 10^5}$	$\frac{\Delta p \dot{m}}{(34,320)(\text{SG})} \times \frac{g}{g_c}$	$\frac{\Delta p \dot{V}}{550}$
W in $\frac{\text{ft-lbf}}{\text{lbm}}$	$\frac{WQ(\text{SG})}{3956}$	$\frac{W\dot{m}}{550}$	$\frac{W\dot{V}(\text{SG})}{8.814}$

[(a)]Table 18.2 is based on $\rho_{\text{water}} = 62.4$ lbm/ft^3 and $g = 32.2$ ft/sec^2. (Multiply horsepower by 0.7457 to obtain kilowatts.)

AFFINITY LAWS

Most parameters (impeller diameter, speed, and flow rate) determining a pump's performance can vary. If the impeller diameter is held constant and the speed varied, the following ratios are maintained with no change of efficiency:

$$\frac{Q_2}{Q_1} = \frac{n_2}{n_1} \quad 18.30$$

$$\frac{h_2}{h_1} = \left(\frac{n_2}{n_1}\right)^2 = \left(\frac{Q_2}{Q_1}\right)^2 \quad 18.31$$

$$\frac{P_2}{P_1} = \left(\frac{n_2}{n_1}\right)^3 = \left(\frac{Q_2}{Q_1}\right)^3 \quad 18.32$$

If the speed is held constant and the impeller size varied,

$$\frac{Q_2}{Q_1} = \frac{D_2}{D_1} \quad 18.33$$

$$\frac{h_2}{h_1} = \left(\frac{D_2}{D_1}\right)^2 \quad 18.34$$

$$\frac{P_2}{P_1} = \left(\frac{D_2}{D_1}\right)^3 \quad 18.35$$

ENERGY OF A MASS

The *energy* of a mass represents the capacity of the mass to do work. Such energy can be stored and released. There are many forms that it can take, including mechanical, thermal, electrical, and magnetic energies. Energy is a positive, scalar quantity (although the change in energy can be either positive or negative).

The total energy of a body can be calculated from its mass, m, and the *specific energy*, U (i.e., the energy per unit mass).[1]

$$E = mU \tag{20.1}$$

LAW OF CONSERVATION OF ENERGY

The *law of conservation of energy* says that energy cannot be created or destroyed. However, energy can be converted into different forms. Therefore, the sum of all energy forms is constant.

$$\sum E = \text{constant} \tag{20.3}$$

WORK

Work, W, is the act of changing the energy of a particle, body, or system. For a mechanical system, *external work* is work done by an external force, whereas *internal work* is done by an internal force. Work is a signed, scalar quantity. Typical units are inch-pounds, foot-pounds, and joules. Mechanical work is seldom expressed in British thermal units and kilocalories.

For a mechanical system, work is positive when a force acts in the direction of motion and helps a body move from one location to another. Work is negative when a force acts to oppose motion. (Friction, for example, always opposes the direction of motion and can only do negative work.) The work done on a body by more than one force can be found by superposition.

From a thermodynamic standpoint, work is positive if a particle or body does work on its surroundings. Work is negative if the surroundings do work on the object. (Thus, blowing up a balloon represents negative work to the balloon.) Although this may be a difficult concept, it is consistent with the conservation of energy, since the sum of negative work and the positive energy increase is zero (i.e., no net energy change in the system).[2]

The work done by a force or torque of constant magnitude is

$$W_{\text{constant force}} = \mathbf{F}\cdot\mathbf{s} = Fs\cos\phi \quad \text{[linear systems]} \tag{20.6}$$

$$W_{\text{constant torque}} = \mathbf{T}\cdot\boldsymbol{\theta} = T\theta\cos\phi \quad \text{[rotational systems]} \tag{20.7}$$

The non-vector forms of Eqs. 20.6 and 20.7 illustrate that only the component of force in the direction of motion contributes to work.

[1]The use of symbols E and U is not consistent in the engineering field.
[2]This is just a partial statement of the *first law of thermodynamics*.

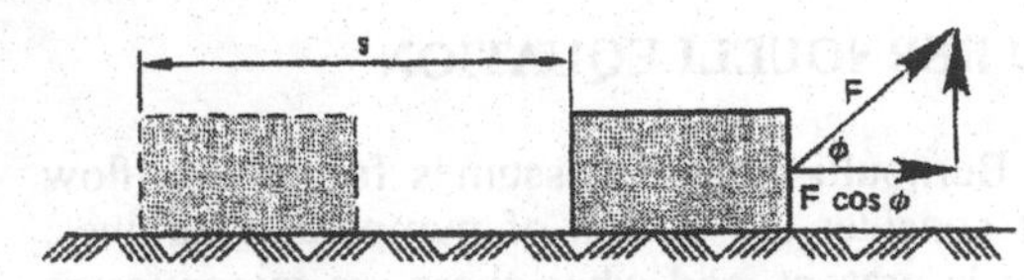

Figure 20.1 Work of a Constant Force

Common applications of the work done by a constant force are frictional work and gravitational work. The work to move an object a distance s against a frictional force of F_f is

$$W_{\text{friction}} = F_f s \tag{20.8}$$

The work done by gravity when a mass m changes from elevation h_1 to h_2 is

$$W_{\text{gravity}} = mg(h_2 - h_1) \quad \text{[SI]} \tag{20.9(a)}$$

The work done by or on a *linear spring* whose length or deflection changes from x_1 to x_2 is given by Eq. 20.10.[3] It does not make any difference whether the spring is a compression spring or extension spring.

$$W_{\text{spring}} = \frac{1}{2}k\left(x_2^2 - x_1^2\right) \tag{20.10}$$

POTENTIAL ENERGY OF A MASS

Potential energy (gravitational energy) is a form of mechanical energy possessed by a body due to its relative position in a gravitational field. Potential energy is lost when the elevation of a body decreases. The lost potential energy usually is converted to kinetic energy or heat.

$$E_{\text{potential}} = mgh \quad \text{[SI]} \tag{20.11(a)}$$

In the absence of friction and other non-conservative forces, the change in potential energy of a body is equal to the work required to change the elevation of the body.

$$W = \Delta E_{\text{potential}} \tag{20.12}$$

KINETIC ENERGY OF A MASS

Kinetic energy is a form of mechanical energy associated with a moving or rotating body. The kinetic energy of a body moving with instantaneous linear velocity v is

$$E_{\text{kinetic}} = \frac{1}{2}mv^2 \quad \text{[SI]} \tag{20.13(a)}$$

According to the *work-energy principle* (see Section 20.9), the kinetic energy is equal to the work necessary to initially accelerate a stationary body, or to bring a moving body to rest.

$$W = \Delta E_{\text{kinetic}} \tag{20.14}$$

A body can also have rotational kinetic energy.

$$E_{rotational} = \frac{1}{2}I\omega^2 \quad \text{[SI]} \qquad 20.15(a)$$

SPRING ENERGY

A spring is an energy storage device, since the spring has the ability to perform work. In a perfect spring, the amount of energy stored is equal to the work required to compress the spring initially. The stored spring energy does not depend on the mass of the spring. Given a spring with spring constant (stiffness) k, the *spring energy* is

$$E_{spring} = \frac{1}{2}kx^2 \qquad 20.16$$

PRESSURE ENERGY OF A MASS

Since work is done in increasing the pressure of a system (i.e., it takes work to blow up a balloon), mechanical energy can be stored in pressure form. This is known as *pressure energy, static energy, flow energy, flow work,* and *pV work (energy)*. For a system of pressurized mass m, the flow energy is

$$E_{flow} = \frac{mp}{\rho} = mpv \qquad 20.17$$

INTERNAL ENERGY OF A MASS

The total internal energy, usually given the symbol U, of a body increases when the body's temperature increases.[4] In the absence of any work done on or by the body, the change in internal energy is equal to the heat flow, Q, into the body. Q is positive if the heat flow is into the body and negative otherwise.

$$U_2 - U_1 = Q \qquad 20.18$$

The property of internal energy is encountered primarily in thermodynamics problems. Typical units are British thermal units, joules, and kilocalories.

WORK-ENERGY PRINCIPLE

Since energy can neither be created nor destroyed, external work performed on a conservative system goes into changing the system's total energy. This is known as the *work-energy principle* (or *principle of work and energy*).

$$W = \Delta E = E_2 - E_1 \qquad 20.19$$

[4]The *thermal energy*, represented by the body's enthalpy, is the sum of internal and flow energies. A more detailed discussion of thermal energy appears in the following chapters.

CONVERSION BETWEEN ENERGY FORMS

Conversion of one form of energy into another does not violate the conservation of energy law. However, most problems involving conversion of energy are really just special cases of the work-energy principle. For example, consider a falling body that is acted upon by a gravitational force. The conversion of potential energy into kinetic energy can be interpreted as equating the work done by the constant gravitational force to the change in kinetic energy.

POWER

Power is the amount of work done per unit time. It is a scalar quantity. (Although power is calculated from two vectors, the vector dot-product operation is seldom needed.)

$$P = \frac{W}{\Delta t} \qquad 20.20$$

For a body acted upon by a force or torque, the instantaneous power can be calculated from the velocity.

$$P = Fv \quad \text{[linear systems]} \qquad 20.21$$

$$P = T\omega \quad \text{[rotational systems]} \qquad 20.22$$

Typical basic units of power are ft-lbf/sec and watts (J/s), although *horsepower* is widely used. Table 20.1 can be used to convert units of power.

Table 20.1
Useful Power Conversion Formulas

1 hp	= 550 $\frac{\text{ft-lbf}}{\text{sec}}$	= 33,000 $\frac{\text{ft-lbf}}{\text{min}}$	= 0.7457 kW	= 0.7068 $\frac{\text{BTU}}{\text{sec}}$
1 kW	= 737.6 $\frac{\text{ft-lbf}}{\text{sec}}$	= 44,250 $\frac{\text{ft-lbf}}{\text{min}}$	= 1.341 hp	= 0.9483 $\frac{\text{BTU}}{\text{sec}}$
1 $\frac{\text{BTU}}{\text{sec}}$	= 778.26 $\frac{\text{ft-lbf}}{\text{sec}}$	= 46,680 $\frac{\text{ft-lbf}}{\text{min}}$	= 1.415 hp	

EFFICIENCY

For energy-using systems (such as cars, electrical motors, elevators, etc.), the *energy-use efficiency*, η, of a system is the ratio of an ideal property to an actual property. The property used is commonly work, or power, or, for thermodynamic problems, heat. When the rate of work is constant, either work or power can be used to calculate the efficiency. Otherwise, power should be used. Except in rare instances, the numerator and denominator of the ratio must have the same units.[5]

$$\eta = \frac{P_{ideal}}{P_{actual}} \quad [P_{actual} \geq P_{ideal}] \qquad 20.23$$

For energy-producing systems (such as electrical generators, prime movers, and hydroelectric plants), the *energy-production efficiency* is

$$\eta = \frac{P_{actual}}{P_{ideal}} \quad [P_{ideal} \geq P_{actual}] \qquad 20.24$$

The efficiency of an *ideal machine* is 1.0 (100 percent). However, all *real machines* have efficiencies of less than 1.0.

PHASES OF A PURE SUBSTANCE

Thermodynamics is the study of a substance's energy-related properties. The properties of a substance and the procedures used to determine those properties depend on the state and the phase of the substance. The *thermodynamic state* of a substance is defined by two or more independent thermodynamic properties. For example, the temperature and pressure of a substance are two properties commonly used to define the state of a superheated vapor.

The common *phases* of a substance are solid, liquid, and gas. However, because substances behave according to different rules, it is convenient to categorize them into more than these three phases.[1]

Solid: A solid does not take on the shape or volume of its container.

Subcooled liquid: If a liquid is not saturated (i.e., the liquid is not at its boiling point), it is said to be subcooled. Water at one atmosphere and room temperature is subcooled, as the addition of a small amount of heat will not cause vaporization.

Saturated liquid: A saturated liquid has absorbed as much heat energy as it can without vaporizing. Liquid water at standard atmospheric pressure and 212°F (100°C) is an example of a saturated liquid.

Liquid-vapor mixture: A liquid and vapor of the same substance can coexist at the same temperature and pressure. This is called a two-phase, liquid-vapor mixture.

Saturated vapor: A vapor (e.g., steam at standard atmospheric pressure and 212°F [100°C]) that is on the verge of condensing is said to be saturated.

Superheated vapor: A superheated vapor is one that has absorbed more heat than is needed merely to vaporize it. A superheated vapor will not condense when small amounts of heat are removed.

Ideal gas: A gas is a highly superheated vapor. If the gas behaves according to the ideal gas laws, it is called an ideal gas.

Real gas: A real gas does not behave according to the ideal gas laws.

Gas mixtures: Most gases mix together freely. Two or more pure gases together constitute a gas mixture.

Vapor/gas mixtures: Atmospheric air is an example of a mixture of several gases and water vapor.

PROPERTIES OF A SUBSTANCE

The thermodynamic *state* or condition of a substance is determined by its properties. *Intensive properties* are independent of the amount of substance present. Temperature, pressure, and stress are examples of intensive properties. *Extensive properties* are dependent on the amount of substance present. Examples are volume, strain, charge, and mass.

In this chapter, and in most books on thermodynamics, both lowercase and uppercase forms of the same characters are used to represent property variables. The two forms are used to distinguish between the units of mass. For example, lowercase h represents *specific enthalpy* (usually just called "enthalpy") in units of BTU/lbm or kJ/kg. Upper case H is used to represent the *molar enthalpy* in units of BTU/lbmole or kJ/kmol.

MASS: m

The mass of a substance is a measure of its quantity. Mass is independent of location and gravitational field strength. In thermodynamics, the customary U.S. and SI units of mass (m) are pound-mass (lbm) and kilogram (kg), respectively.

TEMPERATURE: T

Temperature is a thermodynamic property of a substance that depends on energy content. Heat energy entering a substance will increase the temperature of that substance. Normally, heat energy will flow only from a hot object to a cold object. If two objects are in *thermal equilibrium* (are at the same temperature), no heat will flow between them.

The scales most commonly used for measuring temperature are the Fahrenheit and Celsius scales.[3] The relationship between these two scales is:

$$T_{°F} = 32° + \left(\frac{9}{5}\right) T_{°C} \qquad 21.1$$

The *absolute temperature scale* defines temperature independently of the properties of any particular substance. This is unlike the Celsius and Fahrenheit scales, which are based

[1]Plasma, cryogenic fluids (cryogens) that boil at temperatures less than approximately 110K, and solids near absolute zero are not discussed in this chapter.

[3]The term *centigrade* was replaced by the term *Celsius* in 1948.

on the freezing point of water. The absolute temperature scale should be used for all calculations.

In the customary U.S. system, the absolute scale is the *Rankine scale*.[4]

$$T_{°R} = T_{°F} + 459.67° \qquad 21.2$$

$$\Delta T_{°R} = \Delta T_{°F} \qquad 21.3$$

The absolute temperature scale in the SI system is the *Kelvin scale*.[5]

$$T_K = T_{°C} + 273.15° \qquad 21.4$$

$$\Delta T_K = \Delta T_{°C} \qquad 21.5$$

The relationships between temperature differences in the customary U.S. and SI systems are independent of the freezing point of water.

$$\Delta T_{°C} = \frac{5}{9}\Delta T_{°F} \qquad 21.6$$

$$\Delta T_K = \frac{5}{9}\Delta T_{°R} \qquad 21.7$$

PRESSURE: p

Customary U.S. pressure units are pounds per square inch (psi).

DENSITY: ρ

Customary U.S. density units in tabulations of thermodynamic data are pounds per cubic foot (lbm/ft^3). Density is the reciprocal of specific volume.

$$\rho = \frac{1}{v} \qquad 21.8$$

SPECIFIC VOLUME: v AND V

Specific volume, v, is the volume occupied by one unit mass of a substance. Customary U.S. units in tabulations of thermodynamic data are cubic feet per pound (ft^3/lbm). Specific volume is the reciprocal of density.

$$v = \frac{1}{\rho} \qquad 21.9$$

[4]Normally, three significant temperature digits (i.e., 460°) are sufficient.

[5]Normally, three significant temperature digits (i.e., 273°) are sufficient.

INTERNAL ENERGY: u AND U

Internal energy includes all of the potential and kinetic energies of the atoms or molecules in a substance. Energies in the translational, rotational, and vibrational modes are included. Since this movement increases as the temperature increases, internal energy is a function of temperature. It does not depend on the process or path taken to reach a particular temperature.

ENTHALPY: h AND H

Enthalpy (also known at various times in history as *total heat* and *heat content*) represents the total useful energy of a substance. Useful energy consists of two parts—the internal energy, u, and the *flow energy* (also known as *flow work* and *p-V work*), pV. Therefore, enthalpy has the same units as internal energy.

$$h = u + pv \qquad 21.12$$

Strictly speaking, the customary U.S. units of Eqs. 21.12 are not consistent, since flow work (as written) has units of ft-lbf/lbm, not BTU/lbm. (There is also a consistency problem if pressure is defined in lbf/ft^2 and given in lbf/in^2.) Strictly speaking, Eq. 21.12 should be written as

$$h = u + \frac{pv}{J} \quad \text{[U.S.]} \qquad 21.15$$

The conversion factor, J, in Eq. 21.15 is known as *Joule's constant*. It has a value of 778.17 ft-lbf/BTU. Joule's constant is often omitted from the statement of generic thermodynamic equations, but it is always needed with customary U.S. units for dimensional consistency.

ENTROPY: s AND S

Absolute entropy is a measure of the energy that is no longer available to perform useful work within the current environment. Other definitions used (e.g., the "disorder of the system," the "randomness of the system," etc.) are frequently quoted. Although these alternate definitions cannot be used in calculations, they are consistent with the *third law of thermodynamics* (also known as the *Nernst theorem*). This law states that the absolute entropy of a perfect crystalline solid in thermodynamic equilibrium is (approaches) zero when the temperature is (approaches) absolute zero.[6] Equation 21.16 expresses the third law mathematically.

$$\lim_{T \to 0K} s = 0 \qquad 21.16$$

[6]A molecule with zero entropy exists in only one quantum state. The energy state is known precisely, without *uncertainty*.

SPECIFIC HEAT: c AND C

An increase in internal energy is needed to cause a rise in temperature. Different substances differ in the quantity of heat needed to produce a given temperature increase. The ratio of heat, Q, required to change the temperature of a mass, m, by an amount ΔT is called the *specific heat* (*heat capacity*) of the substance, c. Because specific heats of solids and liquids are slightly temperature dependent, the mean specific heats are preferred for processes covering a large temperature range.

$$Q = mc\Delta T \qquad 21.22$$

For gases, the specific heat depends on the type of process during which the heat exchange occurs. Specific heats for constant-volume and constant-pressure processes are designated by c_v and c_p, respectively.

$$Q = mc_v \Delta T \quad \text{[constant-volume process]} \qquad 21.25$$

$$Q = mc_p \Delta T \quad \text{[constant-pressure process]} \qquad 21.26$$

RATIO OF SPECIFIC HEATS: k

For gases, the *ratio of specific heats, k*, is defined by Eq. 21.28.

$$k = \frac{c_p}{c_v} \qquad 21.28$$

QUALITY: x

Within the vapor dome, water is at its saturation pressure and temperature. There are an infinite number of thermodynamic states in which the water can simultaneously exist in liquid and vapor phases. The *quality* is the fraction by weight of the total mass that is vapor.

$$x = \frac{m_{\text{vapor}}}{m_{\text{vapor}} + m_{\text{liquid}}} \qquad 21.30$$

PROPERTIES OF SATURATED VAPORS

Properties of saturated vapors can be read directly from the saturation tables. The vapor's pressure or temperature can be used to define its thermodynamic state. Enthalpy, entropy, internal energy, and specific volume can be read directly as h_g, s_g, u_g, and v_g, respectively.

PROPERTIES OF LIQUID-VAPOR MIXTURES

When the thermodynamic state of a substance is within the vapor dome, there is a one-to-one correspondence between the saturation temperature and saturation pressure. Knowing one determines the other. The thermodynamic state is uniquely defined by any two independent properties (e.g., temperature and quality, pressure and enthalpy, entropy and quality, etc.).

If the quality of a liquid-vapor mixture is known, it can be used to calculate all of the primary thermodynamic properties. If a thermodynamic property has a value between the saturated liquid and saturated vapor values (i.e., h is between h_f and h_g), any of Eqs. 21.53 through 21.56 can be solved for the quality.

$$h = h_f + xh_{fg} \qquad 21.53$$

$$s = s_f + xs_{fg} \qquad 21.54$$

$$u = u_f + xu_{fg} \qquad 21.55$$

$$v = v_f + xv_{fg} \qquad 21.56$$

PROPERTIES OF SUPERHEATED VAPORS

Unless a vapor is highly superheated, its properties should be found from a superheat table. Since temperature and pressure are independent for a superheated vapor, both must be known in order to define the thermodynamic state.

EQUATION OF STATE FOR IDEAL GASES

An *equation of state* is a relationship that predicts the state (i.e., a property such as pressure, temperature, volume, etc.) from a set of two other independent properties.

Avogadro's law states that equal volumes of different gases at the same temperature and pressure contain equal numbers of molecules. For one mole of any gas, Avogadro's law can be stated as the *equation of state for ideal gases*. (Temperature, T, in Eq. 21.57 must be in degrees absolute.)

$$\frac{pV}{T} = R^* \qquad 21.57$$

In Eq. 21.57, R^* is known as the *universal gas constant*. It is "universal" (within a system of units) because the same value can be used with any gas. Its value depends on the units used for pressure, temperature, and volume, as well as on the units of mass.

Table 21.5
Values of the Universal Gas Constant, R^*

units in SI and metric systems
8.3143 kJ/kmol·K
units in English systems
1545.33 ft-lbf/lbmole-°R
1.986 BTU/lbmole-°R

The ideal gas equation of state can be modified for more than one mole of gas. If there are n moles, then

$$pV = nR^*T \qquad 21.58$$

The number of moles can be calculated from the substance's mass and molecular weight.

$$n = \frac{m}{(\text{MW})} \qquad 21.59$$

Additional Information

Steam

Steam is used to convey energy to space heating or process heating equipment. The advantages of steam include:

- Temperature can be accurately adjusted by controlling steam pressure using simple valves;
- Steam carries a relatively large amount of energy in a small mass;
- When steam condenses back into water, a high rate of energy flow is obtained, which means that the equipment using the heat need not be unduly large. (*See the Appendix of this book for steam-related terms and definitions.*)

Steam Quality: The Steam Tables (shown in the Appendix) show the properties of "dry steam," or steam which is completely evaporated, and therefore contains no liquid water. Steam used for process or heating should be as dry as possible, a condition achieved using steam separators and traps. Steam "dryness fraction" is the term used to describe this aspect of steam quality.

Steam is generated by a combined furnace and boiler. The combustion heat that passes through a stack is directed to feed water. Steam generator capacities are expressed as *mass of steam per hour.* The amount of combustion energy released in a steam generator represents the boiler efficiency, ranging from 75-90%.

Boilers and their associated firing equipment should be properly sized and designed for efficient operation. A boiler's efficiency is reduced if it has to operate with a peak load higher than its maximum continuous rating. In such cases, the pressure may drop and the boiler will be unable to provide the desired quality of steam at the right time and at the correct pressure. By the same token, if the boiler has to work at a small percentage of its rating, efficiency will drop since there will be significant radiation losses. In situations where the steam load tends to be variable, two or more boilers may be used (e.g., a large boiler for winter use, and a smaller one for summer). The firing equipment must respond to the load, while maintaining the proper fuel to air ratio. To reduce heat losses, all potential sources of heat loss in a steam system (including piping) should be lagged or insulated. Lagging not only conserves fuel, but also reduces heat losses that produce condensate inside the line. Reducing that moisture helps to keep the steam quality high. Steam leaks must also be eliminated to ensure efficiency in the system.

Pulleys

Pulleys are used to provide a mechanical advantage in the movement of a tensile force. A block and tackle is a series of pulleys. Pulleys may be attached or free-moving; free pulleys may be attached to a load.

Basic problems involving pulleys normally assume that the ropes approaching and leaving the load-carrying pulley are parallel. The number of ropes is the key factor, not the diameters of the pulleys. A loss factor may be used to address the issue of rigidity of the rope. Loss factors may vary between 1.03-1.06 for wire ropes and chains with contact of 180 degrees.

Free Body Diagrams

A free body, such as a beam, is generally not connected to supports. This allows evaluation of the effect of the forces on it. A free body diagram shows the body in equilibrium, with applied forces, moments and reactions (identified on the diagram) resulting in zero. According to the *Engineer-In-Training Reference Manual* by Michael Lindeburg, PE (Professional Publications), the following method can be used to analyze a three-dimensional structure.

- Step 1: Establish the (0,0,0) origin for the structure.
- Step 2: Determine the (x, y, z) coordinates of all load and reaction points.
- Step 3: Determine the x-, y-, and z-components of all loads and reactions. This is easily accomplished by the use of direction cosines calculated from the (x, y, z) coordinates.
- Step 4: Draw a coordinate free-body of the structure for each of the three coordinate axes. Include only forces, reactions, and moments that affect the coordinate free-body.
- Step 5: Apply the basic two-dimensional equilibrium equations.

Moment of Inertia

Moment of Inertia is the ratio of the torque applied to a rigid body that is free to rotate around a given axis to the angular acceleration around that axis. Formulas to calculate moment of inertia for a cube, a cylinder and the dynamic moment of inertia can be found in *Engineering Formulas* by Kurt and Reiner Gieck (McGraw-Hill Inc.). *The Engineer-In-Training Reference Manual* also provides formulas for mass moments of inertia for a slender rod, thin rectangular plate, rectangular parallel piped, thin disk, solid cylinder and cone, sphere, and hollow cylinder.

Moment of Momentum, or Angular Momentum

Momentum is addressed briefly in the first section of this chapter, under "Statics and Dynamics." Angular Momentum is a vector quantity. It is a measure of the intensity of rotational motion equal to the product of the angular velocity of a rotating body and its moment of inertia relative to the rotation axis and directed along the rotation axis. Calculating for angular momentum involves factors of distance x force x time. Angular momentum is calculated in the same direction as rotation vector, using the right-hand rule.

$$h_o = \frac{r \times mv}{g_c}$$

Strain Equation

According to the *Engineer-In-Training Reference Manual*, strain is elongation expressed on a fractional or percentage basis. It may be expressed as in/in, mm/mm, and percent, or with no units. A strain in one direction is accompanied by strains in orthogonal directions. Dilation is the sum of the strains in the three coordinate directions. The energy stored in a loaded member is equal to the work required to deform the member. Below the proportionality limit, the total strain energy is:

$$U = \frac{1}{2} F\delta = \frac{F^2 L_o}{2AE}$$

Stress

Stress is force per unit area, or F/A. Stress is measured in thousands of pounds per square inch (lbf/in^2, ksi) and MPa. Normal stress represents an area that is normal to the force that is carried. Shear stress means that the area is parallel to the force. (See "Basic Principles of Structural Engineering" in Chapter 3 for more on stress forces.)

The following section is reprinted with permission from *Plant Engineering Magazine.* The authors are Arthur A. Chambers and F. Rowland Dube.

Vacuum Pumps and Systems

Selection, Installation and Maintenance Considerations

The pressure requirements of any vacuum application can generally be satisfied by more than one type of pump. But pressure is only one criterion of pump selection. Certain design features and characteristics often make one type of vacuum pump more suitable than another depending on the specific application. Factors affecting pump selection, installation, and maintenance are discussed here.

Condensable vapors are best handled by liquid-ring and liquid-jet vacuum pumps. These pumps offer functional, as well as economic, advantages in vacuum processes that liberate liquid slugs or appreciable quantities of condensable vapors or dust particles. These units act as direct-contact condensers which increase the pumping speed for vapor without appreciably impairing the pumping speed for air. They often eliminate the need for supplementary condensing equipment. On these applications, oil-sealed rotary pumps depend on water-cooled condensers or refrigerated traps to maintain efficiency.

Initial and operating costs of liquid-ring and liquid-jet vacuum pumps are higher than those of many other types. Installation, however, is simple and inexpensive. Typical applications for liquid-ring and liquid-jet pumps are:

- Casting, molding, forming
- Cooling, chilling, filtering
- Food preparation and packaging
- Filling
- Impregnation, deodorization
- De-aeration, degasification
- Dehydration
- Gas scrubbing

Condensable or volatile gases, pumped in large volumes and at low pressures, can best be managed with vapor-handling systems. These packaged systems are simply a combination of an oil-sealed, triplex, rotary-piston pump. The liquid-ring maintains low pressure in the oil reservoir into which the rotary pump discharges process gases or vapors. Thus, such systems prevent oil contamination in the mechanical pump, corrosion of pump parts, and the need to reclaim oil.

Water vapor requires the high pumping speeds produced by a mechanical-booster/liquid-ring or a mechanical-booster/liquid-jet system. Booster/liquid-sealed pumping systems can be modified to meet space requirements and are easy to install. Typical applications for booster/liquid-sealed systems include:

- Drying, freeze-drying, dehydrating
- Food preparation, chilling
- Liquid oxygen handling
- Distillation
- Central vacuum systems

Other problems such as water pollution or excessive water and energy consumption can be controlled with rotary-piston pumps.

Since they are valved compressors, rotary-piston and rotary-vane pumps do not require much power and consume little water. Cooling water for water-cooled pumps does not contact the process gas and, therefore, does not become contaminated. Where electricity or water is scarce, unitized systems that operate on demand to reduce energy consumption, or air-cooled systems that require no water, can be used. Also, the exhaust from rotary piston pumps can be made relatively clean.

Although the pumping effectiveness of any oil-sealed mechanical vacuum pump is reduced in the presence of condensable vapors, pumps fitted with gas ballasts will, within limits, prevent condensation of insoluble vapors in the pump. In addition, some pumps are designed to handle condensable vapors when used with one or more of the following devices or techniques: centrifuges, cold traps, combination pump units, condensers, oil-reclamation units, special sealants, air stripping, decanting, and high-temperature operation.

Rotary-piston pumps are standard when pump speed must be a variable over a wide pressure range; they provide high volumetric efficiency throughout most of the range. Installation is also simple and inexpensive for small and medium (300 cfm and under) unbalanced pumps and for all sizes of balanced pumps. For large unbalanced pumps, installation costs are high.

Rotary-vane pumps are used in laboratories and for light industry. Power requirements are low, since they are fitted with discharge valves, and they have excellent vibrational characteristics. Exhaust is relatively clean, and installation is simple. The rotary-vane pump is much more sensitive to particulate contamination than is the rotary-piston pump.

Typical applications for oil-sealed rotary pumps include:

- Degassing, leak detection
- Evacuating, sealing, packaging
- De-aeration, drying, freeze drying, evaporating
- Laboratory studies
- Melting, sintering
- Annealing
- Heat treating, brazing

Pump selection can be straightforward or involved, depending on the operating conditions. The following selection criteria are considered important and relevant to nearly every pump application.

Operating pressure determines whether a single-stage or compound, oil-sealed pump or a water-sealed pump is needed and whether staging with an air ejector or booster pumping system is the best solution. Typical pressure ranges of pump types and combinations are shown in Figure 7.1. (The dry-lobe pump (booster) is included for comparison only. Because the pressure range of this pump is very limited, it is seldom used alone.)

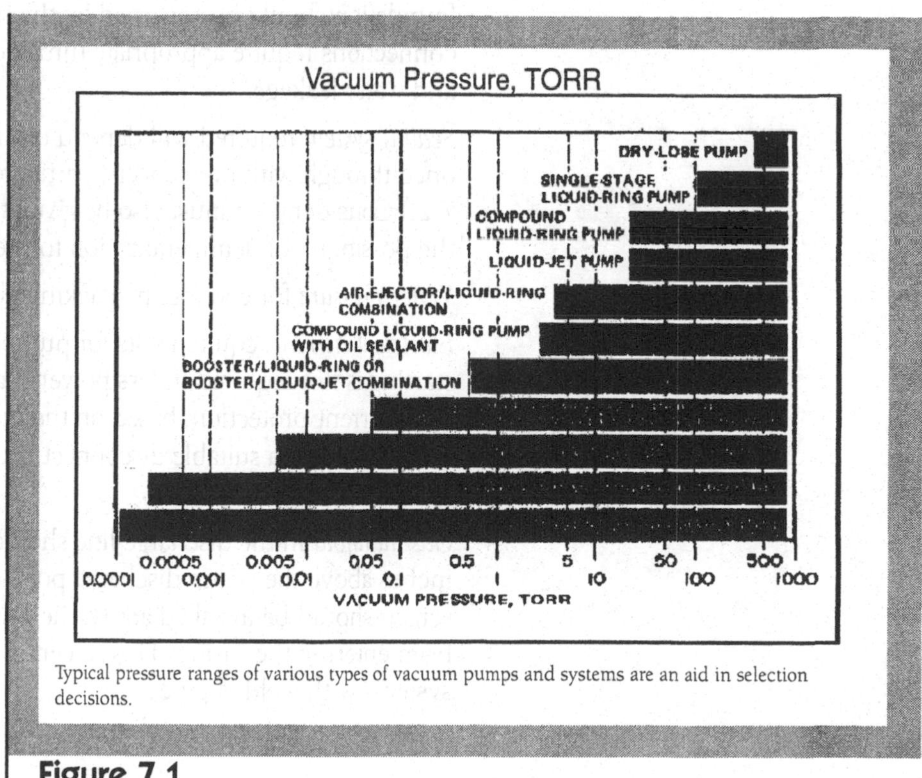

Typical pressure ranges of various types of vacuum pumps and systems are an aid in selection decisions.

Figure 7.1

Pumpdown time required to bring the system from initial pressure to the final desired pressure is important for establishing average pumping speed.

Volume of the system to be evacuated is a factor in the pumpdown time.

Gas load must include gasses that evolve from the process and those that leak into the chamber as a result of its not being fully sealed.

Vacuum manifold configuration, including length, cross-sectional area, entrance size and shape, and fittings, will affect pump speed. In general, the manifold should be at least as large as the pump inlet and as short as possible. With manifolds that are very short, the entrance loss is a prime consideration.

Economic factors may require utilization of existing equipment or attention to process recovery capabilities (non-contamination of air and water). Noise control may also be an economic factor.

Installation and operating costs will depend on the type of pump selected, manpower requirements, the pump's duty cycle (hours per day, days per week, weeks per year) and utility costs (electricity and sealing/cooling water).

Installation considerations may also affect the type of pump chosen. Most vacuum-pumping systems are self-contained and require little, if any, additional equipment to improve their overall performance. But certain accessories can be provided for ease of installation, system protection, economy of operation, and operator convenience.

Liquid-ring pumps should be installed where the sealing liquid will not be exposed to freezing temperatures. Small and medium-size pumps (300 cfm or less) can be set on wooden floors or skids and larger pumps on metal frames or concrete slabs. The pumps are essentially vibrationless, so the rigidity of the foundation should be governed by the structure of the unit base. All piping connections require appropriate thread compound and gaskets to prevent air and water leakage.

Sealing system required will depend on the type of system selected; that is, once-through with no recovery, partial recovery, or full recovery. (See Figure 7.2.) Consideration must also be given to the mineral content of the water and the possibility of demineralization to prevent excessive scaling.

Water pressure for cooling the packing glands must be 5 psig.

Electrical current requirements for pump starting are low. Reduced-voltage starting is not required unless power usage is restricted by the plant. Overcurrent protection, based on the full-load current stamped on the motor nameplate and a suitable disconnect between the controller and power supply should be installed.

Gas manifold in the discharge line should not be positioned more than 39 inches above the pump discharge port to prevent motor overload. A temporary screen should be installed across the pump inlet to prevent abrasive particles from entering the casing. This screen is especially important in new piping systems with welded pipe.

Three Sealing Systems

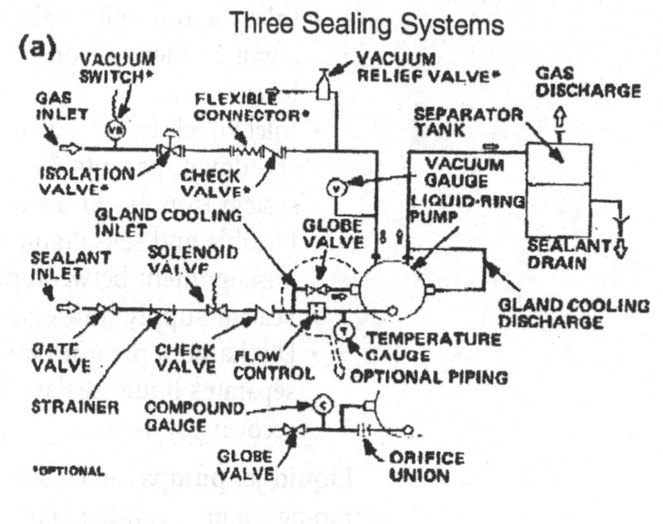

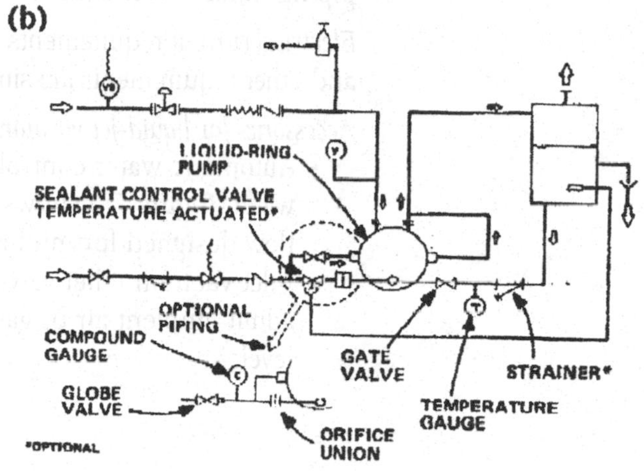

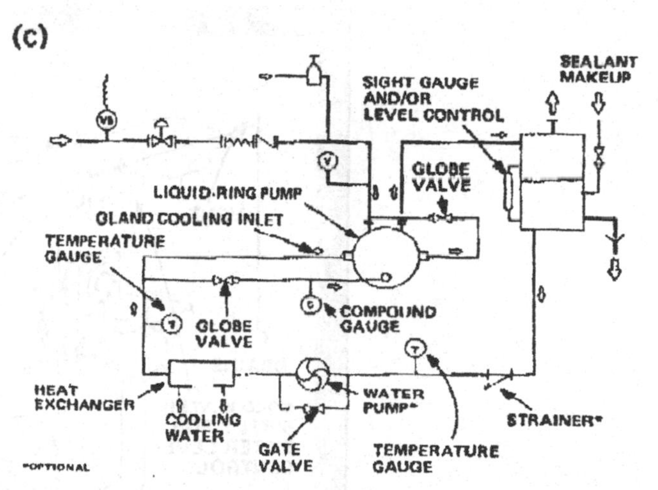

Three systems for liquid-ring vacuum pump piping arrangement are: (a) No sealant recovery; water is taken directly from the main, circulated, and discharged to drain. (This arrangement is most common for small pumps or where water consumption is not a problem.) (b) Partial sealant recovery; partial recirculation is provided to conserve sealant. (c) Full sealant recovery; full recovery is provided to conserve sealant in applications where suitable or compatible sealant is not available from an outside source.

Figure 7.2

Accessories for liquid-ring vacuum pumps include:

- Inlet vacuum relief valve: controls system vacuum. The valve opens to admit ambient air or gas to the pump inlet to control the vacuum level.
- Inlet check valve: isolates the pump from the system automatically at shutdown, permits the system to remain at vacuum, and protects the system from backflow of air and sealant.
- Flexible and self-aligning connectors: compensate for motion or misalignment between pump and system.
- Sealant supply flow control: establishes sealant flow rates.
- Discharge separator tank: collects gas-liquid discharge from pump, separates liquid sealant from gas, and provides storage for sealant-recovery system.

Liquid-jet pumps can be bolted to wooden floors, skids, or concrete slabs. All piping requires leak-tight connections.

Electrical current requirements for starting liquid-jet vacuum pumps are low, and other requirements are similar to those for liquid-ring pumps.

Accessories for liquid-jet vacuum pumps include:

- Automatic water control: eliminates excessive consumption of cooling water. Figure 7.3 shows a pump installation with intermittent water flow designed for minimum water consumption.
- Inlet vacuum relief valve: controls system vacuum. The valve opens to admit ambient air or gas to the pump inlet to control the vacuum level.

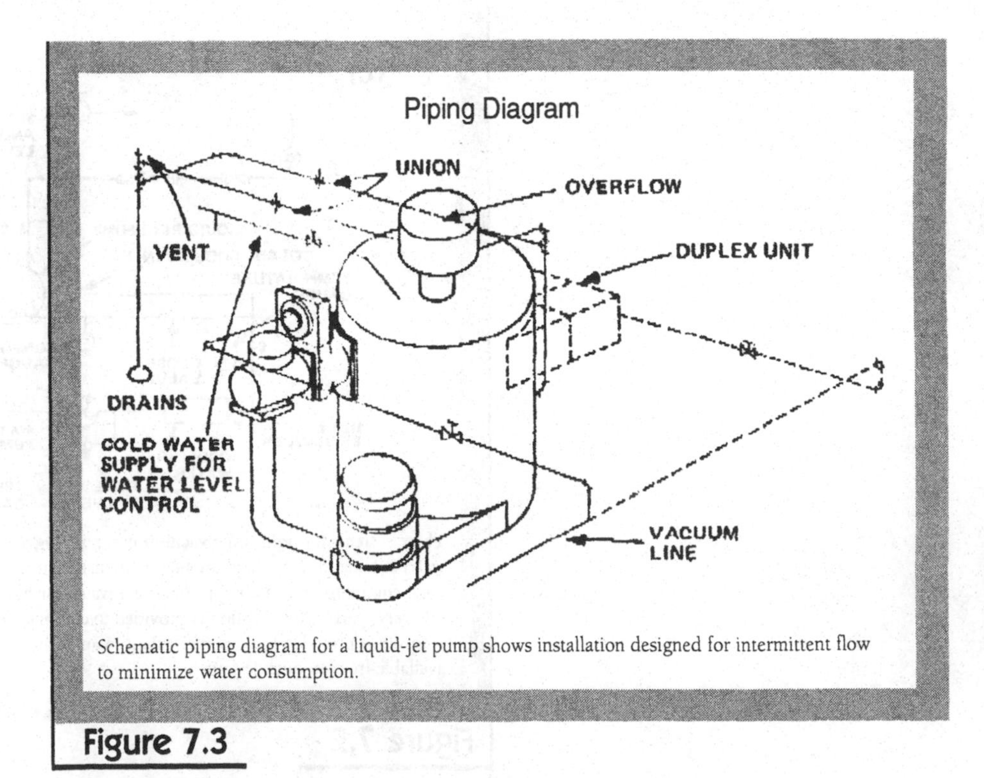

Schematic piping diagram for a liquid-jet pump shows installation designed for intermittent flow to minimize water consumption.

Figure 7.3

Rotary-piston vacuum pump installation considerations are generally the same as for other types. Unrestricted access to the pump and components should be provided for service and maintenance. It is good practice to locate the electrical control station near the pump.

Vibration isolation requirements depend on whether the pump is inherently balanced, partially balanced, or unbalanced. (See Figure 7.4.) However, when a pump is mounted on vibration isolators, deflections during starting and stopping can be significantly larger than during steady-state operation. Therefore, all connections to the pump must be sufficiently flexible to withstand maximum deflections.

Inlet connections must be designed to prevent (1) gas flow restrictions, (2) entry of pump fluids into the process chamber, and (3) ingestion of particulate matter. Since oil-sealed mechanical vacuum pumps produce oil splash at the inlet, the manifolding must be designed to prevent oil from traveling upstream.

Solid contaminants can be kept out of the pump with a simple wet intake trap. For large dust loads, bag-type inlet filters should be used. These should be rated for a pressure drop of no more than 10 percent at pressures above 1 torr, and 50 percent at pressures below 10 militorr (10 microns).

Gauge connections are necessary for process monitoring and system troubleshooting. They should be located on both sides of the main isolation valve to permit independent checking of pressures.

Guidelines for Vibration Isolation of Vacuum Pumps

	INHERENTLY BALANCED PUMPS		PARTIALLY BALANCED PUMPS			UNBALANCED PUMPS		
	Small Vane Pumps to 50 cfm	Triplex piston pipes, all sizes	Medium vane pumps to 130 cfm	Duplex piston pumps to 300 cfm	Simplex piston pumps with 2 flywheels to 50 cfm	Duplex piston pumps over 300 cfm	Simplex piston pumps, all sizes	Large vane pumps over 130 cfm
Frequency of Vibration	Low	High	Low	Low	Low	Very Low	Very Low	Very Low
Amplitude of Vibration	Very Low	Very Low	Low	Low	Low	High	High	High
Foundation Required	Floor sufficient to support pump weight		Solid floor or heavy frame			Concrete foundation with anchor bolts and grouting		
Base Vibration Mounts	Coil springs or air mounts	Elastomer mounts, coil springs or air mounts	Coil springs or air mounts			Not recommended		
Inlet Connector	Shallow, convoluted, light-gauge metal bellows or shallow, convoluted elastomer bellows compatible with pump fluid, or vacuum hose compatible with pump							
Discharge Connector	Flexible connector compatible with pump fluid and gas temperatures of 220 F							

Figure 7.4

Discharge manifold connections must prevent oil loss and oil mist in the discharge gases. A properly designed manifold will take care of these requirements. Oil mist can be eliminated only with in-line or umbrella-type exhaust filters. On multi-pump installations, it is often economical to exhaust into a central system having a filtering capacity based on the total gas load to be handled.

Isolation/air admission valves are essential in a well-designed vacuum system. Since a vacuum pump is a source of leakage when stopped, closing the isolation valve before shutdown will prevent a pressure rise in the chamber. An air-inlet valve is used to break vacuum at the pump inlet at shutdown.

Gas ballasting will help keep the pump oil clean. The most common cause of contamination of mechanical vacuum pumps is water vapor. The condensate can be manually drained from the separation tank or automatically decanted.

Accessories for rotary-piston vacuum pumps include:

- Flexible connectors: to compensate for motion or misalignment between the pump and piping system.
- Special pump oils and sealants.

Mechanical booster systems depend upon the type of backing pump used—liquid ring, liquid jet, or rotary piston. In general, however, the requirements discussed for each of these pumps can be considered relevant to the installation of package booster systems.

Vacuum system pollution controls are often required to prevent discharge of significant quantities of oil sprays and aerosols (smoke) with the exhaust gases. Oil sprays can be reduced to reasonable levels with a well-designed gas-oil separator. The screen decelerates and disperses the oil particles so that they can drain into the reservoir and allow relatively clean exhaust air to escape.

Aerosols cannot be removed by mechanical baffles. They must be removed by coalescing filters that reduce particles to droplets which will then drain into an area upstream of the filter element. Because the filter efficiency is not much better than 100 ppm, pumps with coalescing filters should not exhaust into closed areas when operating under constant, high gas loads. Vacuum pumps can also be equipped with two-stage exhaust filters that use zeolite, activated alumina, or activated charcoal in the final stage. Charcoal is generally preferred because it effectively removes odors.

Electrostatic precipitators in central exhaust systems are useful for smoke removal but offer no protection against residual oil spray. Condensed oil from the precipitator is usually clean enough to be returned to the pump. However, water vapor, which could constitute a significant portion of the gas load, must first be separated from the condensate by decanting.

With water-sealed vacuum pumps, liquid-gas separation may become troublesome if water separators are too small and operating pressures are very high. A closed-loop system should be used for handling contaminating vapors so that the condensate becomes the sealing medium. When light oils are used for sealing, exhaust fumes may become as objectionable as those emitted by the oil sealed mechanical pumps. In such cases, coalescing filters can be used in conjunction with charcoal filters to eliminate aerosols and odors.

Noise generated by most oil-sealed vacuum pumps is caused mainly by the sudden change in the direction of the oil flow at the exhaust port. This noise may be completely eliminated by injecting air into the pump during part of the compression stroke. Injection is done automatically by gas ballasting, a standard feature on most pumps. If the gas-ballast noise suppressor on a single stage pump creates an undesirable increase in inlet pressure, a two-stage (compound pump) should be used. Gas ballasting has a negligible effect on the inlet pressure of a compound pump because of the pump's very high compression ratio.

Cavitation will cause noise in excess of 100 dBA in liquid-ring vacuum pumps sealed with water or high-vapor-pressure fluids. When these pumps operate close to the vapor pressure of the sealing fluid, air must be added at the pump inlet of single-stage pumps or at the interstage of compound pumps to prevent collapse of vapor pockets.

Liquid-jet pumps produce noise by both water hammer and cavitation but at a total level of less than 90 dBA. In such cases, noise suppressors are seldom necessary. Instead, the noise can be eliminated by increasing the gas load on the pump or by operating with low-vapor-pressure fluids.

Maintenance of vacuum pumps is minimal. A vacuum pump that is sized, installed, and operated according to the manufacturer's recommendations requires relatively little care. Only a few routine maintenance procedures are necessary. Replacing parts and overhauling pumps can be done without excessive downtime.

In general, vacuum-pump maintenance involves keeping the pumps clean, inspecting them for leaks, and, with rotary-piston pumps and mechanical blowers, checking the oil level and condition of the oil. Liquid-ring pumps require bearing lubrication approximately every 3000 hours.

All accessories and system piping must be maintained in good condition, and process equipment must be kept free of contaminants.

The following section is reprinted with permission from *Plant Engineering Magazine*. The author is Peter Y. Burke, PE.

Understanding Air Compressor Ratings

Plant engineers often evaluate compressor proposals and make purchasing recommendations. This process is made difficult by the variation in ratings and terminology used in descriptions. To avoid any misunderstanding when comparing positive-displacement-type air compressors, plant engineers need to decipher a number of common terms.

- **Free air**: air at ambient conditions of temperature, humidity, and atmospheric pressure.
- **CFM**: cubic feet of air per minute; volume rate of airflow.
- **ICFM**: inlet cubic feet per minute; cfm flowing through the inlet filter or inlet valve under rated conditions.
- **SCFM**: standard cubic feet per minute; flow rate of free air measured at some point (typically the discharge flange of the compressor package) and converted to standard inlet conditions (usually 14.7 psig, 60°F, and 0% relative humidity).
- **ACFM**: actual cubic feet per minute; flow rate of free air measured at some reference point, usually the discharge flange of the compressor package, based on actual conditions at the reference point.
- **Displacement**: amount of air (in cfm) displaced by the compressor piston or rotor under no load, discharging directly to the atmosphere.
- **BHP**: brake horsepower, usually horsepower delivered to the output shaft of the drive motor. For total package bhp, sum of all motor shaft outputs including the compressor and cooling fans.
- **Total package input power**: total electrical power input to the compressor package, including drive motor, cooling fan motors, control transformers, etc., expressed in kilowatts.
- **Rated capacity**: volume rate of airflow at the rated pressure referenced to a specific point; for example, 110 psi at the discharge flange of the compressor package.
- **Required capacity**: flow of free air, in cfm, at the inlet to the user's system.
- **PSI**: pounds per square inch; force per unit area exerted by compressed air.
- **PSIA**: pounds per square inch absolute; pressure above absolute vacuum. Atmospheric pressure is stated in psia.
- **PSIG**: pounds per square inch gauge; pressure at some point as measured with a gauge and dependent on atmospheric pressure.
- **PSID**: pounds per square inch differential; pressure difference between two points, such as across an aftercooler.
- **Rated discharge pressure**: pressure produced at a reference point, such as the discharge flange of the package, at which compressor performance is rated.
- **Required discharge pressure**: air pressure required at the user's system inlet.

Required capacity and discharge pressure are easy to understand, but rated capacity and pressure are not as clearly understood. There is no accepted standard by which compressor manufacturers rate their products.

A unit rated for a capacity of 300 scfm at 110 psig at some point other than the package discharge means the compressor delivers 300 cfm at standard conditions compressed to 110 psig at that point. This rating does not indicate capacity and pressure delivered to a system, because it does not consider ambient conditions or pressure loss within the package. (See Figure 7.5.)

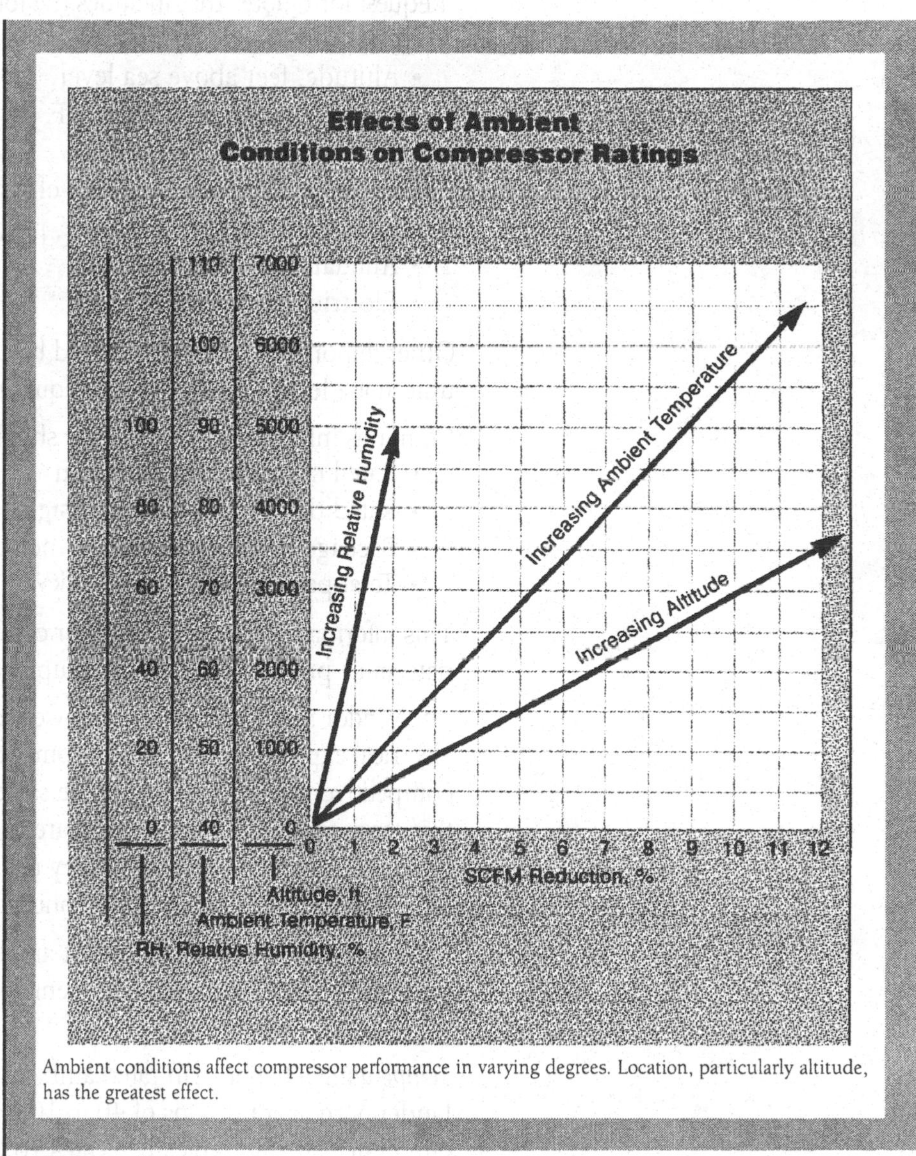

Ambient conditions affect compressor performance in varying degrees. Location, particularly altitude, has the greatest effect.

Figure 7.5

When ratings are considered, ask these questions: Where are these ratings measured—at the package discharge, compressor discharge, or somewhere in between? What are the ambient conditions under which the rating is determined? Does the bhp shown include all motors in the package? (Be sure the bhp rating is for motor shaft output. Compressor or fan input ratings do not consider power transmission losses.)

Do not confuse ambient temperature with inlet temperature. They may be different. For example, the compressor may be in a heated room but draws outside air through a duct into the inlet. Be sure to include a complete description of the equipment being evaluated, including any optional accessories such as heavy-duty intake filters, compressed-air aftercoolers, and special cooling fans or ductwork.

Perhaps the easiest way to compare various products is at system inlet conditions. To ensure an accurate comparison, provide each vendor a simple "Request for Quote" that includes the following information:

- Location
- Altitude, feet above sea level
- Average inlet temperature, °F
- Average relative humidity, %
- Required capacity (free air), cfm
- Required discharge pressure, psig
- Annual operating hours
- Electricity cost, $/Kwh

Other important points that should be considered are size and weight of the unit, noise level, cooling system requirements, starter type, and maintenance.

From this information, the vendor should provide:

- Rated capacity (free air), cfm
- Rated discharge pressure, psig
- Package brake horsepower, including all motors, bhp
- Total package input power, kW

This information from each vendor can be readily compared and helps determine proper sizing of the equipment.

The vendor may also provide a power cost analysis to help in evaluating the electrical expense in operating a compressor. Some vendors even evaluate competitive products to ensure the same rating methods are used. Although this service makes it easy to compare values, carefully examine all the performance data to be certain they match what has been quoted and that estimates are based on identical conditions.

It is helpful to know how different ambient conditions affect performance ratings. The most important ambient conditions are temperature, atmospheric pressure, and humidity.

Temperature. Higher than rated ambient temperatures slightly reduce capacity. Under normal conditions of 40 to 105°F, temperature does not affect a rating in cfm of free air, and typically creates ±6% variation in scfm ratings.

Humidity. Normal fluctuations in humidity produce slight variations in compressor performance. As with temperature, this variation should be well within the reasonable safety margin used in selecting a compressor.

Pressure. Reduced ambient atmospheric pressure due to altitude affects the performance of some compressor package components. If a facility is high altitude (Denver, for example), ratings should be adjusted to the location.

Manufacturers determine compressor ratings for specific package configurations. The type of enclosure and cooling method used in these packages affect the performance ratings of their compressors.

Heat removal is critical to compressor performance, and operating temperatures affect how much power an electric motor can produce. Open compressors release heat and noise into the surrounding area. Enclosed compressors direct heat away from the unit and suppress generated noise.

Aircooled compressors may incorporate package cooling fans. Their ratings should include fan brake horsepower requirement in the overall package ratings.

Watercooled compressors require a cooling system for a continuous supply of water/coolant. Their ratings should include minimum water flow, gal/min; water pressure required, psig; water pressure drop, psig; and maximum inlet water temperature, F.

Noise ratings for packaged air compressors are not typically furnished by manufacturers unless specifically requested. But most manufacturers can provide compressor noise ratings. When evaluating noise level ratings, be certain they are for the exact compressor configuration required. Cooling fans, type of enclosure, and optional features can dramatically affect these ratings.

While maintenance requirements are not air compressor ratings, they should be a part of any comparison. Based on the operating hours and manufacturer's recommended maintenance schedule, annual costs can be estimated. Consider also the ease of maintenance and the amount of time required to perform routine tasks, such as an oil change.

The following section is reprinted with permission from "Cameron Hydraulic Data," edited by C.C. Heald, Ingersoll-Rand, Inc.

Hydraulics

The velocity head energy component is used in system head calculations as a basis for establishing entrance losses, losses in valves and fittings, losses at other sudden enlargements and exit losses by applying the appropriate resistance coefficient K to the $V^2/2g$ term.

In system head calculations for high head pumps the velocity head will be but a small percentage of the total head and is not significant. However, in low head pumps it can be a substantial percentage and must be considered.

When total heads on an existing installation are being determined from gage readings then the velocity head valves as calculated must be included; i.e., the *total suction lift* will be the reading of a vacuum gage or mercury column at the suction flange, corrected to the pump centerline elevation minus the velocity head at point of gage attachment. The *total suction head* and *total discharge head* will be the readings of gages at the flanges corrected to the pump centerline elevation plus the velocity heads at the points of the gage attachments.

Total system head (H)—formerly total dynamic head—is the total discharge head (h_d) minus the total suction head (h_s) if positive or plus if a suction lift: $H = h_d - h_s$ (head) or $H = h_d + h_s$ (lift).

Pump Head—Pressure—Specific Gravity

In a centrifugal pump the head developed (in feet) is dependent on the velocity of the liquid as it enters the impeller eye and as it leaves the impeller periphery and therefore is independent of the specific gravity of the liquid. The pressure head developed (in psi) will be directly proportional to the specific gravity.

Head and *Pressure* are interchangeable terms provided that they are expressed in their correct units. In English Units to convert from one to the other use:

$$\text{Liquid Head in feet} = \frac{\text{psi} \times 2.31}{\text{sp gr}}$$

or

$$\text{Liquid Head in feet} = \frac{\text{psi} \times 144}{W}$$

$$\text{Pressure in psi} = \frac{\text{Head in feet} \times \text{sp gr}}{2.31}$$

or

$$\text{Pressure in psi} = \frac{\text{Head in feet} \times W}{144}$$

Where W = Specific weight in pounds per cubic foot of liquid being pumped under pumping conditions; For Water W = 62.32 lb. per cu. ft. at 68 degrees F (20 degrees C).

A column of water 2.31 ft. high will exert a pressure of one (1) psi based on water at approximately 65 degrees F.

Figures 7.6 and 7.7 are included to help visualize the head pressure relationships of centrifugal pumps when handling liquids of varying specific gravities.

Figure 7.6 illustrates three identical pumps, each pump designed to develop 115.5 ft. of head; when pumping water with a specific gravity of 1.0 (at 68°F) the pressure head will be 50 psi (115.5 ft. divided by 2.31); when pumping liquids of other gravities, the head (in feet) will be the same, but the pressure head (psi) will be proportional to the specific gravities as shown; to avoid errors, it is advisable to check one's calculations by using the above formulas.

Figure 7.7 illustrates three pumps, each designed to develop the same pressure head (in psi); consequently the head (in feet of liquid) will be inversely proportional to the specific gravity as shown.

In these illustrations, friction losses, etc., have been disregarded.

Net Positive Suction Head

The Net Positive Suction Head (NPSH) is the total suction head in feet of liquid (absolute at the pump centerline or impeller eye) less the absolute vapor pressure (in feet) of the liquid being pumped.

It must always have a positive value and can be calculated by the following equations. To help in visualizing the conditions that exist, two expressions will be used: the *first expression* is basis a suction lift-liquid supply level is below the pump centerline or impeller eye; the *second expression* is basis a positive suction (flooded), where the liquid supply is above the pump centerline or impeller eye.

For Suction Lift: $NPSH = h_a - h_{vpa} - h_{st} - h_{fs}$

For Positive (Flooded) Suction: $NPSH = h_a - h_{vps} + h_{st} - h_{fs}$

where

h_a = absolute pressure (in feet of liquid) on the surface of the liquid supply level (this will be barometric pressure if suction is from an open tank or sump; or the absolute pressure existing in a closed tank such as a condenser hotwell or deareator).

h_{vpa} = the head in feet corresponding to the vapor pressure of the liquid at the temperature being pumped.

h_{st} = static height in feet that the liquid supply level is above or below the pump centerline or impeller eye.

h_{fs} = all suction line losses (in feet) including entrance losses and friction losses through pipe, valves and fittings, etc.

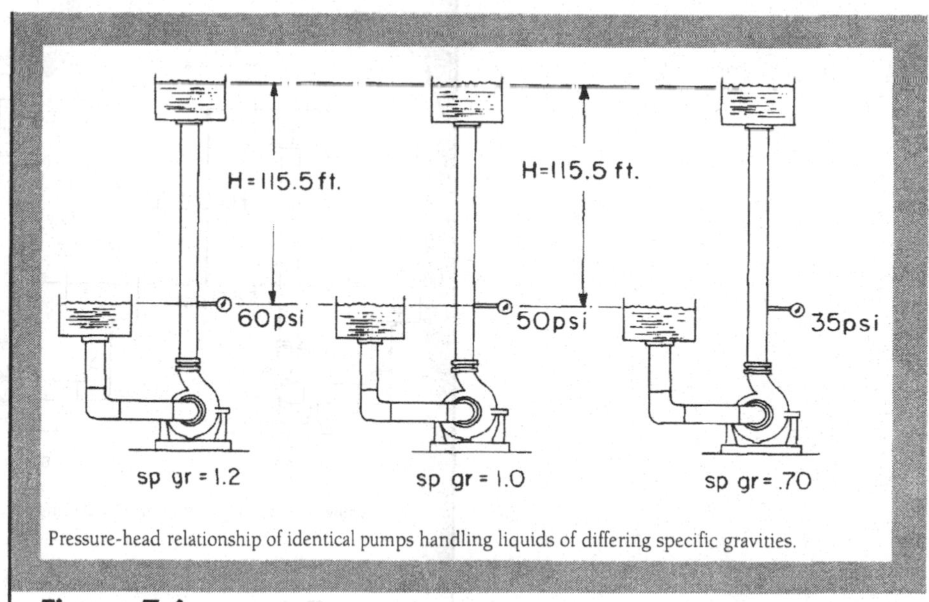

Pressure-head relationship of identical pumps handling liquids of differing specific gravities.

Figure 7.6

Two values of net positive suction head must be considered; i.e., Net Positive Suction Head Required (NPSHR) and Net Positive Suction Head Available (NPSHA).

The NPSHR is determined by the pump manufacturer and will depend on many factors, including type of impeller inlet, impeller design, pump flow, rotational speed, nature of liquid, etc. NPSHR is usually plotted on the characteristic pump performance curve supplied by the pump manufacturer. The Net Positive Suction Head Available (NPSHA) depends on the system layout and must always be equal to or greater than the NPSHR.

The vapor pressure of the liquid at the pumping temperature must always be known to calculate the NPSHA. On an existing installation the NPSH available would be the reading of the gage at the suction flange converted to feet of liquid absolute and corrected to the pump centerline elevation less the vapor pressure of the liquid in feet absolute plus the velocity head in feet of liquid at point of gage attachment.

The following examples show the importance and influence of vapor pressure. In all cases, for simplicity, the same capacity will be used; also, the following suction line losses will be assumed in all cases.

Friction loss through suction pipe and fittings	2.51 ft.
Entrance loss (assume equal to one-half velocity head)	0.41
Total losses	2.92 ft.

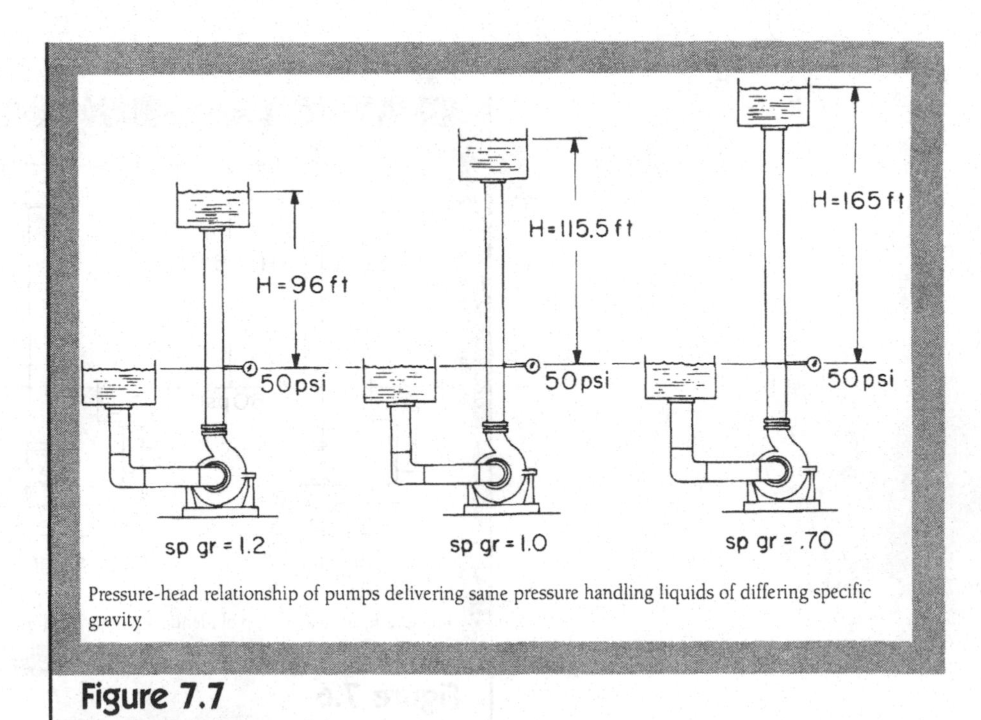

Pressure-head relationship of pumps delivering same pressure handling liquids of differing specific gravity.

Figure 7.7

System Curves

A centrifugal pump always operates at the intersection of its head-capacity curve and the system curve which shows the head required to make the liquid flow through the system of piping, valves, etc. The head in a typical system is made up of three components:

1. Static head
2. Pressure head
3. All losses; i.e., friction, entrance and exit losses

To illustrate, take the typical system shown in Figure 7.8 where the total static head is 70 feet, the pressure head is 60 feet (2.31 x 26) and the friction head through all pipe, valves, fittings, entrance and exit losses is 18.9 feet at the design flow of 1500 gpm; total system head at design flow is 70 + 60 +18.9 = 148.9 feet.

In drawing the system curve (see Figure 7.9), the static head will not change with flow so it is represented by the line AB; the pressure head will not change with flow, so it is added to the static head and shown by the horizontal line CD. The friction head through a piping system, however, varies approximately as the square of the flow, so the friction at 500 gpm will be

$$\left(\frac{500}{1500}\right)^2 \times 18.9 = 2.1 \text{ feet (Point E);}$$

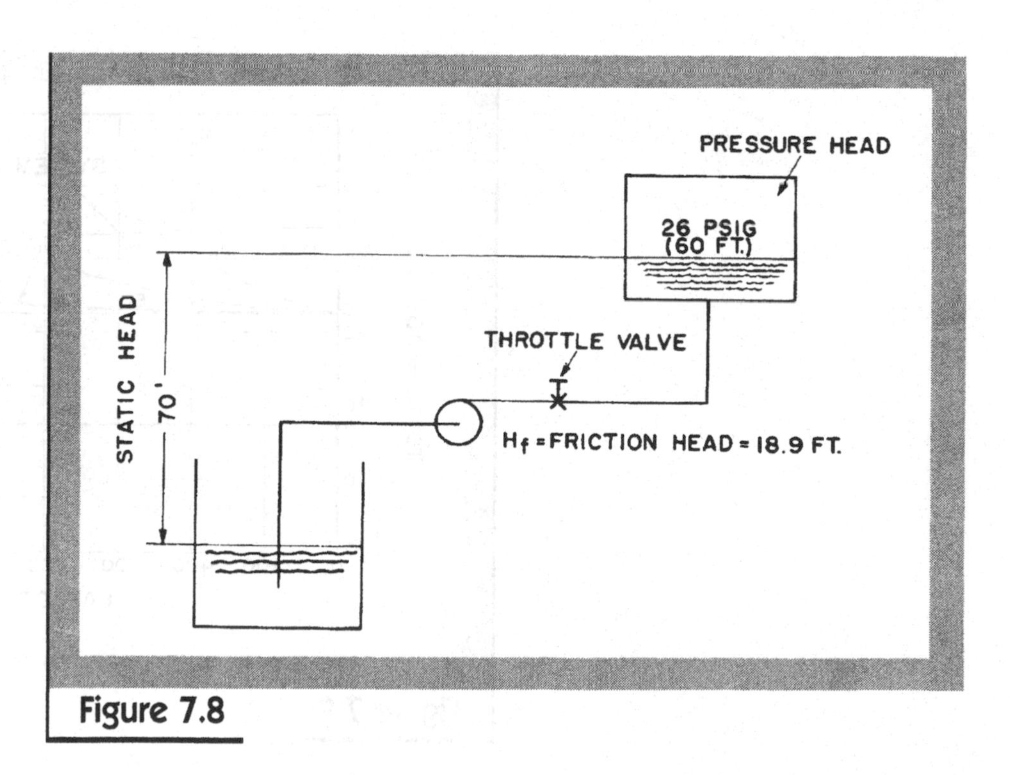

Figure 7.8

likewise the friction at 1000 gpm will be 8.4 feet (Point F); these determine the final curve CEFG. In Figure 7.9 the pump head capacity curve has been superimposed on the system curve. Unless something is done to change either the head capacity curve or the system curve, the pump will operate at 1500 gpm (Point G) indefinitely.

If the throttle valve in the pump discharge line is closed partially it will add friction to the system and the pump can be regulated to operate at Point H (1000 gpm) or at Point J (500 gpm) or at any other point on its curve; this is changing the system curve by throttling and in this case is known as "throttling control."

Parallel and Series Operation

With properly selected Head-capacity (H-Q) curves (preferably curves with continuously rising characteristics) and subject to certain hydraulic and mechanical considerations, parallel and/or series operation of centrifugal pumps can be employed to meet a wide range of service requirements.

In considering multiple pump operation, a system head curve for the entire capacity and head requirements must be made available. The individual pump H-Q curves must be superimposed on the system curve.

For parallel operation of two or more pumps the combined performance curve is obtained by adding horizontally the capacities of the same heads.

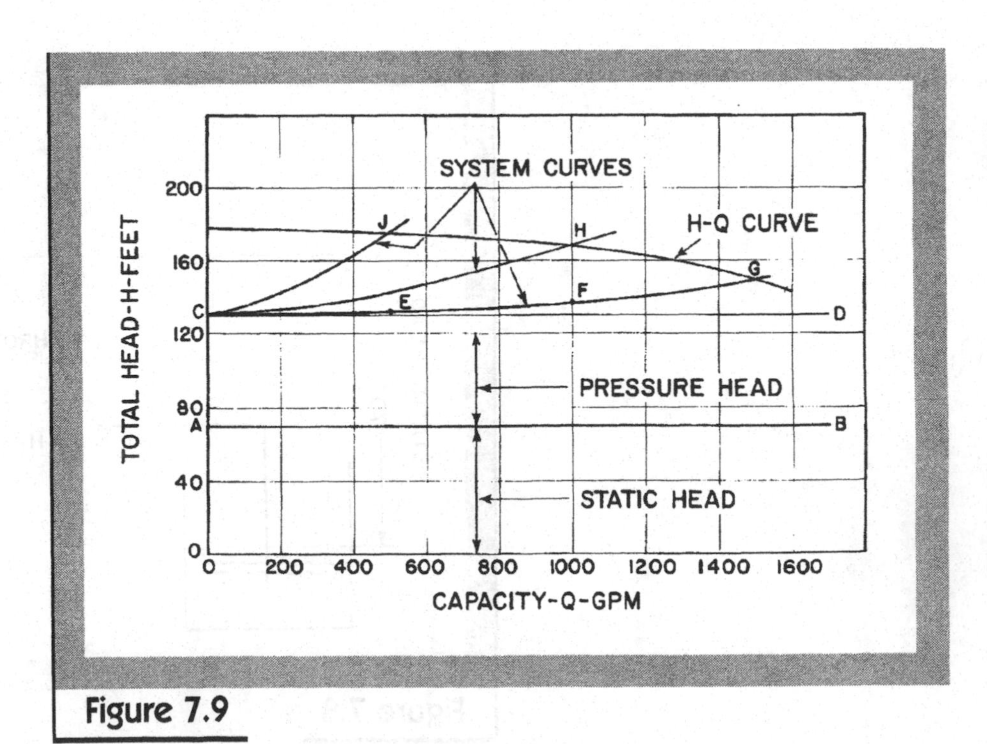

Figure 7.9

For series operation the combined performance curve is obtained by adding vertically the heads at the same capacities.

Since a centrifugal pump always operates at the intersection of its H-Q curve with the system curve, superimposing the system curve on the pump performance curve clearly indicates what flow can be expected and at what heads each pump or its combination will operate.

Figure 7.10 illustrates a two pump series-parallel operation; this could be expanded to a three or more pump combination if desired. In this illustration a typical H-Q curve for a single pump designed for 1000 gpm at 60 ft. head is shown; two pumps in series will deliver 1000 gpm at 120 ft.; or two pumps in parallel will deliver 2000 gpm at 60 ft. system curves have assumed as shown. If the pumps have "run-out capabilities" (dotted portion), they will operate at the intersection of the H-Q curves with the system curves. Note that in this illustration the system curve is based on the assumption that there is 20 ft. static head.

Figure 7.11 offers a simple worksheet to figure head loss.

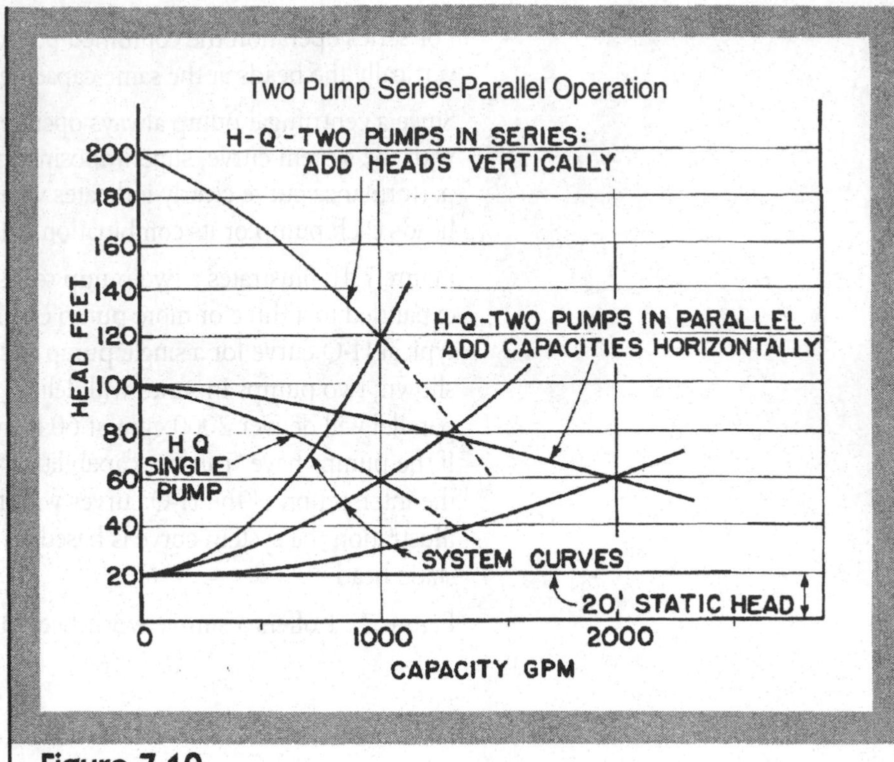

Figure 7.10

Head Loss Calculation Worksheet ΔP Tabulation

SECTION								
Flow Rate (GPM)								
Pipe, Nominal Diameter (inches)								
Straight Pipe (feet)								
Equivalent Lengths (feet)								
Ells								
Tees								
Valves (e.g., gate, ball)								
Total Equivalent Length								
Friction "H_L"								
Other Pressure Drops								
Flow Meter								
Filter								
Pressure-Regulating Valve								
Throttling Valve								
Terminal "H_L" Required								
Total Calculated Head Loss								

Total Pump Head = Total Calculated Head Loss + Difference in Elevation

Figure 7.11

Frequently Used Formulas and Equivalents

Velocity:

$Q = AV$ or $V = Q/A$

$$V = \frac{0.4085 \text{ gpm}}{d^2} = \frac{0.2859 \text{ bph}}{d^2} = \frac{0.0028368 \text{ gpm}}{D^2} = \frac{0.001985 \text{ bph}}{D^2}$$

where:

V = velocity of flow—ft/sec
Q = capacity—ft^3/sec
A = area of pipe or conduit—ft^2
d = diameter of circular pipe or conduit—inches
D = diameter of circular pipe or conduit—ft
gpm = U S gallons per minute
bph = barrels per hour (42 gals—oil)

Velocity equivalents:

ft/sec × 0.3048 = m/sec
m/sec × 3.2808 = ft/sec

Fluid flow—any liquid:

U S gpm × 0.002228 = ft^3/sec
U S gpm × 0.2271 = m^3/hr
U S gpm × 0.227 = metric tons per hour (water at approximately 60°F)
mgd (U S) × 694.5 = U S gpm
ft^3/sec × 448.83 = U S gpm
m^3/hr × 4.403 = U S gpm
U S gallon × 0.8327 = Imperial gallons
Imperial gallons × 1.201 = U S gallons

$$\text{gpm} = \frac{\text{lb per hr}}{500 \times \text{sp gr}} \text{ (water at 60°F)}$$

gpm = boiler hp × 0.069

Fluid flow—oil:

barrels × 42 = U S gallons
barrels per hour × 0.7 = U S gpm
barrels per day × 0.02917 = U S gpm
U S gpm × 1.4286 = barrels per hour
U. S gpm × 34.286 = barrels per day

Head and pressure: (For water at normal temperatures (60°F))

$$\text{Head in feet} = \frac{\text{Head in psi} \times 2.31}{\text{sp gr}}$$

$$\text{Head in psi} = \frac{\text{Head in feet} \times \text{sp gr}}{2.31}$$

Pumping power—

horsepower × 550 = ft-lb/sec
× 33000 = ft-lb/min
× 2546 = BTU/hr
× 745.7 = watts
× 0.7457 = kilowatts
× 1.014 = metric horsepower

$$\text{Brake hp} = \frac{\text{gpm} \times \text{H (in feet)} \times 1.08}{3960 \times \text{efficiency}} \text{ (centrifugal terminology)}$$

$$= \frac{\text{bph} \times \text{H (in feet)} \times \text{sp gr}}{5657 \times \text{efficiency}} \text{ (centrifugal terminology)}$$

$$= \frac{\text{gpm} \times \text{psi}}{1714 \times \text{eff}} \text{ (reciprocating terminology)}$$

$$= \frac{\text{bph} \times \text{psi}}{2449 \times \text{eff}} \text{ (reciprocating terminology)}$$

Note: To obtain the hydraulic horsepower from the above expressions assume a pump efficiency of 100%.

In the above expressions:

gpm = U S gallons per minute delivered (one gallon = 8.338 lbs at 60 Deg F.
bph = barrels (42 gallons) per hour—delivered = 0.7 gpm
H = total head in feet of liquid—differential
psi = lb per sq in—differential
sp gr = specific gravity
eff = efficiency expressed as a decimal

$$\text{Electrical hp input to motor} = \frac{\text{pump bhp}}{\text{motor efficiency}}$$

$$\text{KW input to motor} = \frac{\text{pump bhp} \times 0.7457}{\text{motor efficiency}}$$

Torque—

$$\text{Torque in lb-ft} = \frac{\text{bhp} \times 5250}{\text{rpm}}$$

bhp = brake horsepower

rpm = revolutions per minute

Specific speed—

$$\textit{Impeller specific speed} = N_s = \frac{\text{rpm}\sqrt{\text{gpm}}}{H^{3/4}}$$ (See page 1-20)

where

gpm = design capacity at best efficiency point
H = head per stage at best efficiency point
rpm = speed

$$\textit{Suction specific speed} = S = \frac{\text{rpm}\sqrt{\text{gpm}}}{(\text{NPSHR})^{3/4}}$$ (See page 1-21)

where

gpm = design capacity at best efficiency point for single suction first stage impellers, or one half design capacity for double suction impellers.

Affinity laws

At constant impeller diameter:—(Variable speed)

$$\frac{\text{RPM}_1}{\text{RPM}_2} = \frac{\text{gpm}_1}{\text{gpm}_2} = \frac{\sqrt{H_1}}{\sqrt{H_2}}$$

At constant speed:—Variable impeller diameter)

$$\frac{D_1}{D_2} = \frac{\text{gpm}_1}{\text{gpm}_2} = \frac{\sqrt{H_1}}{\sqrt{H_2}}$$

Frequently Used Formulas and Equivalents (cont.)

Miscellaneous

Temperature equivalents:

	Kelvin	Degrees Rankine	Degrees Celsius	Degrees Fahren-heit
Absolute zero	0	0	−273.15	−459.67
Water freezing point: (14.696 psia 101.325 KPa)	273.15	491.67	0	32
Water boiling point: (14.696 psia 101.325 KPa)	373.15	671.67	100	212

Celsius/Fahrenheit conversions:

$$\text{Deg C} = 5/9\ (°F - 32)$$
$$\text{Deg F} = 9/5\ °C + 32$$

Reynolds Number (R): (see page 1-4)

$$R = \frac{VD}{v}$$

V = Average velocity—ft/sec
D = Average internal diameter—ft
v = Kinematic viscosity of the fluid—ft²/sec (For pure fresh water at 60°F. v = 0.000 0 012 16 ft²/sec.)

Darcy-Weisbach (see page 3-3)

$$h_f = f\,\frac{L}{D}\,\frac{V^2}{2g}$$

Hazen and Williams (see page 3-7)

$$h_f = 0.002083\ L\left(\frac{100}{C}\right)^{1.85} \times \frac{gpm^{1.85}}{d^{4.8655}}$$

Resource Publications

Gieck, Kurt and Reiner Gieck, *Engineering Formulas, Sixth Edition*. New York, NY: McGraw-Hill, 1990.

Heald, C.C., *Cameron Hydraulic Data*. Woodcliff Lake, NJ: Ingersoll-Rand, Inc., 1988.

Lindeburg, Michael L., *Engineering-In-Training Reference Manual*. Belmont, CA: Professional Publications, Inc., 1992.

National Council of Examiners for Engineering and Surveying., *Fundamentals of Engineering Discipline Specific Reference Handbook*. Clemson, SC: National Council of Examiners for Engineering and Surveying, 1996.

Spirax Sarco, Inc., *Design of Fluid Systems: Steam Utilization*. Allentown, PA: Spirax Sarco, Inc., 1991.

Chapter 8

Instrumentation & Controls

Chapter 8 Instrumentation & Controls

Instrumentation and controls play an indispensable role in the safe and economical operation of plant, facility, and industrial operations everywhere. Instrumentation permits operators and managers to monitor the current state of the operation, while process control systems regulate and adjust various tasks, from the simplest (such as self-contained pressure regulators and room thermostats) to the most advanced (like digital systems that control the entire manufacturing process). Facility managers and engineers should be familiar with the vocabulary and basic principles of process control in order to recognize the opportunities and evaluate the benefits of control systems, and to effectively communicate with systems engineers, vendors and co-workers.

This chapter begins with a review of some of the specialized terminology used in the instrumentation and process control field. The section that follows reviews the principles of feedback control loops and includes a discussion of basic and combined modes of control. The chapter concludes with the function and operation of electromechanical relays.

The following section is reprinted from *Principles of Industrial Measurement for Control Applications*, by Ernest Smith, Instrument Society of America.

Terminology

As with every other subject, instrument engineering has its own terminology. Some of the terms have subtle meanings, and a misunderstanding can lead to a completely wrong impression of the performance of a system. The following definitions are intended as an introduction to the use of the terminology.

Range. Every sensor is designed to work over a specific *range*. While an electrical output may be variable and adjusted to suit the application, this is not usually practical with mechanical transducing elements. The design ranges of the mechanisms are usually fixed, and if exceeded, result in permanent damage to or destruction of a sensor.

It is customary to use transducing elements over only the part of their range where they provide predictable performance and often enhanced linearity. This practice also provides a greater safety factor against overloading. To compensate for the accompanying loss of signal (or sensitivity) the electrical output is often amplified and adjusted to a convenient size. Unfortunately, there is a limit to which amplification can be applied and still retain the enhanced performance of the transducer, as the enhancement is offset by amplifier errors, and *noise*. Noise is an interference signal superimposed on the measurement signal. It may be "picked up" from external sources, or be caused by equipment instability.

Zero. When making a measurement it is necessary to start at a known datum, and it is often convenient to adjust the output of the instrument to zero at the datum. The output of a centigrade thermometer is zero at the freezing point of water; the output of a pressure gage may be zero at atmospheric pressure. Zero, therefore, is a value ascribed to some defined point in the measured range.

Zero Drift. One of the problems experienced with sensors occurs when the signal level varies from its set zero value. This introduces an error into the measurement equal to the amount of variation, or *drift* as it is usually termed.

Zero drift may occur for a number of reasons. Changes in ambient temperature, causing physical changes in the sensor, or electrical changes in the amplifier, are common causes, but aging of components or mechanical damage may also be responsible. It is a particularly troublesome problem when the sensor is used for long-term measurements and may not be returned to its zero position for periodic checking.

All sensors are affected by drift to some extent and it is sometimes specified in terms of *short-term* and *long-term drift*. Short-term drift is usually associated with changes in temperature or electronics stabilizing. Long-term drift is usually associated with aging of the transducer or electronic components.

Sensitivity. *Sensitivity* of a sensor is defined as the change in output of the sensor per unit change in the parameter being measured. The factor may be a constant over the range of the sensor or it may vary. Such devices are described as having a linear or a nonlinear output, respectively.

Sensitivity depends on a number of factors which are variable. The mechanical properties of a transducer may vary with temperature, causing a variation in sensitivity, but often it is the electrical part of the sensor which is responsible for the greatest changes. An amplifier may change its gain because of temperature effects on components or variations in power supplies, or even faulty operation. In addition, many sensors must be provided with a voltage from a power source before they give an output. Often the sensor output is directly proportional to this excitation voltage, and any changes in it during operation will introduce corresponding errors. It may be argued that the worst effects of sensitivity changes do not occur when the sensor fails completely, but when the sensitivity changes sufficiently to cause an error, without the error being easily recognizable.

Resolution. *Resolution* is defined as the smallest change that can be detected by a sensor. It is not a universally accepted term, as some people claim that it is a parameter with no absolute value and no precise definition. Sensors that use wire-wound potentiometers or digital techniques to provide their electrical output have finite resolution, which without doubt is meaningful and serves to describe the quality of the instrument. Frequently, however, manufacturers describe their sensors as having infinite resolution. There is no such device. Apart from restrictions incurred by electrical instability, the mechanical construction of a sensor imposes its own restriction, below which changes in output are meaningless.

Response. The time taken by a sensor to approach its true output when subjected to a step input is sometimes referred to as its *response time*. It is more usual, however, to quote a sensor as having a flat response between specified limits of frequency. This is known as the *frequency response*, and it indicates that if the sensor is subjected to sinusoidally oscillating input of constant amplitude, the output will faithfully reproduce a signal proportional to the input. Beyond the limiting frequencies the output may rise or fall, depending on the construction of the device.

Linearity. The most convenient sensor to use is one with a linear transfer function. That is an output that is directly proportional to input over its entire range, so that the slope of a graph of output versus input describes a straight line. This allows a single conversion factor to be applied over the range. In practice, this is never quite achieved, although most transducers exhibit only small changes of slope over their working range. To these curves, a *best* straight line is fitted whose error is usually well within the tolerance of the measurement. Some sensors, particularly those using inductive transducing principles, demonstrate considerable changes in the slope of their output versus input graph and may even reach a point where, regardless of change of input, there is no change of output. The working range of such a sensor is restricted and must be limited to where the graph is relatively linear, or alternatively a different factor must be applied to each reading.

Hysteresis. *Hysteresis* becomes apparent when the input to a sensor is applied in a cyclic manner. If the input is increased incrementally to the sensor's maximum and returned to its zero datum in a similar manner, the calibration may be seen to describe two curves that meet at the maximum. In returning to zero input, the calibration has not returned to its original datum. If now the calibration is continued in the negative direction of input, two further curves will be produced which are a mirror image of the previous ones. Further cycling will eventually link these two halves into one complete loop, which will then be repeatable with every cycle.

This loop is normally referred to as the *hysteresis loop* of the sensor, although it contains any of the other nonlinearity effects which may be present. Consequently, it is usual when specifying a sensor to quote nonlinearity and hysteresis as one parameter.

Calibration. If a meaningful measurement is to be made, it is necessary to measure the output of a sensor in response to an accurately known input. This process is known as *calibration*, and the devices that produce the inputs are described as *calibration standards*.

It is usual to provide measurements at a number of points of the working range of the sensor, so that a ratio of output may be determined. This ratio may be obtained by plotting a graph and calculating its slope, or it may be determined from the measured points by calculation. Such a ratio is described as a *calibration factor*.

The ratio output to input is not always a constant over the range of a sensor, and the calibration graph describes a curve. In these instances a *best straight line* may be fitted through the points and the errors accepted, or a different calibration factor must be provided for every measurement.

Accuracy. The *accuracy* of a sensor is a term causing some confusion, possibly because it is made up of a number of independent quantities, each of which is quoted separately, and possibly because it is usual to quote the accuracy of a measurement as a percentage of *full-scale output* (% FS).

The strict meaning of accuracy when quoted as a percentage of full-scale output is that of a *value of uncertainty* which is applied to converted sensor outputs throughout the entire range of measurement. For example, a measurement with an accuracy of 1± FS and with a range of 0–100 units has a value of uncertainty of ± 1 part in 100 or ± 1 unit which applies to every measurement. A measurement of 50 units would be made with a value of uncertainty of ± 1 unit and an uncertainty of ± 2% of the value. Although the word accuracy is still widely used the word uncertainty is becoming commonplace.

Manufacturers do not normally quote an overall figure of accuracy for a sensor, since some of the variations may not be relevant to certain measurements. A sensor being used at constant temperature would not be affected by errors associated with temperature. Its uncertainty would be largely influenced by its linearity and hysteresis. So that users may have some control in estimating the performancy of their sensors under their own operating conditions, it is usual to provide a breakdown of parameters, which may include:

- Linearity and hysteresis ± %FS
- Residual output at zero ± %FS
- Zero drift with temperature ± %C
- Repeatability ± %FS
- Sensitivity change with temperature %/^{0}C

This list is not exhaustive, nor is it standard, and this can make comparison of specifications difficult. If a calibration over the operating limits of a sensor is not practical, then the overall uncertainty of the sensor signal can only be determined by summing up the manufacturer's specifications. Absolute uncertainty is the algebraic sum of the sensor. It is very unlikely, however, that in practice all the variations would be additive, and the *root-sum-square* method of summation is commonly used to determine a practical figure. This method is derived from probability theory and is summarized in the formula:

$$a = \pm \sqrt{a_1^2 + a_2^2 + a_3^2 + \ldots}$$

where a represents the overall uncertainty to a probability of about 70, and a_1, a_2 and a_3, represent the parameter uncertainties expressed as a percentage of full scale.

In making a measurement, the accuracy or uncertainty of power supplies, amplifiers, and recorders also contributes to the overall value. Some instrumentation engineers treat all these quantities as a "measuring chain" and do not attempt to break them down, arguing that the accuracy of the measurement can only be the accuracy of the chain.

Another way of estimating the overall uncertainty of a measuring chain is to take the algebraic sum of the component values and divide this figure by the square foot of the number of components. Although this method is valid, according to statistical theory, and approximates the root-sum-square answer, it is not so widely used.

Basic Concepts

Block Diagrams. A schematic depicting the major elements in a control system and how they are connected. A block diagram illustrates power and information flow as well as interconnections. In their strictest form, the mathematical relationships between inputs and outputs are given in the blocks of the diagram.

Compensation. The filter or device (and the accompanying design procedure) that compensates for one or more poor response areas in the frequency domain of a process to be controlled.

Controller. The name given to that element of the control system that operates on the comparison of the desired and actual output of the system and provides inputs to the process actuator to bring the output closer to the desired response. The controller incorporates the *intelligence* of the control system. Synthesis in a control system is primarily concerned with defining the controller.

Dynamic System. A process that requires differential equations for its model with time as an independent variable. A dynamic system has energy storage which is out of phase with its motion. In order to synthesize a control, then, information on the time relationship of the energy storages and the motions must be understood. This knowledge is embodied in the solution of time-independent differential equations.

Disturbance. An influence on a process or control that is generally not predictable except statistically. A control system compensates for such disturbances by either measuring them directly or detecting their influence on the output of the system and countering their effects through the controller and actuator.

Error. The difference between the desired value of an output of a controlled system and the actual value. The resulting signal is most often used as an input to the controller, whose output is used to correct the error.

Feedback. Measuring the process output and transmitting back to the originating input location the results of a controller and actuator and the disturbances changing the behavior. This is often described as "closing the loop." An open-loop system is one where there is no feedback. Negative feedback is the common type; positive feedback is used occasionally for increasing the gain. The feedback signal is basic to automatic control, and the accuracy of the required measurement transduction is the limiting factor in the accuracy of an automatic control system.

Filter. The electric, fluidic, or mechanical circuit that selectively reduces or amplifies certain frequency ranges. It is used to reduce noise, compensate for poor response of a process, or smooth the response of a system. It is usually part of the controller or feedback loop.

Gain. The multiplier of the error in a controller to produce the required unit of actuation per unit change in the error. High gain, then, denotes a quick-responding system.

Inputs. The variables that are the causal factors in changing a system's outputs. Inputs vary with time in the same frequency range as the system's response.

Linear and Nonlinear Systems. Systems described by linear and nonlinear differential equations, respectively. Linear response is characterized by a proportional relationship between inputs and outputs. There is a whole body of control theory developed for linear systems, which allows a methodical and complete synthesizing procedure for control system development. Nonlinear systems are all of those systems that are not linear. Natural physical processes are almost all nonlinear. Fortunately, their response in the neighborhood of an equilibrium state can usually be described adequately by a linear system of equations.

Measurement. The determination of the value of variables in a control system. This is generally accomplished by transducers that transform physical variables to signals that can be monitored or read. Measurement of the output and other variables in a controlled system is central to its accurate performance. Measurements may be either continuous or discrete in time.

Model. The representation of a process or system by a set of mathematical equations or a smaller-scale model. This depiction predicts the process or system performance in its more important characteristics. It is used to simulate the process or system so that a controller or other means of varying the performance of the system can be synthesized with less cost and risk than by experimenting iteratively with the actual process. A model needs to be close to but not necessarily equivalent to an actual process. There are always many more variables than can be reasonably modeled, and even those that can be modeled are accurately described over a limited range and by simple interactions with other variables.

On-Off Control. A controller that turns the actuator full on or full off (or full positive or negative). It is the common control system in heating, ventilating, and air-conditioning systems, where the heating or cooling source is either on or off. The controller continuously monitors the error signal and determines when the actuation should be off or on. On-off systems are sometimes called *band-bang* control systems.

Outputs. The variables defined as those of primary interest to be measured and controlled. These variables are defined as part of the development of the model. Experiments with the actual system will indicate which of the many variables are important and essential in characterizing the system's response.

PID Control. Proportional plus integral plus derivative control. This term denotes a control common to the process industry. In fact, hardware implementing PID control has been available for many years. Measurement signals and set points are inputs to the PID control. The output to the actuator is the sum of a proportional constant times the error, an integral (or reset) constant times the integration of the error signal, and a rate constant times the derivative of the error. Proportional control is the most used controller. The error will not become zero in a proportional controller; if it were zero, there would not be any signal to the actuator to hold the process output to the desired value. Integral control, however, can yield output to the actuator with zero error. Integral control adds another energy storage element, however, and will tend to destabilize the overall system. Derivative control embodies a prediction of the output variation and tends to stabilize the system unless there are high-frequency noise sources.

Process Control. Automatic control as applied in the process industries. These are the chemical, metal, glass, paper, and other basic processing industries. They are characterized by the conversion of raw materials to basic products that become the material for industries that produce items more closely related to consumer products. The control problems and solutions in these industries have some intrinsic commonality. Flows, temperatures, chemical reactions, pressure, and chemical and material composition are common characterizing variables. Massive equipment and large capital investments are common. Automatic control is essential to lowering risk and producing consistently high quality. As a result, automatic control in the industries has a long history. Computer control was first widely used in these industries.

Servomechanism. A power-amplifying element of a control system derived from the French term meaning servant. It is a general concept describing the class of automatic controls that embody an intermediate power amplifier. The controller and comparator are often incorporated in servomechanism devices.

Set Point. A process industry term denoting the desired output value for a controlled system. The set point temperature in a chemical reactor is the value the controller and actuator would attempt to maintain in the reactor.

Simulation. The procedure of representing an actual process and hypothesized control system by mathematical or scale physical models and iteratively solving the equations or running the models to characterize the behavior of the actual system from the model. Simulation implies digital or analog computers that embody the solution of the differential equations representing the process to be controlled and its control elements.

Stability. A system characteristic such that for small disturbances or inputs, the output will remain close to its initial value. Unstable systems will not remain in equilibrium and will destroy themselves or move to a new state that is unacceptable. Much of control theory and simulation is concerned with stability since feedback itself can be a destabilizing factor.

Statistical Process Control. A theory and procedure to bring discrete manufacturing processes under control. The procedure relies on a mixture of human and machine interaction with data taken from the controlled process output. The statistical part of the control is the sampling of a relatively small number of products of the process and inferring the behavior of the entire process by averaging sample results. Feedback is performed by an operator relying on the control charts developed from the statistical data. *Process capability* defines the ability of a process to produce products within specifications. *In control* defines a process that is capable and is being controlled so that the output meets the requirements. This theory was developed and codified in the United States by Deming but was first implemented successfully in Japan.

Supervisory Control. Control of the set points in a multivariable, interacting control system. It implies change in the set points to optimize the system configuration, assuming that the dynamic response is handled by a single controller.

Synthesis. The steps leading to development of a controlled system primarily in choosing the measurements and defining the controller. Modeling and simulation are one step in synthesis. The term was first used in the electronics industry, where a control, or filtering, circuit was synthesized to accomplish some desired objective.

Transfer Functions. The key mathematical elements in a block diagram denoting the transformation of input to output for that block. Transfer functions are usually Laplace or Fourier transforms of linear differential equations relating inputs and outputs.

Advanced Concepts

Adaptive Control. A controller that can change its character to adapt to functional changes in the process being controlled. It was first developed for prime mover controls, where the response characteristics of the engine varied significantly with load and speed. Changes in the controller were needed to compensate for varying engine characteristics. The difficult part of implementing the concept has always been identifying the change in the process while the engine is in operations. This is accomplished by adding an additional small input to the system and measuring the correspondingly small response of the system to infer the change.

Computer Control. A control system where a digital computer is the primary means of control and communication. Computer control was first used widely in the process industries, where there are many processes to be controlled, hundreds of measurements, and the need for centralized setpoint control and measurement communication. It is now widely used in most modern control systems.

Cybernetics. A term first coined by Norbert Wiener to denote intelligent control systems, where human reasoning might be embodied in the system.

Discrete Systems. Systems where sampling theory is used to model the process and synthesize the control. It is often used to described computer control systems where analog signals must be discretized to work with a computer and the resulting actuation commands converted into continuous variables by digital-to-analog conversion.

Distributed Systems. Systems where partial differential equations are necessary to describe the behavior and response, most frequently where spatially independent variables are important in addition to the time variable. Examples are acoustics and traveling waves, heat transfer and chemical reacting systems, and geological systems. Partial differential equations are often approximated by ordinary differential ones. This is entirely appropriate since all systems are distributed in nature.

Identification. Denotes the many methods and procedures for determining the differential equations or models for systems from measurements of inputs and outputs. It also denotes the actual online procedure of determining the equations or models. Much effort has been expended to develop identification tools since the knowledge of a system model is essential to the development of a control system.

Interacting Controls. A control system where there are two or more outputs and corresponding inputs that are interrelated. The controller then must compute the inputs to the actuators that account for this interaction and be able to control each process output separately.

Large-Scale Systems. Systems with a very large number of interrelated variables and great complexity. The telephone system and the economy are examples. Theories for simulating and controlling such systems are incomplete.

Man-Machine Systems. Systems where a person is an integral part of the control system, either as a controller or measurement transducer. Examples include airplanes, where the pilot is in the control loop as the primary controller; statistical process control, where the person takes measurements, computes the controls, and implements the actuation change; and most economic systems, where decisions by experts based on economic trends actuate change.

Multivariate Systems. Systems with two or more inputs and outputs that are interrelated.

Optimal Control. Denotes that the controller has been synthesized to minimize (or maximize) some performance measure, usually a time integration of a function of the error signals. This concept has been at the heart of the advanced work in the control and systems engineering field for almost 40 years.

Performance Criteria. A function whose value describes the overall optimality of a controlled system. It is usually a time integration of a linear combination of squares of the differences between the desired values for system state variables and their actual values.

Robustness. The ability of a controlled system to operate after the loss of one or more measurement or actuator.

State Examination. The determination of the *state* of a process from differential equations relating the inputs and outputs and measurements of portions of the inputs and outputs. Kalman was the first to develop the theory to its full potential. It has always been an important concept in stochastic control and filtering.

State Variables. The variables equivalent to the initial conditions for differential equations. This variable set, if known at any point in time, is sufficient (with the describing equations) to predict the future behavior of the system. State variables can be used as feedback variables to optimize linear systems. State estimation is the technique whereby state variables are estimated from those variables that can actually be measured.

Stochastic Control. Control derived from a statistical model of the process. It is also the body of theory associated with the synthesis of control systems using statistical and stochastic techniques.

Control System Elements

Actuator. The element that implements the controller signal. It receives its input from the controller, and its output goes to the process to be controlled. Its power supply will determine the system capability to actuate the process.

Comparator. The element in a control system that compares the desired output with the measured output. It is often a subtracting element with a gain.

Control Computer. The computer, usually digital, that is the controller of a process. In the process industries, the control computer is the central controller, which receive all measured outputs, computes all control outputs, and monitors the system.

Instrumentation. The set of transducers, amplifiers, wiring, and configuration for the measurements in a system. It implies the analysis of errors, noise effects, signal levels, accuracy, and a knowledge of the various transduction elements and available hardware for measurements.

Interfaces. Impedance matching devices that are placed in the signal paths of control and other systems. For example, the analog-to-digital interface for a computer conditions the analog signals and discretizes them for digital processing.

Power Supply. Device or devices that supply controlled power to the actuator in a control system. It is most often electrical power for a motor, pneumatic pressure for a valve, or hydraulic pressure for a motor or linear actuator. Rocket engine gases are more exotic examples.

Process. The physical device or system whose output is to be controlled. One of the strengths of systems engineering is that vastly different processes can be approached with a standard set of concepts and tools.

Transducers. Devices that transform one physical variable into another. Examples include thermocouples for temperature, moving coils and quartz crystals for pressure and motion, magnetic fields for motion, and heated wires for flow.

This section is reprinted with permission from *HVAC Systems Design Handbook*, by Roger W. Haines, Tab Books, Inc.

HVAC Automatic Controls

HVAC systems are sized to satisfy a set of design conditions, which are designed to generate a near-maximum load. Because these design conditions prevail during only a few hours each year, the HVAC equipment must operate most of the time at less than rated capacity. The function of the control system is to adjust the equipment capacity to match the load. Automatic control, as opposed to manual control, is preferable for both accuracy and economics; the human as a controller is not always accurate and is expensive. A properly designed, operated, and maintained automatic control system is accurate and will provide economy in the operation of the HVAC system. Unfortunately, not all control systems are properly designed, operated, and maintained.

Control systems do not operate in a vacuum. For any air conditioning application, it is first necessary to have a building suitable for the process or comfort requirements. The best HVAC system cannot overcome inherent deficiencies in the building. Then, the HVAC system must be properly designed to satisfy the process or comfort requirements. Only when these criteria have been satisfied can a suitable control system be applied.

Fundamentals

All control systems operate in accordance with a few basic principles. These must be understood as a background to the study of control devices and system applications.

Control Loops: Figure 8.1 illustrates a basic *control loop* as applied to a heating situation. The essential elements of the loop include a *sensor*, a *controller*, and a *controlled device*. The purpose of the system is to maintain the controlled variable at some point, called the *set point*. The *process plant* is controlled to provide the heat energy necessary to accomplish this. In the figure, the process plant includes the air-handling system and heating coil, the controlled variable is the temperature of the supply air, and the controlled device is the valve that controls the flow of heat energy to the coil. The sensor measures the air temperature and sends this information to the controller. In the controller, the measured temperature T_m is compared with the set point T_s. The difference between the two is the *error signal*. The controller uses the error, together with one or more *gain* constants, to generate an *output signal* which is sent to the controlled device, which is thereby repositioned, if appropriate. This is a *closed loop* system, because the process plant response causes a change in the controlled variable, called *feedback*, to which the control system can respond. If the sensed variable is not controlled by the process plant, the control system is *open loop*. A common example of this is the use of outdoor air temperature as the sensed variable.

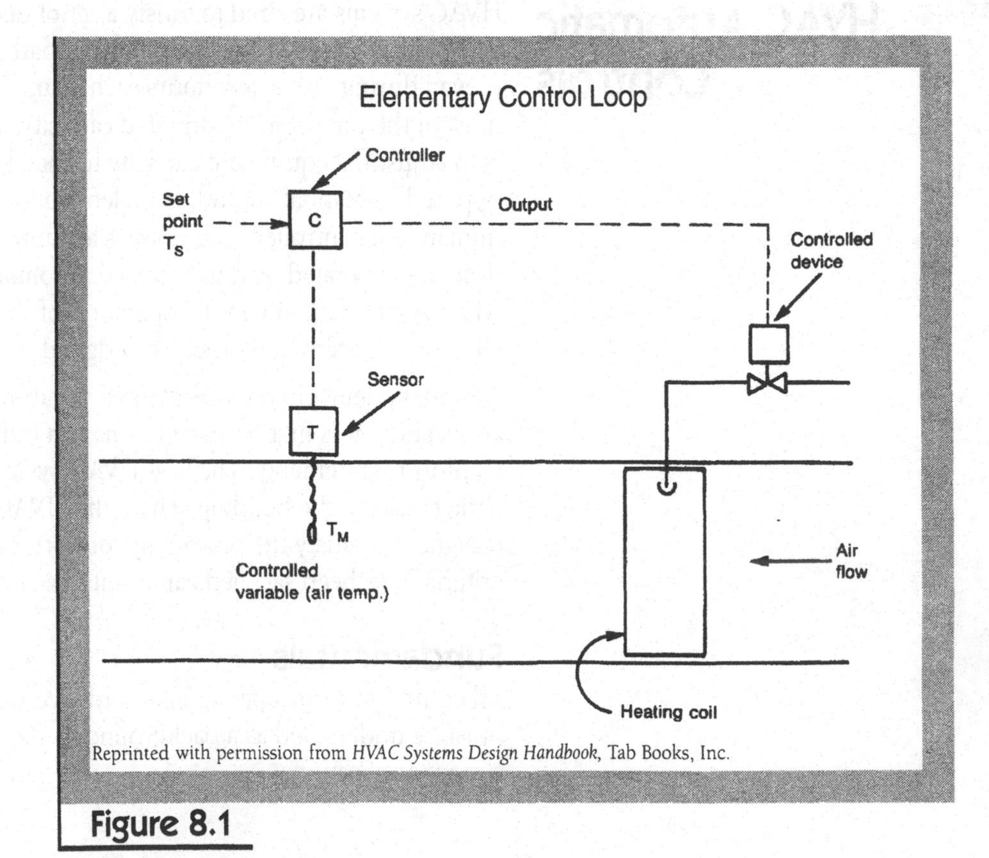

Reprinted with permission from *HVAC Systems Design Handbook*, Tab Books, Inc.

Figure 8.1

Many control systems include other elements, such as switches, relays, and transducers for signal conditioning and amplification. Many HVAC systems include several separate control loops. The apparent complexity of any system can always be reduced to the essentials described above.

Energy Sources: Several types of energy are used in control systems. Most older HVAC systems use pneumatic devices, with low-pressure compressed air at 0 to 20 psig. Many systems are electric, using 24 to 120 volts or even higher voltages. The modern trend is to use electronic devices, with low voltages and currents—e.g., 0 to 10 vdc, 4 to 20 ma, or 10 to 50 ma. Hydraulic systems are sometimes used where large forces are needed, with air or fluid pressures of 80 to 100 psi or greater. Some control devices are self-contained, with the energy needed for the control output derived from the change of state of the controlled variable, or from the energy in the process plant.

Control Modes: Control systems can operate in several different modes. The simplest is two-position, in which the controller output is either "on" or "off." When applied to a valve or damper, this translates into "open" or "closed." Figure 8.2 illustrates two-position control. In order to avoid too rapid cycling, a control differential must be used. Because of the inherent time and thermal lags in the HVAC system, the operating differential is always greater than the control differential.

If the output can cause the controlled device to assume any position over its range of operation, then the system is said to *modulate*. (See Figure 8.3). In modulating control, the differential is replaced by a *throttling range* (sometimes called the *proportional band*), which is the range of controller output necessary to drive the controlled device through its full cycle ("open" to "closed" or "full speed" to "off").

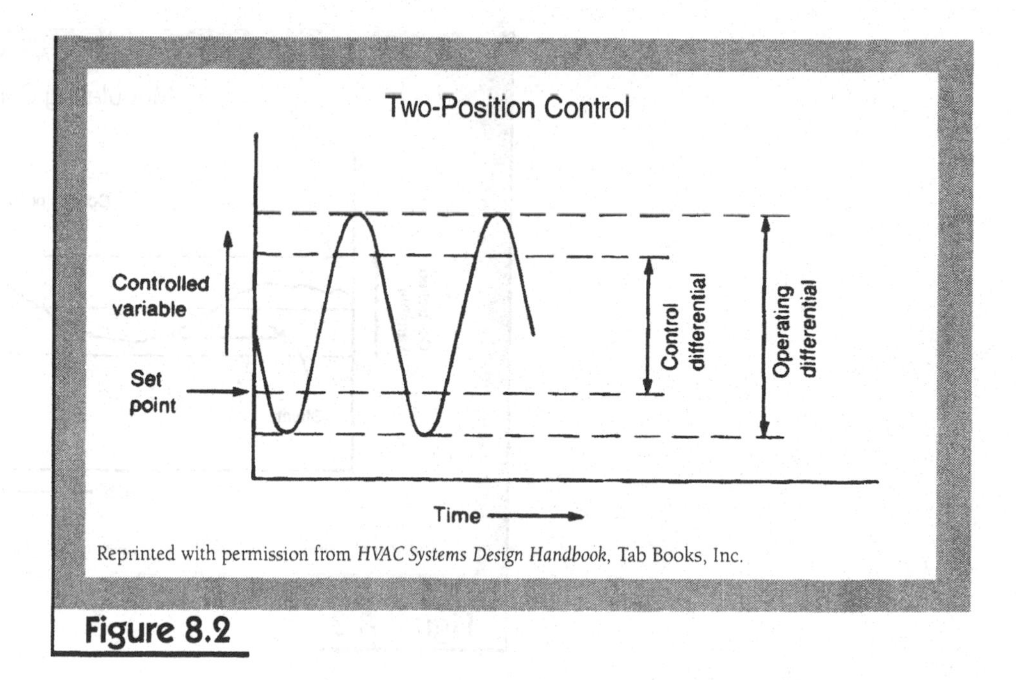

Reprinted with permission from *HVAC Systems Design Handbook*, Tab Books, Inc.

Figure 8.2

Modulating controllers may use one or a combination of three modes: proportional, integral, or derivative.

Proportional control is common in older pneumatic control systems. This mode may be described mathematically by the equation

$$O = A + K_p e$$

where

O = controller output

A = a constant equal to the controller output with no error signal

e = error signal

K_p = proportional gain constant

The gain governs the change in the controller output per unit change in the sensor input. With proper gain control, response will be stable; that is, when disturbed it will level off in a short time if the load remains constant. (See Figure 8.4.)

However, with proportional control, there will always be an *offset*, a difference between the actual value of the controlled variable and the set point. This offset will be greater at lower gains and lighter loads. If the gain is increased, the offset will be less, but too great a gain will result in *instability* or *hunting*, a continuing oscillation around the set point. (See Figure 8.5).

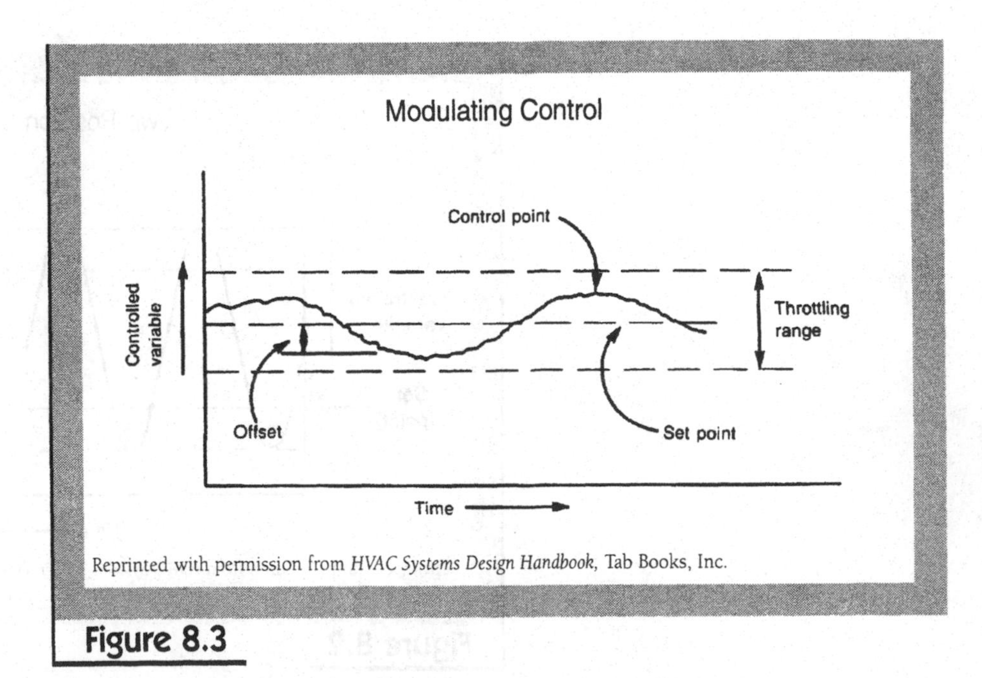

Reprinted with permission from *HVAC Systems Design Handbook*, Tab Books, Inc.

Figure 8.3

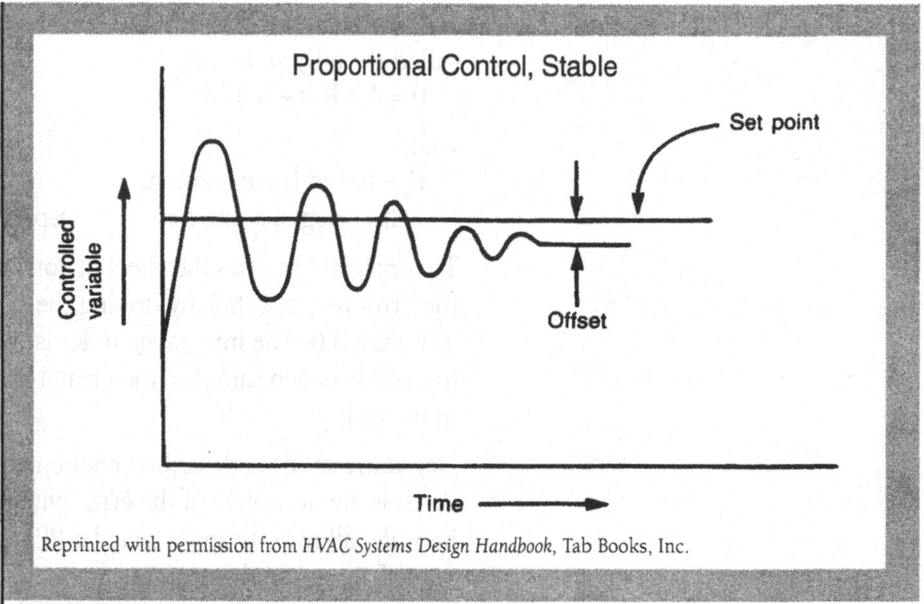

Reprinted with permission from *HVAC Systems Design Handbook*, Tab Books, Inc.

Figure 8.4

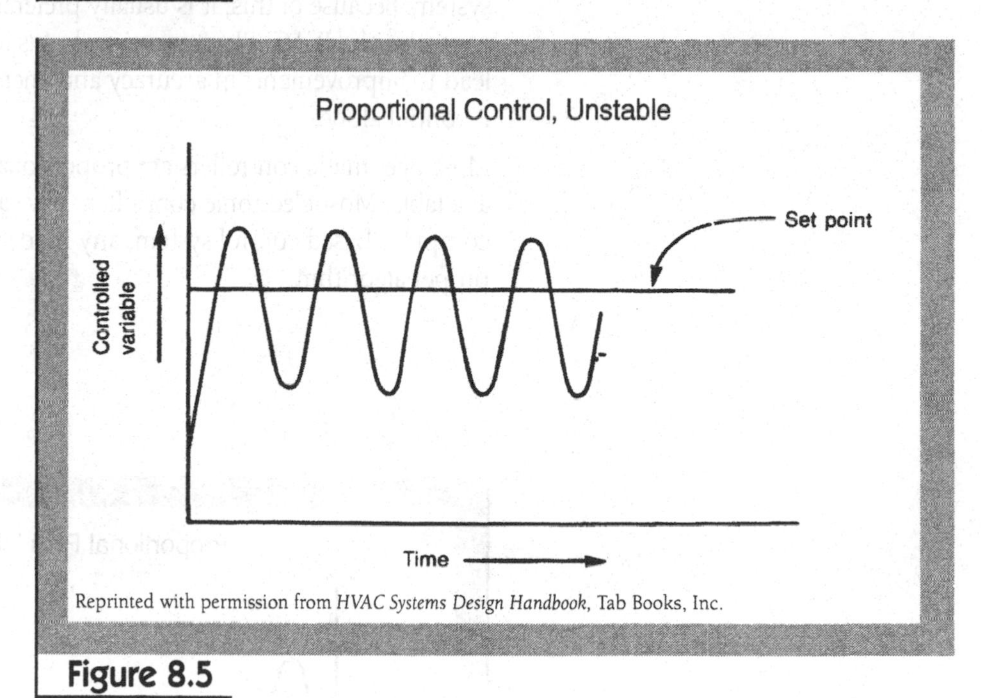

Reprinted with permission from *HVAC Systems Design Handbook*, Tab Books, Inc.

Figure 8.5

To eliminate the offset, it is necessary to add a second term to the equation, called the *integral* mode:

$$0 = A + K_p e + K_i \int e\, dt$$

where

K_i = integral gain constant

e dt = integral of the error with respect to time

The integral term has the effect of continuing to increase the output so long as the error persists, thereby driving the system to eliminate the error, as shown in Figure 8.6. The integral gain, K_i, is a function of time; the shorter the interval between samples, the greater the gain. Again, too high a gain can result in instability.

Derivative mode is described mathematically by the term K_d (de/dt), where de/dt is the derivative of the error with respect to time. A control mode which includes all three terms is called a *PID mode*. The derivative term describes the rate-of-change of the error at a point in time, and therefore promotes a very rapid control response—much faster than the normal response of an HVAC system. Because of this, it is usually preferable to avoid the use of derivative control with HVAC. PI (proportional plus integral) control is preferred and will lead to improvements in accuracy and energy consumption when compared to P control alone.

Most pneumatic controllers are proportional only, although PI mode is available. Most electronic controllers have all three modes available. In a computer-based control system, any mode can be programmed by writing the proper algorithm.

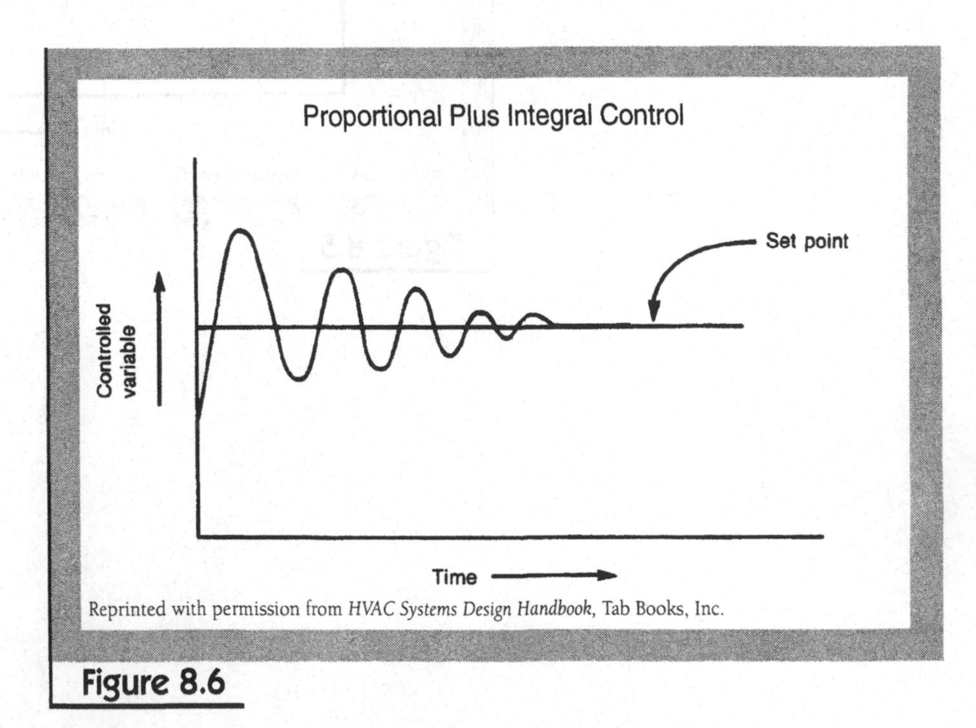

Reprinted with permission from *HVAC Systems Design Handbook*, Tab Books, Inc.

Figure 8.6

The following section is reprinted from *HVAC: Design Criteria, Options, Selections*, by William H. Rowe III, R.S. Means Company, Inc. We strongly recommend the book's chapter on controls. In addition to the information excerpted below, it provides excellent overviews of zoning and control diagrams and sequences.

Control Applications

Controls are manual or automatic devices used to regulate HVAC systems. A bathroom light switch that also operates an exhaust fan is an example of a manual control; a time clock on an office air conditioner is an automatic control. The simplest controls turn equipment on or off. Advanced controls may allow the equipment to modulate in response to demand.

Types of Operation

The control systems most frequently used today are electric, electronic, pneumatic, direct digital control (DDC), and self-contained control systems. The choice of system often depends on the owner's requirements, budget, and the technical level of the owner's operating personnel.

Electric Controls: *Electric controls* are used in residential and small commercial buildings where there are few pieces of equipment to control. Electric controls operate on line or low voltage (see Figure 8.7). Line voltage can be single or three-phase, 60 Hz, 120, 208, 240, 277, or 480 volts ac. Low voltage operates at below 30 volts—usually it is at 24 volts. Electric systems provide control by starting and stopping the supply of electric current and by varying the voltage or current by use of a rheostat or wheatstone bridge circuit. Electric control circuits use mechanical means such as bimetallic strips and bellows to actuate a switch or position a potentiometer; some circuits also incorporate relays and switches.

Electronic Controls: *Electronic controls* are used in small and medium-sized buildings. Electronic systems operate on low voltages and currents and use solid-state components to amplify signals to perform control tasks such as positioning an actuator or switching a relay. (See Figure 8.7). Electronic circuits provide fixed control sequences based on the logic of the solid-state components. Typical characteristics of an electronic system include a low voltage solid-state controller, inputs of 0 – 1V dc, 0 – 10V dc, and 4 – 20 mA, resistance-temperature sensors, thermistor, thermocouples, and outputs of 2 – 10V dc or 4 – 20 mA. Electronic systems use four control modes: two-position, proportional, proportional-integral, and step.

The sensors and output devices used in an electronic system are usually the same type used in a DDC system, which is described below.

Programmable Logic Controllers: *Programmable logic controllers* use a programming language called *ladder logic*. The language uses components that are similar to those used in a line diagram format to represent hard-wired control. *Ladder logic diagrams* include a left vertical line that displays the power or energized conductor. The output element or instruction represents the neutral or return path of the circuit. The right vertical line represents the return path on a hard-wired control line diagram and is omitted. A network may have several control elements, but only one output coil.

Pneumatic Controls: *Pneumatic controls* are used in medium-sized and large buildings and utilize low pressure (0 – 15 psi, or 100 – 200 kPa) compressed air. (See Figure 8.7). The sensor causes a change in air pressure, causing an actuator or motor to react or move. Air pressure acts against spring-loaded pistons or mechanisms to cause motion in the actuator.

Direct Digital Controls: *Direct Digital Controls* (DDC) are used in medium-sized to large buildings and in facilities comprised of a number of buildings. DDC systems are micro-processor based—a signal from a sensor is measured, an algorithmic control routine is performed in a software program within a microprocessor-based controller, and a corrective signal is sent to an actuator or controlled device. (See Figure 8.8).

DDC controllers consist of digital and analog inputs and outputs, converters, a power supply, and software to perform pre-programmed HVAC and energy-management routines. Microprocessor-based controllers can be used as stand-alone controllers that perform a few local tasks. They can also be incorporated in an automated building management system utilizing a personal computer to provide additional enhancement.

A *control loop* consists of a process wherein a sensor reads a condition and sends a signal to a controller; the controller positions an actuator or device to accommodate the sensor's reading. Smaller controllers generally have 8 – 10 control loops; however, some controllers are large enough for 30 – 40 loops.

Controllers in a DDC system are networked together and can share functions and readings. For example, only one outdoor air thermometer sensor is required to control the economizer sections of two or more air handling units. Alternatively, the occupied/unoccupied times may be programmed in the PC and read by all controllers.

DDC control systems generally utilize electronic sensors and electronic, electric, or pneumatic actuators. The electronic input signals must be converted to digital and analog inputs for the controller to understand; the digital and analog outputs from the controller must be converted to electric or pneumatic signals for the actuators to operate. Digital signals are two-position signals: for example, "on/off," "start/stop," "freezing/not freezing," and "open/closed." Analog signals are varying or continuously changing signals. These signals are used for measuring space temperature or humidity or for positioning a valve at a certain point. For example, the valve is 54% closed, the temperature is 74°F, and the humidity is 63%.

DDC systems are more expensive than other systems but give very accurate control, can easily incorporate energy-management programs and graphic system displays, and can be tied into most building automation systems such as fire alarms, security, and lighting controls.

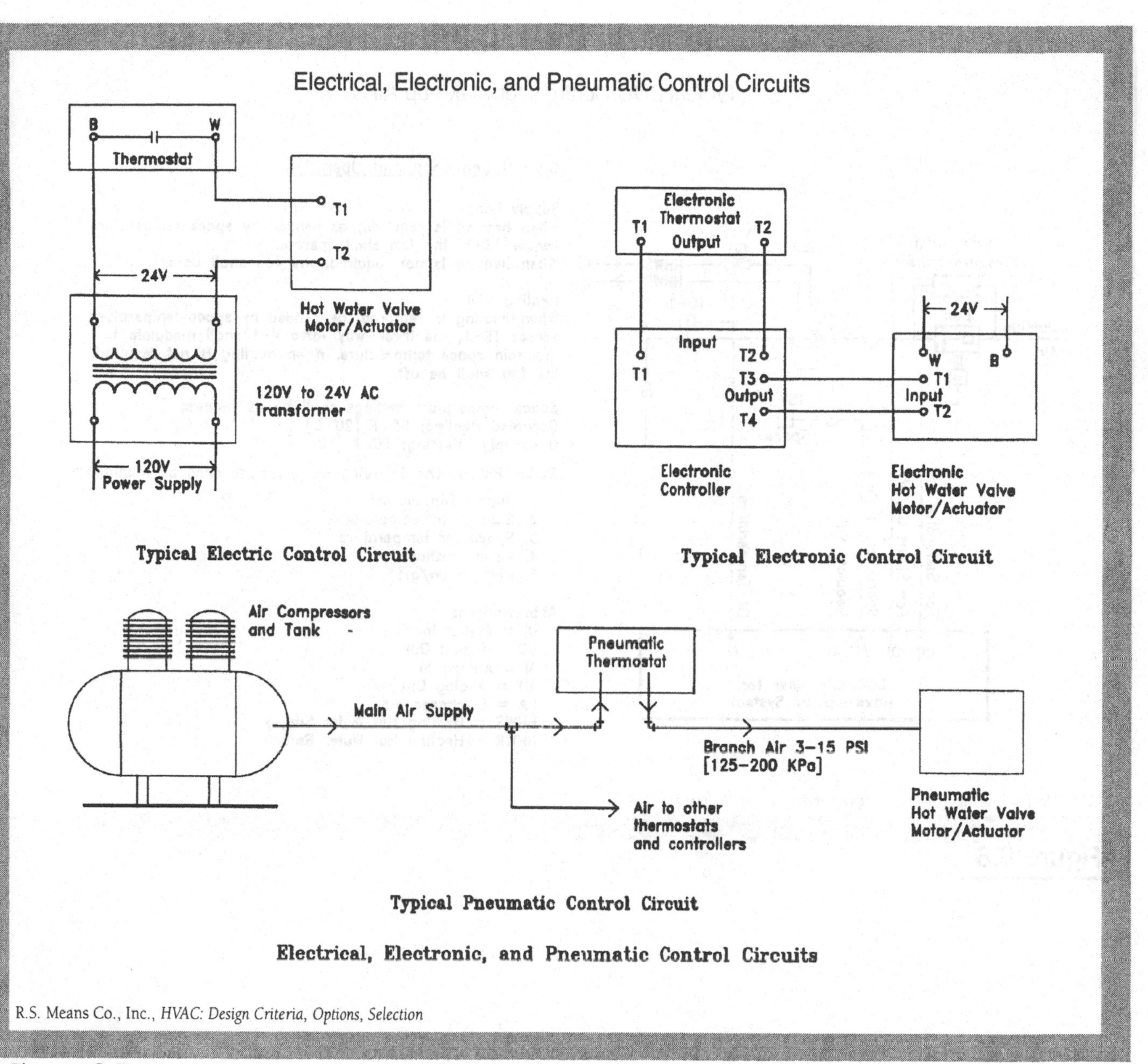
Electrical, Electronic, and Pneumatic Control Circuits
B
W
Thermostat
T1
T2
24V
Hot Water Valve
Motor/Actuator
120V to 24V AC
Transformer
120V
Power Supply
Typical Electric Control Circuit
Electronic
Thermostat
T1
T2
Output
Input
T1
T2
T3
Output
T4
24V
W
B
T1
Input
T2
Electronic
Controller
Electronic
Hot Water Valve
Motor/Actuator
Typical Electronic Control Circuit
Air Compressors
and Tank
Pneumatic
Thermostat
Main Air Supply
Branch Air 3–15 PSI
[125–200 KPa]
Air to other
thermostats
and controllers
Pneumatic
Hot Water Valve
Motor/Actuator
Typical Pneumatic Control Circuit
Electrical, Electronic, and Pneumatic Control Circuits
R.S. Means Co., Inc., HVAC: Design Criteria, Options, Selection

Figure 8.7

Typical DDC Controls for Make-up Air

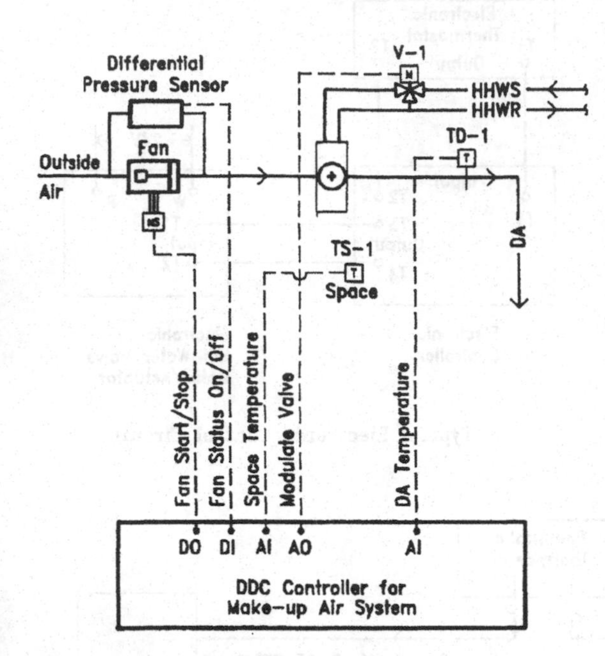

Control Sequence and Operation

Supply Fan:
When heating is required, as sensed by space temperature sensor TS-1, the fan shall operate.
When heating is not required, the fan shall be off.

Heating Coil:
When heating is required, as sensed by space temperature sensor TS-1, the three-way valve V-1 shall modulate to maintain space temperature. When heating is not required, the fan shall be off.

Space temperature settings shall be as follows:
Occupied Heating: 68° F [20° C]
Unoccupied Heating: 50° F [10° C]

Status Points: The following status points shall be monitored:

1. Supply fan on/off
2. Supply air temperature
3. Space air temperature
4. Valve positions
5. Heating on/off

Abbreviations:
DI = Digital In
DO = Digital Out
AI = Analog In
AO = Analog Out
DA = Discharge Air
HHWS = Heating Hot Water Supply
HHWR = Heating Hot Water Return

R.S. Means Co., Inc., *HVAC: Design Criteria, Options, Selection*

Figure 8.8

Self-Contained Controls: *Self-contained controls* are control systems in which the sensor controller and the actuator are part of a single package, such as thermostatic valves. These self-contained controls incorporate sensor, controller, and controlled device in one package. No external source is required. An example of a self-contained control is one where a gas or fluid in a bulb expands and retracts to operate the controlled device.

Sensors, Controllers, and Actuators

Control systems can be divided into three categories: *sensors* sense the controlled medium; *controllers* read the sensor's signal and convert it to a response signal based on predetermined settings; and *actuators* receive the controllers signal and actuate a piece of equipment to cause a change in the controlled medium.

Each process that senses, reads, and responds (sensor – controller – actuator) is also called a *control loop*. A time clock detects the "on" or "off" time and a thermostat detects temperature—these events activate a motor or valve. Examples of other control loops are shown in Figure 8.9.

Examples of Control Loops

Simple Control Loops

Sensors	Controllers	Controlled Devices
Thermostat	⟶	VAV damper/motorized valve/zone pump
Pressure sensor	⟶	Fan motor, space heater, bypass valve
High temperature limit	⟶	Oil/gas burner
Duct smoke sensor	⟶	Equipment shutdown, damper motor
Duct freeze thermostat	⟶	Equipment shutdown
Outdoor air sensor	⟶	Boiler start-up
Time clock	⟶	Pump motor, package unit
Humidistat	⟶	Steam valves/humidifiers
Aquastat	⟶	Oil/gas burner, fan motor
Enthalpy sensor	⟶	Return, exhaust, and intake dampers

R.S. Means Co., Inc., *HVAC: Design Criteria, Options, Selection*

Figure 8.9

The following section is reprinted with permission from *Managing Controls: The Complete Book of Motor Control Application and Maintenance*, by Richard L. Nailen. Copyright© Barks Publications, Inc.

Electromechanical Relays

Mechanical Motor Starters

What is a motor starter? For most of us the answer is simple—it's that box on the wall with the "start" and "stop" pushbuttons on the cover.

Right now that simple device is caught up in a sweeping change in US electrical design and manufacturing practice. How well that change is understood may eventually affect the reliability of motor circuit operation and protection throughout industry.

Although it's widely used around the world, NEMA (National Electrical Manufacturers Association) does not actually define the term "starter." An IEEE (International Electronics & Electrical Engineering) standard defines a motor starter as "an electric controller for accelerating a motor from rest . . . and to stop the motor."

In the US, the expression is usually used to describe a "combination starter" consisting of four basic elements within a single convenient enclosure. (See Figure 8.10.) Referring to the diagram in Figure 8.11:

- *Element A* is the means of isolation—positive separation of the entire motor circuit from the power source, not subject to accidental or automatic reclosure. Physically, this is normally a switch or circuit breaker. (Some changes in breaker design have taken place recently but are not of great concern to us here.)
- *Element B*, circuit protection, is not intended to safeguard the motor itself. Rather, its function is to minimize damage to the conductors and devices upstream from the starter, or within the starter itself, from a downstream short-circuit or accidental ground. This fuse or breaker protection is usually associated with (or part of) Element A.
- *Element C* is the "motor controller" itself, the on-off contact-making device used to routinely start and stop the motor. It's normally a magnetic contactor, energized by a starting circuit, dropping out when the stop button is depressed to remove power from its operating coil.
- *Element D* protects the motor itself from the sustained heat of an overload or sudden stall. Depending upon the degree of sophistication, this "overcurrent protection" may be sensitive to frequent starting, to phase unbalance, or to the loss of a phase, as well as to horsepower overload on the motor. It may or may not compensate for differences in ambient temperature between the starter and motor locations.

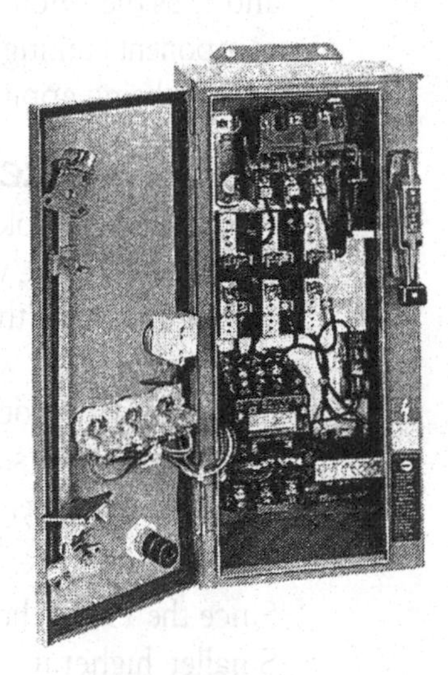

A complete combination starter combining all of the elements shown diagrammatically in Figure 8.11. At the top of the enclosure is the disconnect switch, with fuse block directly beneath. The contactor is at bottom, with control transformer and its fuse. The remaining control circuit components are mounted on the door at left.
Reprinted with permission from *Managing Controls*, Barks Publications, Inc.

Figure 8.10

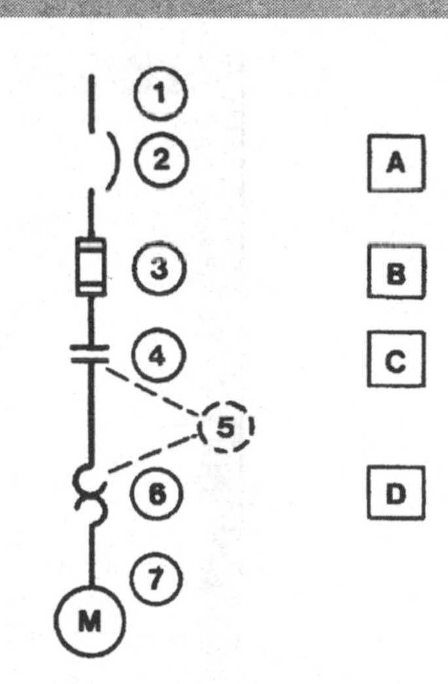

Components of a combination motor starter (circled) include (1) line connections; (2) isolation (disconnect) means; (3) short-circuit protection; (4) routine on-off switching means; (5) control circuit (pushbuttons, etc.); (6) motor overcurrent protection device ("overload"); (7) load connections. The four elements required by NEC Article 430 are Nos. 2, 3, 4, and 6, emphasized by the letters A, B, C, and D in squares.
Reprinted with permission from *Managing Controls*, Barks Publications, Inc.

Figure 8.11

In Article 430, the National Electrical Code (1996) considers only Elements C and D as the "motor controller." (See Figures 8.12 and 8.13.) These are the components taking on a new appearance today. Controllers are getting smaller and are being applied under different rules.

From NEMA Rerates to Adoption of IEC Standards: All of this is similar to what took place in motor design years ago. For a long time the trend was to reduce size, weight, and cost of standard induction motors through "rerate" programs that took advantage of new materials and design methods. The most recent result was the 1964 NEMA T-frame motor line. Without going into the merits of that design, we can recognize that one of the most common complaints about such motors has been that they don't have enough margin. They've been widely perceived as less capable of withstanding abuse than their predecessors.

Since the 1970s, however, that motor rerating trend has been reversed. Smaller, higher temperature ratings have given way to larger machines boasting higher efficiency. Today, operating cost, rather than motor price, tends to govern purchasing decisions.

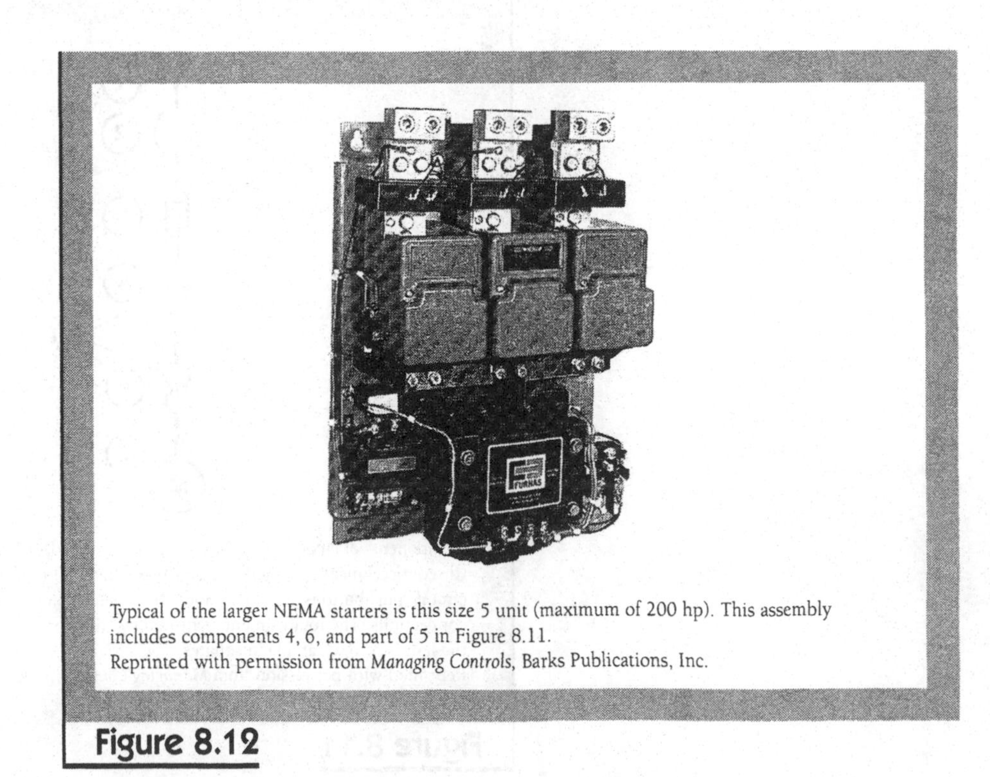

Typical of the larger NEMA starters is this size 5 unit (maximum of 200 hp). This assembly includes components 4, 6, and part of 5 in Figure 8.11.
Reprinted with permission from *Managing Controls*, Barks Publications, Inc.

Figure 8.12

On the other hand, efficiency is not an issue with starters. Losses are insignificant. Hence the economic pressure in this market continues to force decreases in unit size. Rather than a NEMA rerate, as was true for motors in 1964, such decreases are resulting from tacit acceptance in the US of long-standing European design practices expressed in International Electrotechnical Commission (IEC) standards.

Here's how J.T. O'Rourke, president of Allen-Bradley Co., explained the US-versus-overseas philosophies in 1985:

> "What NEMA did was to write standards that yielded general-purpose, robust, easy-to-apply starters. Outside North America, material were hard to come by and more expensive, so IEC standards were more definite-purpose in nature, requiring a significant amount of work to select a starter for a specific purpose."

The NEMA design approach suits the starter to a variety of application requirements, giving most units a sizable reserve capacity for many uses.

Manufacturer interchangeability also underlies the NEMA approach. Added G.R. Hunter, vice president of Square D, "US equipment is larger . . . to service a broad industrial base out of distributor stocks. The idea was to limit the number of devices a distributor had to carry on his shelf." But European manufacturers, dealing more directly with end-users, provided a more tailor-made product.

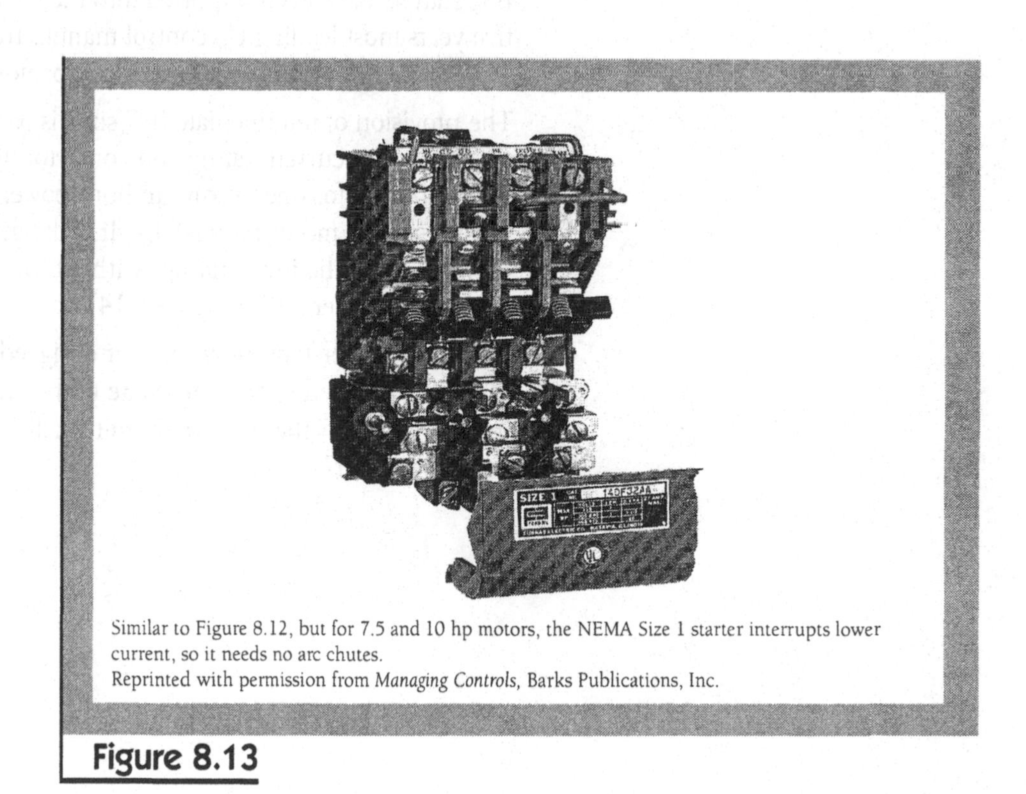

Similar to Figure 8.12, but for 7.5 and 10 hp motors, the NEMA Size 1 starter interrupts lower current, so it needs no arc chutes.
Reprinted with permission from *Managing Controls*, Barks Publications, Inc.

Figure 8.13

In the design of a NEMA starter, contacts are matched to a single value of continuous current, which in turn sets the maximum motor horsepower for that size contactor. (For a limited time period, the rating allows occasional jogging or plugging.) For example, a NEMA Size 2 starter is intended for maximum 460 volt horsepower of 25. The next larger standard motor, 30 hp, would require the next larger standard starter, NEMA Size 3, designed for top horsepower of 50. But suppose the motor is only 15 hp. A Size 2 starter is still required because the next smaller unit (Size 1) is rated for no more than 10 hp.

The NEMA starter range includes ten separate items, Size 00 through 8, to span the motor range from 1 through 900 hp at 460 volts. Depending upon the manufacturer, the IEC starter line offers as many as 15 separate starters for the same horsepowers. Table 8.1 shows the relationship of some of those "intermediate size" IEC devices to the NEMA sizes. The intermediates may be designated by fractions (such as "Size 1-3/4" or "Size 2-1/2"); by NEMA numbers unrelated to NEMA designations.

Table 8.1

Motor Horsepower at 460 Volts	Amperes	Standard NEMA Starter Size	Typical IEC Starter Designation
10	14	1	0+
20	27	2	1-3/4
40	52	3	2-1/2 or 2+
250	302	6	5-1/2 or 5+

This table compares NEMA and IEC starter size designations for a few typical motor ratings. Note that the numbering structure indicates in each case an IEC starter of a little smaller size than the NEMA version, which is true for many motors. Not standardized by IEC documents, and not necessarily used in Europe, these IEC size numbers are commonly applied to starters sold in the US so that users can easily make the size comparison.

IEC starters have been imported into the US for some time. But within the last five years most leading US control manufacturers have also begun marketing their own lines of IEC devices as a companion product to NEMA starter sizes.

The provision of intermediate IEC sizes is based on a closer match-up between specific motor current rating and contactor life. Rather than providing acceptable life for one maximum horsepower rating (and much longer life for several smaller motor ratings), the IEC design trend has been to achieve the desired life for the lower ratings without "overdesign" by fitting smaller contactors to them. (See Figure 8.14).

It's easy to understand how a starter designed for only 7-1/2 hp can be smaller than one intended to accommodate both 7-1/2 and 10 hp motors. But why, as in Figure 8.15, is the IEC 10 hp unit smaller than the NEMA starter made for the same motor?

One answer is that creepage paths and air clearances between live internal parts tend to be larger for NEMA devices. But the more important difference is that IEC standards permit higher temperature rise of terminals than do NEMA standards—a 70 °C maximum rise instead of only 50. (If the unit is properly marked to require lead wire rated for 75 °C, UL listings permit terminal rise of 65 °C—below the IEC limit, but still well above the NEMA value.)

Conversely, of course, if a NEMA-sized starter carries the same load as its IEC counterpart, it will run cooler. And if the IEC temperature limits were applied to a NEMA starter, the latter could carry more load. Unfortunately, specific life-temperature relationships are not readily available for starters, as they have been for motor windings.

In general, the small IEC starter can cost 20 to 30 percent less than the NEMA-sized unit. The contactor coil may use 50 to 70 percent less power.

Another major IEC contrast with NEMA practice is the existence of standardized contactor lifespans, allowing a stated minimum number of operations, in terms of the severity of the motor application. Detailed test procedures to establish that capability appear in IEC Publications 292-1, a 91-page standard issued in 1969 and amended several times since (now being superseded by IEC 947).

Two types of combination starters built to IEC standards. The unit at left contains disconnect switch and fuses in top of cubicle. The starter at right combines these two functions within a circuit breaker.
Reprinted with permission from *Managing Controls*, Barks Publications, Inc.

Figure 8.14

Operational life is defined in two ways. First is the "mechanical life" of the contactor operating mechanism itself—the longevity of the springs, pivots, linkages, etc., without regard to current flowing through the contacts. This is established by the number of tested no load operations "before it becomes necessary to service or replace any mechanical parts." During that long-term test, however, cleaning, lubrication, adjustment, or contact replacement (due to wear that could interfere with the mechanism's operation) are permitted.

Typically, standard IEC contactor mechanical life is 10 million operations. To put that in perspective—such a contactor would survive operation once per minute continuously for 19 years. Some makes are rated for 20 million mechanical operations. (Obviously, in an environment where the contactor is exposed to dirt, corrosion, and poor maintenance, some of that life may be lost.) Specific mechanical lifespans are not dealt with in NEMA standards.

The second criterion of IEC contactor endurance is the "electrical life" of the contacts. Prototype tests must show that the contactor completes a proposed minimum number of operations at a certain current within temperature limits and without need for "repair or replacement." Contacts must not exhibit welding or undue erosion, or excessive display of flame.

All these tests are made under laboratory conditions. Neither expressed nor implied in IEC standards is any specific relationship between the conditions of these tests and the operation of a particular motor.

NEMA contactor testing follows UL Standard 508. Such tests include overheating, over-and-under-voltage operation, dielectric withstand, short-circuit, and a much shorter "endurance" test of no more than 6000 make-and-break cycles.

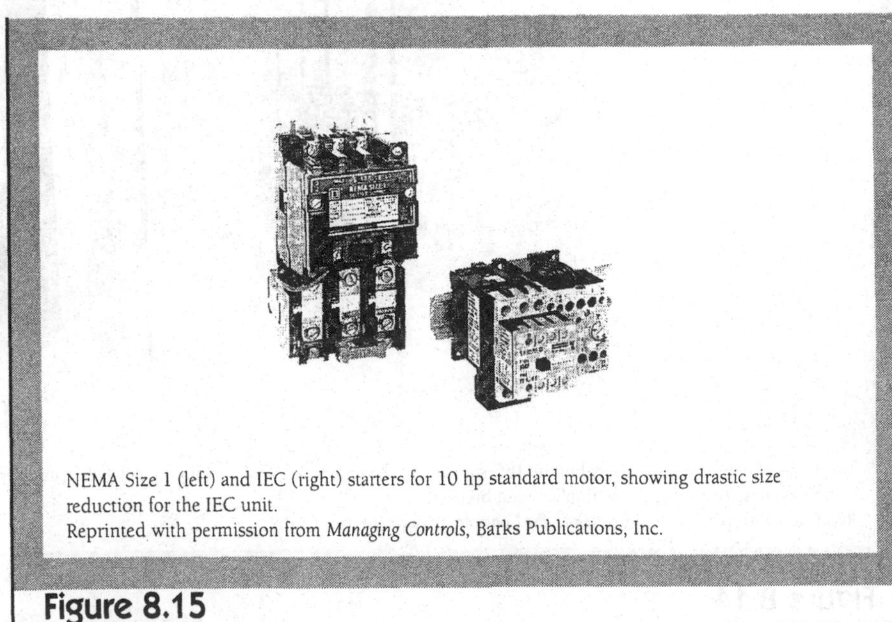

NEMA Size 1 (left) and IEC (right) starters for 10 hp standard motor, showing drastic size reduction for the IEC unit.
Reprinted with permission from *Managing Controls*, Barks Publications, Inc.

Figure 8.15

The IEC electrical life tests—and starter application generally—do recognize the different ways motors are used. Several service categories exist, each associated with a different level of breaking current through the contacts. As far as induction motors are concerned, the three significant categories are:

- *AC2*. Starting of wound-rotor motors.
- *AC3*. Starting cage motors; contacts opened only after motor reaches full speed (hence need not interrupt accelerating current).
- *AC4*. Starting cage motors with jogging, plugging, or reversing.

For AC3 duty, IEC starter contacts must be able to close, or "make," six times rated current, but need interrupt, or "break," only rated current itself. In AC4 service, contacts must break six times rated amperes. Such capability is derived from standard tests at currents 1-1/2 or more times that great.

Most squirrel cage motor applications are considered AC3. To properly select an IEC starter, the user needs to have some idea of how often the motor will be started. Frequency of starts determines the required minimum number of contactor operations, electrically. The relationship between contactor current in amperes and the number of allowable operations is given by published curves like those in Figure 8.16.

Two important precautions apply to the use of such curves. First, be sure of the basis for the horizontal scale if it's in horsepower or kilowatts. Because IEC starter design was originally based on European practice, some current/power relationships have been derived from 50 Hz 380 or 415 volt motor characteristics. Simply ratioing amperes from a 50 Hz to a 60 Hz motor design, in proportion to the voltage ratio (such as 415/460), will not be correct.

Secondly, if the curve scale is in amperes, relate those amperes properly to the rating of the motor involved. If, for example, "standard" full-load amperes are taken from Table 430-150 of the NEC, note that actual motor nameplate current may be quite different. (See, for example, Figure 8.17.) This can be much more important for IEC contactors than for NEMA sizes because the IEC unit will normally be more closely matched to the specific motor—such matching is the purpose of curves like Figure 8.16.

Whatever the applicability of Table 430-150, remember that it stops at 200 hp. Some starter catalogs include tables of "standard motor" currents up to 900 hp, for which no published standard forms any basis. Current-versus-operational life curves will vary with each starter supplier. The IEC standards support only the principle involved, not the numbers. Note the variation between the two solid curves in Figure 8.16, for two different makes of starter typically applied to the same 20 hp motor.

Any application including AC4 duty is troublesome. The dashed line in Figure 8.16 shows how sharply the electrical life of a typical starter is reduced by only a small percentage of AC4 service. To see that more clearly, look at Figure 8.18. For the 20–125 hp range, these curves show the effect of AC4 service in two ways. One is the drop in allowable motor rating for the same total number of starter operations. The other is the reduction in number of allowable operations for the same motor rating. NEMA standards, too, include a derating factor for such application. But the built-in reserve margin is greater, so that factor doesn't have to be invoked as often or as markedly.

Thus, selection of an IEC motor starter involves a determination of the required electrical life and of the way the starter will be used—steps not normally needed when applying NEMA contactors. That's the price paid for getting a device often much smaller, less expensive, quicker to install, and requiring a smaller control transformer than the NEMA-sized equipment.

Choosing Overload Heaters: Some other IEC practices also differ from what has been usual in the US. One of these is overload heater selection and operation.

Neither NEMA not IEC standards prescribe the type of overload to be furnished with any starter. NEMA starters have most often used "melting alloy" overloads. Bimetal thermostatic types are also widely offered. To allow easy interchangeability with varying applications, these have been individually replaceable elements, one per phase. That's the usual way of dealing, for example, with the 150 hp motor controlled by a Size 5 starter designed for maximum horsepower of 200.

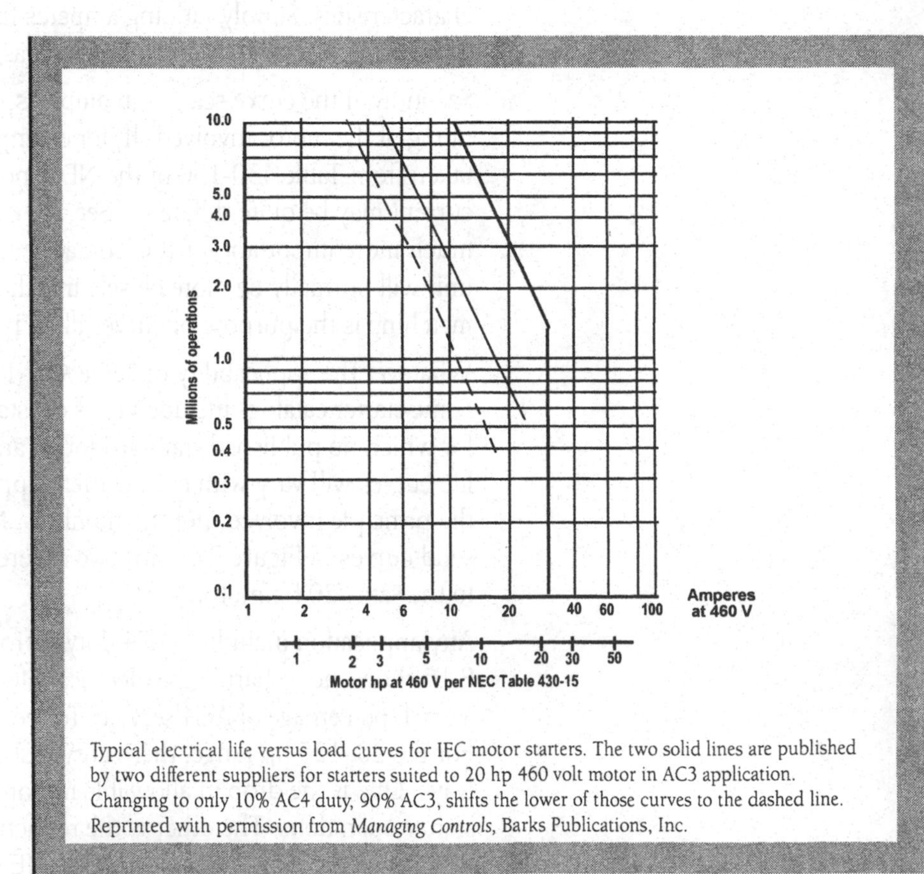

Typical electrical life versus load curves for IEC motor starters. The two solid lines are published by two different suppliers for starters suited to 20 hp 460 volt motor in AC3 application. Changing to only 10% AC4 duty, 90% AC3, shifts the lower of those curves to the dashed line. Reprinted with permission from *Managing Controls*, Barks Publications, Inc.

Figure 8.16

Each individual overload heater responds to its own phase current. Normal motor overload heats all three equally. If a phase is lost, high current through at least one other heater will cause a tripout.

However, that process can take some time. Ideally, severe unbalance or loss of phase would be more directly detectable as a condition different from normal overload, leading to a quicker response. That is possible with the typical IEC starter because, with the greater number of starter sizes permitting less flexibility in overload heater selection, the three overloads are normally supplied as a single unit assembly. The elements aren't individually replaceable. However, the "integral," or "three-pole," assembly normally has a dial adjustment to allow at least ±15% current variation.

These overloads are most often of the heated bimetal type. The three-phase bimetals are interlinked (Figure 8.19a) so that under normal overload all three will cause movement of a contact operating arm to open the contactor after a time determined by the rate of bimetal heating (the severity of the overload). (See Figure 8.19b.) That behavior matches what occurs in a NEMA starter.

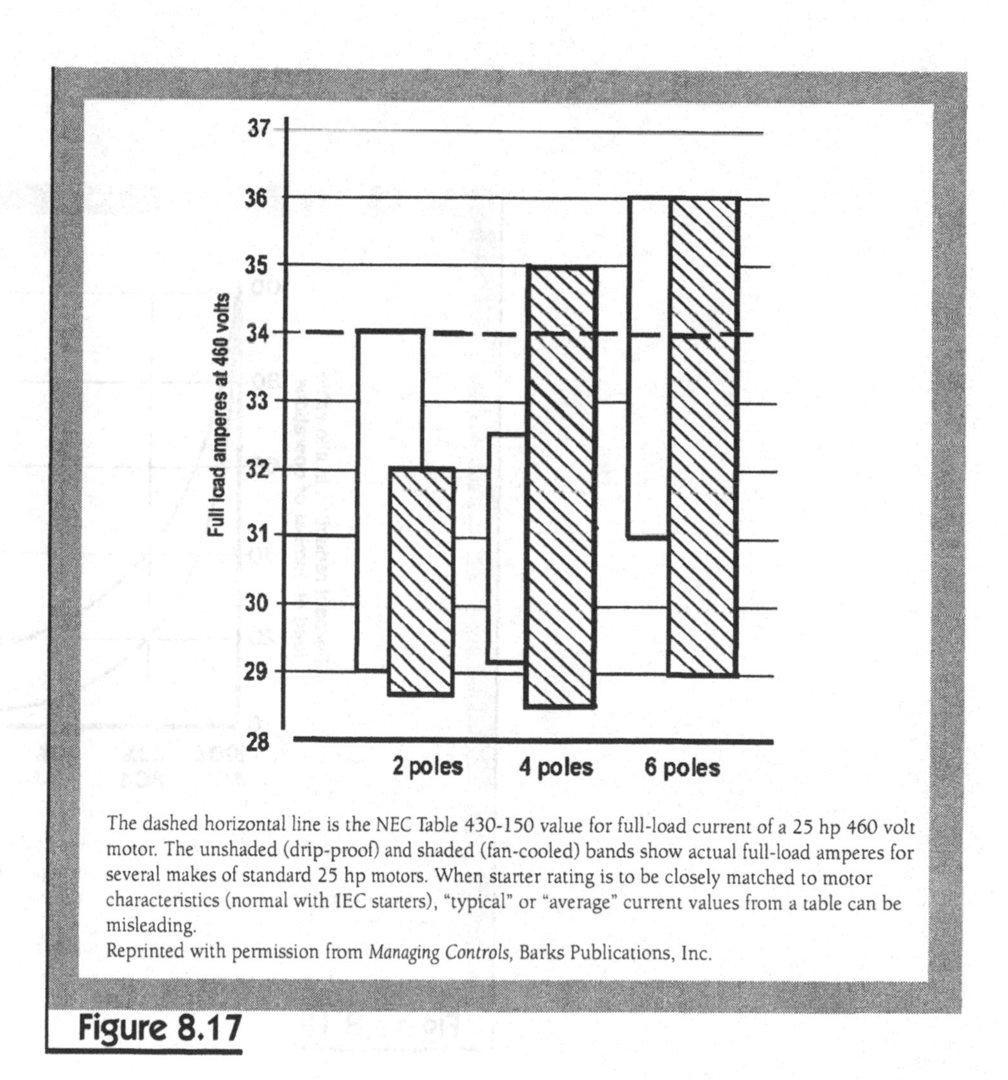

The dashed horizontal line is the NEC Table 430-150 value for full-load current of a 25 hp 460 volt motor. The unshaded (drip-proof) and shaded (fan-cooled) bands show actual full-load amperes for several makes of standard 25 hp motors. When starter rating is to be closely matched to motor characteristics (normal with IEC starters), "typical" or "average" current values from a table can be misleading.

Reprinted with permission from *Managing Controls*, Barks Publications, Inc.

Figure 8.17

In the IEC unit, however, an open phase—causing one of the three bimetals to remain "cold"—increases the speed of contact operation, as show in Figure 8.19c. Hence, the IEC starter is often described as offering "phase loss/ unbalance protection." The term *single-phase sensitivity*, which is probably more descriptive, is also used.

IEC overloads are often convertible from hand to automatic reset operation, whereas such a change with NEMA bimetals requires replacing each element.

Another difference between NEMA and IEC starter standards is in the "protection class." US practice for many years was to supply "Class 20," which in NEMA control standard ICS-2 is required to meet these three conditions:

- At rated current, the overload will trip "ultimately"—that is, no specific time limit applies.
- At 200 rated current, a trip must occur within eight minutes.
- At 600 rated current, a trip must occur within 20 seconds.

That 600 current may be assumed equal to "locked rotor current" for typical motors. However, no such statement is made in any standard. You may see overload relay catalog statements like this: "Class 20 protection (trips in less than 20 seconds at locked rotor amps—600 of FLA)." That sentence clearly

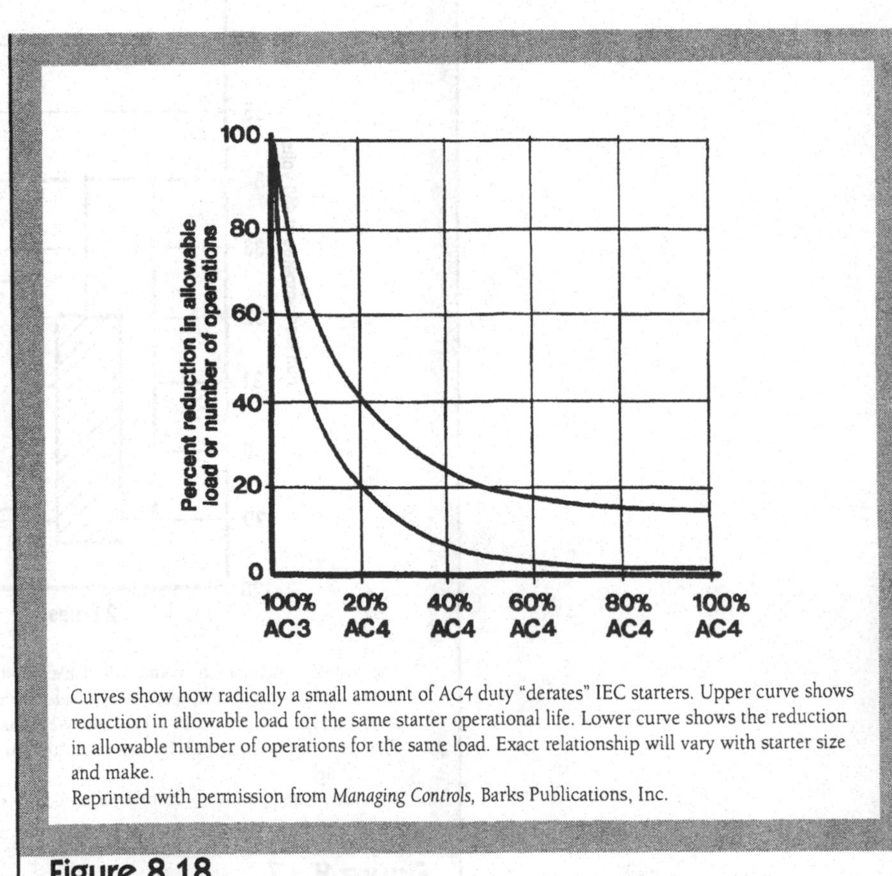

Curves show how radically a small amount of AC4 duty "derates" IEC starters. Upper curve shows reduction in allowable load for the same starter operational life. Lower curve shows the reduction in allowable number of operations for the same load. Exact relationship will vary with starter size and make.

Reprinted with permission from *Managing Controls*, Barks Publications, Inc.

Figure 8.18

implies that "600 of full load current" equals locked rotor amperes. However, depending upon motor size, speed, and type, locked rotor current may be well above or below 600.

Premium efficiency motors tend to have higher locked current than others. So unless the safe-time-versus-current curve for the specific motor is known, the degree of real protection provided by a "Class 20" starter is uncertain. Figure 8.20 shows that this can be cause for concern. Some starter manufacturers now recommend Class 10 protection (the same as Class 20 except that the trip at 600 current must occur within ten seconds) for T-frame motors, but no US standard dictates this.

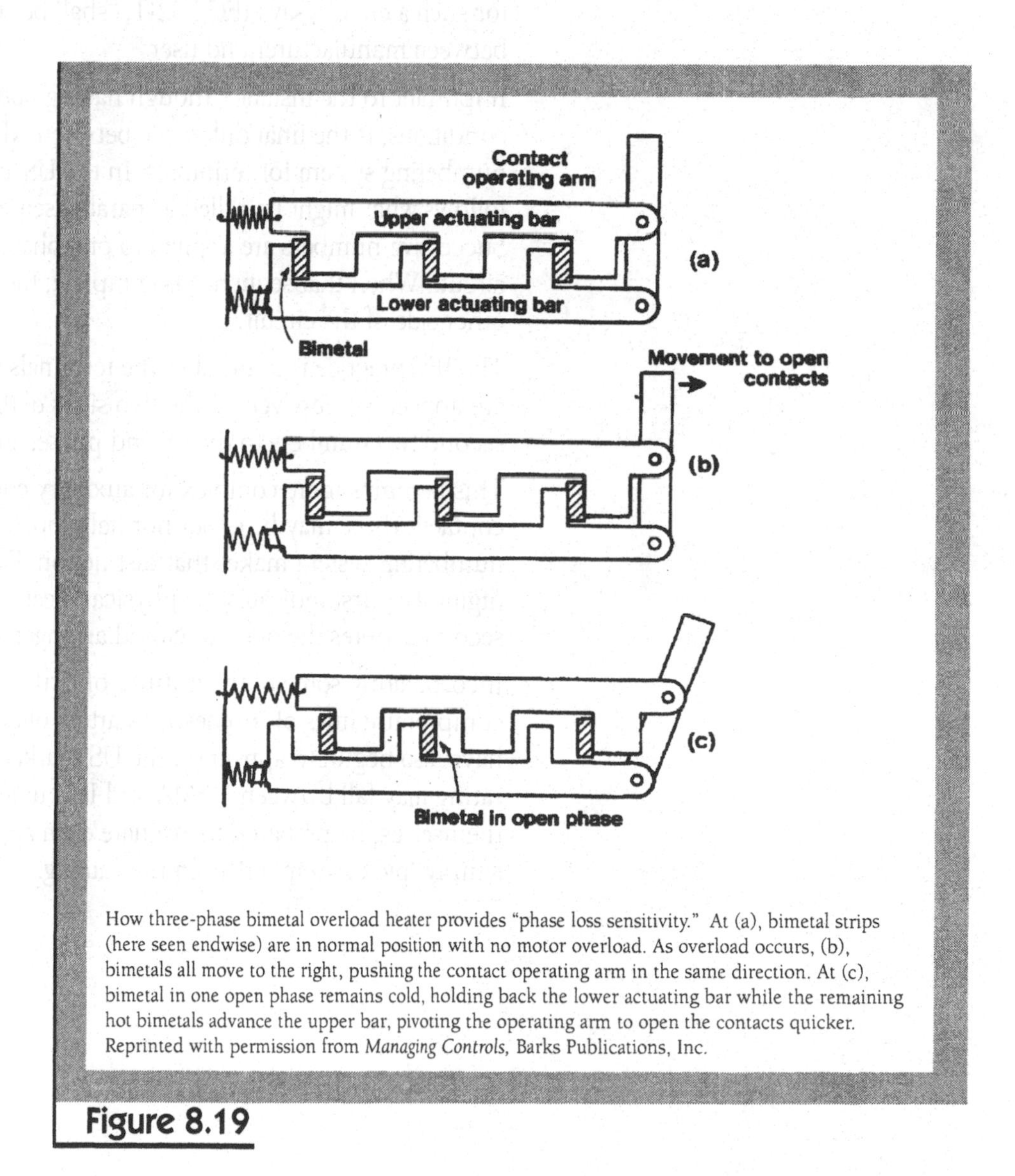

How three-phase bimetal overload heater provides "phase loss sensitivity." At (a), bimetal strips (here seen endwise) are in normal position with no motor overload. As overload occurs, (b), bimetals all move to the right, pushing the contact operating arm in the same direction. At (c), bimetal in one open phase remains cold, holding back the lower actuating bar while the remaining hot bimetals advance the upper bar, pivoting the operating arm to open the contacts quicker. Reprinted with permission from *Managing Controls*, Barks Publications, Inc.

Figure 8.19

Although the broad band of overload heater characteristics in Figure 8.20 merely indicates variations among several suppliers, some individual overload devices themselves exhibit fairly broad tolerances. In one make, for example, a ten-second trip time can occur with currents as low as 4560 or as high as 800. The attempt to match such a characteristic to an actual motor can lead to either over-or under-protection. The IEC starter manufacturers, for whom such matching is normal, usually publish overload heater tripping curves—not so easy to find for standard NEMA starters. And Class 10, not Class 20, is normal protection for IEC equipment.

Some starters are beginning to appear with solid-state overloads, adjustable for various standard protection classes. One make offers the entire range from "Class 3 through Class 30." However, the lowest standard class number is 10; no Class 3 exists.

Starter selection can become more difficult when power factor correction capacitors are connected to the motor terminals, so that the motor starter switches the capacitors as well as the motor. Choosing the right IEC contactor for such a circuit, says IEC 292-1, "shall be subject to special agreement between manufacturer and user."

Important to the installer, though having nothing to do with operating conditions, is the final difference between NEMA and IEC starters—the numbering system for terminals. In the US, connection point designation follows what might be called a "parallel sequence," as in Figure 8.21a. Successive numbers are applied to one phase after another on one side of the circuit. When that sequence is complete, the numbers are continued for the other side of the circuit.

The IEC practice is to number the terminals in series. The first two numbers are applied successively to the two sides of the circuit in the first phase, the second two numbers to the second phase, and so on. (See Figure 8.21b.)

This becomes more complex for auxiliary contact terminals. Unlike the power contacts, these may be either normally open or normally closed. The numbering system makes that distinction. Each terminal is identified by two digits. The first indicates the physical location or sequence of the contact. The second denotes the open or closed arrangement. (See Figure 8.21c.)

Incorporating some of the features of both NEMA and IEC practice, compromise lines of "domestic" starters offering some international advantages have also begun to appear on the US market. Physical size for a given motor rating may fall between NEMA and IEC units. As with the IEC starters themselves, be prepared to evaluate each application carefully rather than to simply "pick a number" from the catalog.

When Off Doesn't Mean Off

"Unless you know the circuit is dead, assume it's live."

You'll see that warning stated many ways in many places. Sometimes, though, we get careless. When a motor is obviously shut down, not running, and the starter is in the off position, it's hard to avoid the confidence that working around the machinery or the circuit will be safe. Unfortunately, three potential hazards exist. Not all of them are explained or even mentioned in published codes and standards.

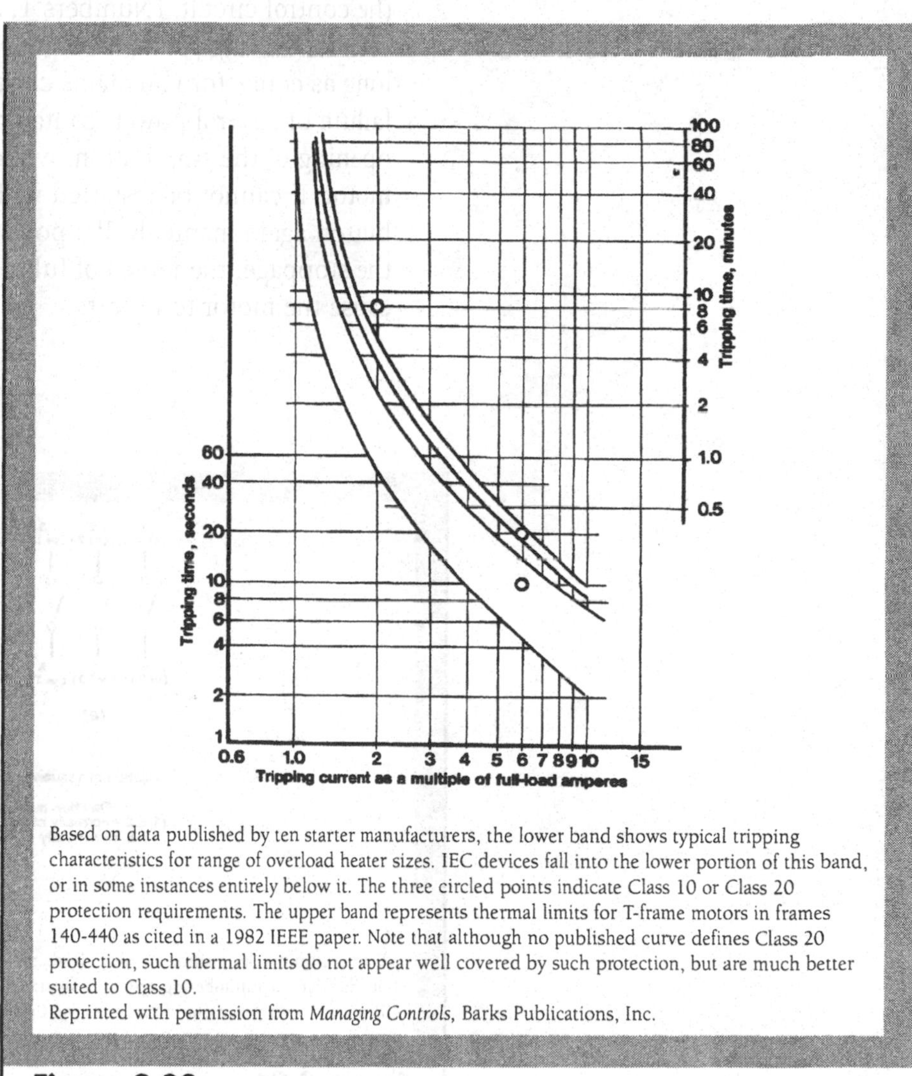

Based on data published by ten starter manufacturers, the lower band shows typical tripping characteristics for range of overload heater sizes. IEC devices fall into the lower portion of this band, or in some instances entirely below it. The three circled points indicate Class 10 or Class 20 protection requirements. The upper band represents thermal limits for T-frame motors in frames 140-440 as cited in a 1982 IEEE paper. Note that although no published curve defines Class 20 protection, such thermal limits do not appear well covered by such protection, but are much better suited to Class 10.

Reprinted with permission from *Managing Controls*, Barks Publications, Inc.

Figure 8.20

Here's what can cause the unwary maintenance worker trouble:

- Unexpected startup of an idle motor. This shouldn't happen, but it does. The danger is twofold. First, someone may be electrocuted. Second, although not in contact with the energized circuit, a worker could be injured by the machinery itself.
- Presence of an unexpected voltage at the motor terminals, even though the starter is "off," and sometimes even when the motor is not running.
- A live circuit to some motor accessory, creating a shock hazard at the motor even though all power to the main winding itself has been disconnected.

Those conditions can arise in several ways. Each has its own remedy.

Standard motor starter circuits prevent an automatic motor restart either following a "stop" command, or following an interruption of supply voltage. Such "three wire control for undervoltage protection" involves two basic features:

- Three wires connect the start-stop pushbuttons to the remainder of the control circuit. (Numbers 1, 2, and 3 in Figure 8.22.)
- The connecting patch between leads 1 and 3 remains closed only as long as contactor C remains closed. Operation of the starter overloads, failure of control power, contactor coil burnout, or of course the opening of the stop button, will drop out the contactor and stop the motor. It cannot be restarted without at least operating the start button again manually. If a power interruption or voltage dip caused the stoppage, the return of full voltage to the circuit will not of itself cause the motor to restart.

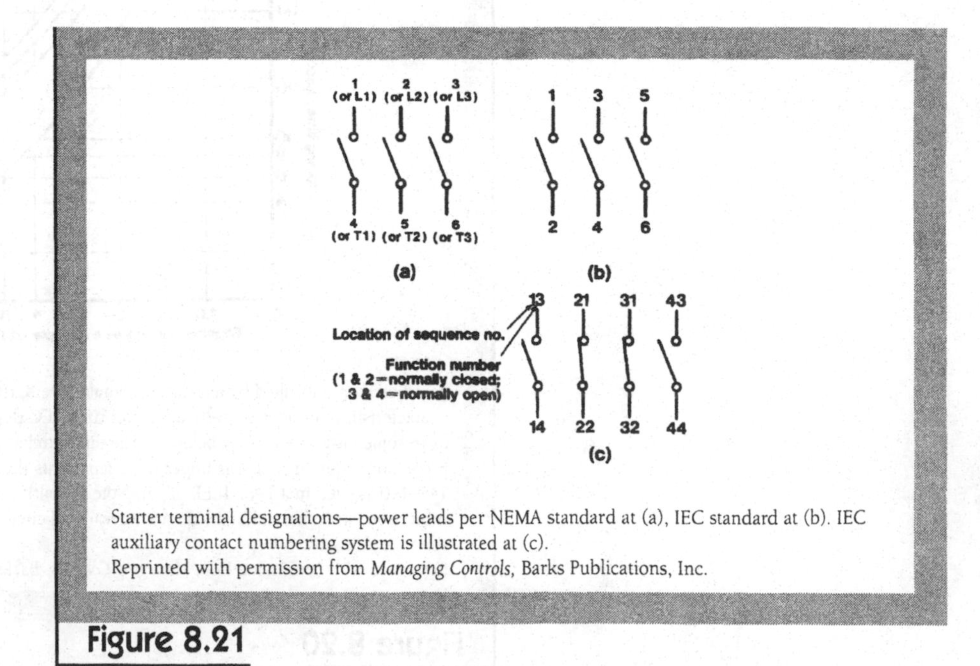

Starter terminal designations—power leads per NEMA standard at (a), IEC standard at (b). IEC auxiliary contact numbering system is illustrated at (c).

Reprinted with permission from *Managing Controls*, Barks Publications, Inc.

Figure 8.21

Automatic Restart: However, several situations may cause an idle motor to start unexpectedly. First, the control scheme itself may be "automatic," in the sense that motor starting results at once when some process variable unpredictably reaches a certain limit. A simple example is a plant air compressor motor, which starts whenever system pressure falls below a pre-set level.

To avoid that hazard while servicing the equipment, a worker's proper safety precaution is to open the disconnect switch or circuit breaker in the motor feeder, upstream from the starting contactor itself. That prevents motor starting regardless of what the control system does.

Accordingly, ANSI C2, the National Electrical Code, calls for motors subject to automatic restart to be capable of being locked out. NEMA Standard MG2, however, the "guide for selection, installation, and use of electric motors and generators," does not mention the hazard of automatic restart. In Section 430-43, the National Electrical Code stipulates that "A motor that can restart automatically after shutdown shall not be installed if its automatic restarting can result in injury to persons." That's a broad provision. Almost any motor driving any useful machine could fall within its terms.

Furthermore, both published and unpublished sources within the engineering profession confirm that a serious and fast-growing new hazard is arising from automation of industrial machinery. Sophisticated electronic controls now govern the operation of many drives. False commands to start (or stop) a drive can originate from transient disturbances on a power line, from electromagnetic interference picked up inductively by a badly shielded or poorly routed control signal, or from a malfunction on a microchip caused by static discharge damage.

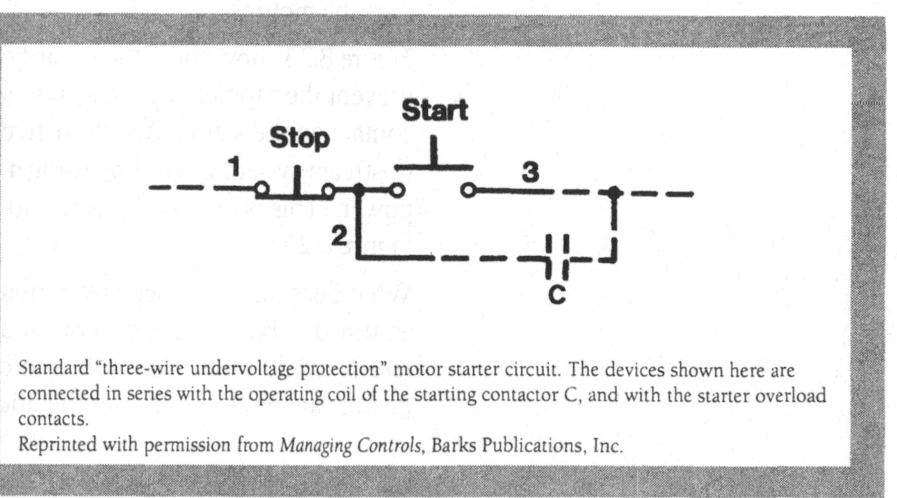

Standard "three-wire undervoltage protection" motor starter circuit. The devices shown here are connected in series with the operating coil of the starting contactor C, and with the starter overload contacts.

Reprinted with permission from *Managing Controls*, Barks Publications, Inc.

Figure 8.22

Yes, the NEC requires an upstream disconnect. But the necessity of opening that disconnect before working on a machine is much more urgent now because the simple nearby starter that is easily locked out is no longer the exclusive provider of start/stop instructions to the motor.

Here's another example of how modern control systems introduce an unexpected danger. In a large chemical processing plant, one master motor control room contains the starter cubicles for most of the plant drives. Each has its own local disconnect. However, a process computer located elsewhere in the building sends start and stop signals to each motor controller via a single "data highway" cable tapped to individual motor control center cubicles. Each cubicle had to be labeled to warn electricians that opening the local circuit disconnect does not interrupt that computer signal, which could introduce a hazardous voltage into the cubicle without warning, even though the motor starting contactor itself is open.

A serious problem can arise even in the absence of such sophisticated control. An electrician or mechanic sent to do work on the motor, control, or driven machinery will make sure the upstream disconnect switch is first opened and supervised (such as by a lockout or tag out procedure) so that it will not be re-closed while work is in process. Location of such a disconnect, visible from the motor unless it can be locked open, is stipulated by NEC Section 430-102(b).

However, when the equipment is not shut down for repairs or adjustment, the process operator will naturally leave the disconnect closed. Knowing that no automatic control will start the motor unexpectedly, such a worker may proceed to work with the machinery (loading or unloading product, or changing tools, for example) without any concern that the motor might suddenly start.

NEC provisions dealing with this danger do not describe in detail the circuitry involved. But certain types of ground faults in motor control circuits can cause startups without warning. This is recognized by NEC Section 430-73, reads: "Where one side of the motor control circuit is grounded, the circuit shall be so arranged that an accidental ground on the remote-control devices will not start the motor."

Figure 8.23 shows how the hazard can arise. The modifications of Figure 8.24 prevent the problem by using a two-pole start button plus an extra auxiliary contact on the starter. An alternative is to isolate the control circuit from the upstream system ground by using a control transformer to provide control power. (This is the usual practice in 480-volt motor control centers.) See Figure 8.25.

What does the NEC mean by remote-control devices? These are normally assumed to be the start-stop pushbuttons, wherever located. In that sense, *remote* would mean "not at the motor itself." Often, of course, these switching devices are built into the combination starter or control center cubicle.

Factory-installed and well-protected, the associated wiring is less likely to develop ground faults than a long run of premises wiring installed in the field. But where safety is concerned, the designer or installer is well advised not to quibble about such detail.

In Figure 8.26a, control power is obtained neither from a local control transformer nor from the motor supply lines themselves. Instead, it comes from a separate panelboard circuit that may supply several control systems. Opening the motor disconnect switch must also open that control power circuit for a least that particular motor (NEC Section 430-74). But during normal drive operation, with the disconnect closed, a ground fault as shown at the stop pushbutton could cause accidental motor starting. The correct principle is to make sure the on/off switching controls are in the *ungrounded* side of the control power circuit. (See Figure 8.26b.)

Medium-voltage motor controllers, also involving control transformers, can likewise present an "unexpected start" hazard unless care is taken to ensure that ground faults cannot bypass the shutdown circuit. The scheme of Figure 8.27 resulted in the deaths of two maintenance workers when a large ball mill accidentally started without warning. Only one ground fault occurred here, rather than the two faults required in a circuit like that of Figure 8.25. It developed at a solenoid, S, located on the mill itself and normally energized whenever the mill was running. Such devices, subject to vibration, contamination, and handling during maintenance operations, are more likely to develop faults than are similar components within the controller assembly.

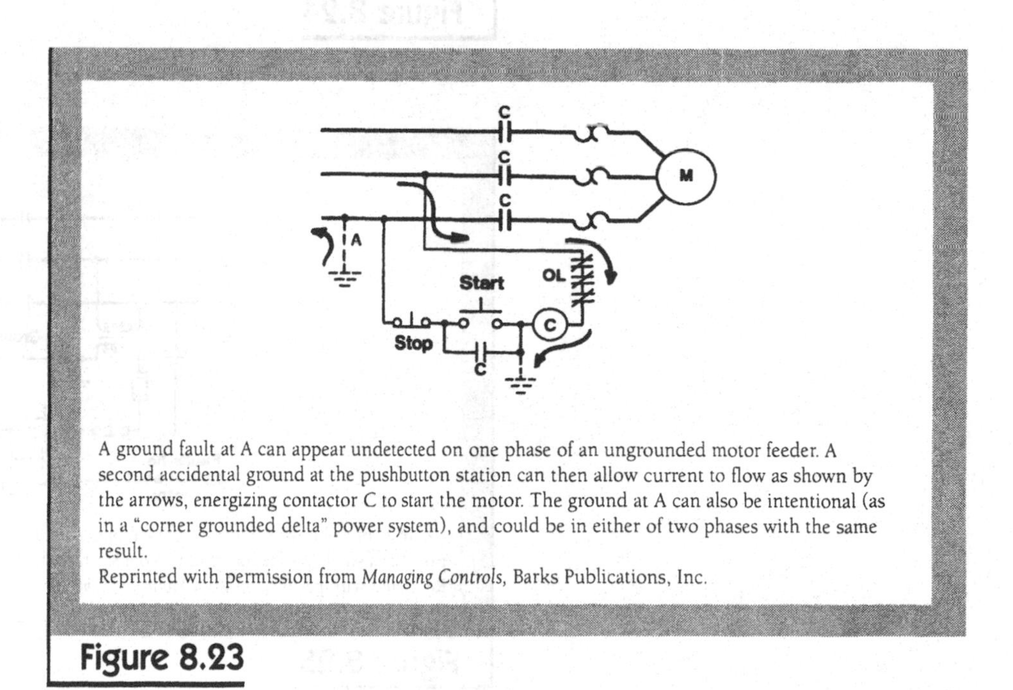

A ground fault at A can appear undetected on one phase of an ungrounded motor feeder. A second accidental ground at the pushbutton station can then allow current to flow as shown by the arrows, energizing contactor C to start the motor. The ground at A can also be intentional (as in a "corner grounded delta" power system), and could be in either of two phases with the same result.

Reprinted with permission from *Managing Controls*, Barks Publications, Inc.

Figure 8.23

Note that the contactor C was intended for operation at 220 volts. The faulted circuit applied only half that voltage to the coil; nevertheless, the contactor closed to start the motor. Again, an essential safety precaution when maintenance work is scheduled is to be sure the motor circuit main disconnect is open. As already pointed out here, though, a danger can be present during normal operation of some drives, when the disconnect would not normally be open.

Users accustomed to the behavior of single-phase motors may raise a question here: What about the "automatic restart" feature of centrifugal (or solid-state) starting switches that control such a motor's auxiliary or starting winding? Supposedly this capability is an advantage in some applications.

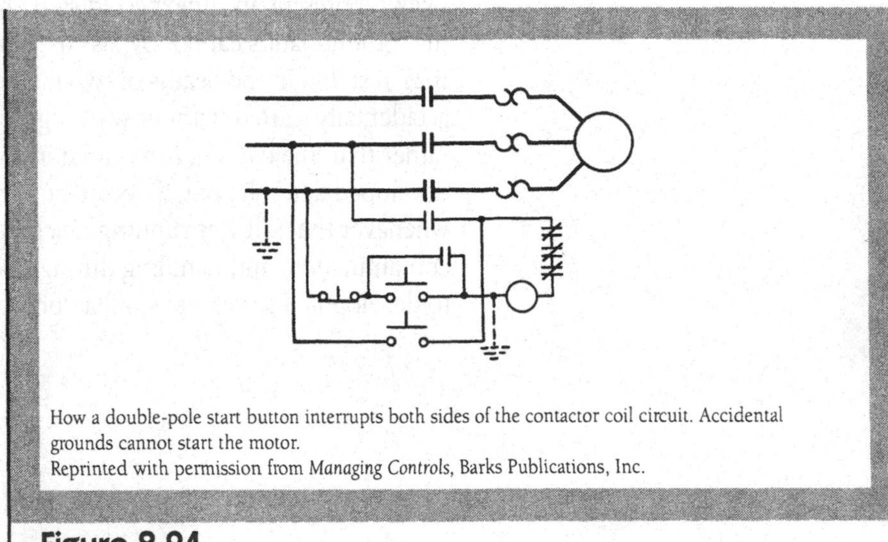

How a double-pole start button interrupts both sides of the contactor coil circuit. Accidental grounds cannot start the motor.
Reprinted with permission from *Managing Controls*, Barks Publications, Inc.

Figure 8.24

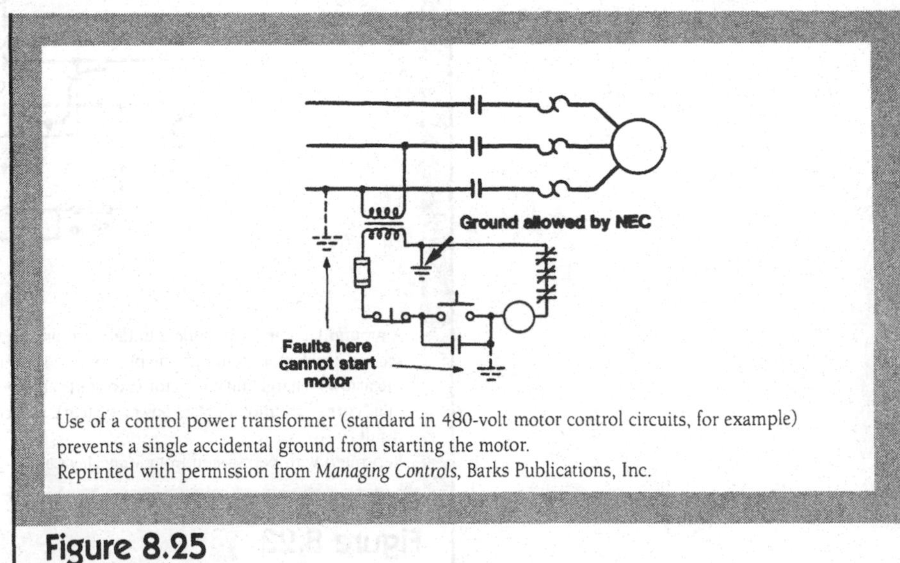

Use of a control power transformer (standard in 480-volt motor control circuits, for example) prevents a single accidental ground from starting the motor.
Reprinted with permission from *Managing Controls*, Barks Publications, Inc.

Figure 8.25

The seeming contradiction is resolved this way. The automatic start (or restart) capability of a three-phase motor, as discussed here thus far, is a feature of the motor's external control system. Either through ground faults or separate control devices, it results in the previously de-energized machine being unexpectedly brought up to full speed. But for the single-phase motor, restart is perhaps better termed "re-energization of the starting winding." In the usual instance, a momentary overload (with the starting winding out of the circuit) has slowed the motor down enough that it hasn't stopped but cannot exert enough torque to re-accelerate. In a power saw, for example, that situation is undesirable. So the restart feature acts to put the starting winding back into the circuit upon such a slowdown, allowing the machine to come back up to speed and complete a cut.

If a single-phase motor were shut down intentionally, an automatic, unexpected startup could be just as dangerous as with a three-phase drive, and is equally subject to the Code restrictions. How the motor's internal switching circuit behaves is normally not relevant.

Any retrofitting operation that replaces an existing motor starter with a variable-frequency drive unit also could be subject to an unexpected restart after power has been interrupted. Some drives are made to shut down and remain de-energized when an outage occurs. Others may be programmed to restart automatically. The typical motor circuit of NEC Diagram 430-1 implies that an electronic inverter would be considered the "supply," and that

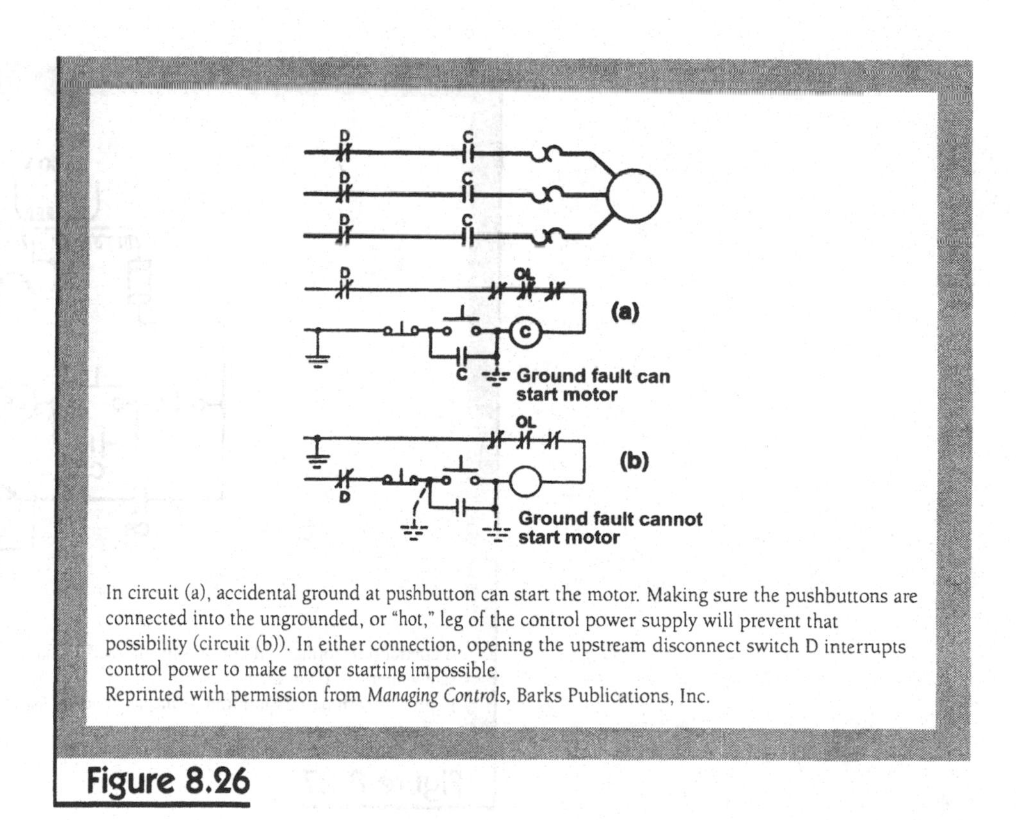

In circuit (a), accidental ground at pushbutton can start the motor. Making sure the pushbuttons are connected into the ungrounded, or "hot," leg of the control power supply will prevent that possibility (circuit (b)). In either connection, opening the upstream disconnect switch D interrupts control power to make motor starting impossible.
Reprinted with permission from *Managing Controls*, Barks Publications, Inc.

Figure 8.26

downstream from the inverter would be a complete set of motor starting and overload protection equipment. But in practice, such equipment is usually either upstream from the inverter, or included within it, or both. Safe contact with the motor itself in any drive installation demands thorough knowledge of how the upstream circuit works and exactly where and how it may need to be locked open while downstream maintenance work is carried out.

Failure of a motor starter to turn a drive off on command can be as dangerous as an unexpected start. This, too, may occur in remote control a-c circuits. The reason? Capacitance—always present between a-c conductors.

Any two conductors separated by insulation constitute a capacitor. The larger the conductor surfaces—length, diameter, or both—and the closer together the conductors are, the greater will be the capacitance. When alternating voltage is applied between the conductors, electrons alternately move toward and away from each conductor. Such a movement constitutes current. We simplify this by saying (though it's not literally true) that "alternating current flows through a capacitor," whereas direct current will not.

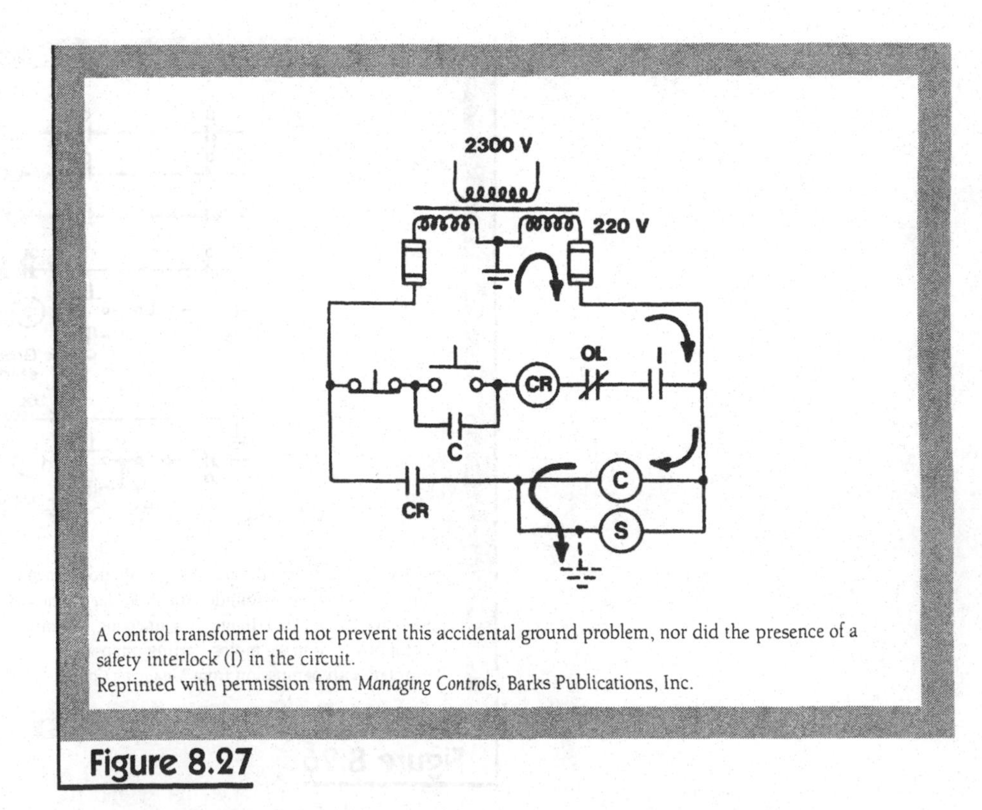

A control transformer did not prevent this accidental ground problem, nor did the presence of a safety interlock (I) in the circuit.
Reprinted with permission from *Managing Controls*, Barks Publications, Inc.

Figure 8.27

Apply that now to the simple motor control circuit of Figure 8.28. If we start the motor by pressing the start button, then redraw the resulting circuit in a simpler form, Figure 8.29 applies. Between wires A and B is a capacitance C. Those two wires will normally be close together, within a single cable or conduit. We don't have to worry about them short-circuiting together, but the insulation preventing that does not get rid of the capacitance.

In Figure 8.29, the intent is that pressing the stop button will open the circuit to de-energize contactor S. Notice, however, that C—through which alternating current "flows"—is connected directly in parallel with the stop button contacts. If C is large enough, which means if wires A and B are long enough, sufficient current will continue to flow through the contactor coil to hold the contactor closed. (This is the contactor's "minimum holding current," usually much less than the normal operating current.) The controlled motor will keep running. In consequence, someone could be injured or killed in an "emergency stop" situation.

Unfortunately, calculating the capacitance is no simple matter. Besides the circuit length, it depends upon the kind of wire insulation; the wire diameter; whether or not the wires are shielded; whether they're in metal conduit, and if so whether or not the conduit may contain water; and on whether wires A and B are separated conductors or part of a multi-conductor cable.

Most a-c control circuits are grounded. Any metallic conduit used must also be grounded. That further complicates the circuit capacitance.

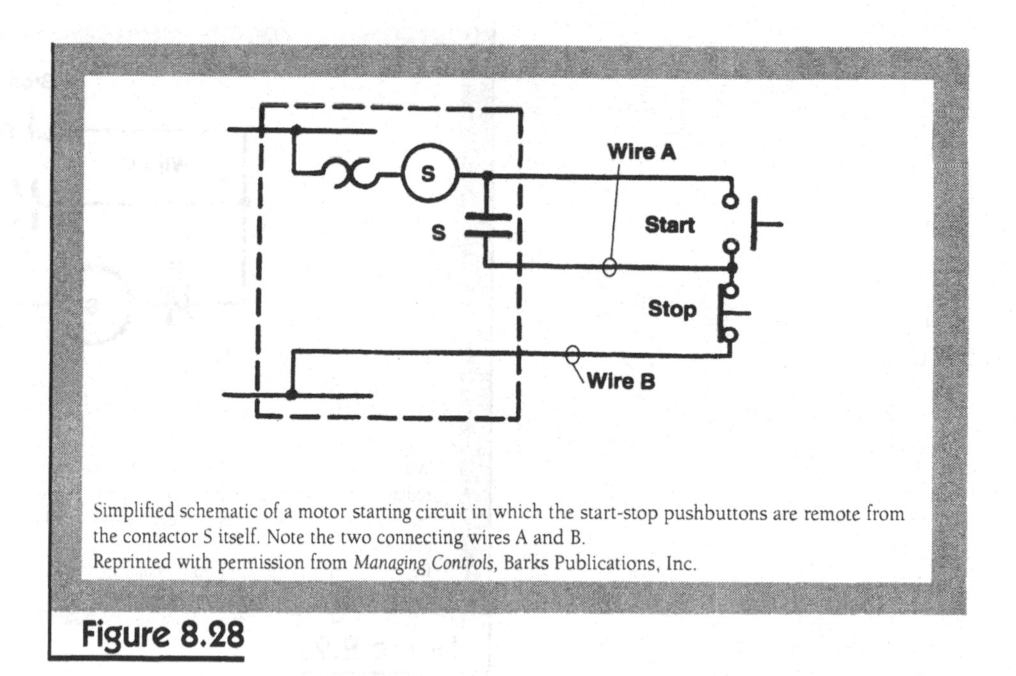

Simplified schematic of a motor starting circuit in which the start-stop pushbuttons are remote from the contactor S itself. Note the two connecting wires A and B.
Reprinted with permission from *Managing Controls*, Barks Publications, Inc.

Figure 8.28

One control equipment supplier has used typical wiring data to develop design guidelines such as those in Table 8.2. Values will vary with the specific starter. Note how the larger wire, with its higher capacitance, reduces rather than increases allowable circuit distances despite lower ohmic resistance.

Table 8.2

Starting Contactor Coil Voltage	Maximum Control Circuit Distance in Feet		
	#14 wire	#12 wire	#10 wire
120	275	230	180
208	90	75	60
240	65	55	45

This table, for a three-wire control scheme as in Figure 8.28, gives the maximum distance between contactor and controlling pushbuttons for a particular type of motor starter, based on water-filled metallic conduit (a "worst case" for underground circuits, resulting in maximum circuit capacitance).

If the circuit cannot be shortened, one solution may be to add a shunt resistor across the contactor coil, as in Figure 8.30 and 8.31. That increases the current through the capacitance when the stop contacts open, raises the voltage drop across the capacitance, and thus lowers the voltage applied to the contactor coil such that the contactor opens more readily. The resistor is out of the circuit when the contactor is open, so it will not affect the closing process. Another scheme uses a double-contact stop button having a grounding contact that short-circuits (discharges) the line capacitance when the button is pressed. A d-c interposing relay will also eliminate the influence of shunt capacitance in the wiring.

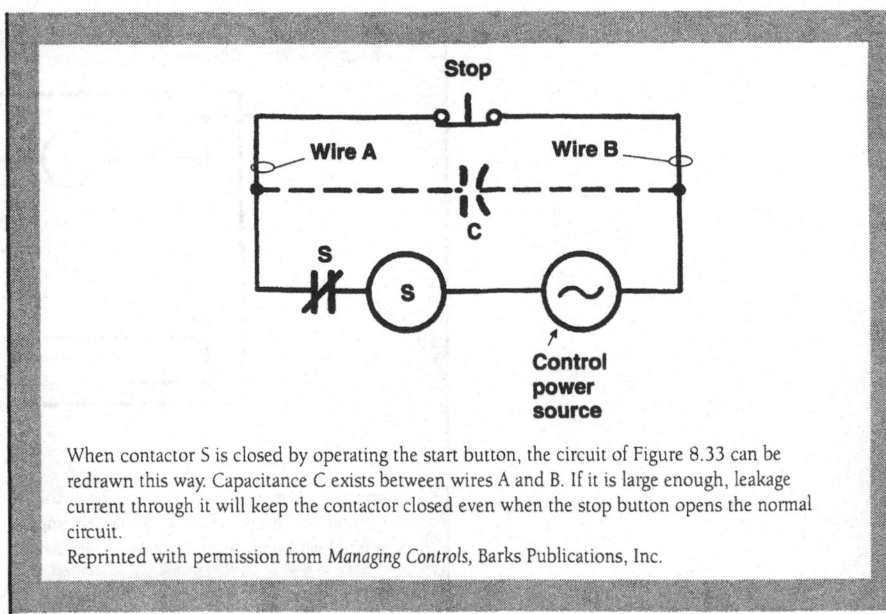

When contactor S is closed by operating the start button, the circuit of Figure 8.33 can be redrawn this way. Capacitance C exists between wires A and B. If it is large enough, leakage current through it will keep the contactor closed even when the stop button opens the normal circuit.

Reprinted with permission from *Managing Controls*, Barks Publications, Inc.

Figure 8.29

For some motors, another type of electrical hazard exists that can endanger personnel even when the motor disconnect switch is open. In one version, potentially dangerous voltage may be present on the motor windings, although incapable of causing machine rotation. In the other version, no such voltage appears, but live wiring and terminals exist elsewhere inside the motor frame or enclosure.

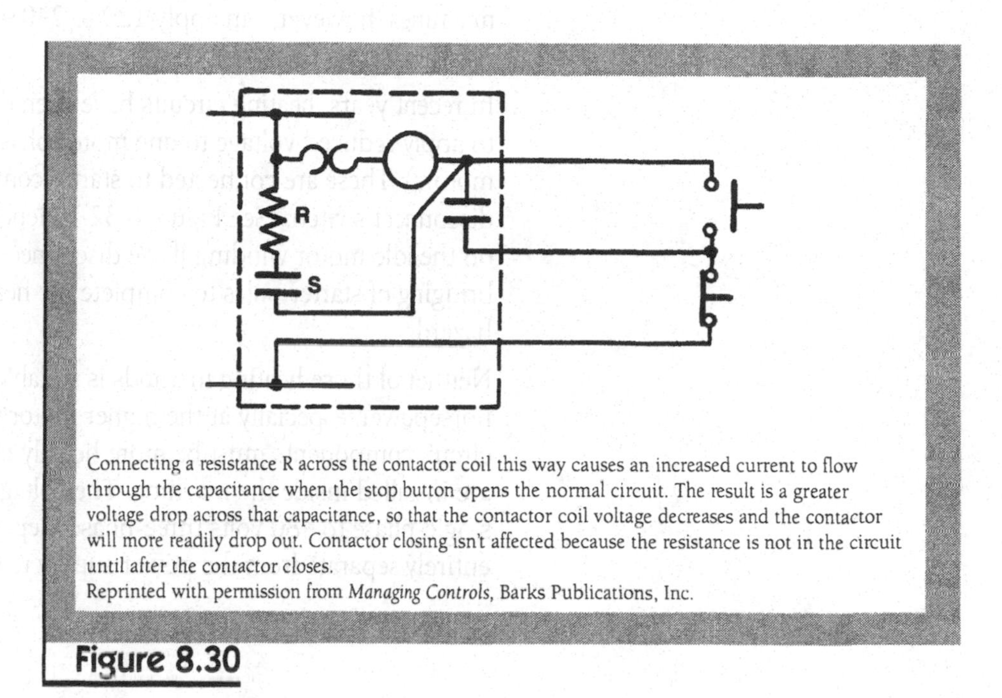

Connecting a resistance R across the contactor coil this way causes an increased current to flow through the capacitance when the stop button opens the normal circuit. The result is a greater voltage drop across that capacitance, so that the contactor coil voltage decreases and the contactor will more readily drop out. Contactor closing isn't affected because the resistance is not in the circuit until after the contactor closes.
Reprinted with permission from *Managing Controls*, Barks Publications, Inc.

Figure 8.30

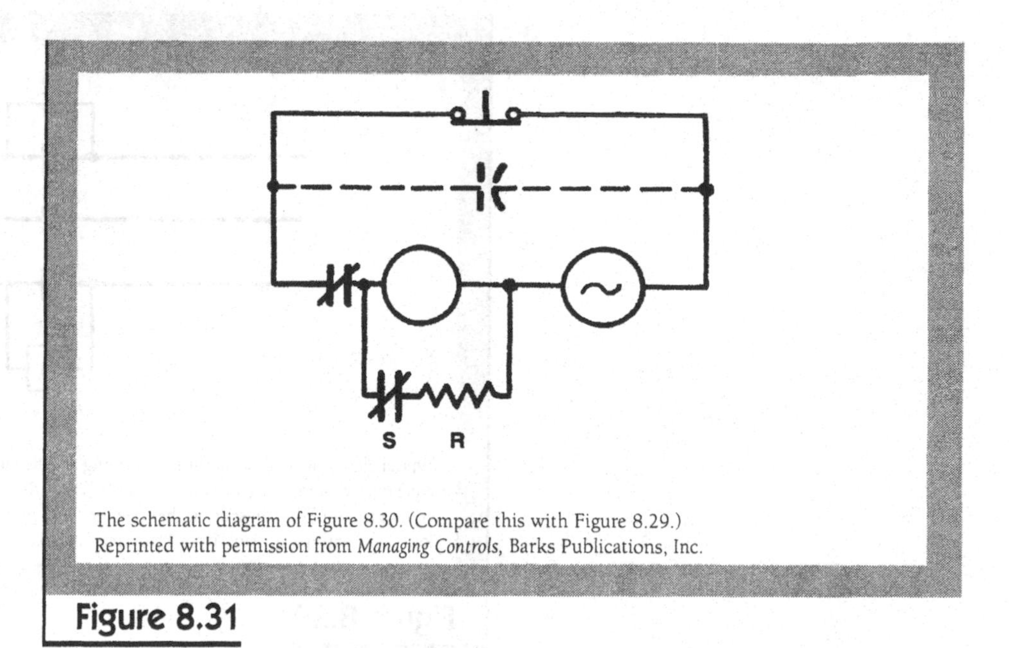

The schematic diagram of Figure 8.30. (Compare this with Figure 8.29.)
Reprinted with permission from *Managing Controls*, Barks Publications, Inc.

Figure 8.31

Those conditions arise when space heating is used to warm the motor interior during idle periods, especially for totally enclosed or large outdoor machines. One heating method applies a reduced single-phase voltage to two of the three winding leads when the motor is shut down. The voltage source may be a control-type transformer fed from the three-phase motor supply itself. Typically, this will be connected upstream from the main disconnect switch, and the circuit will be arranged to apply the heating power automatically whenever the motor starter is de-energized.

For a small motor rated 460 volts or less, the heating voltage will seldom exceed 30 volts. Similar heating arrangements for 2300- or 4000-volt machines, however, can apply 120 to 240 volts single-phase to the "idle" winding.

In recent years, heating circuits have been developed using an electronic device to apply reduced voltage to one motor phase (practical only for low-voltage motors). These are connected to starter contacts, entirely downstream from the disconnect switch. (See Figure 8.32.) Hence, no heating voltage would appear on the idle motor winding if the disconnect were open. But if it is not, the bridging of starter poles to complete the hearing circuit gives rise to a potential hazard.

Neither of those heating methods is suitable for motors above a few hundred horsepower, especially at the higher motor voltage where some of the heating circuit components must be more heavily insulated. Instead, separate heaters are installed inside the machine. The voltage source—ranging from 120 volts single-phase to 480 volts three-phase, depending upon heater wattage—is entirely separate from the power supply to the motor or its control circuit.

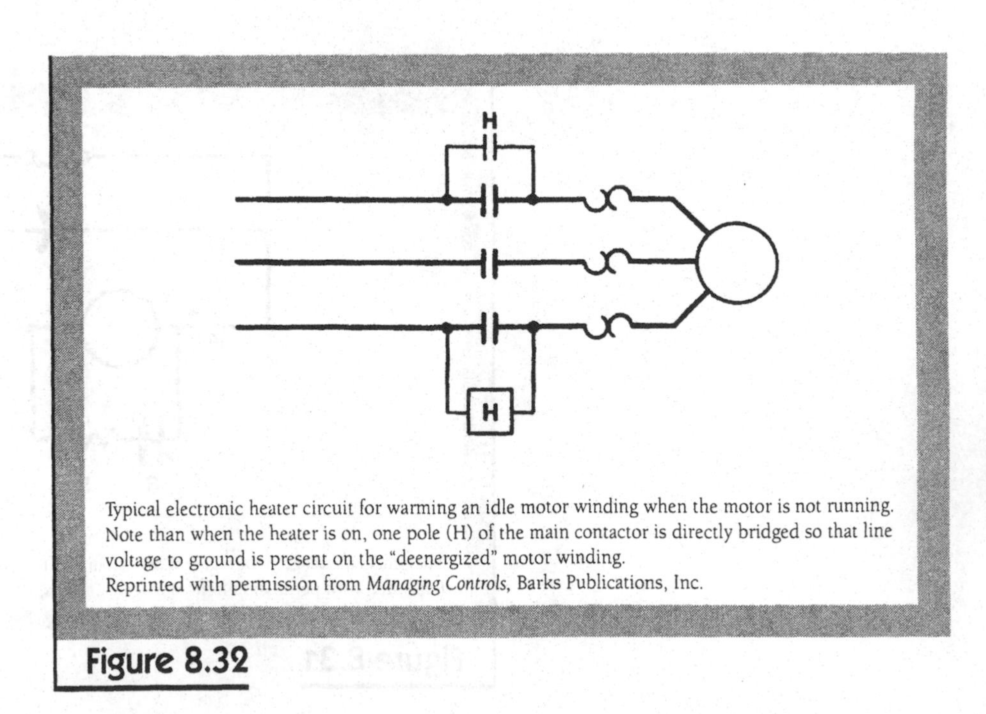

Typical electronic heater circuit for warming an idle motor winding when the motor is not running. Note than when the heater is on, one pole (H) of the main contactor is directly bridged so that line voltage to ground is present on the "deenergized" motor winding.
Reprinted with permission from *Managing Controls*, Barks Publications, Inc.

Figure 8.32

The heaters are usually energized automatically whenever the motor starting contactor opens. That means a potentially lethal live circuit is present within the motor enclosure whenever the motor is off. Disconnecting all power to the motor does not remove that hazard, because the space heater circuit is designed to be energized only when motor power has been switched off.

Code rules governing the connection or use of the motor "controller" don't deal with this hazard. NEC Section 430-71 defines the motor "control circuit" as a circuit that "carries the electric signals directing the performance of the controller." A space-heating circuit of whatever sort does not fit that definition.

Depending upon the motor construction, warning labels may be used to warn workers to avoid or disconnect the heater power, or barriers may be added to prevent contact with the heater wiring. Large heaters necessarily have exposed terminals; smaller ones may be fully insulated, but the insulation can deteriorate over time.

Solid-State Starters: In addition to the electronic heating circuit, a second modern development can also result in the presence of line voltage at the terminals of a non-running motor for which the starter has been turned off. NEC Section 430-73 emphasizes the need for motor control circuits to prevent the starter's being accidentally switched on—but what if the starter remains "off" but still does not fully open the circuit?

That's what can happen with the new electronic solid state, or "soft," starters. The motor voltage is adjusted by variation in SCR (Silicon-controlled rectifiers) firing angle just as in a variable-frequency drive, except that the starter does not vary the frequency but only the voltage.

When a solid-state starter is turned off, voltage is lowered and current reduced below the point where motor rotation is possible. The drive comes to a stop. However, a small leakage current continues to flow through the SCRs. Line voltage thus appears at the motor terminals even though the motor is not rotating. Making sure such a starter is off and cannot accidentally be turned on isn't enough for safe access to the motor connections or winding.

With the so-called *component level starter*, this is not a problem. Such devices are intended for installation downstream (or occasionally upstream) from a conventional magnetic starter or circuit breaker. The contactor in the starter will open to fully isolate the motor from the line whenever the "stop" signal is given. A conventional start-stop pushbutton circuit (as in Figure 8.22) will then keep the circuit open until an operator re-closes it.

Some other makes of solid-state starter include (either as standard or as an option) a built-in isolating contactor on the line side of the SCR assembly. That contactor opens when the motor is switched off, preventing line voltage from appearing at any downstream point.

No published standard governs that isolation option. A survey of 27 solid-state equipment product lines shows that an isolating contactor is optional in nine of them; another 12 are intended for connection in series with a conventional starter. Only one features built-in isolation as standard.

One instruction book for installing solid-state starters reads: "Warning. Equipment is at line voltage when AC power is connected. Pressing 'stop' pushbutton does not remove AC mains potential." Warning labels on the equipment itself may or may not convey a similar message.

Still another electrical hazard of a "de-energized" motor can arise when power factor connection capacitors are connected at the motor terminals. Depending upon the capacitor size, a significant voltage may be present at the machine terminals as long as the speed remains high. A heavily loaded motor will of course come quickly to a stop—often within a second or two—once the winding has been disconnected from the line.

However, if a high-inertia load (or, in the case of an induction generator, a connected engine or turbine) keeps the motor spinning for some time at fairly high rpm, the capacitors will excite the machine windings such that a dangerous voltage may exist at the terminals.

Exposure to such a voltage can be avoided by staying clear of the circuit until the drive has come to a full stop. The alternative is to make certain that no capacitors are on the circuit.

Who is liable when injury results from an unexpected start involving some complex control scheme? Unfortunately, liability issues today are being increasingly handled, as one engineer put it, "by suing everybody who has a nameplate on the equipment"—including the manufacturer of the motor itself. One recent instance of that involved a worker who lost several fingers when a motor-driven machine started unexpectedly during servicing. Most people would rather keep their fingers than win a lawsuit, so understanding what the possible hazards are and how to avoid them is important to anyone working with motors and controls. Basic to that understanding is one simple rule: "Assume nothing."

Resource Publications

Albert, C.L., and Coggan, D.A., eds. *Fundamentals of Industrial Control*. Research Triangle Park, NC: Instrument Society of America, 1992.

Haines, Roger W. *HVAC Systems Design Handbook*. Blue Ridge Summit, PA: Tab Books, Inc., 1988.

Hughes, Thomas A. *Programmable Controllers*. Research Triangle Park, NC: Instrument Society of America, 1997.

Nachtigal, Chester L., ed. *Instrumentation and Control: Fundamentals and Applications*. New York, NY: John Wiley and Sons, 1990.

Nailen, Richard L. *Managing Controls: The Complete Book of Motor Control Application and Maintenance*. Chicago, IL: Barks Publications, Inc., 1993.

Platt, George. *Process Control: A Primer for the Nonspecialist and the Newcomer*. Research Triangle Park, NC: Instrument Society of America, 1988.

Rowe, William. *HVAC: Design, Criteria, Options, Selection*. Kingston, MA: R.S. Means Co., Inc., 1994.

Smith, Ernest. *Principles of Industrial Measurement for Control Applications*. Research Triangle Park, NC: Instrument Society of America, 1984.

For Additional Information

The Institute of Electrical and Electronic Engineers
345 East 47th Street
New York, NY 10017-2394
212-705-7900
www.ieeeusa.org

The IEEE is the world's largest technical professional society, and includes electronic, computer engineering, and computer science professionals. It sponsors conferences, local chapter meetings, and educational programs.

The following web sites are among many Internet resources on instrumentation and controls:

www.ceenews.com – the online version of *CEENews*, a publication for the electrical industry.

www.cfmcontrols.com – the site of a manufacturer's representative for process control.

www-control.eng.cam.ac.uk – the home page of the Control Group at the University of Cambridge; offers a variety of information for the control engineer.

www.drexelbrook.com – a useful site presented by a manufacturer of RF and ultrasonic level controllers.

www.electrical search.com – a fast, direct, and convenient way to access information about instrumentation and controls and other topics in the electrical marketplace.

Chapter 9

Electrical Engineering

Chapter 9

Electrical Engineering

Today, facility and plant engineers must not only possess an understanding of rotating electrical machinery, electronics, the National Electrical Code (NEC), lighting, and industrial control circuitry, but must also keep up with deregulation and its impact on their facilities.

Although all facilities are charged for the energy used in kilowatt-hours, most are also charged for their kilovolt-amp demand. Most facilities have highly inductive loads—that is, the amount of electrical energy they draw from the power company does not match the amount they actually utilize. Therefore, most power companies impose a demand charge on the facility's electrical bill.

To be able to correctly assess a demand charge, facility and plant managers must first understand what inductive loads are and how they cause current to lag in time behind the applied voltage. The first part of this chapter is devoted to explaining this concept, beginning with a review of fundamental units and formulas. Next are sections on power, wave forms, voltage drop, motors, lighting, overcurrent protection, the National Electrical Code, electrical costs, and industrial control.

Units and Formulas

Units of Measurement

The following are the basic electric units of measurement:

- Electrical True Power is measured in WATTS (W)
- Electrical Reactive Power is measured in Volt-Amps Reactive (vARS)
- Electrical Apparent Power is measured in Volt-Amps (vA)
- Electrical Resistance is measured in OHMS (Ω)
- Electrical Potential is measured in VOLTS (V)
- Electrical Current is measured in AMPS (A or I)
- Electrical Inductance is measured in HENRIES (L)
- Electrical Inductive Reactance is measured in OHMS (Ω)
- Electrical Capacitive Reactance is measured in OHMS (Ω)
- Electrical Impedance is measured in OHMS (Ω)
- Electrical Frequency is measured in Cycles per Second or HERTZ (HZ)
- Electrical Capacitance is measured in FARADS (F)
- Electrical Charge is measured in COULOMBS (C)
- Electrical Energy is measured in KILOWATT HOURS (kWh)

Electric Current

Direct current, or DC, flows in only one direction. Alternating current (AC) reverses direction at regular intervals. The chief advantage of AC over DC current is that the former can be more easily transformed to higher or lower voltages, which allows for a safer and more efficient use of power. With the use of transformers, AC power can be adjusted to meet the needs of the power producer, the utility, and the user.

Alternating current has a frequency that is defined by the number of times per second that the current goes through a cycle or sequence. The number of cycles per second is measured in Hertz (Hz). In the United States, most current for domestic use reverses direction at 60 cycles per second—that is, it operates at 60 Hz. Some industrial applications operate at 180 Hz, and fluorescent lighting circuits may operate as high as 400 Hz.

Electrical Formulas

Figure 9.1 lists fundamental electrical formulas.

Ohm's Law

Ohm's Law illustrates the relationship for DC or single-phase AC with a PF of 1.0, as shown in Figure 9.2.

Electrical potential is electromotive force (EMF), measured as voltage (V). EMF means that the electrical pressure (the voltage) is maintained over varying loads. EMF causes free electrons to move in a conductor. This movement is called *electric current,* and is measured in amperage. One amp is equivalent to 6.28×10^{18} electrons moving by one point in one second. An electrical unit of charge is a Coulomb, and is equivalent to 6.28×10^{18} electrons. Therefore, an amp is also equivalent to a Coulomb per second.

Figure 9.1

Electrical Formulas

Ohms Law:

Ohms = Volts/Amperes

Amperes = Volts/Ohms

Volts = Amperes x Ohms

Power — A.C. Circuits:

$$\text{Efficiency} = \frac{746 \times \text{Output Horsepower}}{\text{Input Watts}}$$

$$\text{Three Phase Kilowatts} = \frac{\text{Volts} \times \text{Amperes} \times \text{Power Factor} \times 1.732}{1000}$$

$$\text{Three Phase Volt-Amperes} = \text{Volts} \times \text{Amperes} \times 1.732$$

$$\text{Three Phase Amperes} = \frac{746 \times \text{Horsepower}}{1.732 \times \text{Volts} \times \text{Efficiency} \times \text{Power Factor}}$$

$$\text{Three Phase Efficiency} = \frac{746 \times \text{Horsepower}}{\text{Volts} \times \text{Amperes} \times \text{Power Factor} \times 1.732}$$

$$\text{Three Phase Power Factor} = \frac{\text{Input Watts}}{\text{Volts} \times \text{Amperes} \times 1.732}$$

$$\text{Single Phase Kilowatts} = \frac{\text{Volts} \times \text{Amperes} \times \text{Power Factor}}{1000}$$

$$\text{Single Phase Amperes} = \frac{746 \times \text{Horsepower}}{\text{Volts} \times \text{Efficiency} \times \text{Power Factor}}$$

$$\text{Single Phase Efficiency} = \frac{746 \times \text{Horsepower}}{\text{Volts} \times \text{Amperes} \times \text{Power Factor}}$$

$$\text{Single Phase Power Factor} = \frac{\text{Input Watts}}{\text{Volts} \times \text{Amperes}}$$

$$\text{Horsepower (3 Phase)} = \frac{\text{Volts} \times \text{Amperes} \times 1.732 \times \text{Efficiency} \times \text{Power Factor}}{746}$$

$$\text{Horsepower (1 Phase)} = \frac{\text{Volts} \times \text{Amperes} \times \text{Efficiency} \times \text{Power Factor}}{746}$$

Power — D.C. Circuits:

$$\text{Watts} = \text{Volts} \times \text{Amperes}$$

$$\text{Amperes} = \frac{\text{Watts}}{\text{Volts}}$$

$$\text{Horsepower} = \frac{\text{Volts} \times \text{Amperes} \times \text{Efficiency}}{746}$$

Motor Application Formulas:

$$\text{Torque (lb-ft.)} = \frac{\text{Horsepower} \times 5250}{\text{RPM}}$$

$$\text{Horsepower} = \frac{\text{Torque (lb-ft)} \times \text{RPM}}{5250}$$

Time for motor to reach operating speed (seconds)

$$\text{Seconds} = \frac{WK^2 \times \text{speed change}}{308 \times \text{Ave. Accelerating Torque.}}$$

WK^2 = Inertia of Rotor + Inertia of Load. $(\text{LB-FT})^2$

$$\text{Average Accelerating Torque} = \frac{\frac{(\text{FLT}+\text{BDT})}{2} + \text{BDT} + \text{LRT}}{3}$$

FLT = Full Load Torque, BDT = Breakdown Torque,

LRT = Locked Rotor Torque.

$$\text{Load } WK^2 \text{ (at Motor Shaft)} = \frac{WK^2 \text{ (Load)} \times \text{Load RPM}^2}{\text{Motor RPM}^2}$$

$$\text{Shaft Stress (pds. per sq. inch)} = \frac{\text{HP} \times 321{,}000}{\text{RPM} \times \text{Shaft Diam.}^3}$$

For Pumps:

$$\text{Horsepower} = \frac{\text{GPM} \times \text{Head in Feet} \times \text{Specific Gravity}}{3960 \times \text{Efficiency of pump}}$$

For Fans and Blowers:

$$\text{Horsepower} = \frac{\text{CFM} \times \text{Pressure (pounds/sq. ft.)}}{33000 \times \text{Efficiency}}$$

Speed:

$$\text{Synchronous RPM} = \frac{\text{Hertz} \times 120}{\text{Poles}}$$

$$\text{Percent Slip} = \frac{\text{Synchronous RPM} - \text{Full Load RPM}}{\text{Synchronous RPM}} \times 100$$

Ohm's law states that the voltage drop across a resistance is equivalent to the current passing through the resistance multiplied by the resistance itself. One can get confused because this formula is only really true for DC with an unchanging current value. DC circuits have no inductance or capacitance. Thus, the resistance of a DC circuit is its only opposition to current flow.

In AC circuits, however, the current is sinusoidal—that is, shaped like, or varying according to, a sine curve or a sine wave. This sinusoidal current produces the phenomenon of inductance (L), which adds to the opposition of current flow. The total opposition to current flow in an AC circuit is called impedance (Z). Therefore, in an AC circuit, V = IR, V = IZ.

This changing current produces a magnetic field around a conductor whose polarity is constantly changing. This means that a conductor that has an alternating current passing through it will induce an EMF onto itself that

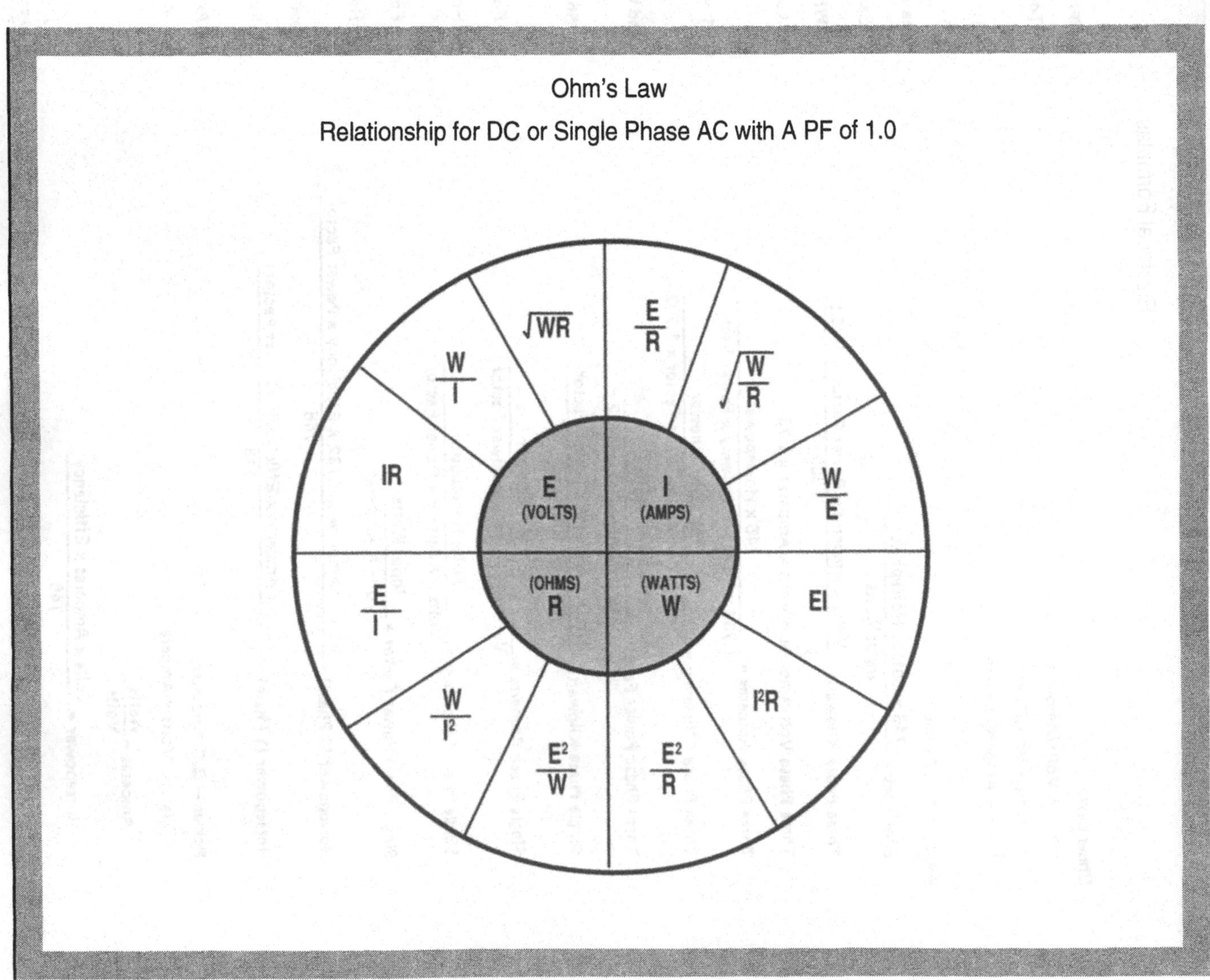

Figure 9.2

opposes any change in the current. Because this EMF is induced via electromagnetic induction, conductors that have an alternating current passing through them are said to possess "self-inductance," or more commonly, "inductance." Inductance is measured in Henries. When a coil is wrapped around a conductor, the value of the inductance is increased. A coil with an inductance of 1 Henry is equivalent to a coil that has a current passing through it, changing at the rate of 1 amp per second, and inducing onto itself a counter EMF of 1 volt.

Inductance, found to some extent in all AC circuits, affects the management of a facility or plant in two principal ways:

1. Because the phenomenon of inductance by definition opposes any change in current (Lenz's Law), it must add to the opposition of current flow in an AC circuit. In fact, it does. The formula $X_L = 2\pi fL$ converts Inductance (L) in Henries to Inductive Reactance (X_L) measured in Ohms.
2. It can be proven that in a purely inductive circuit, the current lags the applied voltage by 90 electrical degrees (i.e., the applied voltage leads the current by 90 electrical degrees). As the current and voltage do not always occur at the same time, there is some wasted energy. The difference in when the current and voltage occur is usually measured in degrees and is always between 0 and 90 degrees. The cosine of this angle is referred to as the *power factor* of a circuit. Thus, low power factors are a direct sign of wasted energy.

Not only is the opposition to current flow greater in comparable AC circuits, with less current flowing due to inductive reactance, but only a portion of this smaller current is actually doing work because some of the current is out of phase with the applied voltage. This manifests itself in excess current and excess heat around electrical machinery, and, ultimately, higher costs.

In order to compensate for this wasted energy, many plant managers have incorporated power factor correcting capacitor banks. It can be proven that in a purely capacitive circuit the current leads the applied voltage by 90 electrical degrees. By adding the right amount of capacitance, the power factor can be brought closer to unity (100%), saving the plant money by allowing it to use less electric current to perform the same work.

What is Power Factor?

Special Electrical Requirements of Inductive Loads

Most loads in modern electrical distribution systems are inductive. Examples include motors, transformers, gaseous tube lighting ballasts, and induction furnaces. Inductive loads need an electromagnetic field to operate.

Inductive loads require two kinds of current: *working power (kW)* to perform the actual work of creating heat, light, motion, machine output, etc.; and *reactive power (kVAR)* to sustain the electromagnetic field.

Working power consumes watts and can be read on a wattmeter. It is measured in kilowatts (kW). (See Figure 9.3.) Reactive power does not do useful "work," but it circulates between the generator and the load. It places a heavier drain on the power source and the power source's distribution system. Reactive power is measured in kilovolt-amperes-reactive (kVAR). (See Figure 9.4.)

Working power and reactive power together make up apparent power. Apparent power is measured in kilovolt-amperes (kVA). (See Figure 9.5.)

A right "power" triangle is often used to illustrate the relationship between kW, kVAR and kVA, as shown in Figure 9.6.

Why Should I Be Concerned About Low Power Factor?

Low power factor means you are not fully utilizing the electrical power for which you are paying. If, for example, a power factor of 80% is used by your boring mill, it would be utilizing only 80% of the power supplied by the utility. That means only 80% of the incoming current is being used to produce useful work.

As the triangle relationships in Figure 9.7 demonstrate, kVA decreases as the power factor increases. At 70% power factor, it requires 142 kVA to produce 100 kW of energy. At 95% power factor, it requires only 105 kVA to produce 100 kW of power. Another way to look at this is that at 70% power factor, it takes 35% more current to do the same work.

Figure 9.8 illustrates the effects of various power factors on an electrical system with 100 kW demand at 480 volts. As the chart shows, the wire size requirement on the system with 100% power factor is No. 1/0. The same system with a 60% power factor requires a No. 4/0 conductor. Actual wire diameters are shown.

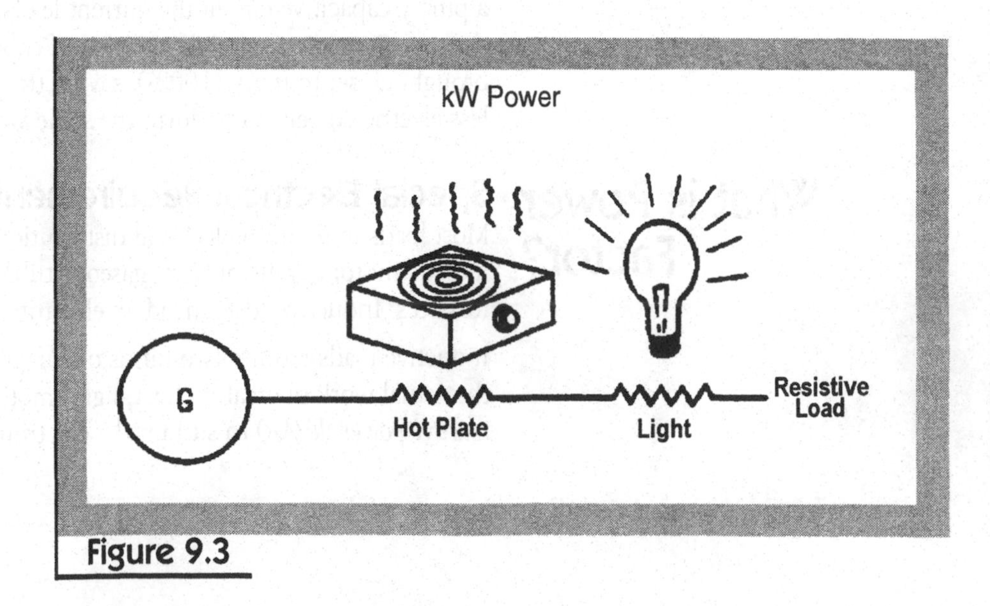

Figure 9.3

What Can I Do to Improve Power Factor?

You can improve power factor by adding power factor correction capacitors to your plant distribution system.

When apparent power (kVA) is greater than true power (kW), the utility must supply the excess reactive current *plus* the working current. Power capacitors act as reactive current generators. By providing the reactive current, they reduce the total amount of current your system must draw from the utility.

Power Quality

Until recently, almost all loads were linear, with the current wave form closely matching sinusoidal voltage wave form and changing in proportion to the load. Lately, non-linear loads—which use current in pulses and at a frequency other than 60 Hz—have increased dramatically.

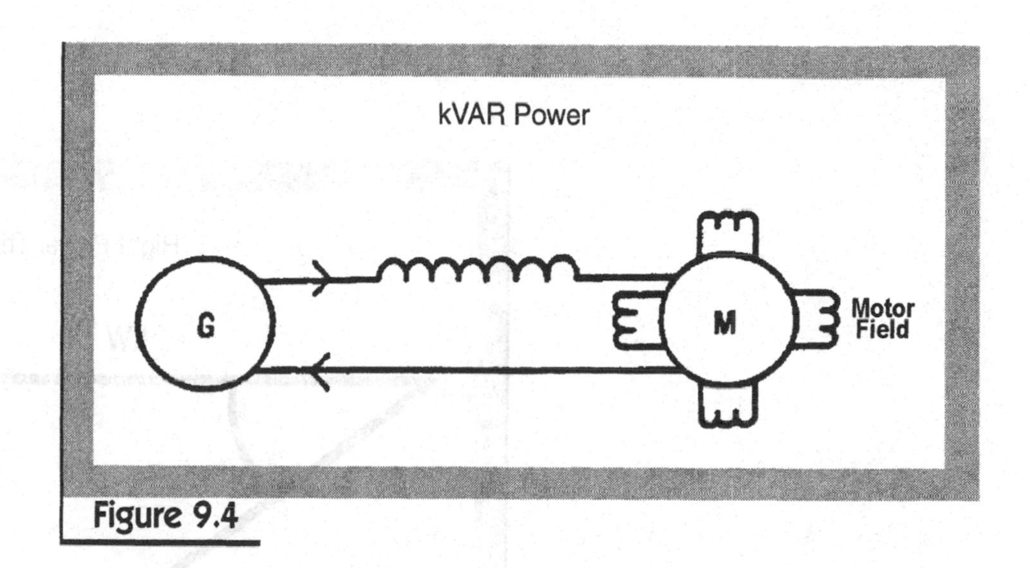

Figure 9.4

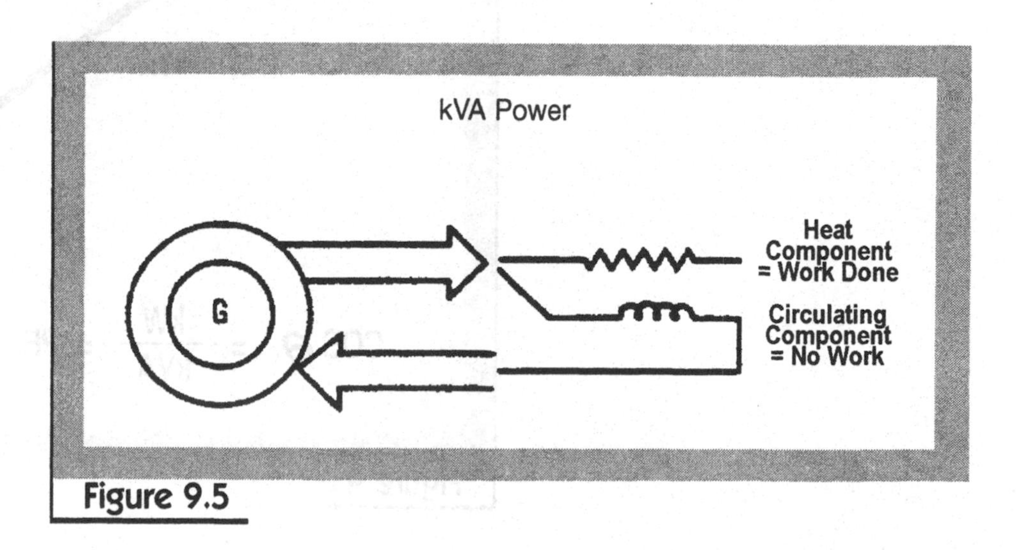

Figure 9.5

Examples of linear and non-linear devices include the following:

Linear Devices

- Motors
- Incandescent lighting
- Heating loads

Non-linear Devices

- DC drives
- Variable frequency drive
- Programmable controllers
- Induction furnaces
- Power supplies
- Personal computers
- Uninterruptible power supplies (UPS)
- Electronic ballasts for fluorescent lighting

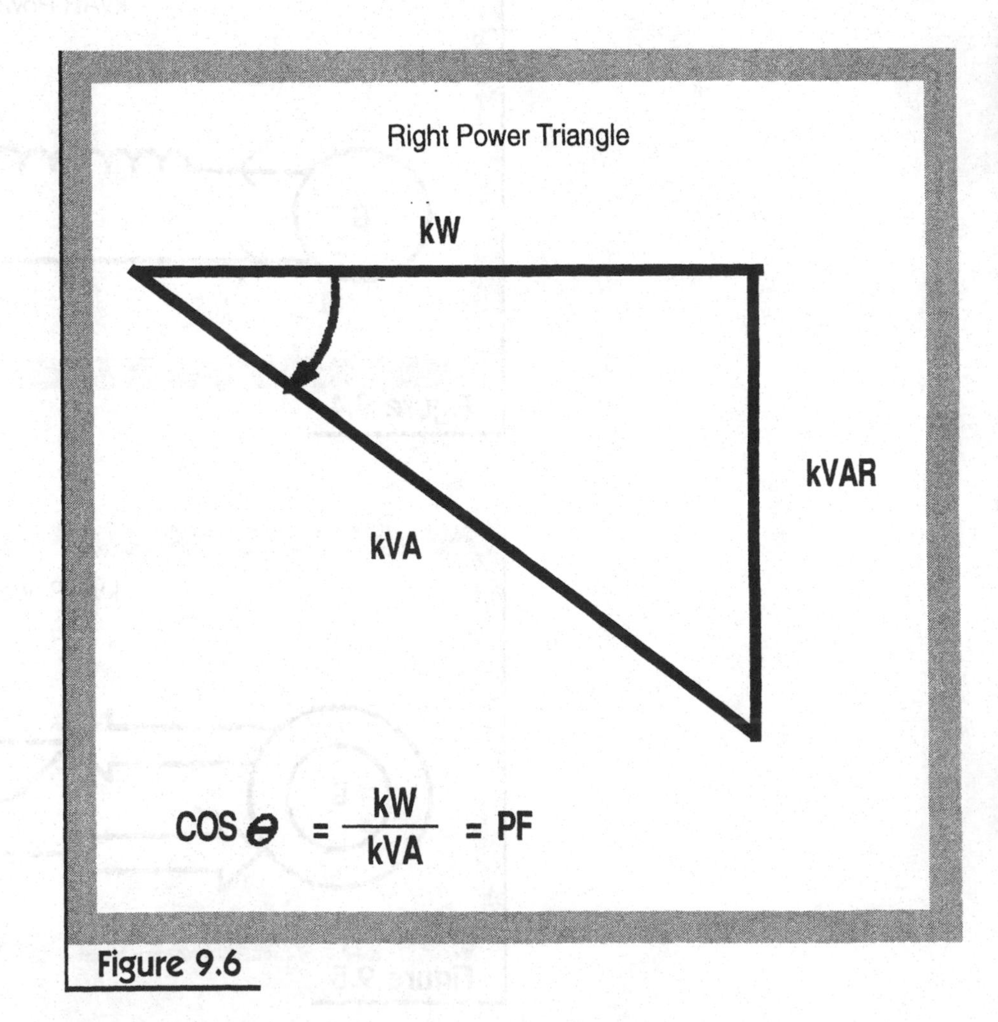

Figure 9.6

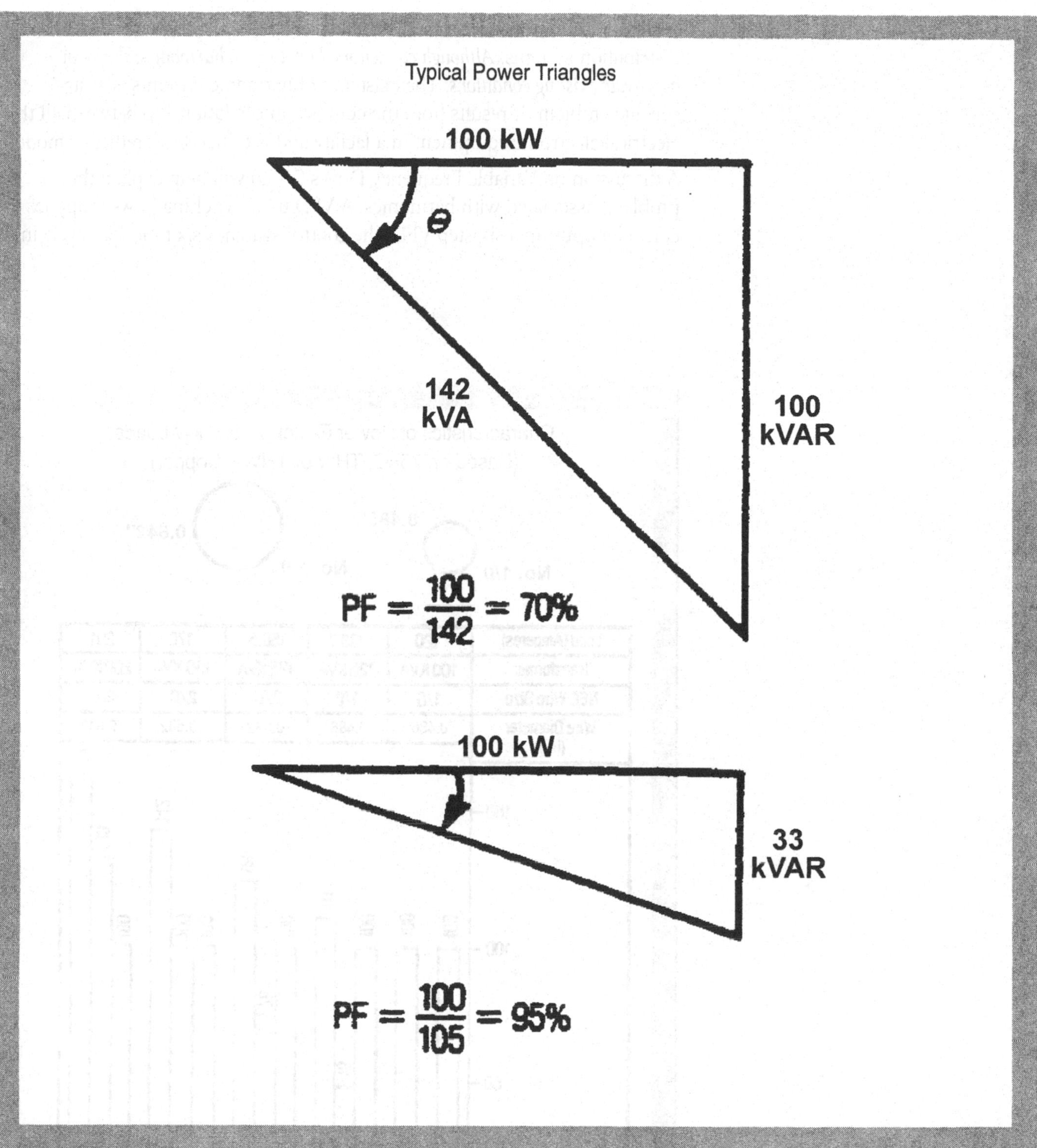
Typical Power Triangles
100 kW
θ
142
kVA
100
kVAR
PF = 100/142 = 70%
100 kW
33
kVAR
PF = 100/105 = 95%

Figure 9.7

The increase in non-linear loads leads to harmonic distortion in electrical distribution systems. *Although capacitors do not cause harmonics, they can aggravate existing conditions.* The existence of harmonic currents is a site-specific problem. It results from the complex interrelationship between all the electrical/electronic equipment in a facility and is difficult to predict or model. A discussion on Variable Frequency Drives (VFD) will help explain the problems associated with harmonics. A VFD uses switching power supplies to control output. In a six-step VFD, the control switches six times per cycle in an

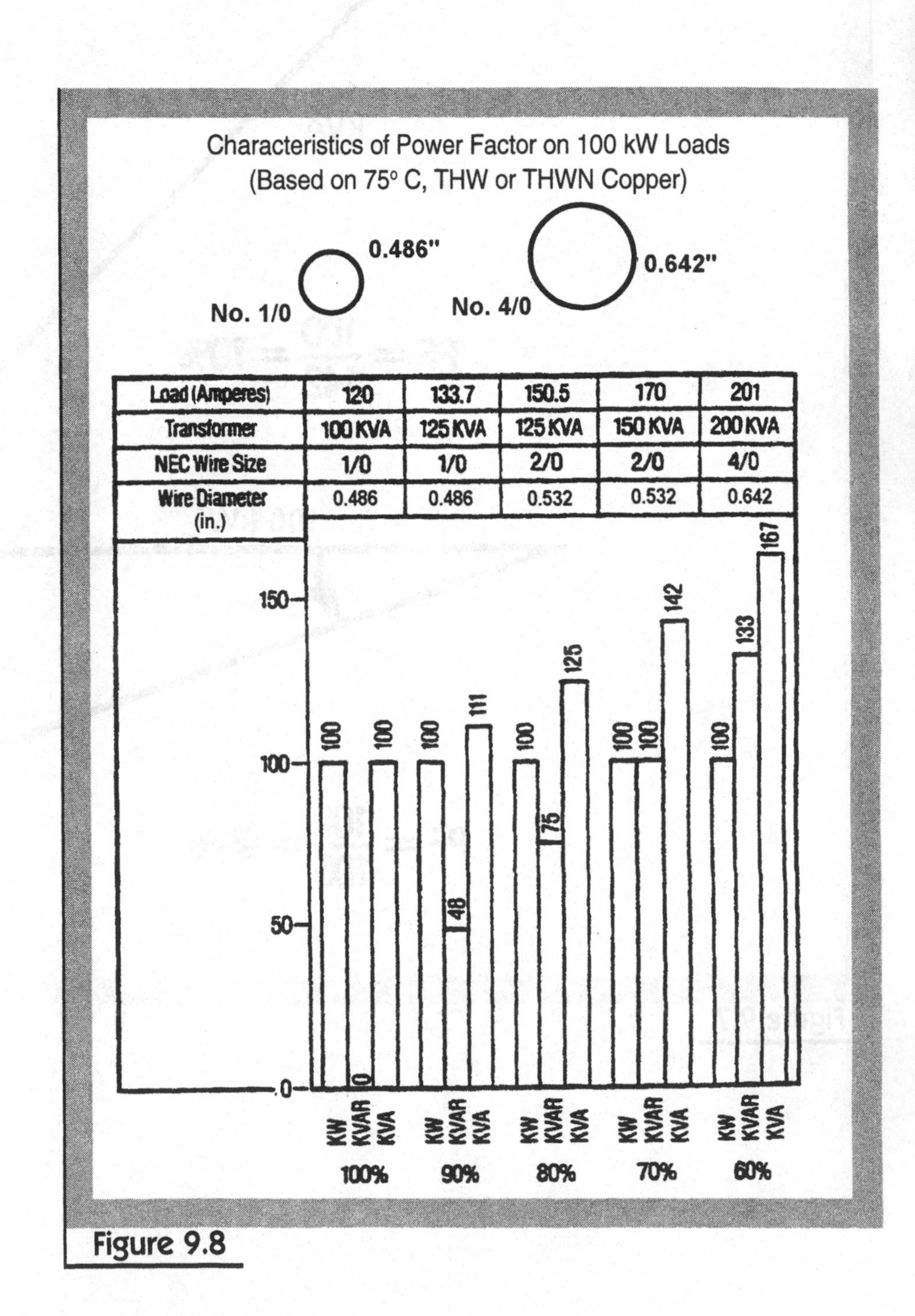

Figure 9.8

effort to simulate a sine wave. As the time between switches changes, the motor receives varying apparent frequencies and changes speed accordingly.

These changes in apparent frequencies lead to two problems: large voltage spikes and distorted current wave shapes. The voltage spikes are short and generally will not affect equipment that does not use a zero crossing of the voltage wave shape for timing. The distorted current wave shape is the "harmonic generator."

Harmonics cause additional noise on a line, which generates heat. The increase in heat can cause nuisance tripping of circuit breakers. Power factor capacitors experience the same problem. Under thermal overload, their fuses blow. Thus, they provide an early warning signal that harmonic currents exist in a facility.

Capacitor Banks and Transformers Cause Resonance:

Capacitors and transformers can create dangerous resonance conditions when capacitor banks are installed at the service entrance. Under these conditions, harmonics produced by non-linear devices can be amplified many fold.

You can estimate the resonant harmonic by using the following formula:

$$h = \sqrt{\frac{kVA_{sys}}{kVAR}}$$

kVA_{sys} = short circuit capacity of the system
kVAR = amount of capacitor kVAR on the line
h = the harmonic number referred to a 60 Hz base

If h is near the values of the major harmonics generated by a non-linear device (i.e., three, five, seven, or eleven), then the resonance circuit will greatly increase harmonic distortion. For example, if the plant has a 1,500 kVA transformer, with a 5-1/2% impedance, and the short circuit rating of the utility is 48,000 kVA, then kVA_{sys} would equal 17,391 kVA.

If 350 kVAR of capacitors were used to improve power factor, h would be:

$$h = \sqrt{\frac{17{,}391}{350}} = \sqrt{49.7} = 7.0$$

Because h falls right on the 7th harmonic, these capacitors could create a harmful resonance condition if non-linear devices were present in the factory. In this case, the capacitors should be applied only as harmonic filtering assemblies.

Reducing Harmonic Distortion: It is possible for non-linear equipment and power capacitors to co-exist. You do not have to give up the benefits of power correction if your plant uses many types of electronic power controls and computers.

To reduce harmonics, avoid odd-number harmonics (e.g., 5, 7, 11, and 13), size reactors to filter site-specific harmonic problems, and use proper filtering and chokes.

Power quality is a major issue for today's facility managers and plant engineers, creating problems in distribution systems that could cause facilities to partly shut down. Regular site analysis will detect potential causes of harmonic distortion. One effective solution to managing distortion is to install dedicated circuits to solve neutral and grounding wire problems. Another solution utilizes a K-rated transformer with an oversized neutral that serves varying degrees of non-linear load without exceeding rated temperature-rise limits. A third method to reduce harmonic problems involves the use of harmonic filters which allow the low-frequency currents to pass but block or impede currents running at higher harmonic frequencies.

Figure 9.9 summarizes the performance features of various types of power conditioning equipment that can be used as a guide for power conditioning technology.

Wave Forms

Ammeter Readings

An ampere of direct current (DC) is defined as that value of current which will deposit 0.001118 grams of silver per second in a standard solution of nitrate of silver and water. Alternating current (AC) will deposit the silver in one half cycle and remove the silver in the other half cycle, and requires another method to define an ampere. An ampere of alternation current is the value that will produce the same heating effect as one ampere of direct current when flowing through the same ohmic resistance for the same period of time.

For a Normal Sine Wave:

$$\text{AMPERES}_{\text{(EFFECTIVE)}} = 0.707\ \text{AMPERE}_{\text{(MAX)}} = \text{AMPERES}_{\text{(RMS)}}$$

For a Distorted Wave:

$$\text{AMPERES}_{\text{(EFFECTIVE)}} = \text{AMPERES}_{\text{(RMS)}}$$

The relationship of .707 x MAXIMUM AMPERE is not true for a distorted sine wave form.

The effective current is determined by the root mean square (RMS) value of the current.

Summary of Performance Features of Various Types of Power Conditioning Equipment

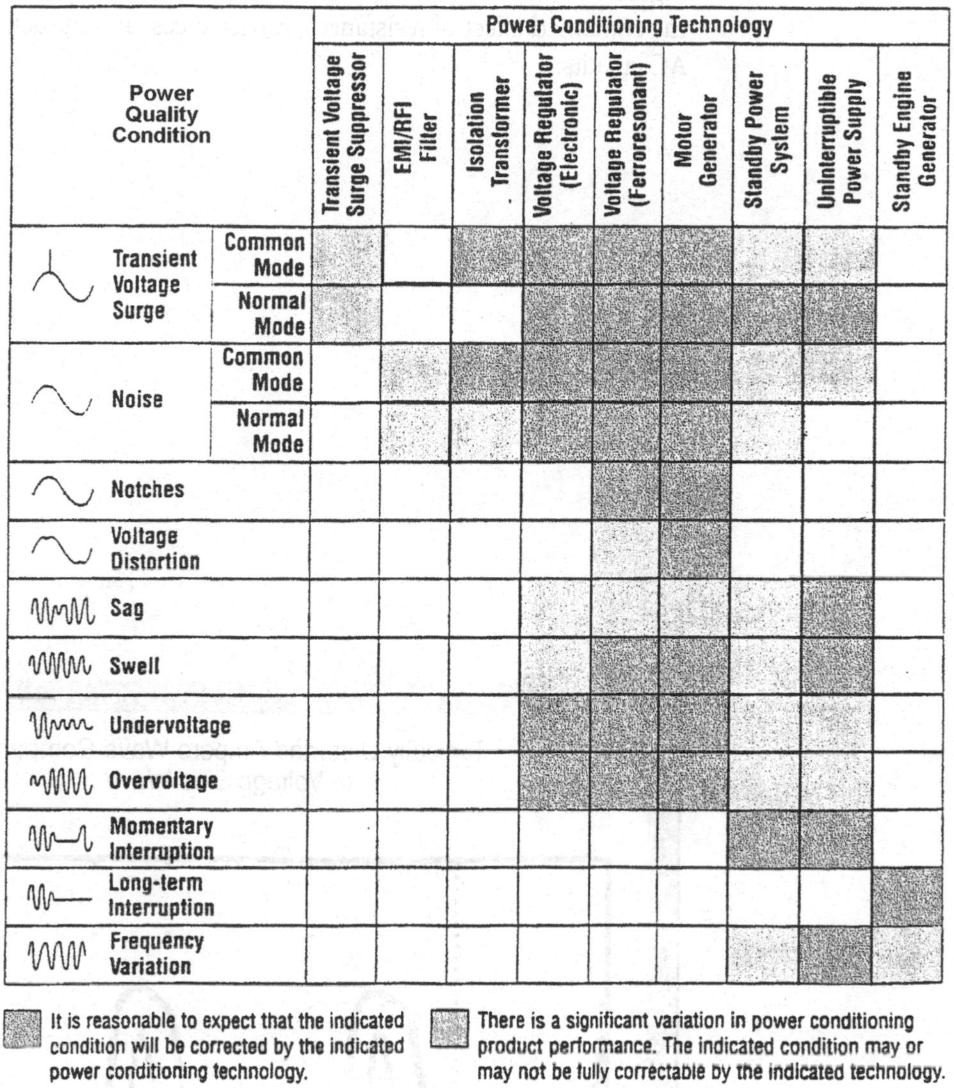

Power Quality Condition		Power Conditioning Technology								
		Transient Voltage Surge Suppressor	EMI/RFI Filter	Isolation Transformer	Voltage Regulator (Electronic)	Voltage Regulator (Ferroresonant)	Motor Generator	Standby Power System	Uninterruptible Power Supply	Standby Engine Generator
Transient Voltage Surge	Common Mode									
	Normal Mode									
Noise	Common Mode									
	Normal Mode									
Notches										
Voltage Distortion										
Sag										
Swell										
Undervoltage										
Overvoltage										
Momentary Interruption										
Long-term Interruption										
Frequency Variation										

It is reasonable to expect that the indicated condition will be corrected by the indicated power conditioning technology.

There is a significant variation in power conditioning product performance. The indicated condition may or may not be fully correctable by the indicated technology.

Courtesy of the Institute of Electrical and Electronics Engineers

Figure 9.9

Causes of a distorted sine wave may include a saturated transformer, poor equipment design, or the presence of harmonics.

Figures 9.10 through 9.16 illustrate various wave forms.

Figure 9.17 illustrates two vector diagrams, which may be used to determine the combined effect of resistances, capacitances, and inductances in AC circuits.

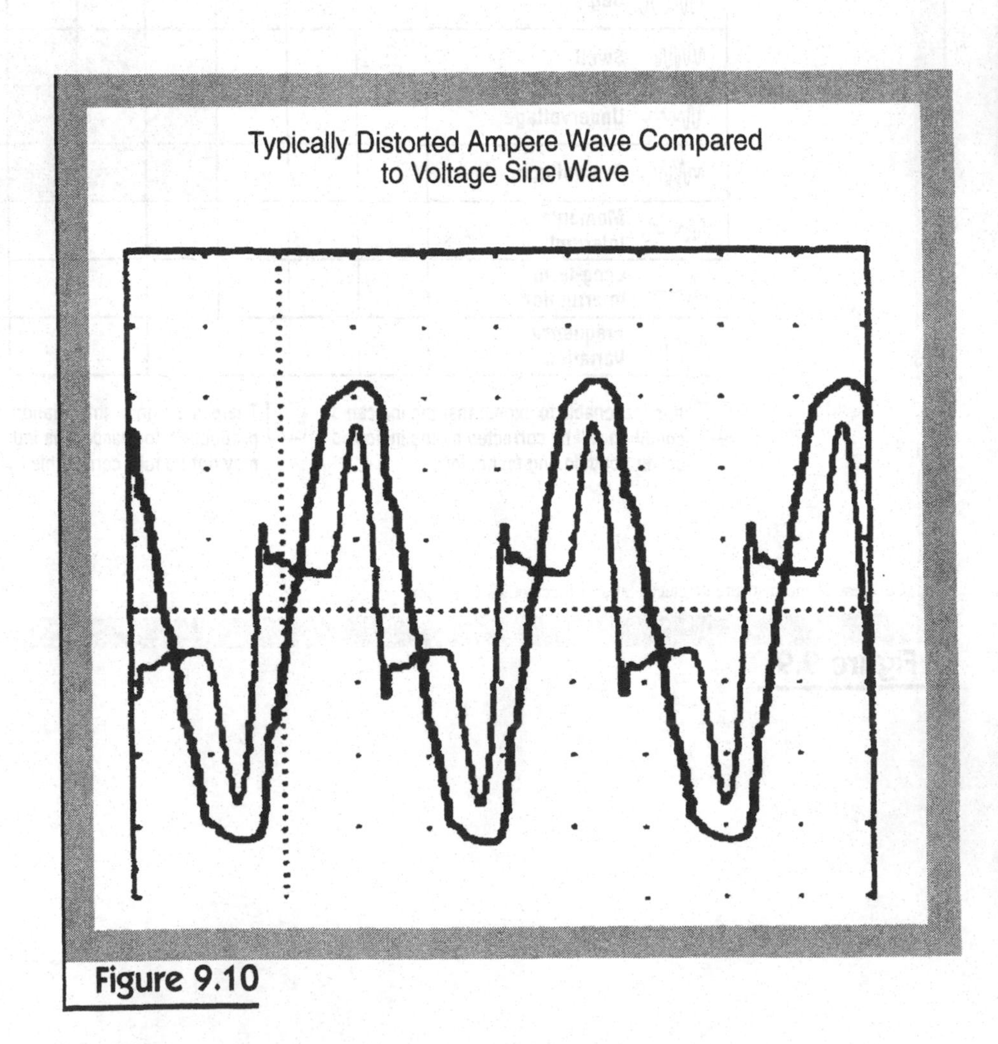

Typically Distorted Ampere Wave Compared
to Voltage Sine Wave

Figure 9.10

RMS Equivalent Heating for Various Wave Shapes

Waveform	Description	Average (RMS calibrated)	Peak (RMS calibrated)	True RMS
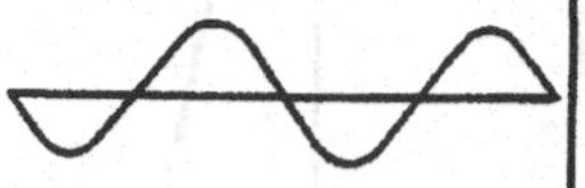	Sine wave	100.0 Amps	100.0 Amps	100.0 Amps
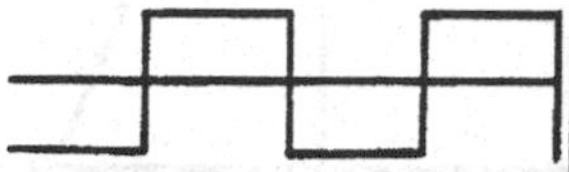	Square wave	100.0 Amps	63.6 Amps	90.0 Amps
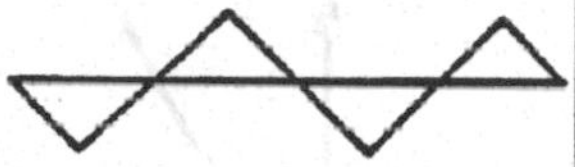	Triangle wave	100.0 Amps	125.7 Amps	103.3 Amps
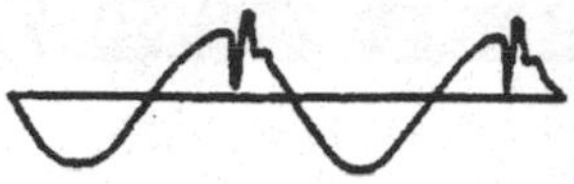	Sine wave with exponentially damped impulse	100.0 Amps	129.2 Amps	100.8 Amps
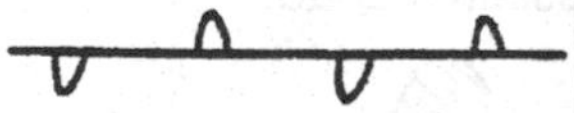	Single-phase electronic load current	100.0 Amps	400.1 Amps	199.4 Amps
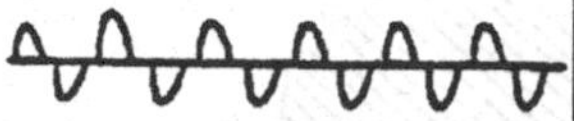	3-phase wye electronic load neutral current	100.0 Amp	156.5 Amps	115.4 Amps
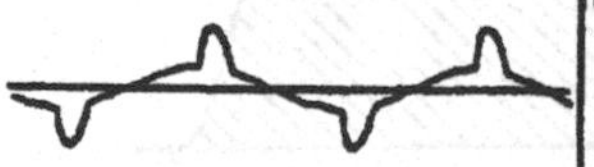	Single-phase electronic load plus 30° linear load	100.0 Amps	201.2 Amps	119.9 Amps

Courtesy of BMI Basic Measuring Instruments.

Figure 9.11

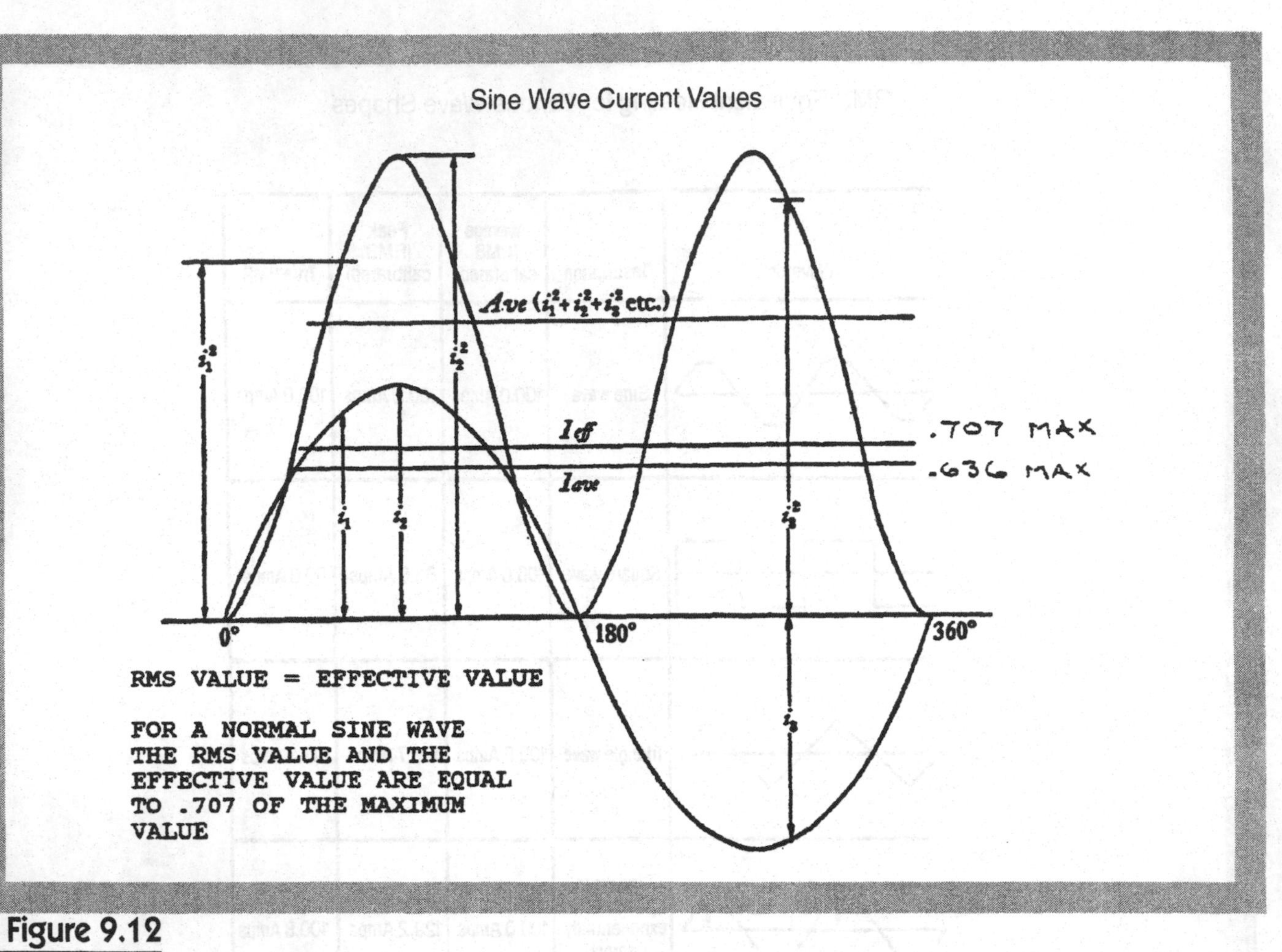

Figure 9.12

Power Wave Form—Resistive Load

P

E

I

Figure 9.13

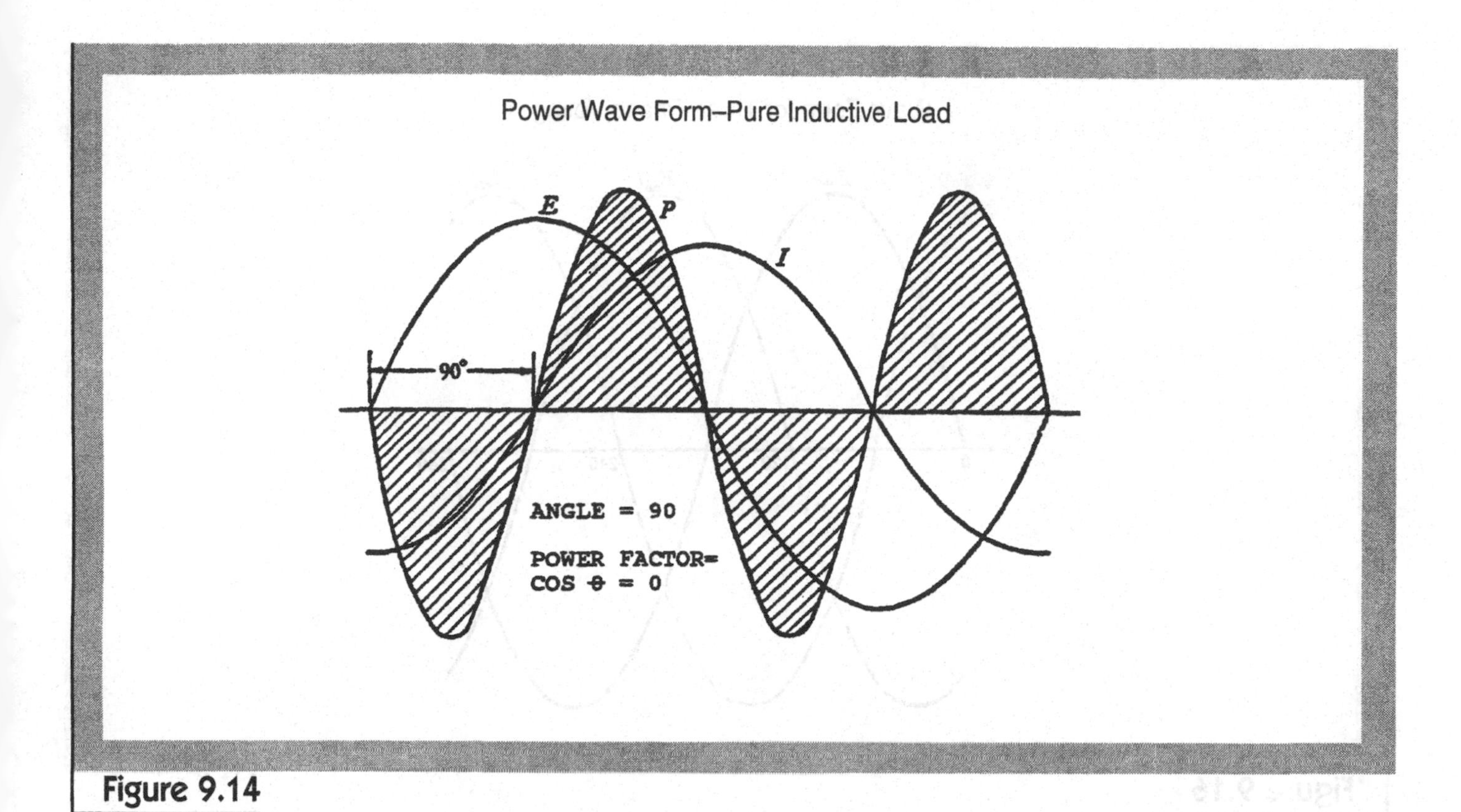

Figure 9.14

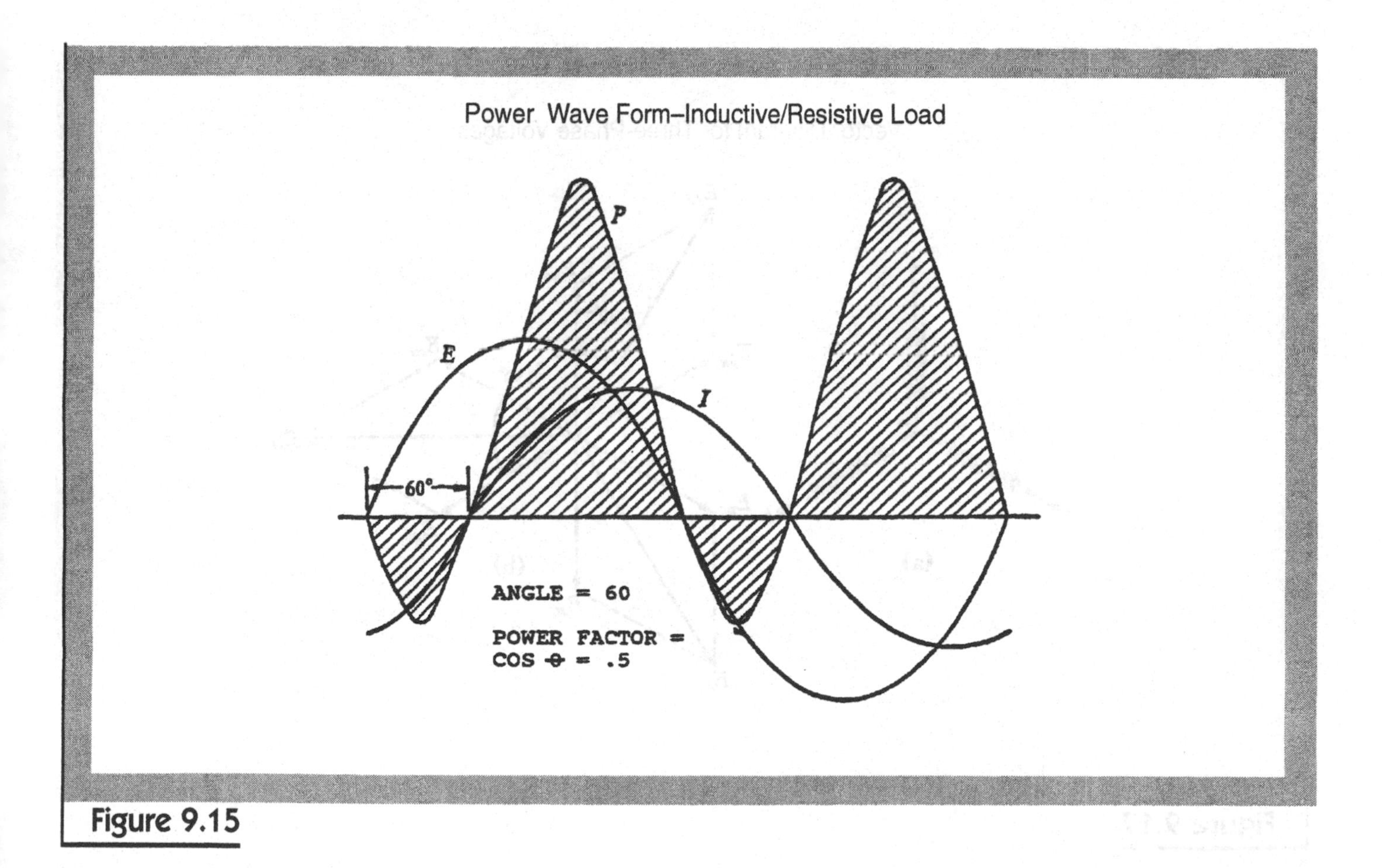

Figure 9.15

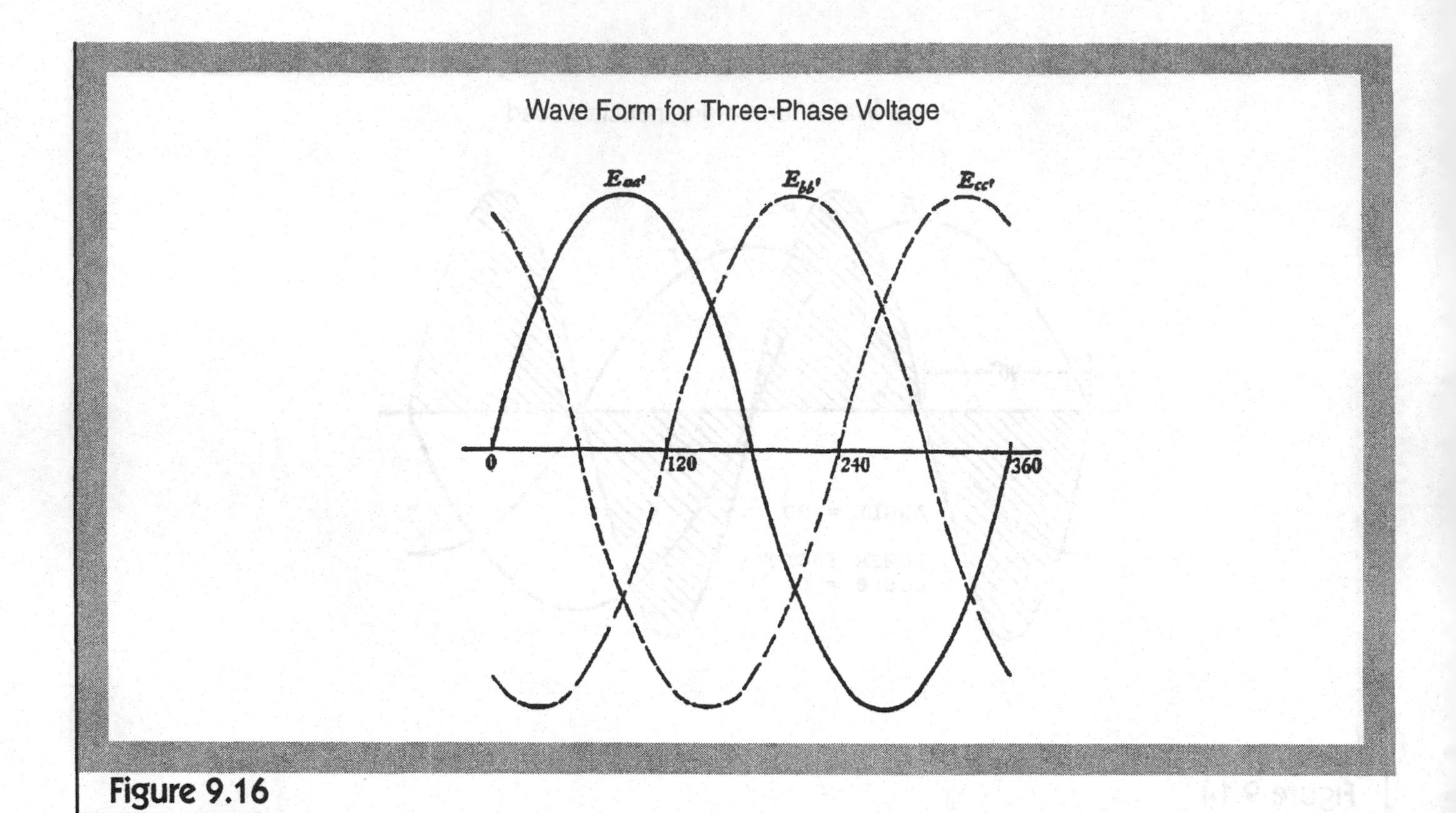

Figure 9.16

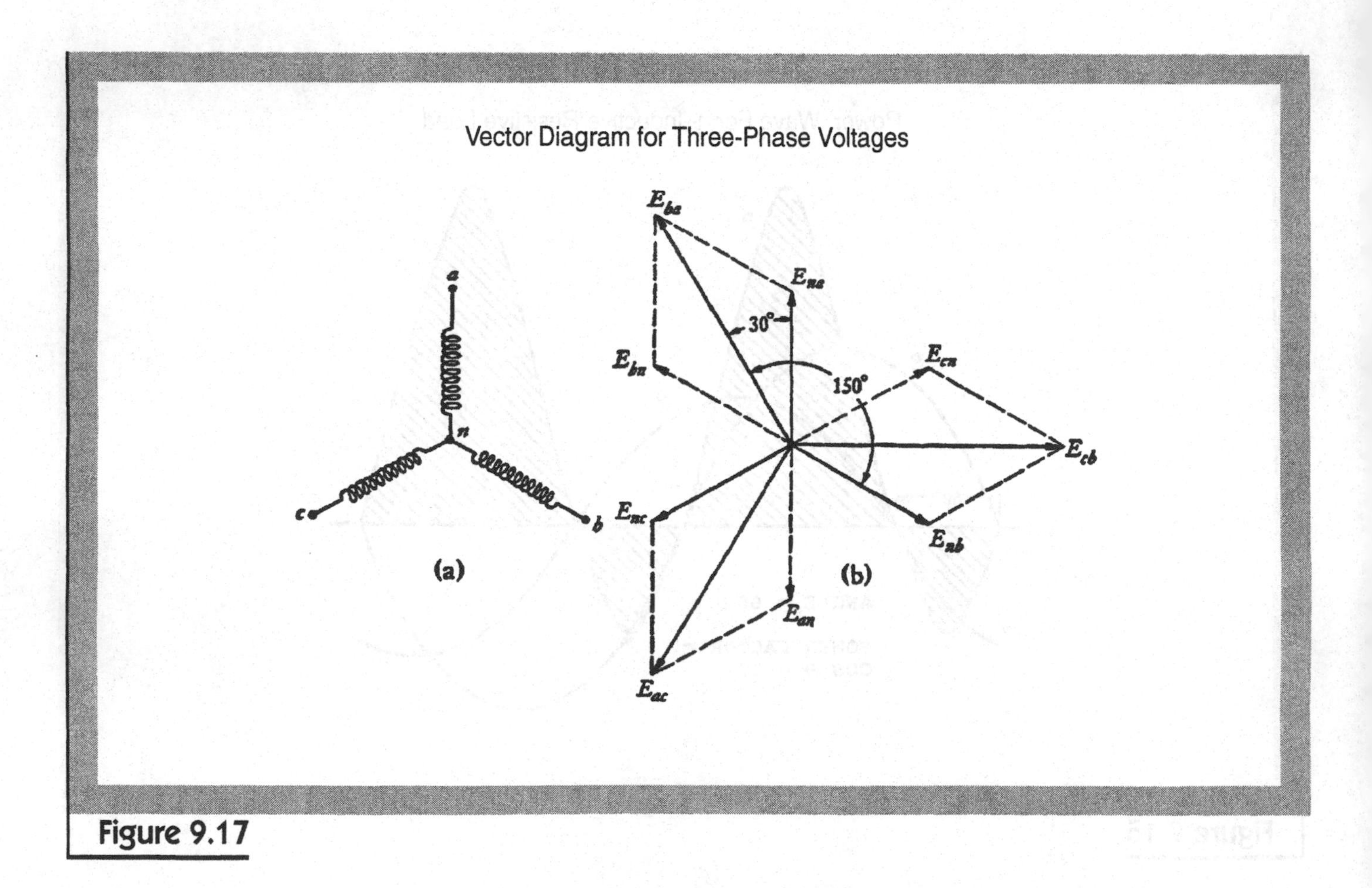

Figure 9.17

Wave forms describe the characteristics of the AC circuit that act exactly like the pulse of the human heart. An electrical engineer can use the shape of the curve and the continuity of the wave form to diagnose the quality and "health" of a power system, in the process making the necessary decisions required to maintain optimum performance of the system. As the demand for higher power quality grows, a plant manager or engineer can increase the power quality of a facility with the installation of transformers with oversized neutrals or harmonic filters.

Resistance, Reactance, and Impedance in Series AC Circuits

Capacitors have capacitance, which is measured in farads (f) or more commonly microfarads (µf). Just as inductance can be converted into inductive reactance, so capacitance can be converted to capacitive reactance. The formula $X_c = 1/(2\pi fC)$ converts capacitance (C) in farads to capacitive reactance (X_c) measured in ohms. Whereas current in a purely inductive circuit lags the applied voltage by 90 electrical degrees it can be proven that current in a purely capacitive circuit leads the applied voltage by 90 electrical degrees. Inductive reactance and capacitive reactance can be thought of as two vectors out-of-phase by 180 degrees, and each is out-of-phase with the circuit resistance by 90 degrees (See Figure 9.18).

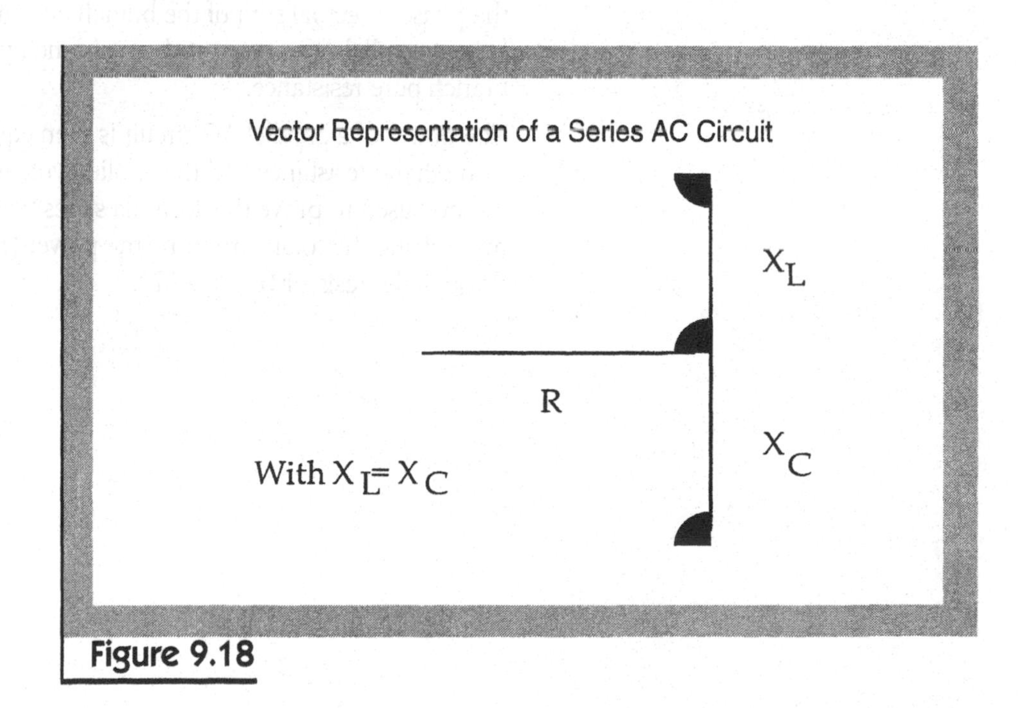

Figure 9.18

The total opposition to current in an AC circuit is called the impedance (Z) of the circuit. In a series AC circuit, the impedance is the vector sum of the resistance, the inductive reactance, and the capacitive reactance. Because the inductive reactance and the capacitive reactance are 180 degrees out-of-phase, their vector sum is the difference between the two vectors. When $X_L = X_C$, a series circuit is said to be in resonance. In a resonant circuit, it is still possible to get high voltage drops across parts of the circuit. Because of this, most power distribution circuits avoid resonance.

A series AC circuit with inductive reactance and resistance can be represented by the vectors in the first picture in Figure 9.19.

The second diagram in Figure 9.19 shows the impedance (Z) of the circuit, which is the vector sum of the resistance (R) and the inductive reactance (X_L). In a series AC circuit the current is equal throughout the circuit. Therefore, we can multiply R, X_L, and Z by I. This is shown in diagram 3 of Figure 9.19. Diagram 4 shows that, in a series AC circuit, it is the voltage drops across the resistance and the reactance of the circuit that are out-of-phase. The power factor of a series AC circuit, then, is the ratio of the resistance to the impedance (PF = R/Z), or the ratio of the voltage drop across the resistance to the voltage drop across the entire circuit.

Power in a single phase series AC circuit is then equal to the product of the voltage drop across the resistance and the current, $P = V_r \times I$. The most common method used to prove this formula states that Power = $V_t \times I \times PF$. By multiplying the total voltage factor, the answer equals the voltage drop across the resistor ($V_r = V_t \times PF$).

In a parallel AC circuit, the voltage drop is equal across all branches of the circuit. Depending on the capacitance, inductance, and resistance of each branch, the current flowing through each branch could be out-of-phase, with respect to each of the other branches. Thus, the total line current is equal to the phasor (vector) sum of the branch currents. Figure 9.20 depicts in phasor form a parallel AC circuit, with one branch pure inductance and the other branch pure resistance.

The power of a parallel AC circuit is then equal to the product of the current through the resistance and the applied voltage ($P = V \times I_r$). The most common method used to prove this formula states that Power = $V \times I_t \times PF$. By multiplying the total current by the power factor, the answer equals the current through the resistor ($I_r = I_t \times PF$).

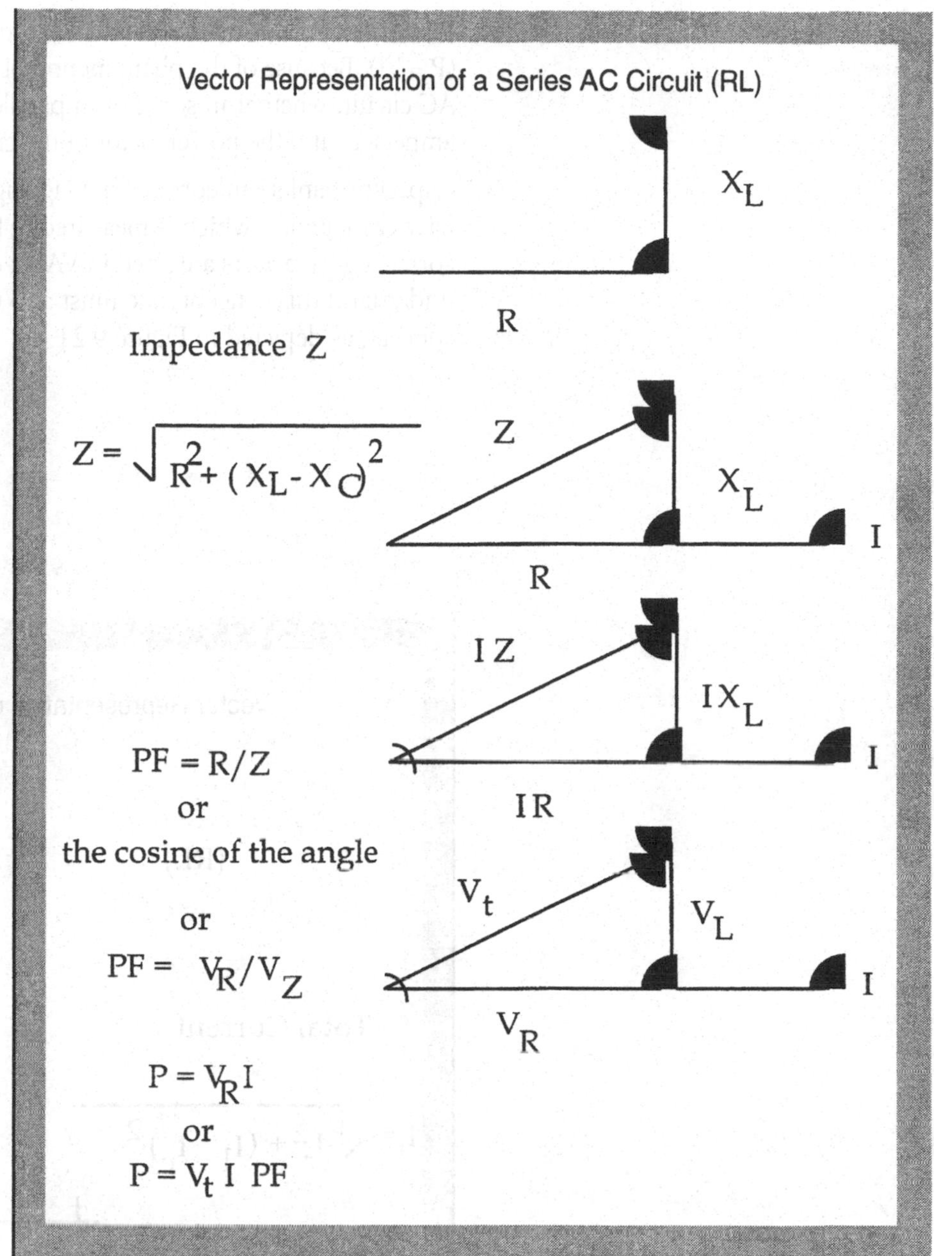
Vector Representation of a Series AC Circuit (RL)
X_L
R
Impedance Z
$Z = \sqrt{R^2 + (X_L - X_C)^2}$
Z
X_L
I
R
I Z
IX_L
I
PF = R/Z
or
the cosine of the angle
IR
V_t
V_L
or
$PF = V_R/V_Z$
I
V_R
$P = V_R I$
or
$P = V_t\ I\ PF$

Figure 9.19

Power in a DC circuit is equal to the product of the voltage and the amperage ($P = VI$). Because of the phenomenon of inductance, power in a single-phase AC circuit, whether in series or in parallel, is the product of the voltage, amperage, and the power factor of the circuit ($P = V_t \times I_t \times (PF)$).

Capacitor banks can correct for a lagging power factor. Although capacitors have capacitance, which is measured in farads, banks of power-factor correcting capacitors are sized in VARS, or more commonly kVARS. To understand this concept, one must also understand the power triangle concept, as depicted in Figure 9.21.

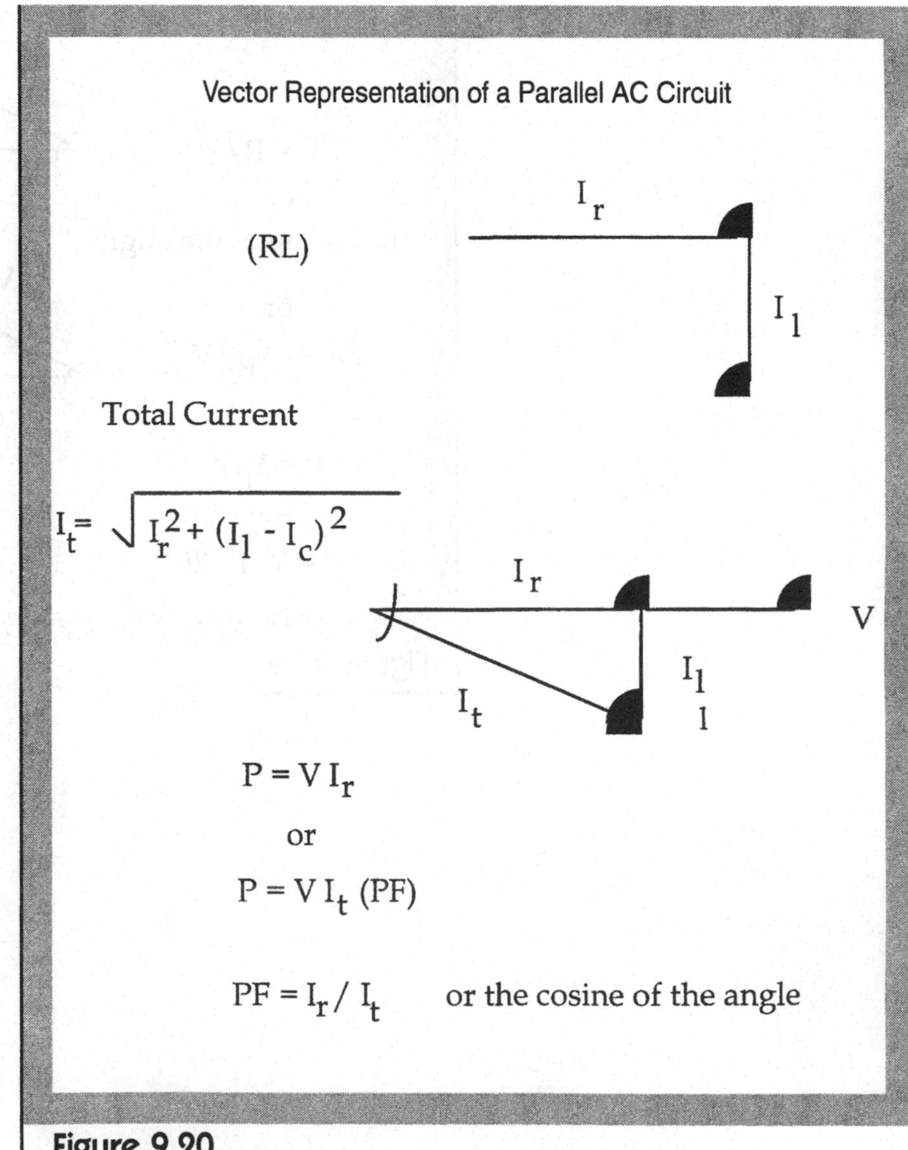

Figure 9.20

In a single-phase AC circuit, the apparent power is equal to the product of the voltage and the amperage, and is measured in Voltamps (VA = V x I). The true power (sometimes referred to as the "active power") is measured in watts and is the product of the voltage, the amperage, and the cosine of the angle (Watts = V x I x cos ø). The reactive power is measured in VARS, and is the product of the voltage, the amperage, and the sine of the angle (VARS = V x I x sine ø). The power factor is also equal to the ratio of the true power to the apparent power (PF = Watts/VA). By adding leading kVARS, (i.e., a capacitor bank), the plant manager can correct for poor power factors due to lagging kVARs.

Three-Phase Power

Most facilities operate on three-phase power because three-phase equipment is more efficient than single-phase equipment. When large facilities are built, it may be somewhat more expensive to install three-phase power, but the savings over time will most likely recoup the added initial costs.

Three-phase power is simply a combination of three single-phase circuits. Thus voltage, amperage, reactance, resistance, impedance and the corresponding true power, reactive power, and apparent power relations of three-phase circuits can be easily understood by viewing the three-phase circuit as three single-phase circuits.

Three-phase power is generated by placing three coils in an AC generator 120 degrees from each other. As the rotor's magnetic field moves past the coils, three alternating voltages are generated, differing in time-phase by 120 electrical degrees. This can be represented by three vectors 120 degrees from each other. Three-phase equipment can be connected in the Delta or in the Wye (star) connection.

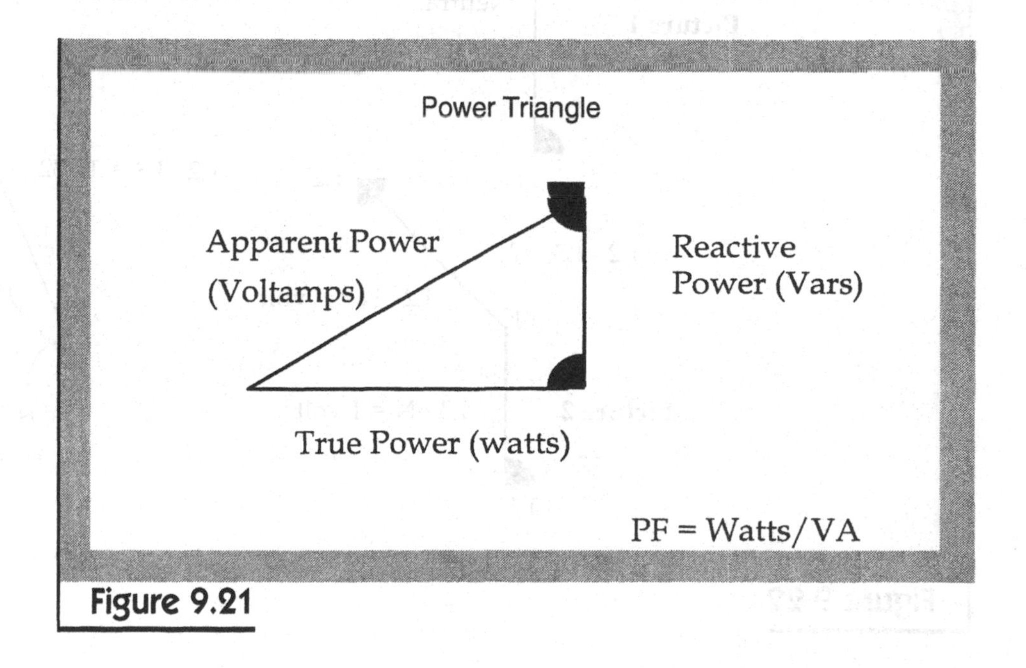

Figure 9.21

When three-phase AC generators are connected in Wye, there is a common connection called the neutral connection. The generator puts out two separate voltages, the phase voltage (V_p), and the line voltage (V_l). As all voltages are a measure of an electrical potential between two points, the phase voltage is equal to the electrical pressure between any one of the lines (L1, L2, or L3) and the neutral connection. The line voltage is equal to the electrical potential between any two of the lines (L1, L2, L3). The phase voltage and the line voltage in a Wye-connected machine differ by a factor of 1.732. To see how this number is derived, refer to Figure 9.22.

Picture 1 in Figure 9.22 depicts the three voltages as vectors 120 degrees apart with the neutral connection. Picture 2 of Figure 9.22 shows two-phase voltages (L2 and L3) of 1 volt each. Since voltage is the potential difference, solving for the line voltage requires subtracting the two vectors. To subtract two vectors rotate any one of them 180 degrees and then add them. In picture 3 of Figure 9.22, L3 was rotated 180 degrees and then added to L2. To add two vectors, put the tail of one to the tip of the other, and draw the resultant vector to solve it. The tail of L3 was put at the tip of L2, and the resultant vector was drawn. The potential difference between L2 and L3 was calculated by using the Pythagorean Theorem: $L2-L3 = (1.5^2 + .866^2)^{.5}$. $L2-L3 = 1.732$ volts.

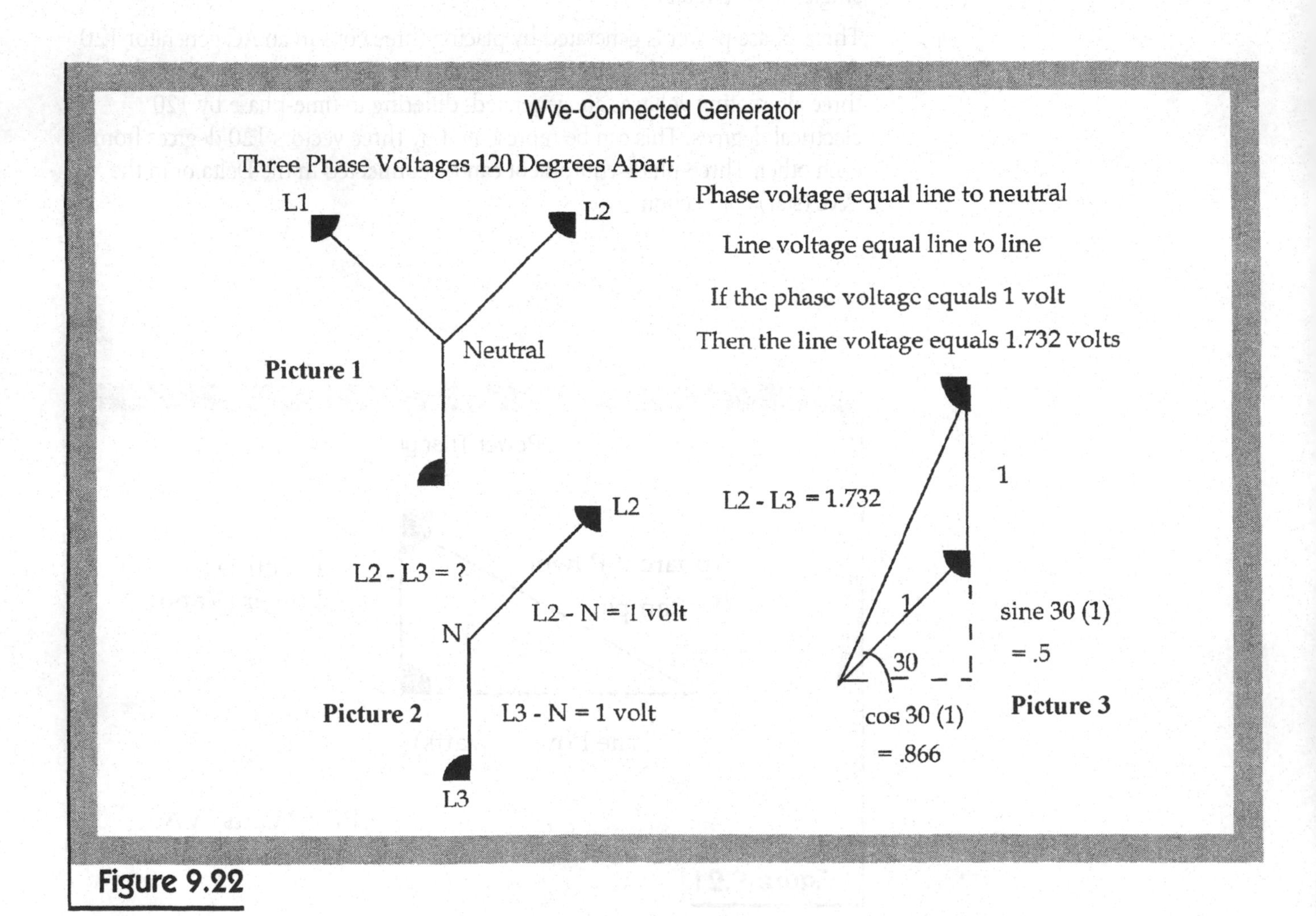

Figure 9.22

In a Wye-connected machine, then, $V_l = 1.732 \times V_p$, and the line current equals the phase current, $I_l = I_p$.

In a Delta-connected machine, it can be shown that the line and phase voltages are equal, $V_l = V_p$, and that the line and phase currents differ in time phase by 120 degrees. Therefore, $I_l = 1.732 \times I_p$.

It is no surprise to the plant manager that the power formulas for three-phase machinery are identical, whether they are connected in Delta or in Wye.

Summary Electrical Power Formulas

		1ø	3ø	DC
True Power	P =	VI cosø	$\sqrt{3}\ V_p I_p$ cosø $1.732\ V_L(I_L)$ cosø	VI
Reactive Power	Vars =	VI sinø	$\sqrt{3}\ V_p I_p$ sinø $1.732\ V_l(I_l)$ sinø	n.a.
Apparent Power	Voltamps =	VI	$\sqrt{3}$ VI 1.732 VI	n.a.

Note: $1.732 = \sqrt{3}$

Voltage Drop

There are several methods to calculate voltage drop. The following method is acceptable for an AC circuit motor load with the conductor size Number 1 or small run in conduit. The inductance is neglected in this calculation. For larger conductors, the solution method must include the power factor and conductor reactance.

Single Phase

$$\text{VOLTAGE DROP} = \frac{2K \times \text{CURRENT} \times \text{LENGTH}}{\text{CONDUCTOR CIRCULAR MIL}}$$

Three Phase

$$\text{VOLTAGE DROP} = \frac{1.732K \times \text{CURRENT} \times \text{LENGTH}}{\text{CONDUCTOR CIRCULAR MIL}}$$

The Constant K varies for conductor type and conductor temperature. The following table may be used for determining K in conductor sizing.

CONDUCTOR TEMPERATURE DEGREES	COPPER OHMS/cmil K	ALUMINUM OHMS/cmil K
20° C	10.6	
25° C	10.8	17
30° C	11.2	
40° C	11.6	
50° C	11.8	
60° C	12.3	
70° C	12.7	

Motors

Understanding the fundamentals of electric motors is vital to the plant manager. Although AC motors are far and away the most common in any plant, DC motors are still found where precision speed control is necessary.

DC Motors

DC motors are designated by how their field circuits are connected, with respect to their armature circuits. A shunt-wound DC motor has its field connected in parallel with its armature; a series-wound DC motor has its field circuit connected in series with its armature; and a compound-wound DC motor has two field circuits, one connected in parallel with the armature circuit and one connected in series with the armature circuit.

A DC shunt-wound motor has a good speed regulation, meaning that for slight load changes there is very little change in speed. A DC series-wound motor has an excellent starting torque. The starter motor in automobiles is a series-wound DC motor. A compound-wound DC motor has good speed regulation and a good starting torque. Interpoles are found on some DC motors. Their purpose is to neutralize the counter EMF of self-induction and thus reduce sparking between the commutator and the brushes. The commutator acts as a rotating rectifier, thereby alternating the current through the armature, thus alternating the polarity of the armature. This provides for the continuing torque of the DC motor.

To reverse direction of a DC motor, either reverse the polarity of the field leads or reverse the polarity of the armature leads. Reversing both the polarity of the field leads and the polarity of the armature leads keeps the rotation in the same direction.

With the advent of solid state devices which can vary the frequency of AC motors, the AC motor can now replace the DC motor, even where speed control is desired.

If, for example, a 440-volt DC motor's output is 30 kW at the pump, and the unit is 90% efficient, what would be the full load-line current in amperage? (Refer to Figure 9.23.) *Note: Most motors list their output in horsepower.*

The answer is P = VI. In order to solve this problem, one must realize that the motor is outputting 30 kW at the pump. Therefore, because the motor is 90% efficient, the kW input to the motor must be 30kW/.9, or 33.33 kVA. Solving for I = 33,333/440, I = 75.75 amps.

AC Motors

AC motors are either single-phase or three-phase. The majority of electric motors, either residential or commercial, are induction motors; a small number are synchronous motors. Induction motors develop a torque because a revolving magnetic field in the stationary portion of the machine induces currents to flow on the rotor bars. When currents flow on the rotor bars, they become magnetized. The interaction between the rotor and stator magnetic fields creates a torque. The rotor can not turn at the same speed as the revolving field in the stator. The difference in the speed of the stator field and the rotor is normally characterized as the percent slip of the machine. Most induction motors operate at between 3 and 5% slip.

Motor Formulas

Figure 9.24 shows common motor formulas.

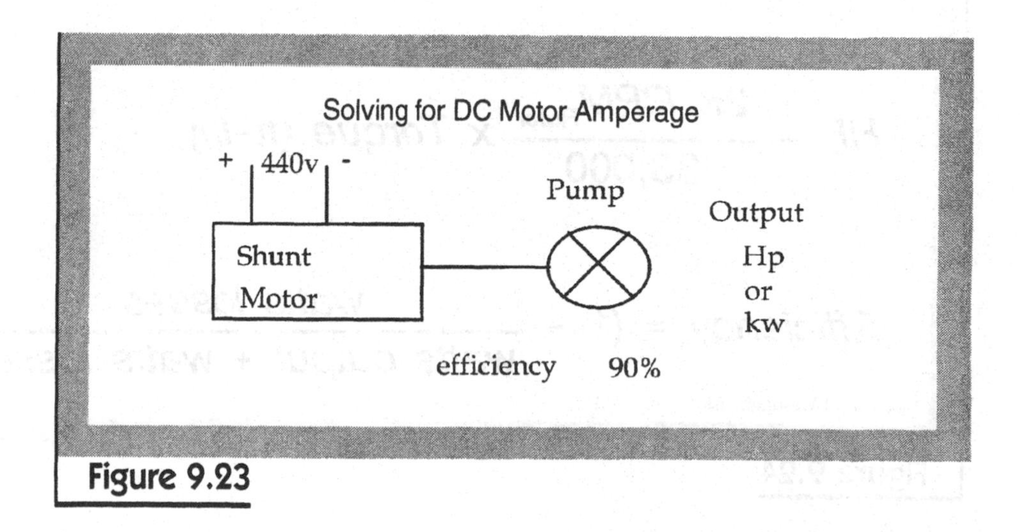

Figure 9.23

Motor Formulas

$$kW\ input = \frac{HP \times 746\ W/HP}{1000\ W/kW \times Efficiency}$$

$$kVA\ input = \frac{kW\ input}{Power\ Factor}$$

$$Amperes = \frac{HP \times 746\ W/HP}{V \times \sqrt{3} \times Efficiency \times Power\ Factor}$$

$$Torque = \frac{Horsepower \times 5250}{RPM}$$

$$Percent\ Slip = \frac{Synchronous\ RPM - Full\ Load\ RPM}{Synchronous\ RPM} \times 100$$

$$Synchronous\ Speed = \frac{Hertz \times 120}{Poles}$$

$$HP = \frac{2\pi\ RPM_{Rotor}}{33{,}000} \times Torque\ (ft\text{-}lb)$$

$$Efficiency = \left(1 - \frac{watts\ losses}{watts\ output + watts\ losses}\right) \times 100$$

Courtesy of Westinghouse

Figure 9.24

Effect of Voltage Variations on Induction Motors

The starting current is proportional to the voltage:

$$\frac{Voltage^1}{Voltage_2} = \frac{Current^1}{Current^2}$$

The starting torque varies with the square of the voltage:

$$\frac{Voltage^2{}_1}{Voltage^2{}_2} = \frac{Torque_1}{Torque_2}$$

Horsepower without overheating varies approximately with the square of the voltage:

$$\frac{Voltage^2{}_1}{Voltage^2{}_2} = \frac{Horsepower_1}{Horsepower_2}$$

Motor Torque:

$$Torque = \frac{Horsepower \times 5250}{RPM}$$

The starting current is proportional to the voltage:

$$Current_2 = \frac{Voltage_2}{Voltage_1} \times Current_1$$

Starting torque varies with the square of the voltage:

$$Torque_2 = \frac{Voltage^2{}_2}{Voltage^2{}_1} \times Torque_1$$

The following section, provided courtesy of Reliance Electric Company, offers information on how to select and apply motors.

Motor Selection & Application

Phase – Motors are designed for either single-phase or three-phase operation. Single-phase motors may be operated on one phase of a three-phase power supply with the rated voltage. Operation of a three-phase motor on a single-phase supply will result in motor damage.

Voltage: Single-Phase Motors – Standard voltages are 115/230.

Three-Phase Motors – Standard voltages are 190/380, 200, 200/400, 230/460, 460 and 575. Specific Definite Purpose motors may have other voltages indicated. Motors rated at 230 or 230/460 volts will operate on a 208 volt supply except that starting and peak running torque at 208 volts will be 20-25% less than at the 230 volt rating. Motors nameplated 208-230/460 will operate satisfactorily at 208 volts.

Voltage Variation – In accordance with National Electrical Manufacturers Association (NEMA) standards, motors will operate with ±10% variation in rated voltage with rated frequency. Performance within this voltage variation will not necessarily be in accordance with performance at rated voltage.

Dual Voltage Motors are easily reconnectable by referring to the information on the motor nameplate.

Frequency – Motors are designed for operation on 50 or 60 Hz. Many motors designed for operation on 60 Hz power may be operated on 50 Hz of a lower voltage rating with resulting decrease in speed and horsepower. In accordance with NEMA standards, motors will operate with ± 5% variation in rated frequency with rated voltage. Performance with this frequency variation will not necessarily be in accordance with performance at rated frequency. See Figure 9.25.

Proper Matching of the Motors to the Load

General information on the torque/speed characteristics of the various NEMA design classes is given in the following section. In addition, typical performance data on motors is given in the engineering data section. The equipment manufacturer and/or user are responsible for selecting the correct motor rating to drive the load.

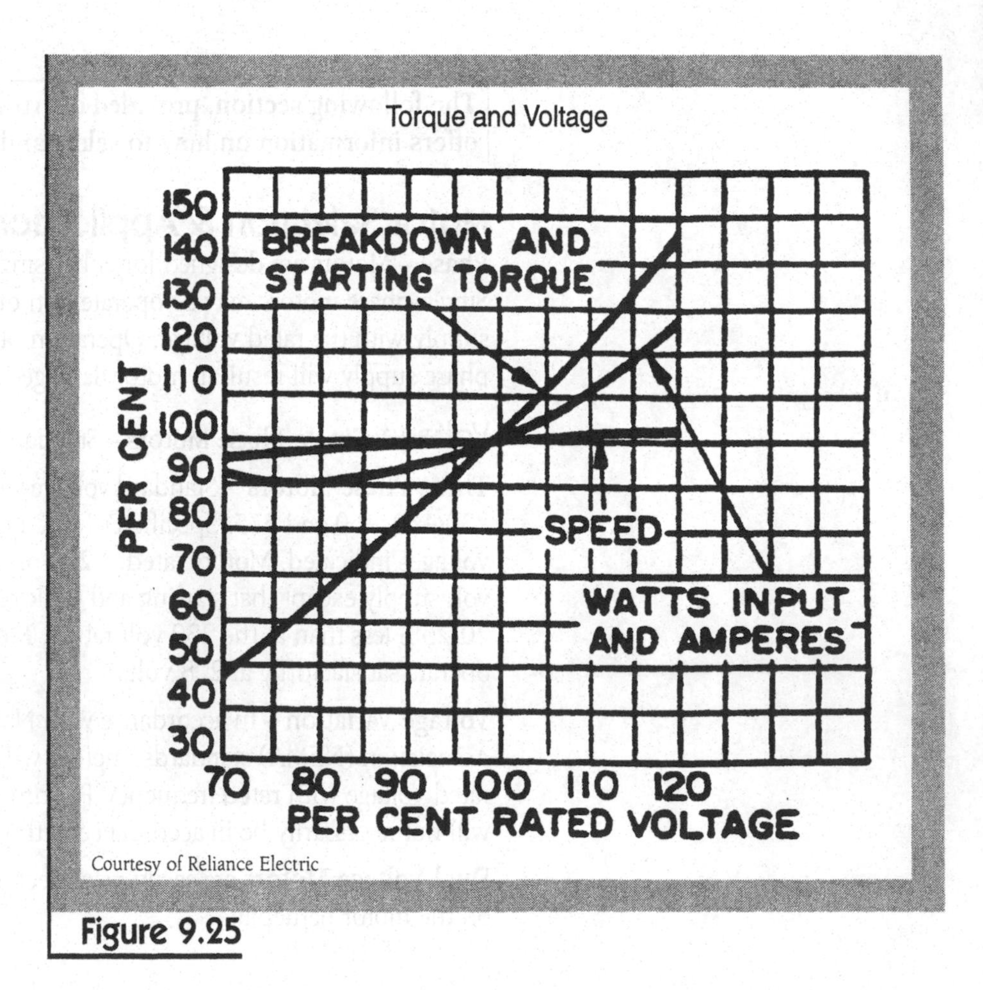

Courtesy of Reliance Electric

Figure 9.25

Torque/NEMA Design Classes: The full load torque of a motor is expressed as follows:

$$\text{Full load torque (ft. lbs.)} = \frac{\text{HP} \times 5250}{\text{Full load speed (rpm)}}$$

Since different loads present different torque requirements at starting (breakaway), minimum (pull-up), breakdown (pull-out) and full load, NEMA has defined four standard design classes A, B, C and D of squirrel cage polyphase induction motors.

The general torque/speed relationship with the four torque points defined is shown in Figure 9.26. Most motors are Design B with a few Design A, C & D. For motors having different characteristics than listed, contact the manufacturer.

Speed/Slip: The synchronous speed of an A-C motor is as follows:

$$\text{Synchronous Speed (rpm)} = \frac{120 \times \text{Power Supply Frequency (Hz)}}{\text{Number of Poles}}$$

The number of poles is a function of motor design. Standard 60 Hz motors are available with synchronous speeds of 3600, 1800, 1200 or 900 rpm.

The actual operating speed of an A-C induction motor is determined by the synchronous speed and the slip. Slip is the difference between the speed of the rotating magnetic field (which is always synchronous) and the rotor speed. Slip generally increases with an increase in motor torque; therefore, actual operating speed generally decreases with an increase in motor torque.

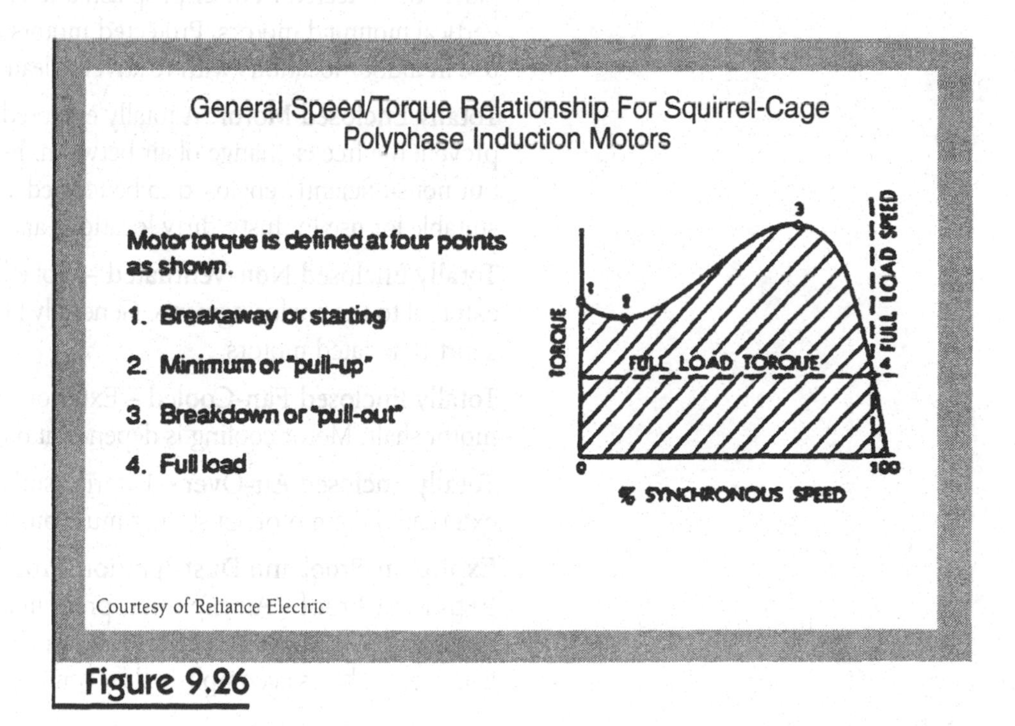

General Speed/Torque Relationship For Squirrel-Cage Polyphase Induction Motors
Motor torque is defined at four points as shown.
1. Breakaway or starting
2. Minimum or "pull-up"
3. Breakdown or "pull-out"
4. Full load
TORQUE
FULL LOAD TORQUE
FULL LOAD SPEED
0
100
% SYNCHRONOUS SPEED

Courtesy of Reliance Electric

Figure 9.26

Service Factor: The service factor of a motor is a multiplier which, when applied to the rated horsepower, indicates a permissible horsepower loading which may be carried under the conditions specified for the service factor.

Safety: The use of electric motors, like that of all other utilization of concentrated power, is potentially hazardous. The degree of hazard can be greatly reduced by proper design, selection, installation and use, but hazards cannot be completely eliminated. Some of the factors that must be considered for the safe application of motors are as follows:

- Enclosure
- Service conditions
- Proper matching of the motor to the load

Information on these factors is presented in this section.

Enclosure:

The selection of the proper enclosure is vital to the successful safe operation of A-C motors. Machine performance and life can be materially reduced by using an enclosure inappropriate for the application. The customer must recognize the specific environmental conditions and specify the correct enclosure. Manufacturers can provide application assistance, but must depend on the customer to provide accurate information on the operating conditions.

Most manufacturers offer motors in the following enclosures:

- Protected
- Totally enclosed (Non-Ventilated, Fan Cooled, Air Over)
- Explosion-Proof
- Dust-Ignition-Proof

Protected Motor: A protected motor is an open machine having all openings limited in size to prevent accidental contact with hazardous parts. Internal parts are protected from drip, splash and falling objects on all horizontal and vertical mounted motors. Protected motors are suitable for general industrial use in indoor locations with relatively clean atmospheres.

Totally Enclosed Motor: A totally enclosed motor is one so enclosed as to prevent the free exchange of air between the inside and the outside of the case but not sufficiently enclosed to be termed airtight. Totally enclosed motors are suitable for use in dusty, dirty locations and in humid environments.

Totally Enclosed Non-Ventilated – Not equipped for cooling by means external to the enclosing parts. Generally limited to low horsepower ratings or short-time rated motors.

Totally Enclosed Fan-Cooled – Exterior surface cooled by external fan on motor shaft. Motor cooling is dependent on motor speed.

Totally Enclosed Air-Over – Exterior surface cooled by a ventilating means external to the motor. Customer must provide sufficient airflow to cool motor.

Explosion-Proof and Dust-Ignition-Proof Motors for Hazardous Locations:
Explosion-Proof – An explosion-proof motor is a totally enclosed machine designed to operate safely in an environment in which flammable gas or vapor is present. This is accomplished by limiting the external surface temperature of

the motor below the ignition temperature of the hazardous gas or vapor surrounding the motor. Hence it is very important that the end user order a motor with the proper temperature code. Also the enclosure is designed to contain an explosion on the inside of the motor which may be caused by arcing, sparking, or abnormal heat. A flame caused by such an explosion will not propagate the motor fits and ignite the gas or vapor surrounding the motor. An explosion-proof motor is suitable for operation in a Class I hazardous environment.

Dust-Ignition-Proof – A dust-ignition-proof motor is a totally enclosed machine designed to operate safely in a flammable dust environment. The motor enclosure prevents the admittance of dust inside the motor. Also, the motor is designed to operate cool enough so as not to ignite any dust, which might accumulate on the external surface of the motor. Should the motor become blanketed with dust and reduce its normal cooling, winding thermostats will take the motor off line before the motor's external surface temperature reaches the ignition temperature of the hazardous dust. A dust-ignition-proof motor is suitable for operation in a Class II hazardous location.

Service Conditions: Standard motors are designed to operate under the following service conditions. Service conditions exceeding these conditions may require special motor selection or re-rating of standard motors.

Altitude – Sea Level to 3300 feet.

Ambient Temperature – 0 °C to 40 °C.

Mounting – Standard motors must be mounted on a rigid mounting surface for satisfactory operation.

Installation – Motor must be in areas or in supplementary enclosures which do not seriously interfere with the ventilation of the motor.

Figure 9.27 provides general information on motor and horsepower rating.

Figure 9.28 features schematics for several types of terminal markings and connections.

General Engineering Data

Frame Designation: Complete frame designation consists of prefix letter(s), followed by the basic frame number and suffix letters(s). In accordance with NEMA standards, the designation for small frames (48, 56) is different from the nomenclature for medium frames (140–440).

Motor Characteristics

PERCENT LOAD VS POWER FACTOR AND EFFICIENCY

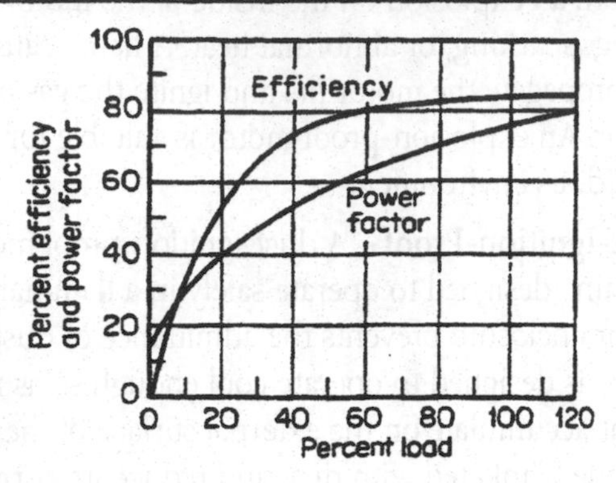

HORSEPOWER VS POWER FACTOR

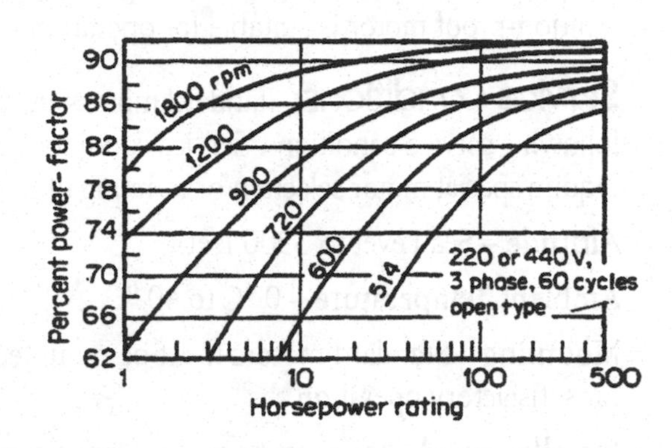

HORSEPOWER VS EFFICIENCY

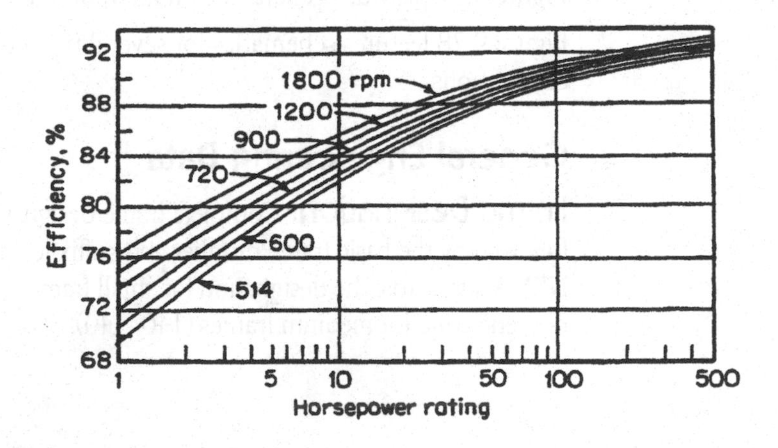

Figure 9.27

Terminal Markings and Connections

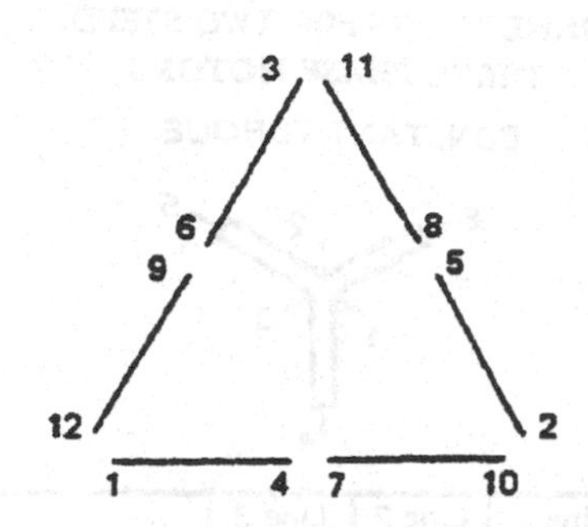

Voltage	Condition	L_1	L_2	L_3	Together
High	Start	1	2	3	4 & 7, 5 & 8, 6 & 9, 10 & 11 & 12
High	Run	1 & 12	2 & 10	3 & 11	4 & 7, 5 & 8, 6 & 9
Low	Start	1 & 7	2 & 8	3 & 9	4 & 5 & 6, 10 & 11 & 12
Low	Run	1 & 6 & 7 & 12	2 & 4 & 8 & 10	3 & 5 & 9 & 11	None

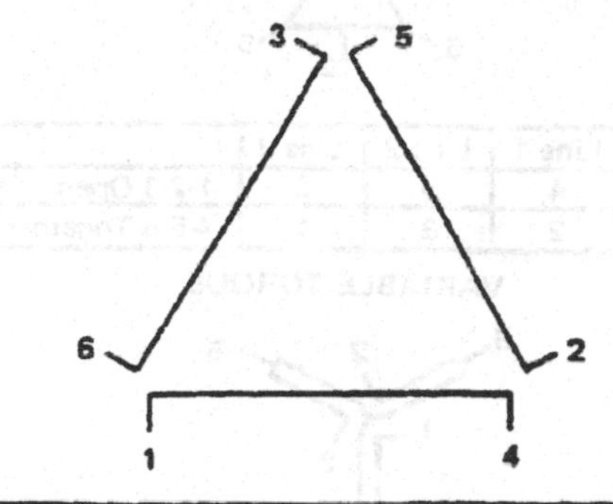

Connection	L_1	L_2	L_3	Together
Run	1, 6	2, 4	3, 5	None
Start	1	2	3	4, 5, 6

CONNECTIONS FOR STAR-START, DELTA-RUN THREE-PHASE MOTORS

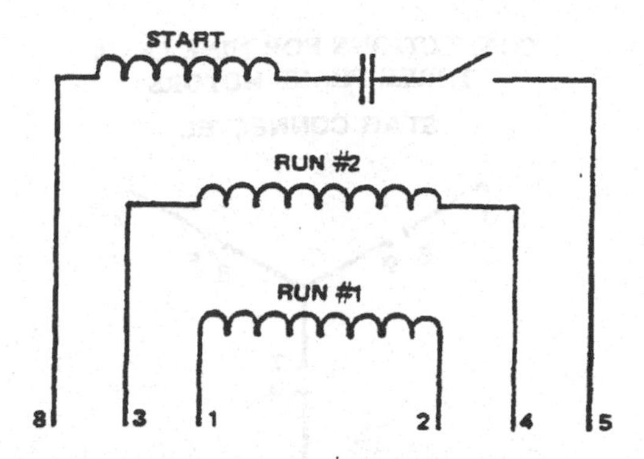

Voltage	Line 1	Line 2	Together
High	1	4, 5	2, 3, 8
Low	1, 3, 8	2, 4, 5	

Counterclockwise rotation facing opposite drive end. For clockwise rotation interchange leads 5 and 8

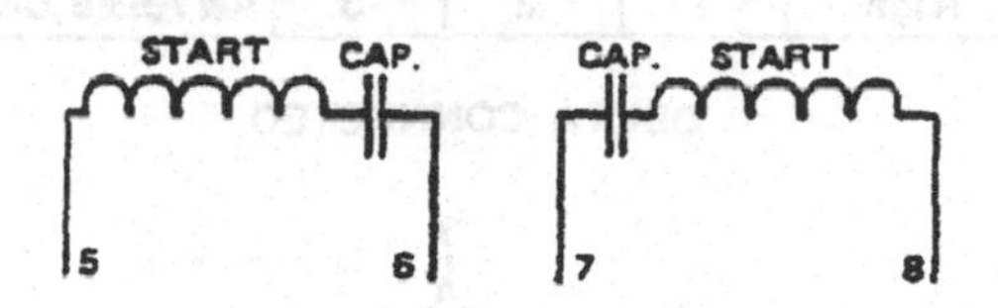

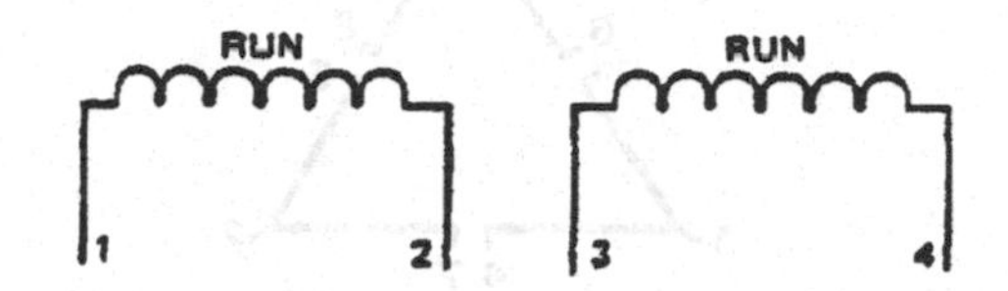

Voltage	L_1	L_2	Together
High	1, 8	4, 5	2 & 3, 6 & 7
Low	1, 3, 6, 8	2, 4, 5, 7	None

Counterclockwise rotation facing opposite drive end. For clockwise rotation interchange leads 5 and 8, 6 and 7.

CONNECTIONS FOR SINGLE PHASE MOTORS

Figure 9.28

Terminal Markings and Connections

CONNECTIONS FOR NINE LEAD, THREE PHASE MOTORS

STAR CONNECTED

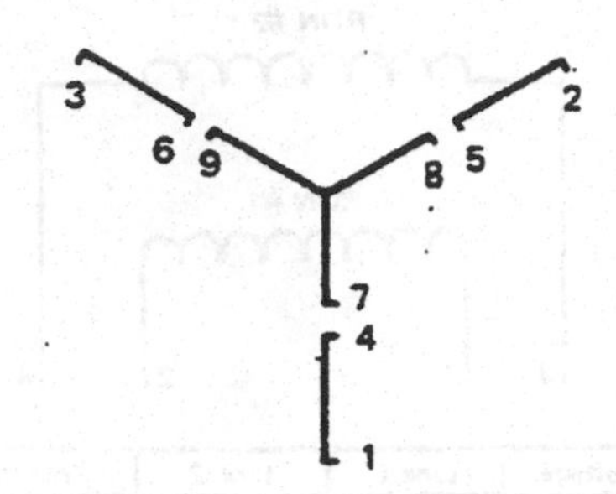

Voltage	Line 1	Line 2	Line 3	Together
Low	1 & 7	2 & 8	3 & 9	4&5&6
High	1	2	3	4&7,5&8,6&9

DELTA CONNECTED

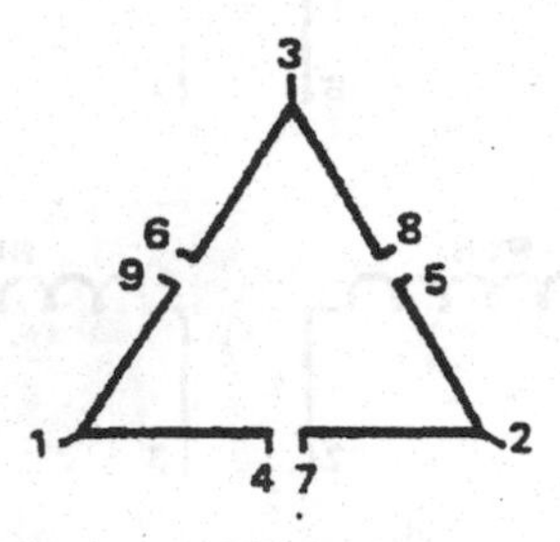

Voltage	Line 1	Line 2	Line 3	Together
Low	1&6&7	2&4&8	3&5&9	None
High	1	2	3	4&7,5&8,6&9

CONNECTIONS FOR TWO SPEED, THREE PHASE MOTORS

CONSTANT TORQUE

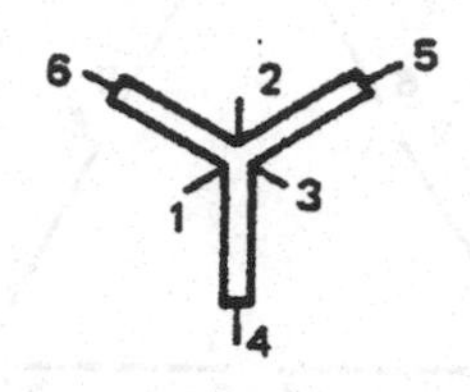

Speed	Line 1	Line 2	Line 3	
High	4	5	6	1-2-3 Together
Low	2	3	1	4-5-6 Open

CONSTANT HORSEPOWER

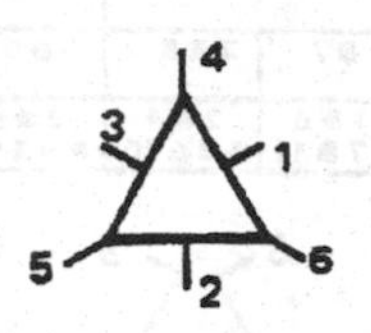

Speed	Line 1	Line 2	Line 3	
High	4	5	6	1-2-3 Open
Low	2	3	1	4-5-6 Together

VARIABLE TORQUE

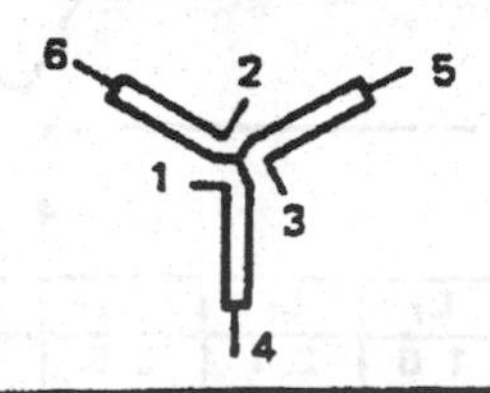

Speed	Line 1	Line 2	Line 3	
High	4	5	6	1-2-3 Together
Low	2	3	1	4-5-6 Open

Figure 9.28 cont.

48 and 56 Frames:

Prefix letters are at the manufacturer's option. Suffix letters are defined by NEMA. Suffix letters for common frames are as follows:

C – NEMA C-Face mounting on drive end

H – NEMA-defined H-base foot mounting

T – NEMA-defined T-frame dimensions

Y – Special mounting dimensions. Recently, 48Y designation has referred to a 48 frame and shaft height motor with a 56 C-face and shaft extension. This terminology has not always been so defined and this definition should not be assumed.

Z – Special shaft extension. Stock 56Z and 56CZ frames designate 56 frame motors with standard 140T frame shaft extensions (7/8" x 2-1/4" for 56Z, 7/8" x 2-1/8" for 56 CZ).

140T–440T Frames:

Prefix letters are at the manufacturer's option. The NEMA standard frame diameters are as follows:

143T, 145T, 182T, 184T, 213T, 215T, 254T, 256T, 284T, 286T, 324T, 326T, 364T, 365T, 404T, 405T, 444T, 445T

In all cases, the first two digits of these frames fix the dimensions D, E, BA, H, U, and N (minus W). For horizontal motors with feet, the first two digits combine to make a number, which is equal to four times the dimension D (height from feet to the center line of the shaft). For example, the "D" dimension of a 286 frame is $28 \div 4 = 7$ inches. The third or remaining digit identifies the value of dimension F.

Suffix letters as defined by NEMA are used in frame designations by most manufacturers.

C – NEMA C-Face mounting on drive end.

LP and LPH – Type P flange-mounting vertical solid-shaft motors having dimensions in accordance with MG 1-18.620. (The letters LP and LPH are to be considered as one suffix and shall not be separated.)

S – Standard short shaft for direct connections (coupled). See dimension tables.

T – Frame designation having standard dimensions.

U – Indicates a frame having standard dimensions established prior to the T-frame.

Y – Special mounting dimensions.

Z – Special shaft extension.

Typical dimensions for estimating purposes are given in the dimension tables included in this section.

Mounting Positions: TFC 48, 56 and 140T frames 3-phase and type CS single-phase have a saddle-mounted conduit box with conduit entrance on either F1 or F2 locations.

Protected TENV 48, 56 and 140T frames and all type CH single-phase motors have F-1 conduit entry.

Frames 180T–440T have F-1 conduit box mounting as standard except for Definite Purpose Oil Well Pumper Motors which have F-2 conduit box mounting as noted. Motors with F-1 conduit box mounting may be converted to F-2 mounting at a manufacturer's modification center.

Evaluating Motor Efficiency

Figures 9.29 and 9.30 show how the motor efficiency and watts loss curves of a high performance motor compare to standard industrial averages.

Ed. Note: These figures, and the text that follows, were developed by Reliance Electric to illustrate the efficiency and cost benefits of its Duty Master XE Motor.

Annual Savings: Actual energy savings depends upon many variables including operating hours, cost of power, duty cycles, and line voltages, to name a few. Tax rates and the future cost of money will also influence the decision. However, a very simple energy savings calculation can be made to estimate the yearly energy savings with a more efficient motor.

Simple Energy Savings Calculation:

This simple payback analysis will assume constant full load operation. Note that this calculation ignores the effects of taxes and the time value of money.

Information you need to know.

HP	=	Motor full load horsepower
Hrs/Yr	=	Number of hours per year the motor is in operation
$/KWH	=	Cost of electricity in dollars per kilowatt hour
HP_{EFF}	=	High performance motor efficiency
STD_{EFF}	=	Standard motor efficiency

If actual motor efficiency is unknown, a manufacturing engineer can help you estimate standard motor efficiencies.

First, calculate the kilowatts saved by using the high performance motor:

$$\text{Kilowatts Saved} = HP \times .746\ [\frac{1}{STD_{EFF}} - \frac{1}{HP_{EFF}}]$$

Then, calculate the yearly energy savings:

Yearly Savings = Kilowatts Saved x Hrs/Yr x $/KWH

Example: Assume that a 25 horsepower motor is going to operate for three 8-hour shifts per day, on a 6-day work week, for 50 weeks per year. Thus, total hours per year are 7200. Also, assume the cost of electricity at $.08 per kilowatt hour. The savings, using a high performance motor with an efficiency of 93%, compared to the 88% which is the industry average for a standard efficient motor, is calculated as follows:

$$\text{Kilowatts Saved} = HP \times .746 \left[\frac{1}{STD_{EFF}} - \frac{1}{HP_{EFF}} \right]$$

$$= 25 \times .746 \left[\frac{1}{.880} - \frac{1}{.930} \right]$$

$$= 1.139 \text{ Kilowatts Saved}$$

$$\text{Yearly Savings} = \text{Kilowatts Saved} \times \text{Hrs/Yr} \times \$/\text{KWH}$$

$$= 1.139 \times 7200 \times .08$$

$$= \$656.36$$

In this case, the yearly savings is $656.36. In other words, if you wanted a one-year payback on your investment, you could afford to spend $656 more for a high performance, energy efficient motor than for a standard efficient motor.

Duty Cycles: Many people question the energy savings of partial load or variable duty cycle applications. The energy saving equations are the same. However, when evaluating a partial load operation, remember to use the horsepower, efficiency, and appropriate hours of operation at the desired load point. A variable duty cycle calculation is simply the summation of several partial load calculations.

Additional Savings: This payback analysis also does not take into account many additional savings you should consider.

- Longer motor life as a result of cooler operating temperatures
- Additional cost savings from improved power factor
- Improved partial load efficiencies

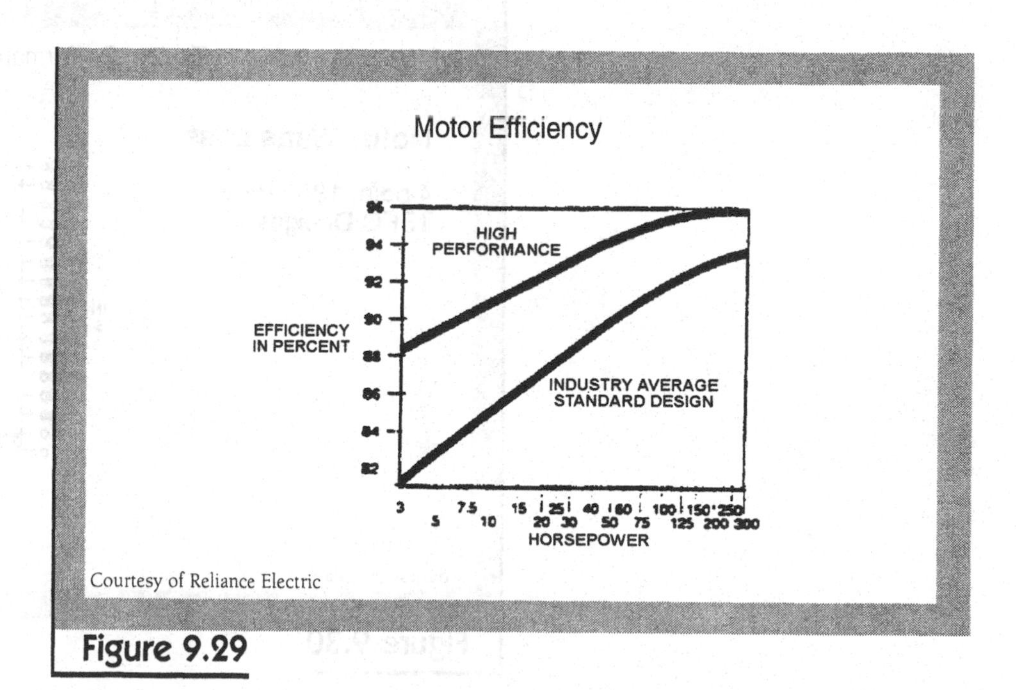

Figure 9.29

- Improved reliability with a more forgiving electrical design
- Quieter operation
- Greater flexibility for use on adjustable frequency power supplies

See Figure 9.31 for detailed motor performance data.

The following section is reprinted with permission from *Cost Planning & Estimating for Facilities Maintenance*, R.S. Means Co., Inc.

Lighting

Lighting systems use the properties of light reflectance and absorption to obtain the most lighting at the lowest energy cost and, at the same time, deliver a specific level of illumination adequate for those working in the area. This illumination should be maintained at recommended levels to conserve eyesight, improve morale, increase safety, improve housekeeping, decrease fatigue, reduce headaches, and increase production, all of which are directly reflected in lower operating costs. In order to achieve these benefits, it will be necessary to consider lamp types, lighting quality and lamp maintenance.

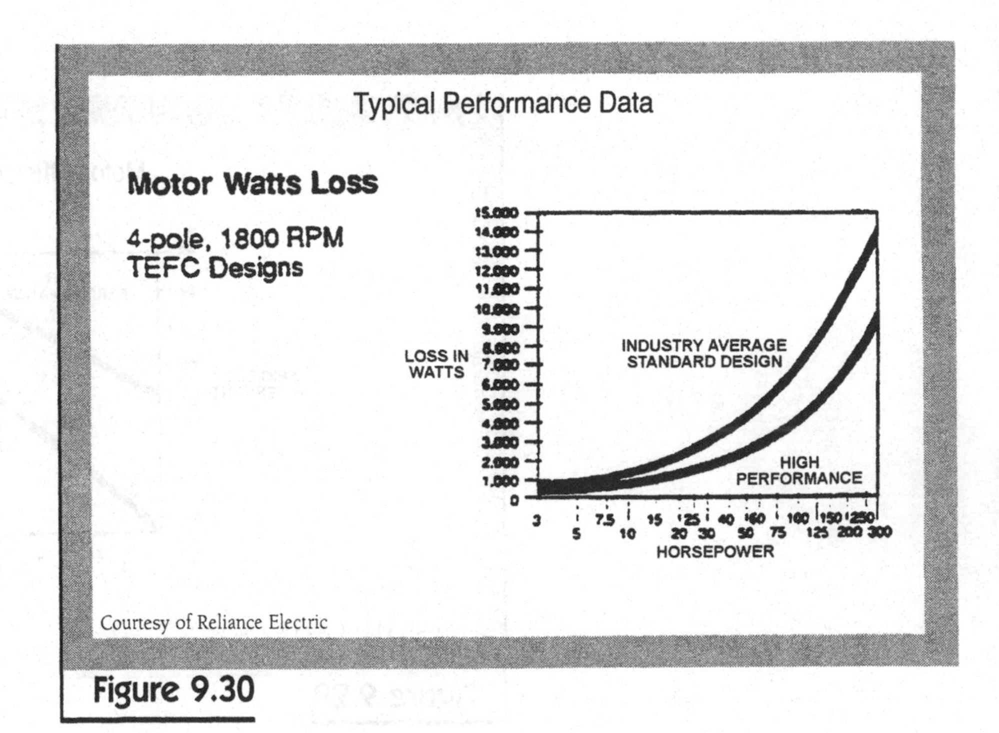

Courtesy of Reliance Electric

Figure 9.30

Typical Performance Data

STANDARD EFFICIENT PROTECTED

HP	FRAME SIZE	FULL LOAD RPM	AMPS @ 460V		TORQUE (FT-LB)			FULL LOAD			NEMA CODE
			FULL LOAD	LOCKED ROTOR	FULL LOAD	LOCKED ROTOR	BREAK DOWN	NOM EFF	MIN EFF	POWER FACTOR	
1	182T	865	2.0	8.4	6.07	9.40	14.8	70.0	66.0	67.9	H
1.5	182T	1155	2.5	14.5	6.82	14.4	18.5	74.0	70.0	75.2	J
	184T	855	2.8	12.4	9.21	15.7	22.1	72.0	68.0	69.5	H
2	184T	1160	3.2	20.8	9.06	22.2	26.4	74.0	70.0	77.1	K
	213T	860	3.8	16.4	12.2	21.2	27.7	70.0	66.0	71.0	H
3	182T	1730	4.6	27.1	9.11	23.2	28.7	78.5	75.5	77.7	J
	213T	1170	5.0	29.1	13.5	26.1	38.5	78.5	75.5	70.9	J
	215T	865	5.4	24.1	18.2	34.1	43.8	75.5	72.0	71.0	H
5	182T	3495	6.8	43.5	7.51	12.3	21.7	81.5	78.5	84.8	H
	184T	1730	7.0	42.7	15.2	32.7	43.8	82.5	80.0	79.9	H
	215T	1160	9.4	54.5	22.6	49.3	63.0	82.5	80.0	72.9	J
	254T	869	7.6	40.0	30.2	50.8	73.5	82.5	80.0	74.0	H
7.5	184T	3480	9.2	60.8	11.3	19.9	30.0	81.5	78.5	90.7	H
	213T	1755	10.5	59.3	22.4	47.1	55.2	84.0	81.5	78.3	H
	254T	1165	10.5	59.4	33.8	63.5	86.8	82.5	80.0	79.7	H
	256T	866	11.0	56.0	45.5	74.0	107	82.5	80.0	76.5	G
10	213T	3514	12.2	73.0	14.9	26.0	38.6	86.5	84.0	87.7	G
	215T	1755	13.5	80.0	29.9	63.9	73.7	85.5	82.5	80.5	H
	256T	1162	13.5	70.0	45.2	88.2	114	84.0	81.5	82.0	G
	284T	872	15.0	72.0	60.2	86.5	134	82.5	80.0	74.8	G
15	215T	3500	17.8	108	22.5	42.0	56.0	87.5	85.5	89.2	G
	254T	1761	20.0	113	44.7	90.3	115	86.5	84.0	80.6	G
	284T	1172	20.0	110	67.2	104	161	86.5	84.0	81.1	G
	286T	872	22.5	115	90.3	154	216	84.0	81.5	73.5	G
20	254T	3512	25.0	141	29.9	52.9	75.9	87.5	85.5	85.4	F
	256T	1753	25.0	131	59.9	110	135	87.5	85.5	85.1	F
	286T	1171	26.0	143	89.7	140	209	87.5	85.5	82.1	G
	324T	879	29.0	140	120	171	256	86.5	84.0	74.7	F
25	256T	3502	30.0	166	37.5	63.4	88.8	88.5	86.5	88.2	F
	284T	1761	32.0	171	74.3	130	164	88.5	86.5	81.9	F
	324T	1177	32.3	170	112	203	250	89.5	87.5	80.7	F
	326T	879	35.0	175	149	221	359	87.5	85.5	76.3	F
30	284TS	3525	36.5	185	44.7	65.0	102	88.5	86.5	86.9	F
	286T	1766	37.0	205	89.3	168	196	89.5	87.5	84.3	F
	326T	1178	37.8	197	134	250	180	90.2	88.5	82.2	F
	364T	877	42.0	210	180	238	399	87.5	85.5	76.1	F
40	286TS	3520	48.0	250	59.7	88.0	131	89.5	87.5	87.1	F
	324T	1769	51.0	260	119	153	256	90.2	88.5	81.4	F
	364T	1177	49.0	280	179	271	402	89.5	87.5	84.8	F
	365T	875	54.0	270	240	360	541	88.5	86.5	78.5	F
50	324TS	3542	60.0	320	74.2	102	169	90.2	88.5	86.4	F
	326T	1770	62.0	340	148	238	334	90.2	88.5	83.1	F
	365T	1177	60.0	350	223	366	481	90.2	88.5	86.1	F
	404T	875	68.0	344	300	470	729	88.5	86.5	77.6	F
60	326TS	3536	72.0	380	89.1	131	201	90.2	88.5	86.0	F
	364T	1769	73.0	396	178	319	394	90.2	88.5	85.2	F
	404T	1179	73.0	416	267	361	627	91.0	89.5	84.5	F
	405T	877	80.0	413	360	584	879	89.5	87.5	78.0	F
75	364TS	3547	85.0	510	111	139	294	91.0	89.5	90.3	F
	365T	1765	91.0	470	223	379	515	90.2	88.5	85.1	F
	405T	1177	90.0	504	335	499	784	91.0	89.5	85.8	F
	444T	882	100	540	447	642	1120	90.2	88.5	77.7	G

This performance data is for standard designs and should be used for estimating purposes only.

Actual ratings may vary slightly from this data.

Courtesy of Reliance Electric

Figure 9.31

Lamp Types

Lighting is a significant factor in the operating costs of a facility and, in the last 25 years, much attention has been given to the development of more efficient lamps. Lamps for interior and exterior fixtures can be incandescent, fluorescent, or high-intensity discharge (HID). HID lamps include mercury vapor, metal halide and high-pressure sodium. (Low-pressure sodium lamps are used principally for outdoor applications because of their strong yellow color.)

Incandescent bulbs are the least efficient of the above, but their low initial cost keeps them attractive. Also, their "warm" (high red content) color spectrum is pleasing for many applications. One significant advantage of incandescent bulbs is that they light up instantly.

Fluorescent bulbs require the use of ballast coils which make the initial fixture costs somewhat higher than the cost of incandescent fixtures. Fluorescent bulbs are, however, more efficient and their color spectrum far more closely imitates natural daylight. A few different fluorescent lamp types are available that offer trade-offs among efficiency, color spectrum and bulb cost.

HID lamps generally are more efficient than fluorescent lamps, but do not generate the same broad light spectrum. HID lamps require special ballasts and require several minutes to warm up before full output is reached.

Mercury vapor lamps put out about 65 lumens per watt and have a blue-green color. Because they have little or no red content in their spectrum, they distort the actual color of objects. They require 3 to 7 minutes after striking to achieve their full brilliance and, since the bulb must cool before it can restrike, it takes about the same amount of time to light up again after a power disruption.

Metal halide lamps are 1.5 to 2 times faster than mercury vapor lamps. Their color is a blue-white and they render color slightly better than mercury vapor lamps. They take longer to warm up or restart—about 15 minutes.

High-pressure sodium lamps have an efficiency of about 110 lumens per watt. They have a golden-white color and can distort colors considerably. About 15 minutes are required for a high-pressure sodium lamp to reach full luminance, but it can restart quickly after a momentary interruption—in only 1.5 to 2 minutes.

Light Quality

Characteristics such as direct glare, reflected glare, brightness ratio, shadows and color are used to describe lighting quality. Both direct glare and reflected glare cause light to shine brightly into the eyes and produce unnecessary strain. Too great a brightness ratio between near and more distant objects can produce fatigue as the eyes shift back and forth to adjust to the varying brightness. Making lighting more diffuse can minimize dark shadows which can annoy and tire the eyes. Color of light and color rendition have a profound effect on light quality.

Lamp burnout clearly reduces the effectiveness of any lighting system, but levels of lighting can begin to deteriorate even before lamp burnout. The amount of illumination initially provided starts to decrease almost as soon as the bulb, or lamp, is put into operation. The rate of decrease is infinitesimal at first, but increases as time passes. This continual decrease of lighting quality is due in large measure to three causes: dirt, lumen depreciation and discoloration. Dirt accumulation on lamps, reflectors, room walls and ceilings will reduce the reflectability of light. Lumen depreciation is the gradual reduction of light output which occurs as lamps age and is caused by heat from within the lamp slowly vaporizing internal components and depositing this material on the inside of the glass bulb. Heat given off by lamps can cause discoloration of reflectors and lenses.

Light Maintenance

Careful observation by those who operate and use a facility is needed to discover and report any evidence of defects in lighting systems. Deficiencies should be prepared promptly to prevent progressive deterioration of the system. Maintain the required illumination intensity by keeping lamps, fixtures and reflective areas in good repair, replacing defective lamps, and keeping the voltage steady. The progressive decrease of light caused by the accumulation of dirt necessitates periodic cleaning of lighting equipment, the frequency of which depends on local conditions. The cleaning schedule for a particular lighting system should be determined by a light meter reading after the initial reading. When subsequent footcandle readings have dropped 20–25 percent, the fixtures should be cleaned again. The exterior surfaces of lighting equipment should be washed with a damp cloth and a non-abrasive cleaner, not just wiped off with a dry cloth. Washing reclaims about 5 to 10 percent more light than dry wiping and reduces the possibility of marring or scratching reflective surfaces of fixtures.

Lamp replacement is a critical portion of light maintenance. Neglected lamp outages reduce illumination and, if burned-out lamps are not replaced, illumination may drop to unsafe footcandle levels in a short time. In some cases it may be satisfactory and more economical to clean lamp surfaces and fixture interiors only at the time of lamp replacement. Lamp replacement is done by either individual fixture or a group of fixtures.

Individual method: With this approach, burned-out lamps are replaced by an electrician on request. To prevent reduced illumination from lamp outages, follow these maintenance procedures:

1. Instruct employees to report burnouts as soon as they appear.
2. Replace blackened or discolored lamps even though they are still burning.
3. Replace fluorescent lamps as soon as they begin to flicker.
4. Replace any lamp with the same type, wattage and voltage as that of the lamp removed. Lamps of higher or lower wattage than called for on the lighting design plans should not be used.

Group method: With this approach, group replacement of lamps before they burn out is considered the most economical method for replacement in large areas. Whenever possible, group replacement should be performed simultaneously with fixture cleaning. Replacement of lamps by this method is accomplished by installing new lamps in all fixtures in the prescribed area after the old lamps have been in service for 70 to 75 percent of their rated life. The rated life of lamps can be obtained from the manufacturer and is the number of hours elapsed when 50 percent of the lamps in a large test group are burned out and 50 percent are still burning. (See Figure 9.32 for typical mortality curves of incandescent and fluorescent lamps.) Proponents of group relamping point out that it may cost as much as ten times more to replace lamps individually than in groups.

The following section is reprinted from "Lighting Fundamentals," Publication Number 430-B-95-003, available from the Environmental Protection Agency.

Determining Target Light Levels

The Illuminating Engineering Society of North America (IESNA) has developed a procedure for determining the appropriate average light level for a particular space. This procedure (used extensively by designers and engineers) recommends a target light level by considering the following:

- the task(s) being performed
- the ages of the occupants
- the importance of speed and accuracy

Then, the appropriate type and quantity of lamps and light fixtures may be selected based on the following:

- fixture efficiency
- lamp lumen output
- the reflectance of surrounding surfaces
- the effects of light losses from lamp lumen depreciation and dirt accumulation
- room size and shape
- availability of natural light (daylight)

When designing a new or upgraded lighting system, one must be careful to avoid overlighting a space. In the past, spaces were designed for as much as 200 footcandles in places where 50 footcandles may not only be adequate, but superior. This was partly due to the misconception that the more light in a space, the higher the quality. Not only does overlighting waste energy, but it can also reduce lighting quality. Within a listed range of illuminance, three factors dictate the proper level: age of the occupant(s), speed and accuracy requirements, and background contrast.

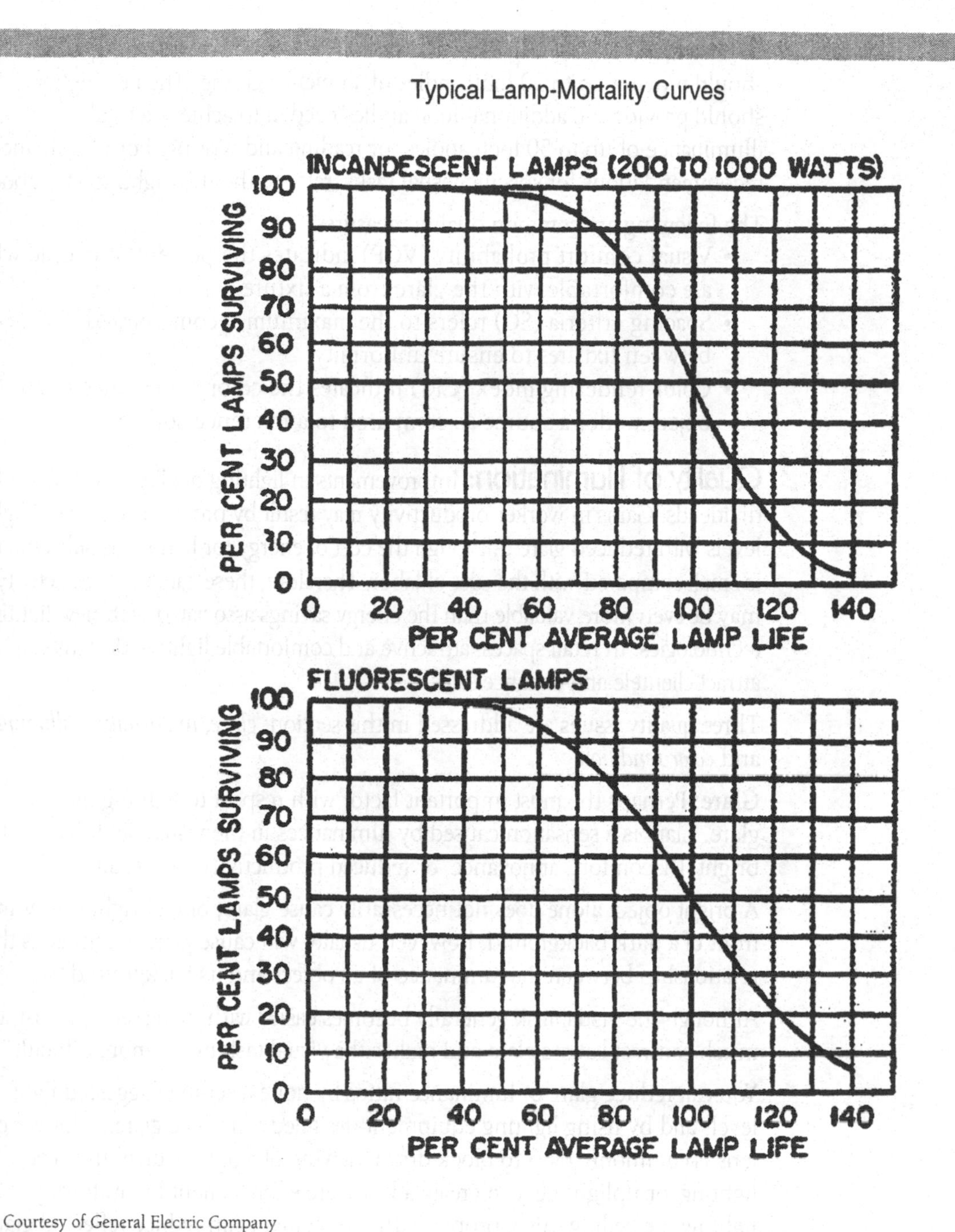

Typical Lamp-Mortality Curves
INCANDESCENT LAMPS (200 TO 1000 WATTS)
PER CENT LAMPS SURVIVING
100
90
80
70
60
50
40
30
20
10
0
0 20 40 60 80 100 120 140
PER CENT AVERAGE LAMP LIFE
FLUORESCENT LAMPS
PER CENT LAMPS SURVIVING
100
90
80
70
60
50
40
30
20
10
0
0 20 40 60 80 100 120 140
PER CENT AVERAGE LAMP LIFE
Courtesy of General Electric Company

Figure 9.32

For example, to light a space that uses computers, the overhead light fixtures should provide up to 30 footcandles of ambient lighting. The task lights should provide the additional footcandles needed to achieve a total illuminance of up to 50 footcandles for reading and writing. For illuminance recommendations for specific visual tasks, refer to the IES Lighting Handbook.

The following are common quality measures:

- Visual comfort probability (VCP) indicates the percent of people who are comfortable with the glare from a fixture.
- Spacing criteria (SC) refers to the maximum recommended distance between fixtures to ensure uniformity.
- Color rendering index (CRI) indicates the color appearance of an object under a source as compared to a reference source.

Quality of Illumination: Improvements in lighting quality can yield high dividends. Gains in worker productivity may result by providing corrected light levels with reduced glare. Although the cost of energy for lighting is substantial, it is small compared with the cost of labor. Therefore, these gains in productivity may be even more valuable than the energy savings associated with new lighting technologies. In retail spaces, attractive and comfortable lighting designs can attract clientele and enhance sales.

Three quality issues are addressed in this section: *glare, uniformity of illuminance* and *color rendition.*

Glare: Perhaps the most important factor with respect to lighting quality is glare. Glare is a sensation caused by luminances in the visual field that are too bright. Discomfort, annoyance, or reduced productivity can result.

A bright object alone does not necessarily cause glare, but a bright object in front of a dark background, however, usually will cause glare. Contrast is the relationship between the luminance of an object and its background.

Although the visual task generally becomes easier with increased contrast, too much contrast causes glare and makes the visual task much more difficult.

You can reduce glare or luminance ratios by not exceeding suggested light levels and by using lighting equipment designed to reduce glare. A louver or lens is commonly used to block direct viewing of a light source. Indirect lighting, or uplighting, can create a low glare environment by uniformly lighting the ceiling. Also, proper fixture placement can reduce reflected glare on work surfaces or computer screens. Standard data now provided with luminaire specifications include tables of its visual comfort probability (VCP) ratings for various room geometries. The VCP index provides an indication of the percentage of people in a given space that would find the glare from a fixture to be acceptable. A minimum VCP of 70 is recommended for commercial interiors, while luminaires with VCPs exceeding 80 are recommended in computer areas.

Uniformity of Illuminance on Tasks: The uniformity of illuminance is a quality issue that addresses how evenly light spreads over a task area. Although a room's average illuminance may be appropriate, two factors may compromise uniformity:

- Improper fixture placement based on the luminaire's spacing criteria (ratio of maximum recommended fixture spacing distance to mounting height above task height)
- Fixtures that are retrofit with reflectors that narrow the light distribution
- Non-uniform illuminance causes several problems:
  - Inadequate light levels in some areas
  - Visual discomfort when tasks require frequent shifting of view from underlit to overlit areas
  - Bright spots and patches of light on floors and walls that cause distraction and generate a low quality appearance

Color Rendition: The ability to see colors properly is another aspect of lighting quality. Light sources vary in their ability to accurately reflect the true colors of people and objects. The color rendering index (CRI) scale is used to compare the effect of a light source on the color appearance of its surroundings. A scale of 0 to 100 defines the CRI. A higher CRI means better color rendering, or less color shift. CRIs in the range of 75–100 are considered excellent, while 65–75 are good. The range of 55–65 is fair, and 0–55 is poor. Under higher CRI sources, surface colors appear brighter, improving the aesthetics of the space. Sometimes, higher CRI sources create the illusion of higher illuminance levels.

Two Approaches to Workspace Lighting

As work stations change in size and we spend increased time in front of our computer monitors, traditional approaches to workspace lighting are being overtaken by new methods. The following are two of the most popular.

Task Lighting: Task lighting involves the use of lighting systems specifically tailored to meet the needs of a particular task. By fine-tuning luminaires in this manner, lighting professionals can help increase worker productivity and improve morale. Moreover, this approach can save energy dollars. For example, a task light using a 13-watt compact fluorescent lamp consumes far less energy than a typical overhead fixture.

Two-Component Lighting: Two-component lighting systems combat the common workspace problems of glare and highly reflective computer screens, as well as the shadows that can come with inappropriate light distribution. Two-component lighting employs indirect lighting alongside portable, individually adjustable task lights. It allows the strongest light to be placed where it is most needed. Softer, more diffuse lighting is used to light surrounding areas. Again, the result is greater worker comfort and reduced energy costs.

Common Terminology

The following are many of the more common terms used in the lighting industry.

- **Arc tube:** A tube enclosed by the outer glass envelope of an HID lamp and made of clear quartz or ceramic that contains the arc stream.
- **Ballast:** A device used to operate fluorescent and HID lamps. The ballast provides the necessary starting voltage, while limiting and regulating the lamp current during operation.
- **Ballast cycling:** Undesirable condition under which the ballast turns lamps on and off (cycles) due to the overheating of the thermal switch inside the ballast. This may be due to incorrect lamps, improper voltage being supplied, high ambient temperature around the fixture, or the early stage of ballast failure.
- **Ballast efficiency factor:** The ballast efficiency factor (BEF) is the ballast factor (see below) divided by the input power of the ballast. The higher the BEF (within the same lamp-ballast type) the more efficient the ballast.
- **Ballast factor:** The ballast factor (BF) for a specific lamp-ballast combination represents the percentage of the rated lamp lumens that will be produced by the combination.
- **Candela:** Unit of luminous intensity, describing the intensity of a light source in a specific direction.
- **Candlepower:** A measure of luminous intensity of a light source in a specific direction, measured in candelas (see above).
- **Coefficient of utilization:** The ratio of lumens from a luminaire received on the work plane to the lumens produced by the lamps alone. (Also called "CU.")
- **Color rendering index (CRI):** A scale of the effect of a light source on the color appearance of an object compared to its color appearance under a reference light source. Expressed on a scale of 1 to 100, where 100 indicates no color shift. A low CRI rating suggests that the colors of objects will appear unnatural under that particular light source.
- **Color temperature:** The color temperature is a specification of the color appearance of a light source, relating the color to a reference source heated to a particular temperature, measured by the thermal unit Kelvin. The measurement can also be described as the "warmth" or "coolness" of a light source. Generally, sources below 3200K are considered "warm;" while those above 4000K are considered "cool" sources.
- **Compact fluorescent:** A small fluorescent lamp that is often used as an alternative to incandescent lighting. The lamp life is about 10 times longer than incandescent lamps and is 3–4 times more efficacious. Also called PL, Twin-Tube, CFL, or BIAX lamps.

- **Daylight compensation:** A dimming system controlled by a photocell that reduces the output of the lamps when daylight is present. As daylight levels increase, lamp intensity decreases. An energy-saving technique used in areas with significant daylight contribution.
- **Downlight:** A type of ceiling luminaire, usually fully recessed, where most of the light is directed downward. May feature an open reflector and/or shielding device.
- **Efficacy:** A metric used to compare light output to energy consumption. Efficacy is measured in lumens per watt. Efficacy is similar to efficiency, but is expressed in dissimilar units. For example, if a 100-watt source produces 9000 lumens, then the efficacy is 90 lumens per watt.
- **Electronic ballast:** A ballast that uses semi-conductor components to increase the frequency of fluorescent lamp operation (typically in the 20–40 kHz range). Smaller inductive components provide the lamp current control. Fluorescent system efficiency is increased due to high frequency lamp operation.
- **EMI:** Abbreviation for electromagnetic interference. High frequency interference (electrical noise) caused by electronic components or fluorescent lamps that interferes with the operation of electrical equipment. EMI is measured in micro-volts, and can be controlled by filters. Because EMI can interfere with communication devices, the Federal Communication Commission (FCC) has established limits for EMI.
- **Fluorescent lamp:** A light source consisting of a tube filled with argon, along with krypton or other inert gas. When electrical current is applied, the resulting arc emits ultraviolet radiation that excites the phosphors inside the lamp wall, causing them to radiate visible light.
- **Footcandle (FC):** The English unit of measurement of the illuminance (or light level) on a surface. One footcandle is equal to one lumen per square foot.
- **HID:** Abbreviation for high intensity discharge. Generic term describing mercury vapor, metal halide, high pressure sodium, and (informally) low pressure sodium light sources and luminaires.
- **High-pressure sodium lamp:** A high intensity discharge (HID) lamp whose light is produced by radiation from sodium vapor (and mercury).
- **Illuminance:** A photometric term that quantifies light incident on a surface or plane. Illuminance is commonly called light level. It is expressed as lumens per square foot (footcandles), or lumens per square meter (lux).
- **Lamp lumen depreciation factor (LLD):** A factor that represents the reduction of lumen output over time. The factor is commonly used as a multiplier to the initial lumen rating in illuminance calculations, which compensates for the lumen depreciation. The LLD factor is a dimension less value between 0 and 1.

- **Light loss factor (LLF):** Factors that allow for a lighting system's operation at less than initial conditions. These factors are used to calculate maintained light levels. LLFs are divided into two categories, recoverable and non-recoverable. Examples are lamp lumen depreciation and luminaire surface depreciation.
- **Low-pressure sodium:** A low-pressure discharge lamp in which light is produced by radiation from sodium vapor. Considered a monochromatic light source (most colors are rendered as gray).
- **Low-voltage lamp:** A lamp (typically compact halogen) that provides both intensity and good color rendition. Lamp operates at 12V and requires the use of a transformer. Popular lamps are MR11, MR16, and PAR36.
- **Lumen:** A unit of light flow, or luminous flux. The lumen rating of a lamp is a measure of the total light output of the lamp.
- **Luminaire:** A complete lighting unit consisting of a lamp or lamps, along with the parts designed to distribute the light, hold the lamps, and connect the lamps to a power source. Also called a fixture.
- **Luminance:** A photometric term that quantifies brightness of a light source or of an illuminated surface that reflects light. It is expressed as footlamberts (English units) or candelas per square meter (Metric units).
- **Lux (LX):** The metric unit of measure for illuminance of a surface. One lux is equal to one lumen per square meter. One lux equals 0.093 footcandles.
- **Mercury vapor lamp:** A type of high intensity discharge (HID) lamp in which most of the light is produced by radiation from mercury vapor. Emits a blue-green cast of light. Available in clear and phosphor-coated lamps.
- **Metal halide:** A type of high intensity discharge (HID) lamp in which most of the light is produced by radiation of metal halide and mercury vapors in the arc tube. Available in clear and phosphor-coated lamps.
- **Par lamp:** A parabolic aluminized reflector lamp. An incandescent, metal halide, or compact fluorescent lamp used to redirect light from the source using a parabolic reflector. Lamps are available with flood or spot distributions.
- **Parabolic luminaire:** A popular type of fluorescent fixture that has a louver composed of aluminum baffles curved in a parabolic shape. The resultant light distribution produced by this shape provides reduced glare, better light control, and is considered to have greater aesthetic appeal.
- **Quad-tube lamp:** A compact fluorescent lamp with a double twin tube configuration.
- **Room cavity ratio (RCR):** A ratio of room dimensions used to quantify how light will interact with room surfaces. A factor used in illuminance calculations.

- **Reflectance:** The ratio of light reflected from a surface to the light incident on the surface. Reflectances are often used for lighting calculations. The reflectance of a dark carpet is around 20%, and a clean white wall is roughly 50% to 60%.
- **Regulation:** The ability of a ballast to hold constant (or nearly constant) the output watts (light output) during fluctuations in the voltage feeding of the ballast. Normally specified as +/- percent change in output compared to +/- percent change in input.
- **Relay:** A device that switches an electrical load on or off based on small changes in current or voltage. Examples: low voltage relay and solid state relay.
- **Spacing criterion:** A maximum distance that interior fixtures may be spaced that ensures uniform illumination on the work plane. The luminaire height above the work plane multiplied by the spacing criterion equals the center-to-center luminaire spacing.
- **Tandem wiring:** A wiring option in which a ballast is shared by two or more luminaires. This reduces labor, materials, and energy costs. Also called "master-slave" wiring.
- **Thermal factor:** A factor used in lighting calculations that compensates for the change in light output of a fluorescent lamp due to a change in bulbwall temperature. It is applied when the lamp-ballast combination under consideration is different from that used in the photometric tests.
- **Tungsten halogen lamp:** A gas-filled tungsten filament incandescent lamp with a lamp envelope made of quartz to withstand the high temperature. This lamp contains some halogens (namely iodine, chlorine, bromine, and fluorine),which slow the evaporation of the tungsten. Also, commonly called a quartz lamp.
- **VCP:** Abbreviation for visual comfort probability. A rating system for evaluating direct discomfort glare. This method is a subjective evaluation of visual comfort expressed as the percent of occupants of a space who will be bothered by direct glare. VCP allows for several factors: luminaire luminances at different angles of view, luminaire size, room size, luminaire mounting height, illuminance, and room surface reflectivity. VCP tables are often provided as part of photometric reports.
- **Very high output (VHO):** A fluorescent lamp that operates at a "very high" current (1500 mA), producing more light output than a "high output" lamp (800 mA) or standard output lamp (430 mA).
- **Work plane:** The level at which work is done and at which illuminance is specified and measured. For office applications, this is typically a horizontal plane 30 inches above the floor (desk height).

The move toward higher efficiency standards for fluorescent and incandescent fixtures has now become a major campaign for most energy managers and plant engineers. The National Energy Policy Act mandates it, as do the lighting accounts for 30% to 40% of commercial buildings. Many projects now call for fluorescent compact lamps, 4-foot T-8, 32-watt lamps with electronic ballasts, and halogen lamps. In addition, lighting retrofits are often rewarded by utility company rebates. However, retrofitting brings new challenges—the disposal of pre-1980 ballasts with hazardous materials, for instance—that plant engineers must face.

Lighting Formulas

Development of an effective and efficient lighting program necessitates an understanding of several fundamental formulas.

Lumen Method: The simplest method to calculate the lighting intensity for an area is the *lumen method.* This method allows for estimation of the average "maintained" footcandles of general illumination for a room or area by using a series of formulas.

The first step is to determine the *total initial lamp lumens*, or the number of lumens per lamp multiplied by the number of lamps in the area. Next, find the *coefficient of utilization* for the installed lamp(s) by contacting the manufacturer(s) or consulting the *IES Lighting Handbook* (Illuminating Engineering Society of North America). Finally, determine the *light loss factor (LLF)*, or the percent of rated output that the lighting system actually produces over time. This factor depends on light design and distribution characteristics, dust accumulation in the area, and the frequency of maintenance, and generally ranges from 40% for poor maintenance conditions to 80% for good ones.

The LLF is actually the product of a number of "recoverable" and "nonrecoverable" factors. Recoverable factors are those that may be corrected, and include *luminaire dirt depreciation (LDD)*, *room surface dirt depreciation (RSDD)*, *lamp lumen depreciation (LLD)*, and *lamp burnout (LBO)*. Nonrecoverable factors usually can not be addressed, and include *luminaire ambient temperature (LAT)*, *voltage variation (VV)*, *ballast factor (BF)*, and *luminaire surface depreciation (LSD)*. The LLF, then, is determined by the following equation:

$$LLF = (LDD \times RSDD \times LLD \times LBO) \times (LAT \times VV \times BF \times LSD)$$

The *IES Handbook* contains detailed information on recoverable and nonrecoverable factors.

Using all of this information, average footcandles may be calculated using this formula:

$$\frac{\text{Total Initial L} \times \text{CU} \times \text{LLF}}{\text{area of room, sq. ft.}}$$

or this one:

$$\frac{\text{Initial L per lamp} \times \text{CU} \times \text{LLF}}{\text{area per lamp, sq. ft.}}$$

where:

L = lamp lumens

CU = coefficient of utilization

LLF = light loss factor

To determine the number of lamps needed for a room or area, use the following equation:

$$\text{Lamps} = \frac{\text{Room area x desired footcandles}}{\text{LLF} \times \text{CU} \times \text{lamp lumens}}$$

Zonal-Cavity Method: The *zonal-cavity method* provides more accurate data then the lumen method because it takes into account the effects of interreflection of light. It is based on the concept that a room or area is divided into three cavities—ceiling, room, and floor—which have effective reflectances with respect to each other and the work plane. Using this method to determine average footcandles requires first calculating the three cavity ratios with the following equation:

$$\text{Cavity Ratio} = \frac{5h(L + W)}{L \times W}$$

where:

h = distance in feet from luminaire to ceiling to determine ceiling cavity ratio
distance in feet from luminaire to work plane to determine room cavity ratio
distance in feet from floor to work plane to determine floor cavity ratio

L = length of room in feet

W = width of room in feet

Next, cavity reflectances must be determined for the ceiling and floor cavities. While it is possible to do so with the use of formulas, the *IES Lighting Handbook* includes a table of reflectances for various reflectance combinations. Finally, use the cavity ratios and cavity reflectances with the manufacturer's information to select a more accurate coefficient of utilization (CU) for the luminaire involved. (Manufacturers provide CU tables with a standard reflectance value, but offer additional information that can be used to make adjustments if your reflectance value is higher or lower.) Use this new CU value with the formula used in the lumen method section to determine a truer average footcandle level.

Overcurrent Protection

The function of system protection is to detect and isolate the affected part of the system from overcurrent or short circuits. The system needs to be coordinated to promptly limit the abnormal condition and limit the effect on the balance of the system.

The system fault study will consider the fault current contributed from several sources, including the utility contribution, site generation, and motors. Two fault current values need to be considered—the symmetrical current and the asymmetrical current. The asymmetrical current is offset from the normal current axis for several cycles after the fault, and asymmetrical current will be higher than the symmetrical current. With the use of high speed interrupting devices that interrupt the current in the first few cycles, the asymmetrical value must be considered in any system study. The fault current at the various locations in the electrical system must be calculated using the conductor types and lengths and the transformer impedances.

When the system fault currents are determined, then the overcurrent device characteristics can be used to provide selective operation of the overcurrent devices to limit the interruption to the smallest portion of the electric system. Coordination curves for fuses and breakers are available from the manufacturers.

Fuses

Low voltage fuses can be grouped into two categories—noncurrent limiting and current limiting. The threshold current is the point where a fuse starts limiting current. The total clearing time is less than a half-cycle. Time delay fuses are available in both noncurrent limiting and current limiting applications. (Time delay typically means that the fuse will not open in less than 10 seconds at five times rated current.) Total clearing time is the time between the beginning of the overcurrent and the final interruption of the current. It is the sum of the minimum melting time plus tolerance and the arcing time. To apply fuses with coordination that will limit the interruption to the smallest portion of the electric system, manufacturers publish selectivity ratio tables.

Breakers

Circuit breakers for low voltage systems have two American National Standards Institute (ANSI) classifications: *molded case breakers,* which feature an integral unit in an insulating housing; and *low-voltage power circuit breakers,* for circuits rated 1000 volts and below. Trip units may be electromechanical-thermal/magnetic, electromechanical-dashpots, or solid-state electronic. Depending on the magnitude of current, the trip unit will initiate an instantaneous response or an inverse-time response. The trip unit characteristics can consist of: (1) continuous current rating, (2) long-time current pickup, (3) long-time delay, (4) short-time current pickup, (5) short-time delay, (6) instantaneous current response, and (7) ground fault. Not all characteristics are available or used in all applications. The instantaneous response of a circuit breaker is basically the time required by the circuit breaker mechanism to open the contacts plus arcing time. The time delay response varies inversely with the current and can vary from fractions of seconds to minutes. The total operational time of a breaker is from the initiation of the fault to clearing of the fault and includes the mechanical time of the breaker and the arcing time.

Medium- and high-voltage fuses are also used. (Medium voltage is 2.4 kV–34.5 kV.) Medium-voltage fuses fall into two categories: *distribution fuse cutouts* and *power fuses*. Medium- and high-voltage breakers are relayed circuit breakers that use an external relay to provide the sensing and trip initiation signal.

Commonly Used Relay Device Function Numbers:

- 12 – Over-speed device
- 21 – Distance
- 25 – Synchronizing
- 27 – Undervoltage
- 32 – Directional power
- 40 – Loss of excitation
- 46 – Phase balance
- 47 – Phase sequence voltage
- 49 – Voltage
- 50 – Instantaneous overcurrent
- 51 – Time overcurrent
- 52 – AC circuit breaker
- 59 – Overvoltage
- 60 – Voltage balance
- 67 – Directional overcurrent
- 81 – Frequency (over/under)
- 86 – Lockout
- 87 – Differential

The coordination of a system requires that not only the fault currents be determined. Equipment and conductor damage curves must also be considered, along with the startup inrush currents. Cable short-circuit protection considerations include the conductor material, insulation material, and cable initial temperature.

There are several computer programs that will do system-fault and coordination studies. Unless you work regularly with these applications, it is wise to contact a firm that regularly performs such studies. It is easy to overlook a key coordination item on a complex system.

Figures 9.33–9.35 illustrate short-circuit currents

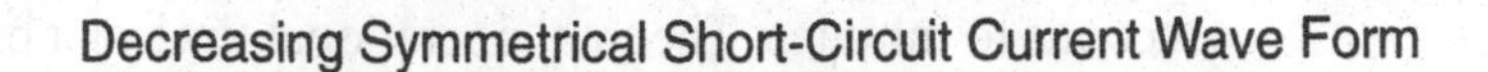
Decreasing Symmetrical Short-Circuit Current Wave Form

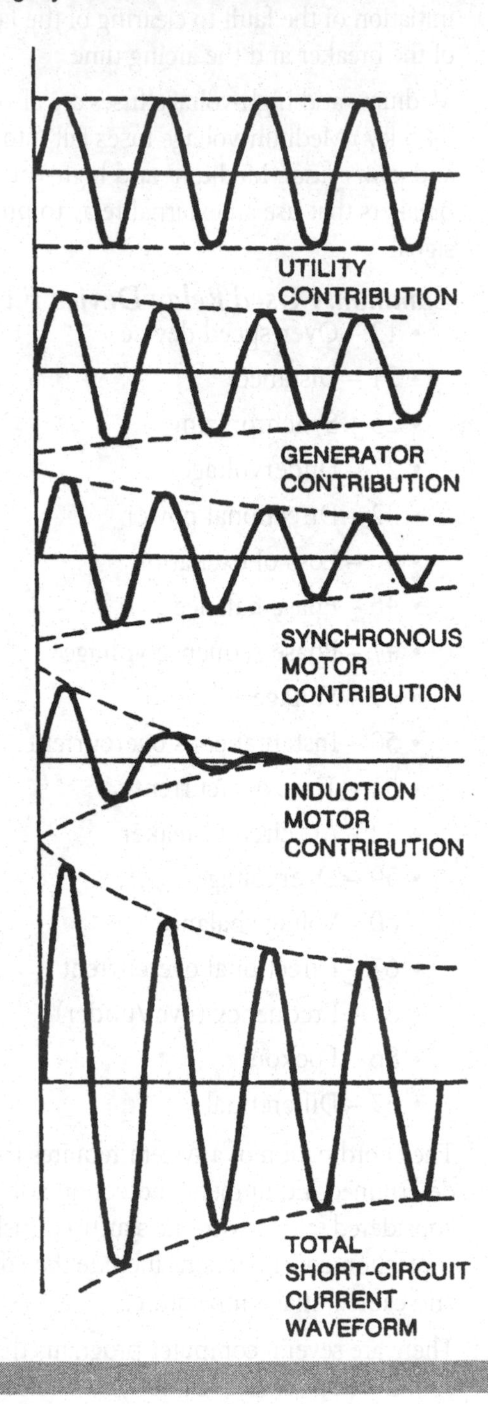

Figure 9.33

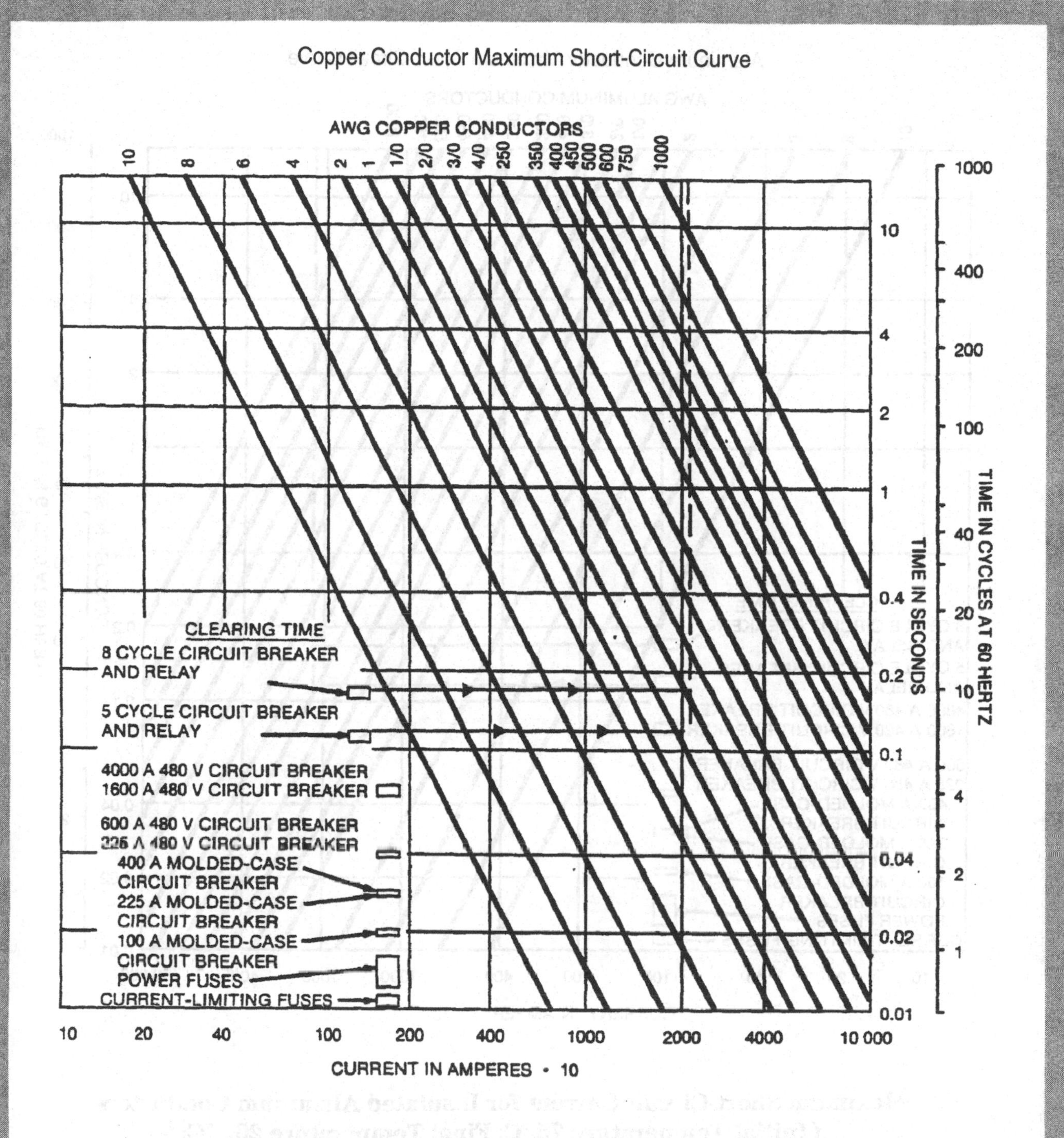

Maximum Short-Circuit Current for Insulated Copper Conductors; Initial Temperature 75 °C; Final Temperature 200 °C;

Figure 9.34

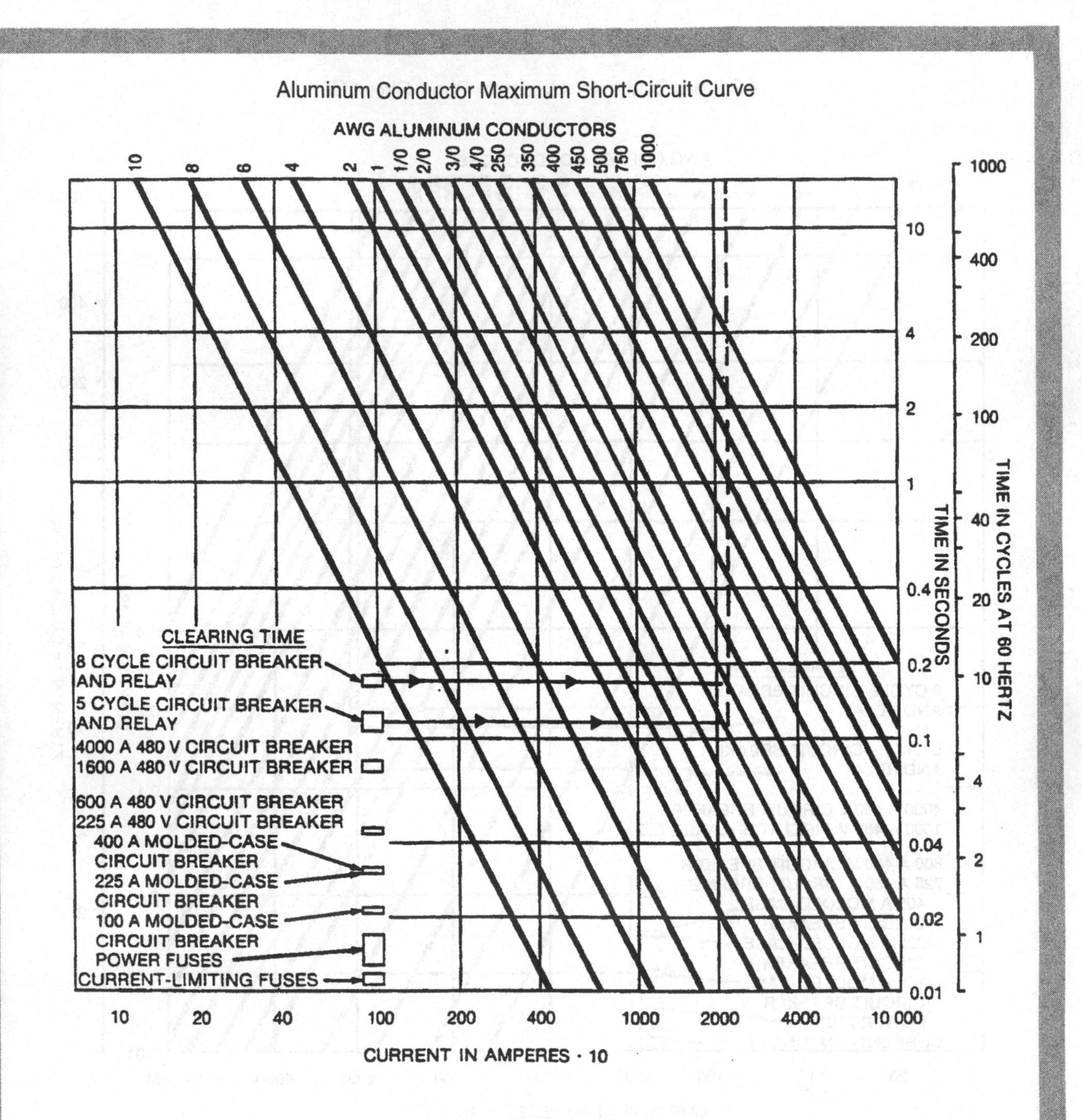

Maximum Short-Circuit Current for Insulated Aluminum Conductors (Initial Temperature 75 °C; Final Temperature 200 °C;

Figure 9.35

National Electrical Code

Published by the National Fire Protection Association, the National Electrical Code (NEC) is revised and re-issued every three years. The most recent edition was released in 1999. The NEC revision process includes a proposal period when suggestions from the public are accepted. Code-making panels review all proposals, and make recommendations that are then reviewed by a correlating committee. After a public comment period, the new code is voted on at an annual meeting for NFPA members.

A companion handbook to the Code is published with each new revision. Serving as a user's guide to the NEC, this handbook includes the complete text of the Code alongside explanatory commentary. The handbook includes sections on code intent and interpretations, historical perspectives, the application of requirements, and supplemental illustrations and tables.

The list of resources at the end of this chapter includes contact information on the National Fire Protection Agency.

Using the Code

The National Electrical Code is an indispensable resource for the facility engineer. The following are just a few ways in which the Code can be put to work.

Service and Feeder Sizing: Employing the appropriate size of electrical conductors or cables is critical to avoid the potential for fire that comes with the overheating of conductor insulation. (Conductors may be referred to as feeders when applied to electrical service entrances, switchboards, or panels.) Conductor size is primarily determined by its connected electrical load. For temperature ratings for a full variety of conductor sizes, see 1999 National Electrical Code (NEC) Tables 310-16 through 310-19.

Raceway Fill: Raceways are channels constructed to house and protect conductors. Because of the heat generated with transmission of current, it is necessary to avoid overloading raceways; in a sense, the wires and cables within need a minimum amount of space to "breathe." The NEC lists the maximum number of conductors and fixture wires that may be enclosed in a number of different raceways. See the first nine tables in Chapter 9 and all of Appendix C in the 1999 NEC.

Panelboards: Panelboards provide circuit control and overcurrent protection, and fall into two basic categories: lighting and appliance, and power and distribution. Panelboards may be main lug only, or may feature a main breaker inside the board. The latter is suitable for use as service equipment. The bussing of a panelboard is copper or aluminum, and will accept plug-in or bolt-on circuit breakers. NEC Article 384 provides an overview of panelboards, including specifications for overcurrent devices, location, and construction.

Determining Load Calculations: A key component in electrical design is estimating power requirements. Electric utilities often require an estimate of load required before new construction. The estimate includes lighting and receptacle loads, HVAC systems, electrical and data processing equipment, and special systems like elevators.

Load calculations must be conducted in accordance with National Electrical Code standards. Consult 1999 NEC Article 220 and Appendix D for calculation guidelines. The following are among the topics addressed.

Lighting Loads:

The general lighting load for a building may be determined by using 1999 NEC Article 220-3(a), which includes a table listing loads by type of occupancy. Unit loads per square foot are measured in volt-amperes, and range from 1 for a church to 3.5 for an office building. Load can be calculated only after the fixtures for a building have been selected.

Motor Loads:

The motor nameplate lists all the data needed to calculate motor load: horsepower, speed, voltage, phase, full-load current, locked rotor current, frame size, and service factor. Motor loads must be calculated in accordance with 1999 NEC Sections 430-24 through 430-26. Tables 430-148 through 430-150 may also be used to determine full-load currents.

Other Loads:

1999 NEC Section 220-3(b) discusses loads for general-use receptacles and outlets not used for general illumination, including fixed multi-outlet assemblies. The Code specifies that single, duplex, and triplex receptacles shall be computed at 180 volt-amperes; quad receptacles count as two duplex receptacles and are computed at 360 volt-amperes. 1999 NEC Tables 220-13 and 220-18 through 220-20 determine feeder loads.

Electric Costs

Electric Rates

Electric rates vary between utilities, especially with the Clinton Administration unveiling the National Electric Restructuring Plan, which could give end users in every state the ability to choose their electricity supplier by January 1, 2003. (This plan also affords individual states the ability to limit consumer options if they believe the local end user would be better off without competition.) Adjoining utilities can have rates that promote different types of customer loads. The facilities engineer needs to review the rate structure of the utility serving the facility. Often there is more than one rate that a utility can apply to the same customer. Again, it is the facilities engineer who needs to check the different rates to obtain the least cost for electric service.

In this section there are typical rate schedules for two adjoining electric utilities. The demand charge between the two utilities is $15.65 compared to $4.40 after the initial rate block. The cost per kilowatt-hour comparison is $.0397 to $.0323 after the initial energy block.

Terminology

The following are some of the more common terms used in discussing energy costs:

Load factor: Ratio of the average load over a designated period of time to the peak occurring during that period.

$$\text{Load Factor} = \frac{\text{kWh / Hours in Period}}{\text{Maximum kW Demand in Period}}$$

Demand factor: Ratio of the maximum demand of the system to the total connected load of the system. (Value is 1 or less.)

$$\text{Demand Factor} = \frac{\text{kW Demand}}{\text{Connected kW}}$$

Diversity factor: Ratio of the sum of the individual maximum demands of the subdivisions of a system to the maximum demand of the system. (Value is 1 or greater.)

$$\text{Diversity Factor} = \frac{\text{Sum of Individual Demands}}{\text{Maximum System Demand}}$$

Electric Tariff Terminology

Billing determinant: An electrical metered or calculated quantity to which a rate is applied for billing purposes.

General terms and conditions: General rules and regulations relative to the provision and billing of electric service.

Grid: The transmission and distribution networks operated by electrical utilities.

Rate: A price or charge per billing unit ($/kW, $/kWh, $/kVAR).

Rate schedule: A rate structure combined with specific stipulations relative to the provision and billing of electric service.

Rate structure: A configuration of rate schedules, customer groupings, demand, and energy charges.

Retail wheeling: The ability of an end user to purchase electricity from a supplier of choice and transmit it over a transmission grid.

Rider: Supplemental schedule that appends or modifies the billing provisions of a standard rate schedule.

Tariff: General terms and conditions; rate schedules; special schedules; riders; clauses; and general description of service territory that utilities are required to file with the state public utility commission.

Billing Demand Terminology:

Contract capacity: A percentage applied to the value of capacity requirements specified in the contract for electric service.

Metered demand: Current month actual value.

Ratcheted demand: Percentage applied to the actual demands of specified months prior to the current bill. (Example: The billing demand in any month shall not be less than 95% of the metered demand established during the previous 11 months.)

Rate Minimum: Absolute minimum value of demand to be used for billing purposes.

The following section is reprinted with permission from *Fundamentals of Electrical Systems and Building Electrical Energy Use,* by Rich Nowak and Rick Coughlin, ASHRAE, Inc.

Understanding an Electric Bill

Figure 9.36 shows the various sections of a typical electric bill. The boldface numbers are added to the bill to identify the sections.

Section 1: Power Company Identification. This section lists the name of the electric utility which, in Figure 9.36, is the Go Nowhere Power Company. Also on the bill is a telephone number, with the appropriate business hours, to call regarding questions about the bill. The utility's address generally does not appear on the bill.

Section 2: Customer Service Address. This is the address where the electric service is provided and where the meter, or meters, record the electric usage and power characteristics. Typically this is the same as the billing address, but it does not have to be.

Section 3: Billing Address. This is the address where the electric bill is sent. It may or may not be the same address where the electric service is provided. In Figure 9.36, the electric usage occurs in Maine, while the electric bills are sent to the organization's headquarters in New York.

Section 4: Account Number. Every electric service has at least one account number. If there is more than one meter that records the electric usage for different parts of the same service address (for example, several different electric services that provide electric power to one large hospital complex), then there will be several account numbers. Generally, one account number coincides with one electric usage meter.

Section 5: Service Period. These are the dates when the meter was read. The electric bill is for the electric usage patterns that occurred during this time period. The service period does not have to be a calendar month and frequently is not. Often the number of days that occurred within the billing period is also given. Service periods reflect when the meter was read and tend to be about 30-day periods. (Note: on alternate read dates, electric meters are sometimes estimated from existing electric usage profiles at the electric utility's office rather than a direct field reading.) In Figure 9.36, the meter was read on August 17, the beginning of the billing period, and then again at the end of the billing period, on September 18. The time period between these two dates is, in fact, 32 days. The term *billing period* is often used instead of *service period*, but both mean the same thing.

Section 6: Service Rate. This is the electric service rate schedule (sometimes referred to as the electric tariff schedule) that applies to the customer. It defines how the customer is billed for the energy usage patterns that occur in a billing period. For the example cited, Happy Home Farms is on an electric service rate schedule designated GS-STOU. To find out what that electric service rate schedule is, you would have to look it up in the utility's rate schedule book.

Section 7: Meter Identification. Every electric meter has an identification number. The electric meter that is read to determine electric bill costs is shown on the bill.

Section 8: Customer Charge. This charge is sometimes referred to as a franchise fee. It is a fixed charge that appears on every electric bill regardless of how much electricity is consumed in the billing period. If Happy Home Farms consumed 0 kWh of electricity between August 17 and September 18, there would still be a customer charge of $135.78. The customer charge is the service period charge for doing business with the utility, regardless of how much electricity the customer uses; it reflects the utility's costs for maintaining the electric service account file, for sending someone to read the meter, for mailing the electric bill, etc.

Section 9: Meter Reads. The values in this area of the bill reflect the actual readings taken from the meter at the beginning and end of the service period. In Figure 9.36, on August 17 (beginning of the billing period), the meter at Happy Home Farms read 3-8-7-5-2-6. At the same time, the meter reader also read the demand part of the meter; but because it does not affect the electric bill (it did affect the previous electric bill), it does appear on this bill. At the end of the billing period on September 18, the meter read 3-8-8-8-0-1 and the demand portion of the meter read 2.53. After reading the demand portion, the dial or counter was reset to zero to get a proper demand reading for the next billing period. Sometimes, actual meter readings are not done every billing period but rather every other billing period; estimated readings are made when actual readings are not. If this is done, it should be indicated on the bill as to which readings are actual and which readings are estimates.

Section 10: Meter Multipliers. The meter multipliers are identified on the electric bill to allow the customer to verify exactly how the usage and demand were determined. The multipliers do not have to be the same value for both energy and demand, but often they are. The multipliers are determined by the internal construction of the meter, and as such, are fixed; they do not change from electric bill to electric bill and are determined by only what meter is being used.

Possible Sections of an Electric Bill

GO NOWHERE POWER COMPANY **1**
TELEPHONE: (800) 548-0036
HOURS: Monday through Friday
8 a.m. through 4 p.m.

4 ACCOUNT NUMBER
294-568000903-3398548-00

CUSTOMER SERVICE ADDRESS:
Happy Home Farms **2**
9 Lefther Ridge Road
Whereyouat, Maine

5 SERVICE PERIOD
8/17/95 TO 9/18/95 32 days

6 SERVICE RATE: GS-STOU

7 METER IDENTIFICATION: 329-623

BILLING ADDRESS: **3**
Farmers Growth, Incorporated
58932 Noplacetopark Avenue
Bigandhot City, New York

8 CUSTOMER CHARGE: $135.78

	SERVICE CHARGE	BEGIN METER READ **9**	END METER READ	MULTIPLIER **10**	USED **11**	RATE **12**	
13	ENERGY	387526	388801	80	102,000 kWh	$0.0876/kWh	$8,935.20
14	DEMAND						
	READ		2.53	80	202 kW		
	BILLED				352 kW	$10.68/kW	$3,759.36

15 MISCELLANEOUS CHARGES, ADJUSTMENTS:

1. Fuel Adjustment,	credit, -$0.00233 per kWh	-$237.66
2. On-Site Power Adjustment,	credit, 18,500 kWh @ -$0.0103 per kWh	-$190.55
3. Environmental Adjustment,	high sulfur fuel use, $0.00032 per kWh	$32.64
4. Power Factor Adjustment,	credit @ -$0.54 per billed kW	-$190.08
5. Rental Charge,	security lights	$184.37
6. Miscellaneous,	nuclear decommissioning @$0.0004 per kWh	$40.80

16 TOTAL CHARGES & ADJUSTMENTS THIS PERIOD $12,469.86

17 TAXES

1. Local Tax,	1.5%	$187.05
2. State Tax,	3.7%	$461.38

18 TOTAL AMOUNT DUE, PAY THIS AMOUNT BY 10/12/95 $13,118.29

19 AFTER THE DUE DATE A 2% LATE CHARGE WILL BE APPLIED

Figure 9.36

Section 11: Used Energy and Demand. This section shows how much energy was consumed during the service period and what the peak demand was for the service period. If there is a ratchet clause in the service rate, then the bill identifies what the actual peak demand is for the billing period and what the billed demand is for the same service period. If the actual peak demand and the billed demand are the same value, then the ratchet adjustment is not in effect; if they are not the same value, then the ratchet adjustment is in effect.

Section 12: Rates for Energy and Demand. Energy rates and demand rates are recorded on the electric bill. If the service rate schedule includes block rates for energy and demand, then there are several rates for both energy and demand. In Figure 9.36, the customer has only one block for both energy and demand; the rate for energy is $0.0876 per kWh and the demand rate is $10.68 per kW of billed demand.

Section 13: Energy Service Charge. Enough information is presented on the bill to allow the customer to verify how the energy charge is calculated. In Figure 9.36, the information is neatly gathered in one section. This will not always be true; sometimes the meter reading values may be at the top of the bill or perhaps on a second page, and the other information may be in a different place or on a different page. In any case, after closely examining the bill, you should be able to decipher how the energy charge is calculated. In Figure 9.36, the energy charge is calculated as follows:

(388,801 – 387,526) usage meter counts x (80 kWh/usage meter count)
= 102,000 kWh

102,000 kWh x ($0.0876/kWh) = $8,935.20

Section 14: Demand Service Charge. As with the energy service charge, each electric bill should contain sufficient information for the customer to verify how the demand charge was calculated. The information may be scattered, but should all be there. In Figure 9.36, the demand charge is calculated.

(2.53 demand meter counts) x (80 kW/demand meter count) = 202 kW, peak demand

The actual peak demand that occurred in the billing period is 202 kW. However, there must be a ratchet clause in the service rate schedule that is in effect for this service period; this is shown by the fact that the actual peak demand is not the same as the billed demand. The ratchet value of demand that is presently in effect must be 353 kW. The ratchet value is greater than the read peak demand and, because the customer is billed for the higher of the two values, the ratchet value becomes the billed demand. The demand service charge is thus based on 353 kW and is calculated as follows:

(353 billed kW) x $10.68/billed kW) = $3,759.36

Section 15: Miscellaneous Charges, Adjustments. In an electric bill, other charges or credits are often included, with some typical ones shown in Figure 9.36. Many of these charges or credits have their origins in federal rules and agreements between a state public utility commission and an electric utility:

- *Fuel Adjustment.* This can be a charge or a credit depending on what the utility had to pay for fuel to operate its electric power plants. If the utility had to pay more than was anticipated for fuel, then the additional cost may be passed along to the consumer as a fuel adjustment charge; if the utility had to pay less than anticipated, then the customers receive a fuel adjustment credit.
- *On-Site Power Adjustment.* This is a credit to customers that generate electricity on-site and return some of that generated electricity to the utility's grid. In essence, this reflects the rate that the utility must pay the on-site power generator if the on-site generator decides to sell power back to the utility.
- *Environmental Adjustment.* This is used to encourage the utility to use more environmental friendly practices. Sometimes, the state public utility commissions may allow the utilities to pass the penalties along to their customers.
- *Power Factor Adjustment.* Depending on the customer's service rate schedule, this adjustment can be either a charge (power factor below a certain threshold value) or a credit (power factor above a certain threshold value). Power factor adjustment is sometimes found as part of the energy or demand charges.
- *Rental Charge.* The utility will often install and maintain such items as security lights, streetlights, traffic lights, etc. When it does, the utility can charge rental fees for the equipment that is used to benefit the customer but owned by the utility. The rental fee is generally a fixed charge that depends on the amount of equipment being rented.
- *Miscellaneous.* State public utility commissions will sometimes allow the utilities to pass along to their customers large unanticipated costs that the utilities incur. One such adjustment charge is for nuclear decommissioning. Under this scenario, the utility is allowed to pass along the costs of decommissioning a nuclear power plant.

Section 16: Total Charges and Adjustments. This is just a totaling of all charges and adjustments pertaining to the bill before taxes. Sometimes this specific section may be missing. In Figure 9.36, this section is the sum of sections 8, 13, 14 and 15.

Section 17, Taxes. Both local and state taxes are added to the totaled number from section 16. If the electric bill is for a pubic entity (such as a public high school), taxes are generally omitted.

Section 18, Total Amount Due. This is the amount that the customer must pay for the electric service during the period identified in the bill. Generally, a due date is also included.

Section 19, Late Charge Statement. After the due date, the utility is allowed to charge a late penalty fee that will appear on the next bill if the customer does not pay the present bill by the stated due date.

An Example of a Utility Bill

Figures 9.37 and 9.38 provide examples of an electric bill and appropriate electric rate schedule. (The service rate in this example is General Service Time-of-Use). In reviewing the bill and the rate schedule, you should be able to determine that:

- This is a time-of-use electric rate schedule that has both seasonal time-of-use rates and daily time-of-use rates.
- There are charges for both electric energy usage and peak demand.
- There is an 80% ratchet clause that windows the previous winter months of December, January, February and March to determine the minimum billed demand.
- There is a hidden power efficiency penalty charge in the form of a reactive demand charge.

The reactive demand charge is invoked any time the on-peak value of the difference (kVAR – 0.50 x kW) is greater than zero. In other words, any time the on-peak power factor falls below 89.4%, the power inefficiency penalty charge will be invoked. (kVAR > 0.5 kW means PF < $\{kW/[(1^2 + 0.5^2)^{0.5}\ kW]\} = 0.894$.) In this case, it is known that the school is all-electric in a northern climate; this means that the largest portion of the load is for resistive electric heat, which has a power factor of one. Rarely, if ever, is the power inefficiency penalty charge involved.

The electric bill in Figure 9.37 is analyzed in Figure 9.39. The different bill sections that were described earlier in this chapter are used as a reference. As you can see from Figure 9.39, not all information that was described earlier is found on a bill. This is to be expected because each utility has its own method of summarizing the electric bill data.

You can see that most of the information described earlier is found someplace, maybe in a different form, on the electric bill in Figure 9.37. However, not all of the information is included, and there is some additional information that was not described earlier. On the Figure 9.37 electric bill, there is an energy comparison report that gives some helpful details regarding how the present energy consumption and energy costs compare with past trends for the same facility. This type of information is relatively new on electric bills but it is becoming more common. Also, an account summary is generally given on a bill to verify the present account balance for the customer. Each utility provides somewhat different information than other utilities.

Sample Electric Bill

Keep This Portion For Your Records

BRIGHT LIGHT POWER COMPANY

For Service Call 1-800-653-8739

Office Hours 8:00 - 5:00

Billing Date
07 31 95

Account Number
546-0120030-015

CUSTOMER SERVICE AT
Happy Home School
8 Lefther Ridge Road
Whereyouat, Maine

General Service, Rate ISTOU

BILLING DETAIL

Prior Balance	767247
Payment	767247CR
Balance Forward	00

ENERGY USAGE

Meter	Date	Reading
718	07/30/95	9189
718	06/28/95	9140

Multiplier 400

kW Hours Used 19600

CURRENT MONTH CHARGES

Customer Charge	10417
Energy Charge	131771

On-Peak
6800 kWh @ 0.071002 / kWh
Shoulder
3200 kWh @ 0.070269 / kWh
Off-Peak
9600 kWh @ 0.063546 / kWh

Demand Charge 421217

On-Peak Measured – 90.40
Billed = 743.04
Shoulder Measured = 86.40
Billed = 86.40
Tax Exempt

CURRENT MONTH TOTAL
552988

NEW BALANCE
552988

PAY THIS AMOUNT
552988

DOES NOT REFLECT PAYMENTS RECEIVED AFTER
07/31/95

ENERGY COMPARISON REPORT

	Number Of Days	Average Cost Per Day	Average kWh Usage Per Day	Total kWh Usage
This Bill	32	143.39	613	19600
Last Bill	29	219.53	1393	40400
One Year Ago	30	133.58	619	18600

Reprinted with permission from *Fundamentals of Electrical Systems*, ASHRAE, Inc.

Figure 9.37

Sample Electric Rate Schedule

ELECTRIC RATE SCHEDULE **BRIGHT LIGHT POWER COMPANY**

RATE IGS-S-TOU
INTERMEDIATE GENERAL SERVICE — SECONDARY — TIME-OF-USE

AVAILABILITY

This rate is available for all general service purposes where service is taken at a secondary voltage and where the customer's maximum measured demand during an on-peak period has exceeded 400 kW but has not exceeded 1,000 kW, subject to the following two paragraphs.

Any customer taking service under this rate whose maximum measured demand during an on-peak period exceeds 1,000 kW twice in the preceding 12 months shall be automatically transferred to the applicable Large General Service rate for secondary service, effective with the next succeeding billing month.

Any customer taking service under this rate whose maximum measured demand during an on-peak period has not exceeded 400 kW in each of the preceding 12 months shall be automatically transferred to the applicable Medium General Service rate for secondary service, effective with the next succeeding billing month.

Electric service must be taken on a continuous year-round basis by any one customer at a single service location where the entire requirements for electric service at the premises are supplied at one point of delivery. Short-term service is not available under this rate.

CHARACTER OF SERVICE

Service will be single or three phase, alternating current, 60 Hz, at one standard available secondary voltage as described in the Company's Standard Requirements.

DEMAND

The monthly on-peak period kW demand shall be the highest 15-minute integrated kW demand registered in any on-peak period during the month as determined by the Company. The monthly on-peak period billing demand shall be the monthly on-peak kW demand, but not less than 80% of the highest monthly on-peak period kW demand occuring in the months of December, January, February or March of the preceding 11 months.

Service connected year-round but used mainly for seasonal business reasons may be split into two periods for on-peak period demand billing purposes, one being the period of high seasonal use and the second being the off-season period of suspended business and very low use. No period of seasonal use will be for less than three consecutive months.

Figure 9.38

Sample Electric Rate Schedule (cont.)

RATE IGS-S-TOU
INTERMEDIATE GENERAL SERVICE — SECONDARY — TIME-OF-USE (cont.)

DEMAND (cont.)

The demands for on-peak period application of the 80% ratchet provision in each separate period will be established independently of the demands in the other period. The 80% ratchet during one period will extend back to the highest monthly on-peak period kW demand occurring in the 11 preceding months, but does not include any demands established in the other period.

The monthly shoulder period kW demand shall be the highest 15-minute integrated kW demand registered in any shoulder period during the month as determined by the Company.

The reactive demand shall be the highest 15-minute integrated kVAR demand registered in any on-peak period during the month.

BASIC RATE PER MONTH	Winter Billing Months December - March	Non-Winter Billing Months April - November
Customer Charge	$104.17	$104.17
Demand Charge		
On-Peak	$10.91/kW	$5.37/kW
Shoulder	$3.14/kW	$2.57/kW
Energy Charge		
On-Peak	$0.087999/kW	$0.071002/kW
Shoulder	$0.083281/kW	$0.070269/kW
Off-Peak	$0.071400/kW	$0.063546/kW

The daily periods for weekdays (Monday through Friday, excluding holidays) shall be as follows:

On-Peak:	7:00 am	to	12:00 pm and
	4:00 pm	to	8:00 pm
Shoulder:	12:00 pm	to	4:00 pm
Off-Peak:	8:00 pm	to	7:00 am

Figure 9.38 cont.

Sample Electric Rate Schedule (cont.)

RATE IGS-S-TOU

INTERMEDIATE GENERAL SERVICE — SECONDARY — TIME-OF-USE (cont.)

DEMAND (cont.)

Winter Saturdays, Sundays and holidays shall be treated as consisting of two shoulder periods, from 7:00 am to 12:00 pm and from 4:00 pm to 8:00 pm, with the remainder to be designated as off-peak. Saturdays, Sundays and holidays during this non-winter period shall be designated as all off-peak.

REACTIVE DEMAND CHARGE

$0.69 per kilovar (kVAR) of reactive demand in excess of 50% of the monthly on-peak kW demand.

MINIMUM CHARGE

The Customer Charge, plus the Demand Charge, plus the Reactive Demand Charge, per month.

Figure 9.38 cont.

Electric Bill Analysis

Ref. Section Number	Ref. Section Description	Applicable Value From Example 2 Bill	Comments
1	Power company ID	Bright Light Power Co.	Located in top left corner of the bill
2	Customer service address	Happy Home School 8 Lefther Ridge Road Whereyouat, ME	Located just below power company identification
3	Billing address	Not applicable	Assume the same as the customer service address unless otherwise indicated
4	Account number	546-0120030-015	Located in top right corner of the bill
5	Service period	6/28/95 - 7/30/95	Determined from meter read dates
6	Service rate	General service, ISTOU	Top portion of the bill; TOU generally indicates a time-of-use rate
7	Meter ID	718	Located in energy usage section of the bill
8	Customer charge	$104.17	Flat charge regardless of energy usage or demand
9	Meter reads	counts (cts) @ end: 9,189 - counts (cts) @ start: 9,140	(9,189 - 9,140) = 49 counts (cts) 49 cts x 400 kWh/cts = 19,600 kWh
10	Meter multipliers	400 for kWh meter	Demand meter probably has the same multiplier, but not certain
11	Used energy & demand	On-peak: 6,800 kWh, 90.40 kW measure, 743.04 kW bill Shoulder: 3,200 kWh, 86.40 kW measure, 86.40 kWh bill Off-peak: 9,600 kWh	Located in billing detail section of the bill; 80% ratchet clause is in effect for on-peak demand - maximum winter demand within the last 12 months was 928.8 kW (= 743.04 kW/0.80)
12	Rates for energy & demand	On-peak: $0.071002/kWh $5.37/kW Shoulder: $0.070269/kWh $2.57/kW Off-peak: $0.063546/kWh	Energy rates are located in the billing detail section of the bill; demand rates are not shown on the bill but can be gotten from the rate schedule
13	Energy service charge	$1,317.71	= ($0.071002/kWh x 6,800 kWh) + ($0.070269/kWh x 3,200 kWh) + ($0.063546/kWh x 9,600 kWh)
14	Demand service charge	$4,212.17	= ($5.37/kW x 743.04 kW) + ($2.57/kW x 86.40 kW)
15	Misc. charges & adjustments	Not applicable	No miscellaneous charges or adjustments applied to this bill
16	Total charges & adjustments	$5,529.88	= $1,317.71 + $4,212.17
17	Taxes	None	School is probably public and therefore tax exempt
18	Total amount due	$5,529.88	= $1,317.71 + $4,212.17
19	Late charge statement	None	The company's late payment policy is found in a separate part of the rate schedule book

Reprinted with permission from *Fundamentals of Electrical Systems*, ASHRAE, Inc.

Figure 9.39

In summary, electric costs are based on how much electric energy has been consumed in the billing period, usually on a monthly basis. Today utility companies promote energy conservation by charging for energy usage with declining block rates. Under this scheme, rates decrease with consumption that is contrary to ascending block rates. With deregulation, facility managers and plant engineers have more freedom to negotiate better rates with utilities or even own their own electricity.

Industrial Control

Electronic Operations

Electrical devices are typically electro-mechanical or electronic. Examples of electro-mechanical devices, which are large and bulky compared to the electronic counterparts, include relays and stepper switches. Examples of electronic devices include diodes and transistors. These items can be put together to form integrated circuits.

Electronics can be analog or digital, with digital the most common because only two states have to be interpreted and systems can operate much faster. Digital electronics use binary numbering. The two binary states can be used to create any number in the binary system.

In binary numbering, the number 2 is raised to a power.

- 2^3 – eight's
- 2^2 – four's
- 2^1 – two's
- 2^0 – one

The number 3 in the binary system is represented as 0011; the number 4 is 0100.

In digital electronics, transistors provide the basics for integrated circuits that contain logic gates. There are many types of logic gates. Some of the types include “and,” “or,” “nand,” and “nor.”

Truth tables and Boolean algebra are used when working with digital logic. Boolean algebra uses three operations to write logic statements: “and,” “or,” and “not.”

The following are examples of typical logic:

The logic for the “and” function is written “AB.”
The logic symbol for the “or” function is written “A+B.”
The logic symbol for the “nand” function is written “$\overline{AB}$.”
The logic symbol for the “nor” function is written “$\overline{A+B}$.”

Control and Industrial Electronics

Control energy sources are either *pneumatic* or *electric*. Control states may be *analog* or *digital*. In an analog system, values vary. Pneumatic controls typically vary the pressure between 3 and 15 pounds per square inch. Electronic controls, on the other hand, usually use one of the following methods:

- Varying voltage between 1 to 5 volts
- Varying current between 4 and 20 milliamperes

Digital systems are simpler: the value is either on or off, high or low, 0 or 1.

Control Devices: Typical pneumatic devices include the following:

- Switches that typically interface to electric systems
- Relays that select the high or low signal
- Controllers that follow logic using one or more inputs for a single output

Typical electric devices include the following:

- Switches that are manually operated
- Relays that are remotely operated
- Controllers that follow a logic sequence as simple as a time clock or as complex as a Programmable Logic Controller (PLC).

Using the "and," "or," and "not" functions from Boolean algebra, several digital logic statements can be written.

Figure 9.40 provides an example using relay contacts to form a digital logic circuit that will turn on a light.

Each "R" contact can operate independently. Any contact that closes will cause the output light to operate. This is an example of an "or" logic circuit: $R_1 + R_2 + R_3 + R_4 = 1$.

Examples of integrated circuits (IC) with "and" and "or" logic gates are shown in Figure 9.41. These integrated circuits are designed with Transistor to Transistor Logic (TTL). This packaging of multiple logic gates onto the IC is referred to as dual in-line packages (DIPs).

Figure 9.42 is an example of an "and" logic gate and truth table.

Figure 9.43 is an example of an "or" logic gate and truth table.

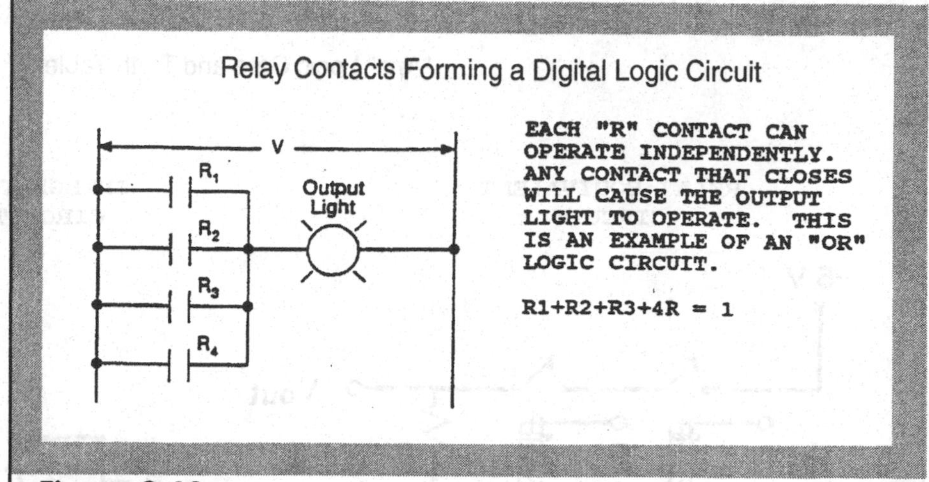
Relay Contacts Forming a Digital Logic Circuit
V
R1
R2
R3
R4
Output Light
EACH "R" CONTACT CAN OPERATE INDEPENDENTLY. ANY CONTACT THAT CLOSES WILL CAUSE THE OUTPUT LIGHT TO OPERATE. THIS IS AN EXAMPLE OF AN "OR" LOGIC CIRCUIT.
R1+R2+R3+4R = 1

Figure 9.40

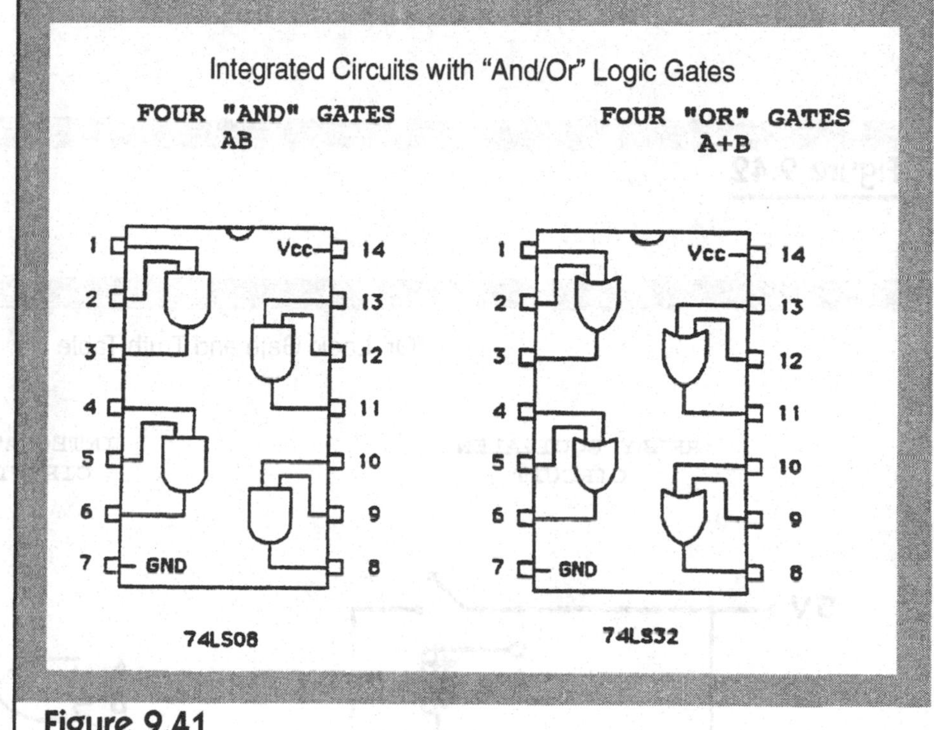
Integrated Circuits with "And/Or" Logic Gates
FOUR "AND" GATES
AB
FOUR "OR" GATES
A+B
Vcc
GND
74LS08
74LS32

Figure 9.41

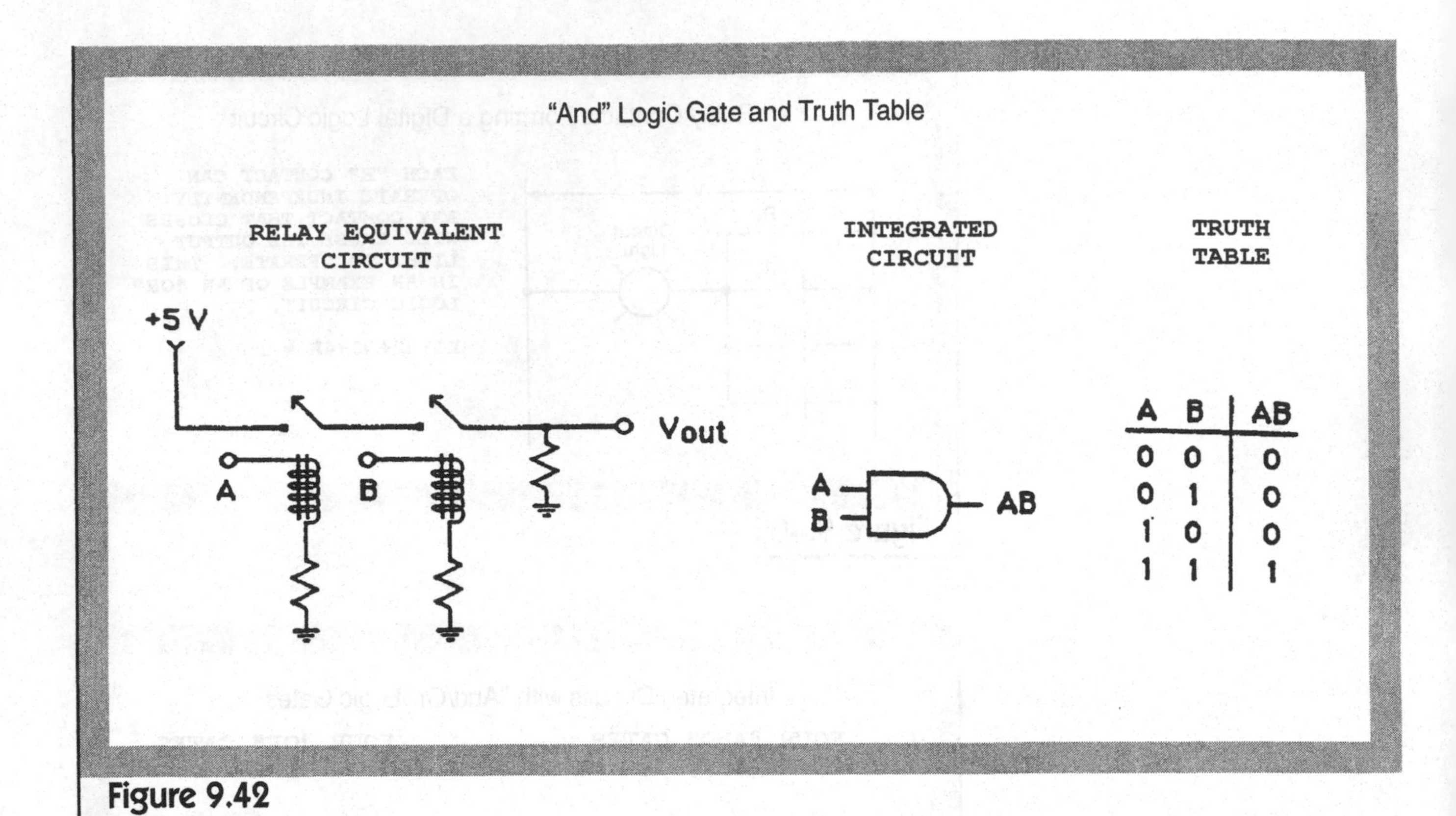

A	B	AB
0	0	0
0	1	0
1	0	0
1	1	1

Figure 9.42

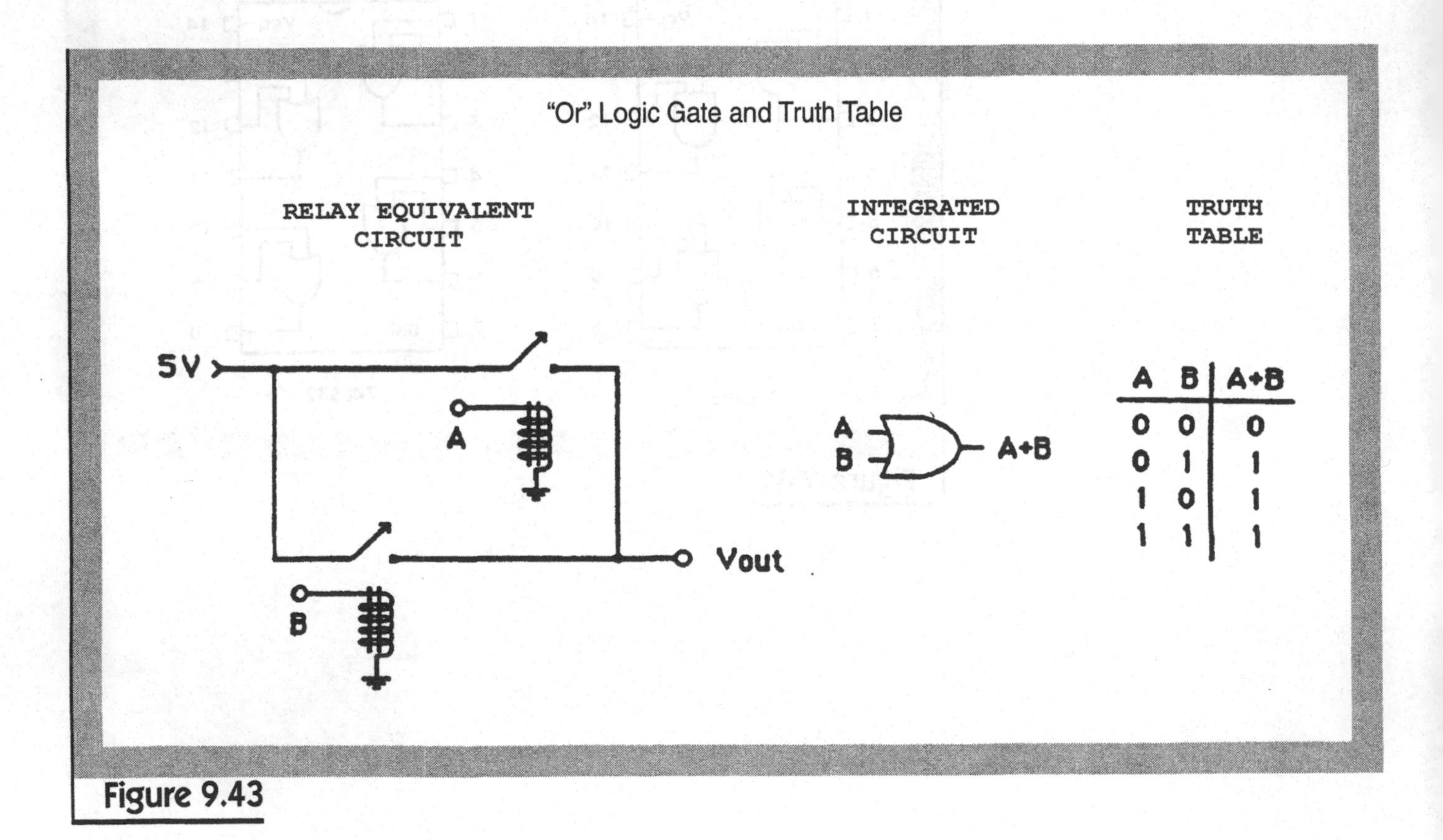

A	B	A+B
0	0	0
0	1	1
1	0	1
1	1	1

Figure 9.43

Resource Publications

Croft, Terrell, and Wilford I. Summers. *American Electricians' Handbook, 12th ed.*. New York: McGraw-Hill Book Company, Inc, 1992.

Fink, Donald G., and H. Wayne Beaty. *Standard Handbook for Electrical Engineers, 11th ed.*. New York: McGraw-Hill Book Company, Inc, 1978.

Higgins, Lindley R., and L.C. Morrow. *Maintenance Engineering Handbook, 3rd ed.* New York: McGraw-Hill Book Company, Inc, 1977.

Illuminating Engineering Society of North America. *IES Lighting Handbook Application Volume*. New York, 1993.

National Fire Protection Association. *National Electrical Code*. Quincy, MA, 1995.

Nowak, Rich, and Rick Coughlin. *Fundamentals of Electrical Systems and Building Electrical Energy Use*. Atlanta: American Society of Heating, Refrigerating, & Air-Conditioning Engineers, Inc., 1997

Power Systems Engineering Committee of the IEEE Industry Applications Society. *IEEE Recommended Practice for Electric Power Systems in Commercial Buildings*. IEEE: Std 241-1990.

Power Systems Engineering Committee of the IEEE Industry Applications Society. *Power and Grounding Sensitivity Electronic Equipment*. IEEE: Std 1100-1992.

Wagner, David L. *Digital Electronics*. Orlando, FL: Harcourt Brace Jovanovich, 1988.

For Additional Information

There are many reference sources that a plant engineer can use for electrical projects. The following is a list of organizations that provide current standards, practices, and guidelines. This list is far from exhaustive, as there are also many excellent books written on electric systems.

American Society for Testing and Materials (ASTM)
100 Barr Harbor Drive
West Conshocken, PA 19428-2959
(610) 832-9500
www.astm.org

American National Standards Institute (ANSI)
11 West 42nd Street
New York, NY 10036
(212) 642-4900
www.ansi.org

American Society of Heating, Refrigerating, and Air-Conditioning Engineers, Inc.
1791 Tullie Circle, N.E.
Atlanta, GA 30329
(404) 636-8400
www.ashrae.com

BOCA International, Inc.
4051 W. Flossmoor Road
Country Club Hills, IL 60478-5795
(708) 799-2300
www.bocai.org

The Institute of Electrical and Electronics Engineers (IEEE)
345 East 47th Street
New York, NY 10017
(212) 705-7900
www.ieeeusa.org

IEEE has a large number of standards that apply to electrical equipment found in industrial and commercial facilities:

IEEE Color Book Series

- IEEE-141 (Red Book): *IEEE Recommended Practice for Electric Power Distribution for Industrial Plants*
- IEEE-142 (Green Book): *IEEE Recommended Practice for Grounding of Industrial and Commercial Power Systems*
- IEEE-241 (Gray Book): *IEEE Recommended Practice for Electric Power Systems in Commercial Buildings*
- IEEE-242 (Buff Book): *IEEE Recommended Practice for Protection and Coordination of Industrial and Commercial Power Systems*
- IEEE-399 (Brown Book): *IEEE Recommended Practice for Industrial and Commercial Systems Analysis*
- IEEE-466 (Orange Book): *IEEE Recommended Practice for Emergency and Standby Power Systems for Industrial and Commercial Applications*
- IEEE-493 (Gold Book): *IEEE Recommended Practice for the Design of Reliable Industrial and Commercial Power Systems*
- IEEE-602 (White Book): *IEEE Recommended Practice for Electric Systems in Health Care Facilities*
- IEEE-739 (Bronze Book): *IEEE Recommended Practice for Energy Conservation and Cost-Effective Planning in Industrial Facilities*
- IEEE-1100 (Emerald Book): *IEEE Recommended Practice for Powering and Grounding Sensitive Electronic Equipment*
- IEEE-P1015 (Blue Book): *IEEE Application Guide for Low-Voltage Circuit Breakers used in Industrial and Commercial Power Systems*

IEEE also has several additional Color Books planned:

- IEEE-P551 (Violet Book): *IEEE Recommended Methods for Calculating AC Short Circuit Currents in Industrial and Commercial Power Systems*
- IEEE-P902 (Yellow Book): *IEEE Guide for Maintaining, Operating, and Safety of Industrial and Commercial Power Systems*

Illuminating Engineering Society of North America (IES)
120 Wall Street, 17th Floor
New York, NY 10005
(212) 248-5000
www.iesna.com

IES publishes various lighting recommended practices and the *IES Lighting Handbook* (IES HB-93: IES Lighting Handbook Reference and Application).

National Electrical Contractors Association (NECA)
3 Bethesda Metro Center, Suite 1100
Bethesda, MD 20814
(800) 888-6322
www.necanet.org

National Electrical Manufacturers Association (NEMA)
1300 North 17th Street, Suite 1847
Rosslyn, VA 22209
(703) 841-3200
www.nema.org

NEMA publishes a wide selection of equipment standards, including:

- NEMA AB1: *Molded Case Circuit Breakers*
- NEMA PB1: *Panelboards*
- NEMA SG2: *High Voltage Fuses*

National Fire Protection Association (NFPA)
One Batterymarch Park
P.O. Box 9101
Quincy, MA 02269-9101
(800) 344-3555
www.nfpa.org

The NFPA publishes the *National Electrical Code* (NFPA-70), along with the following resources:

- NFPA-70B: *Electrical Equipment Maintenance*
- NFPA-70E: *Electrical Safety Requirements for Employee Workplaces*
- NFPA-72: *National Fire Alarm Code*
- NFPA-72E: *Automatic Fire Detectors*
- NFPA-72G: *Installation, Maintenance and Use of Notification Appliances of Protective Signaling Systems*
- NFPA-72H: *Testing Procedures for Local, Auxiliary, Remote Station, and Proprietary Protective Signaling Systems*
- NFPA-75: *Protection of Electronic Computer/Data Processing Equipment*

- NFPA-101: *Life Safety Code*
- NFPA-110: *Emergency and Standby Power Systems*
- NFPA-111: *Stored Energy Emergency and Standby Power Systems*
- NFPA-780: *Lightning Protection Code*

There are many additional NFPA standards that are specific to different industries and special standards. NFPA also publishes many handbooks that expand on the information available in the standards books, including:

- *NEC Handbook*
- *Fire Alarm Signaling Systems Handbook*
- *Electrical Installations in Hazardous Locations*

Underwriters Laboratories, Inc. (UL)
333 Pfingsten Road
Northbrook, IL 60062-2096
(847) 272-8800
www.ul.org

Underwriters Laboratories, Inc. publishes standards for safety of electrical equipment. The standards include the testing procedures to qualify for UL listing. Most of the standards are specific to equipment design but can offer insight to equipment testing requirements.

Chapter 10

Environmental, Health & Safety

Chapter 10 Environmental, Health & Safety

The impact of an organization's Environmental, Health and Safety (EH&S) program can have a substantial effect on the organization as a whole. A "green" organization can attract talented, environmentally conscious employees, as well as venture capital from investors who support the organization's efforts to respect the environment. For the noncompliant organization, its choice to cut environmental corners can lead to poor morale, loss of productivity, and a poor image in the community, as well as substantial fines for violation of regulatory requirements.

Once thought to involve just the plant or manufacturing process, the EH&S umbrella now covers everything from the loading dock to the stack, and from the surrounding neighborhood to the boardroom. While larger organizations may support an environmental administrator, many mid-to-small-size companies continue to include the management of the EH&S program within the facility manager's job description. Considering the sizable impact of noncompliance, it is prudent for the facility manager or engineer to have a broad knowledge of environmental, health and safety regulations and to have the resources available should (or more appropriately, when) expert advice is needed.

This chapter is intended to provide an overview of current environmental, health and safety regulations for day-to-day use by facilities personnel. Source materials have been incorporated by reference and may be consulted should more in-depth information be sought.

The following section is reprinted with permission from *Certified Facilities Environmental Professional Exam Review Book*, Association for Facilities Engineering.

Air Quality and Air Emissions

Because of their importance to public health and environmental quality, the control of air emissions by responsible agencies is one of the most stringently regulated components of plant operation. The conduction of an emissions audit is crucial to both plant process decisions and overall business planning given the level of capital investment associated with air pollution control equipment and associated implications to manufacturing operations in meeting permit conditions. Lack of compliance with applicable air regulations can result in the issuance of consent orders by oversight agencies, fines, or restrictions to plant operations. The air emissions audit consists of the following general steps:

1. Identification of regulations that could potentially be applicable to the plant operation and location;
2. Inventory of air emissions sources for the plant;
3. Determination of which air permits apply to the plant;
4. Evaluation of required air pollution control systems; and
5. Assessment of operational compliance of the plant with permits.

Given the dynamic nature of plant operations, the air emissions audit should be conducted on a periodic basis to ensure that the changes that have occurred have not generated the need for a new permit that was previously not applicable and/or created any violations of existing permit conditions.

Applicable Regulations: In adopting the Clean Air Act Amendments of 1990, Congress created significant new regulatory programs and dramatically altered others. In doing so, Congress built upon the federal regulatory framework created by the Clean Air Act Amendments of 1970 and 1977. Under the 1990 Amendments, a broad range of plant operations that previously did not require air permits are now required to comply with applicable provisions of the regulations.

At the federal level, the key regulations affecting plant operations are provisions of the original Clean Air Act, the 1990 Amendments, and the SARA Title III toxic chemical release reporting requirements. For a number of states, the EPA has delegated the program responsibility for air permitting to the state environmental agency while retaining authority for oversight review as required. It is important to determine which agency has regulatory jurisdiction over the plant, which will depend on the state where the plant is located.

In reviewing the plant operations for compliance with air regulations, particular attention needs to be given to the following:

- Clean Air Act National Ambient Air Quality Standards,
- National Environmental Standards for Hazardous Air Pollutants,
- New Source Performance Standards,
- 1990 Clean Air Act Amendments–Title V Operating Permit
- 1990 Clean Air Act Amendments–Title III Hazardous Pollutants, and
- State-Specific Requirements.

Clean Air Act/Clean Air Act Amendments of 1990

National Ambient Air Quality Standard:

As its name suggests, the primary purpose of the Clean Air Act (CAA) is to improve air quality by limiting the amount of pollution emitted into the atmosphere. Under this legislation, the EPA has established maximum concentration levels for certain air pollutants. These regulatory ceilings are referred to as National Ambient Air Quality Standards (NAAQS). NAAQS have been adopted for six specific types of pollutants, known as criteria pollutants:

- Lead
- Nitrogen oxides (NO_x), measured as NO_2 —These substances are produced when fuel is burned at very high temperatures. These pollutants form one of the principal components of acid rain. NO_x also contributes to the formation of ozone/smog.
- Sulfur oxides, measured as SO_2—These corrosive pollutants are produced when fuel containing sulfur is burned. Sulfur dioxide is another prime component in the formation of acid rain. Industrial plants are a major source for these substances.
- Carbon monoxide (CO)—This odorless pollutant is produced in combustion processes.
- Ozone—An unstable compound of three oxygen atoms, ozone is a major ingredient of smog, a type of urban pollution which can cause eye and respiratory ailments.
- Particulate matter (PM_{10})—These solid or liquid pollutants are produced by fuel combustion and other processes often used by industries.

Each state has enacted a State Implementation Plan (SIP), providing for compliance with NAAQS. The SIP is the central core of the regulatory framework, which includes legally enforceable control requirements. It is important to know the compliance status of the plant location with respect to the above referenced pollutants. The emission standards that a plant must meet are dependent on this status.

Under the CAA, the country is divided into Air Quality Control Regions (AQCR), which have been classified as either attainment areas, nonattainment areas or undesignated areas. Attainment areas are those in which NAAQS have been met and air quality is generally good. Regulations have been enacted to ensure the Prevention of Significant Deterioration (PSD) of the air quality in such attainment areas. These statutes, in other words, protect the desirable status quo. In nonattainment areas, NAAQS for certain criteria pollutants have yet to be met. Areas classified as "undesignated" are assumed to be in compliance.

States have a great deal of emission control responsibility under the CAA regulations. They are responsible for seeing that federal air quality standards are met, but have considerable leeway in deciding how this is done. As a result, the federal government sets attainment goals and provides technology and guidance to assist the states in achieving these goals through implementation of specific regulations and limits on pollution sources, both stationary and mobile. Under the 1990 amendments, states may receive the authority to issue federally enforceable operating permits to major sources of criteria pollutants or air toxics. The type of permit will depend on the attainment status of the area where the source is located, as well as the quantity of pollutants emitted.

National Environmental Standards for Hazardous Air Pollutants: The EPA regulations on National Environmental Standards for Hazardous Air Pollutants (NESHAPs) control pollutants from certain sources which are not covered by NAAQS but which pose significant risk to the public health. Under the regulations, select pollutants have been classified as hazardous. The pollutants with defined standards are radon, beryllium, mercury, vinyl chloride radionuclides, benzene, asbestos, inorganic arsenic, and volatile hazardous air pollutants when associated with certain uses of the pollutant. New and modified plant sources must meet the requirements upon startup. Existing sources that were operating at the time of implementation of the regulations had a period of time to meet compliance guidelines.

New Source Performance Standards: For the industrial categories listed in Figure 10.1, the EPA has established New Source Performance Standards (NSPS) for new and significant stationary sources. The sources falling under these industrial categories need to meet the required technology standards for applicable pollutants, monitor plant performance and submit reports on facility compliance to either the EPA or the site environmental agency with the delegated authority. Applicability of an NSPS to a plant is based on the date which the facility commenced construction, modification or reconstruction.

Clean Air Act Amendments: Although the 1990 Clean Air Act Amendments (CAAA) do not alter the basic regulatory structure of the Clean Air Act, they dramatically increase the number and types of manufacturing activities that are subject to regulations. The titles of the 1990 amendments include:

I. Attainment and maintenance of National Ambient Air Quality Standards

II. Mobile sources

III. Hazardous air pollutants

IV. Acid deposition control

V. Operating permits

VI. Stratosphere ozone protection

VII. Enforcement

VIII. Miscellaneous

IX. Clean air research

X. Disadvantaged business concerns

XI. Employment transition assistance

On an implementation schedule basis, the most critical titles of the act that affect manufacturing plant activities are Title V–Operating Permits and Title III–Hazardous Air Pollutants.

Industrial Categories with New Source Performance Standards (NSPS)

Category	
Aluminum reduction plants	Municipal waste combusters
Ammonium sulfate plants	Natural gas processing plants
Asphalt roofing material manufacturing	Nitric and sulfuric acid plants
Automobile surface coating operations	Non-ferrous smelters
Basic oxygen process furnaces	Nonmetallic mineral processing plants
Battery manufacturing plants	Petroleum dry cleaners
Beverage can coating industry	Petroleum liquid storage vessels
Brass and bronze production plants	Petroleum refineries
Bulk gasoline terminals	Petroleum refinery wastewater systems
Coal preparation plants	Phosphate fertilizer industrial facilities
Electric utility steam generating units	Phosphate rock plants
Fossil fuel-fired steam generators	Portland cement plants
Glass manufacturing plants	Rubber tire manufacturers
Grain elevators	Sewerage treatment plants
Graphic arts plants	Solvent cleaning machine operations
Hot mix asphalt plants	Stationary gas turbines
Industrial/commercial/steam generating units	Steel plant electric arc furnaces
Industrial surface coating operations	Surface coating of metal furniture
Kraft pulp mills	Synthetic fiber production facilities
Lime manufacturing plants	Synthetic organic chemicals industry
Medical waste incinerators	Tape and label surface coating operations
Metallic mineral processing plants	Volatile organic liquid storage vessels

Association for Facilities Engineering, *CFEP Review Book*

Figure 10.1

Required Permits

Title V–Operating Permits: The Clean Air Act Amendments of 1990 contain a variety of new programs designed to reduce emissions of air toxics, improve urban air quality and control the precursors of acid rain. For each major stationary source, the mechanism by which the EPA will integrate all of the federally applicable requirements arising from these programs is the federal renewable Operating Permit under Title V.

The Operating Permit Rule issued on July 21, 1992 implements an entirely new and different air permitting initiative. The rule defines the minimum requirements for state programs, and requires states to develop their programs and seek EPA approval on a tight schedule. Once a state receives approval and publishes regulations to administer its program, companies must apply for their permits within one year. Preparing an application of the magnitude called for in the Operating Permit requires a significant commitment of resources. Producing an application that meets a state's criteria for "completeness" is perhaps the single most important milestone a facility can achieve en route to an acceptable permit.

The key to determining whether a facility needs a federal operating permit is knowing whether or not the facility is a major source. In making this determination, it is necessary to know what regulated pollutants are emitted, the magnitude of these emissions and the air quality status of the region where the facility is located. A source is major if it has the "potential to emit" 100 tons per year (tpy) of a criteria pollutant, 10 tpy of a single Hazardous Air Pollutant, 25 tpy of any combination of Hazardous Air Pollutants or other amounts as defined for nonattainment areas. For example, in regions that are not meeting federal air quality standards, the threshold quantities defining a major source of some pollutants are reduced depending upon the air quality category (i.e., the worse the region's air quality, the lower the threshold for major source determination). Figure 10.2 summarizes the pollutant emission thresholds for major sources.

Before the 1990 amendments, fewer than 20 pollutants were specially regulated under the Clean Air Act. Counting the pollutants covered by all of the Act's new and existing programs and excluding duplicates covered by more than one of these programs, the amended Act now regulates over 380 pollutants in the following categories:

- Criteria pollutants, consisting of SO_2, NO_x, CO, VOC, PM_{10} and lead;
- All pollutants regulated under New Source Performance Standards (NSPS), including H_2S, TRS, H_2SO_4, mist fluorides and Total Suspended Particulates (TSP);
- Hazardous Air Pollutants (HAPs) (189);
- Extremely hazardous substances (138 in final list for coverage by the Accident Release Program) and;
- Class I & II ozone-depleting substances, including CFCs, halons, HCFCs, carbon tetrachloride, methyl chloroform and methyl bromide.

Facilities will need operating permits not only if they are major sources but also if they include sources subject to NSPS acid rain provisions or NESHAPs. Once required, an operating permit must address each regulated pollutant from each source, no matter how small the emission. The permit also will have to account for fugitive emissions, regardless of source categories. States are authorized to set "de minimis" emission thresholds or "insignificant activities" applicable to their own permit programs. De minimis emission quantities are levels below which regulations do not apply.

Major Emission Sources Requiring Operating Permits

Sources	Potential Annual Emissions
Attainment Areas	100 tons
Nonattainment Areas	
Carbon Monoxide	
Moderate	100 tons
Serious	50 tons
PM_{10}	
Moderate	100 tons
Serious	70 tons
Ozone (VOCs and NO_x*)	
Marginal and Moderate	100 tons
Serious and All Transport Regions	50 tons
Severe	25 tons
Extreme	10 tons
Air Toxics (189) Hazardous Air-Pollutants	
One HAP	10 tons+
Two or more HAPs	25 tons

* Does not apply if the EPA finds that NO_x control not required

\+ The EPA retains the authority to reduce the major source threshold under the air toxics program.

Association for Facilities Engineering, *CFEP Review Book*

Figure 10.2

Title III–Hazardous Air Pollutants: Title III of the amendments revamps and revitalizes the federal programs to control hazardous air pollutants. Compliance with Title III is complicated by the fact that the federal program does not replace existing state programs. To the extent that many state programs regulate air toxics by setting acceptable ambient levels, those state programs may continue and compliance with both programs may be required.

The 1990 amendments list 189 hazardous air pollutants (HAPs) that must be regulated under the air toxics program. The EPA must periodically review and, where appropriate, revise this list. Third parties, such as industry and environmental groups, can petition for listing or delisting.

Title III was structured to regulate HAPs by industry category and subcategory. The EPA was directed to publish a list of HAP source categories by November 1991. The EPA estimates that the list will ultimately identify over 750 categories and subcategories.

The 1990 amendments establish for the first time a threshold for sources to be regulated under the air toxics program. For purposes of Title III, the amendments define a major source as one with the potential to emit greater than 10 tpy of any one of the listed HAPs, although the EPA has the authority to reduce the major source threshold. Sources which are not major sources are "area sources" and may be regulated when the EPA issues a subsequent round of area source controls.

Similar to the air toxics control program under the 1977 amendments, the EPA will set the new control standard, Maximum Achievable Control Technology (MACT), by industry category or subcategory. It will be based on reductions achieved in practice by sources of HAP emissions. These standards will generally be based upon the maximum level of control achieved in practice within the category. Sources will be required to achieve the MACT standard within three years after the standard is formally established.

The amendments also provide the sources which have achieved an early reduction of HAPs of 90 percent or more from inventoried 1987 levels the opportunity to apply to the EPA or the state for a six-year extension of the applicable MACT deadline. Whether to grant the extension, however, is up to the EPA.

As part of the Title III air toxics program, the amendments include a new and comprehensive accidental air emission release program modeled in part after New Jersey's Toxic Catastrophe Prevention Act. Although several states have similar programs in place, it is not yet clear how the federal program will interact with those programs.

Initially, the EPA must identify a list of at least 100 Extremely Hazardous Air Pollutants (EHAPs). The EPA is required to review substances designated as extremely hazardous on the SARA Title III Community Right to Know List

("Section 313 List") in developing the EHAP list. It is likely that the EHAP list will include many of those substances. The amendments also contain a list of 16 chemicals that the EPA must include on the EHAP list:

ammonia	hydrogen fluoride
anhydrous ammonia	hydrogen sulfide
anhydrous hydrogen chloride	methyl chloride
anhydrous sulfur dioxide	methyl isocyanate
bromine	phosgene
chlorine	sulfur trioxide
ethylene oxide	toluene di-isocyanate
hydrogen cyanide	vinyl chloride

The amendments require that facilities which manufacture, use or handle any EHAPs perform detailed engineering analyses, establish potential exposure scenarios, undertake comprehensive hazard assessments and develop release prevention and control programs for each of the EHAPs. The amendments do not specify a de minimis amount below which the program does not apply, although the EPA is required to set a threshold for each EHAP as it develops the EHAP list.

Ed. Note: At publication the list of Regulated Toxic Substances included 77 items. A complete list of these substances and threshold quantities for accidental release may be found at www.epa.gov/swercepp/rules/listrule.htm/.

In addition to the above regulations, the EPA is also required to develop regulations establishing release prevention, detection and correction measures including:

- Use, operating, repair, replacement and maintenance requirements for equipment to monitor, detect and control releases;
- Procedures and measures for emergency responses after an accidental release; and
- Requirements for storage of EHAPs as well as for use in operation.

The risk management plan (RMP) to detect and prevent or minimize accidental releases must include:

- An assessment of the potential effects of an accidental release of an EHAP;
- A program for prevention of accidental releases; and
- A response program including specific actions to be taken after an accidental release.

Finally, the {CAA} amendments require protection of employees from potential exposure to hazardous chemicals in the workplace. After finalizing a list of highly hazardous chemicals, the Department of Labor, through the Occupational Safety and Health Administration, must set standards requiring that all employers:

- Develop and maintain written safety information;
- Perform a workplace hazard assessment;
- Consult with and make records available to employees on the assessment;
- Establish a system to respond to identified problems;
- Periodically review the assessment and response system;
- Compile written operating procedures;
- Provide written procedures to employees;
- Ensure contractors have proper training;
- Train employees;
- Establish a quality assurance program;
- Establish a maintenance system for critical equipment;
- Conduct pre-startup safety reviews of all equipment;
- Establish and implement written procedures to manage change to process chemicals, technology, equipment, and facilities; and
- Investigate every accident.

The following section is reprinted with permission from *Cost Planning & Estimating for Facilities Maintenance,* Chapter 6: "Codes and Regulations," by William H. Rowe III, AIA, PE, R.S. Means Co., Inc.

Indoor Air Quality

A topic of increasing interest and awareness to building users and regulators alike is indoor air quality (IAQ). Modern materials and technology have both increased the potential for building contamination and provided ways to alleviate it. Following are some common factors that affect building air quality in existing facilities.

Maintaining Positive Pressure:

With rare exceptions (such as laboratories and certain infectious disease rooms), buildings should be slightly positively pressured in relationship to the outside. In basic terms, buildings should be "pumped up" with more fresh air from the outside than they exhaust from toilets or general building exhaust. This is most difficult to achieve with variable air volume (VAV) systems, particularly in spring and fall when large quantities of both supply and exhaust air are being handled. Having controls in place that will properly track the air quantities of both the supply and return air while maintaining building pressurization will result in better overall air quality.

Maintaining Chillers During the Winter:

Chillers that are shut down in winter (sometimes as a result of cooling tower shutdown) often result in the inability to cool air on warm "swing" winter days. This causes building temperatures to rise, resulting in greater perception of air quality problems and a general feeling of stuffiness.

Closing of Outdoor Air Damper on Freeze Warning:

In winter, fresh outdoor air dampers will close to protect heating and/or cooling coils from freezing. This results in less fresh air to the occupants. While necessary to protect the coils, other designs such as "run around loops," or at least a quick response time for maintenance staff to manually open dampers helps to maintain a feeling of wellness in the building.

With the advent of sick building syndrome, IAQ has become a major topic in the building industry. Building owners, architects, engineers, building managers, and occupants are now actively seeking to ensure that indoor environments are safe and productive. While concerns over indoor air quality have resurfaced recently, proper ventilation and air filtration has evolved in an attempt to keep buildings habitable and to vent out the fumes and stenches that made people sick. The primary reason for poor air IAQ recently is that modern buildings are more airtight and some modern building materials are more prone to pollute the indoor environment. Safer material alternatives should be investigated, and they need to be vented properly.

The effects of poor IAQ are extensive. Respiratory illness, allergic reactions, headaches, and drowsiness are some of the ill effects it can cause. The result can be poor health and, consequently, lack of production, lost work days, and low morale among workers.

While ventilation standards have been increased in response to IAQ, it is important to realize that ventilation is not the cause, and often is a weak cure to the problem. Realizing that if there were no indoor contaminating pollutants, there would be no indoor air quality problem, goes to the heart of the issue. Poor IAQ is caused by any combination of the following:

Excessive emissions of Volatile Organic Compounds (VOC). These originate in large measure from "wet" building materials, such as:

- Plywoods and particleboards containing urea-formaldehyde resins
- Paints
- Adhesives
- Caulking

And other sources such as:

- Cleaning, waxing, and polishing agents
- Equipment
- People
- Office products

Products in the environment, such as:

- Tobacco smoke
- Nitrogen dioxide
- Carbon dioxide
- Carbon monoxide

- Radon
- Formaldehyde
- Sulphur dioxide
- Ozone
- Asbestos

Lack of adequate fresh air ventilation to the building or space in question

Products of combustion from heating plants and vehicles entering the building through the ventilating system or through windows.

Poor IAQ is often improved by increasing the fresh air supply and exhaust and by controlling emissions from materials. However, extensive investigations and testing are often required to find the actual sources of the poor air quality.

Although increased fresh air ventilation will dilute the air and consequently improve the quality of the air, it will not stop the emissions of the pollutants; therefore, indoor air quality is as much an architectural issue as it is a mechanical or ventilation issue.

Architectural Design and Emissions from Building Materials:

While traditional construction materials, such as masonry and plaster, are relatively benign once installed, many newer materials contain volatile organic compounds (VOCs). Most of the VOCs emitted from building materials are mucus membrane irritants and consequently there is a high rate of related symptoms for people in new, remodeled, retrofitted, or refurnished buildings where VOCs are present.

"Natural Materials"

Selecting building construction materials requires some consideration. While natural materials and sustainable design are desirable, there are disadvantages. Many natural materials will decay. Wood studs rot, and wool carpet attracts moths. To combat this decay, chemicals and preservatives are used that are deadlier than some of the artificial substitutes like metal studs or nylon carpet. There are also some natural materials such as lead, arsenic, asbestos, and formaldehyde that we all would rather not use in any case. Also, while there are newer materials that are less polluting, they are often less durable. Therefore, they will be reapplied more often, while a building is occupied, creating a higher frequency of emissions.

"Toxic-Free Products"

In the early stages of project design, the architect must be aware of the building products, their content, and the emission rates of VOCs. Specified products should have low pollutant emission characteristics. As noted earlier, many adhesives, paints, and sealants have very high VOC emission characteristics. Carpeting, too, can have a high emission of compounds such as formaldehyde and fibers. Manufacturers are producing equipment and materials with lower emission rates and many claim that their products are "low polluting," "non-toxic," and "environmentally safe." The American Institute of Architects is one source of information on environmentally friendly building materials. Currently, there are few regulatory guidelines or limits for indoor air pollutants. The number of chemicals utilized in building construction is vast, and the sensitivity of individuals to any one chemical is widely variant.

Baking VOC During Construction

During construction there are procedures that help to considerably reduce VOC emissions. It is typically necessary to run the HVAC (particularly the heating) system during construction to provide warmth for plasterers and other finish trades. The heat often will "bake" the VOCs, vaporizing them so that the exhaust air system can remove them from the building before occupancy. A building may be deliberately baked for a week at warm temperatures with a high ventilation rate to help alleviate pollutants.

Architectural Layout: Good internal building layout contributes to good IAQ. It is important to separate pollutant-generating activities such as food preparation, graphic arts, physical exercise, smoking rooms, and photography from other areas of the building. These areas should be equipped with 100% exhaust and ventilation air systems to convey contaminants directly outdoors. These areas should be kept under negative pressure with respect to abutting occupied spaces.

Office layout and structure have a considerable effect on IAQ. Most office partitions absorb VOCs when the building or space is initially completed and release a constant rate of VOCs into the environment. In this way, many of the VOCs are not removed by the ventilation system at an early stage of occupancy. Office partitions also interfere with good air distribution and circulation, causing air pockets or voids with stale and contaminated air.

It is important to avoid installing "fleecy" products in which microbiological organisms can reside.

Open shelving with large amounts of paper products are another great source and home for VOCs and microbiological organisms.

Operable Windows: Operable windows in offices and new buildings were phased out in the 1960s in order to rely on mechanical ventilation systems and to control energy consumption. However, studies have suggested that incidences of poor IAQ, building-related illnesses (BRI), and Sick Building Syndrome (SBS) have been more frequent in mechanically ventilated buildings than in naturally ventilated buildings. Including operable windows in buildings could improve the IAQ and give occupants more control over their environment.

Site Layout: Site layout is also an important factor in controlling IAQ. As discussed previously in the ventilation section, locating loading areas, dropoff areas, parking areas, and high vehicular traffic areas away from air intake openings can greatly reduce contamination of the indoor air.

Effect of Mechanical Systems and Ventilation on Indoor Air Quality: Mechanical ventilation systems can be either an effective method of controlling IAQ or a source of poor IAQ. Cleanliness, system components, filtering, air distribution, humidity control, recirculation of air, transfer of air, system layout, and ventilation rates could all have positive or negative effects on IAQ.

Cleanliness:

The cleanliness of the intake louvers, ductwork, intake and recirculation filters, diffusers, and registers is important to maintain clear air circulation of ventilation air. Diffusers and registers are likely locations for dust and particulate matter to build up. A proper housekeeping program can ensure adequate cleaning.

System components such as cooling towers, cooling coils, drain pans, humidifiers, mixing boxes, and man-made fibrous insulations and liners are areas where many microbiological organisms can flourish and where organisms like *Legionella pneumophillia*, the bacteria that causes Legionnaires Disease, may be found. These organisms can be picked up in the airstream and circulated throughout the building. Cooling towers must be kept clean and treated with chemicals to avoid contamination. Water introduced into a duct system is a major cause of mold. Leakage of coils, backed-up condensate line, and steam leaks all warrant priority status on the repair list. Cooling coils and their drain pans should be cleaned regularly, and drain pans should be checked to ensure that they are sloped to drain. Humidifiers should be dry steam-type and not water-spray-type. Unitary portable humidifiers should be maintained and cleaned regularly. Mixing boxes should be inspected and cleaned regularly. Fibrous insulations should not be used as duct liners and should be avoided where possible.

Filtration:

System filters should be designed and selected for recirculation air and for outside air. Low efficiency filters used alone are not recommended.

Ed. Note: Low efficiency filters should be used as pre-filters to extend life of high efficiency filters.

Recirculation filters should be capable of removing all VOCs from the airstream. Outdoor air filters should be selected based on their capacity to sufficiently remove from the air the contaminants that are common in the outdoor environment of that area. City projects require filters adequate for vehicular fumes; specialized or HEPA filters may be required for industrial applications or for special applications in hospitals.

Air Distribution and Exhaust:

Proper air distribution is critical to overall air quality. Not only should the proper quantity of air be delivered to a space, but the distribution of the air to match the space layout and the density of occupants is an essential component of overall IAQ design. Moveable office partitions can cause problems by being too close to the ceiling and blocking distribution of the air to the occupants. Sometimes moveable partitions that worked well in an initial layout do not work well with a new layout.

Air movement and transfer, local exhaust, and dedicated supplies can prevent bad air from circulating throughout the wrong spaces. Kitchens and laboratories should be equipped with hoods and kept under negative pressure; graphic arts rooms and smoking lounges should have dedicated exhaust fans and should be kept under negative pressure. Office areas should be kept under relatively positive pressure. Air should not be recirculated from many areas such as kitchens, bathrooms, cafeterias, laboratories, print rooms, and similar spaces where there is a buildup of air pollutants.

Basements often have a buildup of humidity where microbiological organisms and mold will thrive. As these contaminants build up, they propagate through a building in time. Basements should be dehumidified and should have a source of ventilation.

Ventilation Standards:

When deciding on ventilation rates, an analysis should be made of the building structure and materials. It is important to provide fresh air to compensate for VOC emissions and to provide adequate fresh air for the occupants.

Specific Code Requirements:

Currently the only codes and regulations set to improve IAQ are those codes that set minimum values to the volumes of fresh air and air change rate that a space must receive. In the early 1970s the required fresh air rate per person was considered high (15-20 cfm per person). Following the oil crises in the mid-seventies, this level was reduced considerably to 5 cfm per person. However, because of the resulting poor IAQ, ASHRAE revised their standard (62-1989 Ventilation for Acceptable Indoor Air Quality) to increase the recommended fresh air supply to 15-35 cfm per person for most occupancies. It is expected that most state and local codes will adopt the ventilation rates ASHRAE recommends.

One consequence of increased ventilation standards is increased energy costs for the additional fresh air. Studies have indicated that while there is indeed a cost increase, the total energy impact for most buildings is under 10%. Schools, hotels, and retail buildings will be particularly affected by increased ventilation standards as a result of their high number of occupants. As these ventilation standards are implemented, the volume of fresh air and the energy consumed to condition that fresh air will be evaluated with new technology to improve the IAQ of buildings.

For further information you may call the Indoor Air Quality Information Clearinghouse at (800) 438-4318 to speak to an information specialist. You may make inquiries by fax (703) 356-5386 or via e-mail: iaqinfo@aol.com anytime. The address is:

Indoor Air Quality Information Clearinghouse
IAQ INFO
P.O. Box 37133
Washington, D.C. 20013-7133
Direct line: (703) 356-4020

The following section is reprinted with permission from *Certified Facilities Environmental Professional Exam Review Book*, Association for Facilities Engineering.

Wastewater

Since enactment of the Clean Water Act in 1972, a variety of new regulatory requirements have evolved at both the federal and state levels. These have included regulations pertaining to:

- Identification and control of toxics in wastewater;
- On-site pretreatment of wastewater prior to discharge;
- Application of water quality-based effluent limitations on plant discharges;
- Expansion of permitting requirements to include industrial stormwater discharges; and
- Pollution prevention and waste minimization.

Given the changing regulatory environment and the growing emphasis on the control of toxics in industrial wastewater, increasing importance is being placed on the performance of plant wastewater audits to ensure compliance with applicable standards and operational requirements.

Applicable Regulations

The key operational consideration associated with wastewater systems within a manufacturing facility is the management of the operation so that it meets the effluent limitations specified in the facility's discharge permits. Wastewater sources associated with a plant operation that are subject to environmental regulation include:

1. Process-generated wastewater from manufacturing operations;
2. Boiler blow down from a central utility plant;
3. Cooling tower discharge;
4. Stormwater that comes into contact with on-site industrial activity; and
5. Sanitary discharge.

The Federal Water Pollution Control Act of 1972 established the regulatory framework that governs and regulates industrial pretreatment facilities and controls their discharge into Publicly Owned Treatment Works (POTW). Discharge limits under the act are assigned based on the type of manufacturing process and the corresponding Standard Industrial Classification (SIC). Congress has fine-tuned the 1972 act with the following amendments: Clean Water Act of 1977; Municipal Wastewater Treatment Construction Grants Amendments of 1981; and Water Quality Act of 1987.

The Clean Water Act established the National Pollutant Discharge Elimination System (NPDES) program for controlling existing and new point sources and site stormwater management. The act also established New Source Performance Standards (NSPS) for industrial pretreatment. The key concept embodied in industrial wastewater regulations is that all ultimate discharges into the nation's waters are unlawful unless specifically authorized by a permit.

The Safe Drinking Water Act, enacted in 1974, is the basis for protecting public drinking water systems from contaminants that are harmful to the public health. Since 1974, the act has been amended by Congress through the following actions, which has expanded the scope of actions that are regulated: Safe Drinking Water Amendments of 1977, 1979, 1980, and 1986; and the Lead Contamination Control Act of 1988.

The enforcement of the federal acts can reside with the EPA regional office or be delegated to a state environmental agency if the requisite program administrative criteria are met. States and the municipal POTW also have the latitude to impose more stringent standards on industrial wastewater discharge. Industrial wastewater discharge permits are typically required from the state environmental agency prior to discharge to a POTW.

Clean Water Act/Safe Drinking Water Act

Clean Water Act: The Clean Water Act has as its objective the restoration and maintenance of the chemical, physical and biological integrity of the nation's waters. The act requires industry to use, depending on the plant type, either "best potential control technology (BPT)" or "best available technology (BAT)" that is economically available in controlling the quality of wastewater. The control of toxic pollutant discharges has been a key focus of the water quality programs. In addition to control technology requirements, the 1987 amendments require states to implement control strategies for identified "toxic hot spots" waters. These are waters that, because of previous contamination, are expected to remain polluted by toxic chemicals even after industrial discharges have installed wastewater cleanup technologies.

The EPA has issued regulations containing BPT and BAT effluent standards applicable to categories of industrial sources. Certain responsibilities have been delegated by the EPA to the states when required program criteria have been met. In determining the wastewater regulations applicable to a plant, it is important to determine the lead responsible agency.

The environmental regulations associated with the Clean Water Act affect a plant's operations from a number of perspectives. Under the Clean Water Act, an NPDES permit is required if an industrial facility is planning to discharge wastewater through a point source to surface waters that either directly or ultimately connect to navigable waters of the United States. Prior to obtaining the NPDES permit, the applicant is required to demonstrate that water quality standards will be met and that compliance monitoring through analytical testing will be performed. The permit typically specifies which pollutant may be discharged and in what concentrations.

The management of stormwater on industrial sites is also controlled by the NPDES permitting process. Stormwater that comes in contact with industrial operations needs to be controlled and managed with respect to pollutant concentrations. NPDES permits are also required for non-contact stormwater that is released through a point-source discharge to navigable waters of the United States.

Under the Clean Water Act, industrial operations are subject to general pretreatment standards if discharging wastewater to a POTW. The EPA has established categorical pretreatment standards applicable to specific industrial categories. The industrial facility is required to monitor compliance with the pretreatment standards.

To protect water resources, the Clean Water Act also imposes Spill Prevention Control and Countermeasure (SPCC) plan requirements on industrial facilities. The SPCC plan applies to oil and hazardous substance storage tanks, pipelines, oil retention drainage systems and loading/unloading areas.

Safe Drinking Water Act: The primary objectives of the Safe Drinking Water Act and subsequent amendments are to develop:

1. National primary drinking water regulations that incorporate maximum contaminant levels and treatment techniques;
2. Underground injection control regulations to protect underground sources of drinking water; and
3. Groundwater protection programs for sole-source aquifers and wellhead protection areas.

The drinking water regulations established by the act establish maximum contaminant levels for chemicals, radionuclides, microbes and turbidity. The EPA is adding to the list of contaminants on a regular basis. The primary enforcement responsibility for provisions of the act lies with the states.

The implications of the Safe Drinking Water Act to plant operations are:

1. The wastewater discharges to water bodies must demonstrate that no violations of drinking water standards will occur;
2. As drinking water standards become more stringent, increased attention will be given to industrial wastewater discharges;
3. Trends to move towards zero discharge of pollutants will continue; and
4. Regulatory initiatives to encourage in-plant pollution prevention will continue to evolve.

Wastewater Effluent Emissions Inventory

Water Balance: The initial step in preparing a wastewater effluent emissions inventory is to prepare a general water balance for the plant operations. The water balance traces the flow of raw water supply through the plant, accounting for use in product manufacturing, evaporation, fugitive losses and industrial wastewater effluent generation. With the facility water balance completed, the inventory of wastewater emissions from an industrial facility and the review of water pollution control systems can be accomplished through the following audit procedure:

1. Review the NPDES discharge permit and identify all permitted waste streams and outfalls.
2. Assemble and review NPDES Discharge Monitoring Reports and confirm compliance with numerical effluent limitations.
3. For visible outfalls/discharges, verify compliance with general permit criteria (i.e., that the discharge is not causing nuisance conditions in the receiving water associated with color, oil sheens or floating material).
4. Inspect the facility's process/sanitary collection system to ensure that there are no unpermitted connections/waste streams. Depending on site-specific collection system characteristics, verification activities may include one or more of the following: visual inspection of facilities/grounds; review of available collection system/utility mapping; discussions with plant personnel; or performance of limited testing activities, such as dye tests.
5. Visually inspect pretreatment/industrial wastewater treatment system components and review operational characteristics with on-site environmental staff. Discuss potential sources/causes of prior process train upsets and controls which have been or could be implemented to eliminate or mitigate future treatment train upsets.

6. Verify that adequate operational and management controls are in place to minimize the potential for an accidental release to the environment. (i.e., If acids or caustics are used for pH adjustment, materials handling, metering and storage protocols should be consistent with industry standards.)
7. Review internal data collection and management procedures/protocols to ensure that routine monitoring and process control objectives are being satisfied.
8. If problems are identified, determine if funding constraints or budget limitations have precluded implementation of collection or treatment system improvements which would limit future permit violations.
9. If pretreatment or industrial wastewater treatment systems are in operation, verify that operations personnel have required training and/or certifications.
10. Check and confirm that periodic monitoring reports showing compliance with permit conditions have been properly filed with the applicable environmental agency.

The following section is reprinted with permission from *Certified Facilities Environmental Professional Exam Review Book*, published by the Association for Facilities Engineering.

Hazardous Waste

Hazardous wastes are solids, liquids, sludges and contained gaseous materials which could pose short- and long-term health and environmental hazards unless they are properly managed and disposed. The EPA generally defines hazardous waste as that which is toxic, corrosive, ignitable or reactive. Figure 10.3 summarizes the characteristics that result in a waste being classified as hazardous.

The EPA currently lists an estimated 450 wastes and waste streams as hazardous. Figure 10.4 provides examples of hazardous wastes commonly generated by industrial plant operations. Wastes that are mixed with hazardous wastes become hazardous. Under current regulations, plants are required to control hazardous waste from the point of generation to disposal at an approved facility.

Applicable Regulations

The Resource Conservation and Recovery Act (RCRA) was enacted in 1976 as an amendment to the Solid Waste Disposal Act. The most important aspect of RCRA is its requirement for the management and tracking of hazardous waste from generator to transporter to treatment, storage and disposal. RCRA also regulates the cleanup of contamination on active industrial sites.

In order to regulate hazardous waste under RCRA, the EPA first had to determine which specific wastes are hazardous. Since there are thousands of wastes that can be hazardous for many different reasons, this was not a simple task. The definition of hazardous waste has important economic ramifications to industry. The EPA's efforts resulted in a methodology to identify hazardous waste based on the four defining characteristics presented and the approximately 450 listed wastes and waste streams.

The Hazardous and Solid Waste Amendments (HSWA) of 1984 reauthorized and amended RCRA. Provisions of the HSWA regulations imposed new and far-reaching requirements on the management of hazardous waste. The 1984 amendments provided substantial new protection by mandating the EPA to develop regulations to:

- Minimize wastes by reducing, recycling and treating them;
- Ban unsafe, untreated wastes from land disposal;
- Require that land disposal facilities are designed, constructed and operated according to stringent standards; and
- Require corrective action for releases of hazardous waste into the environment.

State laws may also regulate hazardous waste management. These laws either complement the federal regulations or are essentially an adoption of the federal regulations. The EPA, while retaining oversight responsibility, can delegate authority to a state to operate a hazardous waste regulatory program.

Hazardous Waste Defining Characteristics

Characteristic*	Definition
Ignitability	Ignitable wastes can create fires under certain conditions.
Corrosivity	Corrosive wastes include those that are acidic and those that are capable of corroding metal.
Reactivity	Reactive wastes are unstable under normal conditions. They can create explosions and/or toxic fumes, gases and vapors when mixed with water.
Toxicity	Toxic wastes are harmful or fatal when ingested or absorbed. Toxicity is identified through a laboratory procedure called the Toxicity Characteristic Leaching Procedure (TCLP)

*EPA regulations require that all waste generators evaluate their wastes to determine if any of the four hazardous characteristics are exhibited.

Association for Facilities Engineering, *CFEP Review Book*

Figure 10.3

Examples of Hazardous Waste Generated by Industry

Waste Generators	Waste Type
Chemical manufacturers	Strong acids and bases Spent solvents Reactive wastes
Vehicle maintenance shops	Heavy metal paint wastes Ignitable wastes Used lead acid batteries Spent solvents
Printing industry	Heavy metal solutions Waste inks Spent solvents Spent electroplating wastes Ink sludges containing heavy metals
Leather products manufacturing	Waste toluene and benzene
Paper industry	Paint wastes containing heavy metals Ignitable solvents Strong acids and bases
Construction industry	Ignitable paint wastes Spent solvents Strong acids and bases
Cleaning agents and cosmetics manufacturing	Heavy metal dusts Ignitable wastes Flammable solvents Strong acids and bases
Furniture and wood manufacturing and refinishing	Ignitable wastes Spent solvents
Metal manufacturing	Paint wastes containing heavy metals Strong acids and bases Cyanide wastes Sludges containing heavy metals

Association for Facilities Engineering, *CFEP Review Book*

Figure 10.4

RCRA/HWSA

The primary objectives of RCRA and subsequent amendments are to protect human health, groundwater and the environment from the effects of hazardous wastes; to reduce waste and conserve energy and natural resources; and to reduce or eliminate the generation of hazardous waste as expeditiously as possible.

When first passed, RCRA focused on those companies that generated the largest quantities of waste. The regulatory quantity threshold was more than 2,200 pounds (1,000 kilograms), or about five full 55-gallon drums per month. The regulated community included about 15,000 companies that produced approximately 90 percent of the nation's hazardous waste. Small-quantity generators were initially exempt from the regulations. In the 1984 Hazardous and Solid Waste Amendments (HSWA), Congress closed this gap in generator size by requiring the EPA to regulate those small-quantity generators that produce between 220 (100 kilograms) and 2,200 pounds (1,000 kilograms) of hazardous waste in a calendar month.

Under current guidelines, RCRA applies to a plant operation if one of the following activities are occurring:

1. The generation of more than one kilogram/month of acute hazardous waste or 100 kilograms/month of hazardous waste;
2. The transport, treatment, storage or disposal of hazardous waste;
3. The management of recyclable hazardous materials in a manner constituting disposal; and
4. The use of underground storage tanks to store hazardous substances.

Figure 10.5 lists the major provisions of RCRA. Of the provisions listed, Subtitles C, D and I set forth the framework for the EPA's comprehensive waste management programs and are generally found to be the most applicable to plant operations. The primary focus of these subtitles are:

- Subtitle C establishes a system requiring plants to control hazardous waste from generation to ultimate disposal;
- Subtitle D establishes a regulatory framework for managing nonhazardous solid wastes; and
- Subtitle I regulates toxic substances and petroleum products stored in underground storage tanks.

In addition to reducing the RCRA generator size threshold to 220 pounds per month, HSWA expanded the hazardous waste management purview of RCRA by the following actions:

1. Established programs to regulate small-quantity generators;
2. Restricted land disposal of hazardous wastes;
3. Established minimum technology standards for land disposal units; and
4. Initiated listing of new materials as hazardous wastes.

HSWA also gave the EPA expanded authority to require facilities operations that treat, store and/or dispose of hazardous wastes to clean up releases of hazardous wastes to the environment. A key provision of the RCRA/HWSA regulations is the enforcement authority provided the EPA. One of the program's most important tools is the site inspection of all treatment, storage and disposal facilities at least once every two years. During an inspection, regulatory agency personnel can review the company records, take waste samples and assess the facility's operating methods. If a facility is not complying with the RCRA regulations, the EPA or the state can take enforcement action.

Enforcement under RCRA may include civil and criminal penalties, orders to correct the violations, and fines. For minor violations, the EPA typically issues a letter notifying the plant of the noncompliance situation and specifying the actions required to remedy the condition within a certain time period. For severe or recurrent violations, the EPA can levy a penalty of up to $25,000 per day, suspend the facility's permit to operate and bring a legal suit. RCRA is one of a series of laws regulating plant hazardous materials. Figure 10.6 provides an overview of the major environmental regulations governing hazardous materials and the relationship to RCRA.

Permits and Other Requirements

The management of hazardous wastes at a plant location under RCRA/HWSA requires strict adherence to regulations that address waste identification and labeling, handling, storage, inspection, manifesting and, ultimately, proper disposal. It is important to recognize that generators are responsible for their wastes through final disposal.

RCRA Statute Subtitles

Subtitle	Description
A	General Provisions
B	Office of Solid Waste - Authorities of Administrator
C	Hazardous Waste Management
D	State or Regional Solid Waste Plans
E	Duties of the Secretary of Commerce in Resource Recovery
F	Federal Responsibilities
G	Miscellaneous Provisions (i.e., employee protection, citizen suits)
H	Research Development, Demonstration and Information
I	Regulation of Underground Storage Tanks
J	Demonstration of Medical Waste Tracking Program

Association for Facilities Engineering, *CFEP Review Book*

Figure 10.5

Major Environmental Laws Controlling Hazardous Substances

Regulation	Lead Federal Agency	Potential Regulatory Application to Plant Operation
1. Clean Air Act	EPA	Emission of hazardous air pollutants
2. Clean Water Act	EPA	Discharge of hazardous pollutants into the nation's surface waters
3. Resource Conservation and Recovery Act	EPA	Regulates hazardous waste generation, storage, transportation, treatment and disposal
4. Safe Drinking Water Act	EPA	Regulates contaminant levels in drinking water
5. Occupational Safety and Health Act	OSHA	Worker exposure to hazardous substances in the workplace
6. Toxic Substances Control Act	EPA	Manufacture, use and disposal of chemical substances
7. Hazardous Materials Transportation Act	DOT	Transportation of hazardous materials
8. Comprehensive Environmental Response, Compensation and Liability Act	EPA	Provides for the cleanup of inactive and abandoned hazardous waste sites
9. Atomic Energy Act	NRC	Regulates nuclear waste disposal
10. Surface Mining Control and Reclamation Act	Dept. of Interior	Environmental aspects of mining and land reclamation

Association for Facilities Engineering, *CFEP Review Book*

Figure 10.6

Waste Identification: The initial step in determining the applicability of RCRA is the classification of the wastes generated by the plant operation. Wastes that have been determined by the EPA to be hazardous are listed in 40 CFR Part 261, "Identification and Listing of Hazardous Wastes." The EPA list can be organized into three broad categories of wastes:

Source Specific Wastes:
This list includes wastes from specific industries such as petroleum refining and wood preserving.

Generic Wastes:
This list identifies wastes from common manufacturing and industrial processes. An example of this type of waste is used solvents.

Commercial Chemical Products:
This list includes specific commercial chemical products such as creosote and certain pesticides.

All listed wastes are presumed to be hazardous regardless of their concentrations and must be handled according to the EPA's RCRA Subtitle C hazardous waste regulations if generated in quantities above the threshold amounts. If a company can demonstrate that its specific waste is not hazardous, the waste may be delisted and is no longer subject to Subtitle C requirements. Wastes that exhibit one or more of the characteristics of ignitability, corrosivity, reactivity or toxicity are also classified as hazardous. It is the EPA's policy under RCRA that wastes mixed with hazardous wastes are classified as hazardous.

It is important that the most current listing of hazardous wastes regulated by RCRA be obtained from both the EPA and the authorized state agency. Wastes can be added to the listing as well as delisted if found not to be hazardous.

Wastes generated in small quantities are excluded from the majority of the RCRA regulatory requirements. The volume threshold for RCRA is the generation of more than one kilogram/month of acute hazardous waste or 100 kilograms/month of hazardous waste.

Waste Generator Identification Number: A generator of hazardous waste is required to obtain an EPA identification number for each site at which hazardous waste is generated. If the state has been delegated authority, a state identification number may also be needed. Identification numbers can be applied for at the EPA regional office by completing Form 8700-12, Notification of Regulated Waste Activity. The generator identification number is required before hazardous wastes can be treated, stored, shipped or disposed of.

On-Site Waste Handling: Plant operations that generate between 100 kilograms and 1,000 kilograms per calendar month are classified under RCRA as small-quantity generators. Small-quantity generators are allowed to accumulate hazardous waste on site for up to 180 days provided the total amount of accumulated waste does not exceed 6,000 kilograms. If the permitted disposal site for the wastes is greater than 200 miles from the plant operations, wastes can be stored up to 270 days provided the 6,000 kilogram limit is not exceeded. State regulations for small-quantity generators should be reviewed to determine if they are more stringent than the EPA requirements.

Large-quantity generators are allowed under RCRA to store wastes up to 90 days without a permit. If hazardous wastes are stored longer than 90 days, the plant must obtain an RCRA permit or be operating under interim status as a storage facility.

A generator under 40 CFR 262 can accumulate up to 55 gallons of hazardous waste or one quart of acutely hazardous waste at the point of generation without a time limitation. Once the container is filled, it must be moved to a centralized storage area. The 90-day period begins on the day that the container is moved to centralized storage.

Whether it is a small- or large-quantity generator, a plant must meet the storage, testing, labeling and record-keeping requirements of the RCRA regulations. Waste storage must follow the container management practices specified by the RCRA regulations. Containers need to meet the specifications of the U.S. Department of Transportation. Containers must be labeled as hazardous waste along with a description of the contents, such as the chemical name. The container labeling needs to include the generator's name and address and manifest document number.

Tracking System: A manifest document must be prepared for each shipment of hazardous waste. The manifest must accompany any transported waste from the point of generation to the point of final disposal. The manifest contains the following essential information:

- Waste generator identification number,
- Type and quantity of wastes,
- Plant location generating waste,
- Disposal facility to which the waste is being shipped,
- Certification that the generator is minimizing the amount and toxicity of its waste, and
- Signature of each individual handler of the waste.

When the waste reaches its destination, the owner of that facility returns a copy of the manifest to the generator to confirm that the waste has arrived. Generators must retain copies of the manifest for three years after shipment. If a manifest is not returned to a generator, the plant has the responsibility of notifying the EPA or responsible state agency. Up-to-date records documenting the management and employee training procedures must be kept on site.

Figure 10.7 provides a listing of the information that should be kept on site and included in the calendar-year waste summary report, submitted to the EPA regional office or the authorized state agency on a biennial basis. The reporting dates of the responsible agency should be identified for the state within which the plant is located.

Hazardous Waste Transport: A plant that generates hazardous waste is liable for the wastes from the point of generation through ultimate disposal under RCRA. It is, therefore, imperative that the plant conducts a due diligence review of both the licensed transport contractor and the disposal facility. Liability attributable to a transporter or disposal facility can extend back to the original generator of the waste.

Transporters must carry copies of the completed manifests and must put proper symbols on the transport vehicle to identify the type of waste being transported. Like generators, transporters must obtain an EPA identification number.

Agency Inspection: Under RCRA, a regulatory agency has the right to inspect a plant's operations for hazardous waste compliance. When a regulatory inspector visits the plant, he/she should always be accompanied by a plant representative who is knowledgeable of the plant's hazardous waste management activity. The results of the inspection should be reported to the plant engineer and environmental health and safety coordinator.

The following section is reprinted with permission from *Certified Facilities Environmental Professional Exam Review Book*, Association for Facilities Engineering.

Solid Waste

Solid waste is defined as including those wastes that are nonhazardous. Solid waste can comprise solid, semi-solid, liquid or contained gaseous material that is being discarded and is not covered by the hazardous waste definitions of the Resource Conservation Recovery Act or the Hazardous and Solid Waste Amendments.

The management of solid waste has become an important priority for industry due to a number of factors and trends. These include:

Disposal Capacity:

The amount of solid waste being generated is increasing while the disposal capacity is decreasing.

Disposal Costs:

With a growing shortage of landfill capacity, the disposal costs for solid waste are escalating rapidly.

Waste Stream Control:

Solid wastes that are mixed with hazardous wastes become hazardous. The potential for this happening is directly correlated to the plant management of solid waste.

Regulatory Initiatives:

The passage of RCRA in 1976 gave the EPA regulatory responsibilities in the management of solid waste. EPA policies on shifting from solid-waste generation and disposal to source reduction and recycling have implications for plant operations.

Liability of Contamination:
Leaking solid-waste landfills are a major source of groundwater contamination. Companies using a landfill facility can be held liable if environmental cleanups are required.

Applicable Regulations

Historically, state and local governments have controlled solid wastes that encompass industrial and municipal wastes that are not defined as hazardous. With the passage of RCRA in 1976, the EPA under Subtitle D–Solid or Regional Solid Waste Plans, imposed a broader regulatory framework on the use of landfills for the disposal of solid waste. The Hazardous and Solid Waste Amendments (HSWA) of 1984 broadened EPA's technical regulatory controls over solid-waste disposal facilities nationwide. RCRA Subtitle D and HSWA affect the disposal of industrial solid waste by imposing more stringent regulations on the use of municipal solid-waste landfills, municipal waste incinerators and incinerator ash. These tighter regulations on disposal facilities, in turn, resulted in more stringent waste management requirements on plant operations. The Pollution Prevention Act of 1990 reinforces the EPA's environmental management options hierarchy with respect to solid waste (both hazardous and nonhazardous). Pollution prevention focuses on the use of materials, processes or practices that reduce or eliminate the creation of pollutants or wastes at the industrial source. Pollution prevention, as defined by the EPA and the state regulatory authorities, includes practices that reduce

Plant Operations Hazardous Waste Summary Report Contents
(Note: Report prepared on a calendar-year basis.)

- EPA identification number, name and address of the generator
- The year covered by the report
- The EPA identification number, name and address for each off-site processing storage or disposal facility within the United States to which the waste was shipped during the year
- The name and EPA identification number of each transporter used during the year
- The EPA hazardous waste number, United States Department of Transportation (DOT) hazard class and quantity of each hazardous waste shipped off site
- Description of the efforts undertaken during the year to reduce the volume and toxicity of waste generated
- A description of the changes in volume and toxicity of waste actually achieved during the year
- Sampling and testing data of plant wastes
- A copy of all manifests
- Certification signed by the generator representative

Association for Facilities Engineering, *CFEP Review Book*

Figure 10.7

the use of hazardous and nonhazardous materials, energy, water or other resources. Through achievement of this objective, natural resources can be preserved through conservation or more efficient use and the environment protected through the reduction of solid wastes and the more rigorous control of their disposal. At the state and municipal authority levels, regulatory demands are being placed on industry to both reduce the quantity of solid waste generated and manage more efficiently those wastes that require disposal through recycling, source reduction, source separation and on-site treatment. The pressures driving the more efficient management of solid wastes include concern over environmental impacts, particularly ground water pollution, reduced landfill capacity and escalating disposal costs.

EPA Regulations

The EPA management of nonhazardous solid waste under RCRA Subtitle D and HSWA focuses primarily on the disposal aspect of solid waste and not the generation of waste directly. By establishing minimum federal standards for state and waste management plans, Subtitle D has served as a regulatory driver for states to exert control over solid waste generation and disposal. Regulations implemented under authority of Subtitle D set guidelines for defining what constitutes a sanitary landfill. The basic criterion is that operations of the facility should create no reasonable probability of adverse effects on public health or the environment. Under RCRA and HSWA, the EPA's expanded coverage in solid waste disposal incorporates standards for sanitary landfill siting, design and operation, corrective action and closure. Groundwater monitoring requirements were also imposed on landfills to identify if releases of specific indicator parameters of contamination were occurring. In addition to regulating solid waste landfills that may be operated on site by an industrial plant or used for off-site disposal, Subtitle D established regulations that prohibit practices which constitute open dumping of waste. Compliance schedules were established under RCRA which require open dumps to be either upgraded to meet the requirements of a sanitary landfill or closed within a reasonable time (not to exceed five years). Requirements also are placed on the operators of open dumps to remediate any damage to the environment.

The potential responsible party (PRP) provisions of CERCLA and SARA can also impact a company if facilities where nonhazardous solid waste are being sent for disposal are not carefully audited. A large number of landfill sites have been identified by either EPA or state authorities for corrective action due to groundwater contamination. If a plant used such a facility for solid waste disposal and did not keep comprehensive records documenting the nonhazardous nature of wastes sent to the facility, then liability could be imposed based on the need for corrective action due to environmental contamination. To avoid this occurrence, annual audits should be conducted of the operational practices of both the licensed solid waste transporters and the disposal facility itself. Records documenting the quantities, types and composition of solid wastes disposed off site should also be filed at the plant to avoid future liability actions.

Figure 10.8 illustrates the Environmental Management Options Hierarchy of the EPA as reinforced by the Pollution Prevention Act (PPA) of 1990. Pollution prevention applies to the use of both hazardous and nonhazardous materials. In addition to disposal cost considerations, a plant can decrease the risk of liability by reducing the volume of solid waste generated. Since the trend is toward more stringent toxicity definitions and environmental controls at disposal facilities, reducing the volumes of wastes in all categories is a sound long-term management policy.

Provisions of the Pollution Prevention Act go beyond wastes designated as hazardous. It encourages the maximum possible elimination of wastes of all types. The act emphasizes that the preferred method of preventing pollution is to reduce the volume of waste generated at the source and that reuse should be performed whenever possible. The developing policies of the EPA and states are viewing treatment and disposal as last resort measures.

A number of states have enacted legislation that requires pollution prevention, waste minimization and recycling. These regulations pertain to both hazardous and nonhazardous solid waste. Regional solid-waste authorities have implemented programs requiring source separation at the point of collection. Both the state and regional authority programs have implications for how solid waste is managed by plants.

Permits and Other Requirements

Solid Waste Disposal: While Subtitle D of RCRA focuses primarily on the disposal of nonhazardous solid waste, it does have implications for plant operation. It is important that the sanitary landfill used by a plant as the ultimate point of nonhazardous solid-waste disposal has the proper state operating permits that implement the Subtitle D requirements of EPA with respect to facility design, operation, closure and groundwater monitoring. An annual audit of the sanitary landfills, used by the licensed transporters to dispose of a plant's solid waste, should be conducted.

If open dumps were used on a plant site, they should be in the process of being upgraded to a sanitary landfill or scheduled for closure with required remedial measures. Industrial solid-waste landfills on a plant site need to meet the same state permitting requirements and performance criteria as commercial landfill facilities. Sanitary landfill permits typically require renewal on a periodic basis and submittal of monitoring reports on a frequency specified in the operating permit issued for the facility. The operating permit requirements of the state should be reviewed on an annual basis to ensure that landfill facilities operated on the plant site are in compliance with applicable regulations.

Solid Waste Recycling: State laws need to be reviewed to determine how they impact the generation, management and disposal of solid waste. A trend in a number of states is to require source separation and mandatory recycling goals. A number of states have established regulations requiring a minimum 25 percent recycling of solid waste.

Method	Example Activities	Example Applications
Source Reduction (Highest Priority)	• Environmentally Friendly Design of New Products • Product Changes • Source Elimination	• Modify Product to Avoid Solvent Use • Modify Product to Extend Coating Life
Recycling	• Reuse • Reclamation	• Solvent Recycling • Metal Recovery from a Spent Plating Bath • Volatile Organic Recovery
Treatment	• Stabilization • Neutralization • Precipitation • Evaporation • Incineration • Scrubbing	• Thermal Destruction of Organic Solvent • Precipitation of Heavy Metals from a Spent Plating Bath
Disposal	• Disposal at a Permitted Facility	• Land Disposal

Environmental Management Options Hierarchy

Association for Facilities Engineering, *CFEP Review Book*

Figure 10.8

Solid Waste Record Keeping: The composition and quantity of solid waste generated by a plant should be recorded and filed. Information documenting the licensed contractors used to transport solid waste from the plant should also be recorded along with confirmation of the sanitary landfill, municipal incinerator or recycling facility where the waste was disposed. An audit should be conducted annually by the plant to ensure that disposal facilities have the required state operating permits and are in conformance with permit conditions. Monitoring data for the facility should be reviewed to ensure that no groundwater contamination is occurring.

The above actions will significantly help to minimize the future liability of the plant under state and federal (CERCLA, SARA) regulations should groundwater pollution result from the landfill operations.

Pollution Prevention: When Congress passed RCRA in 1976 and HSWA in 1984, it recognized that state and local governments have primary responsibility for nonhazardous solid waste management. The two acts did give the EPA regulatory and technical assistance responsibilities in solid waste management and disposal. With the passage of the Pollution Prevention Act of 1990, the EPA was given an expanded mandate to encourage industry to reduce the quantity of both hazardous and nonhazardous solid wastes through source reduction and recycling. Plants should recognize this management trend on the part of the EPA, states and regional solid waste authorities and initiate programs to reduce the quantity of solid waste requiring disposal through means such as source reduction and recycling.

The following section is reprinted with permission from *Means Environmental Remediation Estimating Methods* by Richard R. Rast, R.S. Means Co., Inc.

Underground Storage Tanks

Introduction

To address a nationwide problem of leaking underground storage tanks (USTs), Congress established a prevention, detection, and cleanup program through the 1984 RCRA amendments and the 1986 Superfund Amendments and Reauthorization Act (SARA).

The 1984 RCRA amendments created a federal program to regulate USTs containing petroleum and hazardous chemicals to limit corrosion and structural defects, and thus minimize future tank leaks. The law directed the EPA to set technical standards for tank design and operation, leak detection, reporting, corrective action, and tank closure. The UST program (new RCRA Subtitle I) is administered primarily by states and requires registration of most USTs, bans the installation of unprotected tanks, sets federal technical standards for all tanks, coordinates federal and state regulatory efforts, and provides for federal inspection and enforcement.

In 1986, Congress created a petroleum UST response program by amending Subtitle I of RCRA through SARA (P.L. 99-499). Before SARA, the EPA lacked explicit authority to require cleanup of contamination from leaking petroleum USTs, as Congress has officially excluded petroleum products (although not petrochemicals) from the Superfund law. The new provisions authorized the federal government to respond to petroleum spills and leaks, and created a Leaking Underground Storage Tank Trust Fund to clean up leaks from petroleum USTs. The money in the fund is derived primarily from a 0.1 cent-per-gallon federal tax on fuels and several other petroleum products.

The 1986 amendments directed the EPA to establish financial responsibility requirements for UST owners and operators to cover costs of taking corrective action and to compensate third parties for injury and property damage caused by leaking tanks. The law required the EPA to issue regulations requiring tank owners and operators selling petroleum products to demonstrate financial responsibility insurance coverage of $1 million.

The following section is reprinted with permission from *Certified Facilities Environmental Professional Exam Review Book*, Association for Facilities Engineering.

Leaking Underground Storage Tanks

As many as 25 percent of all the underground storage tank (UST) systems on plant sites in the United States containing petroleum or hazardous chemicals may be leaking based on studies conducted by the EPA. In addition to the threat to public health and safety and the environment that leaking USTs create, the owner or operator is financially responsible under current federal law to: (1) clean up a spill site, (2) correct environmental damage and (3) compensate others for personal or property damage. The cost for remediation programs can run into the thousands of dollars. Tanks out of service for repair also impact plant operations.

The failure of older USTs is due to a number of factors. Most of the UST systems already in the ground have tanks and piping made of bare steel. When unprotected steel is buried in the ground, it is subject to corrosion and eventual leaking. Tanks and piping also leak if they are not put in the ground properly. Studies conducted by the EPA show that the majority of leaks result from piping failure. Piping has less structural integrity than tanks. As a result, piping is much more susceptible to the effects of installation mistakes, excessive surface loads, stress of underground movement and corrosion. It is important to remember that federal and state regulations apply to the entire UST system, including tanks and piping.

Applicable Regulations

Underground storage tanks (USTs) have been identified by the EPA and the state environmental agencies as a major environmental concern based on the contamination that has occurred due to leaking tanks and piping. Leaking USTs have been found to be a major cause of contamination to groundwater. In addition, leaking USTs can present a threat to public safety due to fire or explosion hazards. Under the Resource Conservation and Recovery Act (RCRA), the EPA imposed regulations under Subtitle C for USTs containing hazardous wastes and Subtitle I for USTs containing petroleum or hazardous substances as defined by the Comprehensive Environmental Response, Compensation and Liability Act of 1980 (CERCLA). A UST under RCRA is defined as storing regulated substances and having 10 percent or more of its volume underground. The volume calculation in determining the underground percentage includes the volume of tanks and piping.

A number of states have either adopted the UST regulatory framework of RCRA or implemented their own regulations. The regulations require submittal of information on the structural integrity of existing USTs and impose requirements on the construction of new USTs. At the municipality level, additional tank registration and permitting are typically required by the local fire department or board of health.

The specific objectives of the EPA, state and municipal regulations relating to USTs (including tanks, piping and detection systems) are to:

1. Prevent leaks and spills;
2. Find leaks and spills;
3. Correct environmental problems created by leaks and spills;
4. Make sure that owners and operators of USTs can pay for corrective actions; and
5. Ensure that each state has a regulatory program in force for USTs that is as strict or more strict than the EPA standards.

RCRA/HSWA

The 1984 Hazardous and Solid Waste Amendments (HSWA) to RCRA, enacted under Subtitle I, require that the EPA develop regulations for the control of USTs that are used to store chemical and petroleum substances. Regulated substances addressed by Subtitle I consist of:

1. Hazardous chemical products regulated by CERCLA, and
2. Petroleum products (including crude oil and refined products that are liquid at standard conditions of temperatures and pressure).

Subtitle I, for the first time, placed USTs that are used for the storage of products as well as wastes under the control of RCRA. The UST provisions of Subtitle I regulate over 300 hazardous materials and products. The use of USTs for the storage of hazardous wastes is regulated under Subtitle C of RCRA.

RCRA Subtitle I grants the EPA the authority for developing and enforcing the federal programs. Under the EPA regulations, each state is required to designate a state agency or local agencies to which UST notifications must be sent by plants. States have discretionary authority to develop their own UST regulations that are either as strict or more strict than the EPA regulations. The state regulations within which the plant is located should be reviewed to determine if they are the same as the EPA's regulatory program or are more stringent.

RCRA Subtitle I regulates both existing and new USTs. Requirements for existing tanks relate primarily to control of the potential for leaking and the proposed response to a leak if it occurs. Specific EPA regulations impose the following requirements on existing tanks:

1. Leak detection or inventory control system and tank testing;
2. Record keeping and reporting;
3. Corrective action if a leak to the environment should occur;
4. Financial responsibility for corrective action; and
5. Proper closure of tanks.

Certain types of tanks are excluded from the hazardous chemicals and regulations. The most frequently occurring types of tanks excluded are:

1. Farm and residential tanks holding 1,100 gallons or less of motor fuel used for noncommercial purposes;
2. Tanks storing heating oil used on the premises where they are stored;
3. Tanks on or above the floor of underground areas, such as basements or tunnels;
4. Septic tanks and systems for collecting storm water and wastewater;
5. Flow-through process tanks;
6. Tanks holding 110 gallons or less; and
7. Emergency spill and overflow tanks.

Other storage areas that might be considered "tanks" are also excluded, such as surface impoundments and pits.

The Superfund Amendments and Reauthorization Act (SARA) was passed in 1986. SARA both extended and expanded the provisions of CERCLA. Section 205 of SARA applies to USTs. Owners and operators of USTs are required to clean up releases from subsurface tank systems that pose threats to the environment and public health. Under SARA, the EPA established financial responsibility standards for owners and operators of USTs. SARA authorizes the EPA and/or states operating under cooperative agreements with the EPA, to require the owners or operators of USTs that have leaked to conduct the necessary environmental cleanup. If the owners or operators do not take action, SARA empowers the EPA to take corrective action and sue to recover the costs of the cleanup.

SARA also created a Leaking Underground Storage Tank Trust Fund that is available for use when the owner or operator of a tank that has leaked cannot be identified or the cleanup costs exceed the financial resources of the owner or operator. SARA also requires owners or operators of large-volume tanks to produce information demonstrating their financial capability, if required, of covering the cleanup costs up to a minimum $1 million dollars per occurrence.

Required Permits and Operational Procedures

The EPA has estimated that the number of USTs that may now be leaking could number in the thousands *(Ed. Note: Possibly tens of thousands.)*. The potential for leaking increases dramatically for tank systems that are over 10 years old and are not protected against corrosion. The cleanup of a UST is costly from the dual perspective of environmental corrective action requirements and the tank being out of service operationally, which impacts plant operations.

The following major points of the UST regulations apply to a plant operation. It is important to remember that the regulations apply to the entire UST system, both tanks and piping.

USTs Installed Before December 1988: By December 1998, USTs that were installed before 1988 must have:

- Corrosion protection for steel tanks and piping, and
- Devices that prevent spills and overflows

Tanks installed prior to December 1988 were also required by the EPA to have leak detection installed by December 1993.

USTs Installed After December 1988: If a plant installed a UST after December 1988 or is planning yet to install one, the plant is required to meet the requirements concerning correct installation so that it meets industry codes, utilizes spill and overflow prevention, and has corrosion and leak protection. Figure 10.9 provides information on the most relevant codes and standards for UST systems.

Corrective Action in Response to Leaks: The response to confirmed leaks and spills comes in two stages: short term and long term. In the short term, action should be taken to stop and contain the leak or spill. The regulating authority should be notified within 24 hours of the leak or spill. Petroleum spills or overfills of less than 25 gallons do not have to be reported if they are immediately contained and cleaned up. Upon containment, the plant EHS department should ensure that the leak or spill poses no immediate hazard to public safety due to exposure, explosion or fire hazards. Contaminated soil during the cleanup process should be properly handled according to EPA and state requirements. The progress in responding to the leak or spill along with collected field information should be reported to the regulating authority no later than 20 days after the event occurred. A report documenting the results of the impact to the general environment is required within 45 days of the event. If groundwater contamination has occurred, the report needs to address subsequent steps to be taken.

Based on the information provided, the responsible regulating authority will decide if additional action is required in the long term at the site. Depending on the extent of the contamination, a Corrective Action Plan could be required that will specify the cleanup actions for the site.

Underground Storage Tanks Relevant Codes and Standards Reference Guide

Regulatory Requirement	Relevant Codes and Standards
Installation	API Publication 1615, 1987, "Installation of Underground Petroleum Storage Systems," Recommended Practice, 4th Edition PEI RP-100-90, 1990, "Recommended Practices for Installation of Underground Liquid Storage Systems"
Tank Filling Practices	API Publication 1621, 1977, "Recommended Practice for Bulk Liquid Stock Control at Retail Outlets," 3rd Edition (A revised edition is now available.) NFPA 385, 1985, "Standard for Tank Vehicles for Flammable and Combustible Liquids"
Corrosion Protection	API Publication 1632, 1987, "Cathodic Protection of Underground Petroleum Storage Tanks and Piping Systems," Recommended Practice, 2nd Edition NACE RP-0169-83, 1983, " Recommended Practice: Control of Corrosion on Underground or Submerged Metallic Piping Systems" NACE RP-0285-85, 1985, "Recommended Practice: Control of External Corrosion on Metallic Buried, Partially Buried or Submerged Liquid Storage Systems"
Closure	API Bulletin 1604, 1987, "Removal and Disposal of Used Underground Petroleum Storage Tanks," Recommended Practice, 2nd Edition
Lining	API Publication 1631, 1987, "Interior Lining of Underground Storage Tanks," Recommended Practice, 2nd Edition NLPA Standard 631, 1990, "Spill Prevention: Minimum 10 Year Life Extension of Existing Steel Underground Storage Tanks by Lining Without the Addition of Cathodic Protection"

Association for Facilities Engineering, *CFEP Review Book*

Figure 10.9

Closure Requirements for Tanks: USTs can be closed either permanently or temporarily. Requirements for a tank that is not protected from corrosion and is expected to remain closed for more than 12 months are:

1. The regulating authority needs to be notified 30 days before the plant closes the UST.
2. A determination of whether leaks have occurred needs to be made.
3. If leaks have occurred, a corrective action plan needs to be prepared and followed.
4. The UST, when closed, can be either removed or left in the ground. The tank is required to be emptied and cleaned. If the UST is left in the ground permanently closed, it should be filled with a harmless, chemically inactive solid, like sand.

The requirements for permanent closure may not apply to USTs at a plant location if one of the following conditions is met:

- If the UST meets the requirements for a new or upgraded UST, then it can remain "temporarily" closed indefinitely as long as it meets the requirements below for a temporarily closed UST.
- The regulatory authority can grant an extension beyond the 12-month limit on temporary closure for USTs unprotected from corrosion.
- The contents of the UST can be changed to an unregulated substance, such as water. Before making this change, the plant must notify the regulatory authority, clean and empty the UST and determine if any damage to the environment was caused while the UST held regulated substances. If there is damage, then corrective actions must be taken.

Tanks not used for three to 12 months must follow requirements for temporary closures:

1. If the UST has corrosion protection and leak detection, the plant must continue to operate these protective systems.
2. If a leak is found, the plant will have to respond just as if it was a leak from an active UST.
3. If the UST is empty, however, the plant does not have to maintain leak detection.
4. The plant must cap all lines, except the vent line, attached to the UST.

Reporting: RCRA requires a plant to report on a UST system at the following milestones: installation, leak or spill release and closing. The key reporting requirements are:

- When installing the UST, the plant needs to fill out a notification form, which is available from the state. This form provides information about the UST, including a certification of correct installation. The notification form should have already been submitted for existing USTs.
- The plant must report confirmed releases to its regulatory authority. The plant must also report follow-up actions that are planned or have been taken to correct the damage caused by the UST.
- The regulatory authority needs to be notified 30 days before the plant permanently closes the UST.

Plants need to check with their regulatory authority about the particular reporting requirements in their area, including any additional or more stringent requirements than those noted above.

UST Inventory: A complete inventory of USTs at a plant location needs to be maintained. Records should be kept for operating, permanently closed and temporarily closed USTs. Engineering drawings documenting the tank location on the plant site, the physical layout of the tank and piping system, construction details and date of installation should be maintained.

Records should be kept for each UST demonstrating the compliance status in four major areas:

1. Leak detection
   - Results of the most recent year's monitoring results, including tightness test performed,
   - Copies of performance claims provided by leak detection manufacturers, and
   - Records of recent maintenance repair, and calibration of leak detection equipment installed on site;
2. Corrosion protection
   - Records of corrosion protection for each tank, and
   - Records of last two inspections of corrosion protection system;
3. Repair or upgrading of UST system
   - Records documenting date and description of repairs or upgrading, and
   - Professional engineer certification of work performed;
4. Closing
   - Record of actions taken to close UST, and
   - Results of site assessment at time of closing.

The following section is reprinted with permission from *Means Environmental Remediation Estimating Methods* by Richard R. Rast, R.S. Means Co., Inc.

CERCLA, SARA, RCRA

Environmental Remediation Regulations

Environmental remediation is controlled by numerous federal, state, and local laws and regulations. Although these regulations are in a constant state of review and change, basic remediation processes are governed primarily by two federal laws: the *Comprehensive Environmental Response, Compensation and Liability Act* (also known as *CERCLA* and/or the *Superfund Law*), and the *Resource Conservation and Recovery Act (RCRA)*. These two laws guide most remediation processes performed in the United States. This book is not intended to provide detailed analyses of these laws; instead, it provides background information and an overview of CERCLA and RCRA.

CERCLA, also known as "Superfund," is the basis for a national program for responding to releases of hazardous substances to the environment. CERCLA created a national emergency response program, provided for cleanup of abandoned sites and active facilities that are not regulated under the RCRA corrective action program, established a trust fund to pay for cleanup of abandoned sites, and created a National Priority List (NPL) of the nation's most contaminated sites. CERCLA establishes the procedures for identifying and cleaning up contaminated sites. CERCLA was amended in 1986 by the *Superfund Amendments Reauthorization Act (SARA)*. SARA reauthorized the trust fund to fund government response actions, extended CERCLA, and added new provisions. References made to CERCLA in this section should be interpreted as meaning "CERCLA as amended by SARA."

The Environmental Protection Agency (EPA) is the federal government agency with the lead responsibility for implementing CERCLA. The EPA's response powers include cleanup of not only hazardous substances but also pollutants and contaminants that may present an imminent and substantial danger to public health or welfare. The government's response authority includes short-term "removal" actions to mitigate immediate threat to public health, and longer term "remedial" actions that are consistent with a permanent remedy. The circumstances under which removal or remedial actions are appropriate are identified in the National Contingency Plan (NCP) and summarized on the following pages.

National Contingency Plan: The National Contingency Plan (NCP) is the regulatory blueprint for implementing the statutory requirements of CERCLA. It is EPA's primary tool for issuing its interpretations of policy and guidance to meet the requirements of Superfund.

The NCP is in Title 40 of the Code of Federal Regulations (40 CFR), Part 300. The methods and criteria for determining the extent of hazardous substance response actions are in 40 CFR 300, Subpart E. The NCP also covers issues such as hazardous substance and extremely hazardous substance release reporting requirements, establishment of national and regional response teams, development of contingency plans to respond to releases, and response phases for cleanup of oil discharges in violation of the Clean Water Act. This section addresses only the hazardous substance response actions.

CERCLA provides two major response authorities—removal actions and remedial actions—to respond to releases of hazardous substances and pollutants or contaminants. The NCP specifies a different set of requirements for removal and remedial actions, as discussed in the following paragraphs.

Removal Actions:

Removal actions are short-term actions in response to a release of hazardous substance(s) that may present an imminent and substantial danger to human health and welfare or the environment. The EPA's removal actions are limited to 12 months or $2 million to ensure proper use of the removal action process. Under CERCLA Section 104 and Executive Order 12580, the U.S. government is authorized to conduct removal actions when the release is on or from a government installation. The 12 months or $2 million limits do not apply to removal actions conducted by the U.S. Government.

The decision to perform a removal action can be made any time during or prior to implementation of the remedial action process, even when the site is not included on the National Priorities List (NPL). The regulations governing removal actions are in 40 CFR 300.415. A removal action is appropriate when:

- A release of hazardous substances creates an actual threat to human health and welfare or the environment.
- A release of hazardous substances has the potential for direct contact with humans, animals, or sensitive environments.
- A known hazardous substance has the potential to migrate to humans, animals, or sensitive environments.
- A fire or explosion hazard is present.
- A release may occur as the result of the integrity of hazardous waste storage or bulk storage containers.
- The release or migration of hazardous substances may be caused or hastened by weather conditions.
- Other appropriate federal or state response mechanisms are unavailable to respond to a release of hazardous substances.
- Other factors may pose a threat to public health and welfare or the environment.

In simplest terms, whenever there is an actual or potential threat of a hazardous substance migrating to, or directly contacting, human or other biological receptors, a removal action is taken to separate or maintain a separation between the receptor and the hazardous substances.

A removal action is not always a permanent solution. A removal action may simply eliminate the immediate problem so the more time-consuming remedial action process may properly characterize and more effectively remediate the contaminated site. Therefore, a removal action should always be consistent with any anticipated future remedial actions.

There are two basic types of removal actions:

- **Emergency removal:** Used when site cleanup or mitigating procedures must be implemented within six months after determining that a removal action is appropriate.
- **Removal with a 6-month planning period:** Used when the planning period for site cleanup or mitigating procedures may exceed six months after determining that a removal action is appropriate.

The following is a list of some examples of potential removal actions. The list does not include every possible option.

- Sealing and removing drums that pose a fire and explosion hazard.
- Installing security fencing and warning signs to prevent direct contact with humans and animals.
- Removing debris or other materials that might attract people to enter the site.
- Evacuating an area and containing the spilled or released material.
- Discontinuing use of contaminated groundwater and supplying an alternate source of drinking water.
- Excavating, consolidating, and removing highly contaminated soils.
- Draining lagoons and providing cover to prevent runon or runoff.
- Capping a disposal area to minimize infiltration of precipitation.
- Constructing drainage controls to prevent (divert) runon and minimizing or capturing runoff from a highly contaminated area before the rainy season.
- Applying absorptive or damping materials to minimize the spread of contamination.
- Stabilizing and maintaining the integrity of existing structures.

Remedial Actions:

Remedial Actions (RA) are permanent remedies taken instead of, or in addition to, a removal action in response to a release or threatened release of hazardous substances. The objective is to ensure that the hazardous substances do not migrate, causing substantial danger to present or future public health and welfare or the environment.

The RA process begins once it is established that a removal action is not warranted or has been completed. The lead agency performs a remedial site evaluation (40 CFR 300.420), which includes conducting a remedial *preliminary assessment (PA)* and a *remedial site inspection (SI)*. The goal of these pre-remedial activities is to gain a better understanding of the nature of the threat. For sites that do pose a threat, the goal is to collect the necessary data to score the site using the *Hazard Ranking System (HRS)* and to identify sites that require immediate response. The HRS provides a numerical ranking that estimates the risk posed by the site. The PAN determines whether a site has ever handled hazardous substances and whether it has released, or has the potential to release, hazardous substances to the environment.

The detailed site investigation and cleanup alternative selection process is referred to as the *Remedial Investigation/Feasibility Study (RI/FS)* process (40 CFR 300.430(a)(2)). An RI/FS includes activities such as project scoping, data collection, risk assessment, treatability studies, analysis of alternatives, and remedy selection. The RI covers site assessment activities to determine site conditions and the nature and extent of contamination. The FS develops and evaluates the cleanup options on which a remedy is based. The goal of the remedy selection process is to select remedies that protect human health and the environment, that maintain protection over time, and that minimize untreated waste. The detailed analysis of alternatives for selecting a remedy involves assessing individual alternatives against nine evaluation criteria listed in 40 CFR 300.430(e)(9)(iii).

The selected remedy is documented in a *Record of Decision (ROD)*, along with supporting documentation. The ROD is signed by the site owner, other responsible parties, the EPA, and other regulators and affected agencies. The regulations governing remedial actions are given in 40 CFR 300.420 through 300.435.

State "Superfund" Laws: Superfund does not preempt states from enacting their own mini-Superfund laws and imposing additional requirements with respect to cleanup of hazardous substance releases. Many states now have such laws. CERCLA Section 120(a)(4) provides that state laws regarding removal and remedial action are applicable to federal facilities not on the NPL. (RAs taken at NPL sites must comply with the substantive portions of state environmental and facility siting laws, provided the laws are applicable or relevant and appropriate.)

The following section is reprinted with permission from *Certified Facilities Environmental Professional Exam Review Book,* Association for Facilities Engineering.

CERCLA–1980

Comprehensive Environmental Response, Compensation and Liability Act (Superfund)

Key Aspects:

1. EPA authorized by CERCLA
2. The list of hazardous substances was expanded to more than 700 substances.
3. Oil and petroleum are *not* included.
4. Reporting and response for containment/cleanup are required when reportable quantity (RQ) is exceeded.
   - The RQ varies with substance, from a minimum of 1 lb. to a maximum of 5,000 lbs.
   - A release is one which is not a federally permitted release, or is not in compliance with a federal permit.
   - "Federal permit" includes permits issued by a federal, state or local authority.

- A release beyond RQ to the air, land or water requires reporting.
- A report is required within 24 hours to the United States Coast Guard at the National Response Center (NRC) (800) 424-8802.
- The largest penalties will be imposed for a failure to notify (report) within the required time frame.

Cleanup Actions

Removal actions:

- Short-term actions are intended to abate imminent threats to public health or the environment.
- Short-term actions are limited to $2,000,000 and one a year per site, if the action is federally funded.
- Sites may be any category presenting imminent threats.

Remedial actions:

- Longer-term actions are taken only after extensive studies are made, and are intended to permanently clean up the site.
- Remedial actions are not limited in cost or duration.
- Federal Superfund funds for remedial actions are limited to only National Priority List (NPL) sites.

Process for Remediation

CERCLA: Any site may be included when the site is alleged to have released hazardous substances in an uncontrolled, unpermitted manner. Once included, the site stays on the list forever. National Contingency Plan (NCP) controls the process for evaluation of CERCLIS and remediation of sites. The EPA regional project remediation manager will be designated. There are four phases to the CERCLIS assessment:

1. Preliminary Assessment (PA)/Site Investigation (SI)
    - PA: A records search is completed.
    - SI: An on-site inspection is always conducted with the sampling.
2. Remedial Investigation (RI)/Feasibility Study (FS)
    - RI: A sampling is done to quantify the extent of contamination.
    - FS: Possible engineering alternatives are assessed to permanently remediate site. There is a public hearing on the FS, then an ROD (record of decision) is issued.
3. Remedial Design (RD)/Remedial Action (RA)
    - RD: Specifications are prepared to implement the ROD.
    - RA: Construction of the RD is completed.
4. Continued operations and maintenance (O&M) are required to preserve the RA.

Enforcement action by the EPA to identify potentially responsible parties (PRPs) begins at RI/FS as well as the requirement that PRPs pay for the remediation process.

If PA/SI identifies a significant threat to human health or the environment, then a site evaluation (SE) is initiated.

- If the SE demonstrates a Hazard Ranking System (HRS) of greater than or equal to 28.5, then the site is added to the NPL.
- NPL sites remain on the NPL only until remediation is completed; once remediated, the site is then delisted.

States cannot administer CERCLA; only the EPA can.

Consequently, many states have adopted parallel state legislation to CERCLA.

CERCLA is a strict-, joint- and several-liability statute; any PRP may be held completely accountable for the entire cost of remediation.

PRPs may be the current owner and/or operator, all prior owners and/or operators at the time of disposal or any transporter who selected the site.

PRPs are liable for all costs of removal or remedial action incurred by the federal or state government, consistent with NCP, the cost of the response given by private parties, consistent with the NCP, damages to natural resources and costs of health assessment/effects studies.

PRPs who do not comply with orders pursuant to the NCP may pay three times the expense in damages plus the cost of removal.

SARA–1986
Superfund Amendments and Reauthorization Act

Key Aspects:

Title I - Response and Liability Permanent cleanup on site is preferred to off-site treatment.

Defenses to liability:

- Release was not a hazardous substance (e.g., petroleum).
- Release was allowed by a permit.
- Release was the result of an:
    1. Act of God,
    2. Act of war or
    3. Act of omission by a third party (innocent landowner).
        - To apply the innocent purchaser or landowner defense, the purchaser must conduct with due care (diligence), all appropriate inquiry (consistent with good practice) of previous ownership and uses of at time of property acquisition to identify possible contamination.
        - The property was acquired by the government via a tax lien or by eminent domain.
        - The property was acquired through an inheritance or bequest.

Title III–Emergency Planning and Community Right-to-Know Act (EPCRA)

Sections 301-303–Emergency Planning

- This act creates a new list of 406 extremely hazardous substances (EHS).
- Facilities must notify the State Emergency Response Commission (SERC) and Local Emergency Planning Committee (LEPC) when the facilities maintain an inventory of one or more EHSs beyond the Threshold Planning Quantity (TPQ) for the substance.
  1. Notification must occur within 60 days of when the TPQ of an EHS is exceeded.
  2. "Facilities" may include any industrial, institutional, municipal, private, public or agricultural facilities where the EHS exceeds the TPQ.
  3. The LEPC is comprised of representatives of facilities who must report EHS to emergency responders (fire, police, emergency medical squads), medics and the public at large.
  4. The SERC, LEPC and facilities are all required to develop emergency response plans in order to offer a coordinated response to a release of an EHS.

Sections 311-312–Community Right-to-Know

- Affected facilities are all those that have hazardous materials as defined by the OSHA Hazard Communication Standard (i.e., posing either physical or health hazards) in amounts:
  1. Of 10,000 pounds or more or
  2. That exceed the TPQ if they constitute an EHS.
     - Facilities must submit either a Material Safety Data Sheet (MSDS) or a list of those regulated hazardous chemicals to SERC, LEPC and the local fire department(s) in compliance with Section 311.
     - Facilities must submit an annual inventory of their hazardous chemicals (exceeding either the TPQ if an EHS, or more than 10,000 pounds for other hazardous materials) by March 1 to comply with Section 312.
  3. The inventory is submitted either on Tier I or Tier II forms to SERC, LEPC and the local fire department(s).
     - Inventory information includes chemical names, annual and daily chemical amounts and location of chemicals at the facility.
     - The purpose of the Community Right-to-Know sections is to develop and improve emergency response plans at the state and local levels and at facilities to better coordinate their responses to the release of hazardous chemicals.

Section 304–Emergency Notification

Facilities must report any release of a chemical that exceeds the reportable quantity (RQ) for that chemical when the release is not allowed by a permit.

Note: Reportable quantities are calculated over a 24-hour period.

1. Notification must be made immediately upon discovery of the chemical release.
2. Information required by notification includes:
   - Chemical name;
   - Whether the chemical is extremely hazardous;
   - Amount released;
   - Time of release;
   - Duration of release;
   - Whether the chemical was released to the air, water and/or land;
   - Acute or chronic health effects;
   - Need for evacuation, if any, and
   - Name and telephone number of contact person.
3. A follow-up written report is required if the release was:
   - An EHS that exceeded its RQ and went beyond the facility property boundary, or
   - A Superfund hazardous substance continuously released, or
   - A U.S. D.O.T. hazardous material released in transportation.
4. Facilities must report a chemical release to the following:
   - If the chemical is an EHS, the amount exceeds its RQ and the release has gone beyond the facility property boundary, notification is required to the SERC and LEPC.
   - If the chemical is a hazardous material as regulated by Sections 311/312 and the amount exceeds its RQ, notification is required to the SERC, LEPC and local fire department(s).
   - If the chemical is a Superfund hazardous substance and the amount exceeds its RQ, notification is required to NRC.

Note: A chemical may be regulated by more than one statute. Consult the Title III List of Lists for assistance. Call (800) 535-0202 for more information.

Section 313–Toxic Chemical Inventory Reporting

A new list of about 654 toxic chemicals under Section 313 was established. Facilities are regulated by Section 313 if they:

1. Have a Standard Industrial Classification (SIC) code of 20 through 39,
2. Have 10 or more full-time employees or
3. Manufacture or process 25,000 pounds of a Section 313 regulated chemical or use more than 10,000 pounds of a Section 313 regulated chemical during the preceding calendar year.

- Facilities meeting these requirements must complete a Toxic Chemical Release Inventory Form (known as a Form R), which describes all releases of Section 313 chemicals to the environment (that occurred as a result of normal operations) during the preceding calendar year.

1. Form R must be completed and submitted to the EPA and the state by July 1 annually.
2. Information required on Form R includes:
   - Name, location, type of business; off-site locations of waste-containing toxic chemicals; whether the chemical is manufactured, imported and/or processed, and categories of use; and an estimated range of the maximum amounts of the toxic chemical at the facility at any time during the preceding year;
   - Amount of the chemical entering the air, land and/or water annually; waste treatment/disposal methods and efficiency of these methods; certification by a senior facility official; amounts of the toxic chemical entering the waste stream prior to recycling, treatment or disposal;
   - Recycling methods used, the percentage recycled and the net change in the percentage recycled;
   - Source reduction methods used; estimates of how much source reduction/recycling will affect the toxic chemical for the next two years; ratio of product production, current year to previous year;
   - Methods used to identify source reduction potentials; amounts of the toxic chemical released due to episodic or catastrophic events; and amounts of the toxic chemical which are treated and the net change from the previous year.
3. Facilities that report 500 pounds or less or manufacture, process or use 1,000,000 pounds or less of a Section 313 chemical may use a shortened reporting form.

Note: Section 313 data is published annually by the EPA and the states. Sections 311/312, the Community Right-to-Know Act, allow the public access to information on the hazardous materials stored or used at a facility. The Freedom of Information Act (FOIA) allows the public access to information on an EHS regulated by Sections 301–303 and to reported releases under Section 304.

Penalties for SARA Title III Violations:

- Failure to comply with various SARA Title III reporting requirements in Sections 301–303, 311/312 and 313 can result in civil and/or administrative penalties ranging from $10,000–$75,000 per violation per day.

Willfully failing to report a release as required by Section 304 (Emergency Release Notification) can result in criminal penalties of up to $50,000 or five years in prison.

TSCA–1976

Toxic Substances Control Act

Key Aspects:

The purpose of TSCA is to reduce unreasonable risks to public health or the environment by regulating chemicals.

The EPA is authorized by TSCA to:

- obtain data from industry on the production, use and health effects of chemicals, and
- regulate manufacturing, processing and/or distribution in commerce, use and disposal of chemicals.

Pertinent Sections

Section 4–Testing

Selected existing chemicals and new chemicals must be tested for persistence, acute toxicity, carcinogenicity, mutagenicity, teratogenicity, behavioral effects and synergistic effects.

Section 5–Premanufacture Notification (PMN)

An initial TSCA inventory of 250 chemicals was published in 1979. Any chemical not listed in the TSCA inventory is "new." The EPA must be given a 90-day advance notification of intent to manufacture or import a "new" chemical—PMN. The EPA may extend the review period an additional 90 days.

PMN includes the following information: chemical identity, molecular structure, categories of use, amount being manufactured, imported or processed, byproducts of manufacture, processing, use and disposal, worker exposure data and any test data on health and environmental effects.

The EPA may approve the production or ban the manufacture after its review. High molecular-weight polymers and low-volume chemicals (i.e., produced at less than 1,000 kg per year) are exempted from PMN.

Section 5.2–Significant New Use Rule (SNUR)

Even if the EPA has approved a chemical for manufacture, it may declare a use of a chemical as a "significant new use" and restrict the use or production of that chemical.

Section 7–Imminent Hazards

The EPA may ask the U.S. District Court for seizure of chemicals or products that the EPA believes may present an imminent and unreasonable risk to the public health or the environment.

Section 8–Reporting and Record Keeping

Types of reporting:

General data collection:—Preliminary Assessment Information Rule (PAIR)

About 350 chemicals are assessed: Comprehensive Assessment Information Rule (CAIR)

Health and safety studies: The EPA is allowed to collect health and safety study information from the manufacturer, independent testing labs and on chemicals produced in small quantities that are not offered for sale.

Substantial risk notification: The EPA must be notified by the industry of any evidence of substantial risk posed by a chemical.

Records of significant adverse reactions: Records of adverse reactions to the health or the environment alleged to be caused by a chemical must be maintained. Records must be maintained for 30 years. This applies only to SIC codes 28 and 2911.

Emergency incidents: Manufacturers and processors must immediately notify the EPA of any emergency incident of environmental contamination, such as: carcinogenic or teratogenic effects, death, serious illness to humans and serious impact to nonhuman organisms.

Section 12 and 13–Imports and Exports

The same record keeping and reporting requirements apply to imports and to exports if a chemical poses an unreasonable risk to the public health or the environment.

Section 6–Chemical Posing Unreasonable Risks

The EPA may prohibit or limit the manufacturing, processing, distribution in commerce, use or disposal of a chemical or chemical mixture if the EPA determines that any of these activities present an unreasonable risk to the public health or the environment.

The EPA may use any of the following measures to limit the impact of a chemical that poses unreasonable risks: ban the manufacturing of the chemical, ban the distribution in commerce of the chemical, restrict the use of the chemical, require warning labels on the chemical containers, issue public warnings about the chemical, require notification dependent on the chemical use, require record keeping or require reporting.

Thus far, the EPA has only regulated four substances under Section 6:

Asbestos: Asbestos is commonly used as insulation for both hot and cold equipment and piping, for fireproofing and for its tensile strength in many building construction materials, including surfacing, fireproofing, soundproofing, pipe covering, boiler insulation, felts, mastics, shingles, sheets, floor tile, ceiling tile, paints and coatings. Its use began in the early 1900s. In 1972/1973 the EPA banned the manufacture and distribution of spray-applied insulation and fireproofing used in construction that contained more than 1 percent asbestos.

In 1976 the EPA banned the manufacture of most other interior construction materials containing more than 1 percent asbestos; distribution continued until about 1980. In 1982 the EPA required public and most private schools (K-12) to identify friable asbestos on the interior of school buildings.

In 1986 TSCA Title II, or the Asbestos Hazard Emergency Response Act (AHERA), required public and private schools (K-12) to identify friable and non-friable asbestos on the interior of school buildings by using trained and certified building inspectors; develop asbestos management plans using trained and certified management planners; and accomplish asbestos abatement by using trained and certified project designers, contractor/supervisors and worker/handlers.

In 1990 the Asbestos School Hazard Amendments and Reauthorization Act (ASHARA) required that individuals who work as asbestos building inspectors, management planners, project designers, contractor/supervisors or worker/handlers in non-school public buildings, including apartment buildings with 10 or more units, must be trained and certified according to AHERA. In 1990 the EPA banned the manufacture of asbestos roofing and flooring felt products, effective in 1992. In 1996 the EPA banned the manufacture of asbestos paper products; the distribution ban took effect in 1997.

TCDD (tetrachlorodibenzodioxin): It is a byproduct which is found in the herbicide 2, 4, 5-T/Silvex, best known from Agent Orange and Love Canal.

The EPA required an advance notice of disposal methods in the Jacksonville, Arkansas, pesticide plant.

CFC (chlorofluorocarbon): Potential damage to the ozone from freon (CFC) led the EPA to restrict use of CFCs in aerosol cans.

PCBs (polychlorinated biphenyls): Widely used as a heating/cooling dielectric, PCBs were used in: electrical distribution transformers, large high-voltage and low-voltage capacitors, railroad transformers, hydraulic systems, heat transfer systems, carbonless copy paper, electric motors with hydraulic fluid, pigments, electromagnetic switches, voltage regulators, compressors, circuit breakers, reclosers, optical liquids, microscopy mounting media, immersion oil and fluorescent light ballasts. The most common health effects resulting from acute PCB exposure are skin irritation and chloracne.

The EPA has established three categories for classification and regulation of PCBs.

- Items containing less than 50 ppm are considered non-PCB items.
- Items containing 50 ppm to 499 ppm are classified as PCB-contaminated.
- Items containing 500 ppm or more are classified as PCB-laden.

Since 1985, the EPA has allowed continued use of PCBs in equipment when:

- They are totally enclosed.
- Equipment containing PCBs and access to areas where such PCB equipment is located have posted warning labels.
- Regular inspections are made of PCB-laden transformers and apparatus.

- PCB network-type transformers were required to be removed from service in or near commercial buildings by October 1, 1990.
- Visual inspections of PCB transformers are required at least every three months, unless there is a potential risk to food or feed, then an inspection is required once a week to identify leaks.
- PCB-laden equipment is registered with the local fire department, and no combustible materials are stored near PCB equipment.

In 1987, the EPA established a PCB Spill Cleanup policy, which requires:

- Immediate (i.e., less than 24 hours) notification to the EPA of any release of PCBs exceeding 50 ppm or of untested mineral oil, when sensitive areas are affected, and for all spills greater than 10 pounds;
- Immediate response to contain and clean up PCB releases;
- Cleanup levels based upon weight of PCBs and PCB-containing materials; and
- Daily inspections are required until release cleanup is finished.

In 1989, the EPA added a manifest tracking requirement for the disposal of PCBs. The EPA now requires that:

- PCB generators or treatment, storage or disposal facilities (TSDFs) must obtain an EPA identification number.
- All PCB shipments must be done with an RCRA Uniform Hazardous Waste Manifest and follow hazardous waste manifesting requirements.
- Commercial PCB treatment, storage and disposal facilities must receive specific EPA approval, which is contingent upon closure, and demonstrate financial assurance to accomplish closure.

Note: No liquids containing more than 50 ppm may be disposed in a landfill. High-temperature destruction of PCBs via incineration or boilers is a common method of disposal. Waste oil containing any detectable amount of PCBs cannot be used as a fuel, a sealant, coating or dust-control agent.

TSCA Record Keeping

- 30 years for adverse health effects
- Five years for all other TSCA records except for:
  - PCB equipment inspection and maintenance records and PCB disposal manifests, which must be kept for three years.

TSCA Penalties

- Civil penalties of up to $25,000 per day per violation are possible.
- Criminal penalties, where knowing or willful violations are proven, carry penalties of up to $25,000 per day per violation and/or imprisonment for up to one year. As with CERCLA and SARA, the most severe penalties are applied for failures to notify, report and maintain proper records.

The following section is reprinted with permission from *The Facilities Manager's Reference* by Harvey H. Kaiser, Ph.D., R.S. Means Co., Inc.

Safety and Health Regulations

Safety

A comprehensive safety department monitors an organization's occupational health and safety, to minimize health hazards and risk of injury. The physical well being of personnel should be the primary consideration of a safety program. In addition, organizations are legally responsible for a growing number of federal, state, and local laws and regulations. An organization will benefit from a sound occupational health and safety program by realizing financial benefits through the prevention of injury, illness, loss of property, or interruption of normal operations and procedures.

A safety program should cover all aspects of an organization's activities. Appropriate standards should be established for each operation and facility, and procedures adopted to assure compliance with these standards.

The scope of an occupational health and safety program may vary widely but there should be provisions for the following:

- Policies and organizational structure
- Occupational health and safety standards
- Surveys and inspections
- Reporting, investigating, and record keeping
- Accident prevention
- Emergency procedures

Policies and Organizational Structure:

A facilities management policy statement on occupational health and safety should be prepared and distributed to all employees. The policy should outline goals and objectives of the program and require the cooperation of all employees to adhere to applicable rules, regulations, codes, and standards.

A health and safety officer is responsible for administering the policy. This officer usually chairs a committee comprised of other department heads or designated representatives. In a large organization, staff may be assigned to the officer to help carry out inspections, investigating incidents, and maintaining records.

Hazards should be minimized in the planning phase rather than after construction or purchasing is complete. Regular coordination meetings should be held with representatives from security, property management, design and construction, maintenance and operations, purchasing, and risk management departments. This ensures that appropriate standards or codes are specified and that safeguards are included in construction, repairs, and purchase of property and equipment.

Occupational Health and Safety Standards:

An occupational health and safety manual should be developed as a guide to safe practices and procedures. The manual should be approved by the organization's chief executive officer and regularly updated. It should include: the policy statement, references to applicable legislation and agency regulations, instructions for illness and injury reporting, and instructions for building evacuation and other emergency procedures.

Fire safety measures should be specified in the manual, such as placing maps on all building floors that outline routes of escape in case of fire. The manual should also include instructions on the storage and disposal of toxic materials and hazardous wastes, where applicable.

Government and industry standards should be considered minimums: it may be advisable for certain conditions to exceed these requirements. For example, higher rates of air exchange may be required to protect books or other print material. Higher HVAC standards are often required to protect sensitive material. Building standards should meet the needs of a particular facility, while providing a barrier-free environment, ventilation, illumination, fire and life safety, and first aid or emergency needs.

Standards for training in the safe use of equipment, the use of protective equipment, and correct emergency procedures should also be specified.

Surveys and Inspections:

Facilities and equipment should be inspected upon acquisition and regular surveys of existing conditions taken on a periodic basis. Surveys should determine compliance with federal, state, local laws, and any special guidelines of the organization. Qualified personnel should conduct the surveys and record their findings in reports to the environmental health and safety officer. Copies should be forwarded to the supervisors of the area surveyed and to maintenance and operations for immediate corrective action. Surveys may vary from annual inspections to daily checks of hazardous areas.

Reporting, Investigating, and Record Keeping:

The occupational health and safety department should establish procedures for reporting, investigating, and record keeping of accidents, injuries, and unsafe conditions and practices. Reporting enhances accident prevention efforts; corrective actions can be developed from reports to eliminate unsafe conditions and practices.

Records of accidental injuries and illnesses provide information which is useful in identifying and analyzing ongoing problems and for program planning. The organization also protects itself in future legal actions by keeping records. Records can be used to determine the strengths and weaknesses of an accident prevention program.

Unsafe conditions are the primary focus of OSHA and EPA standards. Guidelines for identifying and reporting the presence of a hazardous condition and using hazardous materials should be developed and distributed to department heads and supervisors.

Accident Prevention Program:

An accident prevention program serves an organization's objectives by identifying potential problems from reports and records. A program includes many of the items previously described: an occupational health and safety manual, standards, informational materials, and presentation. All personnel must be made aware of hazards throughout the workplace. Warning signs, signals, and notices should be understood and supervisors assigned to transmit such information to employees.

Safety in the environment requires safe working techniques and responsible behavior.

Emergency Procedures:

Numerous emergency conditions—ranging from natural disasters and fire to threats of civil disturbance—can beset an organization and each one requires a special procedure. Emergency procedures should be developed which provide contingencies for various events. Such plans cannot prevent an emergency but can help mitigate its effects and minimize harm to life and damage to property. An organization's emergency plans should be coordinated with local authorities and appropriate agencies. Regardless of size, organizations controlling their own facilities should follow some basic principles for impending emergencies. A chain of command should be established to define the authority and responsibilities of individuals. Responsible and knowledgeable persons should be selected to make decisions in emergencies. Such individuals should have the authority to make modifications in procedures to respond to changing conditions. A control center for maintenance and operations should be established with a direct-line telephone bypassing the switchboard. Emergency power should be connected to the communications system. Equipment locations for items such as portable power generators, lights, and first-aid should be identified and regularly checked for seasonal conditions such as hurricanes, floods, or snowstorms.

An emergency manual should be written, and should include policies for the following conditions:

- Fire
- Hazardous environment
- Civil disturbances
- Hurricanes and floods
- Explosion
- Power loss
- Labor problems
- Snowstorms

Ed. Note: An emergency manual should also cover:

- *Loss of utilities of heating and cooling*
- *Gas leaks*
- *Contamination of water supply*

Building evacuation plans are required for fires, explosions, hazardous atmospheres, and certain natural disasters. An evacuation plan should identify the following:

- Responsibility for activating the plan
- Training and practice requirements
- Special needs of handicapped persons
- Routes of evacuation and assembly points

Communications systems are an important part of emergency procedures. Periodic testing of alerting systems is required to ensure operational readiness and familiarity by all personnel. Alarm systems should be connected to emergency power systems. A careful survey should be made of all operating equipment requiring emergency power. Switching and alternate power sources are expensive and costs can be controlled by careful selection of equipment and circuits to be provided emergency power.

Ed. Note: At the time of publication OSHA has proposed a new rule that all employers covered by the OSHA and the General Duty Clause be required to set up a safety and health program. To quote from their document, Draft Proposed Safety and Health Program Rule, 29 CFR 1900.1, Docket No. S&H-0027: "The purpose of this rule is to reduce the number of job-related fatalities, illnesses, and injuries. The rule will accomplish this by requiring employers to establish a workplace safety and health program to ensure compliance with OSHA standards and the General Duty Clause of the Act (Section 5(a)(1))." The core elements each program must have are:

- *Management leadership and employee participation;*
- *Hazard identification and assessment;*
- *Hazard prevention and control;*
- *Information and training; and*
- *Evaluation of program effectiveness.*

The proposed rule does have a grandfather clause. "Employers who have implemented a safety and health program before the effective date of this rule may continue to implement that program if:

1. *The program satisfied the basic obligation for each core element; and*
2. *The employer can demonstrate the effectiveness of any provision of the employer's program that differs from the other requirements included under the core elements of this rule."*

Further information on this proposed rule may be found on the OSHA web site at www.osha-slc.gov/SLTC/safetyhealth/nshp.html.

The following material on Personal Protective Equipment is excerpted from U.S. Department of Labor Fact Sheet No. OSHA 92-08. It is currently available at http://www.osha-slc.gov/OshDoc/Fact_data/FSNO92-08.html

Personal Protective Equipment

Hard hats, goggles, face shields, earplugs, steel-toed shoes, respirators. What do all these items have in common? They are all various forms of personal protective equipment. OSHA standards require employers to furnish and require employees to use suitable protective equipment where there is a "reasonable probability" that injury can be prevented by such equipment. The standards also set provisions for specific equipment.

While use of personal protective equipment is important, it is only a supplementary form of protection, necessary where all hazards have not been controlled through other means such as engineering controls. Engineering controls are especially important in hearing and respiratory protection which have specific standards calling for employers to take all feasible steps to control the hazards.

Head Protection:

Cuts or bruises to the scalp and forehead occurred in 85% of the cases, concussions in 26%. Over a third of the cases resulted from falling objects striking the head.

Protective hats for head protection against impact blows must be able to withstand penetration and absorb the shock of a blow. In some cases hats should also protect against electric shock. Recognized standards for hats have been established by the American National Standards Institute (ANSI).

Foot and Leg Protection:

Sixty-six percent of injured workers were wearing safety shoes, protective footwear, heavy-duty shoes or boots and 33%, regular street shoes. Of those wearing safety shoes, 85% were injured because the object hit an unprotected part of the shoe or boot. For protection against falling or rolling objects, sharp objects, molten metal, hot surfaces and wet, slippery surfaces, workers should use appropriate footguards, safety shoes or boots and leggings. Safety shoes should be sturdy and have an impact-resistant toe. Shoes must meet ANSI standards.

Eye and Face Protection:

Injured workers surveyed indicated that eye and face protection was not normally used or practiced in their work areas or it was not required for the type of work performed at the time of the accident.

Almost one-third of face injuries were caused by metal objects, most often blunt and weighing one pound or more. Accidents resulted in cuts, lacerations, or punctures in 48% of the total, and fractures (including broken or lost teeth) in 27%.

Protection should be based on kind and degree of hazard present and should: 1) be reasonably comfortable, 2) fit properly, 3) be durable, 4) be cleanable, 5) be sanitary, and 6) be in good condition.

Ear Protection:

Exposure to high noise levels can cause irreversible hearing loss or impairment. It can also create physical and psychological stress. Preformed or molded ear plugs should be individually fitted by a professional. Waxed cotton, foam or fiberglass wool earplugs are self-forming. Disposable earplugs should be used once and thrown away; non-disposable ones should be cleaned after each use for proper maintenance.

OSHA has promulgated a final rule on requirements for a hearing conservation program. Information on the program is available from the closest OSHA office.

Arm and Hand Protection:

Burns, cuts, electrical shock, amputation and absorption of chemicals are examples of hazards associated with arm and hand injuries. A wide assortment of gloves, hand pads, sleeves and wristlets for protection from these hazards is available.

The devices should be selected to fit the specific task. Rubber is considered the best material for insulating gloves and sleeves and must conform to ANSI standards (copies available from ANSI, 1430 Broadway, New York, NY 10018).

Torso Protection:

Many hazards can threaten the torso: heat, splashes from hot metals and liquids, impacts, cuts, acids, and radiation. A variety of protective clothing is available: vests, jackets, aprons, coveralls, and full body suits.

Fire retardant wool and specially treated cotton clothing items are comfortable, and they adapt well to a variety of workplace temperatures. Other types of protection include leather, rubberized fabrics, and disposable suits.

Respirator Protection:

Information on the requirements for respirators to control occupational diseases caused by breathing air contaminated with harmful dusts, fogs, fumes, mists, gases, smokes, sprays, and vapors is available in 29 CFR 1910.134. Proper selection of respirators should be made according to the guidance of ANSI Practices for Respiratory Protection.

Using personal protective equipment requires hazard awareness and training on the part of the user. Employees must be aware that the equipment alone does not eliminate the hazard. If the equipment fails, exposure will occur.

For Copies of OSHA Standards or Clarification: Check your phone book under the U.S. Department of Labor listing for the OSHA office nearest you. This is one of a series of fact sheets highlighting U.S. Department of Labor programs. It is intended as a general description only and does not carry the force of legal opinion. This information will be made available to sensory impaired individuals upon request. Voice phone: (202) 523-8151. TDD message referral phone: (800) 326-2577.

Ed. Note: See Figure 10.10 for a PPE equipment chart.

PPE	Protection Provided	Limitations
Hard Hat	Impact injury to head	No protection against chemical splashing, gases, fumes, vapors, or mists
Hood	Chemical splashing	No protection against impact, gases, mists, fumes, or vapors
Sweat Band	Minimal protection of eyes from perspiration-borne irritants	No real protection of any kind provided
Safety glasses	Large dust particles, chips, and projectiles. Laser protection with proper lenses	Do not fit tightly, no protection from dust clouds or side impact projectiles
Goggles	Dust, projectiles, chips	Not good for liquid splashes, gases, or mists
Full Face Mask	Higher projectile protection, good against splashes and sprays	Need hood for full protection against splashes and sprays. Need respirator to protect against mists, gases, vapors or fumes
Ear Plugs	Noise	Inconvenient to use
Ear Muffs	Noise	Bulky to wear (but improving)
Apron	Splashes and spills, good for working with acids, corrosives, etc.	Only cover clothes in certain areas; no protection of outside edges
Coveralls	Dust, spills	Not good for liquids, gases, vapors, or fumes
Rubberized Gear	Liquid sprays, mists, and spills	Bulky, hot to wear, cumbersome to work in
Gloves	Hand protection from chemical contact, specialty gloves for high heat, etc.	Only protect to top of glove. Material must be compatible with chemical being handled
Sleeves	Increase protection from gloves	May give false sense of security
Special Shoes	Steel-toed protect against crushing injury to toes. Some provide puncture protection. Special shoes provide chemical protection	Not all shoes are the same. Special shoes needed for special situations
Boots	Protect feet from chemical spills and splashes	Must use care to avoid spills to inside of boot
Dust Mask	Good general protection from dusts	Not effective against gases, mists, vapors, or fumes
SCBA	Excellent against all types of air-borne contaminants	Air supply is self-contained, limiting work time
SAR	Excellent against all types of air-borne contaminants	Mobility limited by air supply hose
APR	Excellent against air-borne contaminants for which designed	Special filters needed depending on chemical in atmosphere

R.S. Means Co., Inc., *Hazardous Material and Hazardous Waste*

Figure 10.10

The following section is reprinted with permission from *Cost Planning & Estimating for Facilities Maintenance*, Chapter 6: "Codes and Regulations," by William H. Rowe III, AIA, PE, R.S. Means Co., Inc.

Occupational Safety and Health Act of 1970 (OSHAct)

This legislation is best summarized by quoting the "general duty clause," which states that "...each employer shall furnish to each of his employees employment and a place of employment which are free from recognized hazards that are causing or are likely to cause death or serious physical harm to his employees." OSHAct outlines the procedures and penalties that are involved in compliance. Workplaces are sources of possible health hazards. The presence of toxic chemicals, the use of machinery, or working in an environment that is stressful to the body because of excess noise can all pose real dangers to individuals if not properly controlled. Exposure to harmful effects in the workplace generally are expected to be "as low as reasonably achievable" (known by the acronym ALARA) and applies to all workplace environments. For managers of facilities maintenance and repair, the OSHAct regulations have had an impact on cost as well as overall procedures.

OSHAct has specific requirements for a wide variety of on-site activities that may expose individuals to hazardous materials. Some of these requirements are included in Figures 10.11a-g. Each of the specific OSHAct provisions is written to require that an employer assess for potential dangers, provide information to workers on the hazards likely to occur, provide adequate signage to alert persons to locations where hazardous materials or conditions exist, provide protective equipment or procedures to minimize the dangers, and have in place emergency and evacuation procedures in the event of an accident or injury.

The act has contributed to an increase in overall safety in the workplace. Initially, facility managers absorbed some significant costs to establish procedures and implement systems to comply. There is an overall awareness that some of the current administrative procedures that have been required are burdensome and ineffective and that the system and authority of OSHAct inspectors needs some improvement. Facility managers are nonetheless increasingly aware of both the responsibilities and costs of providing improved safety for employees in the workplace. Figures 10.11a-g are an overview of those aspects of OSHAct that are of primary significance to building managers, and that can influence the costs or procedures that are a part of facilities' operations.

Outline Summary of Selected OSHAct Regulations		
Regulation	**Title**	**Outline of Compliance Requirements**
1910.25 1910.26	Portable Wood Ladders Portable Metal Ladders	Applies to ladders used in construction. Uniformity of steps and overall length for specific trades, such as painters and masons, are specified.
1910.66	Powered Platforms for Building Maintenance	Applies to units permanently installed and dedicated to interior or exterior building maintenance (construction scaffolding is covered elsewhere). Building owners are responsible for determining that equipment meets OSHAct standards before each use and providing assurance of this to employers of workers who will use the equipment. The criteria and methods of assurance, an emergency plan, and review by a professional engineer with expertise in these matters are required.
1910.67	Vehicle Mounted Elevating and Rotating Work Platforms	General safety requirements established to assure that such units are stabilized, workers are safe, and hydraulic and electrical components are in proper working order.
1910.95	Occupational Noise Exposure	Implement a hearing conservation program; maintenance shops and construction areas. Provide the following: • Audiometric testing of employees • 8 hour time weighted average 85 decibels (DBA) • Written notification of hazards • Personal protective equipment
1910.96 1910.97	Ionizing Radiation Nonionizing Radiation	Prevent overexposure to radiation such as X-rays, ultraviolet or infrared light in restricted areas. Certain facilities such as hospitals have radiation safety officers, who identify potential sources of radiation and train staff on proper procedures for managing the location, movement, and disposal of radioactive materials. Also, procedures for monitoring the levels of radioactivity, signage, provision of restricted areas, evacuation procedures, and information on health effects are provided.
1910.101 1910.102 1910.103 1910.104	Compressed Gases Acetylene Hydrogen Bulk Oxygen Systems	Procedures for storing and handling compressed gases in cylinders such as hydrogen, oxygen, acetylene, nitrous oxide are discussed. Additional information available from the Compressed Gas Association.
1910.11	Flammable and Combustible Liquids	Regulates bulk, container, and portable tank storage of flammable and combustible liquids. Requirements for design and construction of flammable containers and rooms, storage requirements, office areas, maintenance storage, and ventilation standards are provided. Emphasis on containers of liquids (including aerosols, gasoline, paints, thinners) not exceeding 60 gallons. Liquids such as paints, oils and varnishes used for painting or maintenance, but kept for less than 30 days, are exempt from the regulations.
1910.12	Hazardous Waste Operations and Emergency Response	Standards for clean-up of existing hazardous waste sites and emergency response to release of hazardous substances. Facilities with hazardous materials present are required to have policies and procedures in place to respond to emergencies involving release of hazardous materials.

R.S. Means Co., Inc., *Cost Planning & Estimating for Facilities Maintenance*

Figure 10.11a

Outline Summary of Selected OSHAct Regulations (cont.)		
Regulation	**Title**	**Outline of Compliance Requirements**
1910.132 1910.133 1910.134 1910.135 1910.136 1910.137 1910.138	Personal Protective Equipment Eye and Face Protection Respiratory Protection Head Protection Foot Protection Electrical Protection Hand Protection	Employers are required to assure that personal protective equipment (PPE) for workers is provided, used and maintained in a safe and sanitary condition. Such equipment shall be used to minimize dangers from: • Impact • Penetration • Compression – rollover • Chemicals • Heat • Harmful dust • Light (optical) radiation The employer shall conduct an assessment of the workplace – usually a walkthrough to observe the following potential sources of hazards: • Motion or movement in personnel that could result in collision with stationary objects. • High temperatures that can result in burns, eye injury, or ignition of equipment. • Chemical exposures. • Harmful dust. • Light radiation; i.e., lasers, welding, brazing, cutting, furnaces, high intensity lights, and similar sources. • Falling objects or potential for dropping objects. • Sharp objects that might pierce the feet or cut the hands. • Rolling or pinching objects that could crush the feet. • Layout of the workplace and location of co-workers. • Electrical hazards. The employer should review past injury and accident data to identify problem areas. PPE appropriate to the anticipated hazards (e.g., to head, hand, foot, eye) are to be provided.
1910.14	General Sanitation	Proper sanitation in toilet facilities, vermin control, waste disposal, showers, food handling, eating and drinking areas should be provided. Food consumption in toxic areas is prohibited, and there are requirements for washing facilities and lavatories.
1910.14	Safety Color Code for Marking Physical Hazards	Proper marking of physical hazards is established: red for danger, fire protection, and stop; yellow for caution and for marking falling and tripping physical hazards.
1910.15	Accident Prevention Signs and Tags	Specifications for accident prevention signs are required at specific hazards, especially where failure to do so may lead to accidental injury. The basic categories of signs are as follows: • Danger • Caution • Safety instruction • Biological hazards • Accident prevention Signs must contain a signal word and a major message and must be readable at five (5) feet or greater as warranted.

R.S. Means Co., Inc., *Cost Planning & Estimating for Facilities Maintenance*

Figure 10.11b

Outline Summary of Selected OSHAct Regulations (cont.)		
Regulation	**Title**	**Outline of Compliance Requirements**
1910.15	Permit-Required Confined Spaces	Applies to certain spaces such as manholes, crawl spaces, vaults, and similar spaces not commonly occupied and also where workers are subject to dangers of hazardous air, collapse, or other significant dangers. In particular, a space is considered a "confined" space if it: • is not ordinarily inhabited by people and is large enough for an individual to enter. • has limited or restricted entry or egress. • is not designated for continuous employee occupancy. A space is considered a "Permit-Required Confined Space" if the above conditions exist and if further the space: • contains or has potential to contain a hazardous atmosphere. • contains a material substance with the potential to engulf the entrant. • has an internal configuration such that an entrant could be trapped by inwardly conveying walls or by a floor that slopes downward and tapers to a smaller cross section. Or: • contains any other recognized serious safety or health hazard. When a permit-required confined space is identified, the following procedures are required: • Hazard identification – the identity and severity of each hazard in the space must be determined and characterized. • Hazard control – procedures and practices that provide for safe entry into the space must be established and implemented. • Permit system – a written permit must be prepared, issued, and implemented. Upon completion of the task, physical barriers such as caution tape, and posting of signs and barriers, prevent unauthorized persons from entering. • Employee training – employees who enter the space, serve as standby attendants, or issue permits must have completed the confined spaces training program. • Equipment – appropriate equipment such as air sampling devices, retrieval equipment, respirators, and ventilation blowers must be provided. • Rescue/emergency procedures for personnel and equipment must be established and implemented. • External hazard protection – physical barriers, "men working" signs, and similar devices to control potential hazards posed by pedestrians and vehicles must be provided. • Written plan – above items must be incorporated with implementation.
1910.15	The Control of Hazardous Energy (lockout/tagout)	Requires employers to establish a program that will enable workers to disable equipment, particularly when the unexpected energization of the equipment or release of stored energy could cause injury to persons working on the equipment. It applies specifically to the control of energy during servicing. The following are required: • Written program. • Employee training and certification tagout devices. • Periodic inspection of procedures. • Coordination with outside contractors and shift/personnel changes. • Lockout or tagout devices, which are affixed to equipment in conjunction with a written program of maintenance procedures and training sessions, and allow workers to minimize dangers of bodily harm from electrical or mechanical activation of machinery and equipment.

R.S. Means Co., Inc., *Cost Planning & Estimating for Facilities Maintenance*

Figure 10.11c

Outline Summary of Selected OSHAct Regulations (cont.)		
Regulation	**Title**	**Outline of Compliance Requirements**
1910.16	Fire Brigades	Requires that facilities establish a written policy that recognizes the existence of a fire brigade for dealing with interim fires. A training program, PPE, emergency response plans, and respiratory protectors are to be provided.
1910.157	Portable Fire Extinguishers	Provides for the placement, use, maintenance, and testing of portable fire extinguishers: • Inspect visually monthly. • Travel distance for class A and D fire extinguishers is 75 feet or less. • Travel distance for class B fire extinguishers is 50 feet or less.
1910.158-1910.163	Fire Extinguishing Systems	Provide training annually. Provides for specifics on fire-related systems.
1910.176 1910.178 1910.179 1910.184	Handling Materials – General Powered Industrial Trucks Overhead and Gantry Cranes Slings	Provides for handling of materials; related to secure storage, housekeeping, and clearance limits to prevent damage to materials, spills and the like. Operators of power trucks who move products are required to undergo training.
1910.21	General Requirements for All Machines	Requires that guards be affixed to machines such as grinders, portable power tools, and power saws. In some cases, manufacturers of equipment may need to be consulted to determine how to retrofit older equipment.
1910.21	Woodworking Machinery Requirements	Provides for safety-related procedures in carpentry shops or other work areas where wood is cut or finished.
1910.22	Abrasive Wheel Machinery	Abrasive wheel machinery may be used only if properly guarded when the wheel is greater than 2 inches in diameter. Specific requirements are provided for flanges, work rests, guard exposure angles, and bench and floor stands.
1910.22	Mechanical Power Transmission Apparatus	Standards on motor-driven equipment such as supply and exhaust fans and compressors; deals with both direct and fan belt drives, the condition of the pulleys, and the drive mechanisms. Levels of required illumination and standards for space clearance and restricted access are provided.
1910.24	Guarding of Portable Powered Tools	Requires that equipment be properly guarded, inspected, maintained, and handled. Generally: • Identify the types and applications of tools. • Identify specific compliance requirements. • Ensure compliance with manufacturers' specifications and requirements.
1910.252	Welding and Welding Operator Oxygen Fuel Gas Welding and Cutting Arc Welding and Cutting Resistance Welding	Provides for specific regulations regarding welding by employees. Covered topics include oxygen-fuel, arc and electric resistance welding. General requirements include specifications on: • Fire prevention and protection • Ventilation • Confined spaces • Personal protection equipment (PPE) • Signage • Health hazard training • Gas storage • Arc shielding

R.S. Means Co., Inc., *Cost Planning & Estimating for Facilities Maintenance*

Figure 10.11d

Outline Summary of Selected OSHAct Regulations (cont.)		
Regulation	**Title**	**Outline of Compliance Requirements**
1910.301-1910.399	Electrical	Provides a comprehensive set of standards regarding electrical systems. Topics include initial design, installation, and maintenance of systems. Much of the standard is devoted to building construction and renovation. For every major replacement, modification, repair or rehabilitation, the standards include regulations on: • Examination, installation, and use of equipment • Splices • Arcing parts • Marking • Identification of disconnecting means • Guarding of live parts • Protection of conductors and equipment • Location in or on premises • Arcing or suddenly moving parts • 2-wire DC and AC systems to be grounded • AC systems 50 to 1,000 volts not required to be grounded • Grounding connections • Grounding path • Fixed equipment required to be grounded • Grounding of equipment connected by cord or plug • Methods of grounding fixed equipment • Flexible cords and cables, uses and splices • Hazardous locations • Entrance and access to workspace (over 600 volts) • Circuit breakers operated vertically • Circuit breakers used as switches • Grounding of systems of 1,000 volts or more supplying portable or mobile equipment • Switching series capacitors over 600 volts • Warning signs for elevators and escalators • Electronically controlled irrigation machines • Ground fault interrupters for fountains • Physical protection of conductors over 600 volts
1910.1	Air Contaminants	OSHAct regulates exposure to over 700 compounds. Examples include solvents, caustic gases, pesticides, asbestos, mercury, antineoplastic drugs, waste anesthetic gases, and welding fumes. These may be present as gas, mist, vapor, particle, or fume and may or may not be discernible. Therefore, management is responsible for alerting workers to potential hazards. Plumbers working on laboratory sinks, and sheet metal workers repairing exhaust ductwork from a toxic gas facility or an infectious disease room, are examples of simple tasks that may have hazardous substances associated with them. MSDS sheets are used to assess the materials that may be encountered.
1910.1	Inorganic Arsenic	**Regulates exposure to inorganic arsenic.**

R.S. Means Co., Inc., *Cost Planning & Estimating for Facilities Maintenance*

Figure 10.11e

Outline Summary of Selected OSHAct Regulations (cont.)		
Regulation	**Title**	**Outline of Compliance Requirements**
1910.1	Lead	Exposure to lead is regulated. Baseline sampling is required in potentially contaminated areas. If lead levels greater than 30 micrograms per cubic meter of air are detected, intervention is required. This includes training, blood tests, and sampling.
1910.1	Bloodborn Pathogens	Healthcare workers who deal directly with patients are most at risk. Hospital workers who come in contact with medical wastes that spill into the environment, repair work in patient rooms, waste drain systems, research facilities in HIV and HBV, and similar risks are required to be notified of the potential for such risk and provided with PPE and policies and procedures for dealing with the substances and avoiding harm.
1910.1	Ethylene Oxide	Provides expanded regulations for this highly toxic gas used in the sterilization of hospital equipment. Health hazard training, monitoring, respiratory protection, emergency evaluation, and PPE standards are regulated.
1910.1	Formaldehyde	Provides expanded regulations for handling exposure. The principal areas of concern are training, PPE, and monitoring during cleanup.
1910.12	Hazard Communication	Requires that information on hazardous materials be made available and is the main focus of OSHAct. The Material Safety Data Sheets (MSDS) provide essential information for use of products. The basic contents of a hazardous communication program will include the following: • General information on health and physical hazard determination • Hazardous chemical lists • Labels and other warning signs • MSDS sheets • Employee information and training • Contractor's responsibilities • List of facility contacts and responsibilities • Methods of MSDS procurements • MSDS locations in facility • Training programs • Program effectiveness audits All hazardous materials must be located and inventoried at each site to determine: • Product name • Manufacturer • Container size • Container type • Number of containers • Location within facility The above information is required to be kept current, usually with a data management system and verified with periodic inventories. Training programs on the hazardous materials (hazmat) must be in place and shall provide for the following: • Location of facility's MSDS • Location of facility's hazard communication program • Location of hazmat inventory lists • Designated coordinators of the facility and their responsibilities • Labeling • Methods to detect the presence of a chemical (odor, appearance, color) • Emergency treatment • Discussion of air sampling data

R.S. Means Co., Inc., *Cost Planning & Estimating for Facilities Maintenance*

Figure 10.11f

Outline Summary of Selected OSHAct Regulations (cont.)		
Regulation	**Title**	**Outline of Compliance Requirements**
1910.12		• Potential physical effects • Potential acute chronic health effects • PPE, work practices, spill/emergency actions
1926.1 through 1926.1148	Safety and Health Regulations for Construction	This part of the code details the requirements to be met for all construction related activity. The sections covered are divided into the following subparts: • General • General Interpretations • General Safety and Health Provisions • Occupational Health and Environmental Controls • Personal Protective and Life Saving Equipment • Fire Protection and Prevention • Signs, Signals, and Barricades • Materials Handling, Storage, Use, and Disposal • Tools—Hand and Power • Welding and Cutting • Electrical • Scaffolding • Floor and Wall Openings • Cranes, Derricks, Hoists, Elevators, and Conveyors • Motor Vehicles, Mechanized Equipment, and Marine Operations • Excavations • Concrete and Masonry Construction • Street Erection • Underground Construction, Caissons, Cofferdams, and Compressed Air • Demolition • Blasting and Use of Explosives • Power Transmission and Distribution • Rollover Protective Structures, Overhead Protection • Stairways and Ladders • Diving • Toxic and Hazardous Substances
1926.58	Asbestos	Provides for protection of workers who encounter asbestos or are exposed to asbestos during construction. Existing facilities in which asbestos materials may be present generally have an industrial hygienist conduct an inventory to identify the location of such materials. During subsequent renovations, materials that might be encountered or disturbed by the renovations are usually removed by firms specializing in removal of these materials. Workers don entire suits to protect them, showers are provided to wash off fibers, and the rooms are subject to negative pressure with HEPA filters to the outside to protect surrounding spaces. Friable asbestos, such as on pipe insulation, plaster, or boilers, is considered the most dangerous. Asbestos siding and vinyl asbestos tile have the fibers encapsulated and the procedures for the removal of these materials is less stringent. These can often be removed without special notification and with wetting procedures.
1926.62	Lead	Lead was banned in paints after 1978. Construction work in pre-1978 facilities will involve protection of workers, protection of adjoining spaces, and proper collection and disposal of the materials.

R.S. Means Co., Inc., *Cost Planning & Estimating for Facilities Maintenance*

Figure 10.11g

The following section is reprinted with permission from *Certified Facilities Environmental Professional Exam Review Book,* Association for Facilities Engineering.

OSHA Key Aspects

The Occupational Safety and Health Administration (OSHA) was created as a bureau within the U.S. Department of Labor.

OSHA only applies to private employers.

OSHA requires that private employers furnish each employee with a place of employment free from recognized hazards which may cause death or serious harm. This requirement is commonly referred to as the "General Duty" clause.

Private employers also must comply with the standards and regulations established by OSHA.

OSHA requires that private employees comply with all the standards and regulations which regulate their actions and conduct.

Pertinent OSHA Requirements to Control Hazards: Control of hazards to employees involves the following measures, in order of preference:

Administrative controls: These are changes to personnel assignments which result in a less hazardous situation to employees. An example of these is allowing only properly trained employees to work in certain areas.

Engineering controls: These are the use of equipment to manage the hazard. Examples of these are the use of local exhaust ventilation, such as a fume hood, to control hazards and the substitution of a less hazardous chemical for the more hazardous chemical currently used.

Work practices: These are the establishment and implementation of procedures that reduce hazards. Examples are lockout/tagout and confined-space entry procedures.

Personal protective equipment: When administrative and/or engineering controls and work practices are inadequate in reducing hazards, personal protective equipment (PPE) is then used. PPE includes respirators, gloves, suits, boots and protective eye wear.

Training: Employees should be trained to understand all the controls and the PPE used to minimize hazards in a process.

Exposure monitoring: The due method for assessing the effectiveness of the hazard controls is to sample for the chemicals in the employees' breathing zones. Exposure monitoring is required by OSHA for certain toxic chemicals.

Medical surveillance: Biological screening and medical examination of employees are other methods of determining employees' exposure to chemicals. They are somewhat less satisfactory measures of assessing exposure, since a biological effect has, to some extent, already occurred. OSHA requires medical surveillance for certain chemicals, such as asbestos and lead.

Recordkeeping: OSHA requires that employee-exposure monitoring results and medical surveillance records be maintained by employers for at least 30 years. OSHA standards also require additional specific records to be maintained by employers, such as for injuries/illnesses that are occupationally related.

OSHA Standards and Regulations

OSHA standards are published in Chapter 29 of the Code of Federal Regulations (29 CFR) and address four major business categories:

General Industry - 29 CFR Part 1910: This applies to all private employers who are not covered by other OSHA or other federal programs.

Construction Industry - 29 CFR Part 1926: This applies to that industry where work for construction, alteration and/or repair, including painting and decorating, is conducted.

Maritime Industry

Agriculture: OSHA does *not* apply to:

- Self-employed persons,
- Farms where only immediate members of the farm employer's family are employed, and
- Workplaces covered by other federal agencies and statutes, such as
  - Mine Safety Health Administration (MSHA).

OSHA standards are of four types:

1. Design standards, which require a specific engineering design to be used,
2. Performance standards, which state the objectives to be achieved but allow flexibility in meeting these objectives,
3. Vertical standards, which apply to particular operations within an individual industry, and
4. Horizontal standards, which apply to all workplaces.

The following section is from the U.S. Department of Labor Program Highlights Fact Sheet No. OSHA 93-26. It is currently available on the Internet at www.osha-slc.gov/OshDoc/Fact_data/FSN093-26.html.

OSHA Hazard Communication Program

Hazard Communication Standard

Protection under OSHA's Hazard Communication Standard (HCS) includes all workers exposed to hazardous chemicals in all industrial sectors. This standard is based on a simple concept—that employees have both a need and a right to know the hazards and the identities of the chemicals they are exposed to when working. They also need to know what protective measures are available to prevent adverse effects from occurring.

Scope of Coverage: More than 30 million workers are potentially exposed to one or more chemical hazards. There are an estimated 650,000 existing hazardous chemical products, and hundreds of new ones are being introduced annually. This poses a serious problem for exposed workers and their employers.

Benefits: The HCS covers both physical hazards (such as flammability or the potential for explosions), and health hazards (including both acute and chronic effects). By making information available to employers and employees about these hazards, and recommended precautions for safe use, proper implementation of the HCS will result in a reduction of illnesses and injuries caused by chemicals. Employers will have the information they need to design an appropriate protective program. Employees will be better able to participate in these programs effectively when they understand the hazards involved, and to take steps to protect themselves. Together, these employer and employee actions will prevent the occurrence of adverse effects caused by the use of chemicals in the workplace.

Requirements: The HCS established uniform requirements to make sure that the hazards of all chemicals imported into, produced, or used in U.S. workplaces are evaluated and that this hazard information is transmitted to affected employers and exposed employees.

Chemical manufacturers and importers must convey the hazard information they learn from their evaluations to downstream employers by means of labels on containers and material safety data sheets (MSDS's). In addition, all covered employers must have a hazard communication program to get this information to their employees through labels on containers, MSDS's, and training. This program ensures that all employers receive the information they need to inform and train their employees properly and to design and put in place employee protection programs. It also provides necessary hazard information to employees so they can participate in, and support, the protective measures in place at their workplaces. All employers in addition to those in manufacturing and importing are responsible for informing and training workers about the hazards in their workplaces, retaining warning labels, and making available MSDS's with hazardous chemicals.

Some employees deal with chemicals in sealed containers under normal conditions of use (such as in the retail trades, warehousing and truck and marine cargo handling). Employers of these employees must assure that labels affixed to incoming containers of hazardous chemicals are kept in place. They must maintain and provide access to MSDS's received, or obtain MSDS's if requested by an employee. And they must train workers on what to do in the event of a spill or leak. However, written hazard communication programs will not be required for this type of operation.

All workplaces where employees are exposed to hazardous chemicals must have a written plan which describes how the standard will be implemented in that facility. The only work operations which do not have to comply with the written plan requirements are laboratories and work operations where employees only handle chemicals in sealed containers.

The written program must reflect what employees are doing in a particular workplace. For example, the written plan must list the chemicals present at the site, indicate who is responsible for the various aspects of the program in that facility and where written materials will be made available to employees.

The written program must describe how the requirements for labels and other forms of warning, material safety data sheets, and employee information and training are going to be met in the facility.

The following section is reprinted with permission from *Hazardous Material and Hazardous Waste* by Francis J. Hopcroft, P.E., David L. Vitale, M.Ed., and Donald L. Anglehart, Esq., R.S. Means Co., Inc.

Hazard Communication Program

The effective management of hazardous wastes begins with the management of hazardous materials. All wastes generated originate with the materials used; the nature of those materials determines how hazardous the wastes will be. The way the materials are managed determines the way in which the wastes must be managed.

OSHA Hazard Communication Standard: *Employer Responsibility:*

The management of hazardous materials in any workplace, including a construction site, is regulated by the Occupational Safety and Health Administration's (OSHA) Hazard Communication Standard, also known as *the Standard.* The Standard requires all employers to:

- Inventory the hazardous materials on the job site
- Obtain a Material Safety Data Sheet (MSDS) for each material
- Post both the inventory list and the MSDS forms in a location accessible to all employees
- Create a written Hazard Communication Program
- Establish a training program

The MSDS Form is prepared by the manufacturer of the material. It informs the reader of the physical and chemical properties of the material, and any health hazards associated with its use. The MSDS also provides guidance on the personal protective equipment to be used when handling the material. Other information includes a list of symptoms resulting from exposure, and what to do in the event of the exposure.

After the inventory and MSDS forms are collected, the employer is required to develop a written **Hazard Communication Program.** The purpose of this program is to inform each employee of the health and safety risks associated with the material(s) that the employee is required to use.

Finally, the employer must provide a ***training program*** to teach each employee how to read an MSDS, what to do in an emergency, and how to protect him or herself in the workplace. The employer must ensure that each employee is provided with the proper personal protective equipment for the job being done, and that he or she knows how and when to use the equipment.

Employee Responsibility:

The Standard also imposes the following responsibilities on the employee.

- To pay attention during the training sessions
- To use the personal protective equipment provided
- To follow appropriate safety guidelines
- The employer and employees must work together to ensure safety in the workplace

The following section is from the U.S. Department of Labor Program Highlights Fact Sheet No. OSHA 93-02.

OSHA Inspections

Inspecting for Job Safety and Health Hazards

The Mandate: The Occupational Safety and Health Act of 1970 seeks to "... assure so far as possible every working man and woman in the Nation safe and healthful working conditions."

As one way to promote worker protection, the Act authorizes the Occupational Safety and Health Administration (OSHA) to set and enforce safety and health standards. The agency conducts inspections to make sure these specific standards are met and that the workplace is generally free from recognized hazards likely to cause death or serious physical harm.

The CSHOs: OSHA calls its inspectors compliance safety and health officers (CSHOs). They are experienced professionals whose goal is to help employers and workers reduce on-the-job hazards.

Inspection Priorities: Not all of the 6 million workplaces covered by federal and state OSHA's can be inspected regularly. The most hazardous conditions need attention first.

Imminent Danger:

Imminent danger situations have top priority. An imminent danger is a hazard that could cause death or serious physical harm immediately, or before the danger could be eliminated through normal enforcement procedures. When compliance officers find imminent danger conditions, they will ask for immediate voluntary correction of the hazard by the employer or removal of endangered employees from the area. If an employer fails to do so, OSHA can go to the nearest Federal District Court for appropriate legal action.

Catastrophes and Fatal Accidents:

High priority is also given to investigation of job fatalities and accidents hospitalizing five or more employees. Such accidents must be reported to OSHA within 48 hours.

Complaints:

OSHA investigates written and signed complaints by current employees or their representatives of hazards that threaten serious physical harm to workers. Complaints, other than imminent danger, received from anyone other than a current employee or employee representative, or unsigned by a current employee, or received anonymously, may result in a letter from the agency to the employer describing the allegation(s) and requesting a response. OSHA will not reveal the name of the person filing the complaint, if so requested.

Programmed Inspections:

OSHA routinely conducts safety and health inspections in high-hazard industries, like manufacturing or construction. The agency develops its general schedule for inspecting the most hazardous industries based on various statistical data, such as job injury/illness rates, worker compensation, and other information.

After entering a workplace in a high-hazard business, OSHA inspectors consult and verify the log of injuries and illness which most employers with more than 10 employees are required to keep.

About five percent of OSHA programmed inspections focus on firms in low hazard manufacturing industries and an additional five percent in nonmanufacturing industries.

Follow-up Inspections:

The agency may reinspect firms cited for imminent danger conditions, or for willful, repeat or serious violations to ensure the correction of cited hazards. OSHA may also conduct follow-up inspections to check the progress of long-term hazard correction programs by employers.

The Inspection:

CSHO "Homework": To prepare for an inspection, compliance officers become familiar with the history of the establishment, the operations and processes in use, and the standards most likely to apply. They gather all equipment necessary to test for health and safety hazards.

At the Worksite:

When an OSHA inspector arrives, he or she displays official credentials and asks to see the employer. Employers should always insist upon seeing the compliance officer's U.S. Department of Labor credentials bearing their photos and serial numbers which can be verified by the nearest OSHA office. Employers have the right to require OSHA to obtain a warrant before permitting entry.

Opening Conference:

The compliance officer will explain the nature of the visit, the scope of the inspection and the applicable standards. Information on how to obtain copies of the OSHA regulations will be furnished. A copy of any employee complaint (edited, if requested, to conceal the employee's identity) will be provided. The employer will be asked to select an employer representative to accompany the compliance officer during the inspection. An authorized representative of the employees, if any, also has the right to go along. The compliance officer will consult with a reasonable number of employees.

Walkaround Inspection:

After the opening conference, the compliance officer and the representatives go through the workplace, inspecting for workplace hazards. When talking with workers, compliance officers will try to minimize work interruptions. The Act prohibits discrimination in any form by employers against workers because of anything they say or show the compliance officer during the inspection or for any other OSHA protected safety-related activity. The compliance officer will discuss any apparent violations noted during the walkaround and if asked, will offer technical advice on how to eliminate hazards.

Closing Conference:

The compliance officer reviews any apparent violations with the employer and discusses possible methods and time periods necessary for their correction. The compliance officer explains that these violations may result in a citation and a proposed financial penalty, describes the employers rights and responsibilities, and answers all questions.

Citations:

OSHA is required by law to issue citations for violations of safety and health standards. The agency is not permitted to issue warnings. Citations include:

1. A description of the violation;
2. The proposed penalty, if any; and
3. The date by which the hazard must be corrected.

In most cases the citations are prepared at the OSHA Area Office and are mailed to the employer. Employers have 15 working days after receipt to file an intention to contest OSHA citations before the independent Occupational Safety and Health Review Commission.

Settlement Agreements:

If an employer believes OSHA's citations are unreasonable, or wishes for any reason to discuss the OSHA enforcement action, he or she may request an informal conference with the Area Director to discuss any citations issued. The agency and the employer may work out a settlement agreement to resolve the dispute and to eliminate the hazard.

The following section is from The Occupational Safety and Health Act of 1970 (OSHAct) as described in the *U.S. Department of Labor Small Business Handbook* from Internet site www.dol.gov/dol/asp/public/programs/handbook/osha.html.

OSHA Violations and Penalties

Workplace Inspections: To enforce its standards, OSHA is authorized under the Act to conduct workplace inspections. Every establishment covered by the Act is subject to inspection by OSHA compliance safety and health officers (CSHOs) who are chosen for their knowledge and experience in the occupational safety and health field. CSHOs are thoroughly trained in OSHA standards and in the recognition of safety and health hazards. Similarly, states with their own occupational safety and health programs conduct inspections using qualified state CSHOs. OSHA conducts two general types of inspections: programmed and unprogrammed. There are various OSHA publications and documents which describe in detail OSHA's inspection policies and procedures. Unprogrammed inspections respond to fatalities, catastrophes and complaints, the last of which is further detailed in OSHA's complaint policies and procedures. The following are the types of violations that may be cited and the penalties that may be proposed:

Other-Than-Serious Violation:

A violation that has a direct relationship to job safety and health, but probably would not cause death or serious physical harm. A proposed penalty of up to $7,000 for each violation is discretionary. A penalty for an other-than-serious violation may be adjusted downward by as much as 95 percent, depending on the employer's good faith (demonstrated efforts to comply with the Act), history of previous violations, and size of business. When the adjusted penalty amounts to less than $50, no penalty is proposed.

Serious Violation:

A violation where there is substantial probability that death or serious physical harm could result and that the employer knew, or should have known, of the hazard. A mandatory penalty of up to $7,000 for each violation is proposed. A penalty for a serious violation may be adjusted downward, based on the employer's good faith, history of previous violations, the gravity of the alleged violation, and size of business.

Willful Violation:

A violation that the employer intentionally and knowingly commits. The employer either knows that what he or she is doing constitutes a violation, or is aware that a hazardous condition exists and has made no reasonable effort to eliminate it.

The Act provides that an employer who willfully violates the Act may be assessed a civil penalty of not more than $70,000 but not less than $5,000 for each violation. A proposed penalty for a willful violation may be adjusted downward, depending on the size of the business and its history of previous violations. Usually no credit is given for good faith.

If an employer is convicted of a willful violation of a standard that has resulted in the death of an employee, the offense is punishable by a court-imposed fine or by imprisonment for up to six months, or both. A fine of up to $250,000 for an individual, or $500,000 for a corporation [authorized under the Comprehensive Crime Control Act of 1984 (1984 CCA), not the OSHAct], may be imposed for a criminal conviction.

Repeated Violation:

A violation of any standard, regulation, rule or order where, upon reinspection, a substantially similar violation is found. Repeated violations can bring a fine of up to $70,000 for each such violation. To be the basis of a repeat citation, the original citation must be final; a citation under contest may not serve as the basis for a subsequent repeat citation.

Failure to Correct Prior Violation:

Failure to correct a prior violation may bring a civil penalty of up to $7,000 for each day the violation continues beyond the prescribed abatement date.

Additional violations for which citations and proposed penalties may be issued are as follows:

- Falsifying records, reports or applications can bring a fine of $10,000 or up to six months in jail, or both;
- Assaulting a compliance officer, or otherwise resisting, opposing, intimidating, or interfering with a compliance officer in the performance of his or her duties is a criminal offense, subject to a fine of not more than $250,000 for an individual and $500,000 for a corporation (1984 CCA) and imprisonment for not more than three years.
- Citation and penalty procedures may differ somewhat in states with their own occupational safety and health programs.

Ed. Note: There is an appeal process for employers and for employees, in the event the inspection was initiated by a complaint. Further details may be found in the document cited or by contacting OSHA.

Further information about OSHA may be found by contacting:

OSHA Office of Information & Consumer Affairs, Room N3637
U.S. Department of Labor
Washington, DC 20210
(202) 219-8151

The following is reprinted with permission from ***Means ADA Compliance Pricing Guide*, a collaboration between Adaptive Environments Center, Inc., and R.S. Means Engineering Staff, R.S. Means Co., Inc.**

ADA Overview

What Is ADA?

The Americans with Disabilities Act (ADA) is a federal civil rights act enacted in 1990 prohibiting discrimination against people with disabilities. There are five sections, or "titles," which cover different aspects of discrimination: Title I, Employment; Title II, State and Local Government; Title III, Public Accommodations and Commercial Facilities; Title IV, Telecommunications; and Title V, which covers miscellaneous provisions of the law. Titles I, II, and III all have sections that deal with construction: Titles II and III proactively, Title I in response to specific requests.

What Is the Process for Complying with ADA?

There is a basic process for complying with ADA:

1. Learning about the requirements of ADA and how they apply to a facility or program;
2. Conducting a survey to identify barriers;
3. Establishing a list of potential modifications for barrier removal, including changes to policies, facilities, and cost estimates;
4. Removing existing barriers.

Although the ADA Standards (ADAAG) are written like a building code, unlike other codes there is no formal sign-off for ADA compliance. This means that facility owners need to know how their facilities are covered under ADA and which modifications might be required. The following is a brief overview of ADA requirements for different facility types to assist in the costing process.

What Are ADA's Standards for Accessible Design?

The ADA Standards for Accessible Design are the enforceable standards issued by the Department of Justice as part of the Final Rule for Title III. The Architectural and Transportation Barriers Compliance Board (ATBCB) developed the ADA Accessibility Guidelines, or ADAAG, to serve as minimum guidelines for the Department of Justice's Standards for Accessible Design. The Standards are often referred to as ADAAG, and in each of the projects in this book, we provide the ADAAG reference citation for ease of use. When altering any building or space, it is important to use the DOJ Final Rule where you will find not only the Standards, but all of the requirements for barrier removal and alterations.

The ADA Standards for Accessible Design established minimum technical requirements for the design and construction of buildings and facilities. They were written with a clear intent: to increase the level of accessibility in the built environment in new construction, alterations, and existing facilities.

The Department of Justice is the agency that enforces ADA Title II and Title III, and ultimately enacts all changes and additions to all aspects of the regulations for both, including the ADA Standards for Accessible Design. Additional sections are being developed and will gradually become incorporated into the ADA Standards. For clarification on Title II and Title III Regulations, including the ADA Standards, contact the Department of Justice. Also, the ATBCB has a number of design-related technical assistance documents.

It is important to remember that since the ADA is an anti-discrimination civil rights act, and not a building code, there is more to compliance than the minimum technical requirements. For full background on the ADA, contact the ADA Technical Assistance Centers.

Ed. Note: ADA Technical Assistance is available at (800) 949-4232 or www.adata.org.

Although the law is complex, the basic provisions for compliance in construction are relatively simple:

- **For new construction and additions:** Both publicly used and employee-only spaces have to comply with the ADA Standards (as defined in the Americans with Disabilities Act Accessibility Guidelines, or ADAAG). Only "non-occupiable" employee spaces, such as elevator pits, are exempt.
- **For renovations:** Any alterations to existing facilities, both public and employee-only common spaces, have to comply with the ADA Standards (ADAAG) unless it is "technically infeasible" to do so. Alterations to an "area containing a primary function" trigger accessibility requirements for modifying the path of travel to the area.
- **For existing facilities:**
  1. *Employee-only facilities or spaces:* Under Title I, employers with 15 or more employees must make "reasonable accommodations" for employees. Reasonable accommodations are made according to individual need and may include architectural modifications, but no modifications are required if not requested.
  2. *Government facilities:* Under Title II, state and local governments must have all their programs accessible. Modifications to existing buildings are required when administrative changes to programs are not sufficient to create access.
  3. *Public accommodations (Privately owned businesses serving the public):* Under Title III, businesses must address use of their services through removing existing barriers where it is readily achievable to do so, that is, "easily accomplished with little expense." Readily achievable modifications are required even when no other renovations are planned.

With such broad requirements, complications arise in deciding exactly what construction complies "to the maximum extent feasible," what are "reasonable accommodations," what is an area of "primary function" and what barriers are "readily achievable" to remove, since each varies greatly depending on particular circumstances. As written, the regulations allow facility owners or managers some leeway in deciding how to comply with the different provisions that apply to their facilities, but it also places responsibility for compliance on them. For complete information refer to the sources of ADA regulations and technical assistance identified in the Resources section at the end of this chapter.

While the ADA Standards are the primary means for meeting the requirements of ADA, ADAAG Section 2.2, Equivalent Facilitation, states "Departures from particular technical and scoping requirements of this guideline by the use of other designs and technologies are permitted where the alternative designs and technologies used will provide substantially equivalent or greater access to and usability of the facility." This is most useful in considering accessibility in relation to projects like children's toilet room fixtures, which is not covered in ADAAG. As stated, the ADA Standards as defined in the ADA Accessibility Guidelines are the minimum requirements that are used to establish accessibility.

Facilities Covered by ADA

All types of facilities are covered under ADA, with the exception of privately owned permanent housing and religious facilities and private clubs, which are exempt from Title III. Most states already have accessibility regulations in place that are enforced as part of the local building codes. The ADA Accessibility Guidelines are based on the American National Standards Institute's *Accessible and Usable Buildings and Facilities (CABO/ANSI 117.1-1992)*, which in its current form is incorporated into all three national model building codes. There are differences, so it shouldn't be assumed that compliance with ANSI 117.1 or a model code is the same as compliance with ADA. ADA does not require any scoping or technical standards to exceed those required by the ADA Standards (ADAAG) but if sections of the local access codes are stricter than ADA, the stricter regulation always governs.

Title I, Employment:

New Construction, Additions, and Alterations: Title I does not specifically address architectural accessibility. Design of employee areas for new construction and alterations is covered in the Title III regulations and ADAAG [also Uniform Federal Accessibility Standards (UFAS) for Title II facilities].

Existing Facilities: Employees with disabilities in companies with more than 15 employees may request that the employer provide "reasonable accommodation" to allow them to carry out their essential job functions. A "reasonable accommodation" may or may not be architectural in nature, depending on the individual's needs. What determines a "reasonable accommodation" is worked out between employer and employee; criteria have been set by the federal Equal Employment Opportunity Commission (EEOC). Reasonable accommodations need not be expensive; examples are raising a desk to allow knee space for a wheelchair user, providing lever door handles, increasing lighting levels, or installing an accessible parking space.

Title II, State and Local Government Services:

New Construction and Additions: Design of new state or local government buildings is covered by ADA and all new construction and additions must be fully compliant. Title II regulations allow the UFAS to be used as an alternative to the ADA Accessibility Guidelines, providing that only one standard is used in any facility. ADAAG is similar to UFAS in its technical requirements, but has more substantial scoping differences.

Alterations: All alterations to spaces used by the public and employee-only areas must comply with the ADA Standards (ADAAG) or UFAS unless it is "technically infeasible" to do so (UFAS has a different formula from ADAAG). If technically infeasible, the alteration must comply "to the maximum extent feasible." "Technically infeasible" is defined as having little likelihood of being done because a major structural member would have to be moved or because an existing physical or site constraint prohibits compliance.

Existing Facilities: Title II requires state and local governments to make their programs accessible. This might not require alterations to an existing facility. For instance, if a town hall is inaccessible, it might be possible to hold meetings in an accessible high school. If, however, a facility is unique in its program (such as one library that serves a given area) the building has to be made accessible in order to achieve program access. As with new construction and additions, either ADAAG or UFAS is acceptable as a design standard for alterations to existing facilities.

Title III, Public Accommodations and Commercial Facilities Operated by Private Entities:

Title III distinguishes between privately-owned businesses that invite the public in to purchase goods and services (public accommodations) and those that don't (commercial facilities). Public accommodations are required to remove barriers in existing facilities where it is readily achievable to do so. Title III regulations list twelve categories of public accommodations:

1. Places of lodging (an inn, hotel, motel, or other place of lodging);
2. Establishments serving food or drink (a restaurant, bar, or other establishment serving food or drink);
3. Places of exhibition or entertainment (a motion picture house, theater, concert hall, stadium, or other place of exhibition or entertainment);
4. Places of public gathering (an auditorium, convention center, lecture hall, or other place of public gathering);
5. Sales or rental establishments (a bakery, grocery store, clothing store, hardware store, shopping center, or other sales or rental establishment);

6. Service establishments (a laundromat, dry cleaner, bank, barber shop, beauty salon, travel service, shoe repair service, funeral parlor, gas station, office of an accountant or lawyer, pharmacy, insurance office, professional office of a health care provider, hospital, or other service establishment);
7. Stations used for public transportation (a terminal, depot, or other station used for specified public transportation);
8. Places for public display or collection (a museum, library, gallery, or other place of public display or collection);
9. Places of recreation (a park, zoo, amusement park, or other place of recreation);
10. Places of education (a nursery, elementary, secondary, undergraduate, or postgraduate private school or other place of education);
11. Social service establishments (a day care center, senior citizen center, homeless shelter, food bank adoption agency, or other social service center establishment);
12. Places of exercise and recreation (a gymnasium, health spa, bowling alley, golf course, or other place of exercise or recreation).

All such establishments have to comply with the requirements for public accommodations; only private clubs and religious establishments are exempt (but any public accommodations leasing spaces from them have to comply).

Business establishments which do not fall into any of the above categories are classified by Title III as "commercial facilities." If part of the facility serves as a public accommodation (such as a tour of a factory), that portion of the facility must be accessible. Commercial facilities must meet the requirements for new construction, additions, and alterations even if a new building or addition will not be open to the public. In new construction, employee common spaces must be fully accessible. Work areas (such as a lab or office floor) must be on an accessible route, and an employee must be able to approach, enter, and exit the area. Individual work spaces are modified on an as-needed basis under Title I, Employment.

New Construction and Additions: All new public accommodations and commercial facilities must be accessible and comply with the ADA Standards (ADAAG). Individual employee-only work spaces (such as a lab station) do not have to comply, but work areas (such as a lab) have to be on an accessible route of travel, and any employee has to be able to approach, enter, and exit the space. Only non-occupiable spaces, such as catwalks and elevator pits, are exempted.

Alterations: All alterations to public and employee-only areas must comply with the ADA Standards (ADAAG) unless it is "technically infeasible" to do so. If so, the alteration has to comply "to the maximum extent feasible."

Alterations to an area containing a primary function trigger additional accessibility requirements in both public accommodations and commercial facilities. A primary function is defined as "a major activity for which the facility is intended," such as a bank's customer service area, or the dining area of a cafeteria. Spaces such as mechanical rooms, entrances, and rest rooms are not areas of primary function, meaning that alterations to these kinds of spaces do not trigger additional accessibility requirements. If an area of primary function is being altered, the path of travel to the area of primary function must be brought into compliance with the ADA Standards (ADAAG) at a cost up to 20% of the total cost of the area's renovation.

For leased places of public accommodation, both the landlord and tenant are responsible for compliance. Allocation of responsibility may be determined by the lease or other contract. ADA does not state who is directly responsible for removing barriers either in existing facilities or during renovations, but places the responsibility for ADA compliance on the contractual agreement between landlord and tenant.

Existing Facilities: For public accommodations where no renovations are planned, all barriers must be removed if it is readily achievable to do so. "Readily achievable" is defined as "easily accomplished with little expense," and this varies depending on the situation. What is readily achievable for a national chain store might not be readily achievable for a single-owner grocery store. Determining what barrier removal measures would be considered readily achievable must be on a case-by-case basis. Factors to consider include the nature and cost of the remedial action; the organization's financial resources; the size of the organization (number of facilities and employees), and type of operation. The Department of Justice regulations to help determine what modifications might be undertaken recommend the following order of priorities in removing barriers:

1. Access into the facility
2. Access to where goods and services are made available to the public
3. Access to rest rooms if provided for public use
4. Other amenities

These recommendations, given in section 36.304(c) of the Title III regulations, can vary in different facilities depending on what service is being provided. Examples of readily achievable modifications are installing grab bars in a rest room, adding a lever to a door knob, or putting up Braille signage. Modifications which might be readily achievable depending on the individual circumstances include installing a ramp up a small flight of stairs, installing a platform lift, or replacing a sink. Installing a new elevator probably would not be considered a readily achievable modification. Barrier removal that is not readily achievable can be addressed at a later date during building modernization, as part of a facility's ongoing obligation to remove barriers.

Surveying the Facility

Surveying the facility is the first step in costing accessibility modifications. ADA regulations do not specify one particular method of identifying barriers, but surveying with an accessibility checklist can be very useful. When surveying a facility for accessibility it is vital to remember that wheelchair access is only one factor to consider. To comply with ADA, people with other mobility impairments such as balance or stamina problems, and people with visual, hearing, or cognitive impairments must also be accommodated: The ADA Standards (ADAAG) include requirements which improve access for people who are blind (raised character and Braille signage), people with hearing impairments (visual fire alarms), people with low fine motor control (lever handles or paddle faucets), control heights for people with limited reach, and many others.

Surveying a facility for accessibility involves identifying spaces, routes of travel, and individual items which are not usable by persons with disabilities and may not be compliant with the ADA Standards (ADAAG). It is important to know the entire list of access issues within a facility to be able to determine which barriers can be removed. Knowing the areas of the facility that need to be surveyed is key: Public areas in public accommodations need to be surveyed to determine barrier removal needs. The existing barriers then need to be prioritized as to severity in impeding access and ease of removal. The Department of Justice recommendations for barrier removal are just that. If an entrance is not fully compliant, but is still usable as an accessible entrance, and there is no accessible rest room in the facility, the rest room would be the higher priority in terms of barrier removal, even though it is listed as a lower priority. The ultimate decision on priorities rests with management.

Involving people with disabilities is critical in setting priorities for barrier identification and removal. To quote directly from the Department of Justice preamble to the ADA Title III Final Rule, discussion of *Section 36.304 Removal of Barriers:* "The Department recommends that this process include appropriate consultation with individuals with disabilities or organizations representing them. A serious effort at self-assessment and consultation can diminish the threat of litigation and save resources ... The Department recommends ... the development of an implementation plan designed to achieve compliance with the ADA's barrier removal requirements before they become effective on January 26, 1992. Such a plan, if appropriately designed and diligently executed, could serve as evidence of a good faith effort to comply."

As stated, this publication is not a survey tool. In some cases, such as an inaccessible entrance that can be ramped, a survey might not be necessary; but even in these cases, other issues could exist (in fact probably do) that should be identified. Again, the most valuable information comes from a user.

Ed. Note: To obtain a copy of the ADA regulations and publications, including the Technical Assistance Manuals for Titles II and III, and information about the technical assistance grant program, call the ADA Information Line at (800) 514-0301 (voice) or (800) 514-0383 (TDD). You may write to the:
Disability Rights Section
Civil Rights Division
U.S. Department of Justice
P.O. Box 66738
Washington, D.C. 20035-6738

Technical Assistance Centers provide resources and technical assistance on the ADA. Their telephone is (800) 949-4232 (voice and TDD). Their website is http://www.adata.org.

The following section is reprinted with permission from *Fire Protection: Design Criteria, Options, Selection* by J. Walter Coon, PE, R.S. Means Co., Inc.

Fire Safety and the National Fire Protection Association (NFPA)

National Fire Protection Association

The National Fire Protection Association is not of itself an authority having jurisdiction, because the NFPA does not approve design plans or conduct tests or inspections of system installations. The NFPA does not approve or certify equipment or materials, nor does the NFPA single out or approve testing laboratories that evaluate equipment, materials, or systems. The authority, or the insurance organization that does have jurisdiction to approve plans and installations may, however, base approval of the plans and the installation on compliance with NFPA Fire Code requirements. As such, the NFPA is the authority, but only within the framework of the local code or the insurance carrier that has jurisdiction. Understanding the origin of the NFPA Fire Codes is important because it explains the reasons and significance of codes in fire protection today.

When automatic sprinklers were first developed beyond the experimental stage, and began to be installed in buildings, several pipe sizing methods were in use. Differences in rules produced installations of all sizes and configurations. There was no proven way to ascertain the effectiveness or reliability of these installations, since there was no clear track record. Insurance companies were providing premium savings for sprinklered buildings, yet with no established standard of installation, they could not establish a foundation of sprinkler installation reliability. As a result, a group of representatives from insurance carriers met in Boston in 1865 to establish a uniform standard for sprinkler installations. Their purpose was to obtain the installation uniformity, system performance reliability, and efficiency needed to establish realistic sprinkler premium rates. This meeting resulted in the first standard based on the consensus of opinions of experienced personnel for the design and installation of automatic sprinkler systems. This first consensus led directly to the birth of the NFPA in 1896.

The small group that established the first technical committee has grown into over 175 NFPA technical committees, consisting of over 3500 volunteer committee members, offering a vast source of experience required to develop, monitor, and periodically update over 270 fire codes. These NFPA technical committees have jurisdiction over three basic NFPA documents: the *Standard*, the *Recommended Practice*, and the *Guide*. NFPA Standards mandate requirements with the use of the word "shall." These requirements are, in most cases, detailed and specific design and installation requirements. NFPA's *Recommended Practice* and *Guide* are exactly what the names imply, and the word "should" is used in these documents in lieu of the word "shall."

A note of caution: The majority of the NFPA Standards have a section at the back of the Standard called the *Appendix*. The Appendix is offered as part of the Standard for information only, and is not part of the body of the Standard. The words "should" and "may" appear in the Appendix. Since Appendix items are not part of the body of the Standard, they are not mandatory requirements, but are for information only, and are used to explain a section of the Standard.

According to the administrative authority of NFPA Technical Committees and the Fire Codes, all Standards, Recommended Practice, or Guides must be reviewed, updated, or reaffirmed in their present content by the Technical Committee in charge of the document every five years, and a new issue of the code is printed each year. It is very helpful to know that a vertical line beside a Section or paragraph or sentence in a Fire Code, Standard, Recommended Practice, or Guide is new wording since the publication of the last document.

The numerous volumes of Fire Codes contain all the current Standards, Recommended Practices, and Guides, but the date of the current issue of the Fire Code volume is not necessarily the latest issue of the particular document being used. For example, the 1989 set of National Fire Codes, with "1989" printed on each volume, may contain Standards dated 1988 or 1987, etc. If the Standard, Recommended Practice, or Guide is being referenced by date in the specification, be sure to check the date of the specific document. This may sound like an unnecessary caution, but changes in an updated document, particularly a Standard, can cause serious problems if the date used in the specification does not coincide with the actual date of the document.

The NFPA Fire Code requirements effectively become the jurisdictional authority when they are used as a basis of acceptance or approval of a system design or installation by the proper authority or the insurance interest having jurisdiction. NFPA Fire Codes actually become law when referenced by number and date of issue in a building code, and can be considered as a "law" when referenced in a specification, by a fire marshal, government agency, local fire prevention bureau, fire department, health department, or any other similar authority or insurance carrier having jurisdiction.

The NFPA may reference other Standards that had some bearing on a particular Standard under consideration. NFPA references these other Standards in the body of their Standard. In such cases, it becomes mandatory to use referenced Standards in the design and installation of the system, if applicable to the design and installation of the particular system.

The following section is from the U.S. Department of Labor Program Highlights Fact Sheet No. OSHA 93-41 from web site www.osha-slc.gov/OshDoc/Fact_data/FSN093-41.html.

Workplace Fire Safety: Fire safety is important business. According to National Safety Council figures, losses due to workplace fires in 1991 totaled $2.2 billion. Of the 4,200 persons who lost their lives due to fires in 1991, the National Safety Council estimates 327 were workplace deaths. Fires and burns accounted for 3.3 percent of all occupational fatalities.

There is a long and tragic history of workplace fires in this country. One of the most notable was the fire at the Triangle Shirtwaist Factory in New York City in 1911 in which nearly 150 women and young girls died because of locked fire exits and inadequate fire extinguishing systems.

History repeated itself several years ago in the fire in Hamlet, North Carolina, where 25 workers died in a fire in a poultry processing plant. It appears that here, too, there were problems with fire exits and extinguishing systems.

When OSHA conducts workplace inspections, it checks to see whether employers are complying with OSHA standards for fire safety.

OSHA standards require employers to provide proper exits, fire fighting equipment, and employee training to prevent fire deaths and injuries in the workplace.

Building Fire Exits: Each workplace building must have at least two means of escape remote from each other to be used in a fire emergency.

Fire doors must not be blocked or locked to prevent emergency use when employees are within the buildings. Delayed opening of fire doors is permitted when an approved alarm system is integrated into the fire door design.

Exit routes from buildings must be clear and free of obstructions and properly marked with signs designating exits from the building.

Portable Fire Extinguishers: Each workplace building must have a full complement of the proper type of fire extinguisher for the fire hazards present, excepting when employer wishes to have employees evacuate instead of fighting small fires.

Employees expected or anticipated to use fire extinguishers must be instructed on the hazards of fighting fire, how to properly operate the fire extinguishers available, and what procedures to follow in alerting others to the fire emergency.

Only approved fire extinguishers are permitted to be used in workplaces, and they must be kept in good operating condition. Proper maintenance and inspection of this equipment is required of each employer.

Where the employer wishes to evacuate employees instead of having them fight small fires there must be written emergency plans and employee training for proper evacuation.

Emergency Evacuation Planning: Emergency action plans are required to describe the routes to use and procedures to be followed by employees. Also procedures for accounting for all evacuated employees must be part of the plan. The written plan must be available for employee review.

Where needed, special procedures for helping physically impaired employees must be addressed in the plan; also, the plan must include procedures for those employees who must remain behind temporarily to shut down critical plant equipment before they evacuate.

The preferred means of alerting employees to a fire emergency must be part of the plan and an employee alarm system must be available throughout the workplace complex and must be used for emergency alerting for evacuation. The alarm system may be voice communication or sound signals such as bells, whistles or horns. Employees must know the evacuation signal. Training of all employees in what is to be done in an emergency is required. Employers must review the plan with newly assigned employees so they know correct actions in an emergency and with all employees when the plan is changed.

Fire Prevention Plan: Employers need to implement a written fire prevention plan to complement the fire evacuation plan to minimize the frequency of evacuation. Stopping unwanted fires from occurring is the most efficient way to handle them. The written plan shall be available for employee review.

Housekeeping procedures for storage and cleanup of flammable materials and flammable waste must be included in the plan. Recycling of flammable waste such as paper is encouraged; however, handling and packaging procedures must be included in the plan.

Procedures for controlling workplace ignition sources such as smoking, welding and burning must be addressed in the plan. Heat producing equipment such as burners, heat exchangers, boilers, ovens, stoves, fryers, etc., must be properly maintained and kept clean of accumulations of flammable residues; flammables are not to be stored close to these pieces of equipment. All employees are to be apprised of the potential fire hazards of their job and the procedures called for in the employer's fire prevention plan. The plan shall be reviewed with all new employees when they begin their job and with all employees when the plan is changed.

Fire Suppression System: Properly designed and installed fixed fire suppression systems enhance fire safety in the workplace. Automatic sprinkler systems throughout the workplace are among the most reliable fire fighting means. The fire sprinkler system detects the fire, sounds an alarm and puts the water where the fire and heat are located.

Automatic fire suppression systems require proper maintenance to keep them in serviceable condition. When it is necessary to take a fire suppression system out of service while business continues, the employer must temporarily substitute a fire watch of trained employees standing by to respond quickly to any fire emergency in the normally protected area. The fire watch must interface with the employers' fire prevention plan and emergency action plan.

Signs must be posted about areas protected by total flooding fire suppression systems which use agents that are a serious health hazard such as carbon dioxide, Halon 1211, etc. Such automatic systems must be equipped with area pre-discharge alarm systems to warn employees of the impending discharge of the system and allow time to evacuate the area. There must be an emergency action plan to provide for the safe evacuation of employees from within the protected area. Such plans are to be part of the overall evacuation plan for the workplace facility.

This is one of a series of fact sheets highlighting U.S. Department of Labor programs. It is intended as a general description only and does not carry the force of legal opinion.

The following section is reprinted with permission from *Cost Planning & Estimating for Facilities Maintenance,* Chapter 6: "Codes and Regulations," by William H. Rowe III, AIA, PE, R.S. Means Co., Inc.

Summary of Procedures for Meeting Code Requirements

With the host of regulatory issues and code requirements, effective facilities management means initiating a system to deal with meeting the standards. The basic steps are:

- Identify the applicable standards. Obtain a copy of the local, state, and federal laws and related professional accrediting societies that affect your facility's operations.
- Determine the specific standards that affect your maintenance program.
- Determine how well current systems meet the applicable standards.
- Identify changes necessary to "meet or beat" the standards.
- Determine costs, manpower, and equipment necessitated by the changes and obtain funding for them.
- Prioritize the needed changes according to health effects, time, and available funding.
- Prepare an action plan showing which items will be done and when.
- Implement the program and achieve compliance.

Resource Publications

Association for Facilities Engineering. *Certified Facilities Environmental Professional Exam Review Book*

Coon, J. Walter, PE. *Fire Protection: Design Criteria, Options, Selection*, R.S. Means Company, Inc.

Hopcroft, Francis J., PE, David L. Vitale, M.Ed., and Donald L. Anglehart, Esq. *Hazardous Material and Hazardous Waste*, R.S. Means Company, Inc.

Kaiser, Harvey H., Ph.D. *The Facilities Manager's Reference*

Rast, Richard R. *Means Environmental Remediation Estimating Methods*, R.S. Means Company, Inc.

Rowe, William H., III, AIA, PE. *Cost Planning & Estimating for Facilities Maintenance*, R.S. Means Company, Inc., Chapter 6, "Codes and Regulations".

For Additional Information

Americans With Disabilities

To obtain a copy of the ADA regulations and publications, including the Technical Assistance Manuals for Titles II and III, and information about the technical assistance grant program, call the ADA Information Line (800) 514-0301(voice) or (800) 514-0383 (TDD). You may write to:

Disability Rights Section
Civil Rights Division
U.S. Department of Justice
P.O. Box 66738
Washington, DC 20035-6738

ADA Technical Assistance (800) 949-4232 or
www.adata.org

Environmental Protection Agency (EPA)
401 M Street, SW
Washington, DC 20460-0003
(202) 260-2090
www.epa.gov

To locate full text of of EPA statutes or laws: www.epa.gov/epahome/laws.html. For a list of the 77 Regulated Toxic Substances and Threshold Quantities for Accidental Release, see www.epa.gov/swercepp/rules/listrule.html.

Indoor Air Quality Information Clearinghouse
IAQ INFO
P.O. Box 37133
Washington, DC 20013-7133
Telephone: (703) 356-4020

For OSHA information: Check your telephone book for the U.S. Department of Labor listing for the OSHA office nearest you, or contact:

OSHA Office of Information & Consumer Affairs
Room N3637
U.S. Department of Labor
Washington, DC 20210
(202) 219-8185

Hazard Communication Standard
www.osha-slc.gov/OshDoc/Fact_data/FSN093-26.html

Personal Protective Equipment:
Telephone: (202) 523-8151
TDD: (800) 326-2577
www.osha-slc.gov/OshDoc/Fact-data/FSN092-08.html

Proposed rule requiring Health & Safety Programs:
www.osha-slc.gov/SLTC/safetyhealth/nshp.html

OSHA Violations and Penalties
OSHA Office of Information & Consumer Affairs
Room N3637
U.S. Department of Labor
Washington, DC 20210
Telephone: (202) 219-8151
www.dol.gov/dol/asp/public/programs/handbook/osha.html

Appendix & Index

Appendix

The pages that follow are offered as a supplement to the material presented in several chapters of this book.

The first spreadsheet provides a template with the formulas needed to create custom loan amortization tables. In this example, a table is created for a three-year loan of $10,000 at 10% interest. The second spreadsheet shows the result, with total interest and principal payments made during each of the three years.

Sample pages from *Means Facilities Maintenance & Repair Cost Data*, a comprehensive source of maintenance information published annually by R.S. Means. The first page, from the book's Maintenance & Repair section, provides cost data and an approximate frequency of occurrence for roofing repairs and replacement. The second page is from the book's section on Preventive Maintenance and offers two equipment maintenance schedules that help establish labor-hours and budgets. The final page lists labor and cost estimates for a number of day-to-day tasks. It is a sample from the book's section on General Maintenance.

B

1% Compound Interest Factors

	Single Payment		Uniform Payment Series				Arithmetic Gradient		
	Compound Amount Factor	Present Worth Factor	Sinking Fund Factor	Capital Recovery Factor	Compound Amount Factor	Present Worth Factor	Gradient Uniform Series	Gradient Present Worth	
n	Find F Given P F/P	Find P Given F P/F	Find A Given F A/F	Find A Given P A/P	Find F Given A F/A	Find P Given A P/A	Find A Given G A/G	Find P Given G P/G	n
1	1.010	.9901	1.0000	1.0100	1.000	0.990	0	0	1
2	1.020	.9803	.4975	.5075	2.010	1.970	0.498	0.980	2
3	1.030	.9706	.3300	.3400	3.030	2.941	0.993	2.921	3
4	1.041	.9610	.2463	.2563	4.060	3.902	1.488	5.804	4
5	1.051	.9515	.1960	.2060	5.101	4.853	1.980	9.610	5
6	1.062	.9420	.1625	.1725	6.152	5.795	2.471	14.320	6
7	1.072	.9327	.1386	.1486	7.214	6.728	2.960	19.917	7
8	1.083	.9235	.1207	.1307	8.286	7.652	3.448	26.381	8
9	1.094	.9143	.1067	.1167	9.369	8.566	3.934	33.695	9
10	1.105	.9053	.0956	.1056	10.462	9.471	4.418	41.843	10
11	1.116	.8963	.0865	.0965	11.567	10.368	4.900	50.806	11
12	1.127	.8874	.0788	.0888	12.682	11.255	5.381	60.568	12
13	1.138	.8787	.0724	.0824	13.809	12.134	5.861	71.112	13
14	1.149	.8700	.0669	.0769	14.947	13.004	6.338	82.422	14
15	1.161	.8613	.0621	.0721	16.097	13.865	6.814	94.481	15
16	1.173	.8528	.0579	.0679	17.258	14.718	7.289	107.273	16
17	1.184	.8444	.0543	.0643	18.430	15.562	7.761	120.783	17
18	1.196	.8360	.0510	.0610	19.615	16.398	8.232	134.995	18
19	1.208	.8277	.0481	.0581	20.811	17.226	8.702	149.895	19
20	1.220	.8195	.0454	.0554	22.019	18.046	9.169	165.465	20
21	1.232	.8114	.0430	.0530	23.239	18.857	9.635	181.694	21
22	1.245	.8034	.0409	.0509	24.472	19.660	10.100	198.565	22
23	1.257	.7954	.0389	.0489	25.716	20.456	10.563	216.065	23
24	1.270	.7876	.0371	.0471	26.973	21.243	11.024	234.179	24
25	1.282	.7798	.0354	.0454	28.243	22.023	11.483	252.892	25
26	1.295	.7720	.0339	.0439	29.526	22.795	11.941	272.195	26
27	1.308	.7644	.0324	.0424	30.821	23.560	12.397	292.069	27
28	1.321	.7568	.0311	.0411	32.129	24.316	12.852	312.504	28
29	1.335	.7493	.0299	.0399	33.450	25.066	13.304	333.486	29
30	1.348	.7419	.0287	.0387	34.785	25.808	13.756	355.001	30
36	1.431	.6989	.0232	.0332	43.077	30.107	16.428	494.620	36
40	1.489	.6717	.0205	.0305	48.886	32.835	18.178	596.854	40
48	1.612	.6203	.0163	.0263	61.223	37.974	21.598	820.144	48
50	1.645	.6080	.0155	.0255	64.463	39.196	22.436	879.417	50
52	1.678	.5961	.0148	.0248	67.769	40.394	23.269	939.916	52
60	1.817	.5504	.0122	.0222	81.670	44.955	26.533	1 192.80	60
70	2.007	.4983	.00993	.0199	100.676	50.168	30.470	1 528.64	70
72	2.047	.4885	.00955	.0196	104.710	51.150	31.239	1 597.86	72
80	2.217	.4511	.00822	.0182	121.671	54.888	34.249	1 879.87	80
84	2.307	.4335	.00765	.0177	130.672	56.648	35.717	2 023.31	84
90	2.449	.4084	.00690	.0169	144.863	59.161	37.872	2 240.56	90
96	2.599	.3847	.00625	.0163	159.927	61.528	39.973	2 459.42	96
100	2.705	.3697	.00587	.0159	170.481	63.029	41.343	2 605.77	100
104	2.815	.3553	.00551	.0155	181.464	64.471	42.688	2 752.17	104
120	3.300	.3030	.00435	.0143	230.039	69.701	47.835	3 334.11	120
240	10.893	.0918	.00101	.0110	989.254	90.819	75.739	6 878.59	240
360	35.950	.0278	.00029	.0103	3 495.0	97.218	89.699	8 720.43	360
480	118.648	.00843	.00008	.0101	11 764.8	99.157	95.920	9 511.15	480

7% Compound Interest Factors 7%

	Single Payment		Uniform Payment Series				Arithmetic Gradient		
	Compound Amount Factor	Present Worth Factor	Sinking Fund Factor	Capital Recovery Factor	Compound Amount Factor	Present Worth Factor	Gradient Uniform Series	Gradient Present Worth	
n	Find F Given P F/P	Find P Given F P/F	Find A Given F A/F	Find A Given P A/P	Find F Given A F/A	Find P Given A P/A	Find A Given G A/G	Find P Given G P/G	**n**
1	1.070	.9346	1.0000	1.0700	1.000	0.935	0	0	**1**
2	1.145	.8734	.4831	.5531	2.070	1.808	0.483	0.873	**2**
3	1.225	.8163	.3111	.3811	3.215	2.624	0.955	2.506	**3**
4	1.311	.7629	.2252	.2952	4.440	3.387	1.416	4.795	**4**
5	1.403	.7130	.1739	.2439	5.751	4.100	1.865	7.647	**5**
6	1.501	.6663	.1398	.2098	7.153	4.767	2.303	10.978	**6**
7	1.606	.6227	.1156	.1856	8.654	5.389	2.730	14.715	**7**
8	1.718	.5820	.0975	.1675	10.260	5.971	3.147	18.789	**8**
9	1.838	.5439	.0835	.1535	11.978	6.515	3.552	23.140	**9**
10	1.967	.5083	.0724	.1424	13.816	7.024	3.946	27.716	**10**
11	2.105	.4751	.0634	.1334	15.784	7.499	4.330	32.467	**11**
12	2.252	.4440	.0559	.1259	17.888	7.943	4.703	37.351	**12**
13	2.410	.4150	.0497	.1197	20.141	8.358	5.065	42.330	**13**
14	2.579	.3878	.0443	.1143	22.551	8.745	5.417	47.372	**14**
15	2.759	.3624	.0398	.1098	25.129	9.108	5.758	52.446	**15**
16	2.952	.3387	.0359	.1059	27.888	9.447	6.090	57.527	**16**
17	3.159	.3166	.0324	.1024	30.840	9.763	6.411	62.592	**17**
18	3.380	.2959	.0294	.0994	33.999	10.059	6.722	67.622	**18**
19	3.617	.2765	.0268	.0968	37.379	10.336	7.024	72.599	**19**
20	3.870	.2584	.0244	.0944	40.996	10.594	7.316	77.509	**20**
21	4.141	.2415	.0223	.0923	44.865	10.836	7.599	82.339	**21**
22	4.430	.2257	.0204	.0904	49.006	11.061	7.872	87.079	**22**
23	4.741	.2109	.0187	.0887	53.436	11.272	8.137	91.720	**23**
24	5.072	.1971	.0172	.0872	58.177	11.469	8.392	96.255	**24**
25	5.427	.1842	.0158	.0858	63.249	11.654	8.639	100.677	**25**
26	5.807	.1722	.0146	.0846	68.677	11.826	8.877	104.981	**26**
27	6.214	.1609	.0134	.0834	74.484	11.987	9.107	109.166	**27**
28	6.649	.1504	.0124	.0824	80.698	12.137	9.329	113.227	**28**
29	7.114	.1406	.0114	.0814	87.347	12.278	9.543	117.162	**29**
30	7.612	.1314	.0106	.0806	94.461	12.409	9.749	120.972	**30**
31	8.145	.1228	.00980	.0798	102.073	12.532	9.947	124.655	**31**
32	8.715	.1147	.00907	.0791	110.218	12.647	10.138	128.212	**32**
33	9.325	.1072	.00841	.0784	118.934	12.754	10.322	131.644	**33**
34	9.978	.1002	.00780	.0778	128.259	12.854	10.499	134.951	**34**
35	10.677	.0937	.00723	.0772	138.237	12.948	10.669	138.135	**35**
40	14.974	.0668	.00501	.0750	199.636	13.332	11.423	152.293	**40**
45	21.002	.0476	.00350	.0735	285.750	13.606	12.036	163.756	**45**
50	29.457	.0339	.00246	.0725	406.530	13.801	12.529	172.905	**50**
55	41.315	.0242	.00174	.0717	575.930	13.940	12.921	180.124	**55**
60	57.947	.0173	.00123	.0712	813.523	14.039	13.232	185.768	**60**
65	81.273	.0123	.00087	.0709	1 146.8	14.110	13.476	190.145	**65**
70	113.990	.00877	.00062	.0706	1 614.1	14.160	13.666	193.519	**70**
75	159.877	.00625	.00044	.0704	2 269.7	14.196	13.814	196.104	**75**
80	224.235	.00446	.00031	.0703	3 189.1	14.222	13.927	198.075	**80**
85	314.502	.00318	.00022	.0702	4 478.6	14.240	14.015	199.572	**85**
90	441.105	.00227	.00016	.0702	6 287.2	14.253	14.081	200.704	**90**
95	618.673	.00162	.00011	.0701	8 823.9	14.263	14.132	201.558	**95**
100	867.720	.00115	.00008	.0701	12 381.7	14.269	14.170	202.200	**100**

10% Compound Interest Factors 10%

	Single Payment		Uniform Payment Series				Arithmetic Gradient		
	Compound Amount Factor	Present Worth Factor	Sinking Fund Factor	Capital Recovery Factor	Compound Amount Factor	Present Worth Factor	Gradient Uniform Series	Gradient Present Worth	
n	Find *F* Given *P* *F/P*	Find *P* Given *F* *P/F*	Find *A* Given *F* *A/F*	Find *A* Given *P* *A/P*	Find *F* Given *A* *F/A*	Find *P* Given *A* *P/A*	Find *A* Given *G* *A/G*	Find *P* Given *G* *P/G*	*n*
1	1.100	.9091	1.0000	1.1000	1.000	0.909	0	0	**1**
2	1.210	.8264	.4762	.5762	2.100	1.736	0.476	0.826	**2**
3	1.331	.7513	.3021	.4021	3.310	2.487	0.937	2.329	**3**
4	1.464	.6830	.2155	.3155	4.641	3.170	1.381	4.378	**4**
5	1.611	.6209	.1638	.2638	6.105	3.791	1.810	6.862	**5**
6	1.772	.5645	.1296	.2296	7.716	4.355	2.224	9.684	**6**
7	1.949	.5132	.1054	.2054	9.487	4.868	2.622	12.763	**7**
8	2.144	.4665	.0874	.1874	11.436	5.335	3.004	16.029	**8**
9	2.358	.4241	.0736	.1736	13.579	5.759	3.372	19.421	**9**
10	2.594	.3855	.0627	.1627	15.937	6.145	3.725	22.891	**10**
11	2.853	.3505	.0540	.1540	18.531	6.495	4.064	26.396	**11**
12	3.138	.3186	.0468	.1468	21.384	6.814	4.388	29.901	**12**
13	3.452	.2897	.0408	.1408	24.523	7.103	4.699	33.377	**13**
14	3.797	.2633	.0357	.1357	27.975	7.367	4.996	36.801	**14**
15	4.177	.2394	.0315	.1315	31.772	7.606	5.279	40.152	**15**
16	4.595	.2176	.0278	.1278	35.950	7.824	5.549	43.416	**16**
17	5.054	.1978	.0247	.1247	40.545	8.022	5.807	46.582	**17**
18	5.560	.1799	.0219	.1219	45.599	8.201	6.053	49.640	**18**
19	6.116	.1635	.0195	.1195	51.159	8.365	6.286	52.583	**19**
20	6.728	.1486	.0175	.1175	57.275	8.514	6.508	55.407	**20**
21	7.400	.1351	.0156	.1156	64.003	8.649	6.719	58.110	**21**
22	8.140	.1228	.0140	.1140	71.403	8.772	6.919	60.689	**22**
23	8.954	.1117	.0126	.1126	79.543	8.883	7.108	63.146	**23**
24	9.850	.1015	.0113	.1113	88.497	8.985	7.288	65.481	**24**
25	10.835	.0923	.0102	.1102	98.347	9.077	7.458	67.696	**25**
26	11.918	.0839	.00916	.1092	109.182	9.161	7.619	69.794	**26**
27	13.110	.0763	.00826	.1083	121.100	9.237	7.770	71.777	**27**
28	14.421	.0693	.00745	.1075	134.210	9.307	7.914	73.650	**28**
29	15.863	.0630	.00673	.1067	148.631	9.370	8.049	75.415	**29**
30	17.449	.0573	.00608	.1061	164.494	9.427	8.176	77.077	**30**
31	19.194	.0521	.00550	.1055	181.944	9.479	8.296	78.640	**31**
32	21.114	.0474	.00497	.1050	201.138	9.526	8.409	80.108	**32**
33	23.225	.0431	.00450	.1045	222.252	9.569	8.515	81.486	**33**
34	25.548	.0391	.00407	.1041	245.477	9.609	8.615	82.777	**34**
35	28.102	.0356	.00369	.1037	271.025	9.644	8.709	83.987	**35**
40	45.259	.0221	.00226	.1023	442.593	9.779	9.096	88.953	**40**
45	72.891	.0137	.00139	.1014	718.905	9.863	9.374	92.454	**45**
50	117.391	.00852	.00086	.1009	1 163.9	9.915	9.570	94.889	**50**
55	189.059	.00529	.00053	.1005	1 880.6	9.947	9.708	96.562	**55**
60	304.482	.00328	.00033	.1003	3 034.8	9.967	9.802	97.701	**60**
65	490.371	.00204	.00020	.1002	4 893.7	9.980	9.867	98.471	**65**
70	789.748	.00127	.00013	.1001	7 887.5	9.987	9.911	98.987	**70**
75	1 271.9	.00079	.00008	.1001	12 709.0	9.992	9.941	99.332	**75**
80	2 048.4	.00049	.00005	.1000	20 474.0	9.995	9.961	99.561	**80**
85	3 299.0	.00030	.00003	.1000	32 979.7	9.997	9.974	99.712	**85**
90	5 313.0	.00019	.00002	.1000	53 120.3	9.998	9.983	99.812	**90**
95	8 556.7	.00012	.00001	.1000	85 556.9	9.999	9.989	99.877	**95**
100	13 780.6	.00007	.00001	.1000	137 796.3	9.999	9.993	99.920	**100**

12% Compound Interest Factors 12%

	Single Payment		Uniform Payment Series				Arithmetic Gradient		
	Compound Amount Factor	Present Worth Factor	Sinking Fund Factor	Capital Recovery Factor	Compound Amount Factor	Present Worth Factor	Gradient Uniform Series	Gradient Present Worth	
n	Find *F* Given *P* *F/P*	Find *P* Given *F* *P/F*	Find *A* Given *F* *A/F*	Find *A* Given *P* *A/P*	Find *F* Given *A* *F/A*	Find *P* Given *A* *P/A*	Find *A* Given *G* *A/G*	Find *P* Given *G* *P/G*	*n*
1	1.120	.8929	1.0000	1.1200	1.000	0.893	0	0	**1**
2	1.254	.7972	.4717	.5917	2.120	1.690	0.472	0.797	**2**
3	1.405	.7118	.2963	.4163	3.374	2.402	0.925	2.221	**3**
4	1.574	.6355	.2092	.3292	4.779	3.037	1.359	4.127	**4**
5	1.762	.5674	.1574	.2774	6.353	3.605	1.775	6.397	**5**
6	1.974	.5066	.1232	.2432	8.115	4.111	2.172	8.930	**6**
7	2.211	.4523	.0991	.2191	10.089	4.564	2.551	11.644	**7**
8	2.476	.4039	.0813	.2013	12.300	4.968	2.913	14.471	**8**
9	2.773	.3606	.0677	.1877	14.776	5.328	3.257	17.356	**9**
10	3.106	.3220	.0570	.1770	17.549	5.650	3.585	20.254	**10**
11	3.479	.2875	.0484	.1684	20.655	5.938	3.895	23.129	**11**
12	3.896	.2567	.0414	.1614	24.133	6.194	4.190	25.952	**12**
13	4.363	.2292	.0357	.1557	28.029	6.424	4.468	28.702	**13**
14	4.887	.2046	.0309	.1509	32.393	6.628	4.732	31.362	**14**
15	5.474	.1827	.0268	.1468	37.280	6.811	4.980	33.920	**15**
16	6.130	.1631	.0234	.1434	42.753	6.974	5.215	36.367	**16**
17	6.866	.1456	.0205	.1405	48.884	7.120	5.435	38.697	**17**
18	7.690	.1300	.0179	.1379	55.750	7.250	5.643	40.908	**18**
19	8.613	.1161	.0158	.1358	63.440	7.366	5.838	42.998	**19**
20	9.646	.1037	.0139	.1339	72.052	7.469	6.020	44.968	**20**
21	10.804	.0926	.0122	.1322	81.699	7.562	6.191	46.819	**21**
22	12.100	.0826	.0108	.1308	92.503	7.645	6.351	48.554	**22**
23	13.552	.0738	.00956	.1296	104.603	7.718	6.501	50.178	**23**
24	15.179	.0659	.00846	.1285	118.155	7.784	6.641	51.693	**24**
25	17.000	.0588	.00750	.1275	133.334	7.843	6.771	53.105	**25**
26	19.040	.0525	.00665	.1267	150.334	7.896	6.892	54.418	**26**
27	21.325	.0469	.00590	.1259	169.374	7.943	7.005	55.637	**27**
28	23.884	.0419	.00524	.1252	190.699	7.984	7.110	56.767	**28**
29	26.750	.0374	.00466	.1247	214.583	8.022	7.207	57.814	**29**
30	29.960	.0334	.00414	.1241	241.333	8.055	7.297	58.782	**30**
31	33.555	.0298	.00369	.1237	271.293	8.085	7.381	59.676	**31**
32	37.582	.0266	.00328	.1233	304.848	8.112	7.459	60.501	**32**
33	42.092	.0238	.00292	.1229	342.429	8.135	7.530	61.261	**33**
34	47.143	.0212	.00260	.1226	384.521	8.157	7.596	61.961	**34**
35	52.800	.0189	.00232	.1223	431.663	8.176	7.658	62.605	**35**
40	93.051	.0107	.00130	.1213	767.091	8.244	7.899	65.116	**40**
45	163.988	.00610	.00074	.1207	1 358.2	8.283	8.057	66.734	**45**
50	289.002	.00346	.00042	.1204	2 400.0	8.304	8.160	67.762	**50**
55	509.321	.00196	.00024	.1202	4 236.0	8.317	8.225	68.408	**55**
60	897.597	.00111	.00013	.1201	7 471.6	8.324	8.266	68.810	**60**
65	1 581.9	.00063	.00008	.1201	13 173.9	8.328	8.292	69.058	**65**
70	2 787.8	.00036	.00004	.1200	23 223.3	8.330	8.308	69.210	**70**
75	4 913.1	.00020	.00002	.1200	40 933.8	8.332	8.318	69.303	**75**
80	8 658.5	.00012	.00001	.1200	72 145.7	8.332	8.324	69.359	**80**
85	15 259.2	.00007	.00001	.1200	127 151.7	8.333	8.328	69.393	**85**
90	26 891.9	.00004		.1200	224 091.1	8.333	8.330	69.414	**90**
95	47 392.8	.00002		.1200	394 931.4	8.333	8.331	69.426	**95**
100	83 522.3	.00001		.1200	696 010.5	8.333	8.332	69.434	**100**

LOAN AMORTIZATION TEMPLATE

	A	B	C	D	E	F	G	H	I
1	Interest	0.1							
2	Period	3							
3	Principle	10000							
4		END YR		PMT		PVIF	INT	PRINC	BALANCE
5		0							
6									
7		1		=PMT(B$1,B$2,B$3)		=(1/(1+B$1))^B7	=B1*B3	=ABS(D7)-G7	=B$3-H$7
8		2		=PMT(B$1,B$2,B$3)		=(1/(1+B$1))^B8	=(B3-SUM(H$7:H7))*B$1	=ABS(D8)-G8	=B$3-SUM(H$7:H8)
9		3		=PMT(B$1,B$2,B$3)		=(1/(1+B$1))^B9	=(B3-SUM(H$7:H8))*B$1	=ABS(D9)-G9	=B$3-SUM(H$7:H9)
10									
11									
12									
13						Total	=SUM(G7:G12)	=SUM(H7:H12)	

LOAN AMORTIZATION EXAMPLE

	A	B	C	D	E	F	G	H	I
1	Interest	10%							
2	Period	3							
3	Principle	10,000.00							
4		END YR		PMT		PVIF	INT	PRINC	BALANCE
5		0							
6									
7		1		($4,021.14)		0.909091	$1,000	$3,021	$6,979
8		2		($4,021.14)		0.826446	$698	$3,323	$3,656
9		3		($4,021.14)		0.751315	$366	$3,656	$0
10									
11									
12									
13						Total	$2,063	$10,000	

Sample page from *Facilities Maintenance & Repair Cost Data*

ROOFING — 5.1 Roof Covering

5.1-235 Modified Bituminous / Thermoplastic

	System Description	Freq. (Years)	Crew	Unit	Labor Hours	1999 Bare Costs				Total In-House	Total w/O&P
						Material	Labor	Equipment	Total		
0300	**MINOR MEMBRANE REPAIRS - (2% OF ROOF AREA)**	1	G-5	Sq.							
	Set up, secure and take down ladder				1		24		24	36.50	43.50
	Clean away loose surfacing				.258		6.20		6.20	9.40	11.30
	Install 150 mil mod. bit., fully adhered				2.597	58	57	11	126	162	189
	Clean up				.390		9.40		9.40	14.25	17.10
	Total				4.245	58	96.60	11	165.60	**222.15**	**260.90**
0500	**FLASHING REPAIRS - (2 S.F. PER SQ. REPAIRED)**	1	2 Rofc	S.F.							
	Set up, secure and take down ladder				.010		.24		.24	.36	.44
	Clean away loose surfacing				.003		.06		.06	.09	.11
	Remove flashing				.020		.48		.48	.73	.87
	Install 120 mil mod. bit. flashing				.019	.56	.41	.08	1.05	1.32	1.53
	Clean up				.004		.10		.10	.15	.18
	Total				.056	.56	1.29	.08	1.93	**2.65**	**3.13**
0600	**MEMBRANE REPLACEMENT - (25% OF ROOF AREA)**	20	G-5	Sq.							
	Set up, secure and take down ladder				.078		1.88		1.88	2.85	3.42
	Remove existing membrane / insulation				4.571		110		110	167	200
	Install 2" perlite insulation				1.481	47	36		83	106	124
	Clean adjacent membrane / patch				1.143		27.50		27.50	41.50	50
	Install fully adhered 150 mil membrane				2.597	58	57	11	126	162	189
	Clean up				1		24		24	36.50	43.50
	Total				10.870	105	256.38	11	372.38	**515.85**	**609.92**
0700	**TOTAL ROOF REPLACEMENT**	25	G-1	Sq.							
	Set up, secure and take down ladder				.020		.48		.48	.73	.87
	Remove existing membrane / insulation				3.501		84.50		84.50	128	154
	Remove flashing				.026		.62		.62	.94	1.13
	Install 2" perlite insulation				1.143	47	28		75	94	110
	Install flashing				.037	1.12	.82	.16	2.10	2.63	3.06
	Install fully adhered 180 mil membrane				2	38	44	8	90	117	137
	Clean up				1		24		24	36.50	43.50
	Total				7.727	86.12	182.42	8.16	276.70	**379.80**	**449.56**

R.S. Means Co., Inc., *Facilities Maintenance & Repair Cost Data*

Sample page from *Facilities Maintenance & Repair Cost Data*

MECHANICAL — PM8.3-310 Deaerator Tank

PM Components	Labor-hrs.	W	M	Q	S	A
System PM8.3-310-1950						
Deaerator tank						
1 Check tank and associated piping for leaks.	.013				✓	✓
2 Bottom - blow deaerator tank.	.022				✓	✓
3 Perform sulfite test on deaerator water sample.	.013				✓	✓
4 Check low and high float levels for proper operation and respective water level alarms.	.337				✓	✓
5 Clean steam and feedwater strainers.	.143				✓	✓
6 Check steam pressure regulating valve operation.	.007				✓	✓
7 Check all indicator lights.	.066				✓	✓
8 Clean unit and surrounding area.	.022				✓	✓
9 Fill out maintenance checklist and report deficiencies.	.130				✓	✓
Total labor-hours/period					.753	.753
Total labor-hours/year					.753	.753

	Description	Labor-hrs.	Cost Each: 1999 Bare Costs: Material	Labor	Equip.	Total	Total In-House	Total w/O&P
1900	Deaerator tank, annually	.753	30	24.50		54.50	64	76
1950	Annualized	1.506	30	49.50		79.50	95.50	115

MECHANICAL — PM8.3-360 Fan Coil Unit

PM Components	Labor-hrs.	W	M	Q	S	A
System PM8.3-360-1950						
Fan coil unit						
1 Check with operating or area personnel for deficiencies.	.035			✓	✓	✓
2 Check coil unit while operating.	.120				✓	✓
3 Remove access panel and vacuum inside of unit and coils.	.468				✓	✓
4 Check coils and piping for leaks, damage and corrosion; repair as necessary.	.077			✓	✓	✓
5 Lubricate blower shaft and fan motor bearings.	.047				✓	✓
6 Clean coil, drip pan, and drain line with solvent.	.473				✓	✓
7 Replace filters as required.	.009			✓	✓	✓
8 Replace access panel.	.023				✓	✓
9 Check operation after repairs.	.120				✓	✓
10 Clean area.	.066			✓	✓	✓
11 Fill out maintenance checklist and report deficiencies.	.022			✓	✓	✓
Total labor-hours/period				.209	1.460	1.460
Total labor-hours/year				.418	1.460	1.460

	Description	Labor-hrs.	Cost Each: 1999 Bare Costs: Material	Labor	Equip.	Total	Total In-House	Total w/O&P
1900	Fan coil unit, annually	1.460	5	48		53	66	81.50
1950	Annualized	3.338	20	109		129	159	196

R.S. Means Co., Inc., *Facilities Maintenance & Repair Cost Data*

Sample page from *Facilities Maintenance & Repair Cost Data*

018 | General Maintenance

		018 500 | Interior General Maint.	Crew	Daily Output	Labor-Hours	Unit	1999 Bare Costs Mat.	Labor	Equip.	Total	Total In-House	Total Incl O&P	
540	0030	Treated cloth	1 Clam	40	.200	M.S.F.		3.22		3.22	3.98	4.97	540
	0040	Hand held duster vacuum		53	.151			2.43		2.43	3	3.75	
	0050	Cannister vacuum		32	.250			4.03		4.03	4.97	6.20	
	0060	Backpack vacuum		45	.178			2.86		2.86	3.53	4.41	
	0070	Washing walls, hand, painted surfaces		2.50	3.200			51.50		51.50	63.50	79.50	
	0080	Vinyl surfaces		3.25	2.462			39.50		39.50	49	61	
	0090	Machine, painted surfaces		4.80	1.667			27		27	33	41.50	
	0100	Washing walls from ladder, hand, painted surfaces		1.65	4.848			78		78	96.50	120	
	0110	Vinyl surfaces		2.15	3.721			60		60	74	92.50	
	0120	Machine, painted surfaces	↓	3.17	2.524	↓		40.50		40.50	50	62.50	
550	0010	CEILING CARE											550
	0020	Washing ceiling from ladder, hand	1 Clam	2.05	3.902	M.S.F.		63		63	77.50	97	
	0030	Machine		2.52	3.175			51		51	63	79	
	0040	Washing acoustical ceiling, hand chemical treatment		4	2			32		32	40	49.50	
	0050	Spray chemical treatment	↓	5.75	1.391	↓		22.50		22.50	27.50	34.50	
560	0010	WINDOW AND ACCESSORY CARE											560
	0020	Wash windows, ground level, w/ sponge squeegee and stepladder											
	0025	Wash windows, trigger sprayer & wipe cloth, 12 S.F.	1 Clam	4.50	1.778	M.S.F.		28.50		28.50	35.50	44	
	0030	Over 12 S.F.		12.75	.627			10.10		10.10	12.50	15.60	
	0040	With squeegee & bucket, 12 S.F. and under		5.50	1.455			23.50		23.50	29	36	
	0050	Over 12 S.F.		15.50	.516	↓		8.30		8.30	10.25	12.80	
	0060	Dust window sill		8,640	.001	L.F.		.01		.01	.02	.02	
	0070	Damp wipe, venetian blinds		160	.050	Ea.		.81		.81	.99	1.24	
	0080	Vacuum, venetian blinds, horizontal		87.27	.092			1.48		1.48	1.82	2.28	
	0090	Draperies, in place, to 64 S.F.	↓	160	.050	↓		.81		.81	.99	1.24	
570	0010	GENERAL CLEANING											570
	0020	Sweep stairs & landings, damp wipe handrails	1 Clam	80	.100	Floor		1.61		1.61	1.99	2.48	
	0030	Dust mop stairs & landings, damp wipe handrails		62.50	.128			2.06		2.06	2.55	3.18	
	0040	Damp mop stairs & landings, damp wipe handrails		45	.178			2.86		2.86	3.53	4.41	
	0050	Vacuum stairs & landings, damp wipe handrails, canister		41	.195			3.14		3.14	3.88	4.84	
	0060	Vacuum stairs & landings, damp wipe handrails, backpack		58.22	.137			2.21		2.21	2.73	3.41	
	0070	Vacuum stairs & landings, damp wipe handrails, upright		31.57	.253	↓		4.08		4.08	5.05	6.30	
	0080	Clean carpeted elevators		48	.167	Ea.		2.68		2.68	3.31	4.14	
	0090	Clean uncarpeted elevators		32	.250			4.03		4.03	4.97	6.20	
	0100	Clean escalator		8	1			16.10		16.10	19.90	25	
	0110	Restrooms, fixture cleaning, toilets		180	.044			.72		.72	.88	1.10	
	0120	Urinals		192	.042			.67		.67	.83	1.03	
	0130	Sinks		206	.039			.63		.63	.77	.96	
	0140	Utility sink		175	.046			.74		.74	.91	1.13	
	0150	Bathtub		85	.094			1.52		1.52	1.87	2.34	
	0160	Shower		85	.094			1.52		1.52	1.87	2.34	
	0170	Accessories cleaning, soap dispenser		1,280	.006			.10		.10	.12	.16	
	0180	Sanitary napkin dispenser		640	.013			.20		.20	.25	.31	
	0190	Paper towel dispenser		1,455	.005			.09		.09	.11	.14	
	0200	Toilet tissue dispenser		1,455	.005			.09		.09	.11	.14	
	0210	Accessories refilling/restocking, soap dispenser		640	.013			.20		.20	.25	.31	
	0220	Sanitary napkin dispenser		384	.021			.34		.34	.41	.52	
	0230	Paper towel dispenser, folded		545	.015			.24		.24	.29	.36	
	0240	Paper towel dispenser, rolled		340	.024			.38		.38	.47	.58	
	0250	Toilet tissue dispenser		1,440	.006			.09		.09	.11	.14	
	0260	Sanitary seat covers		960	.008			.13		.13	.17	.21	
	0270	Partition cleaning		96	.083	↓		1.34		1.34	1.66	2.07	
	0280	Ceramic tile wall cleaning		2,400	.003	S.F.		.05		.05	.07	.08	
	0290	General cleaning, empty to larger portable container, wastebaskets		960	.008	Ea.		.13		.13	.17	.21	
	0300	Ash trays	↓	2,880	.003	↓		.04		.04	.06	.07	

R.S. Means Co., Inc., *Facilities Maintenance & Repair Cost Data*

Laws of Force and Motion

Newton's First Law: The Law Of Inertia:
There is no change in the motion of a body unless a net force is acting upon it.

Newton's Second Law
Whenever a net force acts on a body, it produces an acceleration in the direction of the force, an acceleration that is directly proportional to the force acceleration that is directly proportional to the force and inversely proportional to the mass of the body.

$a \propto F \qquad a \propto 1/M$

$f = kma$

Free fall $\quad W = kmg$

External force and gravity

$$F = \frac{w}{g}a$$

g = gravity = 32 ft/sec/sec

Newton's Third Law
For every force acting on one body there is a force that is equal in magnitude but opposite in direction reacting on a second body.

Velocity and Acceleration

$$\text{Average Speed} = \frac{\text{Distance}}{\text{Time}}$$

$$\overline{V} = \frac{s}{t}$$

$$\text{Average Acceleration} = \frac{\text{Change in Velocity}}{\text{Time}}$$

$$a = \frac{v_2 - v_1}{t}$$

s = distance
a= acceleration
v = velocity
t = time

Acceleration

$v2 - v1 = at$

Acceleration and Distance

$s = v_1t + 1/2at^2$

$2as = v_2^2 - v_1^2$

Steam Terminology, Units & Properties

Enthalpy
This is the term given for the total energy, due to both the pressure and temperature, of a fluid or vapor (such as water or steam) at any given time and condition.

The basic unit of measurement for all types of energy is the British Thermal Unit (BTU)

Specific Enthalpy
Is the enthalpy (total energy) of a unit mass (1 lb). The units generally used are BTU/lb.

Specific Heat Capacity
A measure of the ability of a substance to absorb heat. It is the amount of energy (BTU's) required to raise 1lb by 1°F. Thus specific heat capacity is expressed in BTU/lb°F

The specific heat capacity of water is 1 BTU/lb°F. This simply means that an increase in enthalpy of 1 BTU will raise the temperature of 1 lb of water by 1°F.

Heat transfer is the flow of enthalpy from matter at a high temperature to matter at a lower temperature, when they are brought together.

Sensible Heat (Enthalpy of Saturated Water)
Let us assume that water is available for feeding to a boiler at atmospheric pressure, at a temperature of 50°F, and that the water will begin to boil at 212°F. 1 BTU will be required to raise each lb of water through 1°F Therefore for each lb of water in the boiler the increase in enthalpy is (212 - 50) x 1 = 162 BTU in raising the temperature from 50°F to 212°F. If the boiler holds 22000 lb mass (2638 galls) the increase in enthalpy to bring it up to boiling point is 162 BTU/lb x 22000 lb = 3,564,000 BTU.

It must be remembered this figure is not the sensible heat, but merely the increase in sensible heat required to raise the temperature from 50°F to 212°F. The datum point of the steam tables is water at 32°F, which is assumed to have a heat content of zero for our purposes. (The absolute heat content clearly would be considerable, if measured from absolute zero at minus 459°F). The sensible heat of water at 212°F is then (212 - 32) x 1 = 180 BTU.

Latent Heat (Enthalpy of Evaporation)
Suppose for the moment that any steam which is formed in the boiler can be discharged freely into the atmosphere. When the water has reached 212°F heat transfer between the furnace and water continues but there is no further increase in temperature. The additional heat is used in evaporating the water and converting it into steam.

The heat input which produces a change of state from liquid to gas without a change of temperature is called the "latent heat of evaporation". This latent heat is the difference between the sensible heat of water and the total heat of dry saturated steam.

Absolute Pressure & Gauge Pressure
The theoretical pressureless state of a perfect vacuum is "absolute zero". Absolute pressure is, therefore, the pressure above absolute zero. For instance, the pressure exerted by the atmosphere is 14.7 psi abs. at sea level.

Gauge pressure is the pressure shown on a standard pressure gauge fitted to a steam system. Since gauge pressure is the pressure above atmospheric pressure, the zero on the dial of such a gauge is equivalent to approx. 14.7 psi abs.

So a pressure of 45 psi abs. would be made up of 30.3 psi gauge pressure (psig) plus 14.7 psi absolute atmospheric pressure.

Pressures below zero gauge are often expressed in inches of mercury.

Heat and Heat Transfer
Heat is a form of energy and as such is part of the enthalpy of a liquid or gas.

Total Heat of Steam (Enthalpy of Saturated Steam)
We have established that the steam generated in our boiler contains heat which is described in two ways — sensible and latent. The sum of these is known as the "total heat of steam".

In every one lb mass of steam at 212°F and at atmospheric pressure, the sensible heat is 180 BTU/lb, the latent heat is 970 BTU/lb with the total heat being 1150 BTU/lb. These figures are taken from the steam tables which we will look at in some more detail, later on.

Of course, the proportion of sensible and latent heats remain constant at a given pressure, whatever quantity of steam is involved. For instance, if we were considering a mass of 100 lb of steam rather than 1 lb, each of the figures in the paragraph above would be multiplied by 100.

Density of saturated water and hydraulic pressure.
Note from the above that steam at atmospheric pressure, with a volume of 26.8 ft.^{3} per lb., takes up some 1672 times the volume of the pound of water with which we started. Of course, the density of water itself changes a little with temperature. At 212°F, saturated water has a specific density of about 59.8 lb./ft.^{3} The heated water is a little less dense than is cooler water, nominally taken as 62.4 lb./ft.^{3} at 60°F ambient temperature.

Multiplying the specific volume of saturated liquid (inverse of its density) by 1728 in^3/ft.^{3}, we find that a column of water 28.9 in. high will exert a pressure at its base of 1 lb/in.^{2} This is the theoretical height to which condensate can be elevated by each psi of steam pressure. This also defines the hydraulic pressure at the trap inlet that can be used on modulating service to size a float trap for minimum differential pressure.

PROPERTIES OF SATURATED STEAM

Gauge Pressure In. Hg. Vac.	Absolute Pressure psia	Temperature °F	Heat Content Sensible (h_f) BTU/lb	Heat Content Latent (h_{fg}) BTU/lb	Heat Content Total (h_g) BTU/lb	Specific Volume Steam (V_g) ft³/lb
27.96	1	101.7	69.5	1,032.9	1,102.4	333.0
25.91	2	126.1	93.9	1,019.7	1,113.6	173.5
23.87	3	141.5	109.3	1,011.3	1,120.6	118.6
21.83	4	153.0	120.8	1,004.9	1,125.7	90.52
19.79	5	162.3	130.1	999.7	1,129.8	73.42
17.75	6	170.1	137.8	995.4	1,133.2	61.89
15.70	7	176.9	144.6	991.5	1,136.1	53.57
13.66	8	182.9	150.7	987.9	1,138.6	47.26
11.62	9	188.3	156.2	984.7	1,140.9	42.32
9.58	10	193.2	161.1	981.9	1,143.0	38.37
7.54	11	197.8	165.7	979.2	1,144.9	35.09
5.49	12	202.0	169.9	976.7	1,146.6	32.35
3.45	13	205.9	173.9	974.3	1,148.2	30.01
1.41	14	209.6	177.6	972.2	1,149.8	28.00
Gauge Press. psig						
0	14.7	212.0	180.2	970.6	1,150.8	26.80
1	15.7	215.4	183.6	968.4	1,152.0	25.20
2	16.7	218.5	186.8	966.4	1,153.2	23.80
3	17.7	221.5	189.8	964.5	1,154.3	22.50
4	18.7	224.5	192.7	962.6	1,155.3	21.40
5	19.7	227.4	195.5	960.8	1,156.3	20.40
6	20.7	230.0	198.1	959.2	1,157.3	19.40
7	21.7	232.4	200.6	957.6	1,158.2	18.60
8	22.7	234.8	203.1	956.0	1,159.1	17.90
9	23.7	237.1	205.5	954.5	1,160.0	17.20
10	24.7	239.4	207.9	952.9	1,160.8	16.50
11	25.7	241.6	210.1	951.5	1,161.6	15.90
12	26.7	243.7	212.3	950.1	1,162.3	15.30
13	27.7	245.8	214.4	948.6	1,163.0	14.80
14	28.7	247.9	216.4	947.3	1,163.7	14.30
15	29.7	249.8	218.4	946.0	1,164.4	13.90
16	30.7	251.7	220.3	944.8	1,165.1	13.40
17	31.7	253.6	222.2	943.5	1,165.7	13.00
18	32.7	255.4	224.0	942.4	1,166.4	12.70
19	33.7	257.2	225.8	941.2	1,167.0	12.30
20	34.7	258.8	227.5	940.1	1,167.6	12.00
22	36.7	262.3	230.9	937.8	1,168.7	11.40
24	38.7	265.3	234.2	935.8	1,170.0	10.80
26	40.7	268.3	237.3	933.5	1,170.8	10.30
28	42.7	271.4	240.2	931.6	1,171.8	9.87
30	44.7	274.0	243.0	929.7	1,172.7	9.46
32	46.7	276.7	245.9	927.6	1,173.5	9.08
34	48.7	279.4	248.5	925.8	1,174.3	8.73
36	50.7	281.9	251.1	924.0	1,175.1	8.40
38	52.7	284.4	253.7	922.1	1,175.8	8.11
40	54.7	286.7	256.1	920.4	1,176.5	7.83
42	56.7	289.0	258.5	918.6	1,177.1	7.57
44	58.7	291.3	260.8	917.0	1,177.8	7.33
46	60.7	293.5	263.0	915.4	1,178.4	7.10
48	62.7	295.6	265.2	913.8	1,179.0	6.89
50	64.7	297.7	267.4	912.2	1,179.6	6.68

PROPERTIES OF SATURATED STEAM

Gauge Pressure psig	Absolute Pressure psia	Temperature °F	Heat Content Sensible (hf) BTU/lb	Latent (h_{fg}) BTU/lb	Total (h_g) BTU/lb	Specific Volume Steam (V_g) ft³/lb
52	66.7	299.7	269.4	910.7	1,180.1	6.50
54	68.7	301.7	271.5	909.2	1,180.7	6.32
56	70.7	303.6	273.5	907.8	1,181.3	6.16
58	72.7	305.5	275.3	906.5	1,181.8	6.00
60	74.7	307.4	277.1	905.3	1,182.4	5.84
62	76.7	309.2	279.0	904.0	1,183.0	5.70
64	78.7	310.9	280.9	902.6	1,183.5	5.56
66	80.7	312.7	282.8	901.2	1,184.0	5.43
68	82.7	314.3	284.5	900.0	1,184.5	5.31
70	84.7	316.0	286.2	898.8	1,185.0	5.19
72	86.7	317.7	288.0	897.5	1,185.5	5.08
74	88.7	319.3	289.4	896.5	1,185.9	4.97
76	90.7	320.9	291.2	895.1	1,186.3	4.87
78	92.7	322.4	292.9	893.9	1,186.8	4.77
80	94.7	323.9	294.5	892.7	1,187.2	4.67
82	96.7	325.5	296.1	891.5	1,187.6	4.58
84	98.7	326.9	297.6	890.3	1,187.9	4.49
86	100.7	328.4	299.1	889.2	1,188.3	4.41
88	102.7	329.9	300.6	888.1	1,188.7	4.33
90	104.7	331.2	302.1	887.0	1,189.1	4.25
92	106.7	332.6	303.5	885.8	1,189.3	4.17
94	108.7	333.9	304.9	884.8	1,189.7	4.10
96	110.7	335.3	306.3	883.7	1,190.0	4.03
98	112.7	336.6	307.7	882.6	1,190.3	3.96
100	114.7	337.9	309.0	881.6	1,190.6	3.90
102	116.7	339.2	310.3	880.6	1,190.9	3.83
104	118.7	340.5	311.6	879.6	1,191.2	3.77
106	120.7	341.7	313.0	878.5	1,191.5	3.71
108	122.7	343.0	314.3	877.5	1,191.8	3.65
110	124.7	344.2	315.5	876.5	1,192.0	3.60
112	126.7	345.4	316.8	875.5	1,192.3	3.54
114	128.7	346.5	318.0	874.5	1,192.5	3.49
116	130.7	347.7	319.3	873.5	1,192.8	3.44
118	132.7	348.9	320.5	872.5	1,193.0	3.39
120	134.7	350.1	321.8	871.5	1,193.3	3.34
125	139.7	352.8	324.7	869.3	1,194.0	3.23
130	144.7	355.6	327.6	866.9	1,194.5	3.12
135	149.7	358.3	330.6	864.5	1,195.1	3.02
140	154.7	360.9	333.2	862.5	1,195.7	2.93
145	159.7	363.5	335.9	860.3	1,196.2	2.84
150	164.7	365.9	338.6	858.0	1,196.6	2.76
155	169.7	368.3	341.1	856.0	1,197.1	2.68
160	174.7	370.7	343.6	853.9	1,197.5	2.61
165	179.7	372.9	346.1	851.8	1,197.9	2.54
170	184.7	375.2	348.5	849.8	1,198.3	2.48
175	189.7	377.5	350.9	847.9	1,198.8	2.41
180	194.7	379.6	353.2	845.9	1,199.1	2.35
185	199.7	381.6	355.4	844.1	1,199.5	2.30
190	204.7	383.7	357.6	842.2	1,199.8	2.24
195	209.7	385.7	359.9	840.2	1,200.1	2.18
200	214.7	387.7	362.0	838.4	1,200.4	2.14
210	224.7	391.7	366.2	834.8	1,201.0	2.04
220	234.7	395.5	370.3	831.2	1,201.5	1.96

PROPERTIES OF SATURATED STEAM

Gauge Pressure psig	Absolute Pressure psia	Temperature °F	Heat Content			Specific Volume Steam (V_g) ft^3/lb
			Sensible (h_f) BTU/lb	Latent (h_{fg}) BTU/lb	Total (h_g) BTU/lb	
230	244.7	399.1	374.2	827.8	1,202.0	1.88
240	254.7	402.7	378.0	824.5	1,202.5	1.81
250	264.7	406.1	381.7	821.2	1,202.9	1.74
260	274.7	409.3	385.3	817.9	1,203.2	1.68
270	284.7	412.5	388.8	814.8	1,203.6	1.62
280	294.7	415.8	392.3	811.6	1,203.9	1.57
290	304.7	418.8	395.7	808.5	1,204.2	1.52
300	314.7	421.7	398.9	805.5	1,204.4	1.47
310	324.7	424.7	402.1	802.6	1,204.7	1.43
320	334.7	427.5	405.2	799.7	1,204.9	1.39
330	344.7	430.3	408.3	796.7	1,205.0	1.35
340	354.7	433.0	411.3	793.8	1,205.1	1.31
350	364.7	435.7	414.3	791.0	1,205.3	1.27
360	374.7	438.3	417.2	788.2	1,205.4	1.24
370	384.7	440.8	420.0	785.4	1,205.4	1.21
380	394.7	443.3	422.8	782.7	1,205.5	1.18
390	404.7	445.7	425.6	779.9	1,205.5	1.15
400	414.7	448.1	428.2	777.4	1,205.6	1.12
420	434.7	452.8	433.4	772.2	1,205.6	1.07
440	454.7	457.3	438.5	767.1	1,205.6	1.02
460	474.7	461.7	443.4	762.1	1,205.5	0.98
480	494.7	465.9	448.3	757.1	1,205.4	0.94
500	514.7	470.0	453.0	752.3	1,205.3	0.902
520	534.7	474.0	457.6	747.5	1,205.1	0.868
540	554.7	477.8	462.0	742.8	1,204.8	0.835
560	574.7	481.6	466.4	738.1	1,204.5	0.805
580	594.7	485.2	470.7	733.5	1,204.2	0.776
600	614.7	488.8	474.8	729.1	1,203.9	0.750
620	634.7	492.3	479.0	724.5	1,203.5	0.726
640	654.7	495.7	483.0	720.1	1,203.1	0.703
660	674.7	499.0	486.9	715.8	1,202.7	0.681
680	694.7	502.2	490.7	711.5	1,202.2	0.660
700	714.7	505.4	494.4	707.4	1,201.8	0.641
720	734.7	508.5	498.2	703.1	1,201.3	0.623
740	754.7	511.5	501.9	698.9	1,200.8	0.605
760	774.7	514.5	505.5	694.7	1,200.2	0.588
780	794.7	517.5	509.0	690.7	1,199.7	0.572
800	814.7	520.3	512.5	686.6	1,199.1	0.557

Index

B

C

G

H

I

J

M

N

O

P

Q

R

Z

:inted in the United States
y Bookmasters